Lecture Notes in Mechanical Engineering

Lecture Notes in Mechanical Engineering (LNME) publishes the latest developments in Mechanical Engineering - quickly, informally and with high quality. Original research reported in proceedings and post-proceedings represents the core of LNME. Volumes published in LNME embrace all aspects, subfields and new challenges of mechanical engineering. Topics in the series include:

- Engineering Design
- Machinery and Machine Elements
- Mechanical Structures and Stress Analysis
- Automotive Engineering
- Engine Technology
- Aerospace Technology and Astronautics
- Nanotechnology and Microengineering
- Control, Robotics, Mechatronics
- MEMS
- Theoretical and Applied Mechanics
- Dynamical Systems, Control
- Fluid Mechanics
- Engineering Thermodynamics, Heat and Mass Transfer
- Manufacturing
- Precision Engineering, Instrumentation, Measurement
- Materials Engineering
- Tribology and Surface Technology

To submit a proposal or request further information, please contact the Springer Editor in your country:

China: Li Shen at li.shen@springer.com
India: Dr. Akash Chakraborty at akash.chakraborty@springernature.com
Rest of Asia, Australia, New Zealand: Swati Meherishi at swati.meherishi@springer.com
All other countries: Dr. Leontina Di Cecco at Leontina.dicecco@springer.com

To submit a proposal for a monograph, please check our Springer Tracts in Mechanical Engineering at http://www.springer.com/series/11693 or contact Leontina.dicecco@springer.com

Indexed by SCOPUS. The books of the series are submitted for indexing to Web of Science.

More information about this series at http://www.springer.com/series/11236

Numan M. Durakbasa · M. Güneş Gençyılmaz
Editors

Proceedings of the International Symposium for Production Research 2019

Springer

Editors
Numan M. Durakbasa
IFT, Technische Universität Wien
Vienna, Wien, Austria

M. Güneş Gençyılmaz
Faculty of Engineering
Istanbul Aydın University
Istanbul, Turkey

ISSN 2195-4356 ISSN 2195-4364 (electronic)
Lecture Notes in Mechanical Engineering
ISBN 978-3-030-31342-5 ISBN 978-3-030-31343-2 (eBook)
https://doi.org/10.1007/978-3-030-31343-2

This Springer imprint is published by the registered company Springer Nature Switzerland AG
The registered company address is: Gewerbestrasse 11, 6330 Cham, Switzerland

Foreword

Dear Colleagues and Friends,

It is a great pleasure to welcome you to the "**International Symposium for Production Research - ISPR 2019**" at the TU Wien, Getreidemarkt Campus in Vienna, Austria, from 28 to 30 August 2019, with the theme of "**Cases of Industry 4.0 Applications**". We are particularly pleased with our collaboration with Society for Production Research, İstanbul, Turkey, in hosting this year's symposium again in Vienna.

The purpose of the symposium is to bring together researchers, scientists and experts at universities, companies, institutions, communities, associations and societies in the domain of interest from around the world to share ideas, experiences, research results and vision on production and operations management and technology and to provide a forum for experts and professionals to discuss the actual theme "Cases of Industry 4.0 Applications" and the challenges as well as to advance innovation in this area.

We hope that this gathering will again provide the participants an opportunity for scholarly exchange of ideas and questions and allow them to build or renew friendships with experts, industry and corporate professionals and authorities.

August 2019

Kurt Matyas
Vice-Rector for Academic Affairs

Preface

The 19th International Symposium for Production Research 2019 (ISPR2019) was held at **TU Wien** (Vienna University of Technology) Mechanical and Industrial Engineering Faculty, Vienna, Austria, from 28 to 30 August 2019, with the theme of "**Cases of Industry 4.0 Applications**". This symposium was organized in Vienna, for the third year in a row, by TU Wien and Society for Production Research, İstanbul, Turkey.

The generic theme of "Industry 4.0" was first adopted in the symposium held in 2016 and maintained in the following three symposia in 2017, 2018 and 2019 but each with an emphasis on the relevant developments and progress on the various aspects of this "fourth industrial revolution". In other words, the symposium maintained the same main theme from 2016 but studied a different aspect of it in the subsequent years with the purpose of drawing the attention of the researchers to the cases of Industry 4.0 applications of this new industrial era on the production systems and production management.

The world of science and technology is under increasing influence of the requirements of Industry 4.0. Transition to a new era seems inevitable for every sector of industry. Given the importance of this theme, ISPR2019 hosted numerous distinguished speakers from both the academia and the industry to hear their views on the applications of the Industry 4.0 on the various components of production systems.

A total of over 140 participants attended this year's symposium—academics, practitioners and scientists from 16 countries—who contributed 9 invited talks and 69 papers on the plenary, special, workshop and ordinary sessions. The symposium programme included keynote addresses (opening/closing session), breakout sessions and workshop discussions, a welcome reception, a gala dinner and a closing session on the final day.

This volume of proceedings from the symposium contains 69 refereed papers in 14 categories shown in the contents of the proceedings.

We are very grateful to our host institution, the TU Wien, for its invaluable support and hospitality and for enabling this symposium to be organized, for a third time, in their premises. In particular, we would like to express our gratitude to

Vice-Rector Prof. Kurt Matyas, also the Honorary Chairman of the Scientific Committee of this symposium for his leadership and generous support. We also thank Prof. Christian Bauer, Dean of the Faculty of Mechanical and Industrial Engineering; Prof. Detlef Gerhard, former Dean of the Faculty of Mechanical and Industrial Engineering; and Prof. Friedrich Bleicher, Head of the Institute for Production Engineering and Photonic Technologies, for their interest and support for this symposium.

We would like to thank all the keynote and invited speakers whose contributions enhanced the success of the symposium.

In organizing this event, our colleagues in Vienna and İstanbul contributed endless hours of hard work, energy and wisdom to make this event the success it was. On the Vienna side, our sincere thanks go to the staff of the Research Group Production Metrology and Adaptronic Systems of the Institute for Production Engineering and Photonic Technologies, in particular, to Mr. Erol Güçlü.

On the İstanbul side, we are very grateful to Prof. Çebi, who made available the resources of her department, and to Assoc. Prof. Hatice Camgöz Akdağ, who was involved in every aspect of the organization from the very beginning. Our special thanks go to Ms. Tuğçe Beldek, Mr. Aziz Kemal Konyalıoğlu, our young colleagues, research assistants at İstanbul Technical University and also research assistants at TU Wien, for their hard and dedicated work.

We would like to express our gratitude to the board members of the Society for Production Research in İstanbul for their strong support and dedication to make the symposium a success.

Our very special thanks go to our colleagues, the participants of this symposium. Undoubtedly, they are the core component of this organization.

We would like to recognize and thank our dear colleagues who graciously accepted to join the honorary and scientific committees or who served as peers in this event.

Finally, no such event is possible without the generous support of patrons and sponsors. In this regard, we would like to thank Dr. Michael Ludwig, the Mayor of Vienna, and Ambassador of Turkey, His Excellency Ümit Yardım, for hosting a reception for the participants of this symposium, to all the corporations and individuals who provided invaluable financial and intellectual contributions.

And last but not least, we are grateful to Mr. Antony Doyle, Editor from Springer Nature, for his competent guidance, professionalism and patience.

M. Güneş Gençyılmaz
Numan M. Durakbasa

Production of the Future - Industry 4.0 Applications

The ongoing rapid development of information and communications technology, ICT, digitalization and global networks as well as the convergence of virtual and real worlds are the key factors in modern industrial production. Digitalization and deployment of growing technologies in ICT are transforming the machining and the manufacturing industry worldwide. Production processes will be designed dynamically and efficiently. The production of the future will be characterized by a fully integrated and highly flexible production chain with integrated intelligent systems under the guidance of international standards.

The engineering disciplines working in the industrial sectors according to the Industry 4.0 concept aim for meeting the needs of safe, efficient and cost-effective solutions. The integrated management system in developing high-technology tools and methodology will ensure to achieve this aim by using micro-/nanotechnology methods as a part of next-generation factory strategy.

Industry 4.0 includes a higher priority for customized products, which requires transformations in the design and manufacture of industrial systems and products and their business models. Advanced production in the context of Industry 4.0 requires transformations in the designing, manufacturing, and verification of industrial products and processes based on Geometrical Product Specifications and Verification—GPS. The design system and manufacturing processes are modelled for self-optimization with intelligent measurement processes with the target of practical approach for both the industry and education facilities.

This new concept according to Industry 4.0 can be developed on the basis of advanced production systems and multi-functional integrated factory model as well as extensive use of the IT, AI, CPS, simulation, intelligent quality systems, robotics, sophisticated metrology and advanced engineering data exchange techniques. Moreover, by utilizing advanced information analytics, networked intelligent machines and instruments will be able to perform more efficiently, collaboratively and sustainably, which will enable an agile and optimal industrial production in any kind of industry and especially in up-to-date SMEs towards Industry 4.0.

August 2019

Numan M. Durakbasa
M. Güneş Gençyılmaz

Organization

ISPR2019 was organized by TU Wien, Austria, and the Society for Production Research, Turkey. The symposium took place on the TUtheSky Campus of the TU Wien.

Editors

Numan M. Durakbasa
M. Güneş Gençyılmaz

Co-editors

Peter Kopacek
Ayhan Toraman
Selim Zaim
Jorge Martin Bauer
Serpil Erol
Semra Birgün
Kemal Güven Gülen
Alptekin Erkollar
Mahmut Tekin

Honorary Chair

Kurt Matyas	TU Wien, Austria

Symposium Chairs

Numan M. Durakbasa	TU Wien, Austria
M. Güneş Gençyılmaz	İstanbul Aydın University, Turkey

International Honorary Committee

Organizing Committee

This Book Is Prepared for Publishing by

Scientific Committee

S. Adamczak	Kielce University of Technology, Poland
H. Camgöz Akdağ	İstanbul Technical University, Turkey
K. Altinel	Bosphorus University, Turkey
B. Baki	Karadeniz Technical University, Turkey
Ş. Baray	İstanbul University, Turkey
G. Baş	TU Wien, Austria
N. Başoğlu	İzmir Institute of Technology, Turkey
J. Bauer	National Technological University Buenos Aires, Argentina
A. Baykasoğlu	9 Eylül University, Turkey
E. Bayraktar	The American University of the Middle East, Kuwait
F. Berto	Norwegian University of Science and Technology, Norway
S. Birgün	Fenerbahçe University, Turkey
P. Blecha	Brno University of Technology, Czech Republic
F. Bleicher	TU Wien, Austria
F. Bozbura	Bahçeşehir University, Turkey
I. Budak	Faculty of Technical Sciences in Novi Sad, Serbia
A. Bulgak	Concordia University, Canada
M. Bulu	İstinye University, Turkey
L. A. Crisan	Universitatea Tehnica Cluj-Napoca, Romania
F. Cus	University of Maribor, Slovenia
F. Çebi	İstanbul Technical University, Turkey
D. Delen	Oklahoma State University, USA
H. Demir	Yaşar University, Turkey
M. Demirbağ	University of Essex, UK
B. Dengiz	Başkent University, Turkey
T. Dereli	İskenderun University, Turkey
M. Dinçmen	İstanbul Technical University, Turkey
Á. Drégelyi Kiss	Óbuda University, Hungary
N. Durakbaşa	TU Wien, Austria
B. Durmuşoğlu	İstanbul Technical University, Turkey
A. Erkollar	Sakarya University, Turkey
S. Erol	Gazi University, Turkey
Ş. Esnaf	İstanbul University, Turkey
S. Firat	Marmara University, Turkey
W. Fisher	University of California, Berkley, USA
G. Gencyilmaz	İstanbul Aydın University, Turkey
D. Gerhard	TU Wien, Austria
S. Gözlü	İstanbul Technical University, Turkey
S. Grozav	Universitatea Tehnică din Cluj-Napoca, Romania

K. Gülen	Namık Kemal University, Turkey
A. Güngör	Pamukkale University, Turkey
V. Gyula	University of Miskolc, Hungary
E. Hekelová	City University Bratislava, Slovakia
F. Holesovsky	J. E. Purkyně University in Ústí nad Labem, Czech Republic
Z. Irani	University of Bradford, UK
V. Işler	Hasan Kalyoncu University, Turkey
C. Kahraman	İstanbul Technical University, Turkey
I. Kara	Başkent University, Turkey
T. Katoka	Kindai University, Japan
M. Klumpp	Duisburg University, Germany
D. Kocaoğlu	Portland State University, USA
T. Koç	İstanbul Technical University, Turkey
P. Kopacek	TU Wien, Austria
M. Králik	Slovak University of Technology in Bratislava, Slovakia
L. Kräuter	TU Wien, Austria
C. Kubat	Sakarya University, Turkey
U. Kula	American University of the Middle East, Kuwait
A. Kulakli	American University of the Middle East, Kuwait
J. Kundrak	University of Miskolc, Hungary
A. Kusiak	University of Iowa, USA
V. Majstorovic	University of Belgrade, Serbia
I. Mankova	Technical University of Kosice, Slovakia
A. Markopoulos	National Technical University of Athens, Greece
K. Matyas	TU Wien, Austria
M. Novák	J. E. Purkyně University in Ústí nad Labem, Czech Republic
Y. Omurtag	Robert Morris University, USA
N. Özçakar	İstanbul University, Turkey
D. Özdemir	Bilgi University, Turkey
A. Özok	Okan University, Turkey
E. Öztemel	Marmara University, Turkey
J. Peterka	Slovak University of Technology, Slovakia
D. Pham	University of Birmingham, United Kingdom
M. Pokusová	Slovak University of Technology in Bratislava, Slovakia
S. Popescu	Technical University of Cluj-Napoca, Romania
D. Popescu	Technical University of Cluj-Napoca, Romania
E. Pucher	TU Wien, Austria
A. Sağbaş	Namık Kemal University, Turkey
J. Sanz Juan	Polytechnic University of Valencia, Spain
A. Sharif	University of Bradford, UK
A. Smith	University of Auburn, USA

Ľ. Šooš	Slovakia University of Technology, Slovakia
A. Sorguç	Middle East Technical University, Turkey
A. Soysal	Doğuş University, Turkey
H. Soyuer	Ege University, Turkey
K. Stepien	Kielce University of Technology, Poland
B. Tan	Koç University, Turkey
M. Tanyaş	Maltepe University, Turkey
F. Taşgetiren	Yaşar University, Turkey
H. Taşkın	Sakarya University, Turkey
M. Tekin	Selçuk University, Turkey
A. Toraman	İstanbul Technical University, Turkey
J. Torgersen	Norwegian University of Science and Technology, Norway
O. Torkul	Sakarya University, Turkey
G. Ulusoy	Sabancı University, Turkey
G. Varga	University of Miskolc, Hungary
K. Velíšek	Slovak University of Technology, Slovakia
J. Wang	National University of Singapore, USA
A. Weckenmann	University of Erlangen Nürnberg, Germany
M. Yalçıntaş	İstanbul Commerce University, Turkey
M. Yenisey	İstanbul University, Turkey
M. E. Yurci	Yıldız Technical University, Turkey
S. Zaim	İstanbul Şehir University, Turkey
W. Zębala	Cracow University of Technology, Poland
M. Zerenler	Selçuk University, Turkey

Reviewers

Adrian-Crisan, L.
Aktin, T.
Baray, Ş.
Bayyurt, N.
Bereketli-Zafeirakopoulos, İ.
Beyca, Ö.
Birgün, S.
Bolat, H.
Börühan, G.
Bulut, Ö.
Büyüksaatçi, S.
Camcı, A.
Camgöz Akdağ, H.
Çalışkan, A.
Çebi, F.
Dregelyi-Kiss, Á.
Durakbasa, N. M.
Ekinci, E.
Erkollar, A.
Erol, S.
Ersoy, P.
Esnaf, Ş.
Gençyılmaz, M. G.
Gergin, Z.
Gökçe, M.
Gülen, K.
Kabadurmuş, Ö.
Kandiller, L.
Kazançoğlu, Y.
Kızılay, D.
Kopacek, P.
Kräuter, L.

Namlı, E.
Novak, M.
Öner, E.
Örnek, A.
Özdemir, D.
Özkan, S.
Öztop, H.
Öztürk, Ö.
Öztürkoğlu, Y.
Soyuer, H.
Üney-Yüksektepe, F.
Weckenmann, A.
Yetkin-Ekren, B.
Zaim, S.

Contents

Supply Chain Management

University-Industry Collaboration

Capstone Projects

Artificial Intelligent Applications

Modelling and Optimization of Surface Roughness Parameters of Stainless Steel by Artificial Intelligence Methods

Pavel Kovač, Borislav Savković, Dragan Rodić, Andjelko Aleksić, Marin Gostimirović, Milenko Sekulić, and Nenad Kulundžić(✉)

Faculty of Technical Sciences, University of Novi Sad,
Trg Dositeja Obradovića 6, 21000 Novi Sad, Serbia
{pkovac, savkovic, rodicdr, aleksica, maring, milenkos, kulundzic}@uns.ac.rs

Abstract. The objective of this study is to examine the influence of machining parameters on surface finish in turning of medical steel. A new approach in modeling surface roughness which uses design of experiments is described in this paper. The values of surface roughness predicted by different models are then compared. Used were adaptive-neuro-fuzzy-inference system (ANFIS). The results showed that the proposed system can significantly increase the accuracy of the product profile when compared to the conventional approaches. The results indicate that the design of experiments with central composition plan modeling technique can be effectively used for the prediction of the surface roughness for medical steel difficult to machining. Optimizations of surface roughness parameters was done by use of ant colony method.

Keywords: Turning · Stainless steel · ANFIS modelling · Ant Colony Optimization · Surface roughness

1 Introduction

Increasingly, research in manufacturing processes and systems is evaluating processes to improve their efficiency, productivity and quality. The quality of finished products is defined by how closely the finished product adheres to certain specifications, including dimensions and surface quality. Surface quality is defined and identified by the combination of surface finish, surface texture, and surface roughness. Surface roughness (Ra, Rmax) is the commonest index for determining surface quality [1, 2].

Manufacturing processes do not allow achieving the theoretical surface roughness due to defects appearing on machined surfaces and mainly generated by deficiencies and imbalances in the process. Due to these aspects, measuring procedures are necessary; as it permits one to establish the real state of surfaces to manufacture parts with higher accuracy. To know the surface quality, it is necessary to employ theoretical models making it feasible to do predictions in function of response parameters [3–5].

A lot of analytically methods were also developed and used for predicting surface roughness. An empirical model for prediction of surface roughness in finish turning [6].

N. M. Durakbasa and M. G. Gençyılmaz (Eds.): ISPR 2019, LNME, pp. 3–12, 2020.
https://doi.org/10.1007/978-3-030-31343-2_1

Nonlinear regression analysis, with logarithmic data transformation is applied in development of the empirical model. Metal cutting experiments and statistical tests demonstrate that the model developed in this research produces smaller errors and has a satisfactory result. The mathematical models for modeling and analyzing the vibration and surface roughness parameters in the precision turning with a diamond cutting tool is presented in [7].

Recently, some initial investigations in applying the basic artificial intelligence approach to model machining processes, have appeared in the literature, concludes that the modeling of surface quality in machining processes has mainly used Artificial Neural Networks and fuzzy set theory [8, 9]. Average mean roughness, Ra using neural network (NN) was predicted in [10]. Surface roughness and surface finish have been considered in studies [11–14]. Research of the influence of machining parameters combination to obtain a good surface finish in turning and to predict the surface roughness values using fuzzy modeling is presented in [15]. Also, may notice that the neural network used in the study, where the enabling resolution of the problem that is difficult to define and mathematically modeled. This can be seen in the work where the neural network is based on the face milling machining processes where is aimed to produce the relationship of cutting force versus instantaneous angle φ in [16]. Use of Coolant and Lubricatio agents in Hard Machining is presented in [17].

In this study, cutting speed, feed and depth of cut as machining conditions were selected. An adaptive neuro fuzzy inference system (ANFIS) were developed for modeling these cutting parameters.

2 Material and Experimental Procedure

The aim of the experiment was to carry out the monitoring of machining surface roughness for different cutting regime used. The terms of the experimental study:

(a) Cutting operation was performed on Vertical turning machine for hard turning VL5 (Fig. 1). Machine control unit was Fanuc 18i-TB. Number of Axes: 2 to 3, cutting diameter: 127 to 200 mm, cutting length: 119 mm, tool Stations: 12, spindles: 1, motor power: 20.88 Kw, spindle speed: 4500 rpm.

Fig. 1. Vertical turning machine for hard turning

(b) Tool: The survey was used interchangeable plates tags WNMG 080408-QM, manufacturer SANDVIK Coromant, analogous to the ISO standard (EN ISO 9001:2000, EN ISO 14001:2004) with P30-50. Tool holder for external processing PWLNL 2525 M08 H4 was used.
(c) Equipment for cooling and lubrication was used ALUSOL of Castrol, conc. 6–10%. Castrol Alusol™ RAL BF is a high-performance soluble metalworking fluid which is boron and chlorine free. It contains a unique additive package that works in synergy to enhance machining performance and surface finish, provide excellent product stability, improve bio-resistance properties and lower the overall operations costs.
(d) Workpiece material used in tests was steel IDM 8365 (EN X30CrNi2520+1.3Nb) it has a large degree of elongation, toughness due to the high content of Ni, corrosion resistance due to the presence Cr and Ni, belong to the group of stainless steel and high-alloy steels. Also, mechanical properties of processed materials are given in Table 1. Testing show that the microstructure shall consist of interdendritic eutectic chromium carbides in an austenitic matrix, Fig. 2. This was performed and morphological examination of material microscope, "Leitz" Aristomet QI 01297/"OLYMPUS BH" optical Japan QI 01298 and 01298/1.
(e) For measuring the surface roughness was used Surftest Mitutoyo SJ-301, the needle of the diamond, and the radius of curvature at the top of the needle 5 microns. Path length measured was 3.7 mm. The measured values of: R_a, R_{max}. The measurement results are given in Table 2.

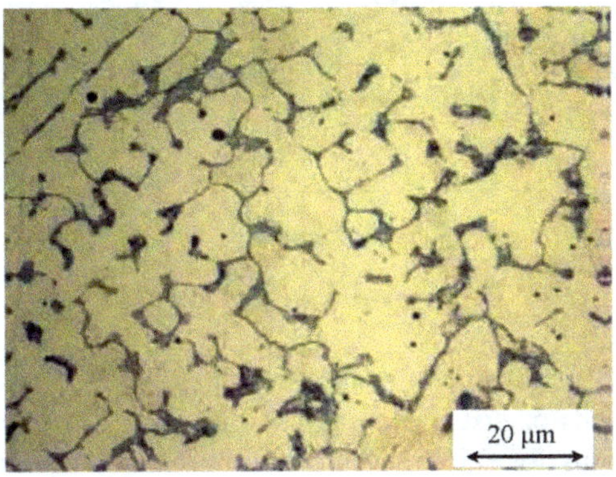

Fig. 2. The microstructure shall consist of interdendritic eutectic chromium carbides in an austenitic matrix

Table 1. Mechanical properties of workpiece material

	Mark:	Unit	Limits	Size
Tensile strength	R_m	MPa	Min 450	486
Prof stress	$R_{p0,2}$	MPa	Min 240	351
Elongation:	A	%	Min 10	10,3
Brinell hardness	HBW 5/750		162–229	207

Table 2. The experimental and modeled values

No.	Factor			Surface roughness			
	v[m/s]	f[mm/rev]	a[mm]	R_a[μm]	R_{max}[μm]	R_a model[μm]	R_{max} model[μm]
1.	3.0	0.12	0.5	0.86	5.22	0.76	5.11
2.	4.33	0.12	0.5	0.88	6.07	0.80	5.78
3.	3.0	0.187	0.5	1.58	10.03	1.86	10.32
4.	4.33	0.187	0.5	1.8	11.06	1.95	11.67
5.	3.0	0.12	1.4	0.97	6.39	0.89	5.46
6.	4.33	0.12	1.4	0.97	6.63	0.94	6.18
7.	3.0	0.187	1.4	2.17	11.04	2.17	11.03
8.	4.33	0.187	1.4	2.4	12.03	2.28	12.47
9.	3.6	0.15	0.84	1.33	6.86	1.32	8.00
10.	3.6	0.15	0.84	1.16	7.58	1.32	8.00
11.	3.6	0.15	0.84	1.17	6.71	1.32	8.00
12.	3.6	0.15	0.84	1.29	7.78	1.32	8.00
13.	2.5	0.15	0.84	1.18	6.87	1.26	7.08
14.	5.22	0.15	0.84	1.25	8.43	1.39	9.06
15.	3.6	0.096	0.84	0.81	4.94	0.54	3.94
16.	3.6	0.234	0.84	3.49	16.26	3.22	16.18
17.	3.6	0.15	0.3	1.105	7.83	1.13	7.48
18.	3.6	0.15	2.34	1.34	8.09	1.54	8.55
19.	2.5	0.15	0.84	1.037	6.72	1.26	7.08
20.	5.22	0.15	0.84	1.082	8.45	1.39	9.06
21.	3.6	0.096	0.84	0.74	5.01	0.54	3.94
22.	3.6	0.234	0.84	3.05	14.92	3.22	16.18
23.	3.6	0.15	0.3	1.34	8.11	1.13	7.48
24.	3.6	0.15	2.34	1.48	8.21	1.54	8.55

3 Implementation of Factorial Experimental Plan

The modeled results by factorial experimental plan are shown in Table 2. In the Table 3 are given results of dispersion analyses, adequacy of models and significance of parameters. Equations for surface roughness modeling determined by central compositional plan design of experiment:

$$R_a = 50.92 \cdot v^{0.135} \cdot f^{2.002} \cdot a^{0.150} \quad (1)$$

$$R_{max} = 106.3 \cdot v^{0.335} . f^{1.585} \cdot a^{0.065} \quad (2)$$

Table 3. Adequacy of models and significance of parameters

Model adequacy		R_a	R_{max}
		$F_a = 3.23658$	$F_a = 3.09891$
Significance	218.57	218.57	218.57
	1.17	1.17	1.17
	375.76	375.76	375.76
	11.38	11.38	11.38

4 Implementation of ANFIS

For training the ANFIS during modeling was used "MatLab" software, which is the most powerful software for technical calculations [18].

Training and testing are the most important characteristics of ANFIS because just training and testing determine its characteristics. To create and train an ANFIS in the MATLAB is used Fuzzy Logic Toolbox. Adaptive neuro-fuzzy inference system is an architecture which is functionally equivalent to a Sugeno type fuzzy rule base [19]. An ANFIS gives the mapping relation between the input and output data by using the hybrid learning method to determine the optimal distribution of membership functions [20]. Both, artificial neural network (ANN) and fuzzy logic (FL) are used in ANFIS architecture [21]. In 1993, Jang first introduced the Adaptive Neuro-Fuzzy Inference System, which was reported as a very efficient system for solving the ill-defined equations involving the automatic elicitation of knowledge expressed only by the if–then rules.

In our case ANFIS is a five-layer neural network that simulates the working principle of a fuzzy inference system. The ANFIS model generated from the membership functions and rules were data-driven by the process data for each mechanical property. Though there are many numbers of membership functions available like triangular, trapezoidal, Gaussian, etc. Each set of process data collected from the extrusions consisted of 30 data points from which 24 and 6 were selected randomly for training and testing, respectively. The models were developed and implemented using 100 epochs. The input and output data sets contained three inputs [cutting speed, feed rate and depth of cut] and one output (surface roughness parameters R_a or R_{max}).

As mentioned before, ANFIS modeling was used for analysis and optimization of surface roughness in turning process. The obtained results of ANFIS model are given in the Table 4, side by side with the obtained experimental results. For reduction of a deviation, is needed to increase the number of input trials.

Calculation of percentual deviation for measured and model surface roughness values was performed according next formula:

$$E = \frac{\left|Ri_{\exp} - Ri_m\right|}{Ri_{\exp}} \cdot 100\% \tag{3}$$

where are: Ri_{exp} - experimental value, Ri_m - model value.

Table 4. Experimental values and values obtained and percentage deviation

No.	Factor			R_i - experimental roughness		R_i - modeled roughness		Deviation E %	
	v[m/s]	f[mm/rev]	a[mm]	R_a[μm]	R_{max}[μm]	R_a[μm]	R_{max}[μm]	R_a[μm]	R_{max}[μm]
1.	3.6	0.15	0.5	1.373	9.321	1.358	9.255	1.105	0.708
2.	3.6	0.15	1.4	1.878	15.950	1.866	15.871	0.643	0.495
3.	3.6	0.12	0.84	1.236	8.413	1.244	8.291	0.643	1.451
4.	3.6	0.187	0.84	2.018	13.138	2.031	13.119	0.640	0.144
5.	3.0	0.15	0.84	2.379	17.754	2.291	17.689	3.841	0.366
6.	4.33	0.15	0.84	2.486	24.334	2.476	23.159	0.404	4.828
	Average deviation							1.213	1.332

In Fig. 3 showed the correlation between the experimental and expected values of surface roughness (R_a, R_{max}). From the diagram can be seen that the regression line is approximately in the direction of the angle of 45° degrees or regression factor for all three sizes is close to the number one (1).

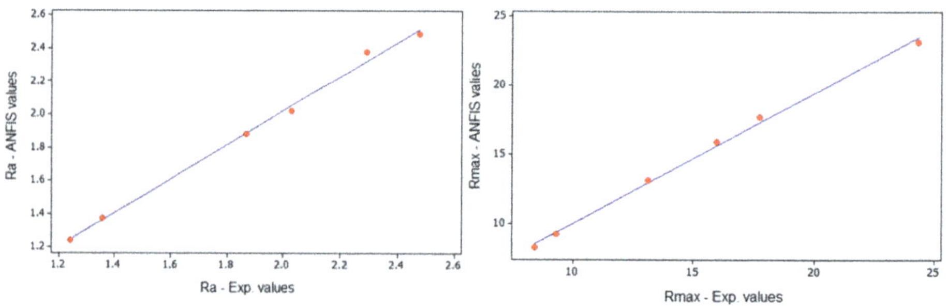

Fig. 3. Correlation between experimental and obtained values of roughness Ra and Rmax

5 Implementation of ANTS Algorithm

The term Ant Colony Optimization (ACO) is then introduced by Marco Dorigo [22], as a result of research on combinatorial optimization approaches. Initially, the algorithm applied to the problem of a commercial passenger and the problem of joining the quadrant. Since 1995 Dorigo and others have worked on various extended versions of the initial idea.

Ants have the ability to find the shortest path that leads from ants to a food source without the use of a visual sign. This ability of the ants is realized by leaving a certain amount of a chemical known as a pheromone by its movement along the ground. In addition, they have the ability to adapt to the new changes around them. For example, when they find the shortest path to a food source, in the event of a barrier on the road or in the concealment of the road, they find a new shortest path, Fig. 4.

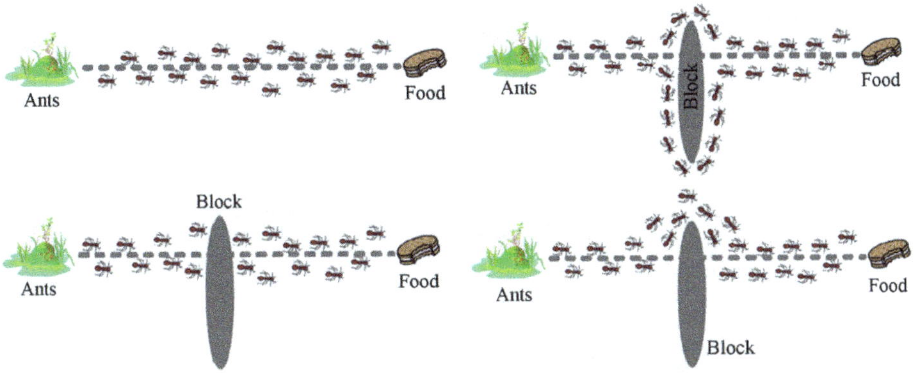

Fig. 4. Finding the shortest path on the road with an obstacle

The ants, although they are blind, can find the shortest path to the food source. This feature of ants is available to solve real problems using certain features and some supplements. Characteristics of ants created by artificial ant are:

- Communication between ants using the chemical pheromone.
- Preferred roads with a higher number of pheromones.
- Faster increase of pheromones on shorter paths than on longer ones.

Characteristics added to real ants:

- Live in an environment where time is calculated discreetly.
- They are completely blind and can access the details of the problem.
- I can retain information to solve a particular problem with a certain amount of memory.

The basic idea of algorithms based on an ant colony is that artificial intelligence agents, using a simple mechanism of communication, can produce solutions to many complex problems [22].

The way in which the algorithm functions is shown is a pseudocode, Table 5.

Table 5. Algorithm: pseudocode – ACO metaheuristic

```
PROCEDURE ACO metaheuristic
        Set parameters
            Initialize pheromone trails
                WHILE (termination condition not met)
                    Construct Ant solutions
                    Deamon actions
                    Update pheromones
                END WHILE
END PROCEDURE
```

The optimization is performed for the target function surface roughness parameters (R_a and R_{max}) minimum, based on the input parameters shown in Table 6 with limit values.

Table 6. Optimal values for R_A and R_{max}

Parameter	Minimal value	Maximal value
Speed v[m/s]	2.5	5.22
Feed f[mm/o]	0.096	0.234
Depth of cut a[mm]	0.3	2.34

The display of the optimization process is shown through the following images. Where are with card acor shown parameters of ant algorithm and problem definition. Until with card criterion are input parameters (v, f and a) with their limited values and input function (R_a and R_{max}) min defined. The results obtained by optimizing the cutting process using an ant algorithm are shown in the Fig. 5. Under the BestSol card, the obtained optimization results are displayed, as well as the value (R_a) min. The results of optimization of surface roughness parameters are shown in Table 7.

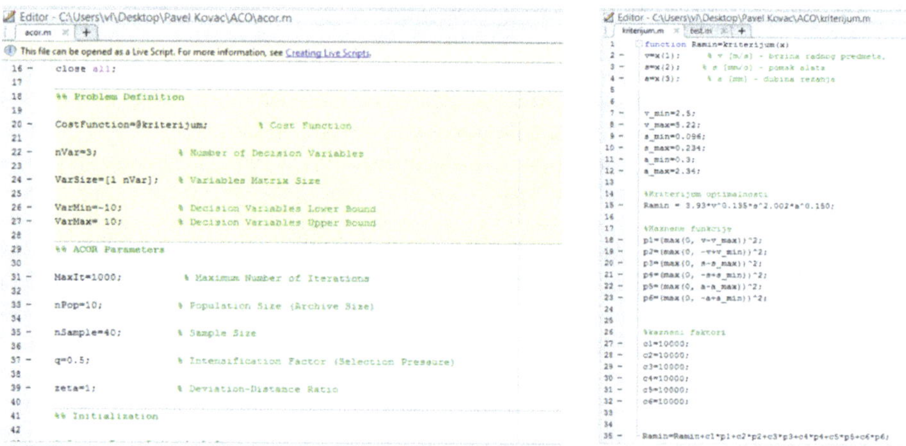

Fig. 5. Display of the optimization ANTS process

Table 7. The results of optimization of surface roughness parameters

Outputs		Input parameters		
Surface roughness parameters [μm]		v[m/s]	f[mm/rev]	a[mm]
R_a	0.440	2.500	0.096	0.300
R_{max}	3.28	2.500	0.096	0.300

6 Conclusion

Intelligent optimization techniques give the influence of cutting conditions on machining surface quality during turning medical steel difficult to machine, are investigated through experimental verification. The investigation results confirm the highly consent of experimental research and intelligent techniques modeling. The intelligent optimization techniques and experimental results show some good information which could be used by future researches for optimal control of machining conditions.

This paper has successfully established ANFIS model, which consist of the pertinent process parameters, for predicting the surface roughness parameters of workpiece. Diagrams on Fig. 3 shows the compared predicted values obtained by experiment and estimated by ANFIS shows a good comparison with those obtained experimentally. The average deviations of models are checked and are found to be adequate. The model adequacy can be further improved by considering more variables and ranges of parameters. Optimal values of surface roughness parameters were successfully determined by ant colony method.

Acknowledgements. The paper is the result of the research within the project TR 35015 financed by the ministry of science and technological development of the Republic of Serbia and CEEPUS project.

References

1. Chen, J.C., Savage, M.: A fuzzy-net-based multilevel in-process surface roughness recognition system in milling operations. Int. J. Adv. Manuf. Technol. **17**, 670–676 (2001)
2. Quintana, G., Garcia-Romeu, M.L., Ciurana, J.: Surface roughness monitoring application based on artificial neural networks for ball-end milling operations. J. Intell. Manuf. **22**, 607–617 (2009)
3. Sivarao, Castillo, Taufik: Machining quality predictions: comparative analysis of neural network and fuzzy logic. Int. J. Electr. Comput. Sci. IJECS **9**, 451–456 (2000)
4. Drégelyi-Kiss, Á., Horváth, R., Mikó, B.: Design of experiments (DOE) in investigation of cutting technologies, Development in Machining Technology DIM, Cracow, pp. 20–34 (2013)
5. Maňková, I., Vrabel', M., Beňo, J., Kovač, P., Gostimirovic, M.: Application of Taguchi method and surface response methodology to evaluate of mathematical models for chip deformation when drilling with coated and uncoated twist drills. Manuf. Technol. **13**(4), 492–499 (2013)
6. Hadi, Y., Ahmed, S.G.: Assessment of surface roughness model for turning process. In: Knowledge Enterprise: Intelligent Strategies in Product Design, Manufacturing, and Management. International Federation for Information Processing (IFIP), vol. 207, pp. 152–158 (2006)
7. Chen, C.C., Chiang, K.T., Chou, C.C., Liao, Y.C.: The use of D-optimal design for modeling and analyzing the vibration and surface roughness in the precision turning with a diamond cutting tool. Int. J. Adv. Manuf. Technol. **54**, 465–478 (2011)
8. Choudhary, A., Harding, J., Tiwari, M.: Data mining in manufacturing: A review based on the kind of knowledge. J. Intell. Manuf. **20**(5), 501–521 (2009)

9. Grzenda, M., Bustillo, A., Zawistowski, P.: A soft computing system using intelligent imputation strategies for roughness prediction in deep drilling. J. Intell. Manuf. **23**, 1733–1743 (2012)
10. Balic, J., Korosec, M.: Intelligent tool path generation for milling of free surfaces using neural networks. Int. J. Mach. Tools Manuf. **42**, 1171–1179 (2002)
11. Pérez, C.J.L.: Surface roughness modeling considering uncertainty in measurements. Int. J. Prod. Res. **40**(10), 2245–2268 (2002)
12. Azouzi, R., Gullot, M.: On-line prediction of surface finish and dimensional deviation in turning using neural network based sensor fusion. Int. J. Mach. Tools Manuf. **37**(9), 1201–1217 (1997)
13. Ho, S.Y., Lee, K.C., Chen, S.S., Ho, S.J.: Accurate modeling and prediction of surface roughness by computer vision in turning operations using an adaptive neuro-fuzzy inference system. Int. J. Mach. Tools Manuf. **42**(13), 1441–1446 (2002)
14. Zębala, W., Gawlik, J., Matras, A., Struzikiewicz, G., Ślusarczyk, Ł.: Research of surface finish during titanium alloy turning. Key Eng. Mater. **581**, 409–414 (2014)
15. Rajasekaran, T., Palanikumar, K., Vinayagam, B.K.: Application of fuzzy logic for modeling surface roughness in turning CFRP composites using CBN tool. Prod. Process. **5**(2), 191–199 (2011)
16. Savković, B., Kovač, P., Gerić, K., Sekulić, M., Rokosz, K.: Application of neural network for determination of cutting force changes versus instantaneous angle in face milling. J. Prod. Eng. **16**(2), 25–28 (2013)
17. Kundrák, J., Varga, G.: Use of coolants and lubricants in hard machining. Tech. Gaz. **20**(6), 1081–1086 (2013)
18. Kovac, P., Rodic, D., Pucovsky, V., Savkovic, B., Gostimirovic, M.: Application of fuzzy logic and regression analysis for modeling surface roughness in face milling. J. Intell. Manuf. **24**(4), 755–762 (2013)
19. Jang, J.S.R., Sun, C.T.: Neuro-fuzzy modeling and control. Proc. IEEE **83**, 378–406 (1995)
20. Ying, L.C., Pan, M.C.: Using adaptive network based fuzzy inference system to forecast regional electricity loads. Energy Conversat. Manag. **49**, 205–211 (2008)
21. Avci, E.: Comparison of wavelet families for texture classification by using wavelet packet entropy adaptive network based fuzzy inference system. Appl. Soft Comput. **8**, 225–231 (2008)
22. Dorigo, M., Maniezzo, V., Colorni, A.: Ant system: optimization by a colony of cooperating agents. IEEE Trans. Syst. Man Cybern. Part B Cybern. **26**(1), 29–41 (1996)

Using Artificial Intelligence for the Improvement of Weapon Efficiency

Recep Benzer[1(✉)], Alptekin Erkollar[2], Birgit Oberer[2], and Semra Benzer[3]

[1] Ostim Technical University, Ankara, Turkey
rbenzer@gmail.com
[2] Swiss Federal Institute of Technology, Zürich, Switzerland
{Erkollar, Oberer}@global.educanet2.ch
[3] Gazi University, Ankara, Turkey
sbenzer@gazi.edu.tr

Abstract. In the defense industry, which is one of the determinants of the political and economic power of the countries, there is a constant need for change, innovation, and modernization in parallel with the technological developments in the world. Therefore, in the field of defense technologies, Military and non-governmental organizations are working to develop high-performance armor systems that will increase mobility by reducing weight and against developing weapon systems. In this paper will be discussed, how artificial intelligence (AI) can be used for the improvement of weapon efficiency.

Keywords: Artificial neural network · Weapon · Efficiency · Defense · Autonomous

1 Introduction

A weapon is any device that can be used with intent to inflict damage or harm. Arms can be used to increase the efficacy and efficiency of activities and not only in the military, e.g., such as hunting, fishing, crime, law enforcement, self-defense, and warfare. Ensuring appropriate human control over weapons, such that they are used effectively and as intended, has been a critical task. Projectiles can miss their targets. Even hand-held edged weapons that are large and unwieldy run the risk of accidents in the hands of an untrained user. Many weapons require skill and experience to be used safely and effectively. As humankind has developed increasingly powerful weapons over time, concern has grown as well about ensuring that weapons are used appropriately. Nuclear weapons, for example, require stringent safeguards to prevent unauthorized use. The two main reasons for having a weapon are for self-defense, as a human being wishing to live a long, fruitful life or a policeman or soldier or security guard to keep someone from taking the right of life from you, and for providing food for yourself and family as a hunter. This can be a simple expression, but the main answer is the tendency of countries to produce weapons and to have power in the world as a gun industry [1]. Autonomous weapon systems might not be categorized based on the system's complexity, but rather on the type of function being executed without

N. M. Durakbasa and M. G. Gençyılmaz (Eds.): ISPR 2019, LNME, pp. 13–22, 2020.
https://doi.org/10.1007/978-3-030-31343-2_2

human intervention (e.g., target selection and engagement) [2]. There is a specific meaning based on tactical, technical parameters in weapon efficiency. Májek and Kacer, [3] define operationally essential elements as the weight of the operation charge, weapon range, projectile maneuvering ability, and system mobility. The United States Navy implemented a program in 1960 to work with dolphins and sea lions to help with defense, mine detection, and design of new submarines and new underwater weapons. "Bottlenose dolphins are better than any machine as far as detecting mines," says Paul Nachtigall, head of the marine mammal research program at the University of Hawaii. These animals can detect mines much faster than a machine might be able to [4, 5]. Every country has to protect its security against threats that may be directed by other countries. These threats can be made using economic, social, political, ideological, or military means. Most of these tools must be supported by a military force to create a severe threat [6]. This feature of the threat assessment requires both the countries with potential risks and their armed forces to be assessed in terms of qualitative and quantitative factors such as firepower, mobility, materials, tools, equipment, and human quality. The determination of the firepower potential and the efficiency of firearms used in the determination of firepower potential constitute a crucial part of this evaluation. In general, the methods developed for the detection of weapon activities are based on the approach of combining the selection and weighting of several weapon characteristics in a formulation for this purpose. Weapon Effectiveness Index and A Technique for Assessing Comparative Force Modernization (TASCFORM) methods are examples of this approach [7]. Weapon activity values showing the degree of effect of weapons to unity firepower and following combat results can be used in different stages of evaluations such as weapon comparison, power comparison, force potential determination, forward force planning. These evaluations refer to the various dimensions of the studies to determine combat power [1]. The Concept of Weapon Activity is based on Combat Force Analysis, Weapon Activity Index, Comparative Force Modernization Assessment Technique, and Operational Lethality Index. The Weapon Activity Index method emerged at the end of a study initiated by the US Army in the 1970s to compare the combat effectiveness of various units. This method is based on the approach of evaluating weapons activities by taking expert opinions with the help of the Delphi method. The part of the technique related to the evaluation of the weapon systems is based on the comparison of performance criteria of a selected base system such as ammunition weight, range, speed, maneuverability, mobility, target detection, target tracking of the weapon and weapon systems with the performance criteria of the selected mode [1, 8].

An autonomous weapon system is a weapon system that can select (search for, detect, identify, track or choose) and attack (use force against, neutralize, damage or destroy) targets without human intervention [9].

Artificial neural networks were developed for use as a potential 'information barrier' technology in the verification of arms control treaty accountable items [10].

2 Literature Review

Weapon activity values showing the degree of effect of weapons to unity firepower and following combat results can be used in different stages of evaluations such as weapon comparison, power comparison, force potential determination, forward force planning. These evaluations refer to the various dimensions of the studies to determine combat power. To better understand the use of weapons efficiency and weapon activity, combat power analysis, and studies in this area should be mentioned. The history of analytical studies to assess the outcome of the battle and to assess the success of the troops in a possible action is not very old. A game is known as the "war game (Kriegspiel)," which started to be played in Germany in the 19th century, can be seen as the first of such studies. The most significant contribution to the studies in this area was made by Lanchester Equations in 1916 [11]. The Weapon Activity Index method emerged at the end of a study initiated by the US Army in the 1970s to compare the combat effectiveness of various units. This method is based on the approach of evaluating weapons activities by taking expert opinions with the help of the Delphi method. The Comparative Force Modernization Assessment Method was developed in 1978 by the Analytic Sciences Corporation (TASC) upon the request of the US Department of Defense to develop a method for quantitative determination of force modernization degree. The technique aims to reveal the force potential of military systems based on the measurable characteristics of the military systems. The method of Operation Lethal Index was developed by Colonel Trevor Nevitt DUPUY, retired from the US army. The method is based on several formulas and principles for use in the calculation of weapon and union activity values by utilizing data from past battles. A weapon system is comprised of four actions, searching for a target, detecting that target, deciding to engage the target, and engaging the target. Developed by Colonel John Boyd, the OODA (**O**bserve, **O**rient, **D**ecide, **A**ct) loop is a model for competitive decision making, comprised of those four actions that define a weapon system: observe, orient, decide, and act. In human operated weapons systems, a human operator completes all of these actions. In a weapon system with autonomy, the human operator relinquishes control over some part of the loop.

2.1 Artificial Intelligence

Artificial Intelligence (AI) is the general name for the modeling of systems that behave similarly to the behavior and behavior of all human beings in nature. Artificial Intelligence can be considered as an interdisciplinary concept. These disciplines are biology, psychology, sociology, computer, mathematics, medicine, and their sub-branches [12–14]. According to the general view, Artificial Intelligence is a system that acts like a human. However, considering that other living things are taken as models, it can be defined as the modeling of behaviors in nature in general. This behavior is six different ways when people are treated like human beings. These; control of a system (robot control, traffic control, etc.), decision making (warfare strategies, diagnosis in medicine, law, expert systems), estimation (finance, traffic, fishing stock review, image process), learning and optimization (the most reasonable solution, shortest path, optimum process). The purpose of artificial intelligence; to provide intelligent systems to help people make decisions and achieve their goals. Along with the development of technology, artificial intelligence appears in the sub-areas of problem-solving, expert systems, robotics, vision, machine learning, and neural networks (Fig. 1).

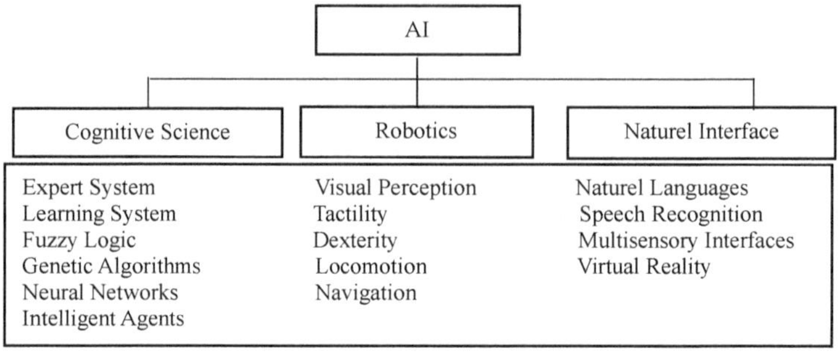

Fig. 1. Overview of artificial intelligence [15]

Considering the potentials of artificial intelligence, Phulera et al. [16] stated that smarter artificial intelligence could replace human jobs, freeing people for other works by automating manufacturing and transportation. Self-writing, self-modifying, and learning software's can help relief programmers of the difficult tasks of specifying functions of the different programs. Artificial intelligence will be used as cheap labor; then there must be an increment in profits for the corporation. Artificial intelligence can make deployment more comfortable and less resource intensive. Compared to traditional programming techniques, expert-system approaches provide the added flexibility (and hence easier modifiability) with the ability to model rules as data rather than as a code. In situations where an organization's IT department is overwhelmed by a software development backlog, rule-engines, by facilitating turnaround, provide a means that can allow organizations to adapt more readily to changing needs.

Furthermore, Phulera et al. mentioned that rapid advances in AI could lead to massive structural unemployment, unpredictable and unseen impacts of new features. An expert system or rule-based approach is not optimal for all problems, and considerable knowledge is required to apply in any of the methods. The ease of rule creation and rule modification can be double-edged. A system can be sabotaged by a non-knowledgeable user who can easily add worthless rules or rules that conflict with existing ones. Reasons for the failure of many systems include the absence of (or neglect to employ diligently) facilities for system audit, detection of possible conflict, and rule lifecycle management (e.g., version control, or through testing before deployment). The problems to be addressed here are as much technological as organizational.

2.2 Autonomous Technology

A weapon system that, once activated, can select and engage targets without further intervention by a human operator [17]. Examples include some 'loitering' munitions that, once launched, search for and attack their intended targets (e.g., radar installations) over a specified area and without any further human intervention, or weapon systems that autonomously use electronic 'jamming' to disrupt communications [18]. State of the art in computing machinery is unlikely to meet all of these requirements within the foreseeable future. Although computers are better at some tasks than humans, humans are better at some tasks than computers (Table 1). Military control and humanitarian impacts are best served by playing to the strengths of both.

Table 1. The differing skills of computers and humans

Computers	Humans
Calculating numbers	Deliberative reasoning
Searching large data sets	Perceiving patterns
Responding quickly to control tasks	Meta-cognition (thinking about thinking)
Simultaneous repetitive routine tasks	Reasoning inductively
Carrying out multiple complex tasks	Applying diverse experience to novel tasks
Sorting data	Exercising meaningful judgment

Computers are also susceptible to some potential problems that make them unpredictable: human error, human-machine interaction failures, malfunctions, communications degradation, software coding errors, enemy cyber-attacks, infiltration into the industrial supply chain, jamming, spoofing, decoys, and other enemy countermeasures or actions. Again, a human in the control loop can determine that a system is displaying aberrant behavior and take appropriate action. There are also other uses of technology that can be seen as being against human dignity. Heyns [19] looks beyond warfare and raises concerns about where we allow machines to make decisions. It was stated that allowing machines to make non-lethal decisions that affect humans is also against human dignity, writing. The notion of 'meaningful human control' should be developed as a guiding principle not only for the use of autonomous weapons but for artificial intelligence in general; not merely focussing on isolated applications of such technologies but the role of technology as such in our future. Allowing technology not only to supplement but indeed to replace human decision-making will undermine the very reason why life is valuable in the first place [19]. Human intelligence generally follows a sequence known as the perception–cognition–action information processing loop, in that individuals perceive something in the world around them, think about what to do, and then, once they have weighed up the options, decide to act. AI is programmed to do something similar, in that a computer senses the world around it, and then processes the incoming information through optimization and verification algorithms, with a choice of action made in a fashion similar to that of humans. Figure 2 illustrates how an autonomous system embedded with AI 'thinks' and makes decisions in this way [20]. While there are many parallels between human intelligence and AI, there are stark differences too. Every autonomous system that interacts in a dynamic environment must construct a world model and continually update that model (Fig. 1). This means that the world must be perceived (or sensed through cameras, microphones and/or tactile sensors) and then reconstructed in such a way that the computer 'brain' has an effective and updated model of the world it is in before it can make decisions. The fidelity of the world model and the timeliness of its updates are the keys to an effective autonomous system.

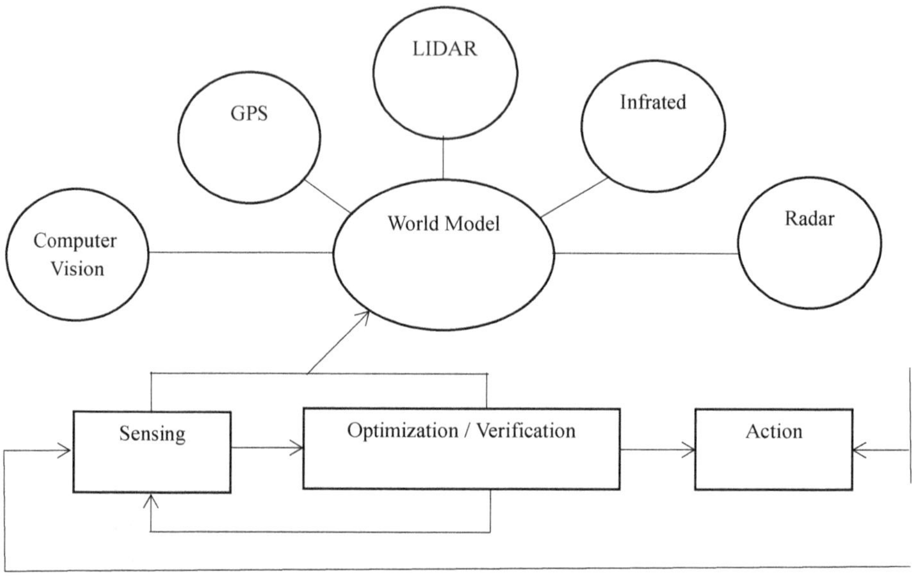

Fig. 2. How AI of an autonomous system Works [20]

3 Lethal Autonomous Weapons Systems

Lethal autonomous weapons systems (LAWS) are weapons systems that apply deadly force independently, with freedom from human control. AI has many subfields, including machine learning, data mining, speech recognition, and image processing. Policymakers frequently refer to any of the subfields of AI as AI. By grouping these various terms under the umbrella of AI, the term is commonly misused by technical experts, policymakers, and the general public [21]. The general idea is that a LAWS, once activated, would, with the help of sensors and computationally intense algorithms, identify, search, select, and attack targets without further human intervention. Whether the human being can still overpower or veto an autonomous weapon's 'decision' for it to be called a LAWS, is also debated. However, military operational necessity precisely seems to require weapons systems that can function once human communication links break down. Furthermore, state of the art research on AI is currently creating software which can 'learn' entirely on its own and even 'learn' to 'learn' on its own. Hence, (precursor) technologies for creating fully 'human-out-of-the-loop' weapons systems already exist [22].

4 Main Challenges for Using AI

Following the assumption of the calculability of the world, usually called digitization and robotization. It is easily overlooked that a human, and thus its actions, is anything but predictable. This is one of the most significant challenges of human-machine interaction, whether cooperative (i.e., the machine provides information to its operator in combat) or uncooperative (i.e., the machine remains intransparent to the opposing side). The artificial Intelligence weapon system, especially learning systems, require a large amount of sensor and training data to function correctly. This creates both quantitative and qualitative challenges. First of all, it is generally questionable whether more sensor technology and thus more data, actually enables more consistent and predictable machine behavior or better human decisions [23]. As Artificial Intelligence, machine learning and deep learning evolves further and moves from concept to commercialization, the rapid acceleration in computing power, memory, big data, and high-speed communication is not only creating innovation, investment and application frenzy. When countries individually and collectively accelerate their efforts to gain a competitive advantage in science and technology, the further weaponization of AI is inevitable [24].

A new report written by a former Pentagon official who helped establish United States policy on autonomous weapons argues that such weapons could be uncontrollable in real-world environments where they are subject to design failure as well as hacking, spoofing, and manipulation by adversaries. And A new report says eight new F-22 fighter jets like these experienced total computer failure when crossing the international date line [25].

The more powerful technology becomes, the more it magnifies design errors and human failures. As society contemplates deploying AI in self-driving cars, in surgical robots, in police activities, in managing critical infrastructure, and in weapon systems, it is creating situations in which errors committed by human users or errors in the software could have catastrophic consequences [26].

5 Restriction (Application Dimension)

Autonomous systems can reduce or eliminate the need for human involvement in some tasks. They are not close to reaching the goal of human-like artificial general intelligence. They are, however, continually becoming more powerful because of developments in machine learning and natural language processing and advances in materials science, networking, energy-storage, and hardware capabilities.

Every expression of daily life, either civil or professional or familial or personal, will be diminished by the iron grip of Artificial Intelligence on the fundamental realities of interpersonal communications. Despite concerns, artificial intelligence technology continues at a dizzying pace. Governments also introduce artificial intelligence strategies. The fact that Russian President Putin has made an ambitious statement that he will control the world that controls the artificial intelligence technology reveals this.

Johnson and Axinn [27] also state that 'To give a programmed machine the ability to 'decide' to kill a human is to abandon the concept of human dignity' since to do so is to treat a rational being as an object. They insist that a machine cannot be moral, but can only follow the values of its programmers. Concluding that arguments based on human dignity are not the best way to argue against AWS is not at all the same as saying that such weapons are acceptable. As explained earlier, there are strong reasons to oppose their use and to argue for a ban. Eventually, the machines will gradually push the humans out of the loop. First, they stand in supervisory roles and finally they'll end up as "killswitch operators" that monitor these autonomous weapons. Machines can be much faster than humans. The act of killing an enemy is based on reflexes, and if soldiers realize that these types of tools can outperform them, they'll eventually come to trust and rely on them. As these systems advance, the ones that rely less on human supervision will dominate. Instead, humans will be given other roles, such as analyzing the behavior of these systems and concentrating on other strategic areas.

There is also rising pressure and incentives to improve these kinds of systems too. All countries are equally pressured to gain superiority, and as such, the inevitability of fully automated, always on orders should be seriously considered in all aspects of AI integration. The best way to avoid catastrophe is by supporting regulation and prohibition of LAWS like chemical and biological weapons, for weaponized AI.

6 Results and Limitations

The decision-making process will be shortened by artificial intelligence-based command control systems, real-time enemy activities can be monitored, and the possible patterns of the enemy can be followed dynamically. In this direction, the decision makers will be able to revise their plans momentarily and will be able to change the speech structures and force structures of the military elements. Weapon systems can be kept under observation by artificial intelligence algorithms against cyber-attacks, and algorithms which can simulate cyber-attacks can detect weak points of system, and immediate corrective intervention can be found.

Considering all the issues discussed above, it can be concluded that the use of artificial intelligence in the military field is necessary. The most crucial factor in deciding is whether the use of artificial intelligence should be performed autonomously or only as a support system. Once a certain maturity has been reached, it will not be reflected in the fact that military operations should be fully autonomous systems. At this point, the target can be applied for alternatives such as target priority determination, target engagement, and even target change according to the current situation, starting from the detection, follow-up, and identification.

Improving collaboration across borders and stakeholder groups; developing policies to assure that development of artificial intelligence will be directed at augmenting humans and the common good; and shifting the priorities of economic, political and education systems to empower individuals to stay ahead in the race with the robots are worrisome potential future [28].

7 Conclusions and Further Steps

Depending on today's weapon requirements, it is required a very complex decision making within a very short time. Additionally, various situations might not be defined in advance. From this point of view, a support for alternative control and evaluation isa very important task for weapon efficency. On the other hand, depending on the situation there can be available big data for decision making, which can not be use directly for manking decisions. Therefore, there is an absolut need for AI support.

While many modern weapons have units for target searching, tracking and seeking without human input, the full-fledged use of autonomous artificial intelligence systems in warfare has prompted concerns among AI and robotics researchers. Alternatively, there are available many negative ideas as well. One special approach is the autonomy of machines.

In summary, as with previous technological advances, AI technology increases the risk that failures in human organizations and actions will be magnified by the technology with devastating consequences. To avoid such catastrophic failures, the combined human and AI organization must achieve high reliability. Work on high-reliability organizations suggests important directions for both technological development and policymaking. It is critical that we fund and pursue these research directions immediately and that we only deploy AI technology in organizations that maintain high reliability.

The next step for this study is a analysis about different countries using AI in the area of weapon efficency.

References

1. Çakır, A., Başdaş, H.E., Leblebicioğlu, K.: Detection of weapon effectiveness values by fuzzy logic method. J. Def. Sci. **5**(1), 22–49 (2004)
2. Scharre and Horowitz (2015)
3. Májek, V., Kacer, J.: Air defence efficiency, according to NATO. Cybern. Lett. (2006). http://www.cybletter.com/index.php?s=2006-II
4. Lee, J.J.: Military whales and dolphins: what do they do and who uses them? Natl. Geogr. Newsl. (2019). www.nationalgeographic.com/news/2019/5/140328-navy-dolphin-sea-lion-combat-ocean-animal-science/
5. Leighton, T., White, P.: Dolphin-inspired target detection for sonar and radar. Arch. Acoust. **39**(3), 319–332 (2014)
6. Jaiswal, N.K.: Military Operations Research: Quantitative Decision Making, vol. 5. Springer, New York (2012)
7. Nor, B., Wand, K.: Weapon scoring using a multilayer perceptron network. J. Battlef. Technol. **IV**(1), 7 (2001)
8. Regan, M.J., Downey, F.M.: A Technique for Assessing Comparative Force Modernization, 6th edn. The Analytic Sciences Corporation (TASC), Virginia (1993)
9. Davison, N.: A legal perspective: autonomous weapon systems under international humanitarian law. Perspect. Lethal Auton. Weapon Syst. **2018**, 5–18 (2017)

10. Aitkenhead, M.J., Owen, M., Chambers, D.M.: Use of artificial neural networks in measuring characteristics of shielded plutonium for arms control. J. Anal. At. Spectrom. **27**(3), 432–439 (2012)
11. Fedyszyn, R., Martel, W.C., Hitchcock, N.E., Hoyler, H.M., Murray, C., Neff, A.J., Perroti, C.S., Sipe, C.R., Sullivan, R.E., Wall, A.N.: Executive Decision Making, Newport, U.S Naval War College (2002)
12. Benzer, S., Benzer, R.: Enterprise business management. In: Erkollar, A. (ed.) Alternative Approaches to Traditional Methods for Growth Parameters of Fisheries Industry: Artificial Neural Networks, Tectum Verlag Germany (2019)
13. Wong, A.S.W., Li, Y., Yeung, P.K.W., Lee, P.W.H.: Neural network predictions of human psychological perceptions of clothing sensory comfort. Text. Res. J. **73**(1), 31–37 (2003)
14. Young, I.A., Krishnamurthy, R., Manipatruni, S., Chen, G.K., Mathuriya, A., Sharma, A., Sumbul, H.E.: U.S. Patent Application No. 16/147,143 (2019)
15. Borana, J.: Applications of artificial intelligence & associated technologies. Science. In: Proceeding of International Conference on Emerging Technologies in Engineering, Biomedical, Management and Science [ETEBMS-2016], 5–6 March 2016
16. Phulera, K., Singh, H., Bhatt, A.: Analytical study on artificial intelligence techniques to achieve expert systems. Int. J. Emerg. Technol. (Spec. Issue NCETST-2017) **8**(1), 137–140 (2017)
17. US Department of Defense: Autonomy in Weapon Systems, Directive 3000.09, 21 November 2012, Glossary, Part II Definitions (2012)
18. ICRC Report, Autonomous weapon systems technical, military, legal, and humanitarian aspects (2014). https://www.icrc.org
19. Heyns, C.: Autonomous weapons in armed conflict and the right to a dignified life: an African perspective. S. Afr. J. Hum. Rights **33**(1), 46–71 (2017)
20. Cummings, M.L.: Artificial Intelligence and the Future of Warfare. International Security Department and US and the Americas Programme (2017). https://www.chathamhouse.org
21. Sagramsingh, R.: Lethal Autonomous Weapons Systems: Artificial Intelligence and Autonomy (2018). http://wise-intern.org
22. Surber, R.: Artificial Intelligence: Autonomous Technology (AT), Lethal Autonomous Weapons Systems (LAWS), and Peace Time Threats ICT4 Peace Foundation and the Zurich Hub for Ethics and Technology (ZHET) (2018)
23. Dahlmann, A., Dickow, M.: Preventive regulation of autonomous weapon systems: need for action by Germany at various levels. SWP Research, Berlin. Stiftung Wissenschaft und Politik German Institute for International and Security Affairs (2019)
24. Pandya, J.: The Weaponization of Artificial Intelligence. Forbes (2019)
25. Markoff, J.: Report Cites Dangers of Autonomous Weapons. N. Y. Times. (2016)
26. Dietterich, T.G.: Robust artificial intelligence and healthy human organizations. Front. Comput. Sci. **13**(1), 1–3 (2019)
27. Johnson, A.M., Axinn, S.: The morality of autonomous robots. J. Mil. Ethics **12**(2), 129–141 (2013)
28. Anderson, J., Rainie, L., Luchsinger, A.: Solutions to address AI's anticipated negative impacts. Pew Research Center (2018)

Big Data and Analytics

Industry 4.0 Score Prediction of Turkish SMEs via Data Classification

Uğurcan Dündar[1(✉)], Fadime Üney-Yüksektepe[1], Zeynep Gergin[1], Doğan Aybars İlhan[1], Mehmet Güneş Gençyılmaz[2], and Ali İhsan Çavdarlı[3]

[1] Department of Industrial Engineering, İstanbul Kültür University, İstanbul, Turkey
{u.dundar, f.yuksektepe, z.gergin, a.ilhan}@iku.edu.tr
[2] Department of Industrial Engineering, İstanbul Aydın University, İstanbul, Turkey
gunes.gencyilmaz@aydin.edu.tr
[3] KOSGEB Small and Medium Sized Enterprises Development Organization, Ankara, Turkey
ali.cavdarli@kosgeb.gov.tr

Abstract. Todays most important industrial concept for companies is to be able to incorporate new technologies into their planning and monitoring systems. These technologies include sensor systems, cloud technologies, automation, predictive/pre-emptive maintenance, 3D printing, smart warehouses, etc. In order to survive in today's competitive business environment, this revolution is very important for Small and Medium Sized Enterprises (SMEs) in Turkey. Hence, in order to analyze the relationship between new technologies and Industry 4.0 score of Turkish SMEs, a data mining study is performed in this research. A survey is performed to gather information on the awareness, readiness and interests of SMEs in new technologies in addition to their Industry 4.0 scores. Aim of this study is to predict whether Industry 4.0 scores of SMEs is low or high by using their technology utilizations. As this is a typical data classification problem, many different data classification methods are applied to determine the best alternative by using WEKA software. Among them, the highest prediction accuracy is 69.11%, obtained by Support Vector Machines. Thus, a Turkish SME's Industry 4.0 score level could be predicted by just investigation of its new technology usage. Therefore, Turkish government could use this approach to determine the current situation of a SME. Moreover, government could determine their supporting programs based the technology usage levels of the SMEs.

Keywords: Industry 4.0 · Data mining · Data classification · Support vector machines · Turkish industry

N. M. Durakbasa and M. G. Gençyılmaz (Eds.): ISPR 2019, LNME, pp. 25–35, 2020.
https://doi.org/10.1007/978-3-030-31343-2_3

1 Introduction

The manufacturing sector is one of the biggest elements in the industry that has a huge impact on economy and focuses on processing raw material. The scope of the manufacturing companies alters with the revolutions in industry. Lately, manufacturing companies should also deal with noteworthy amounts of data processing, geography processing and availability [1]. In order to last their existence in the competition, companies should keep evolving consciously and continuously as a response to needs of market, competition and innovation. These evolutions can be achieved with the information of industry trends and their opportunities.

Industry trends are framed by revolutions occurred. There are three major revolutions that industry had up to last decade. The first major transformation was the usage of steam-powered mechanization at late 1700s. The second one began with electrification of the mechanics and mass production in early 1900s. Then, computers are invented and became widespread also in industry. This situation led the birth of the computer integrated manufacturing concept which is called the third industrial revolution at the end of 1970s. Nowadays, the industry is going through a new transformation which is called Industry 4.0. Even though, there are various descriptions, it roughly, can be explained as establishing a system that enables the interaction between all resources in the environment [2–4]. It includes the following constituents that are called drivers of Industry 4.0 [5]: Big Data and Analytics, Autonomous Robots, Simulation, Horizontal and Vertical System Integration, Internet of Things, Cyber Security, Cloud Systems, Additive Manufacturing and Augmented Reality.

Small and Medium Sized Enterprises (SMEs) are the companies that have less than 250 employees and less than 40 million Turkish liras endorsement or balance [6]. In Turkey, SMEs constitutes the 99.8% of all enterprises [7] and has a huge impact on the country's economy. As a country that is classified with its developing economy, it is important to keep up with the Industry 4.0 for Turkey especially in SMEs since they are the keystone of the Turkish economy.

Gergin et al. [8] surveyed the Industry 4.0 readiness of Turkish SMEs in Marmara Region and scored them. In the survey, the companies are evaluated with a scale of 1.0 to 4.0, and it is applied to 193 SMEs, and the overall score of these SMEs are calculated as 2.69.

With the advancements in technology, the data is created by humanity increasing dramatically. It is estimated that in 2020, each person will produce 1.7 MB data each second [9]. The data generated by industry has a parallel increasing rate with the humanity. So, the Big Data concepts arises which means collection of enormous data and not being able to analyze it with commonly used tools. In Big Data applications the biggest challenge is to gather a valuable information from the bulk data in order to use in decisions to be made [10]. The process of deducing valuable, useful and previously unknown information from data is called data mining [11].

In a strategical manner, data mining applications can be divided into two main categories as supervised learning and unsupervised learning. In supervised learning, training data is analyzed with the knowledge of the instances' classes and attributes in order to create the model of the data which will be used in prediction of the classes of the future instances [12]. However, in the unsupervised learning, instances of the data are grouped according to their similarity based on some context. While Clustering and

Association Analysis are unsupervised learning strategies, Classification and Regression are supervised learning strategies [13].

In this study, we increase the number of instances gathered by [8], with distributing the survey to SMEs that are located in the four biggest industrial cities in Turkey not only in Marmara Region; Istanbul, Bursa, Izmir and Ankara. In total, 513 SMEs took part in the survey from different manufacturing sectors with a number of 75, 147, 226 and 65 from Ankara, Bursa, Istanbul and Izmir, respectively. Industry 4.0 scorecards are prepared according to [8] on each SME while gathering the information of technology instruments that are already being used or planning to be used in the company. Sum of all, utilization information of the technology instruments is used to predict Industry 4.0 level of the SMEs in the scale of Low or High, which is based on the average (2.69) of the Industry 4.0 scores of 513 SMEs that are participated in the survey. Since this is a classical classification problem, several algorithms applied and results are recorded. Results of this study can be used to guide governmental actions to be taken towards SMEs with a quick evaluation of their technology instruments. For further information about general results of the survey [14] can be seen.

The organization of the rest of the paper as follows. In Sect. 2, the theoretical foundation of employed methods is defined briefly. In Sect. 3, the analyzed data is explained step by step as procedure flows from preprocessing to algorithm application and finally to results. At the end, the conclusion and further study opportunities are discussed in Sect. 4.

2 Methodology

Data Mining is the process of obtaining useful information from bulk data by using various computer based learning algorithms. Aim is to discover common and significant characteristics of the data [15]. Due to improved data storage capacities and computer technologies, companies are storing large amounts of data in order to be competitive. Hence, data mining studies become popular while dealing with these huge databases. There exist many data mining studies applied in many different areas such as banking, marketing, stock exchange, insurance, telecommunication, healthcare and quality control [16].

Data mining strategies are grouped into two as supervised and unsupervised learning. In supervised learning, goal is to develop a methodology to predict the class label of an unknown instance. On the other hand, in unsupervised learning, instances with similar behaviour are grouped together. Data Classification and Regression are considered as supervised learning methodologies whereas Clustering and Association Analysis are unsupervised learning methodologies [17]. As data classification is the focus of this study, this topic will be explained in detail in the following subsections.

2.1 Data Classification

In data classification, data is composed of a number of input attributes and a class attribute. Aim of the classification is to find the relationship between the input attributes and the class attribute by proposing an algorithm to predict class values of new instances.

For example, most telecommunication companies are using data classification to predict their loyal and disloyal customers. Past data which includes information of both loyal and disloyal customers are used to develop a method in order to predict a new customer's loyalty. By the help of the proposed data classification method, companies could determine the disloyal customers before they leave. Hence, those customers may be convinced to stay by offering some promotions. This is a popular data classification problem known as churn prediction [18]. Similar classification problems exist in e-commerce, banking, insurance, drug discovery, stock exchange and sports.

A number of different classification algorithms exist in the literature. The popular and widely used ones are explained in the following subsections.

(1) *Decision Tree Algorithms:* In decision tree algorithms, the relationship between input attributes and class attribute is obtained and represented as a tree composed of nodes and branches. Each node is an input attribute and branches are the criteria on that attribute. Depending on the satisfied criteria, the next node will be evaluated. As leaf nodes has class labels, when the user reaches to a leaf node, then the class assignment is obtained [15]. There exist many different decision tree algorithms such as C4.5, NewID, ID3 and CART. They have similar working principle with different attribute selection and branching strategies. As decision tree is easy to understand and easy to apply, they are among the most popular approaches preferred in business life applications [19].

(2) *Bayesian Algorithms:* Main idea of these algorithms depends on the Bayes theorem. Probabilities of each attribute value is calculated by using occurrence frequencies of them for each class. By the help of the independent assumption of attributes, class probabilities of a new instance are calculated by using Bayes theorem. The class with the highest probability is the predicted class for that new instance [17]. Naïve Bayes Algorithm is the popular Bayesian algorithm in literature. Independence assumption should be discussed and criticized before using these classification algorithms.

(3) *Rule Based Algorithms:* Classification rules are constructed in these algorithms. On the left hand side of the rule, one or more attribute conditions exist whereas the class is on the right hand side of the rule. By using these simple rules, the instances, which satisfy the left hand side criteria, are assigned to class on the right hand side. RIPPER, CN2, Holte's 1R are some example of rule based algorithms. In business life, they are preferred due to their simplicity. On the other hand, the rules obtained should be mutually exclusive and exhaustive to get rid of some application problems [15].

(4) *Distance Based Algorithms:* K-Nearest Neighbour is the most popular distance based classification algorithm. The main idea is to calculate similarities of the test instance to the training instances by using a distance measure. The instances with the smallest distance to the test instance are the closest neighbours. Among them, k closest neighbours' class labels are compared and majority class is taken as the predicted class of the test instance. k should be an odd number in order to get rid of the equal frequency class situation. Distance measure used and number of k affects the performance of these algorithms. Moreover, as for each test instance distances should be calculated one by one, the computational time required for

these algorithms will increase with an increase in the number of training instances and number of attributes [17].

(5) *Neural Networks:* These classification algorithms are trying to simulate the neurons in our brains. Hence, the network consists of some nodes and arc carrying some information. If an information is carried through a node, that node changes the information carried. By working on the training set instances, each node determines its best coefficient. Moreover, a threshold value is determined for the final value of the test instance. Thus, a test instance is processed throughout the neural network and a final value is obtained. Comparison of the threshold value and the final value affects the class prediction of the test instance. Although these algorithms have considerably high accuracy performance, learning the training instances is time consuming and the number of nodes in the network is a critical decision [15].

(6) *Support Vector Machines:* Support Vector Machines (SVM) are popular classification algorithms. They are trying to find a hyperplane to separate the classes from each other. The hyperplane that maximized the distance to the closest instances of each class is selected. Hence, if a test instance arrives, its location with respect to the constructed hyperplane affect its predicted class. If the number of classes is more than two, then for each class a hyperplane is obtained by considering the remaining classes together as one class. For the cases without a linear separability, data is transformed to a higher dimension and the procedure is applied. Although they have outstanding high accuracies for two class problems, their efficiency is affected in the multi-class data classification problems [20].

2.2 Performance Evaluation

In order to show the results of a classification algorithm, confusion matrix is used. Figure 1 shows an example of a confusion matrix for two class problem. By using the proposed data classification algorithm, new instances' class labels are predicted and compared with their actual class labels. Hence, the number of correctly classified (TP and TN) and misclassified instances (FP and FN) are counted and inputted to the confusion matrix. As a result, it is easy to analyse the prediction performance of the proposed methodology.

Confusion Matrix		Actual Class	
		Class 1	Class 2
Predicted Class	Class 1	True Positive (TP)	False Positive (FP)
	Class 2	False Negative (FN)	True Negative (TN)

Fig. 1. Example of a confusion matrix

In addition, accuracy and error rate are used to compare the performances of different classification algorithms. Accuracy is the rate of correct prediction and error rate is the rate of misclassification. Equations 1 and 2 gives the accuracy and error rate calculation for the above confusion matrix, respectively. As it can be observed, Error Rate and Accuracy is sum up to 1 and they are correlated [15].

$$Accuracy = \frac{TP + TN}{TP + FP + FN + TN} \tag{1}$$

$$Error\,Rate = \frac{FP + FN}{TP + FP + FN + TN} \tag{2}$$

2.3 Estimation Methodologies

In data classification, generally, a training set is used to develop the model and a test data set is used to test the performance of the model. In order to obtain this data sets, data is randomly partitioned. As the instances in training and test sets affect the proposed model and overall accuracy, this approach is considered as bias. Hence, a more improved estimation methodology, 10-fold cross validation, is popular nowadays. In this approach, data set is randomly divided into 10 mutually exclusive subsets with equal number of instances. The classification model is studied 10 times. Each time, one of the subsets will be tested by using the remaining 9 subsets as training set. Average accuracy of these 10 runs is calculated as the overall accuracy of the model. By the help of these 10 different runs, the bias effect is minimized [17].

3 Implementation and Results

In this section, implementation of classification algorithms and data frames is explained and results are discussed in detail. The steps of the methodology are given in Fig. 2.

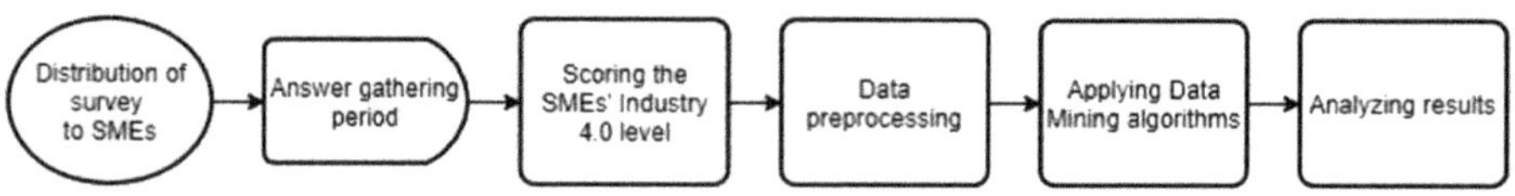

Fig. 2. Methodology steps

3.1 Data Description and Preprocessing

As it is stated earlier in this paper, the data is gathered via survey from SMEs in four industrial cities of Turkey. In total 513 SMEs participated in survey and overall Industry 4.0 score of them is calculated as 2.69. There are 28 questions in the survey, some of them are used to calculate the industry scorecard, some are used to investigate the understanding of Industry 4.0 in SMEs and some are used to investigate the usage of technology instruments. In this paper, those about technology utilization is considered with the scores of each company.

Technology instruments that are questioned in the survey are as follows: (i) 3D Printing, (ii) Sensor Systems, (iii) Artificial Intelligence, (iv) Cloud Technologies, (v) Internet of Thigs, (vi) Additive Manufacturing, (vii) Advanced Robot Technologies, (viii) Automation Systems, (ix) Modelling and Simulation, (x) Virtual Education, (xi) e-Performance Reports, (xii) Big Data, (xiii) Augmented Reality, (xiv) Cyber Security, (xv) Integrated Product Development, (xvi) Predictive Maintenance, (xvii) Productive Maintenance, (xviii) Smart Stocks, and (xix) Mobile Systems.

Each SME that is participated to survey is considered as an instance; while technology instruments are attributes, and class values of instances are their Industry 4.0 level, which is either low or high. Each SME answered the question with choosing one of following choices: "no knowledge about instrument", "not using with knowledge about instrument", "using", and "planned". Hence, there exist 19 input attributes and one class attribute in the data set.

In the data preprocessing phase, the instances that have more than three blank attributes are erased from the data and the ones that has less than or equal to three blank attributes are filled with a value of NaN to represent the empty values. After these procedures there are, 437 instances left and only 90 "NaN" exist. The number of attribute values are decreased to three with converting "no knowledge about instrument" and "not using with knowledge about instrument" to "not using" only. A sample of the data is given in Fig. 3. First column represents the row number and is not used in data mining analysis. On the other hand, Other columns except last column are input attributes. Moreover, the last attribute is the class attribute that will be predicted. In the data set, 219 High class and 218 Low class instances exists.

ID	3D Printing	Sensor Systems	Artificial Intelligence	Cloud Technologies	Internet of Things	Additive Manufacturing	Advanced Robot Technologies	Automation Systems	Modelling and Simulation	Virtual Education	e-Performance Reports	Big Data	Augmented Reality	Cyber Security	Integrated Product Development	Predictive Maintenance	Productive Maintenance	Smart Stocks	Mobile Systems	Level
1	Not Using	Not Using	Not Using	Not Using	Not Using	Not Using	Not Using	Using	Not Using	Not Using	Not Using	Not Using	Not Using	Using	Using	Not Using	Using	Not Using	Using	High
2	Not Using	Not Using	Not Using	Not Using	NaN	Not Using	Not Using	Not Using	Not Using	NaN	Not Using	Not Using	Not Using	Not Using	Not Using	Not Using	Not Using	Not Using	Not Using	Low
.	.	.	.	.	.	.	.	.	.	.	.	.	.	.	.	.	.	.	.	.
.	.	.	.	.	.	.	.	.	.	.	.	.	.	.	.	.	.	.	.	.
.	.	.	.	.	.	.	.	.	.	.	.	.	.	.	.	.	.	.	.	.
437	Not Using	Using	Not Using	Using	Using	Not Using	Not Using	Using	Not Using	Not Using	Not Using	Using	Not Using	Using	Using	Using	Using	Planning	Using	High

Fig. 3. A sample of the data

3.2 Implementation of Classification Algorithms

After the preprocessing phase, the data is studied via WEKA 3.8.2 Data Mining Software. As given in Methodology Section, Decision Tree Bayesian, Rule Based, Distance Based algorithms, Neural Networks and Support Vector Machines are applied to the data set. For each algorithm, a number of different methods are studied. It is observed that the computational time required for each algorithm run is neglectable. So, the selection of the best algorithm is made based on accuracy. The accuracy for each algorithm from each category is given in Table 1.

In each category the algorithms with the best performances are Naïve Bayes, SMO, Kstar, PART, and HoefdingTree with 68.88%, 69.11%, 64.07%, 62.24% and 68.88%, respectively. Error rate values are 100% minus the accuracy values. Hence, the algorithm with the highest accuracy will have the lowest error rate.

Table 1. Accuracy and error rate of algorithms

	Algorithm	Accuracy	Error rate
Bayesian algorithms	BayesNet	68.42%	31.58%
	NaiveBayes	68.88%	31.12%
	NaiveBayesMultinomialText	50.11%	49.89%
	NaiveBayesUpdateable	68.88%	31.12%
Function based algorithms	Logistic	66.59%	33.41%
	MultilayerPerceptron (Neural Networks)	59.95%	40.05%
	SGD	65.90%	34.10%
	SGDText	50.11%	49.89%
	SimpleLogistic	65.90%	34.10%
	SMO (Support Vector Machines)	69.11%	30.89%
	Voted Perceptron	67.51%	32.49%
Distance based algorithms	Ib1	59.95%	40.05%
	Kstar	64.07%	35.93%
	LWL	57.12%	42.88%
Rule based algorithms	DecisionTable	61.56%	38.44%
	Jrip	61.78%	38.22%
	OneR	61.33%	38.67%
	PART	62.24%	37.76%
	ZeroR	50.11%	49.89%
Decision tree algorithms	DecisionStump	57.89%	42.11%
	HoefdingTree	68.88%	31.12%
	J48	64.53%	35.47%
	LMT	65.90%	34.10%
	RandomForest	64.07%	35.93%
	RandomTree	59.73%	40.27%
	REPTree	62.70%	37.30%

The highest accuracy is achieved by SMO which is the Support Vector Machine algorithm of WEKA. Rule based algorithms has the lowest accuracy among all groups. On the other hand, distance based algorithms are better than rule based algorithms. On the other hand, Bayesian, Support Vector Machines and decision tree algorithms have high accuracy values. Hence, those algorithms are much more suitable for this data classification problem.

Table 2. Confusion matrix of SMO algorithm

Confusion matrix of SMO		Predicted	
		High	Low
Actual	High	137	82
	Low	53	165

In Table 2, the confusion matrix for SMO algorithm which has the highest accuracy is given. Among 437 instances, 302 instances are correctly classified. 137 SMEs are actually in High class and the algorithm predict them in High class correctly. Similarly, 165 instances are in Low class and correctly predicted in Low class. Hence total number of correct prediction is 302 which corresponds to an accuracy value of 69.11%. On the other hand, 82 instances are actually in High class but predicted as in Low class and 53 instances are actually in Low class but they are predicted as in High class. Thus, 135 instances are misclassified and error rate is 30.89%. When the class base accuracies are compared, the low class has a higher accuracy with respect to low class.

As a result, SMO algorithm can be used to predict Industry 4.0 class of a SME with 69.11% accuracy. Moreover, the misclassification in Low class prediction is considerably good. Thus, while evaluating a governmental support, the companies that are predicted in Low class may have a prevalence to improve their Industry 4.0 situation.

In addition, attribute selection algorithms are also used and classification algorithms are applied those selected attributes. But the accuracies are lower than the algorithms that are performed with full data set and there was no noticeable computational time reduction. So, the results that are gathered with attribute selections are not presented in this paper.

4 Conclusions

With the advancements in technology and management strategies, it is not enough to produce or serve good quality for today's companies to stay in the competitive stage. Requirements to remain in the competition is to understand the changes in the trends and adjust quickly. These industry trend, especially in manufacturing, are shaped by the transformations or revolutions. After three big transformations, namely first, second and third industrial revolutions, the new era has arisen and called Industry 4.0. The new concept brings the overall interaction and communication between each element in the production environment. For countries that are considered to have a developing economy, keeping up with the trends is one of the most important strategies to make their economy grow continuously.

Report [7] shows that Turkish economy consists of SMEs mostly with 99.8% in total enterprises and categorized in countries that have developing economy. So, for Turkish economy, it is crucial to follow the changes and adapt to them which requires the good understanding of the current situation of SMEs in terms of necessary regulations.

In this study, the survey that is prepared by [8] is used in order to gather the information from SMEs. In total, 513 SMEs are participated from four cities in Turkey. After the preprocessing the number of participations is decreased to 437 and industry level of SMEs are considered as two categorical variables, namely low and high.

Data classification algorithms are deployed in order to conclude valuable and reasonable knowledge from the bulk data. In implementation, 26 different algorithms are applied in WEKA 3.8.2 software that are classified in 5 different categories. The best algorithms that outputs the highest accuracies from each category are furtherly investigated. The best algorithm overall is chosen SMO which is an "Support Vector

Machine" algorithm with 69.11% accuracy rate. It should be noted that computational times are neglected since they are so short.

The results of this study might be used as a guide in making governmental regulations, support programs to SMEs or choosing suitable SMEs to invest in the evolution to Industry 4.0.

In further studies, sector based data mining applications can be practiced with a satisfactory number of instances form each manufacturing sector which requires a bigger distribution network of the survey. Other study can be the employment of other algorithms that are not considered in this paper.

Acknowledgment. This research is made possible with the support of KOSGEB in data collection. The researchers express their gratitude to KOSGEB Management for their support on this study. Authors would also like to thank Istanbul Kültür University [ULEP-2019-U0708/53/07] for funding this researchs's conference attendance.

References

1. McCarthy, I., Anagnostou, A.: The impact of outsourcing on the transaction costs and boundaries of manufacturing. Int. J. Prod. Econ. **88**, 61–71 (2004)
2. Drath, R., Horch, A.: Industrie 4.0: hit or hype? IEEE Ind. Electron. Mag. **8**(2), 56–58 (2014)
3. Hermann, M., Pentek, T., Otto, B.: Design priciples for Industrie 4.0. In: 49th Hawaii International Conference on System Sciences (2016)
4. Siemens: Endüstri 4.0 Yolunda (2016)
5. Turkish Industry & Business Association (TÜSİAD), Industry 4.0 in Turkey as an Imperative for Global Competitiveness: An Emerging Market Perspective, TÜSİAD-T/2016-03/576
6. Ministry of Science, Industry and Technology, Kosgeb, Kobi Stratejisi ve Eylem Planı (KSEP) 2015–2018
7. Turkish Statistical Institute (TÜİK), Small and Medium Sized Enterprises Statistics, 2016, Press release, No: 21540, November 2017
8. Gergin, Z., Üney-Yüksektepe, F., Gençyılmaz, M.G., Aktin, A.T., Gülen, K.G., İlhan, D.A., Dündar, U., Cebeci, Ö., Çavdarlı, A.İ.: Industry 4.0 scorecard of Turkish SMEs. In: Proceedings of the International Symposium for Production Research 2018 (ISPR 2018), Wien, pp. 426–437 (2018)
9. Ahmad, I.: How much data is created per minute? The answer will blow your mind away (2018). https://www.digitalinformationworld.com/2018/06/infographics-data-never-sleeps-6.html
10. Rajaraman, A., Ullman, J.: Mining of Massive Data Sets. Cambridge Univ. Press, Cambridge (2011)
11. Piatetsky-Shapiro, G., Frawley, W.J.: Knowledge Discovery in Databases. AAAI/MIT Press, Massachusetts (1991)
12. Kamber, M., Winstone, L., Gong, W., Cheng, S., Han, J.: Generalization and decision tree induction: efficient classification in data mining. In: Proceedings Seventh International Workshop on Research Issues in Data Engineering. High Performance Database Management for Large-Scale Applications, Birmingham, pp. 111–120 (1997)

13. Han, J., Kamber, M.: Data Mining: Concepts and Techniques. Morgan Kaufmann, San Francisco (2001)
14. İlhan, D.A., Gergin, Z., Üney-Yüksektepe, F., Gençyılmaz, M.G., Dundar, U., Çavdarlı, A. İ.: Comparative analysis of the most industrialized cities in Turkey from the perspective of Industry 4.0. In: International Symposium of Production Research 2019, Wien (2019). (accepted for publication)
15. Roiger, R.J., Geatz, M.W.: Data Mining - A Tutorial Based Primer. Addison Wesley Press, Boston (2003)
16. Giudici, P., Figini, S.: Applied Data Mining for Business and Industry. Wiley, New York (2009)
17. Turban, A., Sharda, R., Delen, D., King, D.: Business Intelligence: A Managerial Approach. Pearson Education Inc., New Jersey (2001)
18. Quinlan, J.R.: Induction of decision trees. Mach. Learn. **1**, 81–106 (1986)
19. Breiman, L., Freidmann, J., Olshen, R., Stone, C.: Classification and Regression Trees. Wadsworth International Group, Monterey (1984)
20. Cortes, C., Vapnik, V.: Support vector network. Mach. Learn. **20**, 273–297 (1995)

The Gender Analysis of Enrolled Students: A Comparison Study of Austrian and Turkish Higher Education

Erol Yücel[1](✉) and Serpil Erol[2]

[1] Department of Technology and Knowledge Management, Baskent University, Ankara, Turkey
erolyucel@baskent.edu.tr

[2] Department of Industrial Engineering, Gazi University, Ankara, Turkey
serpiler@gazi.edu.tr

Abstract. There are more data than ever to analyse and support higher education management decisions and trends nationally and internationally. National higher education systems should rely more on data-based decision making, so on data driven decision support systems (DSS). Using education data provided by Eurostat between 2013–2017, a DSS has been built. By using this DSS, enrolled students related gender data analysed in detail. The data dimensions of the research are enrolled students, gender, education level, fields of education. But, a data-driven DSS for all education levels and fields of European countries will be next research subject in order to provide education intelligence for all education policy makers, managers, teachers, academic staff students and families.

Keywords: Gender · Higher education · Decision analysis · Students · Data visualization · Educational data science and data driven decision support system

1 Introduction

There are more data than ever to analyse and support higher education management decisions and trends nationally and internationally. National higher education systems should rely more on data-based decision making, so on data driven decision support systems (DSS). These systems will surely provide higher education intelligence and performance monitoring urgently needed by the higher education policy makers.

DSS needs comparable data to monitor or benchmark progress of the national education systems, including input and output, nationally or internationally across countries. Fortunately, the United Nations Educational, Scientific and Cultural Organization (UNESCO) in the 1970s developed the International Standard Classification of Education (ISCED) in order to categorise and report both nationally and internationally comparable education statistics [1]. ISCED is first revised in 1997 and its second and current revision has been adopted by UNESCO in 2011. The framework is occasionally updated in order to better capture new developments in education systems worldwide.

N. M. Durakbasa and M. G. Gençyılmaz (Eds.): ISPR 2019, LNME, pp. 36–47, 2020.
https://doi.org/10.1007/978-3-030-31343-2_4

The standards on international statistics on education and training systems are set by the three international organisations jointly administering the annual UOE (UNESCO Institute for Statistics, The Organisation for Economic Co-operation and Development (OECD) and The Statistical Office of the European Union (EUROSTAT) data collection.

The UOE data collection on education statistics covers the 28 EU Member States (Belgium, Bulgaria, Croatia, Czech Republic, Denmark, Germany, Estonia, Greece, Spain, France, Ireland, Italy, Cyprus, Latvia, Lithuania, Luxembourg, Malta, Hungary, Netherlands, Austria, Poland, Portugal, Romania, Slovenia, Slovak Republic, Finland, Sweden and United Kingdom), the EFTA/EEA countries (Iceland, Liechtenstein, Norway and Switzerland), the candidate countries (Albania, North Macedonia, Serbia and Turkey), as well as OECD Member States situated outside Europe (Australia, Canada, Japan, Korea, Mexico, New Zealand, United States) and other countries (e.g. Israel).

The dissemination of education statistics by Eurostat is generally limited to the 28 EU Member States, the EFTA/EEA and Candidate countries. Information on OECD Member States situated outside Europe can be provided by OECD.

Eurostat disseminate education administrative data, besides other categories, from 2013 onwards by using ISCED 2011, even though ISCED 1997 data still are available when needed before 2013. The EUROSTAT education data covers pupils and students – enrolments and entrants, learning mobility, education personnel, education finance, graduates and language learning.

ISCED 2011 (data from 2013 onwards) has nine levels of education, from level 0 to level 8: ISCED 0 (ED 0): Early childhood education, ISCED 1 (ED 1): Primary education, ISCED 2 (ED 2): Lower secondary education, ISCED 3 (ED 3): Upper secondary education, ISCED 4 (ED 4): Post-secondary non-tertiary education, ISCED 5 (ED 5): Short-cycle tertiary education, ISCED 6 (ED 6): Bachelor's or equivalent level, ISCED 7 (ED 7): Master's or equivalent level and ISCED 8 (ED 8): Doctoral or equivalent level.

In this research, ED 6-7-8 levels tertiary education enrolled students [2] will be discussed on gender equity in Austrian and Turkish higher education focusing onto the first level ISCED Fields of Education and Training 2013 (Table 1).

Table 1. ISCED fields of education and training 2013

Code	Fields of education and training
F01	Education
F02	Arts and humanities
F03	Social sciences, journalism and information
F04	Business, administration and law
F05	Natural sciences, mathematics and statistics
F06	Information and Communication Technologies
F07	Engineering, manufacturing and construction
F08	Agriculture, forestry, fisheries and veterinary
F09	Health and welfare
F10	Services

2 Enrolled Students

In this part, first, the students' numbers of two countries by education level, later by fields of education and training, then gender dimension will be added for the analysis on top of ED 6–8 and ISCED 2013 dimensions.

2.1 Enrolled Students' Numbers

Data visualisation of numbers usually explain or show patterns of data for variables better than it is expected. Sometimes charts tell almost all the story about data. Following charts, we believe, will explain the story about the enrolled students of Austria and Turkey with no need of many words. Also, in all the charts, for the sake of not to associate colours with gender, blue is used for female (F) and orange is used for male (M) indicators.

Austria's bachelor's or equivalent level students are increasing yearly for both gender but decreasing slightly for master's or equivalent level and in both levels female dominated. In doctoral or equivalent level, numbers are male dominated and slightly decreasing as seen in Chart 1.

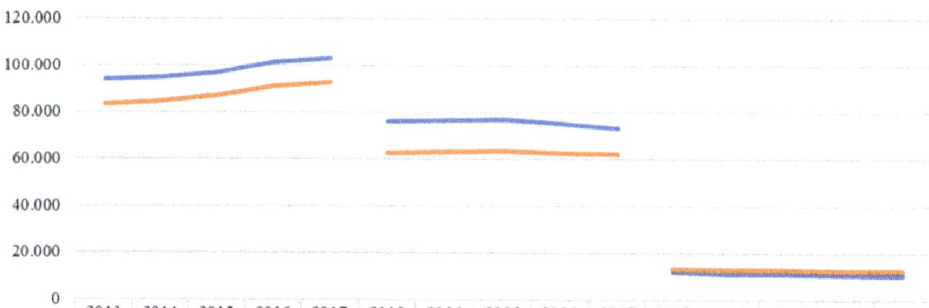

AT	ED6 2013	ED6 2014	ED6 2015	ED6 2016	ED6 2017	ED7 2013	ED7 2014	ED7 2015	ED7 2016	ED7 2017	ED8 2013	ED8 2014	ED8 2015	ED8 2016	ED8 2017
F	94.132	94.737	96.769	101.46	103.05	75.920	76.588	76.940	75.091	73.316	12.014	11.520	11.179	10.877	10.515
M	83.499	84.669	86.998	90.997	92.905	62.478	63.109	63.329	62.661	62.244	13.447	13.059	12.879	12.635	12.353

Chart 1. Austrian enrolled students' number by education level

In Turkey, in every education level, numbers are male dominated and increasing as seen in Chart 2.

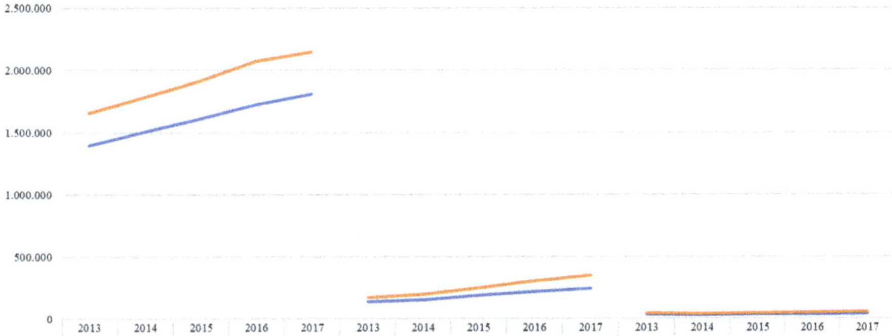

TR	ED6 2013	ED6 2014	ED6 2015	ED6 2016	ED6 2017	ED7 2013	ED7 2014	ED7 2015	ED7 2016	ED7 2017	ED8 2013	ED8 2014	ED8 2015	ED8 2016	ED8 2017
F	1.395.159	1.503.919	1.611.738	1.722.579	1.809.975	138.843	154.348	190.863	222.324	245.083	34.849	28.890	32.688	35.702	37.667
M	1.657.512	1.784.160	1.915.911	2.067.714	2.143.288	175.976	201.445	252.389	305.068	353.448	45.645	38.267	45.535	50.392	53.600

Chart 2. Turkish enrolled students' number by education level

2.2 Enrolled Students' Percentages

In Chart 3, Austrian and Turkish enrolled students' percentage by education level has been shown. ED 6 students' percentage among ED 6–8 in Austria is around %55, but %85 in Turkey, ED 7%38 and %12, ED 8%6.45 and %2 respectively.

Reference [3] states that since the 1970s, the Austrian universities have slowly but steadily become "feminised.", and the widening participation of women in the tertiary sector led to a substantial increase of female participation at the universities as well. As a result, the rate of women increased, for example, among the first year students from 30% in the academic year 1970/71 to 53% in the academic year 2002/03 as seen Chart 3.

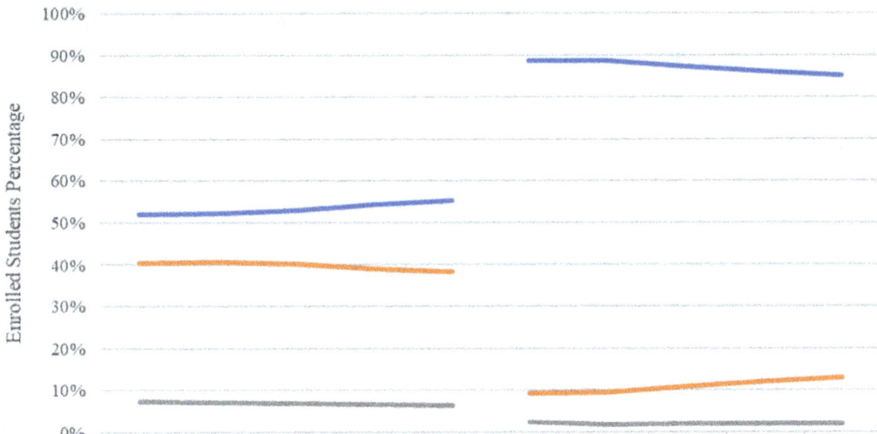

	AT 2013	AT 2014	AT 2015	AT 2016	AT 2017	TR 2013	TR 2014	TR 2015	TR 2016	TR 2017
ED6	52.02%	52.20%	52.79%	54.41%	55.30%	88.53%	88.60%	87.12%	86.07%	85.14%
ED7	40.53%	40.65%	40.30%	38.94%	38.25%	9.13%	9.59%	10.95%	11.98%	12.89%
ED8	7.46%	7.15%	6.91%	6.65%	6.45%	2.33%	1.81%	1.93%	1.96%	1.97%

Chart 3. Austrian and Turkish enrolled students' percentage by education level

2.3 Enrolled Students' Numbers and Percentages Based on ISCED 13 by Gender

In Austria, ED 6–8 students' numbers based on ISCED 13 by gender are female dominated in fields F01–F04 and F08–F09 (six out of ten fields) while F04 Business, administration and law has the most female students and F01 education is the most female-dominated, F07 Engineering is the most male-dominated field. It is observed that the number of students in each ISCED field is proportionally distributed, except F08 and F10 (Chart 4).

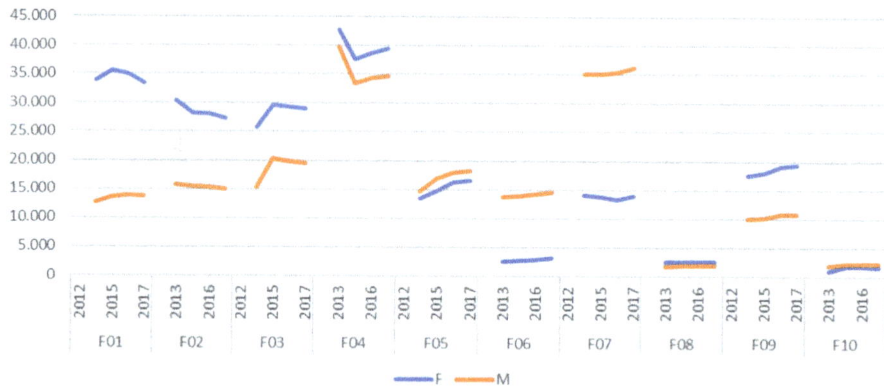

Chart 4. Austrian students' numbers based on ISCED 13 by gender

In Chart 5, we see that Turkish ED 6–8 students' numbers based on ISCED 13 by gender are female dominated in fields F01–F02 and F09 (three out of ten fields) while F04 Business, administration and law has the most male students and F07 Engineering is the second. It is observed that the number of students in each ISCED field is proportionally distributed, except F04.

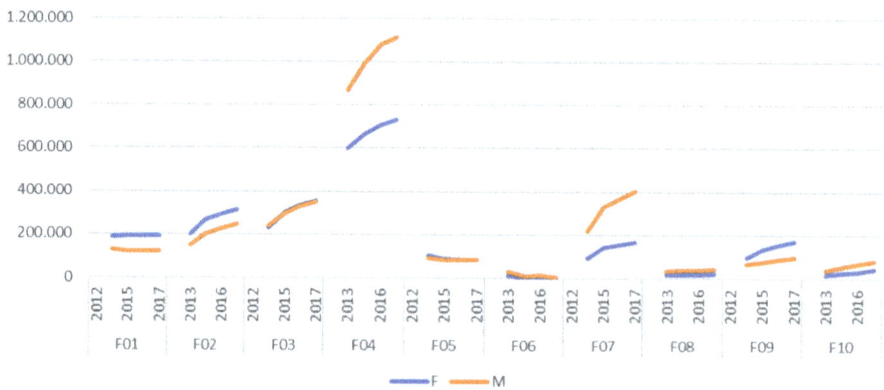

Chart 5. Turkish students' numbers based on ISCED 13 by gender

In Chart 6 we see that in both countries, male students' percentage in ED 8 and ED 7 for Turkey is increasing every year.

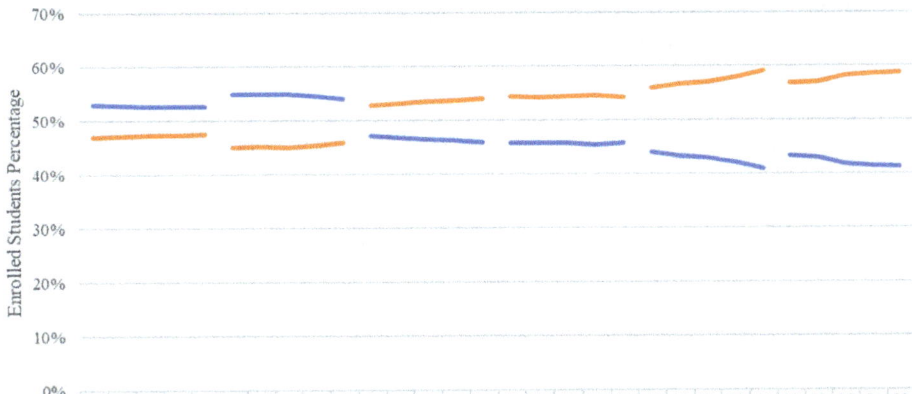

	AT ED6 2013	2014	2015	2016	2017	AT ED7 2013	2014	2015	2016	2017	AT ED8 2013	2014	2015	2016	2017
F	53	53	53	53	53	55	55	55	55	54	47	47	46	46	46
M	47	47	47	47	47	45	45	45	45	46	53	53	54	54	54

	TR ED6 2013	2014	2015	2016	2017	TR ED7 2013	2014	2015	2016	2017	TR ED8 2013	2014	2015	2016	2017
F	46	46	46	45	46	44	43	43	42	41	43	43	42	41	41
M	54	54	54	55	54	56	57	57	58	59	57	57	58	59	59

Chart 6. Austrian and Turkish enrolled students' percentage by education level and gender

2.4 Enrolled Students' Percentages Based on ED 6–8 by Gender and Fields of Education and Training (2014 Data Was Not Available for Turkey, so 2014 Data Is Not Included)

F01 Education is female dominated field of education and training in every education level in both countries while slightly dominated in ED 7–8 in Turkey (Chart 7).

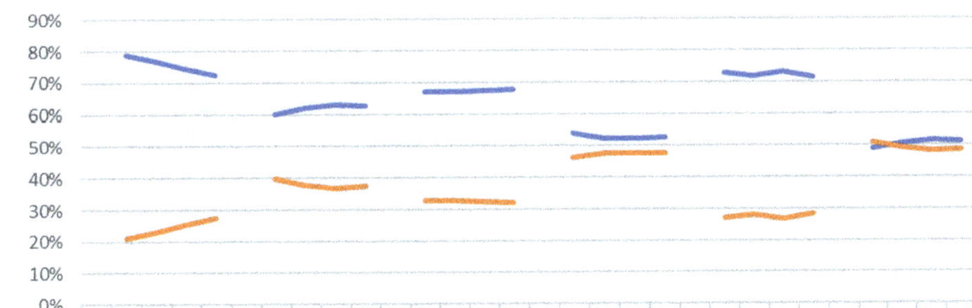

Education F01 ED6	AT 2012	2013	2015	2016	2017	TR 2012	2013	2015	2016	2017
F		79	77	75	72		60	62	63	63
M		21	23	25	28		40	38	37	37

Education F01 ED7	AT 2012	2013	2015	2016	2017	TR 2012	2013	2015	2016	2017
F		67	67	68	68		54	52	52	53
M		33	33	32	32		46	48	48	47

Education F01 ED8	AT 2012	2013	2015	2016	2017	TR 2012	2013	2015	2016	2017
F		73	72	73	72		49	51	52	51
M		27	28	27	28		51	49	48	49

Chart 7. Austrian and Turkish enrolled students' percentage by education level, education (F01) field and gender

F02 Arts and humanities is female dominated field of education and training in every education level in both countries, except ED 7–8 in Turkey (Chart 8).

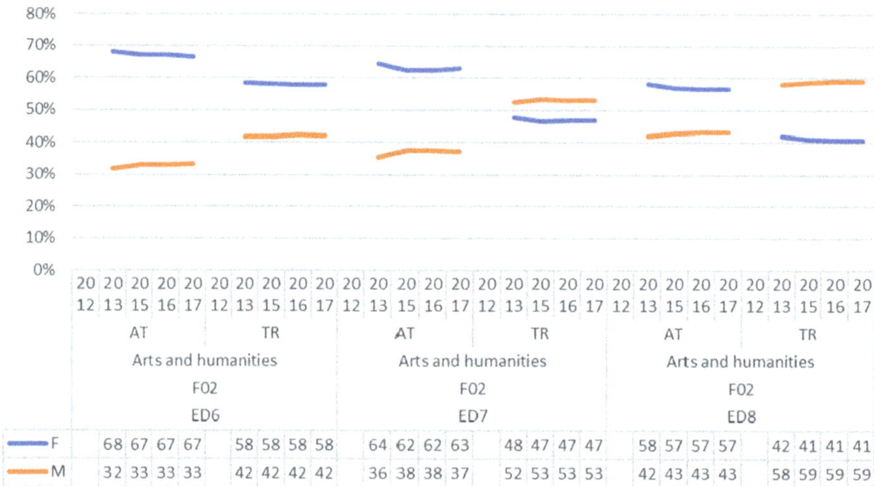

Chart 8. Austrian and Turkish enrolled students' percentage by education level, F02 field and gender

F03 Social sciences, journalism and information is female dominated field of education and training in every education level in Austria, almost equally dominated in ED 6 and male dominated in ED 7–8 in Turkey (Chart 9).

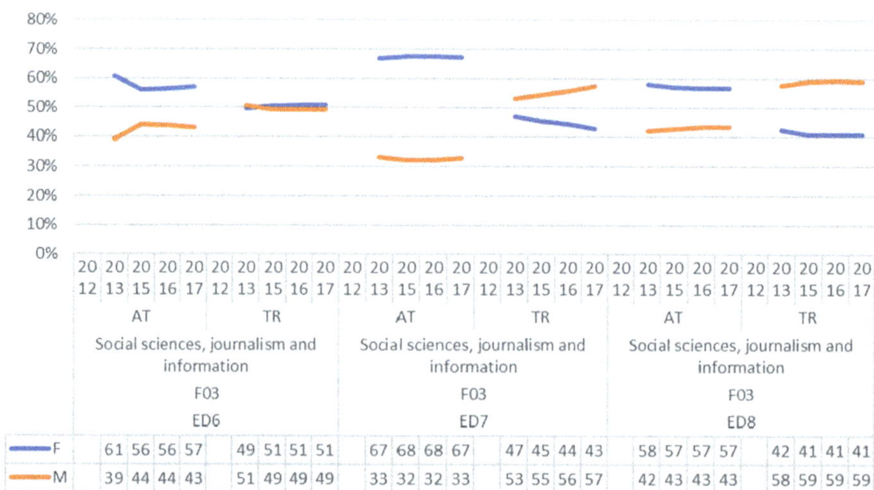

Chart 9. Austrian and Turkish enrolled students' percentage by education level, F03 field and gender

F04 Business administration and law is female dominated field of education and training in every education level in Austria, except ED 8, male dominated in ED 6–8, with an increasing rate in ED 6 in Turkey (Chart 10).

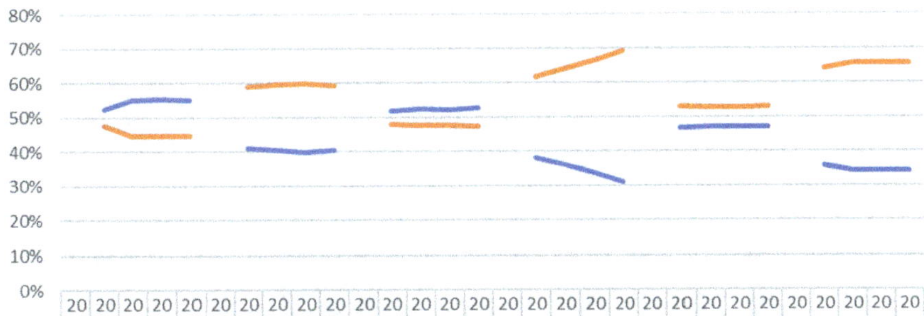

Field	Level	Country	Years	F	M
Business, administration and law F04	ED6	AT	2012 2013 2015 2016 2017	52 55 55 55	48 45 45 45
Business, administration and law F04	ED6	TR	2012 2013 2015 2016 2017	41 40 40 41	59 60 60 59
Business, administration and law F04	ED7	AT	2012 2013 2015 2016 2017	52 52 52 53	48 48 48 47
Business, administration and law F04	ED7	TR	2012 2013 2015 2016 2017	38 36 34 31	62 64 66 69
Business, administration and law F04	ED8	AT	2012 2013 2015 2016 2017	47 47 47 47	53 53 53 53
Business, administration and law F04	ED8	TR	2012 2013 2015 2016 2017	36 34 34 34	64 66 66 66

Chart 10. Austrian and Turkish enrolled students' percentage by education level, F04 field and gender

F05 Natural sciences, mathematics and statistics is equally dominated field of education and training by both gender in every education level in both countries, except ED 8, which is male dominated in Austria (Chart 11).

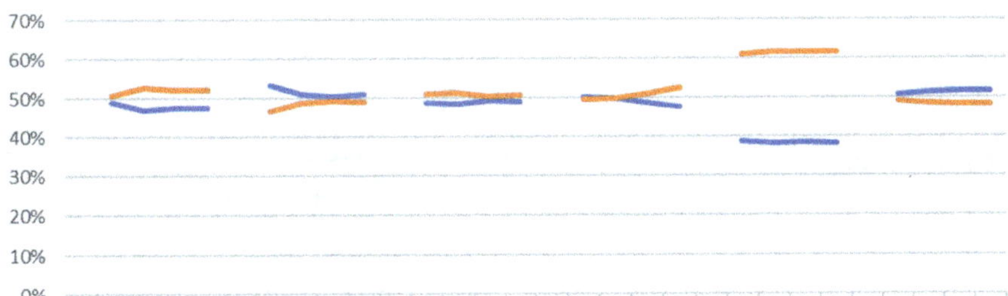

Field	Level	Country	Years	F	M
Natural sciences, mathematics and statistics F05	ED6	AT	2012 2013 2015 2016 2017	49 47 48 48	51 53 52 52
Natural sciences, mathematics and statistics F05	ED6	TR	2012 2013 2015 2016 2017	53 51 51 51	47 49 49 49
Natural sciences, mathematics and statistics F05	ED7	AT	2012 2013 2015 2016 2017	49 49 49 49	51 51 51 51
Natural sciences, mathematics and statistics F05	ED7	TR	2012 2013 2015 2016 2017	50 50 49 48	50 50 51 52
Natural sciences, mathematics and statistics F05	ED8	AT	2012 2013 2015 2016 2017	39 38 38 38	61 62 62 62
Natural sciences, mathematics and statistics F05	ED8	TR	2012 2013 2015 2016 2017	51 51 52 52	49 49 48 48

Chart 11. Austrian and Turkish enrolled students' percentage by education level, F05 field and gender

F06 Information and communication technologies is all male dominated in both countries in every education level with around %80 (Chart 12).

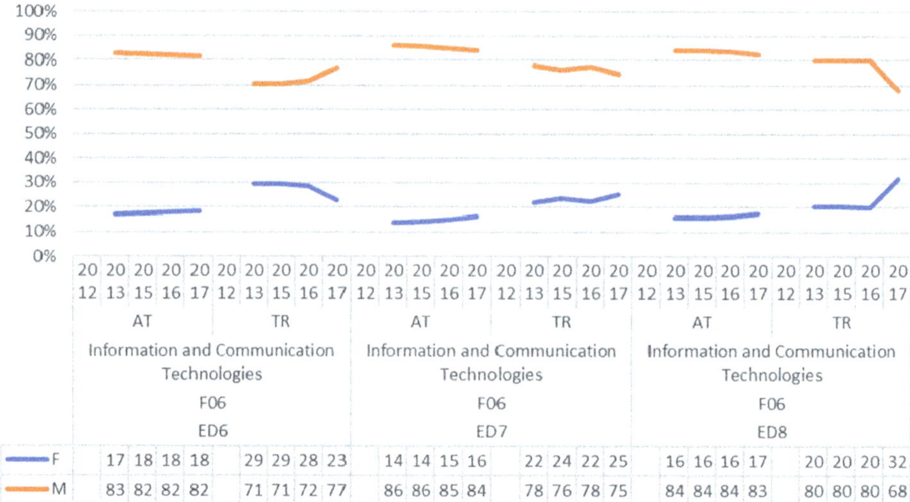

Chart 12. Austrian and Turkish enrolled students' percentage by education level, F06 field and gender

F07 Engineering, manufacturing and construction is all male dominated in both countries in every education level with around %70 (Chart 13).

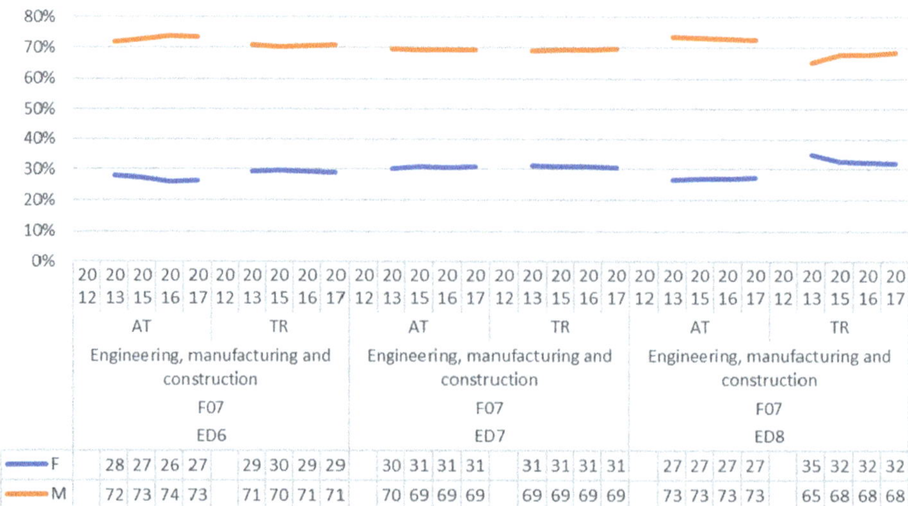

Chart 13. Austrian and Turkish enrolled students' percentage by education level, F07 field and gender

F08 Agriculture, forestry, fisheries and veterinary is male dominated except, ED 7–8 in Austria (Chart 14).

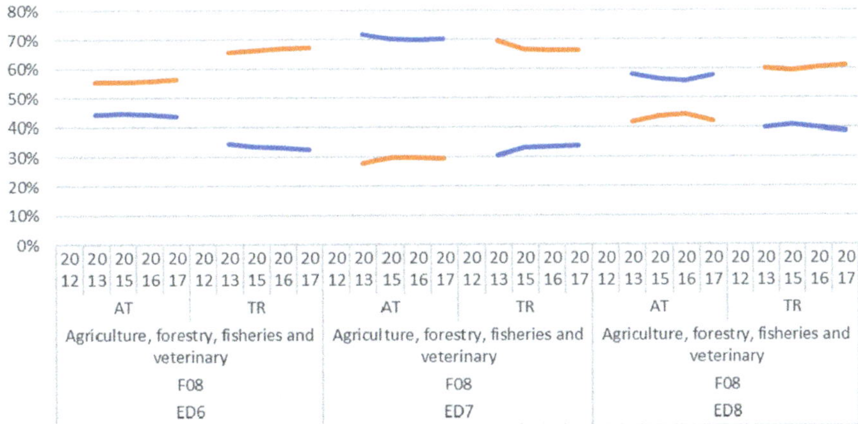

Level	Country	F	M
ED6	AT	44 45 44 44	56 55 56 56
ED6	TR	34 34 33 33	66 66 67 67
ED7	AT	72 70 70 70	28 30 30 30
ED7	TR	31 33 34 34	69 67 66 66
ED8	AT	58 56 56 58	42 44 44 42
ED8	TR	40 41 40 39	60 59 60 61

Chart 14. Austrian and Turkish enrolled students' percentage by education level, F08 field and gender

F09 Health and welfare is a female dominated field of education and training (Chart 15).

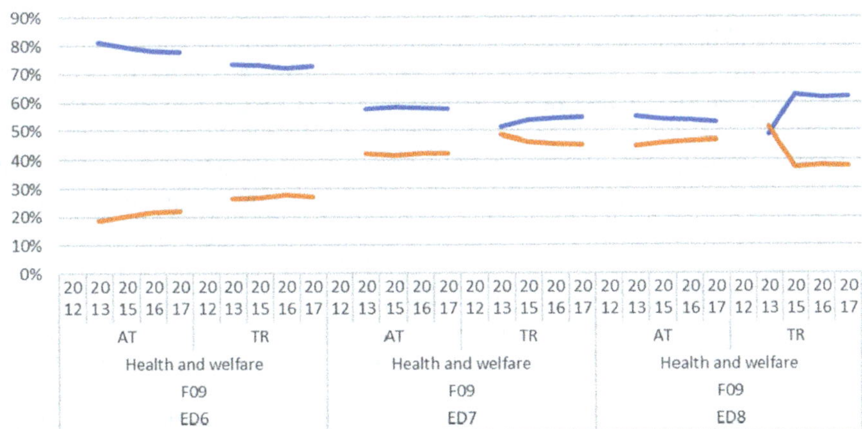

Level	Country	F	M
ED6	AT	81 80 78 78	19 20 22 22
ED6	TR	74 73 72 73	26 27 28 27
ED7	AT	58 58 58 58	42 42 42 42
ED7	TR	51 54 54 55	49 46 46 45
ED8	AT	55 54 54 53	45 46 46 47
ED8	TR	49 63 62 62	51 37 38 38

Chart 15. Austrian and Turkish enrolled students' percentage by education level, F09 field and gender

F10 Services is a male dominated field of education and training (Chart 16).

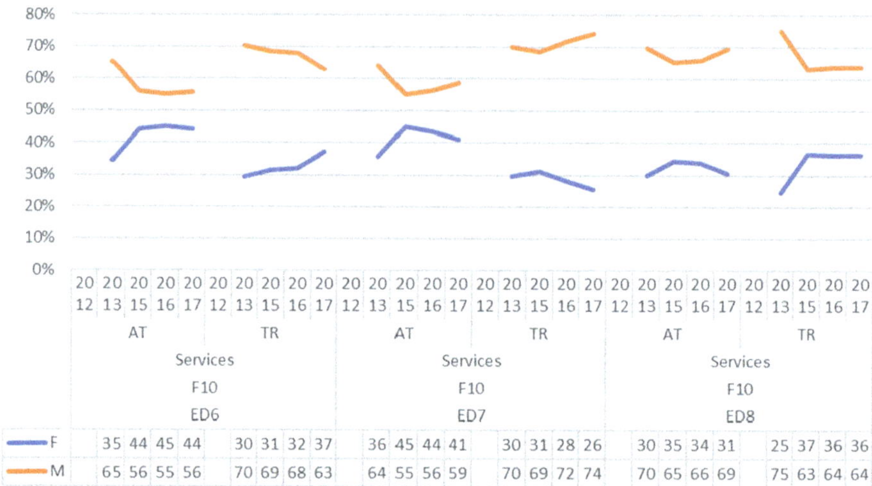

Chart 16. Austrian and Turkish enrolled students' percentage by education level, F10 field and gender

3 Conclusion

The number of enrolled students of Turkey is very high compared to Austria in ED 6–8. But, in Austria, the female students' percentage is higher than male students by almost %5 while it is the opposite in Turkey, the male students' percentage is higher than female students by almost %10.

Reference [4] shows that the integration of Austria into the European Union has helped to reduce gender inequality in universities. If Turkey adapts the European Union policies in gender inequality, the current statistics will favour female students in the next decade.

According to first level ISCED Fields of Education and Training 2013, it is observed that; Education, Health and welfare fields are female while Information and Communication Technologies and Engineering, manufacturing and construction are male dominated. Other fields have mixed results.

Another interesting finding is that in 2017, bachelor's or equivalent level (ED 6) students' percentage is %55,30 in Austria, but, 85,14 in Turkey, master's or equivalent level (ED 7) %38,24 and %12.89, doctoral or equivalent level (ED 8) %6,45 and %1,97 respectively.

A data-driven DSS using EUROSTAT data for all education levels and fields of European countries will be next research subject in order to provide education intelligence for all education policy makers, managers, teachers, academic staff students and families in Europe. In the conference, its protype will be provided.

References

1. https://www.oecd.org/education/isced-2011-operational-manual-9789264228368-en.htm
2. https://ec.europa.eu/eurostat/web/education-and-training/data/database?p_p_id=NavTreeportletprod_WAR_NavTreeportletprod_INSTANCE_LUWsdX8ute5m&p_p_lifecycle=0&p_p_state=normal&p_p_mode=view&p_p_col_id=column-2&p_p_col_count=1
3. Pellert, A., Gindl, M.: Gender equity and higher education reform in Austria. In: Sagaria, M. A.D. (ed.) Women, Universities, and Change. Issues in Higher Education, Palgrave Macmillan, New York (2007)
4. Sagaria, M. (ed.): Women, Universities, and Change: Gender equality in the European Union and the United States. Springer (2007)

Energy Management

Electricity Consumption Forecasting with Artificial Neural Network for Fast-Moving Consumer Goods Sector

Gülfem Yeşil and Bersam Bolat(✉)

Management Engineering Department, Istanbul Technical University, Istanbul, Turkey
gulfemyesil@gmail.com, bolat@itu.edu.tr

Abstract. Nowadays, it is evident that electricity is an indispensable source of energy in the production sectors when industry 4.0 transformation and sustainability become important at the same time. Electricity consumption forecast has crucial importance for effective energy planning in many production sectors. It is important to predict the total consumption of energy consumption and to make a production plan according to it and therefore to make all the functions in the supply chain cost and optimization plans. In this study, Artificial Neural Networks (ANN) method is used for electricity demand estimation for production processes of cold chain product in the fast moving consumer goods sector (FMCG). The impact of the observed independent variables is analyzed on electricity consumption. Estimates in the model are made for the following periods based on the last three years' electricity consumption of the one of the big fast moving goods company located in Turkey.

Keywords: Electricity consumption · Artificial Neural Network (ANN) · Forecasting · FMCG

1 Introduction

Today, one of the most important inputs of developed and developing countries is energy. The level of development of countries is directly proportional to the energy they produce and consume. Although there are many different methods of energy production, all countries need cheap, abundant and clean energy sources. Increasing energy consumption under the condition of developing countries is an indicator of growth. Beyond this, so energy consumption is one of the most important indicators of economic development and social welfare.

Nowadays, it is evident that electricity is an indispensable source of energy in the production sectors when industry 4.0 transformation and sustainability become important at the same time. Electricity consumption forecast has crucial importance for effective energy planning in many production sectors. Most operations in the manufacturing companies are required electrical energy. Some industries such as Fast-Moving Consumer Goods (FMCG) need more energy to manage their supply chain the most effective way, supply high volumes in short times and invest in technology to

N. M. Durakbasa and M. G. Gençyılmaz (Eds.): ISPR 2019, LNME, pp. 51–61, 2020.
https://doi.org/10.1007/978-3-030-31343-2_5

maintain and improve logistics. Fast-moving consumer products are cheap products that are easily consumed or sold in a short period of time and whose usage is limited by day, month or year. This sector includes everything that is consumed quickly on the retailer shelves. Reference [1] shows states that, while the main features of fast-moving consumer goods are low-priced and low loyalty, which are frequently purchased. In terms of firm, it has high sales volume, low profit margin, wide distribution network and fast stock turnover rate. In the sector, a complex supply chain is used which aims to ensure the availability of fast-moving consumer products every day. Given the growth in trade volumes and diversity of products exhibited by the FMCG market, electrical energy plays an increasingly important role on the industry. In addition to being an important input of electricity, companies that want to maintain their shares in the market must add “sustainability”, “green supply chain”, and “energy usage reduction” to their agendas. So, it is important to predict the total energy consumption and to make a production plan according to it and therefore to make all the functions in the sustainable supply chain and optimization plans.

In this study, it is aimed to define a function for the estimation of electric energy consumption with the ANN approach, in the production area of a company in the FMCG sector. This study organized as follows. The second part is dedicated to literature review of ANN model which is frequently utilized for prediction and estimation purpose. In the third part of the study ANN based prediction model is proposed for electricity demand forecasting for especially cold chain production area in a company that operates in FMCG sector. Considering the 2016–2018 periods, annual demand estimation is made by ANN method and electricity demand prediction studies are carried out for future years. And last part is designed for recommendations and future studies.

2 ANN Background

A neural network (ANN) was developed by inspiring from the working principle of the nervous system. ANN is made up of signaling associations called neurons or processing elements. Each neuron receives at least one weighted input. Weights are an expression of connection power with the output as in [2]. Neural networks are nowadays used for recognition of patterns, characters, speech, image compression, process control, process optimization, as well as in science and business world. ANN are typical example of modern interdisciplinary subject, which is used to solve various engineering problems that could not be solved by traditional modeling as in [3, 4]. Many studies that use of ANN for forecasting purpose show the positive effects come from prediction process with this method.

For example, reference [5] presented estimation of gas demand using artificial neural networks. The design and training of the MLP (multi-layer sensor model) was carried out using data describing the actual consumption of natural gas in Szczecin. The model takes into account the calendar (month, month, day of the week, hour) and weather (temperature) factors, which have a significant impact on gas consumption by individual consumers and small industries. It has been found that the model with artificial neural networks can be used successfully to estimate gas consumption at any time of the day and at any time of the year. Another natural gas consumption prediction model is proposed via using data for the production of hot mix asphalt for 2014 for testing purposes. As independent variables in modeling; moisture content, hourly capacity, type of asphalt mixture produced and temperature of produced asphalt. According to the obtained modeling results, it is concluded that ANN can be used successfully in predicting natural gas consumption in the production of hot mix asphalt as in [6].

Reference [7] used ANN in their study. The application and comparison of different artificial neural networks in the prediction of 1-h forward wind speed was investigated. The performance of the model was evaluated according to three measures as mean absolute error, root mean square error and mean absolute percentage error. The results showed that the selection of the type of neural networks for best performance is also dependent upon the data sources.

ANN is a key method to address the issue of nonlinearity of building energy data and the robust calculation of large and dynamic data. Energy consumption is simultaneously affected by several influential independent variables. Reference [8] used solar radiation, outdoor temperature and number of days as inputs and house and heat pump energy usage amount as outputs. The model based on Levenberg-Marquardt and OWO-Newton algorithms promise results of coefficients of determination within 0.87–0.91, which is comparable to prior literature.

Besides forecasting of wind energy and natural gas consumption; ANN has been used for modeling electricity usage in the literature. Reference [9], presented an ANN model to estimate electricity prices. Since the performance of an ANN estimation model depends on appropriate input parameter sets, the focus is on the selection and preparation of basic data, which have a significant impact on electricity prices. The results show that the overall methodology leads to favorable electricity price estimates, while the estimation errors are low or even lower than other estimation models for electricity prices known from the literature. Another electricity prices estimation model is proposed by as in [10] The aim of this study is to examine artificial neural network (ANN) based models for daily price estimation. Training data are clustered in homogeneous groups and a specific estimator is used for each cluster. The models were tested on a dataset of atypical price models and many outliers. This approach makes the price forecasting problem a more challenging task, providing evidence that the proposed models can be considered as useful and robust forecasting tools to the actual needs of market participants, including the traditional generation companies and self-producers, but also the retailers/suppliers and aggregators.

In reference [11] a data-driven forecasting model for the electricity usage of buildings throughout the day is applied. Using variable significance analysis, the key variables were: day type indicator, time of day, HVAC set temperature program, outdoor dry temperature bulb and outdoor humidity. The results show that the proposed model with adaptive training methods can estimate electricity consumption and the highest electricity usage of daily buildings over a period of 15 min, in a test case of a commercial building complex.

In most of study, variety methods have been introduced to forecast long-term electricity demand using macro parameters as in [12]. On the other hand because the electricity is very crucial input for energy cost calculations and yearly production plans for manufacturing firms it is important to forecast the electricity consumption in the whole production processes. In this study, ANN method is utilized for estimation electricity consumption for cold chain product that they are also categorized under the FMCG sector. The food supply chain in cold chain logistics has a special position and a more complex structure in supply chain management. Because of that reason authors proposed ANN based prediction models for electricity usage in cold chain production processes. The model is run with various input combinations while the electrical energy consumption was estimated for each department.

3 Proposed ANN Model and Results

In the proposed model, estimates are made for the following periods based on the last three years' electricity energy consumption and the other inputs which are gathered from the one of the big company that produce cold chain products in Turkey. In cold chain operations, there are points which are very important and critical to complete process that means from production to the deliver to customer. Temperature and air condition is the some of the most important conditions for cold chain. The model is analyzed for 5 different departments of company with 6 different inputs set to see the inputs that effect output for each department. Inputs used in the model are taken from both literature and expert opinions in the company. The inputs used in the model are temperature, wind speed, wet bulb, dew point, volume and number of employees.

The inputs for ANN model can be seen at Table 1 below. The data set consist of 1096 daily data, in below it is shared for 31 days. All of the rest data cannot be shared due to confidentiality of the company.

Table 1. Data set

Day	D1	D2	D3	D4	D5	Temperature °C (°C)	Dew point °C (°C)	Wet bulb °C (°C)	Wind speed - km/h (kmh)	Volume (Liton)	Workers
1	60.852	103.89	19.103	93.04	611.487	4.04	0.46	2.63	16	76197	239
2	8615	21131	27400	27622	189802	6.21	2.96	4.76	15	598102	239
3	9286	18333	7742	8259	105455	8.29	6	6.98	14	97959	239
4	704.5	2099.5	1354	3358	17019.5	9.17	6.79	7.81	17	86973	239
5	8997.5	17389	7401.5	6856	85762	8.29	5.75	6.86	23	16748	239
6	8871.5	17021	6812	5392	73325.5	5.33	0.92	3.69	30	21199	239
7	824	2141.5	3025	4296	19526	3.96	−1.54	2.08	14	110069	239
8	792.5	2148	3175.5	2800	18733.5	5.13	−0.71	2.67	9	129236	239
9	648.5	1988	1519	2588	14785.5	7.75	4.29	5.68	7	132911	239
10	645	2110	2439	3283	18782	8.38	5.5	6.82	15	146168	239
11	799.35	2158.7	1395.9	2425	16281.4	6.79	2.67	4.84	14	110825	239
12	47.749	123.33	22.261	89.78	11717.4	5.04	0.46	3.36	15	116090	239
13	797	2153.5	2655	2629	20381	5.58	−0.21	3.25	12	142635	239
14	4727.1	11818	7575.4	12937	96726.6	6.79	3.54	5.05	10	192957	239
15	685.5	2022	2278.5	2580	18378	10.21	7.83	8.42	12	153027	239
16	653	2038	2338	2954	19970	11.5	8.29	9.39	12	132319	239
17	714	2158	2383.5	2951	19972	10.83	8.71	9.49	12	157367	239
18	897.96	2272.9	3128.2	3224	22024.3	8.92	6.54	7.68	34	156237	239
19	804	2320.5	1994.5	2744	20446.5	2.63	1.08	2.17	23	124544	239

(continued)

Table 1. *(continued)*

Day	D1	D2	D3	D4	D5	Temperature °C (°C)	Dew point °C (°C)	Wet bulb °C (°C)	Wind speed - km/h (kmh)	Volume (Liton)	Workers
20	883.61	2274	3360.1	2797	24331.3	1.71	0.46	1.22	14	139334	239
21	816.64	2213.9	3303.8	2420	22693.4	2.75	1.54	2.32	8	153358	239
22	620.68	1835.9	1022.4	1233	14216.4	6.08	5.13	5.46	10	138270	239
23	1037	2670	3677	3804	28053	8.71	6.29	7.39	12	145666	239
24	1254.5	2931	3121.5	3842	27866	9.88	7.29	8.09	17	122002	239
25	492.5	903	2189	2273	11883	5.58	1.25	3.98	18	123384	239
26	949.5	2172.5	2892.5	3225	21969.5	3.25	−3.08	1.2	14	130922	239
27	951	2355	2312	2364	20507.5	3.33	−2	1.58	9	96605	239
28	763	1638	603	1403	11111	4.67	−1.08	2.66	6	123142	239
29	571.5	1103.5	229.5	234.5	6798	5.13	2.67	4.04	11	52823	239
30	219.5	697.5	80.5	274.5	4946	5.5	2.17	4.11	8	94542	239
31	186.5	617.5	49.5	272.5	4494.5	5.67	2.08	4.19	11	90354	239

70% of inputs are used for learning algorithm, 30% of them are used for testing. For these percentages, literature researches are taken into consideration and by using trial and error approaches as in [13]. The model consist of six inputs and one output for the forecasting for electricity consumption. Figure 1 shows the conceptual ANN model with inputs, outputs and hidden layers in backpropagation form. Output is the electricity consumption value for five functional departments each involved in whole production process to produce the final product. Department 1 (D1) deals with storage and quality control operations, in the Department 2 (D2), Department 3(D3) and Department 4 (D4), the main operations, ancillary production operations and assembly operations are performed respectively. The last Department (D5) serves energy use for other departments.

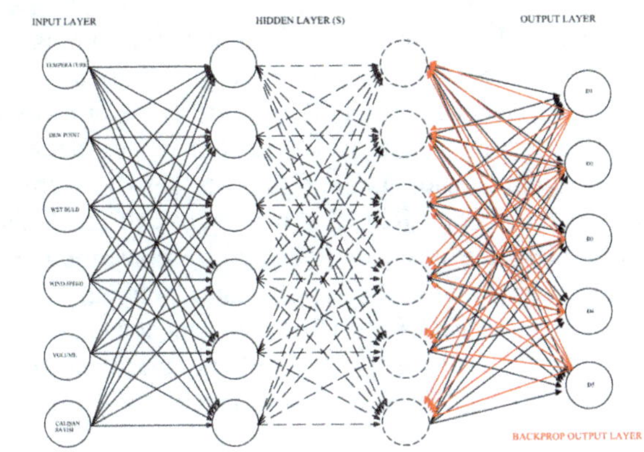

Fig. 1. ANN Model in backpropagation form

There are 4 different models for each department so there are 20 ANN models have been created. Temperature, wet bulb, wind speed, dew point are the constant inputs accordingly experts ant literature. Number of worker and volume of production are added as two other inputs to analyze if they are effect output or not. For first model inputs are temperature, wet bulb, wind speed, dew point. For second model inputs are temperature, wet bulb, wind speed, dew point and number of worker. For third model inputs are temperature, wet bulb, wind speed, dew point and volume of production. For fourth model inputs are temperature, wet bulb, wind speed, dew point, and number of worker and volume of production as seen in Fig. 1.

On the data set, there were missing values with %33 of data for volume of production. For handling this challenge, linear regression method is preferred because of the low percentage of missing value and all the rest data set inputs and outputs are complete. In addition, Root Mean Square Error (RMSE) values are lower than the other imputation method. In Fig. 2, it is shown the volume of production with the imputation of linear regression.

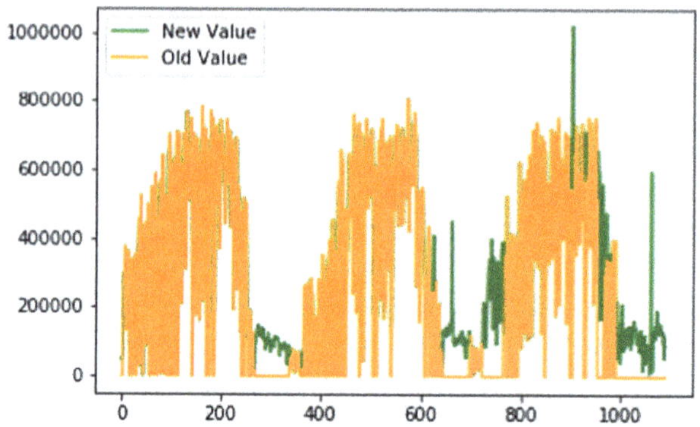

Fig. 2. Production volume prediction

To train and test data and run the model, Python programming language is used thanks to its dynamism and flexibility. The electricity values in data set were drawn between zero and one for easier processing. The performance of ANN model associate with Root Mean Square Error (RMSE). Min RMSE means better performance of training stage of ANN model. Residuals are difference between forecasted and actual data in a regression or ANN model; RMSE shows spread of residuals.

For D1, all inputs effect the amount of electricity consumption. For D1 Fig. 3 shows min RMSE value for training and testing data. Activation function of hidden layer and output layer are chosen as Relu (Rectifier Linear Unit) as in [14] and Selu (Scaled exponential Linear Unit) as in [15]. 1 hidden layers and the learning rate is 0,01. When the model is run for 100 epochs with batch size 10, the results with min RMSE are shown in the Fig. 3 below.

For D2, temperature, wet bulb, wind speed, dew point and number of worker effect the amount of electricity consumption as input. Volume of production does not affect output value. Moreover, Fig. 4 shows min RMSE value for training and testing data for department 2. There are 4 hidden layers and the learning rate is 0,01. When the model is run for 100 epochs with batch size 10, the results with min RMSE are shown in the Fig. 4 below.

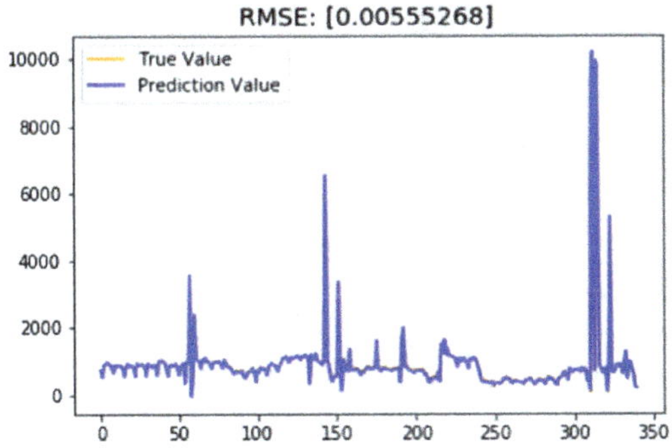

Fig. 3. RMSE value for training and testing data for D1

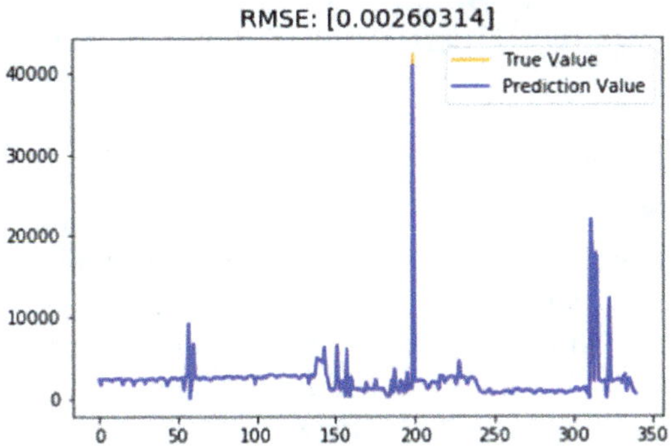

Fig. 4. RMSE value for training and testing data for D2

For Department 3 (D3), temperature, wet bulb, wind speed, dew point and number of worker effect the amount of electricity consumption as input. Volume of production does not affect output value. Moreover, Fig. 5 shows min RMSE value for training and testing data for department 3. Activation function of hidden layer and activation function of output layer are chosen 'Relu' and 'Selu'. There is one hidden layer and the learning rate is 0,01. When the model is run for 100 epochs with batch size 10, the results with min RMSE are shown in the Fig. 5 below.

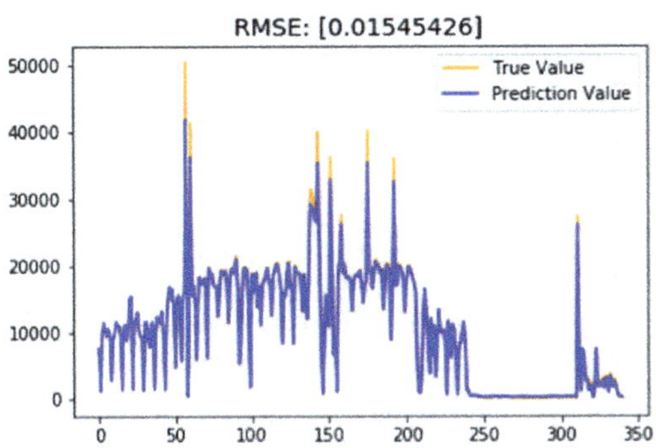

Fig. 5. RMSE value for training and testing data for D3

For D4, all inputs effect the amount of electricity consumption. For D4 Fig. 6 shows min RMSE value for training and testing data. There is one hidden layers and the learning rate is 0,01. When the model is run for 100 epochs with batch size 10, the results with min RMSE are shown in the Fig. 6 below.

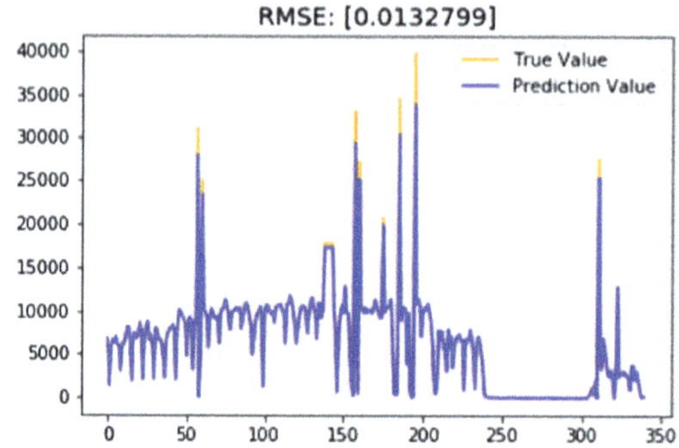

Fig. 6. RMSE value for training and testing data for D4

For D5, temperature, wet bulb, wind speed, dew point and number of worker effect the amount of electricity consumption as input. Volume of production does not affect output value. Figure 6 shows min RMSE value for training and testing data for department 5. There are 2 hidden layers and the learning rate is 0,01. When the model is run for 100 epochs with batch size 10, the results with min RMSE are shown in the Fig. 7 below.

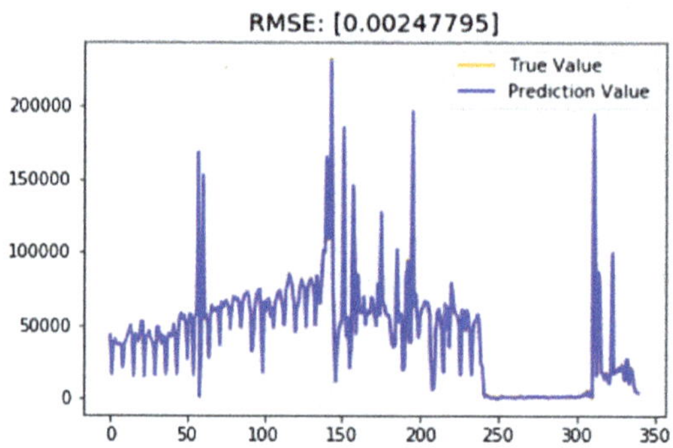

Fig. 7. RMSE value for training and testing data for D5

Belong the results as shown in figures the inputs, temperature, wind speed, dew point, wet bulb and number of workers effect as input each department. Volume of production affect only two departments as input. RMSE values are between 0,0025–0,0155, which means models fit with electricity consumption forecasting.

After training data set, the values are turn back real values to predict future electricity consumption volume. For future prediction long short term memory (LSTM) is used because its flexibility and performance for multiple inputs. In below Table 2, it can be shown some prediction data set for 31 days.

Table 2. Future forecast by ANN

Day	D1	D2	D3	D4	D5
1	66.11	141.35	23.67	69.97	446.22
2	8729.89	22159.53	24124.74	25980.95	119206.2
3	10729.47	19243.79	6367.5	7800.25	97014.3
4	668.48	1981.71	917.64	2908.02	14655.93
5	9117.48	18235.34	6516.64	7979.28	61506.56
6	9136.39	12387.19	9927.35	4593.43	84112.77
7	728.1	1719.72	2334.56	2268.7	12320.16
8	517.9	1363.5	3086.63	2705.1	15366.18
9	704.49	2704.86	1882.13	1946.26	10789.3
10	653.6	2212.7	2147.41	3087.51	11796.15
11	923.6	2265.96	1148.05	2290.41	14978.25
12	45.31	116.41	15.09	77.76	10090.19
13	807.63	2258.31	2337.59	3059.73	14616.79
14	4868.27	8601.11	11039.87	11021.86	110956.47
15	605.72	1623.76	1758.45	1362.49	11595.82
16	426.74	1293.67	2272.57	2853.91	16380.42
17	775.64	2936.16	2953.3	2218.87	14574
18	909.94	2383.54	2754.2	3032.92	13832.5
19	928.98	2435.85	1640.4	2591.11	18810.04
20	838.43	2146.43	2277.21	2422.34	20952.36
21	827.54	2321.64	2908.82	2816.41	16275.21
22	639.22	1336.15	1490.02	1050.75	16307.86
23	916.31	2144.13	2837.75	2008.61	17700.37
24	819.82	1860.53	3034.14	3711.96	22857.12
25	535.02	1228.61	2712.3	1709.37	8671.28
26	962.16	2278.24	2546.7	3032.95	13798.08
27	1098.83	2472.07	1901.53	2232.22	18866.15
28	723.99	1546.1	408.67	1214.74	9567.97
29	579.12	1157.21	202.06	272.92	4875.37
30	238.45	949.01	99.74	206.43	3609.2
31	164.79	495.88	38.2	143.91	2835.86

4 Conclusion and Recommendatıon

This paper presents the production and testing of the ANN model to estimate annual electricity consumption for a cold chain product company. The impact of independent variables of temperature, dew point, wet bulb, number of workers, volume of production and wind speed is observed on electricity consumption in FMCG sector for a cold chain production processes. Estimation of electrical energy consumption is very important for the company because it is one of the basic energy types used.

Production volume is only chosen as input for two departments. For other tree departments it does not improved the RMSE value. The number of workers is an important input for all departments. The following results and recommended improvement can be concluded for the departments which involve to produce cold chain product:

- The reason causing excessive consumption should be identified and corrected.
- Better and less energy consuming lamps should be preferred.
- The consumption on waste department should be analyzed and energies can be reused by a recycling system.
- Baseload times should be analyzed and consumptions during these times can be increased.
- For baseload analysis automation systems can be used.
- The reason causing excessive consumption should be identified and corrected.
- Energy consumption between lines can be reduced by automation systems.
- Since energy is a constantly rotating thing, a system of recycling and reuse of waste energy should be established.
- Pipes and fittings should be examined for reducing losses from transferring energy from one place to another.

For the future study, the model should be tested as new case studies with data of the other companies in the same sector. In order to achieve larger overall performance of the electricity consumption prediction process, it is necessary to increase the number of input data. Also, the other recommended inputs can be added to the model.

References

1. Tiwari, M.K.: An empirical analysis of effect of advertising on marketing of FMCG product. Int. J. Mark. Technol. **2**(2012), 167, 200 (2012)
2. Bodyanskiy, Y., Popov, S.: Neural network approach to forecasting of quasiperiodic financial time series. Eur. J. Oper. Res. **175**(3), 1357–1366 (2006)
3. Flood, I., Kartam, N.: Neural networks in civil engineering II: systems and application. Comput. Civ. Eng. **8**(2), 149–162 (1994)
4. Flood, I., Christophilos, P.: Modeling construction processes using artificial neural networks. Autom. Constr. **4**, 307–320 (1996)
5. Szoplik, J.: Forecasting of natural gas consumption with artificial neural networks. Energy **85**, 208–220 (2015). https://doi.org/10.1016/j.energy.2015.03.084

6. Androjića, I., Dolaček-Aldukc, Z.: Artificial neural network model for forecasting energy consumption in hot mix asphalt (HMA) production. Constr. Build. Mater. **170**, 424–432 (2018). https://doi.org/10.1016/j.conbuildmat.2018.03.086
7. Li, G., Shi, J.: On comparing three artificial neural networks for wind speed forecasting. Appl. Energy **87**(7), 2313–2320 (2010). https://doi.org/10.1016/j.apenergy.2009.12.013
8. Biswas, M.R., Robinson, M.D., Fumo, N.: Prediction of residential building energy consumption: a neural network approach. Energy **117**, 84–92 (2016). https://doi.org/10.1016/j.energy.2016.10.066
9. Keles, D., Scelle, J., Paraschiv, F., Fichtner, W.: Extended forecast methods for day ahead electricity spot prices applying artificial neural networks. Appl. Energy **162**, 218–230 (2016). https://doi.org/10.1016/j.apenergy.2015.09.087
10. Panapakidis, I., Dagoumas, A.S.: Day-ahead electricity price forecasting via the application of artificial neural network based models. Appl. Energy **172**, 132–151 (2016). https://doi.org/10.1016/j.apenergy.2016.03.089
11. Chae, Y.T., Horesh, R., Hwang, Y., Lee, Y.M.: Artificial neural network model for forecasting sub-hourly electricity usage in commercial buildings. Energy Build. **111**, 184–194 (2016). https://doi.org/10.1016/j.enbuild.2015.11.045
12. Ceylan, H., Ozturk, H.K.: Estimating energy demand of Turkey based on economic indicators using genetic algorithm approach. Energy Convers. Manag. **45**, 2525–2537 (2004)
13. Zhang, G., Patuwo, B.E., Hu, M.Y.: Forecasting with artificial neural networks: the state of the art. Int. J. Forecast. **14**, 35–62 (1998)
14. Nair, V., Hinton, G.E.: Rectified linear units improve restricted boltzmann machines. In: Proceedings of ICML, vol. 30, pp. S807–S814 (2010)
15. Klambauer, G., Unterthiner, T., Mayr, A., Hochreiter, S.: Self-normalizing neural networks (2017). arXiv preprint arXiv:1706.02515

Peer Learning Methodology for Sustainable Energy Usage

Frans Folkvord[1,2(✉)], Lutz Peschke[3], Gökçen Baş[4], Simone Vitiello[1], and Nathan Spunda[1,2,3,4]

[1] Open Evidence Research, Barcelona, Spain
{ffolkvord, svitiello}@open-evidence.com, nathan.spunda@gmail.com
[2] Tilburg School of Humanities and Digital Sciences, Tilburg, The Netherlands
[3] Department of Communication and Design, Bilkent University Ankara, Ankara, Turkey
lutz.peschke@bilkent.edu.tr
[4] Vienna University of Technology, Vienna, Austria
goekcen.bas@tuwien.ac.at

Abstract. Humanity must be aware of the possibilities for sustainable energy technologies from their childhood onwards in order to enable a clean and prosperous future. Until now, youth receives only limited and mostly theoretical knowledge about new forms of energy usage, which leaves open doors to the usage of conventional energy sources. It is vital that youth develops a solid understanding of renewable energy's power and its various possible applications, thereby taking into account. Currently, youth are getting increasingly used to learn through transactional forms of communication via their (online) media consumption behaviour and contact with peers through social media. Considering these new communication forms, this study examines if peer-learning methodology (PLM) is an efficient method to train children in increasing their practical knowledge of sustainable energy usage. Nowadays, PLM is used as an educational methodology based on an eclectic integration of multiple theoretical insights from different scientific disciplines, such as developmental psychology, education science and paediatrics, that might be an effective learning methodology. The main idea behind PLM is that in order to educate youth effectively, a communication must motivate the receiver to actively attend to messages and perceive and interpret their content that is provided by peers, include iterative and transactional solicitation of feedback, and activate elaboration of message arguments and counterarguments to encourage individuals to move through the process of learning. In the current study, we investigated how students learn from other peers that have made posters that reflect on new and sustainable forms of energy. In total 14 posters and four movies were shown during an exhibition in both Ankara (Turkey) and in Tilburg (the Netherlands). During the exhibition, 30 pairs of students in Ankara and 12 pairs of students in Tilburg were equipped with a GoPro and they should talk about the exhibition according to the think aloud method. This data will be recoded afterwards by trained researchers in order to establish participants responses.

Keywords: Sustainable energy · Peer-learning methodology · Education · New forms of communication

N. M. Durakbasa and M. G. Gençyılmaz (Eds.): ISPR 2019, LNME, pp. 62–76, 2020.
https://doi.org/10.1007/978-3-030-31343-2_6

1 Introduction

Access to clean, affordable and reliable energy has been a cornerstone of the world's increasing prosperity and economic growth since the beginning of the industrial revolution [1]. For some decades, human activity is overloading our atmosphere with carbon dioxide and other global warming emissions, which trap heat, steadily drive up the planet's temperature, and create significant and harmful impacts on global health, environment, and climate [2, 3]. Therefore, the use of energy by the global population in the twenty-first century and future generations must become sustainable to be able to fulfil upcoming generations' energy-needs, in particular taking into account the increasing energy-need for developing countries [4]. The world needs another energy revolution in which the sources of energy are affordable, accessible, and sustainable [2, 3]. Energy efficiency and conservation, as well as decarbonizing the energy sources, are essential elements in this revolution.

Different initiatives have been launched and agreements have been set in order to increase the usage of sustainable energy. The most apparent and recent initiative is the Paris agreement, that has been ratified globally by practically every country, except the US and North Korea [5]. The agreement, states that society- and economy-wide decisions should be conducted to absolute emission reduction targets. Exposure to air pollutants as a consequence of traditional energy usage, like nitrogen oxides have been identified as a significant risk factor for a number of health conditions including extinction of species, respiratory infections of humans, heart disease, stroke and lung cancer [5, 6]. At the moment, greenhouse gas emissions still remain above agreed levels and would need to fall dramatically in order to meet international agreements [5]. In order to comply with the directives agreed on in the Paris Agreements, participating countries will possibly be obliged to take green initiatives to change behaviour and reduce air pollution in large cities and highly trafficked areas [7].

Solar, wind and water-based energy systems, and engineering of microbes to produce biofuels, are only a few examples that have been developed recently and can replace traditional forms of energy usage, in order to reduce the emissions and have positive benefits on world's energy security and the environment [8]. In addition, an increasing number of studies convincingly shows that renewable energy consumption benefits not only the environment but also for the economic conditions of (developing) countries [4, 9, 10]. Although an increasing number of sustainable energy-forms become apparent, it will take ambitious efforts to increase the actual implementation of these new techniques in everyday life.

Humanity must be aware of their existence, potentials, harvesting, technologies of use and benefits from very childhood onwards in order to enable clean and prosperous future. Until now, youth receives only limited theoretical knowledge about new forms of energy usage, which leaves open doors to continue consuming conventional energy sources. Renewable energies are available everywhere and most of the technologies of application are mature and relatively simple to implement in everyday life. It is vital that youth develops a solid understanding of renewable energy's power and its various possible applications, thereby taking into account effective educational strategies to educate children about sustainable energy usage, both on a theoretical and practical level.

Nowadays, youth are getting increasingly used to more transactional forms of communication through their online behaviour than the more traditional forms of communication, that are considered as one-directional [11]. To adapt the educational system in parallel to this new form of communication, the peer-learning methodology (PLM) has recently been established. In addition to this, there is a need to attract youth for the theme of environmental protection and conservation of nature through the use of renewable energy technologies. This study will address if peer-learning methodology is an effective method to publicly share information and knowledge about renewable energy technologies in an entertaining way. The current study tests if PLM can be used as a great start for youngsters to learn more about all types of renewable energy, in order to grow up more informed about the vital role of renewable technologies in changing world's energy future. Before we explain the methodology of the study and show the results, we will first elaborate more on PLM and about the theoretical underpinnings underlying PLM.

1.1 Peer Learning Methodology (PLM)

Every year new challenges emerge in the field of training and development, like competency development, outsourcing, e-learning, and knowledge management. The common theme of all these educational developments is that businesses, teachers, and political leaders want to assess and see value for their investment in educational programs, not only in improving the theoretical understanding of certain topics by children, but also on the practical understanding of children about what should be done. Core outcomes for educational programs that need to be assessed are if the learning is relevant and immediately applicable to their needs, and if the learning is effective and sustainable.

Different studies have shown that PLM is an effective tool to guarantee a high rate for the use of quality resources and the accuracy of content [12]. Furthermore, peer assessments encouraged feedback to the trainee leading to a learning arena of the collaborative learning. In the current study we test the use of the PLM, which is an innovative learning methodology financed by the European Commission [12]. The main aspect of PLM is that students learn from others (peers) by actively engaging in large trainee groups to experience practical skills, in combination with appropriate and real time feedback in order to improve their deep understanding and learning of real-world aspects. PLM is one the most efficient method among children and young people, therefore our aim is, under this project, to test PLM as an efficient method to reach a bigger number of children and young people and, therefore, to create a higher level of awareness on renewable energy.

1.2 Theoretical Underpinnings

The PLM is based on an selective integration of multiple theories from (social and developmental) psychology, paediatrics and education science. For example, the Self-Determination Theory [13] states that people have three innate needs; (1) competence, (2) feeling of relatedness and (3) autonomy. In addition, the Self-Determination Theory argues that (1) people are proactive with their potential and mastering their inner forces (such as drives and emotions), (2) people have an inherent tendency toward growth development and integrated functioning, and (3) people strive for optimal development and actions, that are inherent in people but they don't happen automatically and should be supported/triggered by the (social) environment. The self-determination theory focuses in particular on the degree to which an individual is self-motivated and self-determined or influenced by other factors. In addition, the Social Cognitive Theory [14] states that seeing or visualizing people (modelling) similar to oneself, like peers, perform successfully typically raises beliefs in observers that they themselves possess the capabilities to master comparable activities. This is called observational/vicarious learning, thereby affecting (1) self-efficacy, (2) outcome expectancies, (3) behaviour capabilities, (4) self-control, (5) managing emotional arousal, and (6) reciprocal determinism. In sum, the main idea behind PLM is that in order to educate youth effectively, a communication must motivate the receiver to actively attend to messages and perceive and interpret their content that is provided by peers, include iterative and transactional solicitation of feedback, and activate elaboration of message arguments and counter-arguments to encourage individuals to move through the process of learning.

The current study uses PLM to increase youngster competence in renewable energy through innovative and creative learning. Furthermore, it aims to train young people to be able to organize their community for launching a local renewable energy initiative with bottom up approach and to train young people's to be able to select the technologically, agriculturally, economically, socially and environmentally best fitting option for their community. As a consequence, it will increase young peoples' competence in renewable energy through innovative and creative learning, and be able to train the local actors to be able to organize their community for launching a local renewable energy initiative with bottom-up approach - train the local actors to be able to select the technologically, agriculturally, economically, socially and environmentally best fitting option for their community. The general aim of the study is to raise awareness about the importance of adopting sustainable habits and behaviours. We aim at making young people and the audience feel more concerned about these issues and make them more eco-responsible citizens. Hereby we hope that the young people realize that some problems concerning sustainability can be dealt with on a local and even personal level. Hence developing a sense of how they can convince others to act in a sustainable way. Partial aims of the study are (1) acquiring practical skills, (2) Obtaining professional skills, (3) accepting environmental values, (4) improving language skills, (5) social and cultural competences.

2 Reception Study in Ankara (Turkey) and Tilburg (the Netherlands) About an Environmental Science Exhibition

Aiming to understand peer learning strategies of young people in the context of environmental science communication, an exhibition about renewable energy and related environmental issues were created and a reception study was executed in Ankara and Tilburg. The study was verified in two parts.

2.1 Creating Popular Science Media

In the first part, exhibits were created within the undergraduate course "Science Writing and Journalism" at Bilkent University, Ankara. Fifteen students prepared text and graphic posters based on data which were provided by Vienna University of Technology and the Provincial Directorate of Environment and Urbanization in Ankara. Additionally, the students searched for other relevant scientific sources. The students were instructed to create media with a high content of storytelling which supports the recipients to understand complex scientific topics in a more sustainable way. The posters could be divided into three different categories. One set of posters dealt concretely with specific topics of energy generation: renewable energies, like geothermal power, wind energy, hydroelectric power as well as nuclear power. Another group of posters presented aspects of power, like how is the renewable energy concept in Arabian countries, used and not used potentials of renewable energy in Turkey, or does Turkey need more energy plants. Other posters discussed topics which are related with climate change and environmental issues, like the relation of urbanization and rainfall, plastic bags and meat consumption as well as the mind experiments about usage of renewable energy on Mars as well as about the question, would Neanderthals live more in harmony with the nature, if they would live today. All posters were accompanied by popular scientific texts.

The posters with specific renewable energies were added with schemes about technical aspects of power plants, which were produced by the project partner iserundschmidt from Bonn/Germany. The methodology of the poster and text creation is based on the approach that media which are received during leisure time will awaken more general and intrinsic interest, but also generate voluntary motivated relevance of youth. Peschke [15] developed a reception model which considers that knowledge can be received in a sustainable way when the topic is connected with own experiences of the life world. It includes Luhmann's theory of social systems [16] and the aspect of Schütz and Luckmann's life world [17].

Schütz's approach connects the everyday life world and the scientific world of knowledge. On the other hand, Habermas [18] understands as the structural components of the life world culture, society and person. The correlation of Habermas' models of the life world with Luhmann's system theory, where science is a sub-system of society, leads to the model where science is a part of the life world. But science is outside of the stock of knowledge which participants of communication use to agree on an object of the world. According to Habermas, this stock of knowledge is located in the culture. As a consequence, media has to contribute to the popularization of science, which is a process of decreasing the distance between science world and life world in order to share knowledge and influence the mass [17] and is the mediatized process of enabling the communicative interaction between science and culture.

The premise of this reception model [15] is, that science as a system has no access to the communicative everyday life practice without medial support. The function of media is, to enable catalytically a compatibility between the scientific object and existing mental images. Thereby, a translation process occurs, which initiates a compatible operation in the system culture – according to Habermas part of the life world. The personal relevance place an important role in this context. In this context, relevance is understood as pertinence in accordance with Foskett [19] and Lancaster [20], regarding the relationship between a medium and an information need. According to the authors, the global information need is higher than the individual information need, because of the big distance between lifeworld and the science world. Since there is no authority which can force an individuum to acquire knowledge, the existence of a voluntarily motivated relevance to deal with a certain scientific topic is needed.

For the reception study arise the following consequence: According to Foskett and Lancaster, the relevance is the relation between information need and medium. If there is an information need according to a topic, the relevance will be evaluated and verified with the help of voluntarily reception of a medium. If there is no information need, there cannot occur a verification of relevance. But a medium is able to generate interest to deal with it. But interest is not sufficient enough to start a process of reception. Otherwise, a medium would have to be only attractive to catch the recipient and the reception would be proceeded automatically, which is not the case. Accordingly, beside the interests, there has to exist an information need to deal with a scientific topic, which means that, a medium is only regarded as relevant with respect of the acquisition of knowledge, if a recipient has an information need which has to be released through awakening interest, dependent on the distance to the everyday life. But this approach is only viable if we assume in accordance with Mizzaro [21] that an information need is not a static and unchanging need but will change over the reception process dependent on the media. Mizzaro calls it a perceived information need. As a consequence, media have potentially impact on the individual perceived information need and can affect relevance, which is in harmony with Hall's representation approach [22].

Considering the approach described above, the students of the Department of Communication and Design at Bilkent University created twelve graphic posters with diagrams, pictures and verbal texts as well as popular scientific text posters [23]. The major types of renewable energy sources that were used were defined as solar, wind, geothermal, hydropower, ocean, tidal, biomass are studied over the last decades for a broad view of feasibility, development and integration in each category and were therefore used in the current study [24]. Students used the relevant literature comprising not only technical knowledge but also environmental relationship, economic and social effects of sustainable energy use. The exhibition additionally consisted of schemes about renewable energies as well as two animation movies about embodied energy and hydropower, which were produced by iserundschmidt – agency for science communication in Bonn, Germany.

3 Methodology

The reception study was conducted with 60 students in Ankara and 24 students in Tilburg. The participants consist of 34 young man and 48 young women. The average age of them was 21,15 years (N = 84; SD = 2,5; Median = 21). The participants were all students that were doing the following studies: Information Technologies (N = 12), Engineering (N = 6), Science (N = 5), Management (N = 18) Social Sciences (N = 11), Fine Arts, Design and Architecture (N = 24) and other subjects (N = 8). Pairs of students were asked to walk through the exhibition and to find out the main information of the different media. The study was conducted in a hallway at the campus of both universities. Participants were equipped with GoPro cameras which opened the opportunity to see how they moved through the exhibition, which details they checked, and the context of communication while decoding the information. Besides the two GoPro cameras, a wide-angle camera was placed which showed additionally the way which the pairs of participants chose in the exhibition. The pairs of participants were informed that they were completely free to decide in which order they analyse the exhibits, how long they wanted to deal with the different media as well as the way and context in which they wanted to discuss the provided topics. The only task which was given by the supervisor was that they should try to find out the main idea of the provided media, that they should talk about what they think, use the English language during the process of communication, and that they should stay always together.

The reception study is based on mainly two approaches. The task that the participants should talk as much as they can about what they think is related to the Think Aloud Method, developed and described by van Someren et al. [25]. The think aloud method is originally developed to observe participants while solving a given problem and to analyse the resulting verbal protocols. It is mainly applied in psychological and educational research on cognitive processes but also for knowledge acquisition (ibid,7). This method has its origin in the psychological research developed from the so-called introspection methods. Introspection is grounded in the idea "that one can observe events that take place in consciousness, more or less as one can observe events in the outside world." (ibid., 29). The advantage is that the produced data are open to analysis and interpretation for anyone. Within the scope of the reception study, the think aloud method is not used in the original way as described by van Someren et al. Instead of given a concrete problem which should be solved during the study, the participants were completely free to decide in which context they want to discuss the certain media. The aim was not to solve a concrete problem but to produce own constructs related with their individual constructed world. This strategy is closely related to Kelly's personal construct psychology. The Personal Construct Psychology is a fundamental work by Kelly [26] which is based on the constructivist approach that the world can only be identified by a human being with help of identification, evaluation and interpretation of certain events in the relation of his/her subjective world. It means that reality is not absolute truth but is based on alternative interpretations, which he/she can choose [27]. The evaluation can differ from human to human, and facts which are generated by these evaluations can be part of alternative constructs as well. Therefore, this philosophical position is described as constructivist alternativism [28]. The relation of the applied methods of the reception study to Kelly's personal construct psychology is generated by the strategy that no concrete question is asked to the participants, and that it is left open what the participants want to discuss. The freedom that the participants can choose the context how they want to discuss the given medium enables to get information about the constructs which are produced by the participants. It avoids that participants deal with constructs which are given by the supervisors. In the worst case the given constructs are not relevant with the life world of the participants.

4 Results

With the help of four examples of the poster exhibitions in Ankara and Tilburg, the effect of PLM will be discussed. The first example is a graphic poster and a text from Seyfafjehi [29] about energy resources on Mars. The text was written in a screenplay style. Professor Neutron and Professor Sun have a fight about the best energy source for living on Mars. Both confront the other the negative sides and disadvantages of solar power, which is defended by Ms Sun, and nuclear power, which is defended by Mr Neutron. The dispute is finished by the interruption of Professor Steam who gives a statement in favour of geothermal energy. The graphic poster visualises the comparison of how solar, nuclear and geothermal power works on Earth and how it would work on Mars with the help of diagrams, schemes and verbal texts.

The participants 1A and 1B remained on that poster 14'19 min which was extraordinary long comparing with the other pairs of participants (almost between 1 and 5 min). The following part of the dialogue shows a typical way of getting access to the information:

1B: So geothermal power is o.k., good. But, how can...
1A: It says here (points the scheme of geothermal energy on the lower left part of the graphic poster)
1B: How can it be renewable?
1A: Because water is off, always water, when you inject water, there, okay, because here in the core, it's after one, I mean.
1B: After a while, it will be down.
1A: Yes, it will be down, but here in the core, there is this energy, that doesn't go that, doesn't go away and when you start, when you inject...
1B: It does not finish...
1A: No, it does not finish. and when you send the water back, here, it heats again and again and again and you always use that steam, I mean, think about the core, it's really big and it's really hot especially on Earth, because, (points to the image on the upper left part of the graphic poster) cooler crusts, Mars has cooler crusts and core.
1B: Yeah, o.k., I see.
1A: And here you have, you don't have to drill so deep, because here is also hots, but on Mars, you have to deep drill, you, like, it stays in here to get hot water.
1B: I see. Actually, on Earth, in Turkey steam comes from the Earth.
1A: Yes, yes, and think about...
1B: Actually, my family lives in Meydan. And there are Geothermal energy stations. Yeah, to produce electric. And it is really close to surface.

The dialogue reveals significant information about the peer learning effect. In this part, participant 1B tries to understand the renewable aspect of geothermal energy. 1A got access to the topic obviously faster and explains it to 1B. It shows the two effects of peer learning: One part gets access to the information with the help of the explanation of the partner. The other partner gets access to the information by explaining step-by-step. Additionally, the two partners demonstrate the strategy of understanding through connecting the topic with stories and experiences from their own life world. Participant 1B obviously understood the topic that Mars has cooler crust and core and you would have to drill deeper to use geothermal power. At that moment, the participant compared this fact with the situation in her home town Meydan where geothermal power can be used in a quite efficient way. In a similar way, when they discussed the comments of Professor Neutron and the potential of nuclear power on Mars, participant 1A talked about a caesium analysis as a high school experiment, around 20 years after the nuclear disaster in Chernobyl.

> *"And I also made a project when I was in high school, I was in Trabzon, and there is a material, there is an element, Caesium, it only can be after a nuclear, nuclear, say, explosion explosion, and after Chernobyl, Caesium came to our land and when I dig the soil in Trabzon and examine it in the laboratory in university we found not high but low Caesium in the soil, so it still affects our soil and we're in Turkey".*

The dialogue and quote indicate and confirm on the one hand the importance of storytelling for the Public Understanding of Science. Stories bridge scientific knowledge with the life world in a catalytic way. From the view of popularisation, it can be understood as the decrease in distance between knowledge from the science world and the experiences of the life world [23]. On the other hand, autonomous decisions of how to deal with the knowledge media reveal information about the constructed world of the participants, especially about their perceived information needs, interests and relatedness to the medium and information itself. Consequently, the peer learning method can be evaluated in the context of reception model based on the perceived information need. Finally, it is obvious that partners which discuss in a self-determined way learn scientific topics more sustainable.

The second example supports this hypothesis. It is a dialogue about a poster which communicates the principle of wind turbines. In the focus of the text and graphics poster is the harmful effect of the wind turbines for birds. Text and graphic poster inform about the fact that every year worldwide 140,000 until 328,000 birds die from the collision with wind turbines, especially which are located on their migration routes. At the same time, it communicates possible solutions to avoid the death of birds [30]. As followed, the dialogue of the participants no. 3 shows a more balanced way of peer learning. The pair remained 5'27 min at that set of posters which is nearly the average of the remaining time of the participants:

> *3B: I didn't know, that Turkey was in any migration route like, you see that?*
> *3A: Yeah, but, but maybe, I don't think there are many ways within Turkey, like, routes of birds inside on, the inside Turkey, so it means that this is not a problem, they should consider, right?*
> *3B: Yeah. (read)*
> *3A: Yeah, exactly. Very important point.*
> *3B: Yeah. but like you have to consider, that would, it affects wind farm's productivity, if they are located away from the migration routes, because migration routes would, the wind be higher on migration routes, or not?*
> *3A: You're right, it helps them, actually you're right.*
> *3B: Yeah.*
> *3A: But it can't be like his that*

3B: I don't think, that is much of the problem, because 328,000 birds die from each..., ok. I don't think this number is not that huge, because it is across the world, I don't think, this is only in Turkey, is it?
3A: No, I don't think so, it's all over the world
3B: 328,000 birds dying is, like, probably one percent of the whole population of the birds dying,
3A: Well it can't be, I think the problem, they just focus on, is this problem
3B: Yeah.
3A: Of that temperature. That the birds are changing their roads and destination just because of their temperature was, which is getting higher
3B: Oh, they're like doing this in seas as well like
3A: Off-shore.
3B: In the oceans.
3A: I've never seen that, to be honest.
3B: I've never seen that since I've never been to the ocean. I've never seen ocean wind farms. that's interesting idea
3A: For... there are some opinions, that we must really destroy the landscape, like, they destroyed the way, we perceive the landscape, like, the natural beauty the natural landscape, but I think it's not a really important factor.

Comparing with the dialogue of participants no.1, the dialogue of the participants no. 3 are more balanced in a way, that there is not one who explains the other the problems and facts which are provided of the poster. The interaction of the two partners is more shaped by a common discovery of the information. The peer learning strategy in the context of self-determination is more significant since the role of the teacher is dissolved. The learning strategy is negotiated between the partners in an emancipated way and less influenced by a concrete problem of one of the partners.

The two other posters of Akıncı [31] and Cansaran [32] had two different kind of eyecatchers which enabled the entrance to the posters. Akıncı created a poster about the comparison of rainfall between the small city Edirne inside of a rural area and the mega city Istanbul. Both cities are located in the same region of Turkey, called Marmara. The distance between both cities is around 240 km. Without anthropogenic influences, the climate conditions should be very similar. But in one region of Istanbul the annual rainfall is extraordinary high caused by mirror-glass equipped skyscrapers and of the density of car traffic. Cansaran analysed the forecasts and real energy demands in Turkey. He reveals that the forecasts are much higher than the real needs. The design of infographics of both posters are very different. Cansaran's poster contains tables, text and pure bar charts. It follows the strategy and demand of Tufte in a consequent way. In contrast, Akıncı presented her topics with embellished elements. Her poster is divided in three vertical sections. The upper section shows statistics about the numbers of cars in Edirne and Istanbul, as well as the number of skyscrapers and high buildings in Istanbul. The lower part visualises the population in Edirne and Istanbul. The diagrams and charts are embellished with symbols. They visualise the meaning of the diagram according to Neurath's strategy. Neurath used for his charts a mixture descriptive and analytical codes. The charts, like bar charts are in general very abstract representations and get a concrete meaning with addition of symbols. Neurath [33] called his mixed codes ISOTYPE (International System of Typographic Picture Education). He wanted to establish an international visual language which can be understood cross-cultural and independent on language and class barriers in a transdisciplinary way. Acıncı used for the representation of car numbers icons of a VW Bug and for the representation of the cities a mosque (Edirne) and a bridge (Istanbul).

The center of Akıncı's poster is the comparative presentation of two maps which shows the rainfall in the "Rural City Edirne" and the "Urban City Istanbul" in 2017. The overlay of the different rain zones on the maps not too detailed. Therefore, the participants understood mainly quite fast the main information is that the rainfall in the European part of Istanbul was extraordinary high. Above and below the maps are charts in the style of Neurath. The density of the information is quite low. In contrast, Cansaran's poster consist of pure diagrams and tables. The eyecatcher is the red-coloured verbal text "*Turkey has no need to build new energy plants, forecasts are mostly wrong*!" in the middle of the poster. In other words, he already gave instructions how to read the tables and diagrams. This had significant consequences for the reception of the posters. As described above, all graphic posters were accompanied with texts which gave more detailed information about the topic. In the case of Akıncı's poster, nearly all participants started the reception with the map of the graphic poster. Most of them did not read the text poster beside the graphic poster. The relevance for the life world of the participants arose from the information about Istanbul. Nearly all Turkish participants have already been in Istanbul. They know the problematic situation regarding the car traffic, in particular in the center of Istanbul near the Bosphorus, where car traffic is seen all day, almost every day. The participants from other countries than Turkey know Istanbul less. Therefore, the perceived information needs of these participants were much more related to information about Istanbul than Edirne. Typically, the participants decoded provided information quite fast and started exchanging own experiences with Istanbul or other big cities ("have you ever traveling with public transportation in Istanbul? […] it is awful […] I don't have space to move inside the metros." (3B from Pakistan); "I always compare a population of Istanbul with the population of Poland, and in Istanbul are 18 [million citizens], and it is like half of Poland" (3A from Poland)). The most of the recipients did not read the text poster, and started discussing own life experiences and media news. Many of them discussed the activities of Greta Thunberg and the Friday for Future activities combined with criticisms of governmental activities and ignorance. In contrast, the participants dealt with Cansaran's poster in a completely different way. Some of them rejected the reception because of "too many numbers" (22A/B) or dealt in a very sketchy way. Many of them were attracted by the headline but before they started decoding of the tables and graphs, they read the text poster. It revealed that the pure information of charts constitutes a threshold which is difficult to get over. As a consequence, many recipients escape from the topics or fall back on text media which they regard as easier to understand. The catchy text "Turkey has no need to build new energy plants, forecasts are mostly wrong!" helps the access to the topic of the poster but at the same time it avoids the deeper decoding activities, since the main idea is already given. The described results are only a little segment of a broader context. But it reveals that environmental communication cannot provide physical reality as objective facts. It needs to connect phenomena with experiences of the lifeworld.

5 Discussion

The current study examined if PLM is an effective method for youth to learn and process information and knowledge about renewable energy technologies. Consequently, planning, management and implementation of the training concept are based on the interrelation of basic elements and guidelines with a focus on high quality.

The dialogues that were recorded between the students revealed significant information about the peer learning effect in the usage of sustainable energy. As the results showed, PLM is an effective learning methodology since youth access and learn the information with the help of the explanation of the peer and because of explaining the information step-by-step to the peer. In addition, participants used stories and experiences from their own life world to explain and elaborate on the information to their peers to demonstrate the strategy of understanding through connecting the topic with. Next, peers also interacted through a common discovery of the information, thereby showing higher self-determination to acquire the presented information since the role of the teacher is dissolved.

Based on these results we think that the dialogues and quotes indicate and confirm on the one hand the importance of storytelling when considering the public understanding of science. As the current study shows, stories have the opportunity to bridge scientific knowledge with the life world in a catalytic way and can be an effective strategy to decrease the distance between knowledge from the science world and the experiences of the life world [15]. The peer learning strategy in the context of self-determination appeared to be more significant since the role of the teacher is dissolved, and more responsibility and determination is activated among students themselves. The learning strategy is negotiated between the partners in an emancipated way, and less influenced by handling a concrete problem or situation that is provided by the teacher, making the learning process intrinsically more determined.

Students learn from others (peers) by actively engaging in large trainee groups to experience practical skills, in combination with appropriate and real time feedback in order to improve their deep understanding and learning of real world aspects. In this case, the important subject of sustainable energy usage [4, 9, 10]. PLM is an effective methodology to reach a higher number of children and young people, with their own communication skills and tools, and, therefore, to create a higher level of awareness on renewable energy. In addition, they also stimulate acquiring practical skills, obtaining professional skills, accepting environmental values, improving language skills, and learn social and cultural competences.

Currently, scientific knowledge is mostly outside of the stock of knowledge of the general population, whereby media can, and should, contribute to the popularization of science, which is a process of decreasing the distance between science world and life world in order to share knowledge and influence the mass to use more sustainable energy [17]. Science as a system has no access to the communicative everyday life practice without medial support, so the main function of media is to enable catalytically the compatibility between the scientific object and existing mental images. Global information need is becoming higher than the individual information need, because of the big distance between lifeworld and science world. Accordingly, beside the interests, there has to exist an information need to deal with a scientific topic, which means that, a medium is only regarded as relevant with respect of the acquisition of knowledge, if a recipient has an information need which has to be released through awakening interest, dependent on the distance to the everyday life. Peers increase the interest and need to acquire certain knowledge as the results showed, so PLM can be a very effective methodology to increase the theoretical and practical knowledge about sustainable energy among youth.

The current study has several strengths. First, most participating students considered it as joyful to participate and reported to have learned much about sustainable energy. This was found also among Dutch and international students studying in Tilburg. Second, testing both Turkish and Dutch students gave us the possibility to compare cultural differences between participants from these countries. Because the posters were created taking into account sustainable energy options in Turkey, it was interesting to see how Dutch students perceived the posters, because it is less directly relevant for them. Third, considering that the study was conducted in a real-life exhibition with multiple technological devices to collect the data, the ecological validity of the study is guaranteed.

In contrast, some limitations also exist. First, only a small group of participants were included in the current study. Conducting a study like this is time and energy consuming, therefore it was difficult to obtain more participants. Second, no control groups were established in the current study, so comparing the PLM to other forms of educational methods is difficult. Because the study was exploratory and we interpreted the results qualitatively, we recommend future studies to experimentally test a larger number of participants with a control group to see potential differences in educational settings.

6 General Conclusion

This study showed that PLM is an effective and joyful education methodology to inform and activate youth to learn more about sustainable energy. The translation from scientific knowledge to the mass is highly needed in order to modify human's energy behaviour and make it more sustainable. Considering the rapid economic and societal developments in developing countries, and its increased energy usage that develops in parallel, we need to rethink our energy usage immensely and educate youth effectively to learn more about sustainable energies and their usage (see also [5, 6]).

Acknowledgment. This research paper was generated within the scope of the project "POWER: Empowerment of Youth on Renewable Energy for Sustainable Societies", which is funded by the Erasmus+ Program of the European Union 2017-3-TR01-KA205-048402.

References

1. Chu, S., Majumdar, A.: Opportunities and challenges for a sustainable energy future. Nature **488**(7411), 294 (2012)

2. Beckage, B., Gross, L.J., Lacasse, K., Carr, E., Metcalf, S.S., Winter, J.M., Howe, P.D., Fefferman, N., Franck, T., Zia, A., Kinzig, A.: Linking models of human behaviour and climate alters projected climate change. Nat. Clim. Chang. **8**(1), 79 (2018)
3. Urry, J.: Climate change and society. In: Why the Social Sciences Matter, pp. 45–59. Palgrave Macmillan, London (2015)
4. Ito, K.: CO2 emissions, renewable and non-renewable energy consumption, and economic growth: evidence from panel data for developing countries. Int. Econ. **151**, 1–6 (2017)
5. United Nations Framework Convention on Climate Change (UNFCCC). https://unfccc.int/process/the-paris-agreement/status-of-ratification. Accessed 18 Apr 2019
6. IPBES: Nature's dangerous decline 'unprecedented' species extinction rates 'accelerating' (2019). http://campaign.r20.constantcontact.com/render?m=1102243211802&ca=2853ad8d-f207-4e95-adda-5fcc9e7bfc88. Accessed 9 May 2019
7. Kushwaha, G.S., Sharma, N.K.: Green initiatives: a step towards sustainable development and firm's performance in the automobile industry. J. Clean. Prod. **121**, 116–129 (2016)
8. Larcher, D., Tarascon, J.M.: Towards greener and more sustainable batteries for electrical energy storage. Nat. Chem. **7**(1), 19 (2015)
9. Inglesi-Lotz, R.: The impact of renewable energy consumption to economic growth: a panel data application. Energy Econ. **53**, 58–63 (2016)
10. Ozturk, I., Acaravci, A.: CO2 emissions, energy consumption and economic growth in Turkey. Renew. Sustain. Energy Rev. **14**(9), 3220–3225 (2010)
11. Barnlund, D.C.: A transactional model of communication. In: Communication Theory, pp. 47–57. Routledge (2017)
12. Moldovan, L.: Innovative method of peer assisted learning by technology and assessment of practical skills. Procedia Technol. **12**, 667–674 (2014)
13. Ryan, R.M., Deci, E.L.: Self-determination theory and the facilitation of intrinsic motivation, social development, and well-being. Am. Psychol. **55**(1), 68 (2000)
14. Bandura, A., Walters, R.H.: Social Learning Theory, vol. 1. Prentice-hall, Englewood Cliffs (1977)
15. Peschke, L.: Infografiken. Visualität und Wissensaneignung in der mediatisierten Welt. Springer VS, Wiesbaden (2019)
16. Luhmann, N.: Soziale Systeme. Grundriß einer allgemeinen Theorie. Suhrkamp, Frankfurt am Main (1987)
17. Schütz, A., Luckmann, T.: Strukturen der Lebenswelt. UVK, Konstanz (2003). [1979]
18. Habermas, J.: Theorie des kommunikativen Handelns. Band 2: Zur Kritik der funktionalistischen Vernunft. Suhrkamp, Frankfurt am Main (1987)
19. Foskett, D.J.: A note on the concept of relevance. Inf. Storage Retr. **8**(2), 77–78 (1972)
20. Lancaster, F.W.: Information Retrieval Systems: Characteristics, Testing and Evaluation. Wiley, New York (1979)
21. Mizzaro, S.: How many relevances in information retrieval? Interact. Comput. **10**(3), 305–322 (1998)
22. Hall, S.: The work of representation. In: Hall, S., Evans, J., Nixon, S. (eds.) Representation. Cultural Representations and Signifying Practices, 2nd edn. SAGE, London u.a. (2013)
23. Peschke, L.: Visuality of the Anthropocene. A Concept for a Poster Exhibition. Book on Demand, Norderstedt (2019)
24. Jackson, T.: Renewable Energy: Prospects for Implementation, 1st edn. Butterworth-Heinemann (1913). ISBN 978-1-4832-5695-5
25. van Someren, M.W., Barnard, F.Y., Sandberg, J.A.C.: The Think Aloud Method. A Practical Guide to Modelling Cognitive Processes. Academic Press, London (1994)
26. Kelly, G.: The Psychology of Personal Constructs, vol. 1–2. Norton, New York (1955)

27. Bonarius, H., Holland, R., Rosenberg, S.: Personal Construct Psychology. Recent Advances in Theory and Practice. MacMillan, London (1981)
28. Viney, L.L.: Social science research in the 1990s: the contribution of constructivism. Int. J. Pers. Constr. Psychol. **5**(3), 295–305 (1992)
29. Seyfafjehi, S.: Energy resources on mars. In: Peschke, L. (ed.) Visuality of the Anthropocene. A Concept for a Poster Exhibition. Book on Demand, Norderstedt (2019)
30. Yılmaz, A.D.: Serial killer wind turbines? In: Peschke, L. (ed.) Visuality of the Anthropocene. A Concept for a Poster Exhibition. Book on Demand, Norderstedt (2019)
31. Akıncı, N.: Urbanized city through rainfall: how does climate change relate to it? In: Peschke, L. (ed.) Visuality of the Anthropocene. A Concept for a Poster Exhibition. Book on Demand, Norderstedt (2019)
32. Cansaran, O.E.: Power generation of Turkey. An analysis on guesses and facts. In: Peschke, L. (ed.) Visuality of the Anthropocene. A Concept for a Poster Exhibition. Book on Demand, Norderstedt (2019)
33. Neurath, O.: Gesammelte bildpädagogische Schriften. In: Haller, R., Kinross, R. (eds.) Hölder-Pichler-Tempsky, Wien (1991)

Flexible Manufacturing Systems

Analysis of a Dedicated Flexible Manufacturing System with Closed Loop Layout: A Case Study in Production of Electro-Mechanical Products

R. Şirin Uyan[1] and Adalet Öner[2(✉)]

[1] Scheneider-Turkiye, Istanbul, Turkey
sirin.uyan@scheneider-electric.com
[2] Industrial Engineering Department, Yasar University, Bornova, Turkey
adalet.oner@yasar.edu.tr

Abstract. This study concerns with a dedicated flexible manufacturing system with closed loop layout. The production system consists of different types of parts with different processing times moving on the closed loop conveyor. An analytical model is proposed to show the dynamics and interactions in the system. Since the model is nonlinear and ignores random machine failures, a detailed simulation model has been developed to be able to make a proper analysis of the system. The objective is to find the best configuration in order to maximize the throughput. A number of scenarios representing different configuration settings have been evaluated and compared with respect to the objective function. Existing optimization methods and tools, which are used along with simulation models, have been addressed and used to find the best solution. The results have been discussed and recommendations have been made for future.

Keywords: Production systems · Simulation · Optimization · Flexible manufacturing systems

1 Introduction

A Flexible Manufacturing System (FMS) is a structure of computer controlled semi-independent workstations, which have connection through an automated transportation system. There are many different FMS configurations which vary with the types of components used in the system such as the types of machine tools, types of material handling system, type of storage areas for in-process inventory and the variety of part types to be processed [1].

This study concerns with a dedicated flexible manufacturing system with closed loop layout. A number of ordered operations are performed on a fixed set of part types. The processing time of each operation is different for each part type. Each operation is assigned and performed on only one machine station for all part types. There is a fixed route for each part through the system. A conveyor is used to move parts between machines. There is local storage areas with limited capacities in between machine

N. M. Durakbasa and M. G. Gençyılmaz (Eds.): ISPR 2019, LNME, pp. 79–90, 2020.
https://doi.org/10.1007/978-3-030-31343-2_7

stations. In fact, the system has a unidirectional cyclic design and operates similarly to a dedicated non-homogeneous transfer line.

This kind of production system is used when the same series of operations are performed on the different parts of a final-product whose parts have different size, shape and material. The computers regulate the machines to conform to the changes in size, shape and material if two consecutive parts are of different types. For example, Schneider Electric uses this system for metal coating and related operations in producing medium voltage switching devices. Basic structure of the system is seen in Fig. 1.

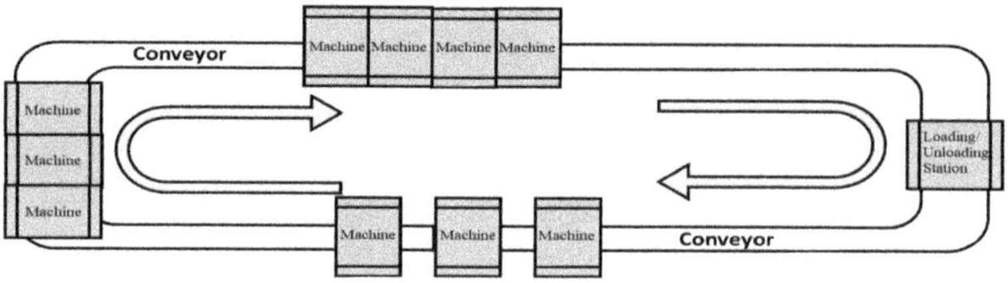

Fig. 1. A dedicated FMS structure with closed loop layout

The parts are transported between machines in load-bearing containers i.e. trolleys which are moved by a conveyor system. The speed of the conveyor may vary at different sections on the layout due to technological requirements. The trolleys have different fixtures for different part types. Therefore, there is a specified trolley type associated with each part type. Trolleys accommodate one or more parts simultaneously depending on the structure of its fixture. Accommodation capacity of each trolley type is fixed and predetermined. The machines process the parts in batches accommodated in the trolley without dropping them off.

Unprocessed parts enter in the system when they are loaded at the loading station in the trolleys that are compatible with that part type. The trolley loaded with parts is moved through the system to visit all stations sequentially until the last operation is performed on the last station. When all operations are completed, the trolley then moves forward to complete a closed loop path and arrives back to the loading/unloading station where parts are dropped off. Empty trolley is then loaded again with new unprocessed part(s) of the same type and it continues to revolve in the system [2].

If a machine at a station is busy, the trolley coming from the previous station waits in a local buffer storage area, which has a limited capacity. If the local storage area between two consecutive stations is full, then the trolley cannot leave upstream station and hence blocks it. The station stays blocked until an unoccupied space is available in the buffer storage area in between.

If someone could observe a time-lapse animation of the system, he/she would see a set of trolleys revolving constantly in a closed loop. This set is comprised of different trolley types and hence we can define disjoint subsets each having a number of trolleys from a different type. Since different parts of a product are processed in the system, the number of subsets should be equal to the number of part types and each subset should have at least one element. The cardinalities of the subsets, i.e. the number of elements in each subset of trolleys should be decided. A great number of combinations may be defined as alternative feasible solutions, however, the problem is to find the best combination, in other words to find the best configuration of the system in order to maximize the efficiency.

2 Problem Definition

In production of industrial electro-mechanical switching devices (SD), the production process can be divided into several stages. During the early stages, the parts of the SD are produced separately and then, they go through a process called "Metallization Process". The last stage is the assembly line in which the parts are combined together and assembled to form the end product, SD.

Metallization process includes some consecutive operations essentially for coating the parts with the layers of special materials. There are two functions of the metallization process. The first one is protection of the surface of the parts from harsh environment conditions such as moisture, dust, chemicals etc. It also includes mechanical protection since the process provides resistance to shocks, which may possibly cause micro cracks and eventual failures and malfunctioning of the parts. The second function is to establish a Faraday cage around the parts by coating it with a conductive (metalized) paint. If the final product is electrically grounded in a proper way, the electrical field generated during the operation of the switchgear is totally kept inside the product, as the metallization layer is fully covering the surface area of the product.

The operations of the Metallization Process are performed on a specially built, carousel like platform, which involves processing machines and conveyor segments connecting the machines. Conveyor segments form a closed layout. The structure of the system is shown in Fig. 2.

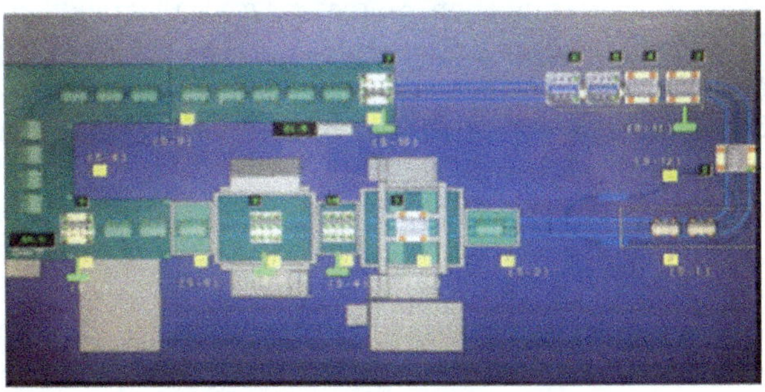

Fig. 2. The layout of the system

There are n types of parts to be processed in the system and the same set of ordered operations are applied to every part type. Each operation is processed by a specific machine. The number of machines (and hence operations) is m.

Since part types have separate shapes and sizes, each part type is carried on a particular type of trolley, which is associated with that part type. Each trolley type has a specific fixture on it in order to accommodate one or more parts simultaneously. A_j indicates number of parts that can be accommodated in the trolley type j. The machines process the parts in batches accommodated in the trolley without dropping them off. The notation p_{ij} represents the processing time of a batch of part type j (a load of trolley type j) on machine i, such that $1 \leq i \leq m$.

The system produces semi-finished parts to be used in finished product (SD). The bill of materials (BOM) determines B_j, which indicates the numbers of each product type in one unit of finished product. The aim is to increase the throughput of the system in terms of final product. Therefore the performance is measured by the quantity of end product that can be produced using the parts processed in the system in a given period of time, for example in a week.

Decision variables are the numbers of elements in each subset of trolleys types. In other words, main question will be "how many trolleys should be used from each trolley types?" The objective is to maximize the quantity of finished product while ensuring the following technological constraints are not violated.

- Total number of trolleys allowed in the system is limited to a specific number R due to physical restrictions.
- Each subset of trolleys should have at least one element. In other words, there should be at least one trolley from each type.
- Pre-emption is not allowed.

3 Literature Review

There are many articles in the literature regarding to flexible manufacturing systems (FMS) and optimization with simulation. Studies have focused on different topics, such as analysis of the flexible manufacturing system, closed loop layout design and some of them focused on optimizing some performance metrics in the system.

Basic definitions and a classification scheme of flexible manufacturing systems is given in [3]. In reference [4], a classical FMS system is studied. Different types of conveyors, transporters and workstations are considered, and a deterministic analytical model (queuing model) is proposed to make an analysis of classical FMS systems. The objective function is minimizing waiting times and queue size at particular crucial points in the system. However, our system is not a classical FMS, rather a specific variant of FMS. Moreover, we do not need manage a queue.

In reference [5], a study is presented to analyse throughput of non-homogeneous transfer line, which has different process times in each machines. The problem settings are similar to our problem. However, its layout is not a closed loop. Each part type is moving through between workstations and when its process on the last machine is finished, it leaves the system. In our closed loop layout, the parts are unloaded after they are done with the last workstation, and then the trolleys are loaded again with the same type of parts and hence sequence of the jobs are determined in a different way in our problem. They proposed different analytical approaches, which delivers approximate solutions. They do not prefer analytical models because they cause significant errors and poor estimate when compared to verified and validated simulation outputs.

Performance analysis of flexible manufacturing systems is studied in [6]. This paper concluded that analytical models are complex and usually nonlinear and therefore they are difficult to solve. They advised to use simulation models since they are more effective to analyze such systems.

The performance analysis of closed loop material handling systems is presented in [7]. In this paper, a manufacturing system is studied which consists of a set of workstations, an inventory system, and a material handling system. Model has developed for analyzing the performance of a manufacturing system consisting of N workstations, one loading station, N unloading stations, and an inventory system linked by conveyors. Incoming parts enter from loading station and recirculate throughout the conveyor, leave the system through the respective unloading station after being processed. Although there are some similarities with our problem settings, there are significant deviations. For example, in that model, the parts have an option to bypass a machine to be processed in others. In our problem, the parts should be processed sequentially in the same order. Furthermore, the parts arrive in a stochastic process, which does not conform to our problem settings. As usual, a simulation model is used to conduct the analysis.

An analysis of the performance measures of classical FMS systems is given in [8]. This study focused on application of Petri nets for measuring performance of FMS. He considers flexible routes for the parts in the system, which indicates a deviation from our problem environment since there is no route flexibility in our model. A simulation model is used as the analysis tool. Additionally, the bottleneck technique (an analytical model) has been developed to compare and verify simulation results. Designing optimal FMS for particular requirements is a complex problem and hence it is hard to develop accurate mathematical models to calculate performance measures. Therefore, simulation models are used for numeric modeling technique for analyzing highly complex systems.

In reference [9], improvement of the resilience of production systems had been studied. A methodology is proposed to assess the performance in face of breakdowns and to identify the level of resilience for a production system. Due to its modular structure, arbitrary production lines have been analyzed. A simulation model is employed and optimization procedures are used along with the simulation model. They use genetic algorithm to find the best configuration of the system to maximize the resilience.

Another simulation model is proposed in [10] to make an analysis of FMS. This study focused on strategic issues like variants of the simulation models to run and analyzing the outputs. In their problem settings, there are a number of machines usually performing different operations, however some of them identical and performs the same operation. The parts may have different routes. In our problem environment, there is only one machine dedicated for each particular operation and the parts have the same route through the machines. The aim of that study is determining the machine mix i.e. the number of machines performing each operation and the number of flexible machines performing any operation.

Almost all the studies use a simulation model as the basic analysis tool since underlying relationships usually lead to the complicated and nonlinear analytical models. We will follow the same path to demonstrate the analysis of our problem setting, i.e., a dedicated FMS with closed loop layout.

4 Mathematical Model

If machine failures ignored, the system can be considered as a deterministic model and therefore an analytical model can be demonstrated. A mathematical model has been developed to find the best composition of the trolleys to maximize throughput of the system. The notation is given as follows.

Parameters:

j: index for trolley types, $1 \leq i \leq n$,

B_j: quantity of part type j needed to manufacture an finish-product (shown in BOM)

A_j: number of parts that can be accommodated in the trolley type j (determined by fixture of that trolley)

R: maximum number of trolleys allowed to operate in the system

Decision Variables:

x_j: number of type j trolleys to be used in the system, (Each part type is represented by an associated trolley type)

y: the number of finished product that can be produced in one complete tour of all trolleys

k: the number of tours that can be completed in a given period

$$\text{Objective function;} \qquad Max.Z = k * y \tag{1}$$

$$\frac{A_j}{B_j} * x_j \geq y, \qquad \forall j = 1, 2, \ldots, n \tag{2}$$

$$x_j \geq 1, \qquad \forall j = 1, 2, \ldots, n \tag{3}$$

$$\sum\nolimits_{j=1}^{n} \mathrm{x}_j \leq R \tag{4}$$

$$k = f\left(p_{jk}, \mathrm{x}_1, \mathrm{x}_2, \ldots, \mathrm{x}_{n,}\right) \tag{5}$$

$$x_j \text{ integers}, \qquad \forall \mathrm{j} = 1, 2, \ldots, \mathrm{n} \tag{6}$$

Notice that the formulation leads to a nonlinear integer-programming problem since the objective function includes a multiplication of two decision variables and each decision variable is restricted to be integers. Furthermore, Eq. (5) indicates that the tour time k is a function of $x_j's$. Although the model (1)–(6) is relatively simple, that function in Eq. (5) creates a great deal of complexity and ambiguity and therefore needs to be elaborated. For this reason, let us adopt some additional notation as follows.

i: index for work stations, $0 \leq i \leq m$, Loading/unloading station is denoted by i = 0
V: total number of trolleys currently used in the system,

$$V = \sum_{j=0}^{n} x_j$$

P_{ij}: operation time of parts carried in trolley type j on station i
g: index for conveyor segments between stations.

Since the system is designed as unidirectional cyclic, conveyor segments are represented by a set of ordered pairs of stations, $G = \{(0,1),(1,2),(2,3),\ldots,(m-1, m),(m,0)\}$

T_g: transfer time on conveyor segment g.

If only one of a particular type of trolley is allowed in the system, in other words, only one x_j is equal to 1,

C_j: total job completion time or the time for completing one tour for trolley type j if only one of trolley of type j is revolves in the system. It is defined as the sum of all processing and transfer times,

$$C_j = \sum\nolimits_{i=1}^{m} P_{ij} + \sum\nolimits_{g \in G} T_g \tag{7}$$

Then the number of tours, k, in a given particular duration of production time, T, can be defined as follows.

$$k = T/C_j = \frac{T}{\sum_{i=1}^{m} P_{ij} + \sum_{g \in G} T_g} \tag{8}$$

On the other hand, if only one of each type of trolley is allowed in the system, in other words, each x_j is less than or equal 1, then the number of tours, k, is given as follows.

$$k = \frac{T}{Max\{C_1, C_{2,}, \ldots C_n\}} \tag{9}$$

If more than one trolley from each type is allowed in the system as required in real life application, i.e., $1 \leq x_j \leq R$, then the number of tours, k, is given as follows

$$k = \frac{T}{Max\left\{\sum_{j=1}^{n} x_j P_{ij}, \forall i = 1, 2, \ldots m\right\}} \tag{10}$$

However, the equation above holds in specific conditions, such as if machine blocking does not occur in the system. If machine blocking occurs, new relations should be investigated.

5 Simulation Model

The ARENA software has been used for creating simulation model for the analysis of the system. An animation model has also been developed to accompany to the simulation model. The framework of the animation model is shown in the below Fig. 3.

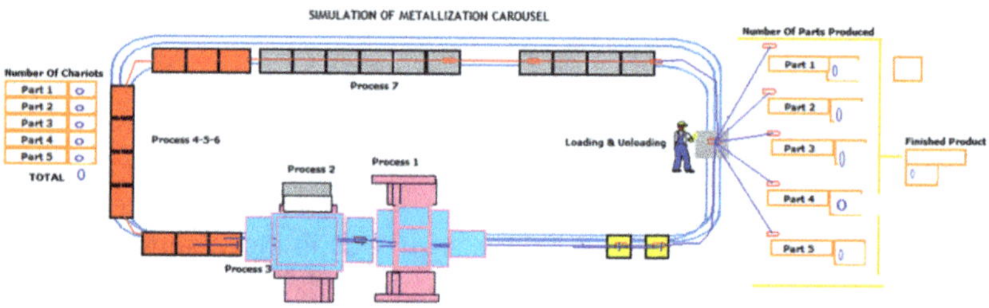

Fig. 3. The framework of the animation model

Two external files have been used in the model, inputs are received from the first one and outputs of the simulation are written to the second file. The input file includes the following data.

- Processing times of each trolley type at each workstation
- Transfer times between stations
- Loading/unloading times of each type of trolley
- The number of parts on each trolley type

The simulation model is essentially a deterministic one except the machine breakdowns. Since it has a stochastics nature, statistical analysis has been conducted on historical data to find the best distributions for Time-To-Failure and Repair times. Acceptance tests has been applied to verify the candidate distributions.

To verify the model, tests have been conducted using the ARENA software-debugging tool described as in [11] for each sub-module in the process of the model development. First, only a single part is allowed to enter into the system and solitary part flow is observed through the system. The same observations are carried for each other part types. Furthermore, especially the part interactions are investigated carefully.

The system tested for many different values of part configurations and processing times. The aim is to create wide variety of different situations where the model logic might fail. A detailed animation model has been developed to accompany the simulation model. It allowed us to track the flow of parts and to view the activities that occurs within the system.

In order to validate the model, i.e. to check if the model has accurate representation of the real system, the outcomes of the simulation model are compared to the observations on the real system. The real system is observed for different configuration of trolley types is on action. Numbers of produced parts of each type are compared to the simulation outputs.

Simulation output has been analysed following the procedures and techniques described in [12]. Initially, the warm-up periods have been determined. Confidence intervals have been calculated for 10 replications for each scenario (a particular configuration of trolleys). It has been observed that the values of half widths of confidence intervals vary depending on the scenario, therefore it is required to standardize half widths and hence the concept of "relative error" is used [12, 13]. Using this concept, numbers of required replications have been determined in order to attain satisfactory relative errors. Each scenario has been run again with the calculated number of replications and it has been observed that relative error of each scenario has been decreased under threshold value.

6 Optimization via Simulation

The simulation model provides us with the results of a particular scenario. In our problem, each scenario represents a particular configuration of trolley types. In other words, each scenario is defined by determining the numbers of trolleys from each trolley type. Since many scenarios can be described, the task is to find the best scenario with respect to the performance measure. Two different approach has been followed in this study. The first one is to list only the promising configurations and hence determine scenarios to be fed into simulation model. It is a filtering process to eliminate the unpromising configurations. The rules of this process can be described as follows.

Step-1: Determine mandatory minimum number of trolleys in the system, V_{min}
V_{min} is equal to the number of part types to be processed.
Assign a trolley for each part type. This is the first and base scenario.
Total number of trolleys in the system, $V = V_{min}$

Step-2: Increase the total number of trolleys in the system, $V = V + 1$

Step-3: Determine number of trolleys for each part type such as $\left\lfloor \frac{A_j}{B_j} \right\rfloor$ and $\left\lfloor \frac{A_j}{B_j} \right\rfloor + 1$

Step-4: Determine the combinations of trolley configurations such that total number of trolleys is equal to V, and the number of trolleys from each type should conform to Step 3.
Each combination represents a separate scenario. Add the scenarios generated into the list.

Step-5: if V is equal to total number of trolleys physically allowed in the system, R, then exit,
otherwise, go to Step-2.

This procedure is useful to eliminate the unpromising scenarios. The size of the list that contains competent scenarios can be reduced considerably. Once the list is prepared, each scenario is evaluated by simulation model. It can be a time consuming process to run each scenario in the computer, however there is a very useful tools in ARENA software to run many scenarios simultaneously.

The other approach is to use special techniques described in [14]. One of the tools suggested is the "OpQuest" tool in ARENA software. It uses a linear combination procedure, suggested in connection with the scatter search methodology.

7 Solution of a Sample Problem

A sample problem has been solved using the two optimization approaches stated above. The sample contains 5 part types and allowable total number of trolleys in the system is 30. The number of parts required in one finished product and the number of parts loaded in one trolley for each part type are given in the following tables respectively (Tables 1 and 2).

Table 1. Number of parts to be used in a finished product, B_j

Part type 1	Part type 2	Part type 3	Part type 4	Part type 5
3	1	1	3	3

Table 2. Number of parts that are loaded simultaneously in one trolley, A_j

Part type 1	Part type 2	Part type 3	Part type 4	Part type 5
6	1	2	12	12

There are seven operations that are executed on machines. Loading and unloading operations are done manually. The processing times are given in table below (Table 3).

Table 3. Processing times, p_{ij}

Operation	Part 1	Part 2	Part 3	Part 4	Part 5
Loading	554	401	501	140	140
Transfer on conveyor to the next station	40	40	40	40	40
Operation 1	406	371	350	319	387
Transfer on conveyor to the next station	14	14	14	14	14
Operation 2	783	325	293	547	282
Transfer on conveyor to the next station	14	14	14	14	14
Operations 3	162	162	162	162	162
Transfer on conveyor to the next station	8	8	8	8	8
Operations 4–5–6	1200	1200	1200	1200	1200
Operation 7	26	26	26	26	26
Transfer on conveyor to the next station	33	33	33	33	33
Unloading	98	297	232	88	85

The facility is assumed to work seven days a week and two-shifts a day (16 h). Time horizon is selected to be 72 days. There may be one or two operators for loading and unloading operations. If the number of operators is one, then the same operator does both operations. Otherwise, one operator is assigned only for loading operation and the other one does the unloading operation. The quantity of throughput by the total number of trolleys are shown in the following figure (Fig. 4).

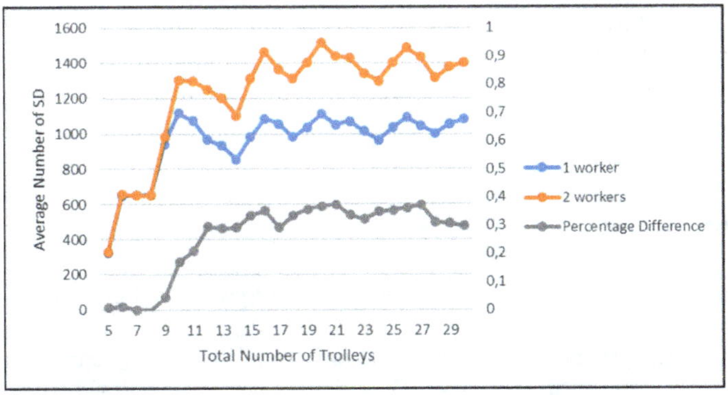

Fig. 4. Maximum number of finished products by the total number of trolleys in the system

The best configurations that yields maximum number of finished products are described in the following table. It can be seen that, if one operator is used to execute both loading and unloading operations, best configuration is comprised of total 10 trolleys. The distribution of those trolleys for trolley types are 2, 4, 2, 1 and 1 respectively. On the other hand, if loading and unloading operations are done by separate operators, then the best configuration is to assign 4, 8, 4, 2 and 2 trolleys for each part types with a total 20 trolleys in the system (Table 4).

Table 4. Best configurations

Number of workers for loading/unloading	Max number of finished product	Total number of trolleys in the system	Distribution of trolleys (pieces)				
			Trolley type 1	Trolley type 2	Trolley type 3	Trolley type 4	Trolley type 5
1	1116	10	2	4	2	1	1
1	1100	20	4	8	4	2	2
1	1085	16	3	6	3	2	2
2	1519	20	4	8	4	2	2
2	1487	26	5	10	5	3	3
2	1466	16	3	6	3	2	2

8 Conclusion

A dedicated flexible manufacturing system with closed loop layout has been studied. An analytical model is proposed to show the dynamics and interactions in the system. It has been observed that such systems tend to be nonlinear w.r.t. mathematical modeling. Since the model is nonlinear and ignores random machine failures, simulation is the only tool to make a proper analysis of the system. However, it is an optimization problem, and the objective is to find the best configuration in order to maximize the throughput. Therefore it is required to employ the optimization techniques using simulation model to solve the problem. A number of scenarios representing different configuration settings should be evaluated and compared with respect to the objective function.

References

1. Tempelmeier, H., Kuhn, H.: Flexible Manufacturing Systems, 1st edn. Wiley-Interscience, New York (1993)
2. Werner, L.: Analysis and design of closed loop manufacturing systems. Master's thesis, Massachusetts Institute of Technology Operations Research Center, USA (2001)
3. Browne, J., Dubois, D., Rathmill, K., Sethi, S.P., Stecke, K.E.: Classification of flexible manufacturing systems. FMS Mag. **2**(2), 114–117 (1984)
4. Koenigsberg, E., Mamer, J.: The analysis of production systems. Int. J. Prod. Res. **20**(1), 1–16 (1982)
5. Dhouib, K., Gharbi, A., Ayed, S.: Simulation based throughput assessment of non-homogeneous transfer lines. Int. J. Simul. Model. (IJSIMM) **8**(1) (2009)
6. Kumar, B.S., Mahesh, V., Kumar, B.S.: Modeling and analysis of flexible manufacturing system with FlexSim. Int. J. Comput. Eng. Res. **5**(10), (2015)
7. Pourbabai, B.: Performance modeling of a closed loop material handling system. Eur. J. Oper. Res. **32**(3), 340–352 (1987)
8. El-Tamimi, A.M., Abidi, M.H., Mian, S.H., Aalam, J.: Analysis of performance measures of flexible manufacturing system. J. King Saud Univ. Eng. Sci. **24**(2), 115–129 (2012)
9. Schattka, M., Puchkova, A., McFarlane, D.: Framework for simulation-based performance assessment and resilience improvement. IFAC-PapersOnLine **49**(12), 289–294 (2016)
10. Standridge, C., Kleijnen, J.P.C.: Experimental design and regression analysis: an FMS case study. Eur. J. Oper. Res. **33**(3), 257–261 (1988)
11. Kelton, W.D., Sadowski, P., Swets, N.: Simulation with Arena. McGraw Hill, Boston (2010)
12. Law, A.M., Kelton, W.D.: Simulation Modeling and Analysis. McGraw-Hill Series in Industrial Engineering and Management Science. McGraw-Hill, New York (2000)
13. Ozturk, G.: Simulation & analysis of Izmir metro transportation system. Master thesis, Yasar University, Izmir, Turkey (2012)
14. April, J., Kelly, J., Glover, F., Laguna, M.: Practical introduction to simulation optimization. In: Chick, S., Sanchez, T., Ferrin, D., Morrice, D. (eds.) Proceedings of the 2003 Winter Simulation Conference, pp. 71–78 (2003)

Optimization of the Machine and Device Layout Solution in a Specific Company Production

Marian Králik(✉), Vladimír Jerz, and Michael Paštéka

Institute of Manufacturing Systems, Environmental Technology and Quality Management, Slovak University of Technology in Bratislava, Bratislava, Slovak Republic
{marian.kralik,vladimir.jerz, michael.pasteka}@stuba.sk

Abstract. The subject of the paper deals with the plant layout optimization of the existing production halls in a specific company producing components. The aim is to suggest a layout based on analysis of the product portfolio, realization of the capacity calculation and proposition of the optimal machine and device configuration. For this purpose, some optimizing calculations with utilization of simulation were realized. The result is a detailed layout including 3D visualization.

Keywords: Manufacturing systems · Production flow analysis · Optimization · Simulation · Visualization

1 Introduction

Appropriate machinery and equipment layout in manufacturing systems within Industry 4.0 has a direct impact on minimizing material handling costs, thereby contributing to better business management. Another way to make production more efficient is to use lean production as a systematic approach that avoids unnecessary waste by continuously improving production processes. To maximize the company's benefits, a combination of these two ways of cost and production system optimization should be applied. There is an increasing number of the software systems on the market that enable the simulation of production systems with interactive 3D display of the chosen solution. Software tools offering comprehensive solutions that fully adapt to customer requirements are becoming increasingly popular. The advantage of the software products on this basis is the possibility of creating a new and more efficient layout solution, or the entire production system in a virtual environment without the need to suspend production. This design is then directly used after its optimization.

N. M. Durakbasa and M. G. Gençyılmaz (Eds.): ISPR 2019, LNME, pp. 91–103, 2020.
https://doi.org/10.1007/978-3-030-31343-2_8

2 Description of the Solution Procedure

The procedure for creating the optimum layout design for the machines and equipment in manufacturing systems is generally known. We want to point out to some of the solutions used in a company in Slovakia, which produces components ranging from small to medium series. In particular, manufacturing technologies such as turning, milling, material cutting, grinding, honing are used in manufacturing of the components. The company controls the quality of products by the means of the control and measuring technology, coordinate measuring machine, surftest and contour measuring system. The machine base is made up of universal center lathes, automatic saws, NC milling machines, center grinders, honing machines, biaxial and three-axial CNC lathes with driven tools up to four-axial machining centers.

2.1 Preparation of Input Data

The first step in solving the task of optimal layout is to collect accurate data. Accuracy and quality of collected data is a key element of optimization. Any error or inaccuracy in this step will negatively affect every subsequent step. Technological process of the production of components gives information about time consumption for the production of components, the need of production machines and equipment, necessary production tools, cutting conditions, requirements for the quality of components, etc. In the company, the technological procedures for the production of components are in the technological procedures folder in .xls formats and in the given structure according to Fig. 1.

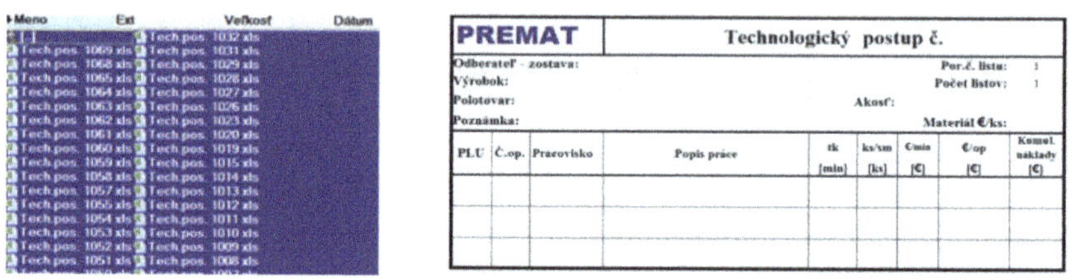

Fig. 1. Technological procedures and their structure

Thus, searching for information about a particular manufactured component requires knowledge about its manufacturer, on the basis of which the designer opens a specific folder with the name of the manufacturer.

2.2 Component Basis Analysis

Components manufactured by the Slovak company include parts for hydraulic systems, parts for bearing reducers, parts of rolling bearings, parts for pneumatic springs clamping, or needle and roller pulleys (Fig. 2).

Fig. 2. Production program of the company

Individual components are used in hydraulic pump or heavy machinery designs such as bulldozers, tractors or harvesters. The P-Q diagram was built on the basis of theoretical knowledge. The "x" axis shows the selected part of the component base and the "y" axis shows the annual production of the individual components. The company produces more than 250 types of components. In the P-Q diagram, 60 components have been selected for clarity to represent a cross-section of the entire manufacturing spectrum, so that all the manufacturing machines available to the company are included (Fig. 3).

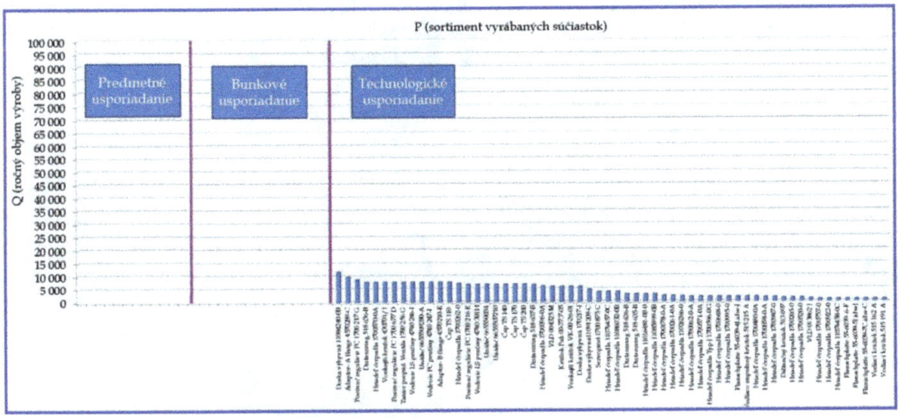

Fig. 3. P-Q diagram of the manufactured components

The P-Q diagram shows that production in the company focuses on a wide range of products, with a maximum annual production of one type of component not exceeding 12,000 pieces. A big problem is updating the data. The company receives "just in time" orders from large customers according to current demand. Production is planned for a maximum of one week in advance, and may additionally modify depending on market requirements. For example, they produce 50 different types of pump shafts and there is no way to determine which type of pump will be the most demanded one in the future.

2.3 Production Segmentation

The segmentation of production is only possible for the components produced in larger quantities. The analysis of the component basis shows that the company does not have products manufactured in large quantities. The company is currently applying for the production of components in larger quantities. The developed procedure will serve as a methodology, if this business plan is successful. For these reasons, the question arose: “Which annual production of the selected component will be economically advantageous for rearranging the machines”? This could happen in practice, provided that the customer would standardize the pumps and at the same time the market demand for them would increase.

A matrix of “machines – components” was constructed upon the company and machine basis data. The individual colors represent the blank from which the components are made (Fig. 4).

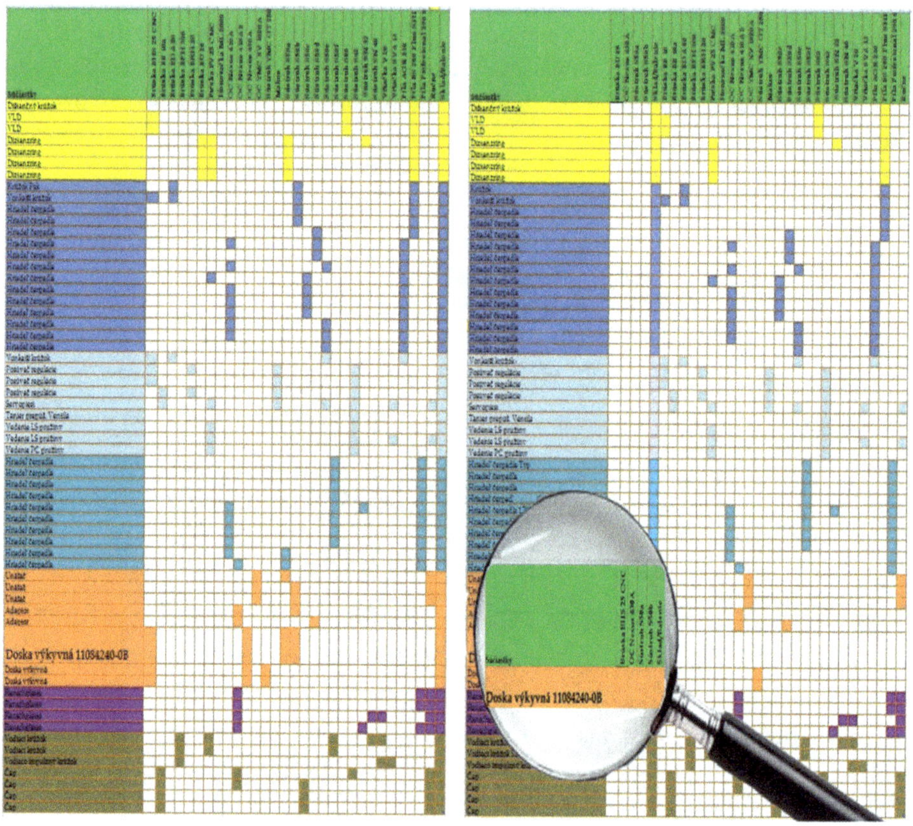

Fig. 4. “Machines – components” matrix – components before and after production segmentation

The need to create a cellular arrangement for the Swinging plate component ("Doska výkyvná") results from segmentation. This autonomous workstation will need at least 4 machines to operate [1]:

- BHS 25 CNC grinder,
- MazakNexus 430A machining center,
- 2 x S50i lathe.

The exact need for machines, workers, or robots results from the capacity calculations.

2.4 Capacity Calculations

As "Swinging plate" is a really manufactured (part) component, the technological procedure (job description) contains all the necessary data. It provides the information about duration of the technological operation and the number of pieces to be produced per change. The standard was set at 390 min of net time, because it was empirically found, that factors, such as:

- the time needed to exchange tools,
- cleaning the workplace,
- handling time (clamping/unclamping a semiproduct or a part),
- interoperable transport,
- quality control,
- machine maintenance,

take 90 min of a single shift time in case of the technological arrangement. Therefore, production times and standards were recalculated to 480 min. By the subsequent capacity calculations (for a limited range of the contribution, we do not mention it), the number of necessary machines was set to 9 with 100% usability, thus verifying the correctness of the calculation formulas.

2.5 Designing of Production Layout

Based on the calculated data from the previous steps, it is possible to design a new production layout. The company produces a wide range of products, so the technological arrangement of the machines is the most suitable for this type of production. Until now, the company has used all the advantages of the technological arrangement of machines:

- changing the production program does not disturb the production,
- ease of implementation of multi-machine operation,
- better machine utilization,
- easier maintenance.

No changes were made after consulting with the management. But with the further design of the manufacturing disposition solution, an increase in the production of "Swinging plate" is envisaged, and the design of the measures should allow for this alternative. The original layout is shown in Fig. 5.

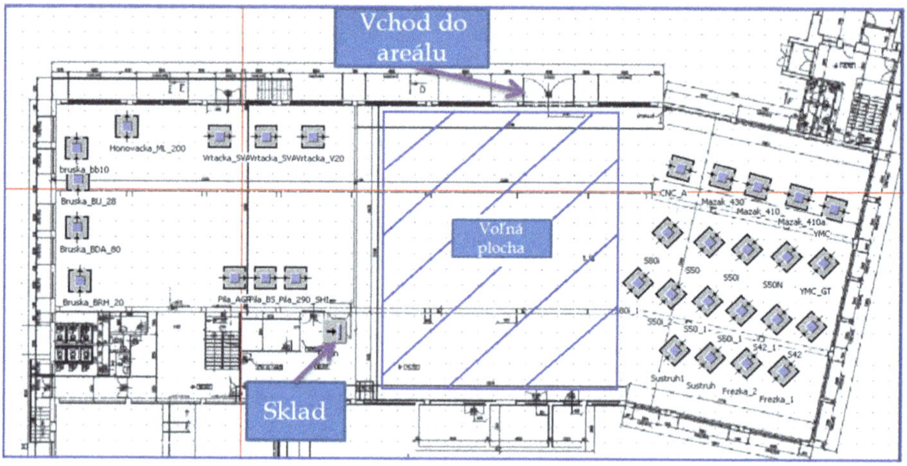

Fig. 5. Current production machinery layout

Capacity calculations have shown that 9 machines are needed to produce a given component. On the basis of capacity calculations and machine times, the exact need of individual machines was determined:

- 6 pcs of S50i lathe,
- 2 pcs of Mazak Nexus 430A machining center,
- 1 pcs of BHS 25 CNC grinder.

The location of the workstation would be best suited to the free area shown in Fig. 5. From among the several variations, the arrangement shown in Fig. 6 was chosen. Creating a new arrangement near the warehouse would minimize the total interoperable transport time. Arranging workplaces with respect to the continuity of operations has created a pyramid arrangement. The advantage of this arrangement is a clearly defined and irreversible material flow.

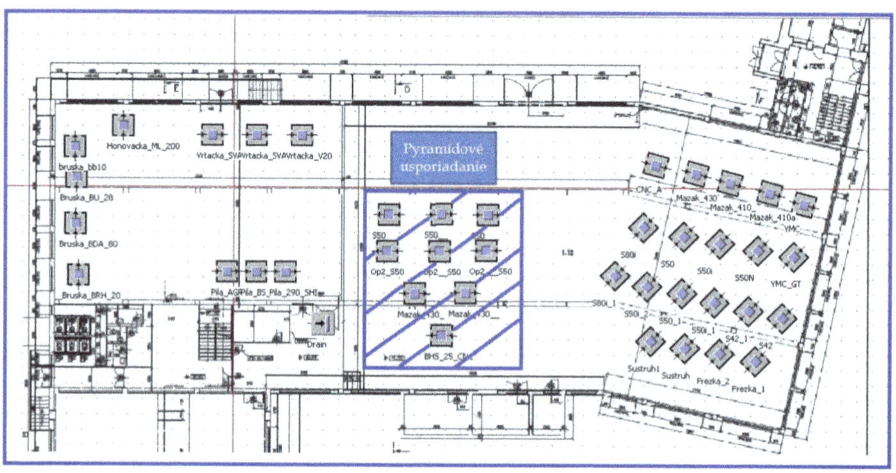

Fig. 6. New production layout design for the component production enhancement

2.6 Simulation

The selected layout had to be verified by simulation. The main task of the simulation was to find out:

- the annual production of the proposed system in three-shift operation,
- percentage use of workplaces,
- verifying the correctness of the current standards (since they have not been calculated but obtained empirically).

Since it was a relatively simple set of nine machines, a manual simulation was performed using calculations. The reason was also the assumption that the designer can design a simpler manufacturing system in this way. Subsequently, the calculations were verified by the Tecnomatix simulation software product, specifically in the Plant Simulation module.

A brief description of creating a simulation model is provided in the text below.

The handling times of the individual workplaces were indicated by the red arrows (Fig. 7). Within this time, removal of the semi-finished product from the pallet, its clamping, unstrapping, and inserting into the pallet were included. Empty palettes were marked in green. The letter "O" represents the casting and the number 117 is the expected production at 100% utilization of the machines. Numbers in other squares express the operation number and the expected production of one change without including handling times, interoperable transport, etc. For example, the designation 2/176 indicates that 176 pieces of "Swinging plate" must be produced on the third machine during one working shift (Fig. 7). Number 2 indicates that the first two technological operations have been carried out on semi-finished products and are ready for the third operation.

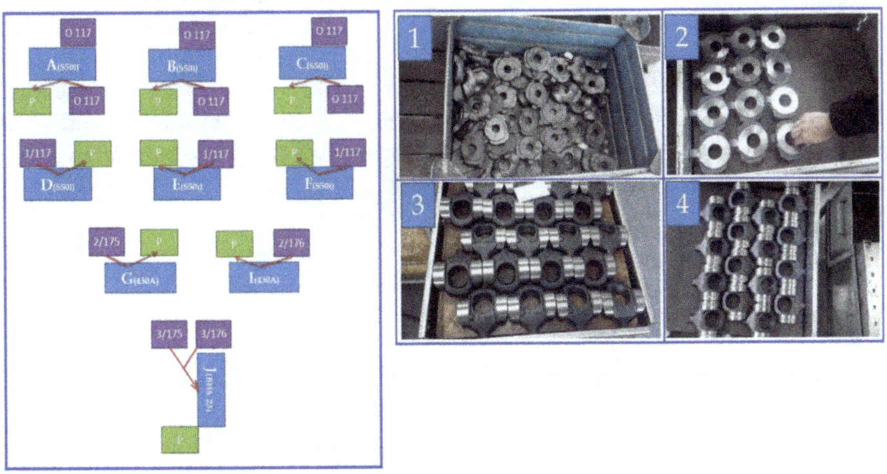

Fig. 7. The contents and location of pallets after the first working shift and the "Swinging plate" component in individual stages of the production process

Subsequently, it was necessary to determine the interoperable transport after each working shift so that the workstation was in the same condition as at the beginning, thus ensuring the repeatability of the cycle (Fig. 8).

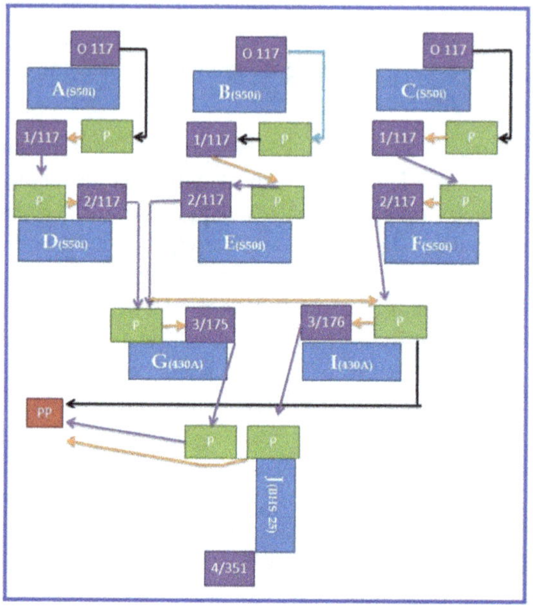

Fig. 8. Contents and location of pallets after the first working shift and the component itself

Figure 8 expresses the proposed interoperable transport. Each worker must perform the exact number of operations in a given order to ensure cycle repeatability. Individual handling operations are performed after meeting the norm of workplace.

Interoperable transport times were estimated by the complexity of the operations and the length of the trajectory. 30 s are added to each worker at the start of the manipulation to find a pallet truck. For example, 170 for worker E originated as follows:

- 30 s to search for a truck,
- 10 s to lift the pallet,
- 10 s to transport to G,
- 120 s to displace 56 pieces to the pallet brought in by a worker D.

Other workers transport the entire daily standard without rehandling.

One-off times had to be deducted from the total working time to set the standard correctly. It included a mandatory break, the time needed to clean the workplace, the interoperable transport, the tool change time, the quality control, the regular machine maintenance regarding the proposed layout. Thus a real number of minutes in norms was obtained when the workplaces can produce, tool change times in machining tools, etc.

It is necessary to include the failure rate of individual workplaces in the calculation of the expected annual production. However, the company does not have statistical data on failures at individual workplaces, nor the times to eliminate failures. A 10% failure rate is considered when planning production of the production systems. This information is relevant when individual machines are connected in series. When creating a new layout, the pyramid arrangement was designed in parallel with the current technological arrangement, so that in the event of failure of any machine, it is possible to move production immediately to another machine. For this reason, the failure rate will only include the time it takes for re-setting of the machine.

In order to verify the results of the manual simulation, in particular the correctness of the calculation formulas, the model created in the Tecnomatix software product was proposed. Modern simulation products allow to insert the building floor plan into the model background, that helps better imagine the layout of the machines (Fig. 9).

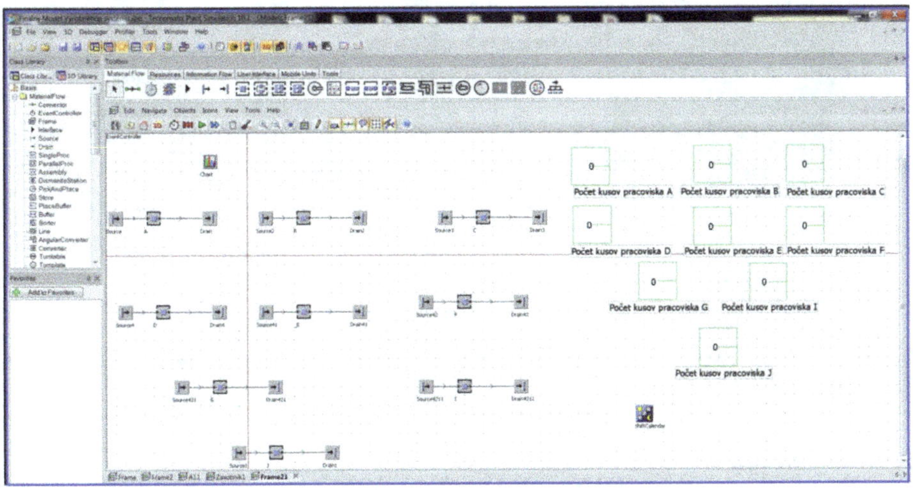

Fig. 9. Basic model of the proposed system

All data was entered into the pyramid model. E.g., to include a mandatory break into the model, the ShiftCalendar (Fig. 10) is used. It enables to insert complete production planning.

In the company, the first shift starts working at 6 a. m. and ends at 2 p. m. with a mandatory break from 10:45 a.m. to 11:15 a.m.

In order to simplify the model and to interpret it using one graph, the times of some operations were joined in the simulation. The term "Elimination of bottlenecks (ÚM)" expresses the time when the machines no longer produce after the standard has been met to avoid overproduction that would impair the cycle repeatability. After all data was entered, the simulation could be started (Fig. 11).

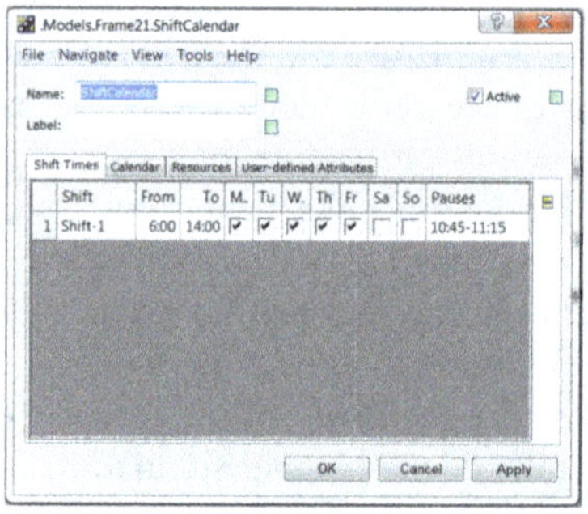

Fig. 10. ShiftCalendar

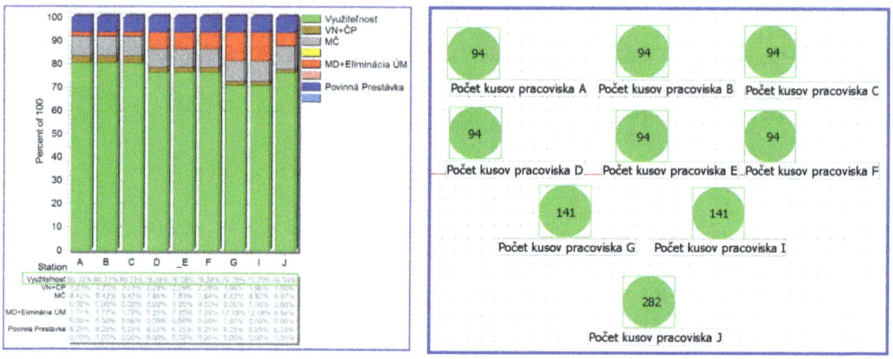

Fig. 11. Graphical representation of simulation study results (workplace utilization and single shift production)

2.7 Visualization

Currently, there is a large number of software products used to simulate manufacturing systems. Apart from standard means of modeled process visualization by 2D images with icons representing the system elements, the latest software products also allow for the use of virtual reality. If we want to bring the simulation model closer to the user with these options, 3D models of individual production machines, handling and transport equipment must be available. This is an illustrative approach to visualization of the static elements layout. For animation of transport, manipulation and technological operations, the models must be composed of several parts, which are then assigned a kinematic structure (e.g. Tecnomatix Process Designer).

Tecnomatix/Plant Simulation software has a basic 3D view that allows a better idea of the manufacturing system (Fig. 12).

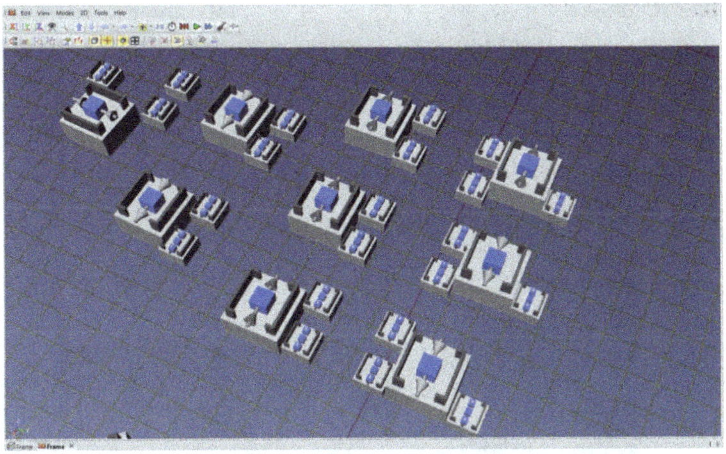

Fig. 12. Basic 3D view of the proposed layout in the Tecnomatix/Plant Simulation software environment

There are many graphic software products on the market to create 3D models. For example, the Solidworks software, uses parametric modelling. Parametric modelling means that first a 2D plan of the desired shape is created and assigned a dimension (parameter). The desired shape is then "drawn" into space. If necessary, adjustments such as removal, chamfering, etc. are made into the "drawn" shape. In this way, the Mazak 430A machining center model was created (Fig. 13).

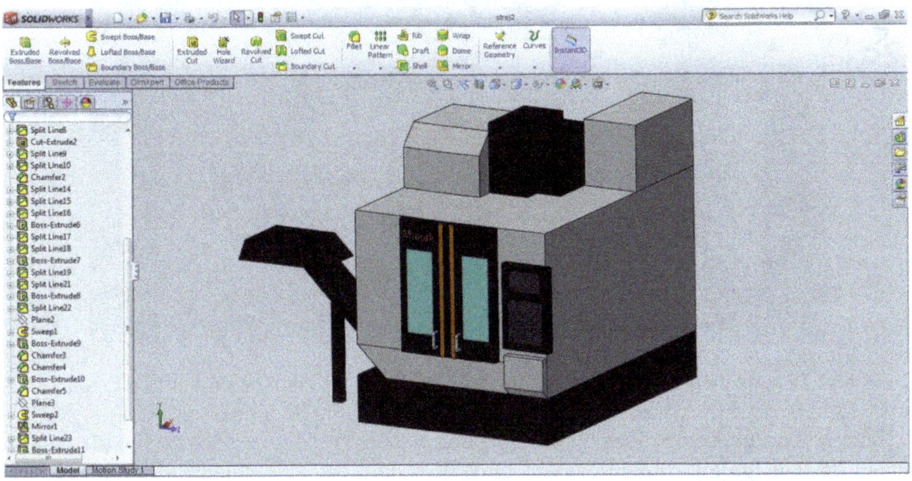

Fig. 13. Lathe model in the Solidworks software environment

Another program for creating spatial models is the Photomodeller which applies so-called digital photogrammetry followed by Photoshop.

Once the 3D models have been created, it is necessary to export them. Each software offers multiple export formats. Tecnomatix software uses its own format with a .jit extension that the used software (Photomodeler, Photoshop) cannot create. For this reason, we have tried to export all types of formats from Solidworks and Photomodeler and then import them into Tecnomatix. Among all formats, WRML (.wrl, Fig. 14) was the most suitable.

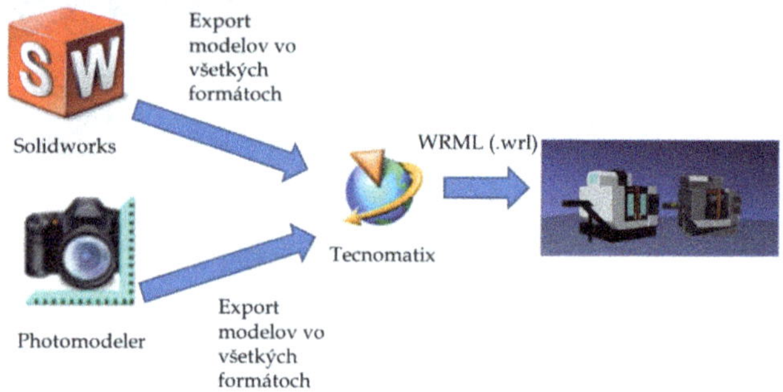

Fig. 14. Importing 3D models into simulation software

After creating the models and importing them successfully, the machines could be arranged according to the simulation model (Fig. 15).

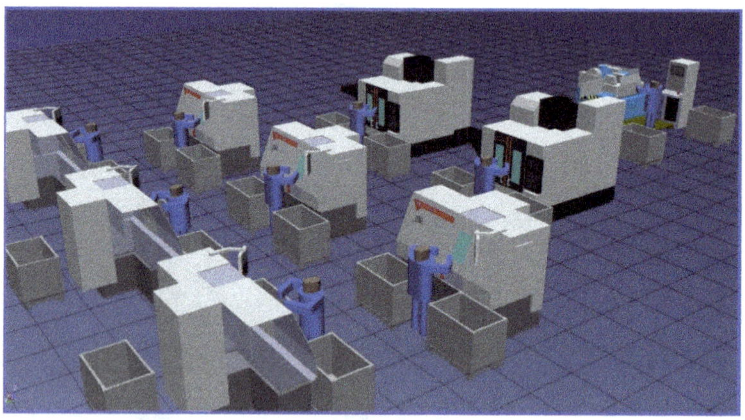

Fig. 15. Visualization of the selected and by means of simulation verified layout

The above mentioned findings, led to the demand for integration of the production design knowledge into one integrated unit. A system for automated production system design containing the data needed for a designer of the production systems has been developed:

- component database,
- technological processes,

- drawing documentation,
- machine database.

A system for automated production system design was created, so called Integrated Design System (ISP). After consideration of the required parameters and consulting with programming experts, the Wampserver 2.4 software package was selected, which includes:

- Apache 2.4.4,
- MySQL 5.6.12,
- PHP 5.4.12.

Due to wide range of issues and the limited scope of the paper, the system will be presented in other professional publications.

3 Conclusions

The paper describes the process of the production project, which can be used in the plants implementing the knowledge and principles of Industry 4.0. The individual described parts of the solutions were subsequently implemented into the system of automated production system design, so-called Integrated Design System (ISP). The created software provides the designer with a better idea of the machine and the component base of the enterprise, thus speeding up the collection process of the necessary data. At the same time, it contributes to more efficient corporate management.

Acknowledgment. The article was written with the help of projects: KEGA 027STU-4/2019.

References

1. Králik, M., Morávek, I.: Programovanie CNC výrobnýchstrojov. Edíciaučebníc. Slovenskátechnickáuniverzita v Bratislave, FX, s.r.o. (2008). ISBN 978-80-89313-36-5
2. Bouyssou, D.: Modelling inaccurate determination uncertainty, imprecision using multiple criteria. In: Locket, A., Islei, G. (eds.) Improving Decision Making in Organizations. Lecture Notes in Economics and Mathematical Systems, vol. 335, pp. 78–87. Springer, Berlin (1989)
3. Azibi, R., Vanderpooten, D.: Construction of rule-based assignment models. Eur. J. Oper. Res. **138**, 274–293 (2002)
4. Slowinsky, R., Greco, S., Matarazzo, B.: Minig decision-rule preference model from rough approximation of preference relation. In: Proceedings of the 26th IEEE Annual International Conference on Computer Softwre & Applications (COMPSAC 2002), Oxford, pp. 1129–1134 (2002)
5. The ISPR2018 website (2018). http://www.ispr2018.org/
6. Kusiak, A.: ISPR2018 homepage (2017). http://www.ispr2018.org/wp-content/uploads/2017/12/Kusiak-Presentation-Smart-Manufacturing-Sept-13-2017_Vienna.pdf
7. Kováčik, R.: Optimization of a layout solution of the machines and devices in production in a specific company. Ph.D. thesis, Slovak University of Technology in Bratislava, Slovak republic (2014)

Healthcare Systems and Management

A Process Design for Rare Diseases Patients in Turkey

Hatice Camgöz Akdağ, Gül Beyza Özuzcan(✉), and Dilay Durucali

Department of Management Engineering,
Istanbul Technical University Management Faculty, Istanbul, Turkey
camgozakdag@itu.edu.tr, beyzaozuzcan@gmail.com,
durucalid@gmail.com

Abstract. This paper is conducted to establish a better pathway for rare disease patients and doctors especially in Turkey. Through the project, contacts made with numerous rare disease patients, geneticists who expertise in this field, academicians, students of medicine department, and many more. After collection and evaluation of the raw data, the analysis is made to implement it for real life improvement scenarios. The analysis technique of this project was value stream mapping, which is broadly used to find the bottlenecks and slacks in the healthcare sector. Afterwards, final suggestions are explained on the last section. To conclude, the patient flow analysis for rare disease patients and especially the detailed flow for cystic fibrosis patients presented that a comprehensive strategy should be designed and plans should be made for rare diseases case in Turkey. This project was the first step for this mission, and also, the first project focusing on the rare diseases field from an engineering perspective.

Keywords: Rare diseases · Cystic fibrosis · Patient flow · Value stream mapping · Engineering analysis method

1 Introduction

Today health is the most essential field for any society as it has a significant effect on the development of humanity. The increased medical treatment and technological solutions are used more and more to cure patients. Especially the actions in the rare diseases field is gaining more importance in Europe and many other societies. Unfortunately, the awareness, actions and resources are very inadequate in Turkey. There is only one source of publication about rare diseases in Turkey, which is also gathering most of the information from foreign resources. Developments should be carried out on this field by starting an engineering analysis of the rare disease patient flow in Turkey.

Almost 9 years ago, it is stated by the Council of the European Union that, necessary plans and strategies should be developed to increase the synergy between local, national and transnational centres [2]. Turkey is way behind in that sense and an action to guide and unite all the rare disease patients should be taken. To begin with, the pathway of the rare disease patients in Turkey is researched and analysed carefully. In this project, a flow analysis is made for current and future states through findings from

N. M. Durakbasa and M. G. Gençyılmaz (Eds.): ISPR 2019, LNME, pp. 107–118, 2020.
https://doi.org/10.1007/978-3-030-31343-2_9

the interviews and surveys together with the focus group with doctors. Finally, a pathway of diagnosis and treatment of rare disease patients in Turkey is created with Value Stream Mapping technique. As a conclusion, recommendations are added accordingly.

2 Literature Review

A rare disease is any disease which is observed very uncommonly worldwide. Although it has many definitions, the main idea is to state that these diseases do not appear commonly and therefore, the physicians might not be able to diagnose and treat the patients. According to EURORDIS, it is called as rare disease, if a disease is observed less than 1 in each 2,000 people [5]. In another words, around 37,000 people is affected by a rare disease in Europe by taking the total population of Europe is 742,855,349 based on the assumptions of United Nations.

Many associations and organizations or non-governmental organizations have been established in order to provide economic, psychological and social support to patients and their families. In addition to this, these associations not only provide assistance but also create awareness about rare diseases. When associations in Turkey examined, it is observed that many associations just represent people with the same disease and unfortunately, there is no common work. Combining associations under a single roof will provide an effective communication as well as the sound of a strong community that cannot be underestimated in our country, in which 23.2% of the consanguineous marriages.

3 Fıeld Research

A field research study is conducted on rare diseases patients and doctors as online surveys, face-to-face interviews and focus group study. Prof. Dr. Melih Bulut and Prof. Dr. Ugur Ozbek helped us through all the stages of our study, who are the field expert doctors in Turkey and in contact with many patient organisations. The total duration of the study was approximately 10 weeks. An average of 80 answers on each survey is collected. Main outcomes of this study can be stated as follows: the diseases, which are observed in Turkey are analysed, the duration of the testing and diagnosis phases are examined, the knowledge of patients about their diseases and similarity of different diseases are analysed, the range of the patients ages and genders are gathered, the examination time of doctors on patients for one visit is gathered, active patient organizations are observed and contacted, the obstacles of the doctors about the diagnosis are analysed, and finally, suggestions and wishes of both patients and doctors are collected and analysed.

After analysing the data pool, the details of cystic fibrosis disease are explained more in detail. The following outcomes are achieved and used for value stream mapping: The average age of diagnosis is gathered, patient waiting times are collected and analysed, the misdiagnosis rate is calculated, the average number of the doctors that a patient visits is analysed, the average approximate time of the testing phase and examinations is gathered, and previous and current situation of free of charge tests provided by public hospitals for CF patients are analysed.

There are some problems with rare diseases issue in Turkey. Firstly, the health data inappropriately to the E-Nabiz platform (Online Health Portal), currently used International Coding of Diseases ICD-10 does not cover many rare diseases and it should be updated with the ICD-11 instead. Secondly, beginning from the Ministry of Health, there is a lack of awareness and care about the rare diseases patients in Turkey. Our aim is to put a preceding step for this trainings as well. Other problems are low budget allocation and no budget plan, non-membership of EU and etc. Survey results can be stated as follows:

- The system is inadequate and the ministry of health needs to establish a policy. In total, there are 8000 diseases without a registry. For instance, asthma entered under the child diseases; albinism entered as a blindness. In our first survey, 70% of the patients stated that there is an inadequate knowledge of healthcare personnel about the causes of misdiagnosis. In this case, it should be handled by the ministry of health to create a programmed education and awareness.
- 10-min examination time is insufficient. It is impossible to diagnose a rare disease in 10 min. Most of the tests required for rare diseases are expensive and not paid by the government.
- The newborn heel-screening test is implemented for 4 kinds of diseases as follows Hypothyroidism, Phenylketonuria (PKU), Biotinidase Deficiency and Cystic Fibrosis respectively. Thanks to newborn screening program when started in 2015, real frequency of the disease in Turkey could be learn [11].
- The frequency of the Cystic Fibrosis in North America and Europe is around 1/2,500. The exact frequency in our country is unknown. However, it is thought to be from 1/3,000 to 1/4,000. According to TUIK 2018 data, Turkey's population of 2018 was 80,810,525 [3, 4]. In this case, from 20,203 to 26,937 people suffer from the Cystic Fibrosis. Such a large number represents a considerable community.
- When symptoms are first realised, patients are on average 1.5 years old. The age range is from 0 to 19.
- After realising the symptoms, patient waits an average of 66 days until the examination. Waiting times range from 0 to 1800 days.
- After make an appointment for the examination, expected time is 16 days and ranges from 0 to 450 days.
- After examination, full diagnosis takes an average of 98 days. This period ranges from 0 to 1800 days.
- The patients were 3 years old at the time of diagnosis. The ages range from 0 to 40 years.
- Patients go to 4.2 doctors, until the diagnosed. Misdiagnosis lead to prolong the duration of the diagnosis approximately 15 months and ranges from 0 to 180 months. 68% of the patients were misdiagnosed.
- When it is asked the patients who had been misdiagnosed, how many times were diagnosed wrongly, the answer is received as 2.2 times. The gender distribution of cystic fibrosis patients participating in the survey were equal.

- When adequacy of resources in Turkey is asked with 1–5 Likert scale, average answer as 2,1 is received.
- When similar diseases with the most common misdiagnosis is asked, 20 kinds of diseases are obtained. For instance, asthma and bronchitis are the ones that have almost inseparable symptoms.

4 Value Stream Mapping

4.1 Introduction

Value stream mapping (VSM) is a flowchart method to illustrate, analyse and improve the steps required to deliver product or service and uses a system of symbols to depict various work activities and information flows. In our study, the VSM technique is used because main problem was the astronomic numbers of cycle and waiting times, backward flowing pathways and loops on patient doctor interactions. For sure, there are other reasons, which are unmeasurable such as inadequate knowledge of patients and healthcare personnel. However, the focus group study with doctors has provide us valuable information about this issue as well. Before designing the VSM, a flow analysis designed for current and future scenarios. In this project, there are 2 current flows for patients born before 1st of January on 2015 and after 1st of January on 2015 due to the heel-screening test mentioned in the previous section. However, the future flow is only one and united flow with the rare diseases hub in Turkey.

4.2 Current Flow Analysis

(1) Current Flow Diagram: The flow is designed for a patient going to a state hospital as many rare disease patients first choice is a state hospital. There are 3 main stages in the patient flow; appointment and examination, testing, results and diagnosis.

Although it seems like the same flow of a regular illness, a flow for the rare disease patients have a huge difference compared to that, due to its start with patients experience of an intense disease symptom, unpredictable length of testing and diagnosis phase and loops created within this patient-doctor cycle until disease is truly diagnosed. This is the critical path for our research as it delays the diagnosis with an unpredictable time and results with fatal consequences for a patient. However, obstacle of diagnosing a disease cannot be assigned only to doctors, because it is quite challenging to make right diagnosis in 10 min appointment and even with available state hospital test results. Even for a normal disease, there is always a misdiagnosis probability. It is harder if disease has similar symptoms or test results with hundreds of many other diseases. This scenario is much more challenging for the rare disease patients because of the following reasons.

1. Inadequate options and resources for specific testings
   a. Lack of testing options in most of the hospitals
   b. High costs and long durations for specific tests
2. The healthcare personnel and doctors are lack of adequate knowledge about rare diseases due to
   a. Inadequate academic lecturing
   b. Underestimation due to lower prevalence rates compared to rare
   c. No trainings designed by the ministry of health

A general flow of the rare diseases patients in below indicated in Fig. 1.

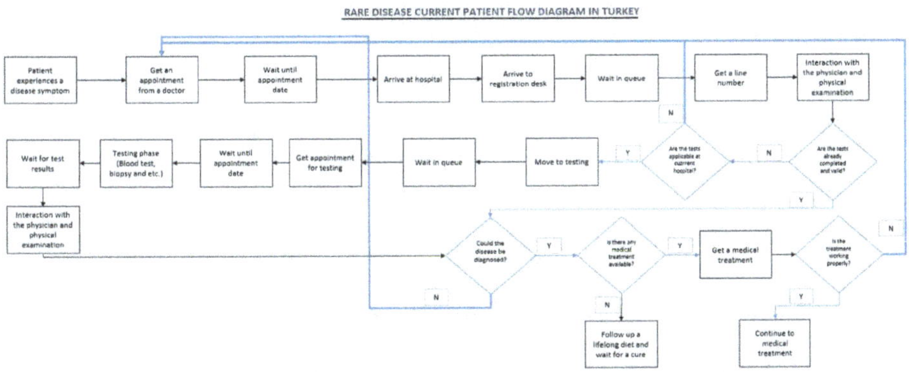

Fig. 1. Rare disease current patient flow diagram

On the figure, it can be clearly observed that the loops of taking an appointment might occur from three points: unavailability of test at the current hospital, undiagnosed disease by the current doctor, or ineffective treatment on the disease.

For the current cystic fibrosis flow of patients born earlier than year 2015, there are 4 backward paths observed. Two of them is the same for the general case of rare disease patients, which are tests unavailability at the current hospital, and the ineffectiveness of the treatment. Other 2 activities are different than the general case. Either patient actually is not a cystic fibrosis patient, or unavailability of medical treatment for cystic fibrosis patients at current hospital.

It can be clearly observed on flow that fully understanding that the patient is a Cystic Fibrosis patient is happening almost at the end of the flow. Therefore, there is a tremendous amount of time loss for these patients who are mainly children under serious health risks. On the other side, there is no backwards loop at the current flow of patients born after year 2015, which is a very good development as seen in Figs. 2 and 3 respectively.

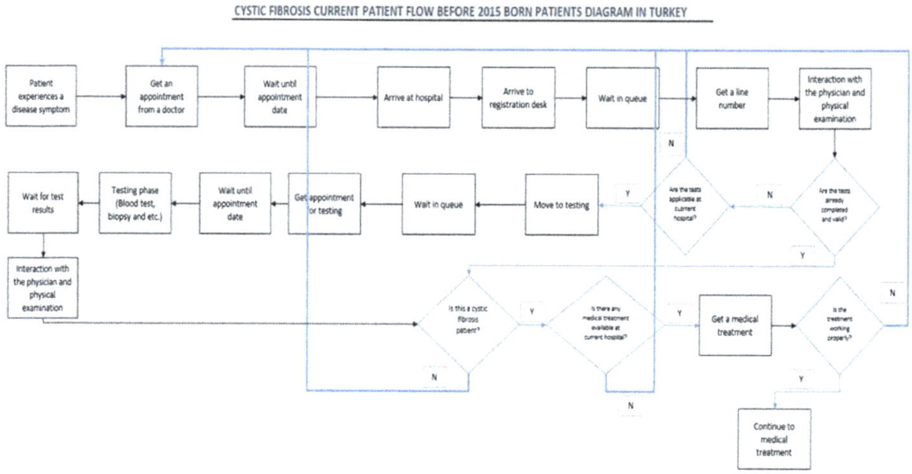

Fig. 2. Cystic fibrosis current patient flow: before 2015 born patients diagram

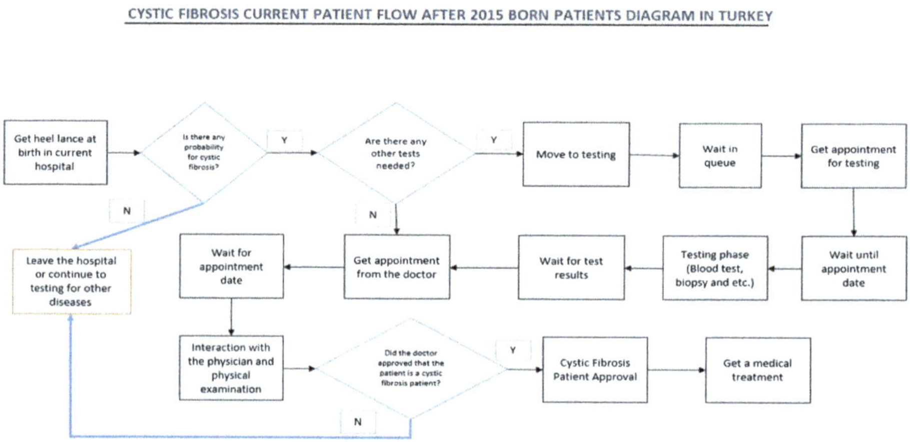

Fig. 3. Cystic fibrosis current patient flow: after 2015 born patients diagram

(2) Current State Map: After the evaluation of the 2 alternatives of CF patient flows a general current state map is created with average time values stated for each step. The value added times are

- Registration desk: The time spent during the patients process on the registration desk of the hospital to give her ID number to the person in charge, the control of personnel on the computer, whether the patient has already booked appointment, and approving appointment and giving a line number to the patient.
- Examination: The time interval of doctor examining a patient (Sector standard is 15 min in Turkey)

– Registration desk: The time of patient taking an appointment for testings stated by the doctor. The patient comes to desk, the personnel searches for the testings entered by the doctor on the system, personnel prints out and gives the line numbers for each testing to the patient.
– Test appointment: The total time of the actual the testing processes.
– Test examination: The real time required for the analysis of a test.
– Examination after testing: The patient's second visit to the doctor for the test results evaluation and examination. Again, the sector standards is the limit here.

The non-value added times are the waiting times in between the value-added activities, which is wanted to minimize as possible if the conditions held on real life. These are for example waiting on the lines, waiting because of unforeseen events such as process delays from previous patient, or time passed during changing clothes, wearing some tools before testing and etc. However, reducing these times are usually not possible due to many procedures, regulations and standards implemented on health sector. Therefore, minimizing them is beyond the topic of this study. Figure 4 shows the Cystic Fibrosis Patients Current Value Stream Map.

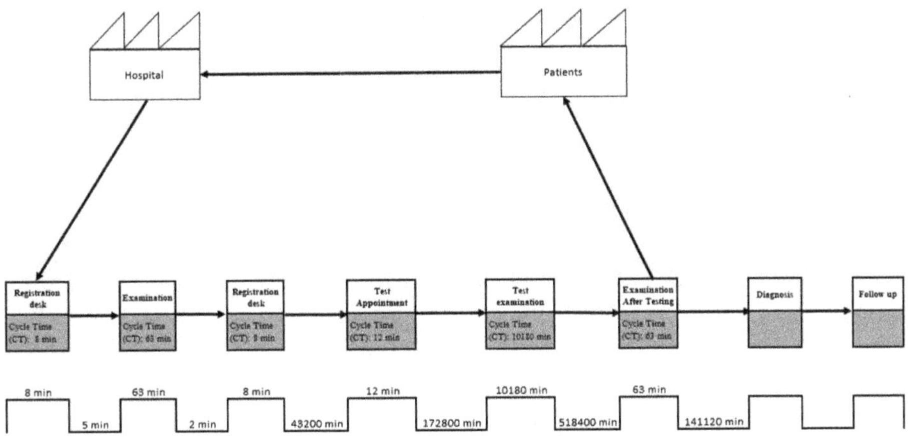

Fig. 4. Cystic fibrosis patients current value stream map

4.3 Future Flow Analysis

(1) Future Flow Diagram: The future flow analysis is designed based on the findings of the selected concept with AHP technique in the next section of this report. The selected concept to improve the rare disease patient flow is establishing a rare diseases hub in Turkey as a rare diseases hospital to serve only for these specific diseases. For the future flow there are some assumptions taken as stated below.

Assumption 1: The cloud based healthcare data storage system in Turkey (E-nabiz) will be improved including family genetics, and ICD-11 codes. ICD-10, current code system used in Turkey does not include rare diseases. ICD-11 codes are the new disease codes announced by WHO including many rare diseases. (*ICD: International Classification of Diseases)

Assumption 2: Doctors and healthcare personnel are started to take trainings about rare diseases.

Assumption 3: The rare diseases hub will serve as a hospital under ministry of health.

Assumption 4: An area of profession for rare diseases will start to work in this hub as an intern or as a doctor.

Assumption 5: The European Reference Networks (ERN) for rare diseases will be followed.

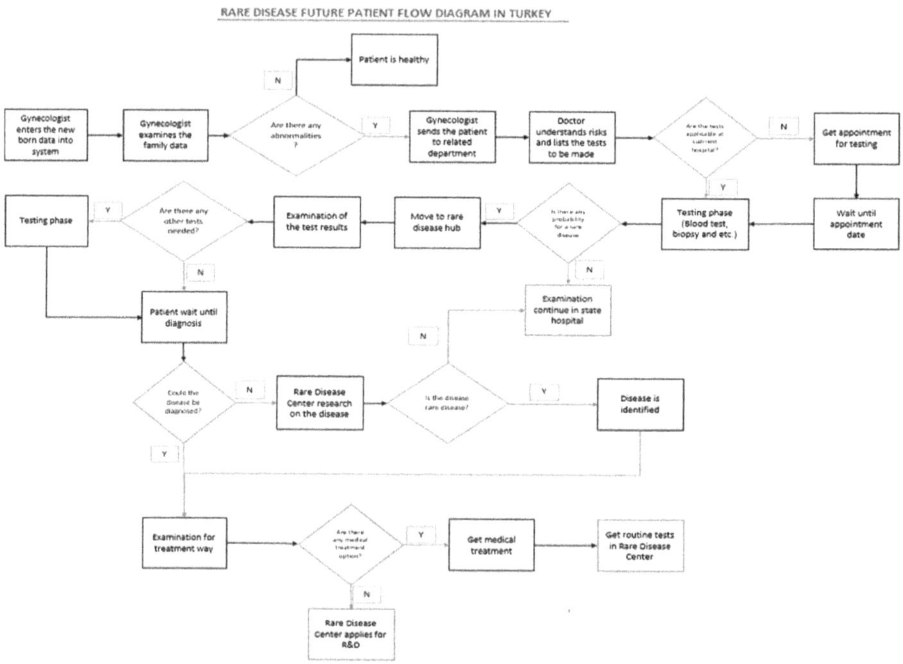

Fig. 5. Rare disease future patient flow diagram in Turkey

In the future flow given in Fig. 5, the steps are different than the current flow as the genetic data of a patient is analysed at the birth and necessary actions started earlier. If there is a rare disease possibility for this patient, the hospital sends the patient to the rare diseases hub and examinations continue at there. The centre makes necessary tests, search for the rare disease, search for treatment options and apply to ministry of health if there isn't any treatment yet. Basically, rare diseases hub will serve to the patients as a special centre such as the cancer treatment centres. After that, if the rare disease diagnosis is made patient gets the treatment and follows up the regular controls at this centre. If not diagnosed, patient goes back to a state hospital for following examinations.

The future flow for a cystic fibrosis patient is also created as special circumstances occur for every different disease. Figure 6 shows the cystic fibrosis patient flow design with the suggested rare disease hub. As it can be observed very clearly, there are no backward flows or loops created. The patient flow is on one and fluent direction with an establishment of rare disease hub, and other assumptions stated.

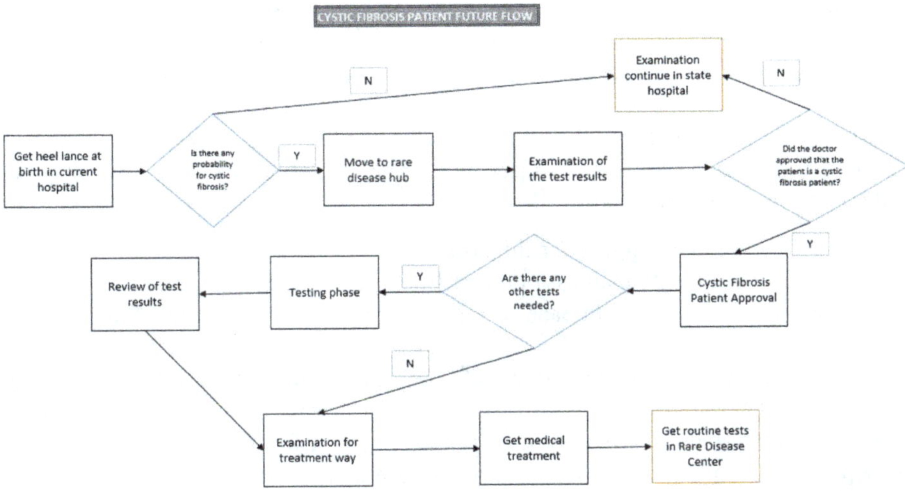

Fig. 6. Cystic fibrosis patients future flow in Turkey

(2) Future State Map: Then, a future state map is generated for cystic fibrosis patients in Turkey. The total value-added average cycle time is reduced from 10,334 min to 10,217 min with this suggested future flow. The cycle time of activities reduced significantly, due to lowering the visiting amount of a doctor to only once. Also, there are other effects such as healthcare personnel's and patient's increased knowledge about rare diseases, more resources availability and increased use of technology. Including all these quantitative and non-quantifiable affects the cycle time is reduced as given in Fig. 7.

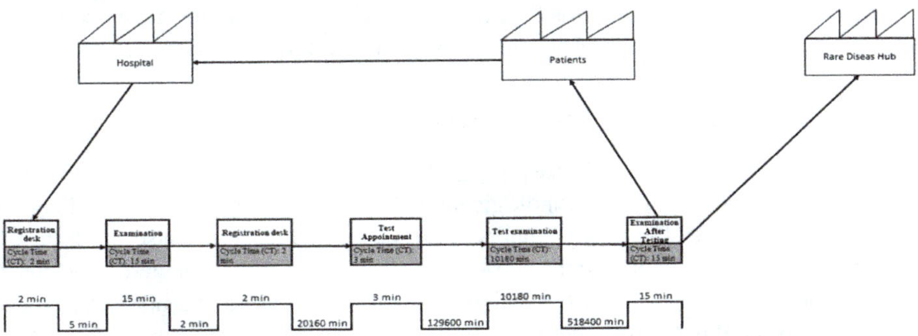

Fig. 7. Cystic fibrosis patients future flow

5 Concept Analysis

5.1 Concept Generation and Concept Selection with AHP Technique

The Analytic Hierarchy Process (AHP) is a structured technique for organizing and analysing complex decisions. In this study, evaluated options are website, app, new hub, department of hospital. When choosing from the options, different criteria are considered which affect not only cost but also the sustainability of hub. Also, articles that is about hospital location selection are considered. We have identified 3 main criteria as an efficiency, feasibility and ease of communication and 7 other sub-criteria beneath them.

The top of the hierarchy is goal of the analysis and then multi criteria that consist of several factors. The last level is the alternatives. The weights of the criteria is calculated by the dual evaluation method.

Step 1. Develop the weights for the criteria by

- developing a single pair-wise comparison matrix for the criteria
- multiplying the values in each row together and calculating the nth root of said product
- normalizing the aforementioned n^{th} root of products to get the appropriate weights
- calculating and checking the Consistency Ratio (CR)

Step 2. Develop the ratings for each decision alternative for each criterion by

- developing a pair-wise comparison matrix for each criterion, with each matrix containing the pair-wise comparisons of the performance of decision alternatives on each criterion
- multiplying the values in each row together and calculating the nth root of said product
- normalizing the aforementioned n^{th} root of product values to get the corresponding ratings
- calculating and checking the Consistency Ratio (CR)

Step 3. Calculate the weighted average rating for each decision alternative. Choose the one with the highest score.

The weights obtained by comparing each criterion are indicated. Then, the scores of the positions were determined by collecting the data obtained by multiplying the weight of the criteria for each position by the weight of that criterion. In accordance with this scoring, new hub establishment has the most consistent results.

5.2 Framework for the Selected Concept

Technically it is possible for Turkey to establish a rare diseases hub from the constructional, administrative and technological levels. However, the lack of knowledge of the healthcare personnel cannot be underestimated in this case, because the hub will work by their helps as well. Firstly, a training design should be considered for physicians. Secondly, conferences for the patients should be organized and interactions with the physicians is aimed to be increased during the pathway. From the technological perspective, data analysis tools can be used to develop the pathway estimations with the each entered data. Later on, if the budget is enough, new technological implementations can be considered.

Since the health budget for the 2019 is still argued at the Grand National Assembly of Turkey, the budget estimation for rare diseases implementations is unknown as well. However, EU health budget plan for 2021–2027 period for ERN is € 413 million. In other words, Europe will allocate this money only for 30 million rare disease patients living in EU [2]. The health budget according to Grand National Assembly estimations was 33.583 billion Turkish lira for 2018 period. The rare diseases hub will be located under Health Institutes of Turkey, whose budget is 53 million Turkish lira in Turkey. Another finding showed that there is a test made at birth for PKU and cystic fibrosis patients, and there is plan for SMA test but the budget is not given for any of them [1].

In Turkey, currently there is no decisions or estimated budgets for such a project. A rare diseases hub as a patient flow guider should be created under the Ministry of Health with necessary budget amount to be calculated. Looking at legal structure of health sector in Turkey shows us that only rare diseases section is only mentioned under the Medical Social Services Unit. Since there is no legal documentation for rare diseases in Turkey, EU legislations will be offered to use.

6 Conclusions and Future Work

There are 4 main parts executed on this study to design a rare disease patient flow in Turkey. These parts are the literature review, the field analysis, value stream mapping implementation and concept analysis for future studies.

The results of this study can be collected on 3 main points, which are informing the community, the health personnel and the rule makers, which are Ministry of Health and government authorities. Informing the society and increasing the public awareness about rare diseases is one of the key points. The world rare diseases day on the last day of February is for that aim a good option to increase the awareness. Commercials, social media postings and videos should be increased. This can be done via including public communications technical personnel on the rare diseases hub. Especially training the healthcare staff and the rule makers are very important as they significantly contribute to the decision making and technological application part. A rare diseases hub will be a starting point for collection of all genetic information and evaluation of them to develop the national understanding on rare diseases. It will also serve as a research centre as it will have the collected rare diseases patients data. All in all, the biggest lesson learned from study was that, this rare diseases field should not only be studied by medicine and genetics & bioengineering students, but also by the management and computer engineering students as well. In that way, different perspectives can be considered for the most effective patient flows.

This study aimed to execute the first analysis on the rare diseases patients in Turkey. Taking this as a key resource, it is suggested that a detailed analysis and a feasibility plan for the rare diseases hub to be carried out. Furthermore, to raise awareness of the community, technological solutions and social media platforms could be used to promote the importance of rare diseases case. The world rare disease day on the end of February is the best opportunity serving to this issue.

As a conclusion, necessary training programs for the doctors and other healthcare personnel should be strictly planned to increase their knowledge level about rare diseases. Besides, comprising a rare diseases expertise for geneticists and doctors would increase the effectiveness of rare diseases hub, and contribute to Turkish health sector by enabling researchers to develop treatment options inside the country.

References

1. Budget Plan: TBMM Planning and Budget Commission (2017). https://sgb.saglik.gov.tr/Shared%20Documents/2018_Yili_Plan_Butce_Sunumu_14_11_2017.pdf
2. Council Recommendation of 8 June 2009 on an action in the field of Rare Diseases: Official Journal of the European Union (2009)
3. New Born Screening Program: Turkish Public Health Institution. https://dosyaism.saglik.gov.tr/Eklenti/11173,259822214447pdf.pdf?0
4. Population Projections: Turkish Statistical Institute (2009). http://www.tuik.gov.tr/HbPrint.do?id=30567
5. What is a Rare Disease? EURORDIS (2014). https://www.eurordis.org/content/what-rare-disease

An Integrated Approach for the Evaluation of Sustainability and Patient Satisfaction in Hospitals: A Case of Turkey

Aziz Kemal Konyalıoğlu, Tuğçe Beldek(✉), Hatice Camgöz Akdağ, and Gizem Deniz

Management Engineering, Istanbul Technical University, Istanbul, Turkey
{konyalioglua, beldek, camgozakdag, denizgi}@itu.edu.tr

Abstract. Improving quality is a very important issue in hospitals. Measuring quality is directly proportional to measuring sustainability criteria. Customer satisfaction is seemingly correlated with the factors which affect sustainability evaluation. Besides, the evaluation sustainability and satisfaction consist of different kinds of uncertainties. In this case, Quality Function Deployment (QFD) is very useful tool to evaluate customer satisfaction integrated with the fact that the weights of QFD are calculated by the method of Analytic Hierarchy Process (AHP). In this study, we propose an integrated approach for the evaluation of sustainability in Turkish Hospitals by using Quality Function Deployment and Analytic Hierarchy Process in order to assess patient satisfaction by considering the most important criterium in hospitals. This study enlightens the relations between patient satisfaction in hospitals and sustainability assessments in the same place.

Keywords: QFD · Quality management · Healthcare · AHP

1 Introduction

Sustainability is now being considered as a very important topic. As energy resources are continuing to decrease, every part of world seeks to develop their sustainability qualifications in sense of water, energy and any kind of energies. Furthermore, by keeping their sustainability qualification high, they need to try their customers to be satisfied. In this case, hospitals are taken into consideration as a type of establishments which are very important for the next generations. Hospitals are supposed to be eco-friendly buildings and establishments.

While developing the sustainability factors in a hospital, meeting customer satisfaction and focusing on patient needs are equally important because in aspect of customer satisfaction, some qualifications of sustainability can be related to customer needs. In Turkey, in this case, as a developing country, sustainability gains importance thanks to the awareness of renewable energy as well customer satisfaction in hospitals are considered while designing a hospital. Even if many of hospitals are opened in Turkey, it is not certain that patient satisfaction in hospitals are interrelated with sustainability qualifications.

N. M. Durakbasa and M. G. Gençyılmaz (Eds.): ISPR 2019, LNME, pp. 119–128, 2020.
https://doi.org/10.1007/978-3-030-31343-2_10

Total Quality Management can be defined as continuous improvement. In hospitals, patient satisfaction can be a part and measure of quality management. Total Quality Management can be also provided by using sustainability tools. Patient satisfaction in a hospital can then be related to some qualifications of green buildings.

In this study, there are five main parts which consist of introduction, literature review, methodology, application and finally results and conclusion. In the introduction part of this study, it is tried to explain the aim and the scope of the study. In the literature review part, the previous studies about sustainability, quality management, quality function deployment (QFD) and Analytic Hierarchy Process (AHP) is tried to be investigated. In methodology part, Analytic Hierarchy process and QFD are explained systematically while in application part, the weights of QFD are calculated by using AHP and QFD approach correlated with sustainability qualifications in general and patient satisfaction in the hospitals of Turkey. Finally, the results of QFD and AHP integrated approach considering sustainability qualifications and patient satisfaction results are discussed.

2 Literature Review

2.1 Analytic Hierarchy Process

The Analytic Hierarchy Process (AHP) provides weight comparisons between different criteria according to experts' opinion. These weights are given within a scale which compare pairs relatively. If one attribute dominates one another, then the appropriate evaluation is given according to the scale. After the selection of a criteria it is important to measure inconsistency if exists (Saaty 2008). AHP helps to solve complex problems which include multiple criteria to give a decision between them. The first step is to draw a hierarchical table that decompose decision criteria. These sub-criteria are compared pairwise to give different levels of importance (Kurttila et al. 2000).

It is possible to combine different methods such as fuzzy approach, linear programming, and quality function deployment and so on, with AHP. This possibility leads with more accurate and detailed results and a better way to reach the aim. AHP also gives chance to evaluate both qualitative and quantitative data according to its numerical scale. If compared attributes are equal than it will be "1", if one another is more important than it will be "1/9" or "9" according to the related one. Basic steps of AHP are:

1. Problem definition
2. Giving more detailed definitions according to the objective
3. The effective criteria identification
4. Drawing hierarchical tree for criteria, sub-criteria and alternatives
5. Making pairwise comparisons which is calculated by $n(n - 1)/2$, where n is the element amount. The diagonal elements are 1 and the others are compared each other according to the importance.

6. Finding maximum Eigen value, CI (consistency index), CR (consistency ratio) and normalized values
7. Taking the decision if these values are at the satisfactory level (Vaidya and Kumar 2006) (Fig. 1).

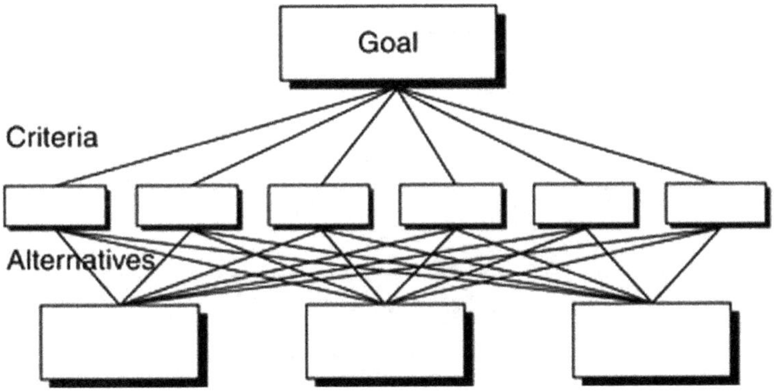

Fig. 1. Analytic hierarchy scheme (Saaty 2013)

2.2 Quality Function Deployment

The most important thing at product and service design is to meet customer requirements. With suitable ergonomic principles and meeting customer needs, high quality of good or service can be reached. In terms of quality, more efficient methods are developed to simplify product and process improvement. These techniques are used in different areas to get benefit from not only quality issues but also ergonomics. Quality Function Deployment (QFD) helps to convert customer requirements to technical properties of the product/service to be developed. These customer needs mostly come out of basic ergonomic principles. Figure 2 shows "House of Quality (HoQ)" which is the matrix for QFD analysis (Bergquist and Abeysekera 1996).

QFD steps can be summarized as follows:

1. Product Planning
2. Product Design
3. Process Planning
4. Process Control (Bouchereau and Rowlands 2000)

Within the first step of QFD, product planning may be done via HoQ that consists of customer requirements. HoQ provides both converting these needs to technical properties and benchmarking between other providers. At last, different weights give a result to prioritize these properties to be focused.

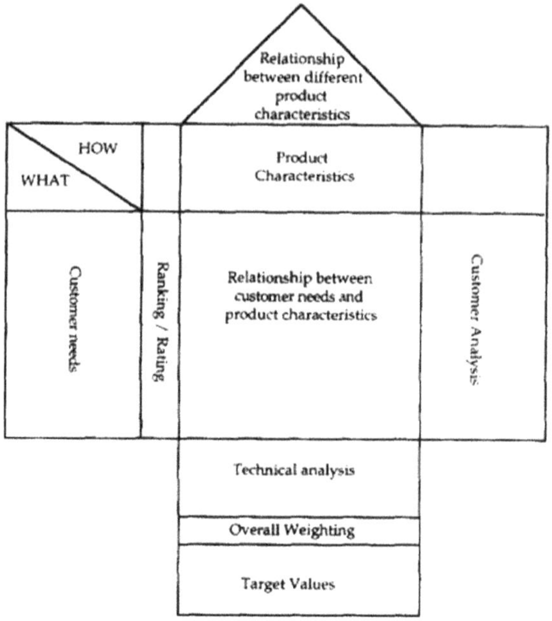

Fig. 2. House of quality (Bergquist and Abeysekera 1996)

2.3 AHP and QFD Integration

There are different examples at the literature that uses AHP and QFD together. Armacost et al. studied about construction industry at United States. There were lack of modernization at the building sites. A development was needed to reach minimum energy consumption and so on. To build a proper wall panel, this study used QFD to define the customer requirements clearly. They also integrated AHP to be able to prioritize customer needs to make decision making (1994). Hua et al. studied about marketing, using Kotler's model for strategic orientation. They used QFD and AHP together to make strategic decision while benchmarking for different types of adoptions (1994).

Bhattacharya et al. studied robot selection with a integrated QFD-AHP model. There are different factors that should be considered while selecting a robot for a specific work. This integrated model leads economic point of view while converting customer requirements into specifications. For this study, seven factors were considered to evaluate different needs (2005). Dai and Blackhurst studied combination of AHP and QFD for supplier selection. There are three different aspects, economic, social and environmental, for meeting sustainability criteria. To compare different suppliers in terms of sustainability there are different tools. In this study QFD was used to define stakeholders' needs and AHP was used for the assessment of different suppliers (2012). Chadawada et al. studied location selection both using QFD and AHP. This combined model allows considering several factors and selecting location from different alternatives (2015).

3 Methodology

3.1 Analytic Hierarchy Process

Analytic Hierarchy Process (AHP), which is a very useful tool in decision making and prioritization, has been developed by Saaty in 1980 and uses pairwise comparison in order to reduce complexity in complex decisions (Wind and Saaty 1980). Furthermore, AHP simply consists of three main steps which includes

1. Calculation of the criteria vectors' weights
2. Calculation of the A matrix which shows the optional scores
3. Rankings the options according to the A matrix.

The calculation of the criteria vectors' weights is the most important part compared to other steps. The comparison matrix is called as A matrix whose a_{jk} shows the importance of the *j*th criterion relative to the *k*th criterion (Saaty 1980). Table 1 shows the relative importance scores.

Table 1. The relative scores based on a_{jk} values (Saaty 1980)

Value of a_{jk}	Interpretation
1	j and k are equally important
3	j is slightly more important than k
5	j is more important than k
7	j is strongly more important than k
9	j is absolutely more important than k

After building *A* matrix, it is possible to construct normalized A matrix by using (1)

$$\overline{a_{jk}} = \frac{a_{jk}}{\sum_{i=1}^{m} a_{lk}} \tag{1}$$

Where $\overline{a_{jk}}$ values are the entities of normalized A matrix. For criteria weighting vector, it is possible to say that the final step is to calculate w's by using (2)

$$w_j = \frac{\sum_{l=1}^{m} \overline{a_{jl}}}{m} \tag{2}$$

According to Saaty, the matrix B, whose entity b_{ih} shows the comparison of ith criterion compared to hth criterion based on jth criterion, should then be calculated as seen in (3). In this case,

$$b_{ih} b_{hi} = 1. \tag{3}$$

Last but not least, in B matrix, sum of each column should be divided by average of this column in order to have S matrix.

The scores v can be found as

$$v = S \cdot w$$

3.2 Quality Function Deployment

The QFD methodology mainly focuses on customers' needs and obligations in order to evaluate their satisfaction levels. A brief QFD methodology is shown in the Fig. 3.

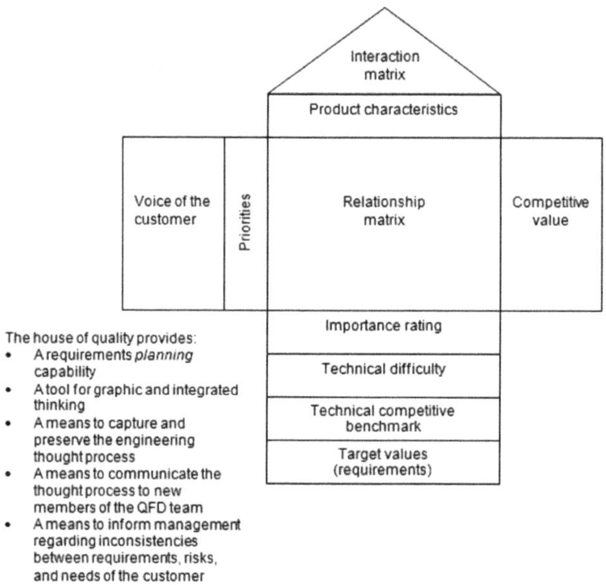

Fig. 3. Methodology for house of quality or quality function deployment (URL1)

In QFD, the weights of Priorities are calculated and the relationship matrix is constructed. Respectively, the QFD methodology consists of

1. Assessing the customer demands
2. Rating of importance of customer demands
3. Evaluation of Customer Evaluations
4. Assessing the technical requirements
5. Understanding the interrelationship between technical requirements in order to assess hows and whats
6. Calculating column weights (Zaim and Şevkli 2002)

4 Application

Firstly, the sustainability criteria in a hospital are weighted by using AHP method. The criteria are taken as

1. Green certification, which is given to fulfill green building requirements
2. Facility Design is for evaluation of the Green buildings
3. Professional Credibility is in order to evaluate if the hospitals are well-known or not
4. Customization is for evaluating if the private hospitals are more sustainable than public ones
5. Regulatory and Law is for measuring if there exist some laws forced to save energy, water and to regulate air conditioning
6. Communication is for being able to understand if there is any better connection between the workers and to satisfy patient needs
7. Technical Services is for measuring how important technical services are in order to be sustainable and satisfy patient needs
8. Admission and Discharging are administrative stuff
9. Workers' Awareness is for understanding if the workers are aware of the sustainability concept or not
10. Indoor quality is for measuring air conditioning, infrastructure of hospitals and hygiene.

These factors have been evaluated by experts who study mainly in sustainability area and the AHP matrix have been calculated in order to measure sustainability criteria pairwise comparison weights in order to put in quality function deployment as seen in the Table 2.

Then the normalized matrix has been calculated based on the comparison matrix as seen in the Table 3.

The weights of each factors of sustainability have been found as respectively in the Table 4. Table 4 shows us that the most important criteria for sustainability in hospitals are Regulatory and Law, Green Certification and Indoor Quality when the least important one is admission and discharging.

The weights have been put in QFD in order to interrelate the patient satisfaction factors and sustainability criteria. The QFD leads to understand which sustainability criteria are interrelated with patient satisfaction factor in hospitals as seen in the Fig. 4.

While building QFD matrix, the patient satisfaction qualifications have been taken as

- Nurse Care
- Housekeeping
- Courtesy and Attitude
- Infrastructure of Hospital
- Safety Indicators
- Administrative Procedures
- Hospital's Image
- Process of Clinical Care

- Hygiene
- Doctors' Experiences

By the view of experts, the highest relative weights are found as Infrastructure of Hospitals, Hospitals' Image and Safety Indicators which are normally the needs that patients desire mostly in a hospital.

Table 2. AHP comparison matrix for sustainability criteria in hospitals in Turkey

	Green Certification	Facility Design	Professional Credibility	Customization	Regularities and Law	Communication	Technical services	Admission and Discharging	Workers Awareness	Indoor Quality – Air condition etc
Green Certification	1	3	4	3	1/2	5	2	4	3	1/2
Facility Design	1/3	1	3	3	1	3	3	3	2	1
Professional Credibility	1/4	1/3	1	1/4	1/3	2	1/2	3	1/4	1/5
Customization	1/3	1/3	4	1	1/4	2	3	5	2	1/3
Regularities and Law	2	1	3	4	1	2	4	6	2	2
Communication	1/5	1/3	1/2	1/2	1/2	1	1/3	2	1/2	1/4
Technical services	1/2	1/3	2	1/3	1/4	3	1	4	1/3	1/2
Admission and Discharging	1/4	1/3	1/3	1/5	1/6	1/2	1/4	1	1/2	1/3
Workers Awarenes	1/3	1/2	4	1/2	1/2	2	3	2	1	1
Indoor Quality – Air condition etc	2	1	5	3	1/2	4	2	3	1	1

Table 3. The normalized matrix for sustainability criteria in Hospitals in Turkey **Cw (Normalized)**

1	0.139	0.367	0.149	0.190	0.1	0.204	0.105	0.121	0.238	0.07
2	0.046	0.122	0.112	0.190	0.2	0.122	0.157	0.091	0.159	0.141
3	0.035	0.041	0.037	0.016	0.067	0.082	0.026	0.091	0.02	0.028
4	0.046	0.041	0.149	0.063	0.05	0.082	0.157	0.152	0.159	0.047
5	0.278	0.122	0.112	0.253	0.2	0.082	0.21	0.182	0.159	0.281
6	0.028	0.041	0.019	0.032	0.1	0.041	0.017	0.061	0.04	0.035
7	0.069	0.041	0.075	0.021	0.05	0.122	0.052	0.121	0.026	0.07
8	0.035	0.041	0.012	0.012	0.033	0.02	0.013	0.03	0.04	0.047
9	0.046	0.061	0.149	0.032	0.1	0.082	0.157	0.061	0.079	0.141
10	0.278	0.122	0.186	0.190	0.1	0.163	0.105	0.091	0.079	0.141

Table 4. Weights of each criteria

1	0.168	16.8%
2	0.134	13.4%
3	0.044	4.4%
4	0.095	9.5%
5	0.188	18.8%
6	0.041	4.1%
7	0.065	6.5%
8	0.028	2.8%
9	0.091	9.1%
10	0.146	14.6%

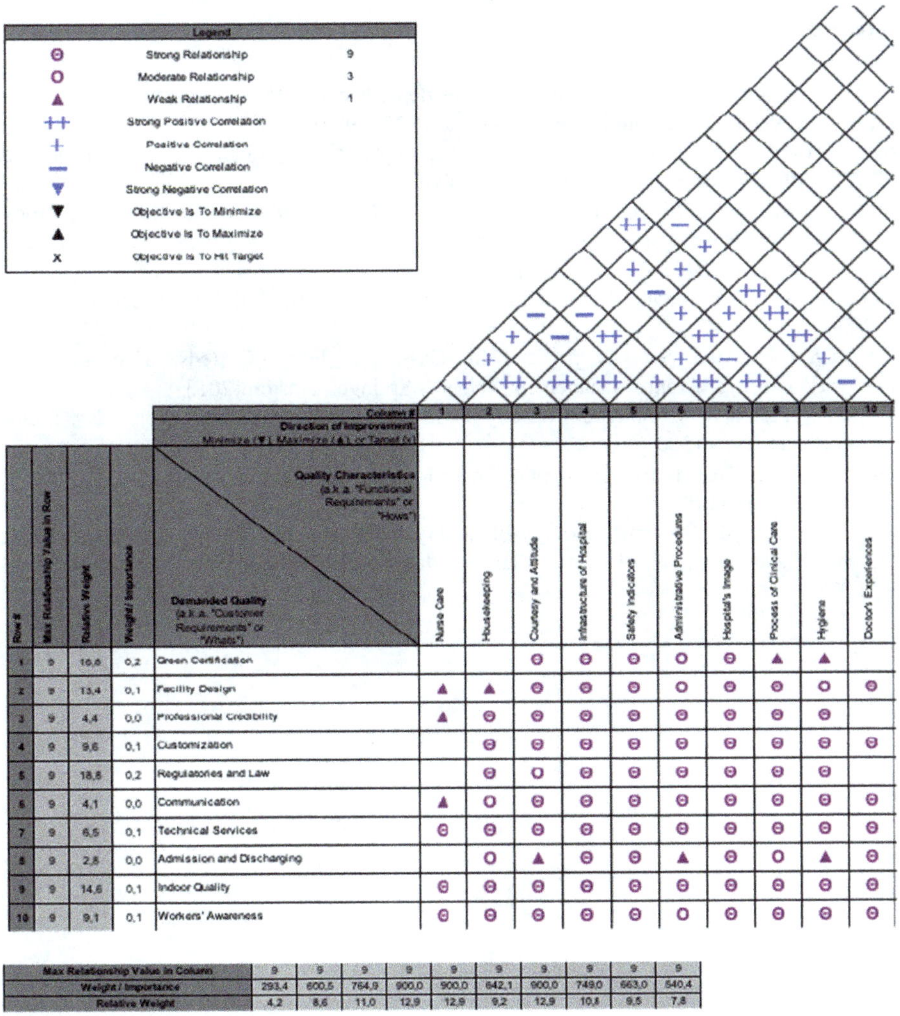

Fig. 4. House of quality for sustainability and patient satisfaction

5 Conclusions

Patient satisfaction is very crucial topic while sustainability is gaining importance. Hospitals are trying to be sustainable when they meet patient needs. In this study, it has been tried to interrelate the sustainability factors and patient needs. It has been seen that regulatory and law and green certification have the highest weights compared to other sustainability factors when Infrastructure of Hospitals, Hospitals' Image and Safety Indicators which are the needs that patients desire mostly in a hospital that are interrelated to the highest weighted sustainability factors. In conclusion, for an improved quality management in terms of sustainability and patient satisfaction, these factors should be improved vice-versa. For further researches, a large scope of sustainability factors and patient satisfaction qualifications should be taken into consideration.

References

Bergquist, K., Abeysekera, J.: Quality function deployment (QFD)—a means for developing usable products. Int. J. Ind. Ergon. **18**(4), 269–275 (1996)

Bouchereau, V., Rowlands, H.: Methods and techniques to help quality function deployment (QFD). Benchmarking Int. J. **7**(1), 8–20 (2000)

Kurttila, M., Pesonen, M., Kangas, J., Kajanus, M.: Utilizing the analytic hierarchy process (AHP) in SWOT analysis—a hybrid method and its application to a forest-certification case. For. Policy Econ. **1**(1), 41–52 (2000)

Saaty, T.L.: Decision making with the analytic hierarchy process. Int. J. Serv. Sci. **1**(1), 83–98 (2008)

Saaty, T.L.: Analytic hierarchy process. In: Gass, S.I., Fu, M.C. (eds.) Encyclopedia of Operations Research and Management Science. Springer, Boston (2013)

Vaidya, O.S., Kumar, S.: Analytic hierarchy process: an overview of applications. Eur. J. Oper. Res. **169**(1), 1–29 (2006)

Wind, Y., Saaty, T.L.: Marketing applications of the analytic hierarchy process. Manag. Sci. **26**(7), 641–658 (1980)

Zaim, S., Şevkli, M.: The methodology of quality function deployment with crisp and fuzzy approaches and an application in the Turkish shampoo industry. J. Econ. Soc. Res. **4**, 27–32 (2002)

Industrial Applications

An Analysis of Methods Allowing Detection of Large Local Irregularities of Surface

Krzysztof Stępień[1(✉)], Dariusz Janecki[2], and Stanisław Adamczak[1]

[1] Department of Manufacturing and Metrology,
Kielce University of Technology, Kielce, Poland
{kstepien, adamczak}@tu.kielce.pl
[2] Centre for Laser Technologies of Metals,
Kielce University of Technology, Kielce, Poland
djanecki@tu.kielce.pl

Abstract. Measurements of form deviations of 3D elements can be conducted with the use of various strategies, differing in a number and distribution of sampling points located on an investigated surface. Low number of sampling points or measured sections can lead to the situation that some surface irregularities are not detected by a measuring system. An application of higher density of sampling points, in turn, results in significant lengthening of measurement time, which is undesirable. This is the reason why authors propose a novel, adaptive measurement strategy consisting of two stages: a preliminary measurement and additional measurements. During the preliminary measurement an investigated area is scanned along preselected trajectory. If measurement results show that there is significant change of sensor readings in a certain fragment of an investigated surface, then we conduct additional measurements. In this paper authors present methods that can be used to localize areas where significant change of signal may occur.

Keywords: Measurement · Form deviation · Adaptive strategy · Rotary elements

1 Introduction

Rotary elements are very significant group of machine parts. Such parts should be usually characterized by high dimensional accuracy. Therefore accurate measurements of form deviations of rotary elements is the fundamental matter. Taking into account requirements of Industry 4.0 it is recommended that such measurements should be conducted in-process. It usually requires application of advanced measuring systems, including coordinate measuring machines.

Measurements of form deviations of 3D elements can be conducted with the use of various strategies, differing in a number and distribution of sampling points located on an investigated surface [1, 2]. Low number of sampling points or measured sections can lead to the situation that some surface irregularities are not detected by a measuring system. An application of higher density of sampling points, in turn, results in significant lengthening of measurement time, which is undesirable, if one takes into

N. M. Durakbasa and M. G. Gençyılmaz (Eds.): ISPR 2019, LNME, pp. 131–137, 2020.
https://doi.org/10.1007/978-3-030-31343-2_11

account requirements of modern manufacturing processes [3–6]. This is the reason why authors have developed a novel, adaptive measurement strategy. Proposed adaptive strategy consists of two stages: a preliminary measurement and additional measurements. During the preliminary measurement an investigated area is scanned along preselected trajectory. If measurement results show that there is significant change of sensor readings in a certain fragment of an investigated surface, then we conduct additional measurements. General scheme of the strategy is presented in diagram shown in Fig. 1.

However, the question is how can we decide what is regarded as the significant change of sensor readings. According to the authors we can do that with the use of two groups of methods. The former is the group of methods based on an analysis of values of sensor readings and the latter is the group of methods based on an analysis of correlation between values in neighbouring profiles.

2 Criteria Based on an Analysis of Values of Sensor Readings

These criteria can be divided into two main groups:

- a criterion based on the value of standard deviation,
- a criterion based on an assessment of risk of exceeding of critical value.

(a) the criterion based on standard deviation

The key value in the former criterion is the value of mean square deviation of the surface, which is generally given by following equation

$$s = \sqrt{\frac{\sum_{i=1}^{n}(x_i - \acute{x})^2}{n-1}} \tag{1}$$

where x_i are subsequent values of variable x, $\acute{x}$ is the mean value of variable x and n is the number of variable x. In proposed concept value of the deviation will be calcuated not with respect to the mean but with respect to the surface of the reference cylinder. It can be determined, for example, by the minimization of the quality indicator described by following equation

$$J(E_{x0}, E_{y0}, E_{xL}, E_{yL}, R_0) = \frac{1}{2}\int_0^{2\pi}\int_0^{L}\left(R(\varphi, z) - R_0 - E_x(z)cos\varphi - E_y(z)sin\varphi\right)^2 dzd\varphi \tag{2}$$

with respect to variables $(E_{x0}, E_{y0}, E_{xL}, E_{yL}, R_0)$.

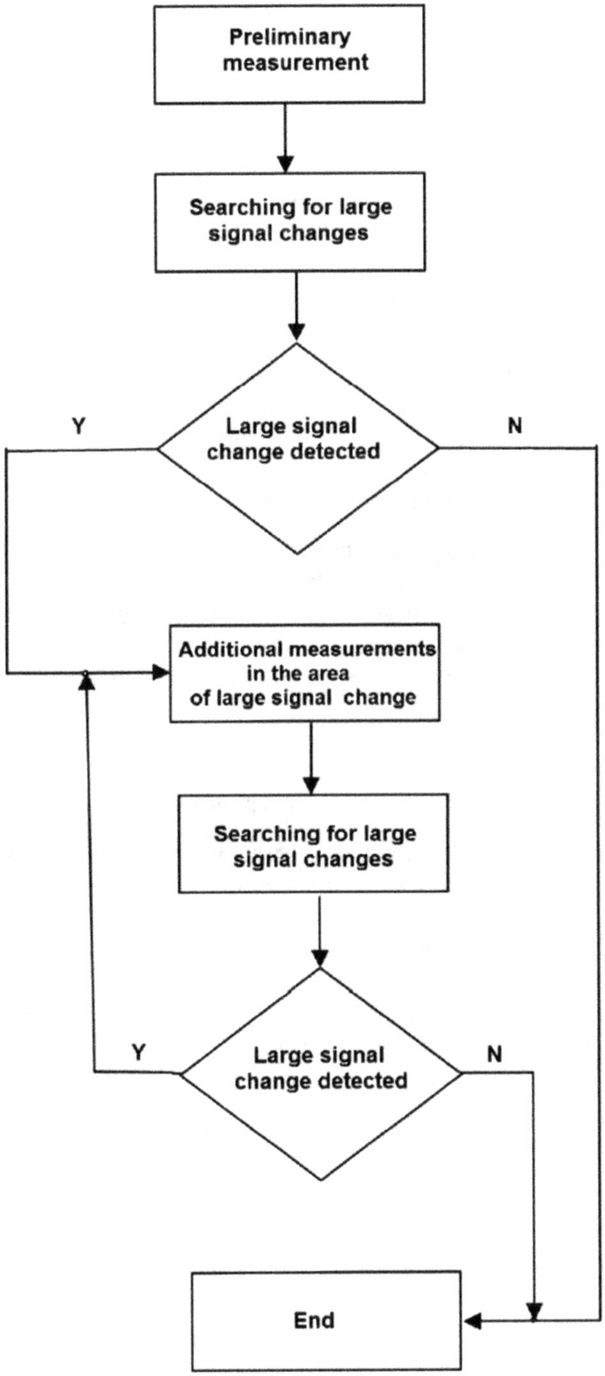

Fig. 1. General scheme of the adaptive strategy proposed by authors

After calculation of the parameters of the reference circle it is easy to calculate deviations of measured surface from the reference according to the Eq. (3)

$$\delta_i = R_i(\varphi, z) - R_0 - E_x(z)cos\varphi - E_y(z)sin\varphi \tag{3}$$

Taking into account Eqs. 1 and 3 one can formulate equation allowing calculation of deviations of the surface from the reference cylinder, which is following:

$$s_{cyl} = \sqrt{\frac{\sum_{i=1}^{n} \delta_i^2}{n-1}} \tag{4}$$

Considering solutions applied in the field of quality assurances, the threshold value that defines large local irregularities will be equal to $3s_{cyl}$. It means that if the distance of given point of the surface from the reference cylinder is larger than $3s_{cyl}$ then this point lies in the area where it is necessary to conduct additional measurements.

Finally, the procedure can be formulated in the following way:

1. We conduct the preliminary measurement.
2. We calculate the parameters of the reference cylinder.
3. We calculate the value of mean square deviation of distances of points of the surface from the surface of reference cylinder.
4. We localize points, whose distance from the reference cylinder is equal or higher than $3s_{cyl}$.

(b) the criterion based on an assessment of risk of exceeding of critical value.

Let us now consider the criterion that is based on assessment of risk that the values of the profiles exceed a certain threshold value. In this criterion the preliminary measurement should be conducted with the use of circumferential or longitudinal sections strategy. Firstly, let us assume that symbols C_{p1}, C_{p2}, C_p and C_{v1}, C_{v2}, C_v denote values of peaks and valleys in two neighbouring profiles and on the whole measured surface. It can be assumed that in additional section defined by variable z lying between z_1 and z_2 peaks or valleys can exceed global values C_p or C_v if the difference $C_{p1}-C_p$ will be small in relation to the absolute value of difference z_1-z_2.

This condition should be formulated in such way that it would be independent of the change of the scale of length and radius deviation. Let us now assume that the coefficient that represents maximum foreseen slope of the function $C_p(z)$ will be denoted by μ.

If for the new potential section defined by the coordinate lying between already measured sections, both conditions are fulfilled, i.e.

$$1 - (C_{p1} + C_{p2})/C_p \leq \mu\ (z_2 - z_1)/(z_{max} - z_{min}) \tag{5}$$

then we assume that the probability of exceeding treshold value by points of the profile is low.

3 Correlation-Based Criteria

Areas of the surface where the significant change of signal may occur can be localized also with the use of correlation methods. Let us assume that we conduct measurements in circumferential or longitudinal sections and we calculate correlation between neighbouring profiles. If the correlation coefficient is high then it can be assumed that there will not be any significant irregularities between the profiles. But if the value of the coefficient is low then the probability that in the area between the profiles significant irregularities occur is high and it is necessary to conduct additional measurements in this area. The correlation can be calculated by three methods: Pearson's correlation, Spearman's correlation or the cross-correlation function.

3.1 Pearson's Correlation

Pearson's correlation coefficient is used to assess linear relationships between variables. Thus, with the use of this coefficient it is possible determination how the change of the former variable will affect the change of the latter one. In general the coefficient of Pearson's correlation can be formulated as follows:

$$r(x, y) = \frac{cov(x, y)}{S_x S_y} \tag{6}$$

where the covariance *cov(x,y)* is calculate with the use of Eq. (7)

$$cov(x, y) = \frac{\sum(x_i - \acute{x})(y_i - \acute{y})}{n} \tag{7}$$

It is obvious that if we apply equations given above to evaluate correlation between profiles, then subsequent values of variables x and y will be the values of profiles in neighbouring sections.

3.2 Spearman's Coefficient

Spearman's coefficient is used to evaluate correlation between sets of data in two cases: if the data are of qualitative character, which makes it possible to set them in a relevant order or if the data are of quantitative character but their number is relatively low. Individual values are attributed by so-called ranges. Spearman's coefficient is defined by Eq. 8.

$$r_s = 1 - \frac{6\sum_{i=1}^{n} d_i^2}{n(n-1)} \tag{8}$$

where d_i are the difference of values of individual ranges and n is the number of all ranges.

In the case of comparison of geometrical profiles, as ranges, for example, values of subsequent amplitudes harmonic components of the profiles can be used. Thus, the application of this method in this case would require using the algorithm of Fast Fourier Transform.

3.3 The Cross-correlation Function

In this method we apply so-called normalized cross-correlation function given by following equation

$$\rho = \frac{2\sum_{k=0}^{N} R_1^k R_2^k}{\sum_{k=0}^{N} \left(R_1^k\right)^2 + \sum_{k=0}^{N} \left(R_2^k\right)^2} \tag{9}$$

where appropriate values R_1 and R_2 denote values of profiles in neighbouring sections, and N is the number of values of profiles in one section.

It is easy to notice that Eq. 9 can be used to calculate correlation coefficient if the profiles are not shifted in phase. If there is a risk that there is a difference in phase (for example the workpiece can be moved between subsequent measurements), the we can modify Eq. 9 and formulated it as a function of the phase shift τ. Then the correlation coefficient will be the maximum value of the function. The function is given by following equation

$$r(\tau) = \frac{2\sum_{k=1}^{N} R_1(\varphi_k + \tau) R_2(\varphi_k + \tau)}{\sum_{k=1}^{N} R_1(\varphi_k) + \sum_{k=1}^{N} R_2(\varphi_k)} \tag{10}$$

where values of profiles in neighbouring sections are denoted by R_1 and R_2 and τ is the phase shift between the profiles.

4 Conclusions

The strategy of measurements of form deviations proposed by authors requires performing additional measurements in areas where significant local irregularities may occur. But the question is what changes we regard as significant and how to localize the area where such changes are possible. Authors have conducted the theoretical analysis of this problem and proposed two approaches to solve these problems. The former approach is based on an analysis of values of profiles and the latter is based on calculation of correlation between neighbouring profiles. Taking into account results of the theoretical analysis and practical aspects of using methods presented in previous sections show that most convenient in practice are following methods: the method based on the value of standard deviation and the method based on the value of the cross-correlation function. Authors are going to use these methods during experimental verification of measurements of rotary elements with the use of adaptive strategy.

Acknowledgment. The paper has been elaborated within the framework of the research project entitled "Theoretical and experimental problems of integrated 3D measurements of elements' surfaces", reg. no.: 2015/19/B/ST8/02643, ID: 317012, financed by National Science Centre, Poland.

References

1. Jermak, C.J., Rucki, M.: Static characteristics of air gauges applied in the roundness assessment. Metrol. Meas. Syst. **23**(1), 85–96 (2016)
2. Żórawski, W., Makrenek, M., Góral, A.: Mechanical properties and corrosion resistance of HVOF sprayed coatings using nanostructured carbide powders. Arch. Metall. Mater. **61**(4), 1839–1846 (2016)
3. Kundera, C., Kozior, T.: Research of the elastic properties of bellows made in SLS technology. Adv. Mater. Res. **874**, 77–81 (2014)
4. Adamczak, S., Bochnia, J., Kundera, C.: Stress and strain measurements in static tensile tests. Metrol. Meas. Syst. **19**(3), 531–540 (2012)
5. Poniatowska, M.: Research on spatial interrelations of geometric deviations determined in coordinate measurements of free-form surfaces. Metrol. Meas. Syst. **16**(3), 501–510 (2009)
6. Humienny, Z., Turek, P.: Animated visualization of the maximum material requirement. Measurement **45**(10), 2283–2287 (2012)

Enhancement of Images from Industrial X-Ray Computed Tomography Systems by Hybrid Approach

Mario Sokac[1(✉)], Zeljko Santosi[1], Djordje Vukelic[1], Marko Katic[2], M. Numan Durakbasa[3], and Igor Budak[1]

[1] Faculty of Technical Sciences, University of Novi Sad, Novi Sad, Serbia
{marios, zeljkos, vukelic, budaki}@uns.ac.rs
[2] Faculty of Mechanical Engineering and Naval Architecture, University of Zagreb, Zagreb, Croatia
marko.katic@fsb.hr
[3] TU Wien (Vienna University of Technology), Vienna, Austria
numan.durakbasa@tuwien.ac.at

Abstract. Application of the computed tomography (CT) within industry has been rising in recent years due to its non-destructive abilities and accuracy. Nevertheless, there are some challenges related to CT scanning, such as presence of artefacts. The aim of this research is to investigate to what extent the application of some advanced algorithms can influence the accuracy of the X-ray CT images. In this paper, after a brief overview of different existing methods used for reduction of different types of artefacts, preliminary research of a new approach for CT image enhancement is presented. It is based on a hybrid methodology using two different methods - Fuzzy Clustering and Region Growing - joined in order to exploit their advantages. Results show that the proposed methodology contributes to CT image enhancement, with borders of segmented objects on CT images more easily extracted.

Keywords: Computed tomography · X-ray CT · Artefacts · Industrial CT images · Image processing

1 Introduction

High demands in the industry for fast and accurate product measurement is pushing the development of new systems that are used for those purposes. One such system is the X-ray computed tomography (CT) and its use in the industry field, especially for the metrology and manufacturing applications [1]. X-ray CT system, due to its non-destructive abilities and high accuracy, found its place in such high demanding field with its constant development and improvement [2]. However, X-ray CT systems are prone to different types of artefacts [3, 4], due to numerous factors included [5]. But, from the hardware side of X-ray CT, there have been numerous research investigating the possibilities of suppression of these artefacts from the software side as well. Today, software support plays an important role as is provides a qualitative and quantitative side of processing and interpretation of X-ray CT results. This allows an accurate

N. M. Durakbasa and M. G. Gençyılmaz (Eds.): ISPR 2019, LNME, pp. 138–146, 2020.
https://doi.org/10.1007/978-3-030-31343-2_12

visualization and application of different types of dimensional measurements to be performed. However, one of the biggest challenges that X-ray CT systems are facings are presence of artefacts on the obtained results as they have a big influence on the quality of generated CT data [6]. They present artificial structures that are found in the X-ray CT results and that don't correspond to the real data [7]. In the field of metrology artefacts can seriously affect, or even prevent acquisition of reliable measurement data. They may appear as variations in shades of gray, or as a deviation in the reconstructed data. The presence of these artefacts can have a great influence the obtained results, i.e. reconstructed 3D models, which can make them unreliable for any further analysis [7].

2 Literature Review

When it comes to application of different methods, in order to deal with artefacts and their influence on the X-ray CT results, there have been many different solutions and research directions. There are numerous factors that influence the CT dimensional measurements such as surface quality, material, acquisition parameters, CT machine setup, etc. [8]. One such parameter is also surface roughness of the material, which is shown to have a strong influence on CT dimensional measurements [6].

In addition to those factors, numerous methods have been developed such as MAR methods (Metal Artefact Reduction), region growing, active contours method and others that reported more or less successful elimination, or at least reduction of negative impacts in order to obtain reliable and accurate measurement results [9, 10]. When it comes to MAR methods, sinogram in-painting technique shows to be one of the preferable techniques used to remove unwanted streaking artefacts caused by metal implants which degrade the image quality. And this technique often consists of interpolating the missing projection data, but the result is often a loss of image quality with additional artefacts in the whole image [11–13]. Many authors also investigated and discussed about industrial X-ray CT accuracy [14] and measurement uncertainty [15]. Also projection-based MAR for industrial 3D X-ray computed tomography was investigated [7], as well as scatter correction methods applied in dimensional X-ray CT [8]. For enhancement and development of new methods for surface extraction new methods have also been developed by implementing graph cuts method [16], multi-level Morse complex [17], or 3D Canny algorithm with sub-voxel resolution [18].

Development and application of various methods for analysis and processing of information from CT images is an active research topic for a long period of time. These methods span from different fields and have been used for different purposes [19, 20]. Among these methods are solutions that deal with specific problems related to the removal/reduction of artefacts, or image enhancement. In spite of some of developed approaches of this kind, there is still space for the development of new improved methods related to reducing the impact of artefacts and improving borders extraction on resulting X-ray CT images. Application of artificial intelligence methods (namely fuzzy logic) for analysis and processing of the generated CT image data is applied for a long period of time in different fields [21, 22]. It aims to improve the 2D image by enhancing the object borders and thus eliminate/reduce negative impacts present on image. In addition to fuzzy logic, application of other methods such as region growing, active contours method and others are all used for their application in image processing and handling [23]. Their application has shown to give elimination or at least reduction of negative impacts in order to obtain reliable and accurate measurement results.

3 Research Methodology

3.1 The Nature of the Research

Fuzzy logic, more specifically Fuzzy C-means (FCM) clustering method proved to be a very popular method when it comes to image processing and analysis by many researchers in various fields, because it is not confined to a single area of application. Although, today there are many different solutions and implementations of this method, there is still a room for the improvement connected to application of this method. This also includes its integration and/or combining with other methods in order to develop hybrid approaches with emphasized positive features of single methods.

3.2 Materials

The need for this research is based on a fact that X-ray CT images that are impaired with presence of artefacts and/or the use of inappropriate parameters are present in today's application of X-ray CT systems. This influences the proper surface extraction and interpretation of results from acquired X-ray CT images. Because of these problems, different methods have been developed that deal with improvement and enhancement of these results, especially those closely related to the field of coordinate metrology and dimensional accuracy. Improvement of processing algorithms for CT image reconstruction, through the integration of FCM and region growing (RG) methods for the purpose of development of a new hybrid approach, has a significant potential. This potential lies in advantages of these methods that can be synergized through their integration, while their disadvantages can be reduced.

3.3 Methods

The main goal of this research is to develop a new method which will feature an improved processing and analysis of generated X-ray CT images, and consequently enable surface reconstruction of dimensionally more accurate surface 3D models. The special focus, during the development of this model, will be on its application on impaired CT images with poorly displayed objects borders on them.

Preliminary research in this area suggests that the development of such model would contribute in improving the quality of boundary extraction by enhancing the contrast (borders) between certain objects present on images. FCM method plays a big role in enhancement of the pixels of interest, while for the segmentation and generation of 2D binary images the RG method is used (Fig. 1). In this way this approach is primarily focusing on the successful extraction of the objects borders on the image, and thus improving accuracy of reconstructed surface 3D models.

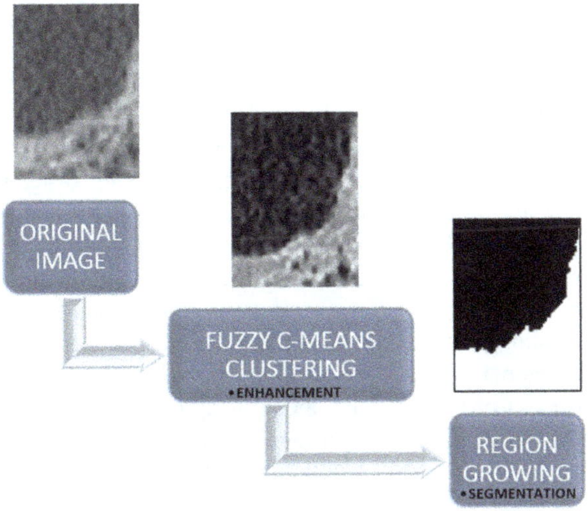

Fig. 1. Basic principle of the proposed approach

The experimental data was acquired using an industrial X-ray CT system NIKON XT H 225 located at University of Zagreb, Faculty of Mechanical Engineering and Naval Architecture in Zagreb, Croatia (Fig. 2).

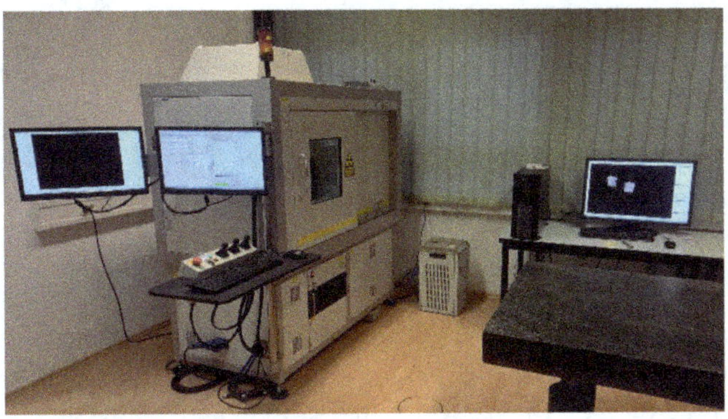

Fig. 2. Industrial CT NIKON XT H 225 located at University of Zagreb, Faculty of Mechanical Engineering and Naval Architecture in Zagreb, Croatia

4 Results and Discussions

Multi-material components (MMCs) are usually parts which are manufactured from several different materials, for example, polymers and metal components, or composite materials with some metal components present. One of the factors which severely limit the application of CT system for the surface reconstruction and analysis of these components is the presence of artefacts. In order to investigate the present implementation of the above algorithm, X-ray CT dataset was obtained by an industrial X-ray CT system with which an MMC part was scanned - a steel hex screw mounted in a plastic housing.

MMC components are a particularly interesting case for research due to the presence of different materials with different densities. However, different thicknesses of materials, and above all, different structure and density of the material lead to a larger or smaller absorption of the X-rays. As such, it strongly influences proper edge detection, resulting with presence of artefacts. These obtained images are not adequate for further analysis and data extraction (surface 3D model) due to poor pixel intensity that influence the visibility of an object on the CT images. Therefore, any subsequent analysis and diagnosis may become invalid in terms of dimensional analysis. In order to achieve improvement of these results, the application of the proposed hybrid model provides adequate processing, extraction and displays real structure of the 3D model from X-ray CT data. A single image containing M8 steel hex screw mounted inside a plastic housing was used for analysis of the proposed algorithm, and it is shown on Fig. 3.

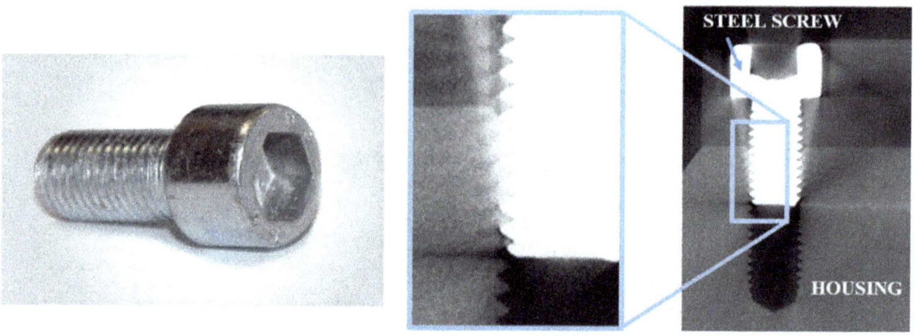

Fig. 3. Steel hex screw placed inside a plastic housing

Steel is a material that has a high density and, within MMC, it can present a problem for the X-ray CT systems due to its higher density and the need for greater penetrating power of the X-rays, while the surrounding plastic has a much lower density.

The first step in this approach is the application of FCM method in order to adequately enhance the contrast of the border between the object of interest (hex screw) and the background (plastic housing). The input parameters which are used for FCM method are number of clusters C and the membership degree (i.e. fuzzification) m. After that, the RG method is used for boundary detection of object present on X-ray CT image (Fig. 4a, b and c).

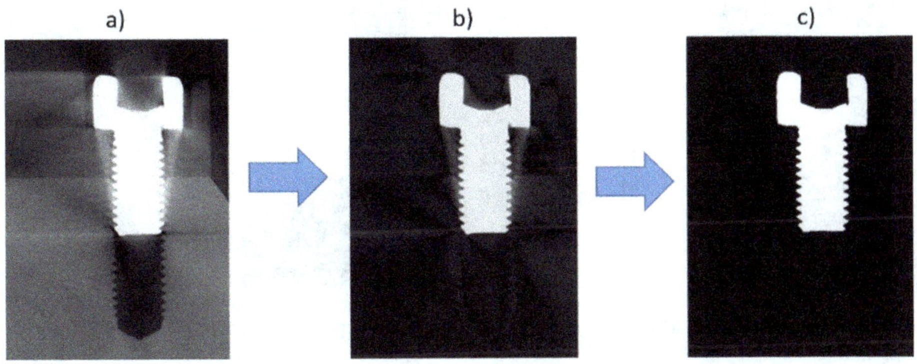

Fig. 4. Working stages of the proposed method showing (a) original image, (b) enhanced image using FCM method and (c) segmented image of hex screw using RG method

In order to perform image analysis, and show influence of application of FCM method, two 2D profile lines on two images were compared (original image and image processed by FCM) in order to determine the differences between them. This 2D profile on the image is drawn along the arbitrary path and is consisting of the segments of the predefined path of the line. Each segment of the line profile is calculated by the number of points (profile points) which correspond to the pixels density of the image.

Profile is calculated along the path of the 2D line (Fig. 5a). This 2D line calculates the pixels intensity along the defined path. It can be observed that there is a normalization of the pixel intensity, as well as an increase in their contrast which is enhanced by applying FCM method. From the graph in Fig. 5b it can also be seen that higher values of contrast have a positive effect on the enhanced visibility, which contributes to an easier and better segmentation using RG method.

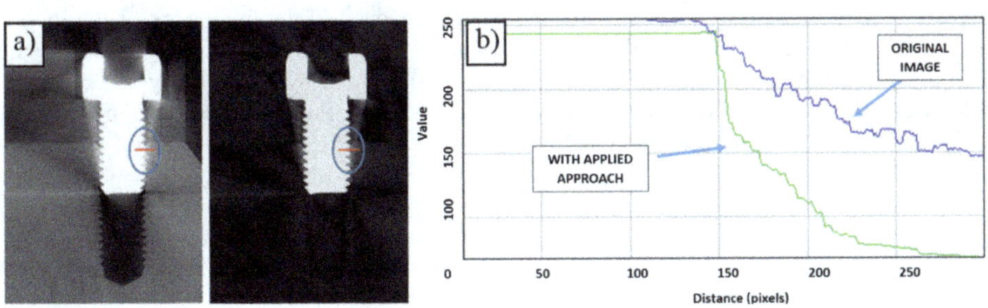

Fig. 5. Showing: (a) path of the 2D profile line and (b) pixel intensity variations before and after FCM method

Figure 6a and b shows the segmentation before and after implementation of the RG method.

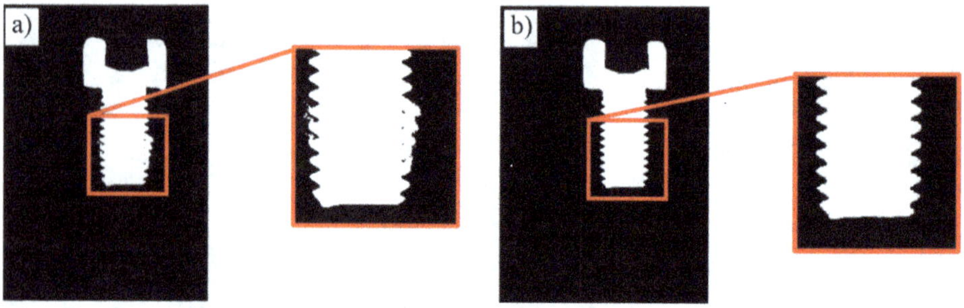

Fig. 6. Segmentation performed (a) before and (b) after implementation of the proposed approach

On the basis of the above-described results of preliminary analysis of the hybrid method which combines FCM and RG methods, it is shown that it shows good results that requires further investigations.

5 Conclusions

The research presented in this paper aims at development of a new methodology to improve the segmentation and analysis of X-ray CT images in the area of industrial metrology. By combining various concepts and methods from the field of visual computer techniques, the additional information is obtained, where the use of these methods aids in accurate reconstruction of objects boundaries. Presented results of preliminary analysis of the proposed hybrid methodology, which combines fuzzy clustering with region growing methods, are very promising and present a good foundation for further investigations.

Acknowledgements. This paper presents the results achieved in the framework of the Project no. 114-451-2723/2016-03 funded by the Provincial Secretariat for Higher Education and Scientific Research, and within the project TR - 35020, funded by the Ministry of Education, Science and Technological Development of Republic of Serbia. Project IKARUS supported parts of presented research (European Regional Development Fund, MIS: RC.2.2.08-0042, Faculty of Mechanical Engineering and Naval Architecture, University of Zagreb).

References

1. Durakbasa, N.M., Bauer, J.M., Bas, G., Kräuter, L.: Towards a sophisticated approach to cost oriented automation and intelligent metrology in the advanced manufacturing. IFAC-PapersOnLine **51**(30), 54–59 (2018)
2. Akdogan, A., Vanli, S., Durakbasa, N.: Re-engineering of manufacturing parts by computed tomography data. In: Proceedings of the 12th International Conference on Measurement and Quality Control - Cyber Physical Issue IMEKO TC 14 2019, pp. 114–121 (2019)
3. ISO 15708-1:2017: Non-destructive testing – radiation methods for computed tomography – Part 1: Terminology (2018)

4. ISO 15708-2:2017: Non-destructive testing – radiation methods for computed tomography – Part 2: Principles, equipment and samples (2018)
5. Townsend, A., Pagani, L., Blunt, L., Scott, P.J., Jiang, X.: Factors affecting the accuracy of areal surface texture data extraction from X-ray CT. CIRP Ann. Manuf. Technol. **66**(1), 547–550 (2017)
6. Aloisi, V., Carmignato, S.: Influence of surface roughness on X-ray computed tomography dimensional measurements of additive manufactured parts. Case Stud. Nondestruct. Test. Eval. **6**, 104–110 (2016)
7. Amirkhanov, C., Heinzl, M., Reiter, J., Kastner, M.E.: Gröller: Projection-based metal-artifact reduction for industrial 3D X-ray computed tomography. IEEE Trans. Vis. Comput. Graph. **17**(12), 2193–2202 (2011)
8. Baer, M., Hammer, M., Knaup, M., Schmidt, I.: Scatter correction methods in dimensional CT. iCT **14**(1), 41–46 (2012)
9. Krumm, M., Kasperl, S., Franz, M.: Reducing non-linear artifacts of multi-material objects in industrial 3D computed tomography. NDT E Int. **41**(4), 242–251 (2008)
10. Bartscher, M., Hilpert, U., Goebbels, J., Weidemann, G.: Enhancement and proof of accuracy of industrial computed tomography (CT) measurements. CIRP Ann. Manuf. Technol. **56**(1), 495–498 (2007)
11. Boas, F.E., Fleischmann, D.: CT artifacts: causes and reduction techniques. Imaging Med. **4** (2), 229–240 (2012)
12. Chen, Y., Li, Y., Guo, H., Hu, Y., Luo, L., Yin, X., Gu, J., Toumoulin, C.: CT metal artifact reduction method based on improved image segmentation and sinogram in-painting. Math. Probl. Eng. **2012**, 1–19 (2012)
13. Heinzl, C., Kastner, J., Georgi, B., Lettenbauer, H.: Comparison of surface detection methods to evaluate cone beam computed tomography data for three dimensional metrology. In: International Symposium on Digital industrial Radiology and Computed Tomography (DIR 2007), Lyon, pp. 21–29 (2007)
14. Dewulf, W., Kiekens, K., Tan, Y., Welkenhuyzen, F., Kruth, J.: Uncertainty determination and quantification for dimensional measurements with industrial computed tomography. CIRP Ann. Manuf. Technol. **62**(1), 535–538 (2013)
15. Gayou, O.: Influence of acquisition parameters on MV-CBCT image quality. J. Appl. Clin. Med. Phys. **13**(1), 1–12 (2012)
16. Haitham Shammaa, M., Ohtake, Y., Suzuki, H.: Segmentation of multi-material CT data of mechanical parts for extracting boundary surfaces. Comput. Aided Des. **42**(2), 118–128 (2010)
17. Nagai, Y., Ohtake, Y., Suzuki, H.: SegMo: CT volume segmentation using a multi-level Morse complex. Comput. Aided Des. **107**, 23–36 (2019)
18. Yagüe-Fabra, J.A., Ontiveros, S., Jiménez, R., Chitchian, S., Tosello, G., Carmignato, S.: A 3D edge detection technique for surface extraction in computed tomography for dimensional metrology applications. CIRP Ann. Manuf. Technol. **62**(1), 531–534 (2013)
19. Budak, I., Mirkovic, S., Sokac, M., Santosi, Z., Puskar, T., Vukelic, D.: An approach to modelling of personalized bone grafts based on advanced technologies. Int. J. Simul. Model. **15**(4), 637–648 (2016)
20. Drégelyi-Kiss, Á., Durakbasa, N.M.: Measurement error on the reconstruction step in case of industrial computed tomograph. In: Proceedings of the International Symposium for Production Research 2018, Vienna, pp. 309–323 (2018)
21. Heil, U., Gross, D., Schulze, R., Schwanecke, U., Schomer, E.: Metal artifact reduction in x-ray computed tomography: inpainting versus missing value. In: IEEE Nuclear Science Symposium and Medical Imaging Conference, Sidney, pp. 2675–2677 (2011)

22. Hiller, J., Hornberger, P.: Measurement accuracy in X-ray computed tomography metrology: Toward a systematic analysis of interference effects in tomographic imaging. Precis. Eng. **45**, 18–32 (2016)
23. Jiménez, R., Comps, C., Yagüe, J.A.: An optimized segmentation algorithm for the surface extraction in computed tomography for metrology applications. Procedia Eng. **132**, 804–810 (2015)

Intelligent Monitoring System for Zero-Defect Production of Irreplaceable Parts

Ildiko Maňková[1(✉)], Marek Vrabeľ[1], Ladislav Vargovčík[2], and Martin Eckstein[3]

[1] Faculty of Mechanical Engineering, Technical University of Košice, Košice, Slovakia
{ildiko.mankova,marek.vrabel}@tuke.sk
[2] Prototyping and Innovation Centre, Faculty of Mechanical Engineering, TU Košice, Košice, Slovakia
ladislav.vargovcik@tuke.sk
[3] MTU Aeroengine Munich, Munich, Germany

Abstract. Hole making process monitoring of irreplaceable parts is crucial in terms of waste production elimination. Defects on machined part do not directly affects its end user. Strict quality control regulations in all production levels ensure the required product quality. However, faulty parts increase production costs proportionally. These costs are included in the sale price for which end user purchases the product. In order to achieve a significant reduction of the waste production a new generation of sensors will be applied to obtain appropriate signals directly from the machine control unit. For the real time and effective signal processing, wavelet transformation and neural network approach will be used to identify changes of the cutting tool or machined surface. Such a proposed system has all necessary assumptions to reduce scrap production in manufacturing chain and thus reduce the price of final product. The paper describes a approach in real-time monitoring for drilling boltholes. In an experimental setup, process data origination from the CNC of a Sinumeric 840D, collected by an OPC-Server had been processed. Comparing OPC data logging, DAU data logging and profibus data logging with respect to data quality, sample rate and real-time behaviour, profibus data logging appears to be the favourable choice.

Keywords: Monitoring · Hole making · Ti6Al4V alloy · Spindle power · Profibus

1 Introduction

Today, production of highly stressed components is intended to make contribution towards a safer production of the holes within manufacturing process of irreplaceable parts made from titanium alloy Ti6Al4V by providing solutions for the extraction and processing of internal controller data for purpose of linking it to tool wear and the integrity of machined workpieces as represented by microstructure, residual stress Aero engines clearly differ from other propulsion systems used in general mechanical engineering and in vehicle construction above all by the materials used and the extreme

N. M. Durakbasa and M. G. Gençyılmaz (Eds.): ISPR 2019, LNME, pp. 147–156, 2020.
https://doi.org/10.1007/978-3-030-31343-2_13

stresses occurring in operation. Boltholes in rotating turbine and compressor disks belong to the most highly stressed geometric features of aero engines. Consequently, it is important for engine manufacturers to assess the quality of these holes in an early stage of the manufacturing chain and not as late as in final inspection [1].

Lauro et al. [2] stated that monitoring applied to machining processes can improve the process through an increase in tool life and surface quality with a simultaneous decrease of electric energy and waste material. According to requirements of end users identified within EU funded project SIMON [3], real production application of monitoring solution for machining processes is highly dependent on its level of integration into the machine tool or CNC unit controller. A high level of integration can be achieved by using internal data, provided by the components of the machine tools themselves. The newest generation of numerical controllers in machine tools can be used to log data from all rotary and translatory axes without additional sensors which may change machine tool properties. The potential of this type of signal, with respect to monitoring surface integrity of the workpiece, has not yet been explored.

Axinte and Gindy [4] in their research concluded that monitoring set-ups based on single sensory signals (e.g. cutting forces, vibration, machine power and acoustic emission) or combinations of these have been employed to identify/characterize tool condition. Despite the costly equipment and extensive creative energy dedicated to these attempts, both academics and researchers should recognize that not much has been implemented in industry. This is due to the lack of robustness of the monitoring methods combined with a reduced flexibility of the sensors to fit to the industrial set-ups. Since most spindle drives of modern computer numerical control (CNC) machine tools allow direct/indirect measurement of spindle power, this could be regarded as an easy solution for tool condition monitoring. Electric drives and spindles provide the mechanical force necessary to remove material from the part. The major advantage of motor related parameters to detect malfunctions in the cutting process is that the measurement apparatus does not disturb the machining process. The capacity to measure power already exists in the drive controller as part of the drive control loop or can be readily retrofitted and is suitable for use in production environments [5]. However, a current literature review clearly reveals that knowledge of using power sensor for TCM (tool condition monitoring) and tool life prediction is still very limited [6].

Corne et al. [7] within their very recent research evaluated the usefulness of spindle power data for real-time tool wear/breakage prediction (TWBP) during hole making of super alloy instead of using force data that are impractical in production floor. They concluded that real time process monitoring and control capability with instant data processing is vital, especially, when drilling Ni-/Ti-based superalloys because catastrophic failure and premature breakage of tools occur in random manner. The essential objective of process monitoring described in literature, is to reliably identify machining process deviations and problems or anomalies occurring during machining such as excessive tool wear, tool fracture, collisions, process instabilities (e.g. chip clogging) at an early stage during manufacturing and consequently being able to initiate adequate actions [8–10].

Nowadays main topics of process monitoring research in hole making in irreplaceable parts can be focused on following [11]:

- To develop intelligent monitoring system operating in a real time aimed for hole making process in machining titanium alloy. This topic is highly relevant in sectors with repeatable production, where high productivity is required under strict requirements for the end user's properties. This system responds directly to applications associated with current changes in CNC technologies that combine digitization, sensor data processing, product quality assurance through process management, decreasing of environmental impacts by reducing waste production and related energy efficiency (also called 4th Industrial Revolution).
- Integration and fusion in process monitoring (power and current signals of spindle motor and machine drives) to predict surface integrity in hole making. In terms of hardware, to identify positions of the advanced sensors placements, determine their cross connection and wiring for transmitting data to processing unit. Develop algorithms for signal filtration and its processing. To elaborate and validate sensor calibration procedures for monitored variables.
- To identify anomalies occurring on machined surface of the hole and determine their correlation with cutting edge tool wear behaviour. To develop algorithms for sensor signals data processing using methods of wavelet transformation and artificial neural networks. To define sensor signals as a source relating to cutting edge tool wear by utilizing procedures related to evaluation of the of information edge sharpness (edge roundness).
- To determine the impact of the cutting tool edge changes (due to e.g. microchipping) on the resulting surface integrity in hole making of the irreplaceable parts by methods of statistical analysis (ANOVA, RSM, PCA etc.). To implement the assembled monitoring system into selected machining centres and to determine procedures for adjusting prediction algorithms, robustness, performance and reliability.

According to above mentioned topics this paper is intended to make contribution towards a safer production of boltholes in highly stressed irreplaceable parts made from titanium alloy Ti6Al4V by providing solution for the extraction and processing of internal controller data for purpose of linking it to tool wear and the integrity of machined workpiece.

2 Evaluation of Sensor Signal Quality

Considerably increased reliability and price availability of monitoring systems notwithstanding, very few of them have been used in actual conditions of machining work-pieces with dedicated use, the irreplaceable components. An essential reason can be relatively complicated and sophisticated use of aforesaid techniques and their operation in industrial practice. One of main challenges in the field of process monitoring is development of the autonomous algorithms to reduce operation severity known in these days. Therefore, there is real need of intelligent production system to gather needed data during machining that subsequently are capable of identifying tool state and machined surface. For production cost of machining titanium alloy Ti6Al4V being high due to strictly hold machined surface integrity, the following task is needed to fulfil [11]:

- reducing of the faulty pieces in production due to timely based identification of the deviations (faults) of tools or work-pieces when machining by drilling in real time
- increasing of the energy based efficiency of producing holes made from difficult-to-cut super-alloy as productivity increasing when maintaining high level of process reliability
- reducing man power requirements to operate machine tools (especially to asses tool life from tool wear viewpoint) due to global lack of work-force to increase efficiency and output capacity of drilling process and that is considering of the actual tool life and not a tool maker's value often applied to produce zero failure of work-pieces (in efficient production, number of tool changes should be hold minimum while tool edge is supposed to be used until its limiting tool life, for instance, maximum tool flank wear etc.)

Proposal of the ways to reach the goals is coming from methodology of obtaining both current and power data from spindle and drive units of CNC machining centre: To define suitable software/hardware open architecture platform to obtain directly internal data from machine tool control unit, or to define other hardware that is to apply in machine tool working area (e.g. Hall's sensing head to measure power that is located in wiring unit chamber of the machine tool). Final choice of the measuring system is assessed from such viewpoints as data quality, frequency of sampling frequency, its applicability in the real industrial plants as well as its efficiency in the real time. Further important aspect is its assessment from viewpoint of costs of the measuring devices. Therefore, various types of applicable inter-connections as TCP/IP, USB port etc., between data units and hardware components shall be investigated to increase data quality and measurement reliability to reduce noise affecting final character of the data.

When designing a system to collect internal data form the CNC machine tool, Siemens Sinumerik 840D controller is considered due to the largest commercial application across industry. Currently, there are several options for obtaining internal signals of the machine tool, but in this paper is only tested three different systems. The first is so called OPC data logging system, which uses the open structure of the user interface of the machine tool to transfer data. Another system is Siemens Tracer and DAU which uses analog port from drive control to transfer data to the evaluation unit. The final applied system is called Fieldbus data logging. The Profibus client is located in the profibus network topology, and the internal data are transferred to the data collection and evaluation unit (TCP/IP).

2.1 OPC Data Logging

OPC is the common abbreviation for "OLE for Process Control". It provides a universal means for the exchange of data between different hardware without the need to know the exact internal design of each device under consideration [12]. This is achieved by providing a common bi-directional abstraction level as outlined in Fig. 1. Since commercial OPC software is not tailored to the needs of cutting process monitoring, a universal OPC server provided by SIEMENS with the symbolic name "OPC-SINUMERIK Machineswitch V2.4.98" could be installed and configured to the requirements of the experimental setup. A corresponding client (laptop) is configured to log both the effective power of the spindle and the effective power of the motor moving the z-axis. In a first step, the data is simply stored on the hard drive of the HMI and subsequently accessed with a USB flash drive. In a second step, "near real-time-access" is achieved with an additional executable programme. This programme can be installed as a background service on the HMI and is configured to read and send OPC data to an external client connected to the HMI via Ethernet. During the tests, the data rate was identified as a limiting factor for subsequent data processing. The maximum data rate was found to be around 10 samples/second. This is in compliance with the data rate of other OPC applications installed in manufacturer automation.

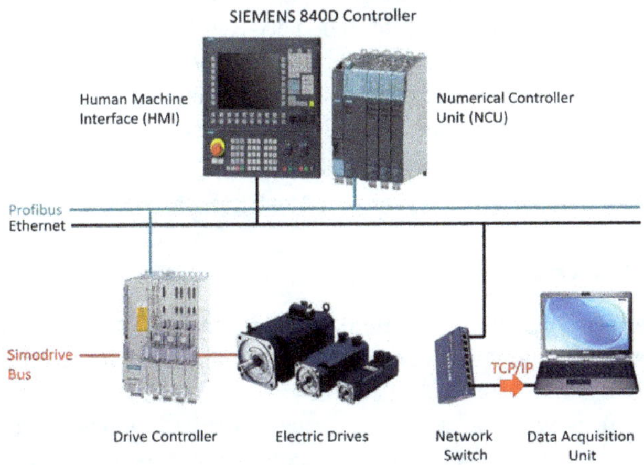

Fig. 1. Test configuration for OPC datalogging

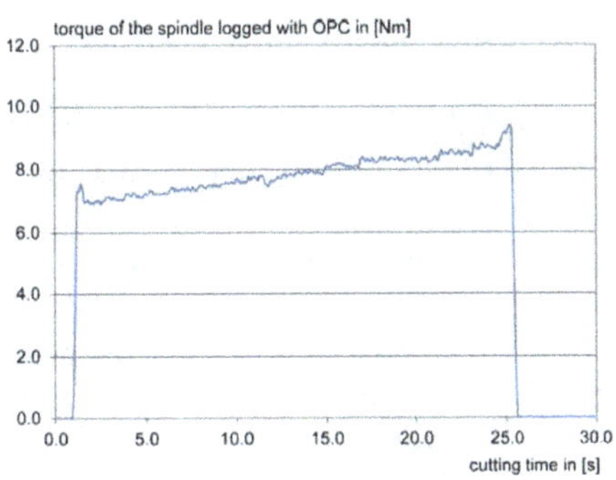

Fig. 2. OPC pattern of a pattern of a twist drill. New tool condition

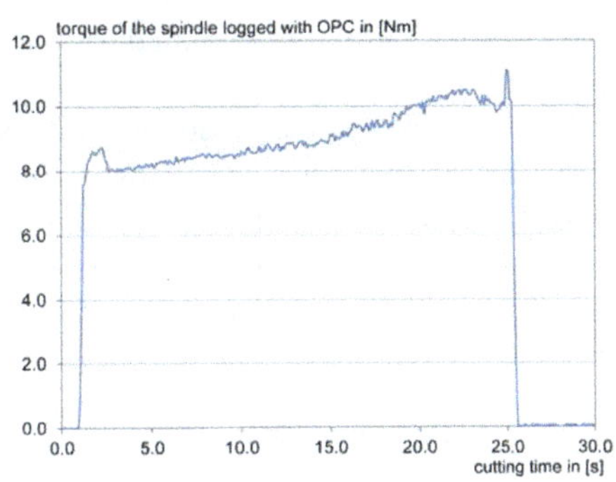

Fig. 3. OPC pattern of a pattern of a twist drill. Worn tool condition

Representative torque data generated during drilling and finishing and logged from the spindle by OPC are outlined in Figs. 2 and 3. These patterns require less interpretation. A significant change in amplitude can be observed between a new and a worn tool for both drilling with a twist drill as well as for finishing with face cutting tools. It has to be considered that the torque logged via OPC is calculated by the machine tool controller. Input data is the effective power and the parameters of the motor(s) installed. Furthermore, it is important to consider that the machine tool applied for the OPC data logging were not calibrated prior to the cutting tests.

Summarizing the most important results from the machining tests with OPC enabled, it can be stated that OPC can be considered a good source for data with respect to spindle torque and torque proportional to the feed force. It has to be taken into account that the data rate with OPC is limited and data originating from the z-axis must be judged with the knowledge about the mechanical set-up of the respective of axis. Under the machining setup chosen for this work, which delivered a data rate of about 10 samples/second, a considerable difference between the OPC pattern created by sharp and dull tools within a drilling and a finishing cycle was observed. It is obvious that this data is not suitable to log signals representing phenomena that occur in a frequency above 5 Hz. From its concept, OPC data logging adds no additional hardware to the machine tool and can be performed independently of the numerical controller's hardware platform.

2.2 Siemens Tracer and DAU

Although numerical controllers communicate digitally with their periphery, they also provide some special analogue ports. These ports are used for setup, diagnosis and service. SIEMENS Sinumerik controllers provide a so-called "DAU" port. In addition, Siemens Sinumerik provides a so called "Trace-Function", which can be configured to log multiple data from rotary and translator machine tool axes in a digital format. The trace function is an excellent tool for post process data analysis up to a sample rate of app. 1000 samples/second. Unfortunately the trace function is not usable for real-time process monitoring. Consequently, within this work only the analogue DAU port was tested for its usability to log data during machining. The configuration for data logging is outlined in Fig. 4.

Example plot obtained from drilling tests are provided in Fig. 5 (Test was done for nickel alloy at that case). To allow real-time data processing during the tests it was necessary to convert the analogue data received from the DAU port to digital data. This data was sent to the USB port of a data acquisition PC applied for data processing. The limitation of this solution is that only one channel (e.g. effective power of the feed drive) can be submitted via DAU. In a series of cutting tests implausible data was recorded within a cutting cycle. During a more detailed investigation of the processing of internal data, it became evident that this port is configured to automatically adjust its measuring range to minimize signal deviations caused by the limited resolution of the 8 bit D/A converter that supplies the analogue signal. Considering the corresponding algorithm, it is possible to calculate back to the original signal.

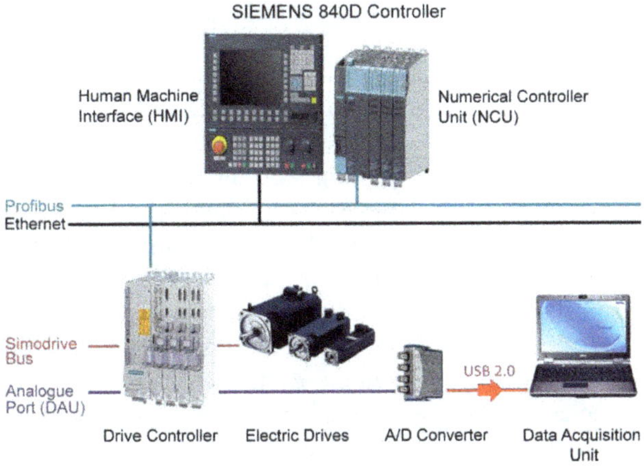

Fig. 4. Test configuration of DAU data logging

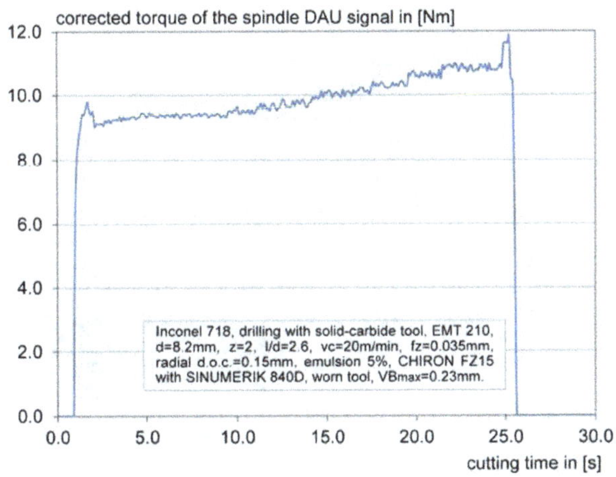

Fig. 5. Reconstructed DAU signal

In summary it can be stated DAU data logging is limited to one recordable channel at a time. This channel can be configured to the individual monitoring need. But it is not possible to switch this pre-selected channel during processing. In theory, the data rate is only limited by the maximum data rate of the internal D/A converter and the external A/D converter. But it has to be kept in mind, that the internal D/A translation are performed with an 8 bit converter. This converter offers $2^8 = 256$ different signal levels. If the machining process causes an elevated level of signal dynamics, DAU data needs to be continuously screened for plausibility and pre-processed. To avoid misinterpretation or perhaps false alarms, the algorithm of switching between measurement ranges must be known and considered for the interpretation of the signals. On the other hand, this method needs only a very limited instrumentation which is completely outside of the cutting zone.

2.3 Fieldbus Data Logging

Fieldbus systems are used to connect different devices on different command levels (e.g. sensors, actors and controllers on the slave level) with numerical controllers and workstations on the master level with one serial connection. The devices within the master level communicate via token passing procedure. The token is a sequence of bits that travels continuously through the master ring topology. The device that holds the token can communicate with the profibus devices. At SIEMENS this token is called compile cycle. SIEMENS already supplies compile cycles for different purposes including monitoring. For this work a compile cycle was used, that is commercially available and can be applied both for the so-called "PowerLine" and the "SolutionLine" series of an 840D controller. This compile cycle is called "Tool and Process Monitoring". This specific compile cycle can read up to 8 configurable addresses within one interpolator cycle. A logging client was installed to the topology of the profibus in the machine tool for the configuration of the machine tool for fieldbus data logging. In order to make data storage and transmission as convenient as possible, a small embedded PC was installed.

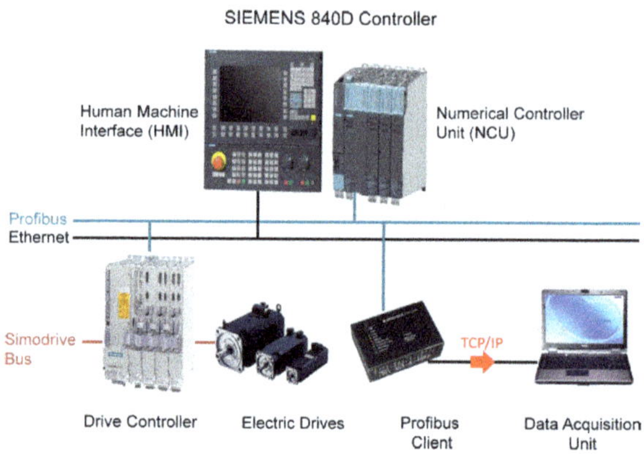

Fig. 6. Test configuration of fieldbus data logging

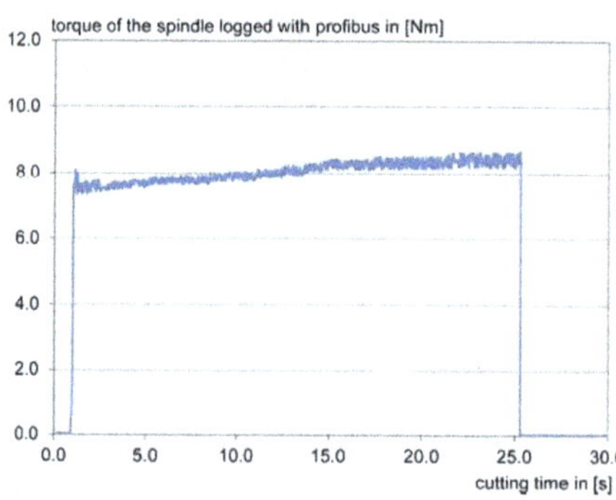

Fig. 7. Profibus pattern of a twist drill. New tool condition

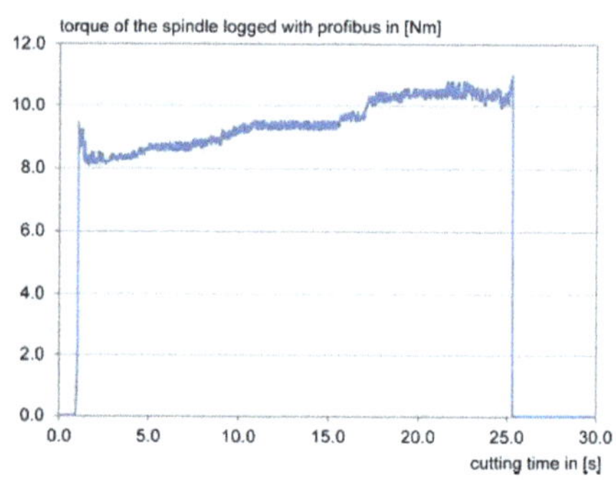

Fig. 8. Profibus pattern of a twist drill. Worn tool condition

Representative plots obtained from the configuration outlined in Fig. 6 are provided in Figs. 7 and 8. The sample rate that can be achieved depends on the interpolator (IPO) cycle. With the hardware installed in the machine tools MICRON UCP1050 approx. 100 samples/second ±5 samples/second was achieved. According to new information provided by SIEMENS, sample rates of up to 2 kHz can be achieved on the newest controllers if the very latest compile cycles are applied. But so far, these new compile cycles have been reserved for SIEMENS applications and tasks.

In order to compare the quality of the fieldbus data with the data obtained from a dedicated force measurement system, an additional KISTLER rotating dynamometer type HS-RCD 9125A was linked to the test machine. Since it is not possible to apply coolant through the RCD, the comparison was made for finishing. This is the most sensitive and important machining step when working on a very low level of cutting forces.

With respect to the capabilities of profibus data logging performed within the conditions of this work, it can be stated that this data can be logged with only small modifications to the machine tool. The data is extracted and transmitted to the monitoring unit in a completely digital form. This rules out conversion losses and minimizes signal interference with other sources of electric and electromagnetic signals near the machine tool. With the compile cycle applied, the sample rate was 100 samples/second. Compared to external sensors, profibus data logging can be installed in a way that does not expose data logging devices to the harsh environment of chip- making processes at all. On the other hand, signal dynamics can only be picked up to a level of 50 Hz, which is much less than the capabilities of many external sensors (e.g. for acceleration, force or acoustic emission). Compared to force data obtained from an RCD, profibus data still has an acceptable quality, especially in the time domain.

3 Conclusions

For this article, SIEMENS Sinumerik controllers qualify themselves as an excellent hard- and software platform with “open controller architecture”. Comparing OPC data logging, DAU data logging and profibus data logging with respect to data quality, sample rate and real-time behaviour, profibus data logging appears to be the favourable choice. Compared to the price of many dedicated external sensors, all three methods to log internal data provide data access with small investments on a high level of integration. Internal data logging, as it is performed within this work, does not change the properties of the metal cutting system.

Acknowledgment. The authors would like to thanks to the Research Agency of Slovak Ministry of Education to support research project VEGA 1/0219/18 and KEGA 070TUKE-41/2017.

References

1. Eckstein, M.: Process Monitoring – Ein innovativer Ansatz zur Steigerung der Zuverlässigkeit höchstbelasteter Bauteile der Fluggasturbine. Deutscher Luft- u. Raumfahrtkongress, Jahrbuch (2004)
2. Lauro, C.H., et al.: Monitoring and processing signal applied in machining processes – a review. Measurement **58**, 73–86 (2014)
3. Sensor Fused Intelligent Monitoring System for Machining-Final Report for European IMS Project EP 26504 (2002)
4. Axinte, D., Gindy, N.: Assessment of the effectiveness of a spindle power signal for tool condition monitoring in machining processes. Int. J. Product. Res. **42**(13), 2679–2691 (2004)
5. Teti, R., et al.: Advanced monitoring of machining operations. CIRP Ann. – Manuf. Technol. **59**, 717–739 (2010)
6. Drouillet, C., et al.: Tool life predictions in milling using spindle power with the neural network technique. J. Manuf. Process. **22**, 161–168 (2016)
7. Corne, H., et al.: Study of spindle power data with neural network for predicting real-time tool wear/breakage during Inconel drilling. J. Manuf. Syst. (2017). http://dx.doi.org/10.1016/j.jmsy.2017.01.004
8. König, W., Klocke, F., Ketteler, G., Memis, F., Rehse, M.: Prozessüberwachung beim Bohren, Fräsen und Schleifen unter Verwendung neuer Auswerte- strategien, VDI-Berich 1995 1179, 91–113
9. Abu-Mahfouz, I.: Drilling Wear detection and classification using vibration signals and artificial neural network. Int. J. Mach. Tools Manuf **43**, 707–720 (2003)
10. Klocke, F., Gierlings, S., Brockmann, M., Sage, C., Veselovac, D.: Adaptive control of manufacturing processes for a new generation of jet engine components. In: CIRP - ICME Capri (2010)
11. Vrabeľ, M., et al.: Development of intelligent monitoring system for zero-defect production of irreplaceable parts. Reserach project VEGA 1/021918-20 (2017)
12. Iwanitz, F., et al.: OPC-Fundamentals, Implementation and Application, Hüthing Fachbuchverlag, 3rd Issue (2006)

Periodical Technical Emission Control by Using an IoT Universal Short-Test Procedure

Ernst Pucher(✉), Andreas Gruber, Andreas Eidmann, and Christian Spitzwieser

Institute for Powertrains and Automotive Technology, Vienna University of Technology, Vienna, Austria
{ernst.pucher, andreas.gruber, andreas.eidmann, christian.spitzwieser}@tuwien.ac.at

Abstract. At least since the diesel gate scandal, the public has become aware of the possibilities of software engineers, and so it is not surprising that inspection personal does not trust the on-board diagnostic functions too much. At this point only a modern tail pipe test can bring the required result. The patented procedure is similar to that exhaust gas test, which has already been carried out for several decades on gasoline engines and three-way catalytic converter. In addition to the known quantities CO2, CO, HC and O2, the determined emission values additionally include NOx, NO2, NO, particle number and average particle size.

A main development goal for this device was the application of state-of-the-art Internet-of-Things (IoT) technologies. The first step was the implementation of wireless communication between the measurement device itself and its control-computer. They are able to communicate via WiFi 5-Protocol. The next step was to create a network-application to upload the gained data and make it accessible for the user through a web-browser based GUI. The final step is linking all those measurement systems together by the use of a Message Queuing Telemetry Transport based data platform. In using this widely acclaimed IoT standard protocol, it is possible to control, collect the data from, and service a huge amount of measurement systems. The investigations showed predominantly good correlations between the emissions during the short test in comparison to RDE measurements. This instrument significantly upgrades the periodic inspection of motor vehicles on the road, as it reflects a real emission behavior at the tailpipe of the exhaust system and does not make use of car specific engine control data.

Keywords: Periodical technical inspection · Internet-of-Things · Universal short-test procedure · Big data and analytics

1 Introduction

Energy production from carbonaceous fuels conveys an additional challenge for manufacturers of vehicles with highly efficient engines. Due to this focus, the pollutants NOx and particle number have become the focus of society. However, this development

N. M. Durakbasa and M. G. Gençyılmaz (Eds.): ISPR 2019, LNME, pp. 157–168, 2020.
https://doi.org/10.1007/978-3-030-31343-2_14

is not entirely new to the experts of the field of exhaust emissions of internal combustion engines. Thus, since some years, a highly exceeding of NO_2 air-quality limits could be observed [1]. One up-to-date monitoring technology is images of the European environmental satellite SCIAMACHY (Fig. 1).

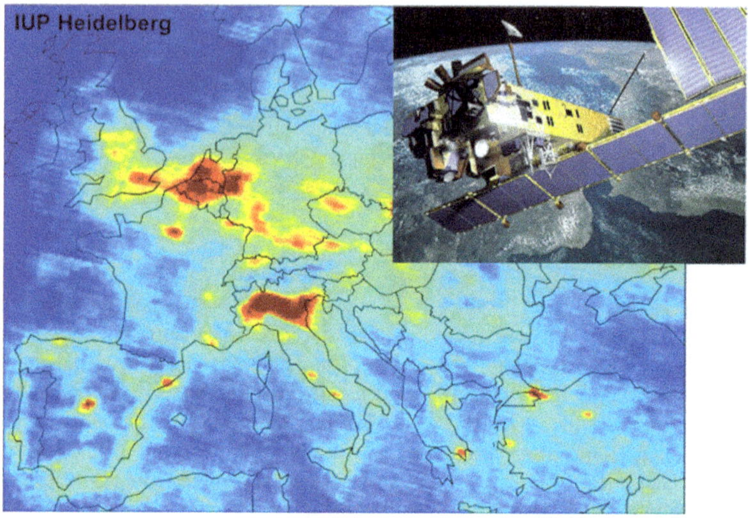

Fig. 1. European NO_2 air pollution map from SCIAMACHY

The good news: Nitric oxide and particulates Emissions are mainly generated locally and can therefore be very well influenced by local measures.

2 Certification Methods for Determining Distance Related Exhaust Emissions of Individual Vehicles

However, the design of an efficient emission reduction should aim at maximum conversion of pollutants at all possible operating points, which also arise in real traffic. In this case, when the vehicle was certified, the exhaust emission behavior was only checked on chassis dynamometers with CVS (Constant Volume Sampling) exhaust gas measurement (Fig. 2) [2]. The possibilities to influence these tests in the legal framework by adapting different parameters were given. So, it is not surprising that on the one hand, the fuel consumption of vehicles in real traffic were hardly achievable, and on the other hand, the emissions exceed the legal limits many times over.

A completely different situation arises in the so-called RDE investigation (Real Driving Emissions). Here a vehicle is examined for its emission behavior in real traffic (Fig. 3) [3]. Of course, the pollutants resulting from the varying real-world dynamic mode of operation are not identical to those of the CVS on the chassis dynamometer.

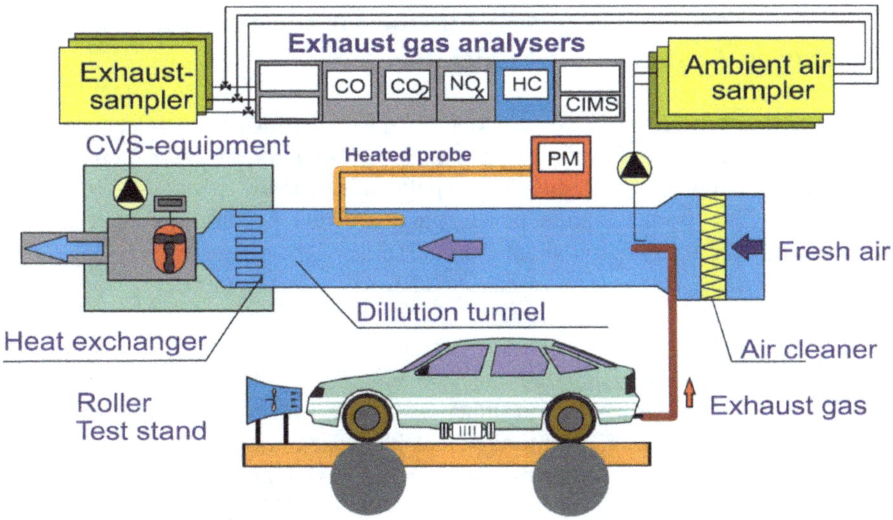

Fig. 2. Roller test bench with constant volume sampling exhaust gas measuring system

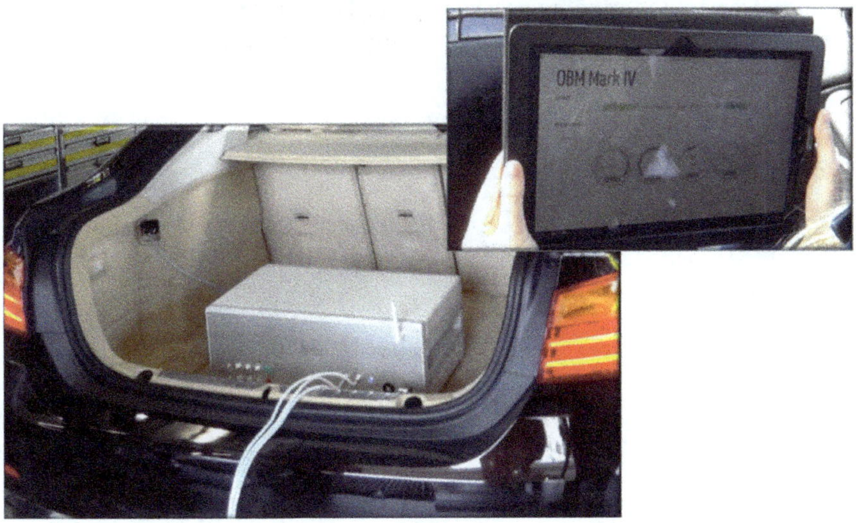

Fig. 3. Real-world testing with OBM (On-board measurement) IoT device

An improvement should be the new Euro 6d emissions legislation, in which the manufacturers must demonstrate compliance with the limit values for passenger cars according to the new WLTP test cycle (Worldwide Harmonized Light Vehicles Test Procedure) and additionally in an RDE test in real traffic. These methods should allow an objective assessment of the emission behavior of the drive systems in the homologation of the vehicles.

3 Current Test Procedures in Europe for the Periodic Technical Inspection of Vehicles

For periodic inspection in Europe, a variety of measurement and maintenance procedures to verify the function of the exhaust emission behavior are used. Depending on the vehicle age, a direct measurement of the pollutants from the exhaust system or a so-called examination of the engine management and the exhaust gas aftertreatment system should give information about the emissions of the vehicle.

For powertrains with gasoline engine and three-way catalyzer, a method for determining the combustion air-fuel ratio Lambda and carbon monoxide concentration (CO) at elevated idle speed is used in tailpipe measurement. This method, which was established in 1988 by the main author of this article, provides information about the engine Lambda control system and function of the catalytic converter. For older Gasoline engines CO and HC measurement at idle speed is used.

Diesel engines have very different strategies in the Member States of the EU. Most use opacity measurements at free engine acceleration mode. However, it can be shown that the methodology, which dates to the seventies, has not been suitable for this since the introduction of particle filters in 2005. The principle of free engine acceleration and turbidity measurement is shown in Fig. 4. In this case, the soot emission is determined at full load acceleration from engine idle speed. For vehicles with Diesel particulate filter however the measurement results are hardly interpretable at the lower measuring range of the unit.

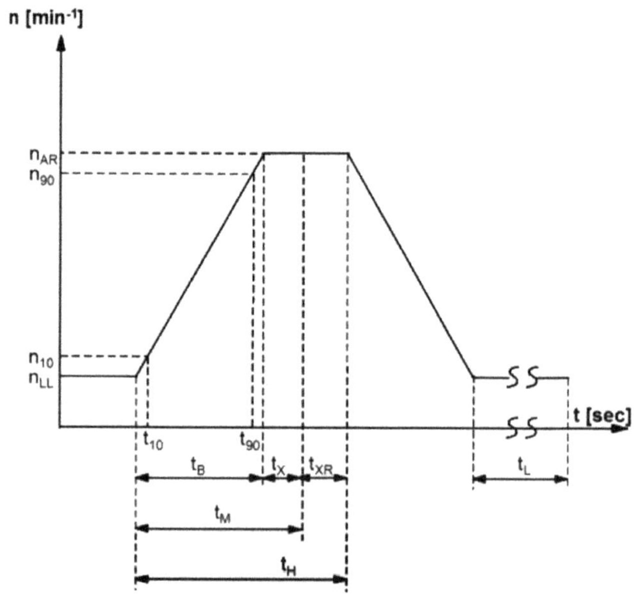

Fig. 4. Diesel - opacity at free acceleration

Some European states are using the on-board diagnostic function to determine the state of the exhaust after-treatment. The interrogation of the diagnostic interface merely shows a standardized protocol which is exchanged between the test devices and the computers of the motor control. As some situation in the last years showed, however, there are often significant differences between the values supplied by the vehicle in direct comparison with real emission. Germany has recognized this problem and has returned during a legislative amendment with the beginning of 2018 back to tailpipe measurement.

Both procedures are anchored in the practice of regular inspection for motor vehicles on the market to check their compliance with the legal framework.

Other issues:

- The most important air-quality related exhaust constituents NOx and ultrafine PN are not measured
- The free acceleration test for Diesel is used only very limited in reality-Specifically with HDV
- We have more than 400 Million vehicles in Europe - How will we handle the deterioration problem
- A significant share of Euro 6 cars, trucks and busses still have highly exceeding emissions

At this point, the newly developed universal short test method address and close the gap in the periodic inspection of emission control systems in modern gasoline and diesel engines.

From our experience of a high number of real-world test on all vehicle categories and evaluation of periodic inspection programs of Diesel and Gasoline vehicles over the last 20 years we learned:

- A direct tail pipe check that is independent of the vehicle itself is required

4 New Universal Short-Test Procedure for Particle and NOx Measurements at Periodic Technical Inspection

The new procedure enables a good transition from the previously known routine for the executing test institutions. It has clear parallels with the exhaust gas test in the gasoline engine, which eliminates costly training for the examiner. For the workshops and test site operators, there would be further benefits to the introduction of the new test procedure [4]. On the one hand, the number of test equipment could be minimized, since a single device can be used to measure the emissions of all combustion engines on the market. Of course, this would also be reflected in the operating and maintenance expenses for the annual calibration of the test equipment. On the other hand, eliminates the sometimes-lengthy conditioning phases, which are required in diesel vehicles with predominantly short-distance operation, to clean the exhaust line for the old opacity measurement. In addition, interference or defects in particle filters can be detected.

Another advantage is the ability to detect built in Ad-Blue emulators, which are installed to reduce the operating costs, by measuring the nitrogen oxide emissions with high-resolution (Fig. 5).

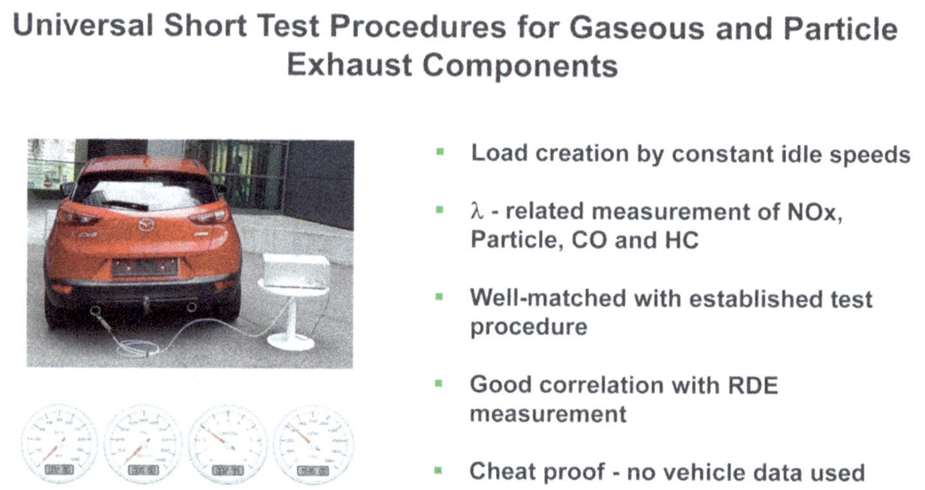

Fig. 5. Universal Exhaust Emission Short-Test procedure for periodic technical inspection (PTI)

The developed emission device software is based on the programming language Go, which was developed by Robert Griesemer, Rob Pike and Ken Thompson in 2012 for the programming of technical systems. A selection criterion for the Go programming language was the threading system, which was included directly in the language scope, with which the connection and evaluation of the gas sensors was implemented. The sequential process communication system (CSP) included in the scope of Go enables efficient and secure communication and synchronization of the different sensor data channels of the emission meter. The scope of the Go system environment includes an extensive software library, which has considerably simplified the connection to the various interfaces (device drivers) of the sensors used.

A central topic of current motor vehicle research and application is the communication possibilities between vehicles (V2V) and between vehicles and stationary infrastructure units (V2I) made possible by technical progress. The emission meter was integrated into a modern V2I system, which was developed in the context of this work. The implementation of the V2I system was based on a cloud infrastructure with a data transfer connection based on the communication standards 3G and 4G. The meter thus has the capability of emission data transmission to remote data analysis units in real time. Thus, a remote evaluation and remote control of vehicle emissions is possible. During the development of the communication architecture, attention was paid to a solution that guarantees complete and error-free data transmission after disconnections. In numerous test drives with onboard measurement, the practicality, reliability and stability of the V2I system were examined and confirmed.

5 IoT Connection of the Emission Measuring Device

A development goal for this device was the application of state-of-the-art Internet-of-Things (IoT) technologies [5]. The first step was the implementation of wireless communication between the measurement device itself and its control-computer. They are able to communicate via WiFi 5-Protocol (former: IEEE 802.11ac). After this was established, the next step was to create a network-application to upload the gained data and make it accessible for the user through a web-browser based GUI. The final step is linking all those measurement systems together by the use of a MQTT (Message Queuing Telemetry Transport) -based data platform. In using this widely acclaimed IoT standard protocol, it is possible to control, collect the data from, and service a huge amount of measurement systems. In Austria alone, if said measurement method is implemented as a standard, there is the demand for several thousand individual measurement devices.

The aim of the developed variant was to use and implement an architecture that meets all requirements for a scalable IoT system. Criteria for the selection continued to be an integrated solution to prevent data loss during a connection interruption, as well as an IoT system, which corresponds to a standard. The use of Open Standards allows for easy system extensibility, future-proofing, and compatible third-party system support.

The final version was developed on the basis of a professional IoT architecture. It was the system IoT Watson I.B.M. selected, which was developed according to the IoT architecture MQTT. MQTT is the most commonly used IoT standard. MQTT defines a publish- and subscribe system with a central broker and the associated protocols for connecting IoT nodes. IoT nodes are on the one hand devices which send the signals of the IoT sensors to the central broker. Data transfer to the broker is mentioned in the MQTT Architecture Event. The counterpart to this is a command, which is sent by the broker to all devices, for example to send a central start command. The second type of defined IoT nodes are applications. On the one hand, these receive the measurement data of the devices and on the other hand can send command to devices.

The logon of the IoT nodes at the central broker is defined in the MQTT protocol. IoT application nodes have the option of defining filters when sending the commands or receiving the events in order to specifically communicate with only a subset of the IoT nodes. The graphics in Fig. 6 shows the integration of the OBM emission meter into the IBM IoT Watson system.

In the software of the decentralized emission measuring device an MQTT device node has been implemented. This is responsible for sending the OBM measurement data to the central broker. As an MQTT device adapter an implementation for the programming language Go was chosen. This adapter offers the possibility to temporarily persist the measurement data in case of interruption of the data communication. Possible is an in-memory persistence or a persistence in the file system. If in-memory persistence is selected, the data will only be saved during runtime of the emissions meter software. Should the emissions meter software be terminated in the absence of a data connection, the metering data would not be sent to the central IoT broker and would be lost. If persisted in the file system, the measurement data would

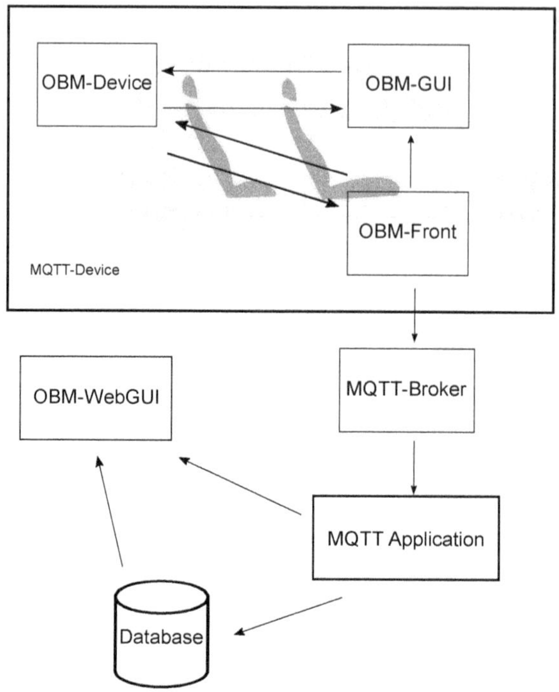

Fig. 6. IoT integration of the short test device

also persist after stopping and starting the emissions meter software. As soon as there is a connection between the emission measuring device and the IoT Broker, the data is sent automatically.

By registering the emission measuring device MQTT Device with the IoT Broker, the respective nodes can be uniquely identified. The MQTT transmission protocol fulfills all requirements for data transmission according to the current safety criteria. Through appropriate authorization, authentication and encryption of the measurement data during transport, these are protected against improper access.

6 Validation of the Short-Test Procedure with Real Driving Emission Tests

The methodology and the validation of the results will be presented to give an overview of the quality of the test procedure [6]. Therefore, the results of LDV (Light Duty Vehicle) NOx and particulate number measurements of the New Short Test Procedure are compared with real driving emission RDE of the investigated vehicles. The most important criterion for the assessment of such a new measurement method is specifically the correlation between measurement results from the short-test procedure and the emission performance at certification tests. Depending on whether the designated field of application is a certified vehicle emission test center or a car repair shop, other relevant criteria are the test time needed to obtain valid results and the reliability under various preconditions. To test the performance of the measurement method, extensive real-world vehicle exhaust emission measurements were carried out to examine the specific emission behavior of the test objects. Then, the results of both the real-world and the short-test measurements were compared against each other and the correlations between them were calculated.

The measurements were performed with the OBM 5.0 real-time emission measurement system. This device can determine emission data second by second and display them live on a handheld device and in a web application via data uplink. Measured exhaust gases are CO2, CO, O2, HC, NO, NO2, NOx and particle number PN (Fig. 7). In consequence the emission behavior of the tested vehicle is examined and it allows detailed trip analyses during and after the measurement.

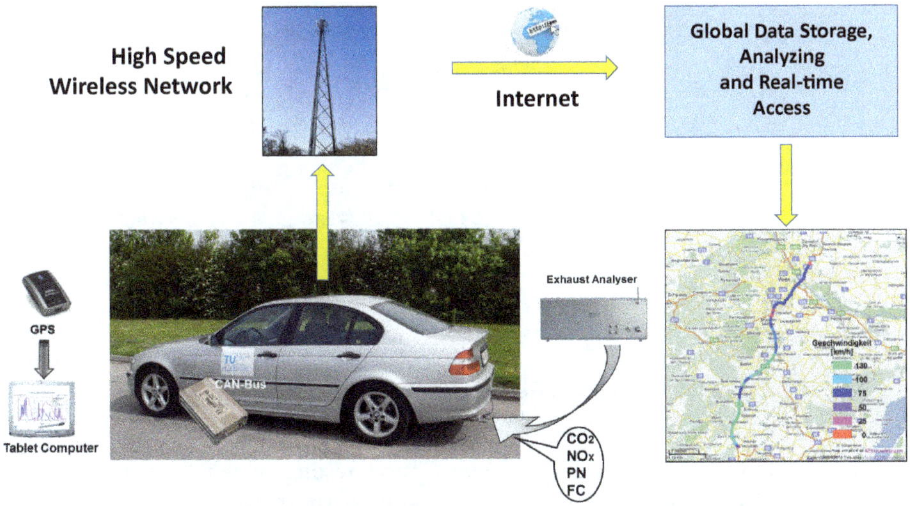

Fig. 7. OBM 5.0 real-time V2I and IoT in-car emission measurement system

The measurement trips consisted of two characteristic parts: An urban cycle with low average speed and frequent stops as well as a highway part with a high average speed and no stops. The urban part was performed in the inner city of Vienna, Austria, where the Ring was driven twice, which adds up to a total trip length of 10 km. The highway drive consisted of three sections at 80 km/h, 100 km/h and 130 km/h average speed with a length of a least 5 km for each part. It was driven twice, once out and once back into the city of Vienna to minimize the influence of road slope and wind. The whole trip length was approximately 70 km and the measurements were performed in around two hours.

The first short-test was performed after a short drive with the vehicle, directly before the inner-city driving cycle. The engine running time was under 10 min and therefore, this recording represents test conditions with low engine and exhaust gas temperatures and correlates to a typical vehicle repair shop situation where the vehicle is tested almost directly after a cold engine start. The second short-test was done after the urban test cycle with a driving time of approximately 30 min. It represents a test with a fully warmed-up engine. The last short-test procedure was completed directly after the highway drive under high, steady load conditions. Engine and exhaust gas temperatures were generally higher there and they represent conditions when a vehicle is tested directly after a long, extensive drive, rather untypical for a normal repair shop environment. During the short-test procedures, also exhaust gas temperatures were recorded to determine threshold values for engine warm-up times. Every short-test procedure recording consists of at least 60 s idling, followed by another 60 s of elevated speed of 2000 rpm and ends with another 30 s of idling. During the procedure, NOx concentrations and the particle numbers are measured and recorded. These results are then multiplied with the current lambda value second-wise (Fig. 8).

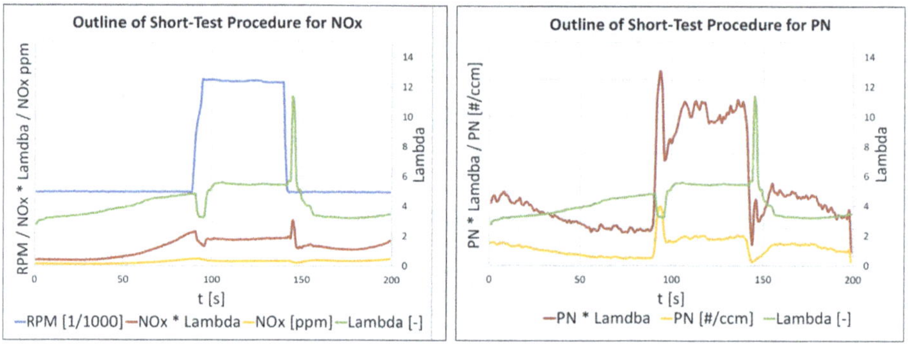

Fig. 8. Outline of short-test procedure for NOx and PN (particle number)

The diagrams in Fig. 9 shows the correlation results of ten Euro 6 light duty vehicles. Satisfying correlations were found between short-test and real- world measurement results for NOx and particulate number PN. In general, the NOx emissions in city driving exceeded significant acceptable values. Particle emissions were well below the limits. But single vehicles had defective filters.

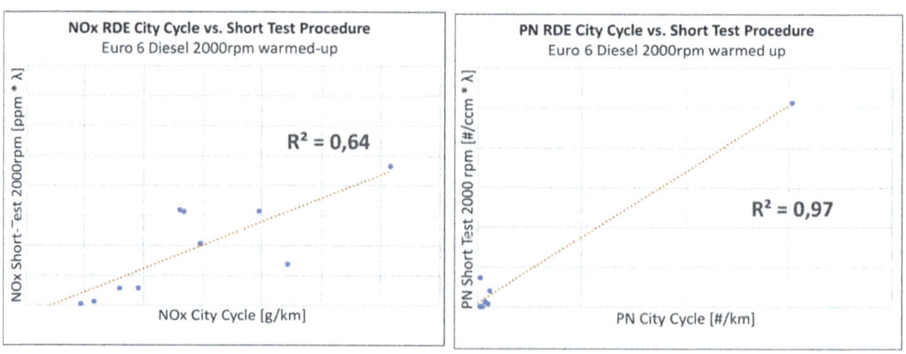

Fig. 9. Correlations for NOx and Particle Number (PN) measurements of Euro 6 cars

7 Summary

The real-world measurements we have carried out over the past 15 years have resulted in the knowledge that motor vehicles of the same emission standards show enormous fluctuations in their pollutant emissions. This should be controlled for by conducting periodic exhaust emission inspections. However, the measuring procedures currently used for so-called short tests are no longer able to keep up with advanced engine technology or measurement devices.

In most global regions no exhaust emissions are measured during the periodic inspection of diesel vehicles. In general, there is no measurement of pollutant components such as oxides of nitrogen (NOx). The current situation is particularly unsatisfactory for the measurement of particle emissions as well. The employed measurement method for exhaust opacity is too insensitive for modern diesel engines with particle filters. Furthermore, new direct-injection combustion in gasoline engines now also cause significant particle emissions due to partial non-premixed operation mode which needs to be subject to an inspection. A vehicle independent direct tail pipe check must be conducted.

Specifically, city driving measurements provide very satisfying correlations between RDE distance related mass emissions measurements and concentration measurements of the short test procedure.

The presented new universal short-test procedure can reliably detect vehicles with exceeding NOx and PN emissions. Thanks to the short test time, it is suited well for the application during periodic technical inspection.

The aim of the developed IoT variant was to use and implement an architecture that meets all requirements for a scalable system. Criteria for the selection continued to be an integrated solution to prevent data loss during a connection interruption. The final version was developed on the basis of the professional IoT Watson I.B.M. and is according to the IoT architecture MQTT - Message Queuing Telemetry Transport.

Acknowledgement. We would like to thank in particular our former colleague Lukas Gruenberger for his cooperation in real-world measurements and short testing.

References

1. Gadjanski, M.: Wasserstoff-Brennstoffzellenfahrzeuge im realen österreichischen Straßenverkehr. Dissertation Technische Universität Wien, Institut für Fahrzeugantriebe und Automobiltechnik (2017)
2. Pucher, E.: Abgasemissionen - Gesetzliche Vorschriften, Abgasmesstechnik, Schadstoffe und ihre Entstehung, Minderung von Schadstoffen. In: van Basshuysen, R., Schäfer, F. (Hrg.) Handbuch Verbrennungsmotor - Grundlagen, Komponenten, Systeme, Perspektiven. herausgegeben von: MTZ Motortechnische Zeitschrift; Vieweg + Teubner GWV Fachverlage GmbH, Wiesbaden, S. 796–813 (2011). ISBN: 978-3-8348-1549-1
3. Pucher, E., Eidmann, A.: Real-world emission measurement of free-flow and congested traffic with a V2I capable PEMS. In: Proceedings of the 22nd ITS World Congress 2015, Paper-Nr. ITS-2868 (2015)

4. Pucher, E.: Short Test procedures for nanoparticles emission and gaseous exhaust components of combustion engines. In: Proceedings of the 20th ETH Conference on Combustion Generated Nanoparticles, Zurich (2016)
5. Eidmann, A.: Vehicle to Infrastructure (V2I) Real-time communication for mobile measurement systems. Doctoral thesis at the TU Wien, Institute for Powertrains and Automotive Technology (2018)
6. Grünberger, L., Pucher, E.: Results of LDV NOx and particulate number measurements of a new short test procedure versus real driving emissions. In: Proceedings of the 28th CRC Real World Emissions Workshop, Garden Grove, California, 18–21, March (2018)

Production Planning at a Chocolate Company: A Two-Phase Approach by Aggregation

Zehra Düzgit[1(✉)], Ayşe Beyza Kuzuoğlu[1], Hazal Kalelioğlu[1], Hazal Kolay[1], Merve Başak Güler[1], Senem Akyol[1], and Ayhan Özgür Toy[2]

[1] Department of Industrial Engineering, İstanbul Bilgi University, Istanbul, Turkey
zehra.duzgit@bilgi.edu.tr, bkuzuoglu@gmail.com, hazalkalelioglu@gmail.com, kolay.hazal@gmail.com, basak.guler02@bilgiedu.net, akyolsenem@hotmail.com

[2] Industrial Engineering Department, Yaşar University, Izmir, Turkey
ozgur.toy@yasar.edu.tr

Abstract. There are some factors to be considered while developing production planning policy of chocolate and chocolate-based products. One of these factors is that chocolate is a perishable food, therefore it has limited shelf life. Another factor is that; its demand is not always easy to forecast. Holidays, mothers' day, teachers' day, valentines' day and new year are examples of the peak periods. For these special days, demand generally follows a seasonal demand pattern where seasonality may also contain trend for some specific product types. Moreover, there are two religious holidays (Ramadan Feast and Feast of Sacrifice) in Turkey whose dates shift each year. This phenomenon makes forecasts challenging. Underestimating demand causes loss of customer goodwill, lost customers and market share whereas overestimating demand causes excess inventory to keep in stock and risk of fat blooming. Accurate forecasting is critical since it provides a fundamental input for the production plan.

The study is conducted in a chocolate company in Turkey. The company does not implement a systematic planning method for chocolate production, instead the planning is based on past experiences with respect to experts' opinions. The objective of this study is to determine the optimal production and ending inventory levels for the period of 2018 so as to minimize total production and inventory holding cost subject to production, inventory and capacity related restrictions. A two-phase optimization method is adopted as a solution method. Firstly, a mathematical model is developed for the monthly aggregate production planning on product group basis. In order to solve the problem, monthly demand for product groups are forecasted based on the sales data of the previous two years. Secondly, a mathematical model for weekly disaggregate production planning is developed for each end item using the outputs of the aggregate planning as input. The disaggregate production plan gives the weekly planned production and inventory levels for end items for year 2018 with minimum deviation from the aggregate plan. Although the proposed solution model is implemented for year 2018, it can be used for the coming years by updating some parameters.

Keywords: Production planning · Demand forecasting · Aggregate production planning · Disaggregate production planning · Seasonality · Shelf life

N. M. Durakbasa and M. G. Gençyılmaz (Eds.): ISPR 2019, LNME, pp. 169–181, 2020.
https://doi.org/10.1007/978-3-030-31343-2_15

1 Introduction

In this study we consider the production planning for a chocolate producing company in Turkey. Annual chocolate consumption per capita in Turkey is reported in [1] as 3.1 kg. There are some peculiarities in chocolate industry effecting production planning factors: the basic raw material, the cocoa bean, is a stock market item, hence that the procurement price is affected by the market conditions; all semi-finished and finished products of chocolate production process are perishable; there are seasonality in the demand. If shelf life of a chocolate-based product is exceeded, the product may display a diagnosis of spoilage named as fat blooming, which is special for chocolate. According to [2], when managing perishables, age dependency forms the main concern. So, shelf life which can defined as the periodic time between manufacturing and retail purchasing of qualified or saleable conditioned product as defined in [3], is one of the concerns that must be taken into account. A product may be perishable, consumable, repairable or recoverable. Therefore, it cannot be ignored in inventory policy [4]. The shelf lives of products are generally being calculated by deducting the product's storage time or reducing inventory held up in stores [5].

A typical chocolate production process has the following stages: *(i)* **Mixing** is the stage of combining the required proportion of ingredients and mixing them until it becomes a homogenous and viscous liquid; *(ii)* **Pre-Grinding** breaks crystal sugars, cocoa mass and solid materials in milk powder into 200–250 microns; *(iii)* **Grinding** further breaks the mixture into 18–20 microns; *(iv)* **Conching** is the stage when the chocolate mixture is first molded, then pounded and lastly aired. Lecithin and vanillin are added during conching. *(v)* **Filtration** turns the mixture to a semi-finished product by filtering it into 300–400 micron. *(vi)* **Pumping** is transferring the liquid chocolate to the 50 °C water-jacketed tanks. *(vii)* **Tempering** is a heat treatment that regulates the crystal structure of chocolate. *(viii)* **Moulding** is draining the liquid chocolate into the required sizes and shapes of molds. *(ix)* **Packaging** is when the chocolates are picked from the molds by robot arms and packaged.

In the company we consider there are 652 different product types which are grouped in 10 product families. Product families are results of similarities in production processes of product types. The 10 product families and the number of product types in each family are depicted in Table 1. Demand forecasting hence the production planning has been done at product family level in the company.

For the production planning purposes, we need to elaborate on the demand seasonality. There are special dates/events where the demand in chocolate-based products is increased. Dates like Valentine's Day, Mothers' Day, Fathers' Day, Teachers' Day, New Year's Day, Religious Holidays are the peak seasons for chocolate industry. Dates for events such as Christmas, Valentine's Day, Mothers' Day, Fathers' Day are traced by the Gregorian Calendar whereas Islamic holidays are determined by Hijri Calendar. Since the lunar calendar lags behind the solar calendar by around ten days every year, Islamic holidays shift every year according to the Gregorian Calendar, which makes demand forecasting for chocolate-based products difficult. Data indicates that annual demand patterns also varies by product families.

Table 1. Product families and product types

Product family ID, f	Product types in the product family	Product type ID, k
1	Couverture, Confiserie, Ganache Chocolate	1, …, 45
2	Madlen Chocolate	46, …, 199
3	Filled Chocolate, Special Chocolate, Mini Chocolate	200, …, 469
4	Chocolate Bar, Compound Chocolate	470, …, 528
5	Drop Chocolate, Chocolate Pieces	529, …, 572
6	Praline	573, …, 579
7	Chocolate Hazelnut Creme	580, …, 584
8	Dragees, Almond Candy, Granule	585, …, 637
9	Turkish Delight	638, …, 649
10	Cocoa Powder	650,651,652

The aim of this study is to determine the production and ending inventory levels of a chocolate manufacturer company to come up with an efficient production plan. The data on hand are the sales, inventory and production levels for previous years. We employ a two-step solution approach. The first step is forming a monthly aggregate production plan based on product families. The demand forecasts are inputs of the aggregate plan. The second step is constructing a disaggregated production plan through the outputs of the aggregate plan. The disaggregate production plan divides the monthly plan into weekly form based on product types. There are some restrictions we consider. The shelf life of chocolate is one of the restrictions; the seasonality factor should be considered if exists for each product family; the production and inventory capacity of the plant should also be considered.

The paper is organized as follows. Problem Statement and Solution Approach are given in Sect. 2. Numerical Study is described in Sect. 3. Conclusion is in the last section.

2 Problem Statement and Solution Approach

The company produces and sells K different product types. For the company we consider there are 652 different product types ($K = 652$). Let k denote each product type, i.e., $k = 1,\ldots, 652$. As we discussed earlier product types are grouped into product families, there are F product families; $F = 10$ for our case. Our goal is to generate weekly production plan to meet the demand, d_{kw}, for each product type k in week w, around the year which minimizes the total annual cost; note that $w = 1, \ldots, 52$ by definition. The annual cost comprises production cost and inventory holding cost. Let x_{kw} and y_{kw} be the production quantity and ending inventory respectively, for the product type k in week w. We assume that c_{kw} is incurred for every unit of product type k produced in week w; we assume that unit production cost is non-increasing in time. h_{kw} is the unit holding cost for product type k, incurred for every unit on hand at the end of week w. We assume that there is no setup related cost for the production. In our model shortages are not allowed, that is all the demand for a week must be satisfied by the production in that week or earlier, i.e. from the stock. As we discussed above, products are perishable. Products in stock at the end of their shelf life, SL_k, are disposed at no or negligible cost. We assume that oldest items are used first to meet the demand, i.e., FIFO. Moreover, as the company policy there are lower and upper bounds on production at every week for product type k, denoted by LB_k and UB_k respectively. The mathematical model for determining the weekly production plan which minimizes the total annual cost is as follows when demand for each week is known for a product type k.

$$\text{Minimize } z = \sum_{w=1}^{52} (c_{kw}x_{kw} + h_{kw}y_{kw}) \tag{1}$$

subject to

$$y_{kw} = y_{k(w-1)} + x_{kw} - d_{kw} \qquad \forall w \tag{2}$$

$$LB_k \leq x_{kw} \leq UB_k \qquad \forall w \tag{3}$$

$$y_{kw} \leq \sum_{j=w+1}^{j=w+SL_k} d_{kj} \qquad \forall w \tag{4}$$

$$x_{kw}, y_{kw} \geq 0 \qquad \forall w \tag{5}$$

In this formulation (1) is the total annual cost, (2) is the inventory balance equations, (3) is company-imposed bounds on production quantity, and (4) is the shelf life constraint. Solving this mathematical model for every product type yields the optimal weekly production plan for the company. However, there are some issues which prevent implementing this mathematical model for the company. We need to develop a solution approach which considers all those issues. We will address those issues one by one. The first issue is that the demand data are not readily available. The company has past two years' sales data. The sales data are monthly and at product family level. Moreover, as we discussed earlier, previous years' data cannot be used as they are, for predicting the demand since Gregorian Calendar dates of Islamic holidays shift every year. Another issue is that lower and upper bounds on production are at product family level, therefore cannot be used directly in the model above for determining the production plan at product type level. Next, we present our approach formulating these issues in obtaining the optimum production plan.

2.1 Forecasting the Demand

As we discussed above the company under consideration has a wide product range, specifically 652 different product types, which are grouped into 10 product families. Past two years' data on monthly sales, production and ending inventory levels for each product family are available. Every product family follows different sales pattern since certain product families are demanded more in certain occasions. Therefore, we analyze the data and forecast the demand for each product family separately.

Our analysis indicates that: *(i)* sales data on product families #5, #8, and #10 do not contain any trend or seasonality. We implement simple exponential smoothing forecasting technique for these product families. In order to determine the smoothing constant, α, the total Mean Absolute Percentage Error (MAPE) is minimized in past two years' data. *(ii)* sales data on product families #1, #6, and #9 display seasonal demand pattern. Seasonal index method is used, and seasonal fluctuations are measured through seasonal indices. *(iii)* sales data on product family #2, #3, #4 and #7 exhibit a seasonal pattern with trend component. Holt-Winter's method is used to forecast the demand for these product families.

Since the lower and upper bounds on production are given monthly at product family level, an aggregate production plan is necessary which is to be decomposed into a disaggregate plan afterwards. Our solution approach comprises forecasting the demand, finding an aggregate monthly production plan at product family level and then disaggregating this plan to obtain the weekly production plan at product type level.

The demand forecasts are taken as input for the aggregate production plan as demonstrated in Fig. 1. Then, the outputs of the aggregate plan are used as input for the disaggregate production plan.

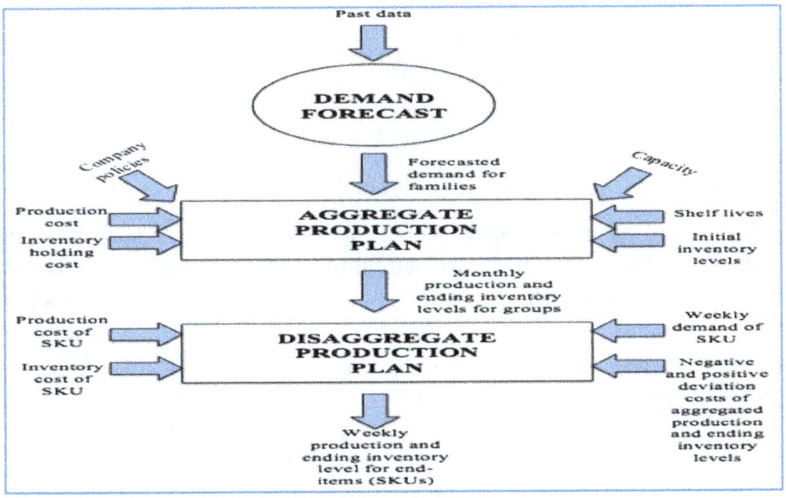

Fig. 1. Production planning diagram

2.2 Aggregate Production Plan

Aggregate production planning deals with matching capacity to changing demand, over a finite time horizon, in order to achieve long run profitability for the company [6].

We construct a linear programming model for the aggregate monthly production plan for each product family. The objective is to minimize the total annual cost, comprising production and inventory holding costs.

Let f represent product families, $f = 1, \ldots, 10$ and m denote months, $m = 1, \ldots, 12$. Unit production cost for family is C_{fm} and unit holding cost is H_{fm} for the product family f in month m. As we discuss above we forecast the demand for product family f, in month m, which we denote by $\tilde{D}_{fm}$. Product family f has a shelf life SL_f. Y_{f0} is the initial inventory for product family f at the beginning of a year. The company imposes lower and upper bounds on production for product family f, in month m, denoted by LB_{fm} and UB_{fm} respectively. The company enforces a certain level of safety stock for product family f, at the end of month m, denoted by SS_{fm}.

The continuous decision variables of the model are X_{fm} and Y_{fm}. X_{fm} represents production level of product family f in month m whereas Y_{fm} represents ending inventory level of product family f at the end of month m.

The following linear programming model is solved for each product family f:

$$Minimize\ z = \sum\nolimits_{m=1}^{12} \left(C_{fm}X_{fm} + H_{fm}Y_{fm}\right) \tag{6}$$

subject to

$$Y_{f(m-1)} + X_{fm} - \tilde{D}_{fm} = Y_{fm} \qquad \forall m \tag{7}$$

$$LB_{fm} \leq X_{fm} \leq UB_{fm} \qquad \forall m \tag{8}$$

$$Y_{fm} \geq SS_{fm} \qquad \forall m \tag{9}$$

$$Y_{fm} \leq \sum\nolimits_{j=m+1}^{m+SL_f} \tilde{D}_{fm} \qquad \forall m \tag{10}$$

$$X_{fm}, Y_{fm} \geq 0 \qquad \forall m \tag{11}$$

The objective of the mathematical model (6) is to minimize annual production cost and inventory holding cost for a product family f. Constraint (7) is the inventory balance equation. Constraint (8) imposes a lower and upper bound on production level. Constraint (9) ensures that a certain level of safety stock is preferred by the company in order not to have any shortage. Constraint (10) regulates the shelf life restriction. Constraint (11) is non-negativity restrictions.

2.3 Disaggregate Production Plan

Once we obtain the aggregate monthly production plan for product families, we disaggregate this plan to obtain the weekly production plan for each product type. When we decompose the monthly aggregate plan at the production family level into the weekly plan at the product type level due to the integrality of units and precision, the disaggregated plan, hence the production and inventory, may deviate from that of the aggregated plan. Our objective is to minimize these deviations.

For the disaggregate production plan, we use the solution of the aggregate model for production quantities and inventory levels as parameters. Other paramcters of the disaggregate production plan are *(i)* the distribution of the product type sales within its product family and *(ii)* weekly demand factor.

Sales distribution (the percentage) of each product type is determined based on past data by simply dividing each product type sales to the total sales of the product family which it belongs to.

For the weekly demand factor we use the following illustrative example, by referring to the 2018 calendar. To form weekly demand for each product type, the special days of 2018 calendar are shown in Fig. 2. As shown in Table 2, a weight factor is assigned to each week which sums to totality for each month. According to the company policy, if there is any special day/week within a month, the previous week is assigned a larger weight compared to other weeks. Weekly forecasted demand for each product type is simply the multiplication of forecasted monthly demand by the weight factor of every week, which we denote by $\tilde{d}_{kw}$.

Fig. 2. Calendar of year 2018

Table 2. Assigned weights of weeks of year 2018

Months	Weeks	Dates	Weights
January	Week 1	1 Jan - 7 Jan	0,25
	Week 2	8 Jan - 14 Jan	0,25
	Week 3	15 Jan - 21 Jan	0,25
	Week 4	22 Jan - 28 Jan	0,25
			1
February	Week 5	29 Jan - 4 Feb	0,2
	Week 6	5 Feb - 11 Feb	0,35
	Week 7	12 Feb - 18 Feb	0,1
	Week 8	19 Feb - 25 Feb	0,35
			1
March	Week 9	26 Feb - 4 Mar	0,1
	Week 10	5 Mar - 11 Mar	0,2
	Week 11	12 Mar - 18 Mar	0,15
	Week 12	19 Mar - 25 Mar	0,25
	Week 13	26 Mar - 1 Apr	0,3
			1
April	Week 14	2 Apr - 8 Apr	0,2
	Week 15	9 Apr - 15 Apr	0,2
	Week 16	16 Apr - 22 Apr	0,25
	Week 17	23 Apr - 29 Apr	0,35
			1
May	Week 18	30 Apr - 6 May	0,15
	Week 19	7 May - 13 May	0,05
	Week 20	14 May - 20 May	0,2
	Week 21	21 May - 27 May	0,3
	Week 22	28 May - 3 Jun	0,3
			1
June	Week 23	4 Jun - 10 Jun	0,3
	Week 24	11 Jun - 17 Jun	0,2
	Week 25	18 Jun - 24 Jun	0,25
	Week 26	25 Jun - 1 Jul	0,25
			1
July	Week 27	2 Jul - 8 Jul	0,2
	Week 28	9 Jul - 15 Jul	0,2
	Week 29	16 Jul - 22 Jul	0,3
	Week 30	23 Jul - 29 Jul	0,3
			1
August	Week 31	30 Jul - 5 Aug	0,35
	Week 32	6 Aug - 12 Aug	0,35
	Week 33	13 Aug - 19 Aug	0,05
	Week 34	20 Aug - 26 Aug	0,1
	Week 35	27 Aug - 2 Sep	0,15
			1
September	Week 36	3 Sep - 9 Sep	0,25
	Week 37	10 Sep - 16 Sep	0,25
	Week 38	17 Sep - 23 Sep	0,25
	Week 39	24 Sep -30 Sep	0,25
			1
October	Week 40	1 Oct - 7 Oct	0,2
	Week 41	8 oct - 14 Oct	0,2
	Week 42	15 Oct - 21 Oct	0,3
	Week 43	22 Oct - 28 Oct	0,3
			1
November	Week 44	29 Oct - 4 Nov	0,1
	Week 45	5 Nov - 11 Nov	0,25
	Week 46	12 Nov - 18 Nov	0,3
	Week 47	19 Nov - 25 Nov	0,1
	Week 48	26 Nov - 2 Dec	0,25
			1
December	Week 49	3 Dec - 9 Dec	0,3
	Week 50	10 Dec - 16 Dec	0,3
	Week 51	17 Dec - 23 Dec	0,3
	Week 52	24 Dec - 30 Dec	0,1
			1

For the disaggregate production plan model we retain the notation introduced for general model and aggregate production model above. For the tractability of the model we introduce the notation *k(f)* to refer to product type *k* in product family *f* (See Table 1); likewise we use the notation *w(m)* to refer to week *w* in month *m* (See Fig. 2).

The outputs of the aggregate production plan, namely X_{fm} (production level of product family *f* in month *m)* and Y_{fm} *(*ending inventory level of product family *f* at the end of month *m),* are used as parameters for the disaggregate production plan. c_{kw} is the unit production cost of product type *k* in week *w*. h_{kw} is the unit inventory holding cost of product type *k* at the end of week *w*. y_{k0} is the initial inventory level of each product type *k*.

Let α^- and α^+ represent the cost of negative and positive deviations from aggregated production level, respectively. Let β^- and β^+ represent the cost of negative and positive deviations from aggregated ending inventory level, respectively.

The decision variables of the model are as follows. Let x_{kw} denote the production level of product type *k* in week *w*. Let y_{kw} denote the ending inventory level of product type *k* at the end of week *w*. δ_{fm}^{1-} is the negative deviation from the planned aggregate production level of product family *f* in month *m;* δ_{fm}^{1+} is the positive deviation from the planned aggregate production level of product family *f* in month *m;* δ_{fm}^{2-} is the negative deviation from the planned aggregate ending inventory level of product family *f* in month *m;* δ_{fm}^{2+} is the positive deviation from the planned aggregate ending inventory level of product family *f* in month *m*.

The following linear programming model is solved for all *f*:

$$Minimize\ z = \sum\nolimits_{m=1}^{12} (\alpha^- \delta_{fm}^{1-} + \alpha^+ \delta_{fm}^{1+} + \beta^- \delta_{fm}^{2-} + \beta^+ \delta_{fm}^{2+}) + \sum\nolimits_{\forall k(f)} \sum\nolimits_{w=1}^{52} (c_{kw} x_{kw} + h_{kw} y_{kw}) \qquad (13)$$

subject to

$$y_{k(w-1)} - y_{kw} + x_{kw} = \tilde{d}_{kw} \qquad \forall k(f), \forall w \qquad (14)$$

$$\sum\nolimits_{\forall k(f)} \sum\nolimits_{\forall w(m)} x_{kw} + \delta_{fm}^{1-} - \delta_{fm}^{1+} = X_{fm} \qquad \forall m \qquad (15)$$

$$\sum\nolimits_{\forall k(f)} \sum\nolimits_{\forall w(m)} y_{kw} + \delta_{fm}^{2-} - \delta_{fm}^{2+} = Y_{fm} \qquad \forall m \qquad (16)$$

$$x_{kw} \geq 0 \qquad \forall k(f), \forall w \qquad (17)$$

$$y_{kw} \geq 0 \qquad \forall k(f), \forall w \qquad (18)$$

$$\delta_{fm}^{1-}, \delta_{fm}^{1+}, \delta_{fm}^{2-}, \delta_{fm}^{2+} \geq 0 \qquad \forall m \qquad (19)$$

The objective of disaggregate model (13) is to minimize the total cost of deviations from the aggregated production plan, production cost, inventory holding cost. (14) implies weekly inventory balance equations. Weekly production deviation constraint (15) adjusts the production level of each product type with both negative and positive deviation from the planned aggregate production level according to the production level of each product family. Weekly inventory deviation constraint (16) controls the ending level of each product type with both negative and positive deviation from the planned aggregate ending inventory level according to the ending inventory level of each product family. (17), (18), (19) are the non-negativity restrictions of the decision variables.

3 Numerical Study

3.1 Parameters of Demand Forecasting

For the product families which exhibit a constant demand pattern (#5, #8, and #10), the smoothing constant, α, must be determined to be used in simple exponential smoothing forecasting technique. α represents the amount of reliance to past data. To decide on the optimal α value, error measurement method Mean Absolute Percentage Error (MAPE) is used to minimize the forecast error, based on past years. The optimized values for α are 0.528 for the product family 5, 0.3 for the product family 8, and 0.5 for the product family 10.

For the product families which exhibit a seasonal demand pattern (#1, #6, and #9), seasonal indices are computed and tabularized in Table 3. If the seasonal factor of the month is higher than 1, this indicates that this month's demand is high; if it is lower than 1, this indicates that this month's demand is low.

Table 3. Seasonal factors

Seasons (Months)	Product family 1	Product family 6	Product family 9
January	1.15	1.10	0.34
February	1.19	0.93	0.51
March	1.19	1.51	0.53
April	0.82	1.08	0.82
May	0.82	0.95	0.81
June	0.82	1.04	3.18
July	0.66	0.67	0.67
August	0.78	1.04	2.18
September	0.95	1.00	0.66
October	1.17	0.98	0.52
November	1.13	0.90	0.71
December	1.32	0.79	1.06

For the product families which exhibit a seasonal and trend demand pattern, i.e., families #2, #3, #4 and #7, smoothing parameters of Holt-Winter's method (α, β, γ) are determined. α value indicates the level value; β stands for trend factor and γ indicates the seasonal factor. The optimum values of three smoothing parameters are shown in Table 4.

Table 4. Optimum α, β, γ Values for Holt Winter's

Product family	Optimum α Value	Optimum β Value	Optimum γ Value
2	0.2	0.1	0.1
3	0	0	0.1
4	0.2	0.1	0.1
7	0.2	0.1	0.1

We present in Table 5 forecasted demand for each product family, calculated by aforementioned methodologies.

Table 5. Monthly demand forecasts of product families for 2018 (in kg)

Product family	Months											
	1	2	3	4	5	6	7	8	9	10	11	12
1	136,936	140,710	141,268	97,038	96,820	97,944	78,906	92,845	112,630	138,885	134,078	156,605
2	9,146	7,623	10,665	17,893	31,415	53,299	1,894	31,063	27,852	4,302	13,322	33,755
3	11,413	15,538	13,269	12,403	18,362	23,049	7,865	29,585	17,892	15,762	16,623	17,810
4	923	957	1,108	1,947	1,365	2,174	1,411	1,142	1,665	1,569	1,647	1,087
5	26,694	32,269	26,484	31,818	30,325	25,880	29,341	28,515	27,912	28,589	28,339	28,280
6	14,770	12,490	20,230	14,525	12,695	13,960	8,960	13,980	13,390	13,100	12,100	10,635
7	4,260	6,106	10,932	8,245	8,942	13,409	10,375	9,934	12,429	7,785	12,441	23,242
8	683	521	932	821	835	698	785	773	752	770	765	762
9	642	954	991	1,529	1,520	5,940	1,249	4,070	1,240	980	1,335	1,983
10	10,426	9,311	3,597	11,780	13,877	5,187	10,281	9,782	8,417	9,493	9,231	9,047

3.2 Parameters of Aggregate Production Plan

Values of the parameters we use in our models are obtained by inquiring the company representatives. These parameters are as follows:

Shelf life of product families are $SL_f = 6$ mo. for $f = 1,2,5,6,7,10$; $SL_f = 5$ mo. for $f = 3,4$; $SL_f = 2$ mo. for $f = 8,9$. Lower and upper bounds on production level of product families are: $LB_{1m} = 59{,}000$ kg. $\forall m$; $LB_{2m} = 3{,}000$ kg. $\forall m$; $LB_{fm} = 0$ for $f = 3, \ldots, 10$ and $\forall m$; $UB_{9m} = 7{,}000$ kg. $\forall m$; $UB_{fm} = \infty$ for $f = 1, 2, 3, 4, 5, 6, 7, 8, 10$ and $\forall m$.

Company-imposed safety stock levels for product families are: $SS_2 = 1{,}000$ for $m = 4$; $SS_9 = 1{,}200$ for $m = 5,7$.

Initial inventory levels are required to be used in inventory balance equation, for $m = 1$. Initial inventory level of product families are (in kg): $Y_{10} = 69{,}835$; $Y_{20} = 9{,}811$; $Y_{30} = 12{,}608$; $Y_{40} = 721$; $Y_{50} = 28{,}876$; $Y_{60} = 3{,}530$; $Y_{70} = 700$; $Y_{80} = 1{,}200$; $Y_{90} = 0$; $Y_{10,0} = 2{,}176$.

The company prefers not to publicize the unit production cost and inventory holding cost.

The outputs of aggregate production plan are monthly production levels and monthly ending inventory levels of each product family. We summarize the optimum aggregate production plan and ending inventories in Tables 6 and 7. Please note that the outputs are rounded to the nearest integer for table formatting purposes. The objective function values, i.e., total annual costs, are not reported due to the company privacy policy.

Table 6. Production level of product family *f* in month *m* (in kg), X_{fm}

Product family	Months											
	1	2	3	4	5	6	7	8	9	10	11	12
1	67,101	140,710	141,268	97,038	96,820	97,944	78,906	92,845	112,630	138,885	134,078	156,605
2	3,010	3,948	10,665	18,893	30,415	53,299	3,010	29,947	27,852	4,302	13,322	33,755
3	0	14,343	13,269	12,403	18,362	23,049	7,865	29,585	17,892	15,762	16,623	17,810
4	202	957	1,108	1,947	1,365	2,174	1,411	1,142	1,665	1,569	1,647	1,087
5	0	30,087	26,484	31,818	30,325	25,880	29,341	28,515	27,912	28,589	28,339	28,280
6	11,240	12,490	20,230	14,525	12,695	13,960	8,960	13,980	13,390	13,100	12,100	10,635
7	3,560	6,106	10,932	8,245	8,942	13,409	10,375	9,934	12,429	7,785	12,441	23,242
8	0	4	932	821	835	698	785	773	752	770	765	762
9	642	954	991	1,529	2,720	4,740	2,449	2,870	1,240	980	1,335	1,983
10	8,250	9,311	3,597	11,780	13,877	5,187	10,281	9,782	8,417	9,493	9,231	9,047

Table 7. Ending inventory level of product family *f* at the end of month *m* (in kg), Y_{fm} (Only non-zero stock levels are specified)

Product family	Months											
	1	2	3	4	5	6	7	8	9	10	11	12
1												
2	3,675			1,000			1,116					
3	1,195											
4												
5	2,182											
6												
7												
8	517											
9					1,200		1,200					
10												

3.3 Parameters of Disaggregate Production Plan

As we explain above for each product family sales distribution (percentage) are computed from the past data. These percentages are {0.52, 0.11, 0.07, 0, 0.15, 0.06, 0.02, 0.04, 0.01, 0.01} respectively for each family. Initial inventory levels are required to be used in inventory balance equation, for $w = 1$. Initial inventory level of product types is (in kg): $y_{10} = 486$; $y_{20} = 2{,}415$; …; $y_{651,0} = 840$; $y_{652,0} = 750$.

The company prefers not to publicize the unit production cost and inventory holding cost.

The outputs of disaggregate production plan are weekly production levels and weekly ending inventory levels of each product type. However, since there are 652 product types and 52 weeks, these outputs cannot be tabularized. Other outputs are the deviations which are represented by a quartet $\{\delta_{fm}^{1-}, \delta_{fm}^{1+}, \delta_{fm}^{2-}, \delta_{fm}^{2+}\}$ in Table 8, instead of negative deviation from the planned aggregate production level of product family *f* in month *m;* positive deviation from the planned aggregate production level of product family *f* in month *m;* negative deviation from the planned aggregate ending inventory level of product family f in month m; positive deviation from the planned aggregate ending inventory level of product family f in month m, respectively.

Table 8. Deviations of product family f at month m $\{\delta_{fm}^{1-}, \delta_{fm}^{1+}, \delta_{fm}^{2-}, \delta_{fm}^{2+}\}\}$(in kg). The outputs are rounded to the nearest integer. * denotes that there is no deviation.

Product family	Months											
	1	2	3	4	5	6	7	8	9	10	11	12
1	{*,6,*,309}	{5,*,*,243}	{*,*,*,*}	{*,*,*,*}	{*,*,*,}	{*,*,*,*}	{*,*,*,*}	{*,*,*,*}	{*,*,*,*}	{*,*,*,*}	{*,*,*,*}	{*,*,*,*}
2	{*,*,*,29}	{*,*,*,14}	{*,3,*,6}	{2,*,*,1}	{*,*,*,1}	{*,*,*,*}	{*,*,*,*}	{*,*,*,*}	{*,*,*,*}	{*,*,*,*}	{*,*,*,*}	{*,*,*,*}
3	{*,2,*,22}	{5,*,*,6}	{*,2,*,2}	{*,*,*,*}	{*,*,*,*}	{*,*,*,*}	{*,*,*,*}	{*,*,*,*}	{*,*,*,*}	{*,*,*,*}	{*,*,*,*}	{*,*,*,*}
4	{*,*,*,4}	{*,*,*,2}	{*,*,*,*}	{*,*,*,*}	{*,*,*,*}	{*,*,*,*}	{*,*,*,*}	{*,*,*,*}	{*,*,*,*}	{*,*,*,*}	{*,*,*,*}	{*,*,*,*}
5	{*,26,*,294}	{32,*,*,111}	{*,*,*,27}	{3,*,*,7}	{*,*,*,2}	{*,*,*,1}	{*,16,*,*}	{*,4,*,*}	{*,36,*,*}	{*,17,*,*}	{2,*,*,*}	{13,*,*,*}
6	{*,1,*,23}	{*,*,*,137}	{*,66,*,*}	{*,*,*,*}	{*,*,*,*}	{*,*,*,*}	{*,*,*,*}	{*,*,*,*}	{*,*,*,*}	{*,*,*,*}	{*,*,*,*}	{*,*,*,*}
7	{*,*,*,28}	{*,*,*,92}	{*,58, *,*	{*,*,*,*}	{*,*,*,*}	{*,*,*,*}	{*,*,*,*}	{*,*,*,*}	{*,*,*,*}	{*,*,*,*}	{*,*,*,*}	{*,*,*,*}
8	{*,*,*,13}	{*,*,*,8}	{*,*,*,3}	{*,*,*,*}	{*,*,*,*}	{*,*,*,*}	{*,*,*,*}	{*,*,*,*}	{*,*,*,*}	{*,*,*,*}	{*,*,*,*}	{*,*,*,*}
9	{*,*,*,*}	{*,*,*,6}	{*,*,*,*}	{*,*,*,*}	{*,*,*,*}	{*,*,*,*}	{*,*,*,*}	{*,*,*,*}	{*,*,*,*}	{*,*,*,*}	{*,*,*,*}	{*,*,*,*}
10	{*,*,*,15}	{143,*,*,90}	{*,*,*,*}	{*,*,*,*}	{*,*,*,*}	{*,*,*,*}	{*,96,*,*}	{*,386,*,*}	{300,*,*,238}	{*,162,*,*}	{448,*,*,85}	{*,*,*,*}

4 Conclusion

We consider the demand forecasting and production planning problem of a chocolate producing company in Turkey. We observe that there is high seasonality in the demand for some product families. Moreover, Islamic holidays that are traced by Hijri Calendar complicates forecasting since Hijri Calendar and Gregorian Calendar lags ten days every year. We modify available forecasting techniques to account this issue. We assign weighting factors for the weeks which help us resolve the problem.

Chocolate is a perishable product therefore our models for production planning takes perishing account.

In the company we consider there are 652 different product types which belong to 10 product families. We want to come up with an efficient weekly production and inventory plan for each product type.

If the demand were known, we would be able to solve a classical lot sizing problem subject to known demand. However, the demand is not known in the company. Thus, demand must be forecasted. The company forecasts demand at product family level. Moreover, the company imposes lower and upper bounds on production and ending inventory levels at product family level. Since lower and upper bounds are at the product family level, they must be transformed into product types.

We suggest a two-step solution methodology. In the first step, we form a monthly aggregate production plan based on product families. The demand forecasts are inputs of the aggregate plan. In the second step, we develop a disaggregated production plan. The outputs of the aggregate plan become inputs of the disaggregate plan. The disaggregate production plan divides the monthly plan into weekly form based on product types.

The proposed solution considers shelf life of chocolate, safety stock requirements, and the seasonality factor if exists. Given the dates of the special dates/events, the proposed solution approach can be employed for any year.

References

1. Türkiye'de Çikolata Pazarının Büyüklüğü. NTV (2017). https://www.ntv.com.tr/ekonomi/turkiyede-cikolata-pazarinin-buyuklugu-5–3-milyar-tl,OUwZnIYp0EmBChpI0dW17Q
2. Geboers, F.: Demand Forecasts and Ordering Implications for the Bread Assortments of Dutch Retailers. MS Thesis, Eindhoven University of Technology, Eindhoven, Netherlands (2010)
3. Ojeda, A.F.A.: Production Planning with Raw Material Shelf-Life Considerations by Mixed Integer Programming. MS Thesis, Concordia University, Montreal, Quebec, Canada (2012)
4. Ziukov, S.: A literature review on models of inventory management under uncertainty. Bus. Syst. Econ. **5**(1), 26–35 (2015)
5. Sarker, B.R., Babu, P.S.: Effect of production cost on shelf life. Int. J. Prod. Res. **31**(8), 1865–1872 (1993)
6. Davizón, Y.A., Martínez-Olvera, C., Soto, R., Hinojosa, C., Espino-Román, P.: Optimal control approaches to the aggregate production planning problem. Sustainability **7**, 16324–16339 (2015)

Structural and Modal Analyses of Naca 66-206 Aircraft Wing Model

Ismail Bogrekci[1], Pinar Demircioglu[1(✉)], H. Saygin Sucuoglu[1], Emrah Guven[1], Neslihan Demir[1], and M. Numan Durakbasa[2]

[1] Mechanical Engineering Department, Aydın Adnan Menderes University, Aydın, Turkey
{ibogrekci,pinar.demircioglu,hilmisucuoglu, neslihan.demir}@adu.edu.tr, 141802016@stu.adu.edu.tr

[2] Department of Industrial Metrology and Adaptronic Systems, Vienna University of Technology (TU Vienna), Vienna, Austria
durakbasa@ift.tuwien.ac.at

Abstract. In this study, NACA 66-206 wing type was modelled and analysed to understand to mechanical behaviour under lift and weight forces. Structural and modal analyses were applied using Ansys Workbench Static Structural and Modal Analysis tools. Three different materials (aluminum 6061, carbon fiber and strong unidirectional epoxy glass) were selected for structure of the wings. The wings were created with three parts as airfoil, rib and spar. The affected forces were calculated and applied to the wings for three motion velocities as 10, 50 and 100 m/s. In the analyses; the lift forces were calculated by the buoyancy equation for the aircraft. The weight forces were also added to get more realistic results. The motion velocity and material effects to the structural strength of the wings were compared according to the results obtained from static and modal analyses. The maximum deformations were observed for the 100 m/s motion velocity with the values of 2.28 mm for aluminum, 0.16 mm for carbon fiber and 7.71 mm for epoxy glass. The similar effects of the motion velocity were found for the strains and equivalent stresses. It can be concluded from these results that it was better to manufacture the NACA 66-206 wings with the carbon fiber material as it had high natural frequency values and caused to lowest weight force.

Keywords: Lift and weight forces for aircraft wing · Maximum deformation · NACA 66-206 aircraft wing model · Natural frequency · Structural and modal analyses

1 Introduction

In order to meet the transportation needs such as land, sea and air transportation, technological developments are increasing more and more. The most effective and fast transportation is aviation. The most common, known and preferred type, airplane, helicopter, hot air balloons, etc. air vehicles have more complex technologies and techniques. The aircraft is any machine that can receive support from the reactions of air in the atmosphere [1]. On the plane, there are wings holding the plane in the air and

N. M. Durakbasa and M. G. Gençyılmaz (Eds.): ISPR 2019, LNME, pp. 182–191, 2020.
https://doi.org/10.1007/978-3-030-31343-2_16

there are four forces to fly. There are forces on the plane that go up, down, move or slow down (Fig. 1).

The four main factors that allow the aircraft to fly are:

1. ***Lift Force;*** The force provided by the wings that allow aircrafts to remain in the air.
2. ***Gravity Force;*** the force pulling down the plane. The whole weight of the plane acts as a single point.
3. ***Drag Force;*** The force of air molecules, which tends to slow down an object, which is opposite to the effect and direction of movement of the aircraft. It is caused by differences in friction and air pressure.
4. ***Thrust Force;*** a force that moves an aircraft in the direction of the motion. It consists of a propeller, jet engine or rocket. Air is pulled in and then pushed out in an opposite direction. One example is a household fan [2].

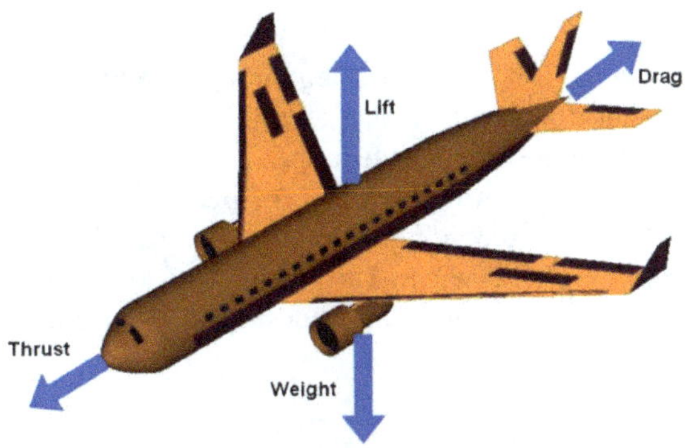

Fig. 1. The four forces of flight for aircraft [3]

During the flight, the lifting of an airplane must be in balance with its weight and go beyond its thrust and friction. An airplane uses engines for thrust and wings for lifting. The drag is reduced by the shape of the plane and its weight is controlled by the materials in which it is built [4].

The wings have emerged as the main part of the elevator which is heavier than the plane. The wings carry some serious loads in the aircraft structure. The wing is the main lift source and must be carefully designed to optimize the lift/drag ratio throughout the flight [5]. When designing the wings, many factors such as size, weight, speed, climbing speed and the use of the aircraft should be considered. The aerodynamics of the wings should be evaluated under the loads related to the many manoeuvres or wing loading [6].

Wings have three types of movement which are yaw, roll and pitch (Fig. 2). A yaw motion is a movement of the nose of the aircraft from side to side. A pitch motion is an up or down movement of the nose of the aircraft. A rolling motion is an up and down movement of the wing tips of the aircraft [7]. The axes are alternatively designated as vertical, longitudinal and transverse. These axes move with the vehicle and rotate relative to the Earth along with the craft [8].

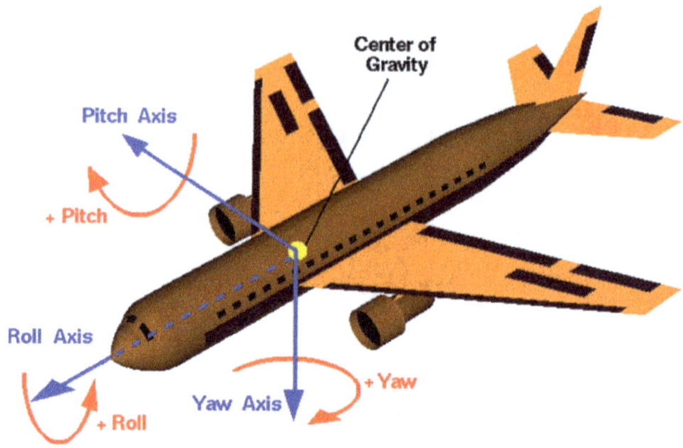

Fig. 2. Roll, pitch and yaw movements of wing [7]

When the wing is in motion, the air passing through the top and bottom of the wing due to its oblique position will not be able to continue equally. As the air flows, the low-pressure area will be formed in the upper part of the wing and changes the direction at the top. A change in the direction of the airflow indicates that the wing applied a force to it (Fig. 3). According to Newton's third rule, the airflow applies an equal and opposite force to the wing. This force tries to lift the wing and push it backwards. Pushing the wing backwards is undesirable because it cuts off the speed of the aircraft.

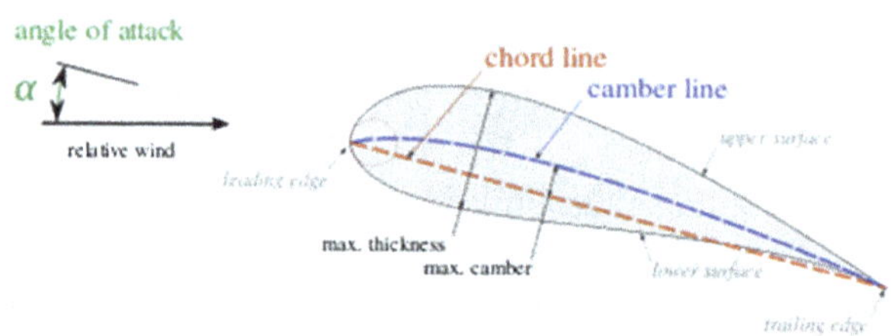

Fig. 3. Airfoil structure

Finite Element Analysis enables complex problems to obtain approximate solutions [9]. Finite Element Analysis is used for model validation and update. The modal analysis provides an experimental evaluation of the dynamic properties of the structure. In addition, modal analysis is generally used to confirm the Finite Element Analysis results [10]. Modal analysis was performed to determine the vibration characteristics (natural frequencies and mode shapes) when designing a structure. Important parameters in the design of a structure for dynamic loading conditions are natural frequencies and mode shapes [11]. Before applying a modal test, there are some features, such as selecting suspension points, driving points and, most importantly, response points for a modal test for model validation [10]. One of the most common analyses is static analysis. Static analysis is used to determine displacements, stresses, stresses and internal forces in structures or components caused by loads that do not generate significant inertia or damping effects [11].

The static, fatigue and modal analysis for behaviour of the wing and finite element analysis for three different materials and velocities were conducted. According to the study, stress values increase with increase in airplane wing speed. Deformation and frequency values are higher for carbon epoxy material [9]. The modal analysis of an aircraft wing as a cantilever beam to obtain the relevant mode patterns of natural frequencies were realized. The results of this study showed that an aircraft wing could be considered as a cantilever beam (excluding gravity) by ignoring all forces in the aircraft [12]. The natural frequency analysis to avoid failure and to avoid resonance in the material to simulate the boundary conditions was investigated. The results showed that by analysing the frequency of the wing structure, it helps to calculate the resonance that should not be equal to or more than the natural frequency. The material used for the aircraft wing is the aluminum alloy which has less weight and does not affect the weight of one third of the steel [13].

In this study, the structural and modal analysis was applied to NACA 66-206 aircraft wing modelled to understand the mechanical behaviour. The analyses were carried out for three different moving speeds of 10, 50 and 100 m/s and three different materials as aluminum, carbon fiber and strong unidirectional epoxy glass (Epoxy S-Glass UD). The forces affected by the aircraft wing were calculated by the buoyancy equation for the aircraft. Weight was added to the affected force to get more realistic results. The speed and material affecting the wing structural strength were compared according to the results obtained from the analyses.

2 Material and Method

The aircraft wing model was prepared with Autodesk Inventor Software to apply the structural and modal analyses. Designed wing was analysed with CAE (Computer Aided Engineering) methods to observe the structural strength with the means of occurred deformation, stress and strain and natural frequency values.

2.1 Design and Material of Wings

Detailed design view, dimensions and weights of the wing were given in Table 1 and Fig. 4, respectively.

Table 1. Dimension and weight of designed wing

Maximum length (mm)	Maximum height (mm)	Thickness of sheet (mm)	Weight with aluminum (kg)	Weight with carbon fiber (kg)	Weight with Epoxy S-Glass UD (kg)
448.7	47	5	63.56	32.12	45.89

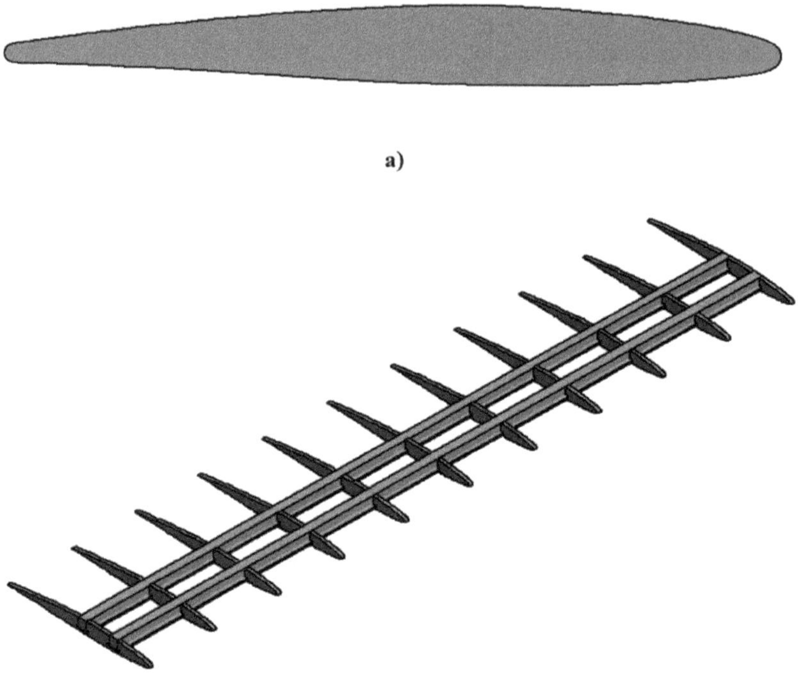

a)

b)

Fig. 4. Designed wing structure (a) Airfoil (b) Spars and ribs

Aluminum is very desirable metal because of its malleability and elastic features. Although aluminum is relatively weaker, its weight is almost one third of the steel. So, the weight to strength ratio of the aluminum is better than steel. The carbon fibers generally have excellent tensile properties, low densities and high thermal–chemical stabilities. In recent years, reinforced carbon fiber usage has been increased to meet the demand of different industries such as aerospace, construction and automobile. Epoxy glass is suitable for a broad range of applications within the electrical and mechanical fields such as spacers, mechanical barriers, electrical insulation components/spacers, etc. It has very low moisture absorption, high heat resistance up to 180 °C without major impact on the mechanical properties and very good dielectric properties.

According the information mentioned above; materials of the wings were selected as aluminum 6061, carbon fiber and epoxy S-glass UD. Mechanical properties of the selected materials are given in Table 2.

Table 2. Mechanical properties of aluminium carbon fiber and epoxy glass [14]

Property	Material		
	Aluminium 6061	Carbon fiber	Epoxy glass
Yield-Tensile strength (MPa)	280	945	1,700
Elastic modulus (GPa)	70	180	50
Poisson's ratio	0.33	0.3	0.3
Density (g/cm^3)	2.71	1.6	2

2.2 Calculation of Forces

The lift force and force caused from the weight of the wing were calculated and applied in the analyses. The lift force was found using the equation given in below.

$$L = \frac{1}{2}\rho C_L V^2 S$$

L	Lift force (N)
ρ	Air density (kg/m^3)
C_L	Lift coefficient
V	Motion velocity (m/s)
S	Wing area

Air density was used as 1.225 kg/m^3. Lift coefficient was selected as 0.5 as angle of attack of the selected wing model was 5° [15]. Three different velocities were used with the values of 10, 50 and 100 m/s. Wing area was directly obtained from design software (1.08 m^2).

2.3 Static Structural Analysis

Static Analysis is a submodule of Ansys Structural Analysis and used to determine responses of a structure under applied loads. This module can be used to determine the displacements, strains, stresses, and forces in a structure. Static Analysis was applied to the designed wing. The main purpose of this analysis was to observe the maximum stress, strain and displacement values and areas (Fig. 5).

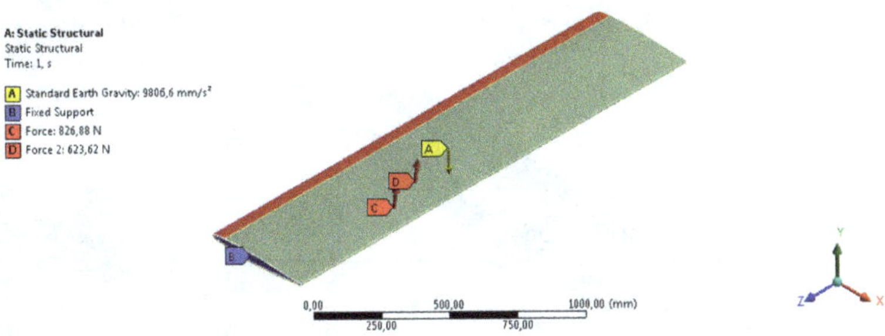

Fig. 5. Static analyses with Ansys workbench

2.4 Modal Analysis

Modal analysis is a method to find the vibration characteristics (natural frequencies and mode shapes) of a designed structure. Ansys Workbench Modal tool was used to modal analysis to wing structures with the different material (aluminum, carbon fiber and epoxy glass). The values of the maximum reached natural frequencies were calculated for three modes.

The applied forces in the analyses were given in Table 3.

Table 3. The applied forces in analyses

Motion velocity (m/s)	Lift force (N)	Weight force (N)	Total (N)
Aluminium			
10	33.075	623.62	656.69
50	826.87	623.62	1,450.49
100	3,307.50	623.62	3,931.12
Carbon fiber			
10	33.075	315.18	348.25
50	826.87	315.18	1142.05
100	3,307.50	315.18	3622.68
Epoxy S-glass UD			
10	33.075	450.26	483.33
50	826.87	450.26	1,277.13
100	3,307.50	450.26	3,757.76

3 Results and Discussion

3.1 Static Analysis

Static analysis was conducted to find the maximum occurred von Misses stresses, strains and deformation on wing structure (Fig. 6). The obtained results were given in Table 4.

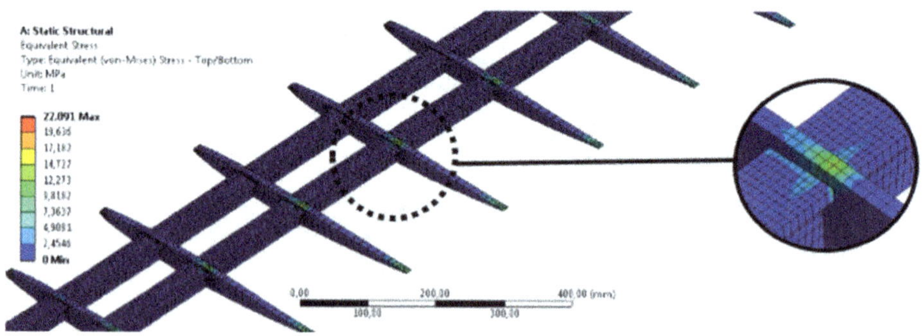

Fig. 6. Static Analyses Results (Max. Equivalent Stress)

Table 4. Static analysis results

Motion velocity (m/s)	Deformation (mm)	Strain (mm/mm)	Equivalent stress (MPa)
Aluminium			
10	0.0204	2.01×10^{-5}	1.4251
50	0.5656	3.13×10^{-4}	22.091
100	2.2800	1.25×10^{-3}	88.617
Carbon Fiber			
10	0.0015	8.60×10^{-7}	0.849
50	0.0401	2.24×10^{-5}	22.143
100	0.1611	8.98×10^{-5}	88.702
Epoxy S-glass UD			
10	0.0686	9.84×10^{-5}	0.781
50	1.9172	1.64×10^{-3}	16.064
100	7.7137	6.60×10^{-3}	64.375

3.2 Modal Analysis

Modal analysis was applied to find the natural frequency values for three modes (Fig. 7). The obtained results are given in Table 5.

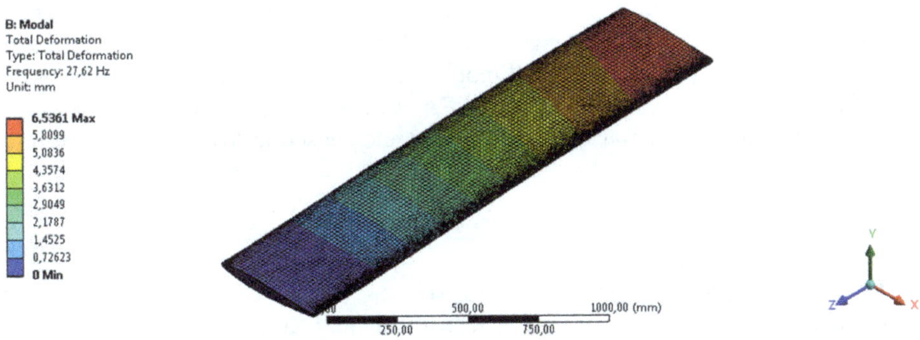

Fig. 7. Modal analysis results

Table 5. Modal analysis results

Material	Natural frequency (Hz)		
	Mode 1	Mode 2	Mode 3
Aluminium	27.62	84.81	91.51
Carbon fiber	146.18	446.78	483.26
Epoxy S-glass UD	17.51	47.79	55.88

It was obtained from static analysis results that the motion velocity had important effects on occurred deformation, strain and stress. While the deformation was only 0.0204 mm for 10 m/s it reached the 2.28 mm for 100 m/s motion velocity for aluminium wing. The same effects were observed for carbon fiber and Epoxy S-glass UD materials. The maximum natural frequency caused to the deformation were found for carbon fiber with the value of 146.18 Hz for mode 1, 446.78 for mode 2 and 483.26 for mode 3. It can be understood these analyses results that the carbon fiber material was the best choice for Naca 66-206 Aircraft Wing Model design and manufacturing.

4 Conclusions

In this study, NACA 66-206 wing type was modelled and analysed to understand to mechanical behaviour under lift and weight forces. Structural and modal analyses were applied for three different materials (aluminum 6061, carbon fiber and strong unidirectional epoxy glass). The affected forces (lift and weight) were calculated and applied to the wings for three motion velocity as 10, 50 and 100 m/s. According to the static and modal analyses results; the maximum deformations were observed for the 100 m/s motion velocity with values of 2.28 mm for aluminum, 0.16 mm for carbon fiber and 7.71 mm for epoxy glass. The similar effects of the motion velocity were found for the strains and equivalent stresses. When the strength of the materials was compared with each other, the minimum stress was occurred on the strong epoxy glass structure. Although, the occurred stresses were so close to each other for aluminum and carbon fiber, carbon fiber caused to low weight force and also the natural frequency was higher than aluminum and epoxy glass for all modes. It can be concluded from these results that it was better to manufacture the NACA 66-206 wings with the carbon fiber material as it had high natural frequency values and caused to lowest weight force.

References

1. International Civil Aviation Organization (ICAO). Operation of Aircraft, Part I International Commercial Air Transport – Aeroplanes (Ninth Edition), Canada (2010)
2. NASA: The Four Forces of Flight (2019). https://www.nasa.gov/audience/foreducators/k-4/features/F_Four_Forces_of_Flight.html
3. NASA: Four Forces on an Airplane (2019). https://www.grc.nasa.gov/www/k-12/airplane/forces.html
4. MIT: Theory of Flight (2019). http://web.mit.edu/16.00/www/aec/flight.html
5. Fielding, J.P.: Introduction to Aircraft Design. Cambridge Aerospace Series (1999)
6. Mohite, P.M.: Lecture Notes – Basic Aircraft Construction (2019). http://home.iitk.ac.in/~mohite/Basic_construction.pdf
7. NASA: Aircraft Rotation (2019). https://www.grc.nasa.gov/www/k-12/airplane/rotations.html
8. Aircraft principles axis (2019). https://www.revolvy.com/page/Aircraft-principal-axes
9. Peruru, S.P., Abbisetti, S.B.: Design and finite element analysis of aircraft wing using ribs and spars. Int. Res. J. Eng. Technol. (IRJET) **04**(i.06), 2133–2139 (2017)

10. Raid, S.Z.: Finite Element, Modal Testing and Modal Analysis of a Radial Flow Impeller. Iran. J. Sci. Technol. Trans. B, Eng. **29**(B2), 157–169 (2005)
11. Abdelal, G.F., Abuelfoutouh, N., Gad, A.H.: Finite Element Analysis of Satellite Structures, Applications to Their Design, Manufacture and Testing. Springer Verlag, London (2013)
12. Demirtas, A., Bayraktar, M.: Free vibration analysis of an aircraft wing by considering as a cantilever beam. Selcuk Univ. J. Eng. Sci. Tech. **7**(1), 12–21 (2019)
13. Saran, V., Jayakumar, V., Bharathiraja, G.: Analysis of natural frequency for an aircraft wing structure under pre-stress condition. Int. J. Mech. Eng. Technol. (IJMET) **8**(8), 1118–1123 (2017)
14. http://www.matweb.com/. 20 May 2019
15. Aerodynamic Lift, Drag and Moment Coefficients. https://aerotoolbox.net/lift-drag-moment-coefficient/

The Effect of Cutting Speed and Feed on Cutting Forces in Face Milling with Square Shaped Insert

János Kundrák[1], Angelos P. Markopoulos[2](✉), Nikolaos E. Karkalos[2], and Tamás Makkai[1]

[1] Institute of Manufacturing Science, University of Miskolc, Miskolc, Hungary
{kundrak, tamas.makkai}@uni-miskolc.hu
[2] School of Mechanical Engineering, National Technical University of Athens, Athens, Greece
{amark, nkark}@mail.ntua.gr

Abstract. Metal cutting operations are usually employed for the manufacturing of various components in the automotive and aerospace industries. Especially, face milling is particularly preferred for creating high quality surfaces for high-end components, due to its versatility and high efficiency. A major challenge of this process is the achievement of essential goals, such as low surface roughness, minimum deformation and workpiece damage, avoidance of unfavorable residual stresses and at the same time the achievement of high productivity, i.e. high material removal rate. For that reason, it is essential to determine the optimum process parameters in order to reach the necessary compromise between productivity and quality of the manufactured product. In the present study, the effect of feed and cutting speed on cutting forces and specific cutting forces during face milling with square shaped cutting insert is investigated and conclusions are drawn on the importance of each process parameter, as well as the recommended values of these parameters in order to achieve the desired goals.

Keywords: Metal cutting · Face milling · Cutting forces · Specific cutting forces · Square insert

1 Introduction

The fabrication of parts for the automotive and aerospace industry, often, involves the creation of perfectly flat surfaces with high dimensional accuracy in order to comply with the requirements of various assembled structures. Flat surfaces can be created by several methods, but throughout the years, face milling has been proven to be among the most effective techniques for that purpose. For face milling, various types of milling inserts and cutters have been developed, both for usual and specialized purposes, such as milling of hard-to-cut materials or milling with high feed. Progress regarding the increase of face milling efficiency can only be conducted with detailed investigations of the effect of process parameters on the surface quality and tool and workpiece damage so as to achieve rapid production without lowering tool life or the expected life of the

N. M. Durakbasa and M. G. Gençyılmaz (Eds.): ISPR 2019, LNME, pp. 192–200, 2020.
https://doi.org/10.1007/978-3-030-31343-2_17

produced part. A relatively easy way to meet the desired goal, is by monitoring the cutting forces during milling and determine the effect of process parameters on them, in order to be able to select their appropriate values.

As face milling is a long established process, a considerable amount of work has already been conducted regarding the influence of feed, cutting speed, depth of cut or even chip cross-section shape on cutting forces [1–4], mainly in cases with usual rectangular inserts. Furthermore, Selaimia et al. [5] conducted experiments on stainless steel specimens in order to determine the correlation between process parameters and outputs such as cutting force, cutting power, specific cutting force and material removal rate during face milling with square inserts. Their findings indicate that surface roughness and specific cutting force are mostly affected by feed; cutting force and cutting power are mostly affected by the axial depth of cut, whereas Material Removal Rate (MRR) is affected by both parameters. Muñoz-Escalona and Maropoulos [6] developed an analytical model for the prediction of surface roughness during face milling with square inserts. The geometric model was validated by experiments on aluminum alloy workpieces. It was shown that the model is capable of predicting surface roughness for various combinations of tool and workpiece materials, although it is not appropriate for cases with increased tool wear. Nath et al. [7] in their work commented that the square insert lacks the advantage of round inserts, as it engages the workpiece at a constant axial depth of cut and not at a variable one.

Kundrák and Felhő [8] created a surface roughness model based on CAD software, in order to predict the surface topography after face milling with various types of inserts mounted on the same milling head. More specifically, they conducted experiments with various combinations of square and dodecagonal inserts and compared the theoretical and experimental surface roughness values. It was shown that the model was capable of predicting surface roughness with a relatively low error, in most cases. The use of two different types of cutting inserts was the most challenging case for the developed model. San Millan Torrecillas et al. [9] investigated the tool wear during face milling of hardened cast iron with square ceramic cutting inserts. It was found that during milling of cast iron the wear on the clearance surface was dominant, the tip radius was considerably altered and roughness increased considerably, as well as the values of force components, which were increased up to five times.

Zhenyu et al. [10] developed an advanced model for the prediction of surface roughness during face milling with square inserts, taking into consideration both static and dynamic effects. After validating the model with results from aluminum milling experiments, they concluded that the model exhibited low error in terms of Ra prediction. In the aforementioned work, the most important factors affecting surface roughness were the run-out errors in the radial and axial directions. Gurdal et al. [11] showed that square cutting inserts exhibit faster wear rates than the high feed or round ones, as uniform chipping occurred during their experiments leading to accelerated notch formation. Wertheim et al. [12] presented a detailed analysis of the function of the cutting inserts according to their geometrical characteristics, including the square milling insert. Li et al. [13] presented a comprehensive milling forces model for various geometries, using multiple input parameters and validated their models with experiments. Rodrigues and Coelho [14] investigated seven different types of chip breakers for square carbide cutting tools and found that the alteration in chamfer characteristics of the insert was able to affect directly cutting forces and specific cutting forces.

In the present study, the effect of feed and cutting speed during face milling with square inserts is investigated in order to determine the relative importance of each parameter in connection to cutting forces and specific cutting forces. A wide range of cutting speeds, ranging from normal cutting speeds, used in industrial practice, up to high speed machining, are investigated. More specifically cutting speeds from 100 m/min to 500 m/min, at equal intervals of 100 m/min step, are considered. Furthermore, five different feed speeds, namely 0.1, 0.2, 0.3, 0.4 and 0.8 mm/tooth, are employed in the analysis. From the subsequent analysis, it is possible to select the appropriate values of process parameters, which can maintain adequate material removal rate. At the same time, with the adoption of the proposed process parameter values, excessive power consumption and cutting forces, which lead to unfavorable results for face milling, can be avoided.

2 Experimental Procedure

In the present work, face milling experiments were conducted on C45 carbon steel specimens under various feed rate and cutting speed conditions. For the experiments, a milling head with a single square cutting insert was employed. The cutting insert is a Dijet SEKN 1203 AFTN JC5030 coated carbide insert with $\kappa_r = 45°$, $\gamma_0 = 0^o$, $\alpha_0 = 20^o$ and a $0.85 \times 45°$ chamfer. The milling head type is Canela 0748.90.063 with a diameter of 63 mm. Furthermore, the width of cut is 58 mm, cutting length 50 mm. The workpiece hardness was measured at HB180. Two sets of experiments were conducted for the purposes of this study. For the first set, a_p and f_z were kept constant at 0.8 mm and 0.4 mm/tooth, respectively, whereas v_c was varied from 100 to 500 m/min, i.e. five different values. For the second set, a_p and v_c were kept constant at 0.8 mm and 200 m/min, respectively, whereas f_z was varied from 0.1 to 0.8 mm/tooth, namely 0.1, 0.2, 0.3, 0.4 and 0.8 mm/tooth.

During the experiments, the three cutting force components were measured using Kistler 9257A dynamometer and subsequently amplified using three Kistler 5011A charge amplifiers. The amplified force signals were recorded on a PC using a CompactDAQ-9171 data collector with four channels. Finally, the recorded data were processed using special measurement software, prepared by LabView programming language. The experimental procedure, the relevant experimental equipment and typical results are depicted in Fig. 1.

Indicative graphs of the cutting force components variation in face milling cases such as the ones conducted in the present work is presented in Fig. 2. Due to the fact that the coordinate system of the dynamometer is fixed on the workpiece, the shape of the force components curves is anticipated to be in the form shown in Fig. 2, on the left; the F_x component, which lies in the direction of feed, exhibits both positive and negative values, as the direction of the force vector changes, when the cutting insert passes from the symmetrical plane, whereas the other two force components remain positive throughout the experiment. The force components, as calculated in a coordination system attached to

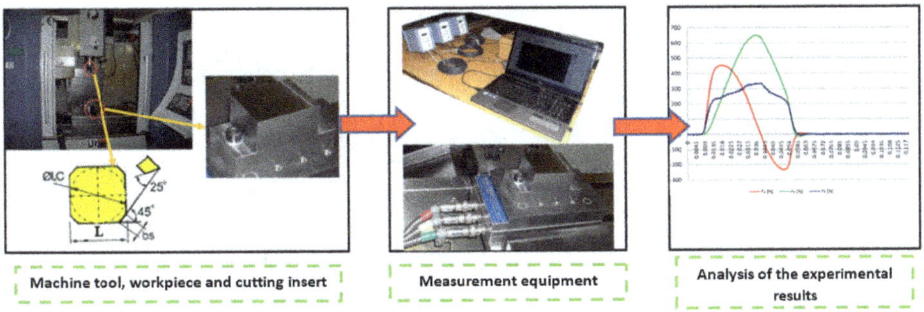

Fig. 1. Experimental procedure, equipment and results

the cutting edge of the tool [1], are depicted on the right of Fig. 2. It can be seen that F_z coincides with F_p, however, the other force components differ in magnitude and direction throughout the process.

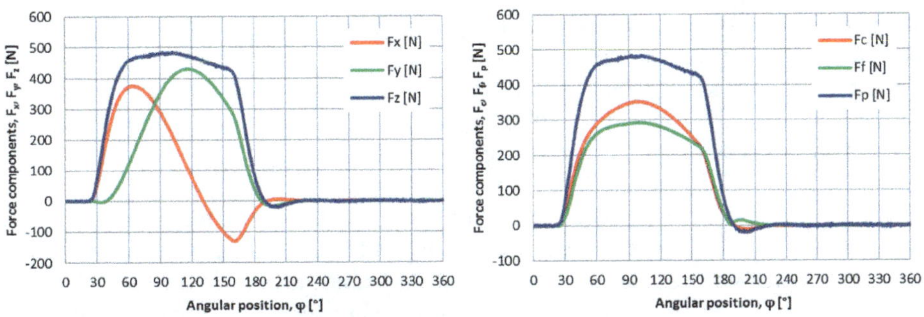

Fig. 2. Typical force components graph for face milling experiments

After analysis of the recorded data, the maximum values of the cutting force components, for both coordination systems, are obtained and summarized in Tables 1 and 2.

Table 1. Cutting force components (F_x, F_y, F_z) values for all experimental cases

v_c (m/min)	f_z (mm/tooth)	F_x (N)	F_y (N)	F_z (N)
100	0.4	452.31	649.20	333.69
200	0.4	452.79	624.19	280.07
300	0.4	421.57	602.96	257.02
400	0.4	399.04	598.98	264.39
500	0.4	369.55	569.58	250.83
200	0.1	198.77	235.84	204.77
200	0.2	375.41	429.94	484.10
200	0.3	469.49	575.88	548.03
200	0.8	808.69	1158.59	602.66

Table 2. Cutting force components (F_c, F_f, F_p) values for all experimental cases

v_c (m/min)	f_z (mm/tooth)	F_c (N)	F_f (N)	F_p (N)
100	0.4	591.91	288.40	333.69
200	0.4	547.99	328.54	280.07
300	0.4	505.21	371.76	257.02
400	0.4	419.74	442.19	264.39
500	0.4	375.38	457.05	250.83
200	0.1	194.69	157.64	204.77
200	0.2	352.65	293.42	484.10
200	0.3	480.70	353.22	548.03
200	0.8	999.13	612.38	602.66

3 Result and Discussion

In this section, results regarding the aforementioned face milling experiments are presented and analyzed. In Fig. 3, the results regarding the variation of cutting force components with the increase of cutting speed are presented. It can be seen that, in most of the cases examined, the cutting force components values are decreased with an increase of cutting speed. This is anticipated, as it is considered that the increase of cutting temperature at increased cutting speeds leads to softening of the workpiece material and reduces its cutting resistance [15]. Values of F_y cutting force components are the higher in every case, followed by F_x and F_z values. However, the increase of cutting speed was found to affect F_z cutting force component to the greatest extend, followed by F_x and F_y. More specifically, a fivefold increase of cutting speed, led to 1.33 times lower F_z, 1.22 times lower F_x and 1.14 times lower F_y force.

When the coordinate system attached to the cutting edge of the tool is considered, it can be seen that F_c decreases while F_f increases, with increasing cutting speed. More specifically, up to 300 m/min, F_c is the highest component, while for higher cutting speeds, F_f is slightly higher than F_c. The decrease of F_c and the increase of F_f are almost linear, for the cutting speed range of the analysis.

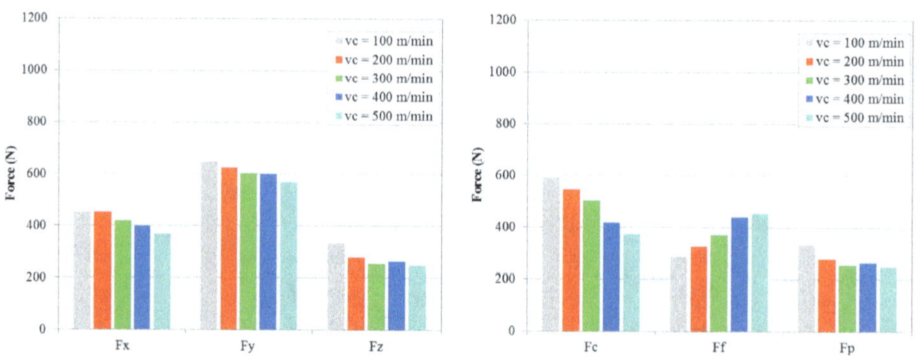

Fig. 3. Variation of maximum cutting force components for different values of the cutting speed

As it can be seen in Fig. 4, an increase in feed rate, results to a clear increase of the values of every force component, for both coordination systems. This is also anticipated, as the engagement of cutting insert with the workpiece is increasing. In most cases, F_y force exhibits the higher values, followed by F_z and F_x values. The increase of feed rate is observed to affect mostly the F_x and F_y force components. More specifically, an eightfold increase of feed rate led to 4.91 times higher F_y, 4.07 times higher F_x and 2.94 times higher F_z.

When F_c, F_f, F_p are considered, it may be observed that the corresponding increase is 5.13, 3.88 and 2.94 times, respectively. F_c is mostly affected by the increase in feed rate, followed by F_f. Comparing this variation of force components values with the variation between cases with variable cutting speed, it is observed that the effect of feed rate on cutting force values is considerably more significant than that of the cutting speed.

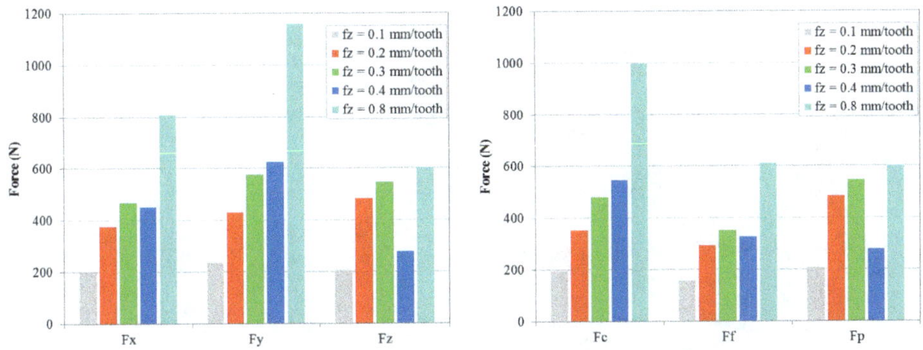

Fig. 4. Variation of maximum cutting force components for different values of feed rate

After the cutting forces are analyzed, it is also important to analyze the variation of specific cutting forces with varying cutting speed and feed rate. The specific cutting forces are calculated as the ratio of maximum cutting forces to the chip cross-section value, which is the product of axial depth of cut and the feed rate value [16]. They are denoted with k_i with subscript i being x, y, z for x, y and z components of cutting forces, respectively. For the cases with variable cutting speed, the chip cross-section does not change, thus the variation of specific cutting forces is similar to the variation of the cutting forces at it can be seen by comparing Figs. 3, 4 and 5.

On the other hand, when variable feed rate is considered, the specific cutting force values are changing due to different feed rate values. In this case, the values of specific cutting force components are presented in Fig. 6. As feed rate increases, the specific cutting forces decrease in almost every case; k_y force values are the higher, followed by k_z and k_x force components.

As far as the k_c, k_f and k_p are concerned, it can be seen from Fig. 6 on the right, that they tend to decrease with increasing feed rate. It can be observed that k_c is higher than k_f in all cases. However, the latter presents a steeper descend than the former; for an eightfold increase in feed rate, k_f is decreased to its half.

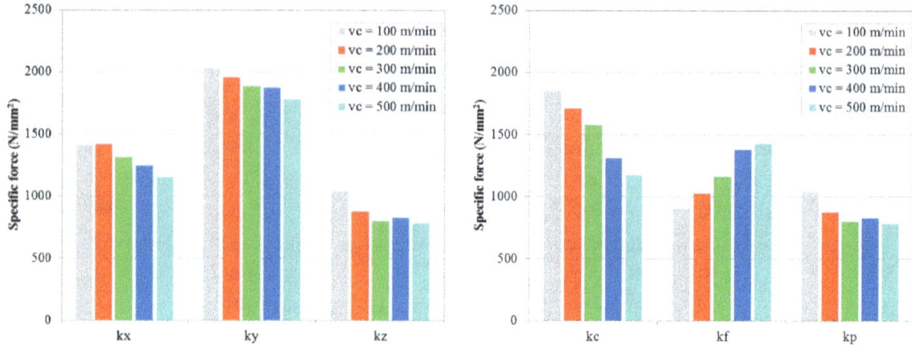

Fig. 5. Variation of specific cutting force components at different cutting speeds

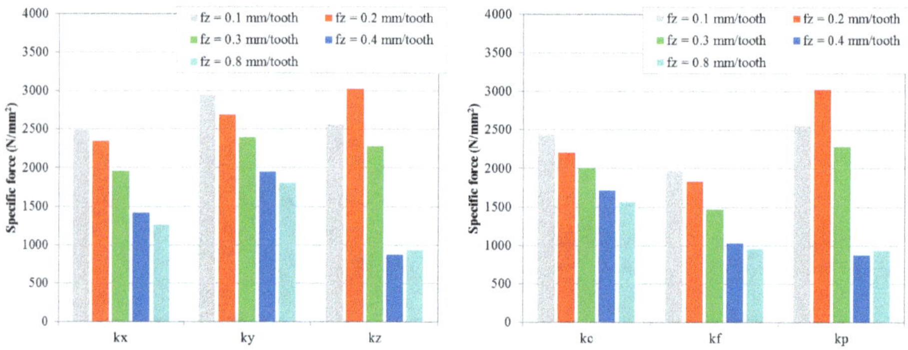

Fig. 6. Variation of specific cutting force components at different feed rates

4 Conclusions

In the present paper, face milling experiments with a milling head with a single square milling insert were conducted, under various cutting speed and feed rate values. From the analysis of the experimental results regarding cutting forces and specific cutting forces, various conclusions can be drawn.

As expected, an increase of cutting speed led to a decrease of cutting force components, due to increased cutting temperature and subsequent softening of the material workpiece. Furthermore, an increase in feed rate led to a clear increase of cutting force values, due to the more intense engagement of the cutting tool and the workpiece. Furthermore, it was observed that in most cases, F_y values are the higher and it was also seen that increase of cutting speed affects mostly F_z whereas an increase in feed rate affects mostly F_y. On the other hand, F_c is the highest force for cutting speeds up to 300 m/min, while F_f takes over for higher speeds. The force component F_c was the one mostly affected by feed rate, followed by F_f. The subsequent analysis of specific cutting forces, regarding cutting speed variation, was in accordance with the trends observed for cutting force components. It is worth noting that in the case of variable feed rates, the trends for specific cutting forces are the inverse of the trends observed for cutting forces, due to increasing chip cross section values.

Finally, as the increase of cutting speed led to much smaller variation of cutting forces compared to the increase of feed rate, the latter was proven to be the predominant factor regarding the control of cutting forces during face milling. Consequently, if it is desired to determine the optimum values leading to increased MRR and avoid excessive power consumption and tool wear, an appropriate adjustment of feed rate should be conducted.

Acknowledgement. The authors greatly appreciate the support of the National Research, Development and Innovation Office – NKFIH (No. of Agreement: OTKA K 116876).

This study was carried out as part of the EFOP-3.6.1-16-00011 "Younger and Renewing University – Innovative Knowledge City – institutional development of the University of Miskolc aiming at intelligent specialization" project implemented in the framework of the Szechenyi 2020 program. The realization of this project is supported by the European Union, cofinanced by the European Social Fund.

References

1. Kundrák, J., Markopoulos, A.P., Makkai, T., Deszpoth, I., Nagy, A.: Analysis of the effect of feed on chip size ratio and cutting forces in face milling for various cutting speeds. Manuf. Technol. **18**(3), 431–438 (2018)
2. Karpuschewsk, B., Kundrák, J., Felhő, C., Varga, G., Sztankovics, I., Makkai, T., Borysenko, D.: Preliminary investigations for the effect of cutting tool edge geometry in high-feed face milling. In: Jármai, K., Bolló, B. (eds.) Vehicle and Automotive Engineering 2, VAE 2018. Lecture Notes in Mechanical Engineering, pp. 241–254. Springer, Cham (2018)
3. Borysenko, D., Karpuschewski, B., Wenzel, F., Kundrák, J., Felhő, C.: Influence of cutting ratio and tool macro geometry on process characteristics and workpiece conditions in face milling. CIRP J. Manufact. Sci. Technol. **24**, 1–5 (2019)
4. Karpuschewski, B., Batt, S.: Improvement of dynamic properties in milling by integrated stepped cutting. Ann. CIRP **56**(1), 85–88 (2007)
5. Selaimia, A.A., Yallese, M.A., Bensouillah, H., Meddour, I., Khattabi, R., Mabrouki, T.: Modeling and optimization in dry face milling of X2CrNi18-9 austenitic stainless steel using RMS and desirability approach. Measurement **107**, 53–67 (2017)
6. Muñoz-Escalona, P., Maropoulos, P.G.: A geometrical model for surface roughness prediction when face milling Al 7075-T7351 with square insert tools. J. Manuf. Syst. **36**, 216–223 (2015)
7. Nath, C., Brooks, Z., Kurfess, T.R.: Machinability study and process optimization in face milling of some super alloys with indexable copy face mill inserts. J. Manuf. Process. **20**, 88–97 (2015)
8. Kundrák, J., Felhő, C.: 3D roughness parameters of surfaces face milled by special tools. Manuf. Technol. **16**(3), 532–538 (2016)
9. San Milan Torrecillas, R., Seleznev, A.E., Gurin, V.D., Peretyagin, P.Y.: Specifics of Wear of Ceramic Cutting tool inserts featuring Al_2O_3-TiC dies when face milling hardened cast iron. Mater. Sci. Forum **876**, 43–49 (2016)

10. Zhenyu, S., Luning, L., Zhanqiang, L.: Influence of dynamic effects on surface roughness for face milling process. Int. J. Adv. Manuf. Technol. **80**, 1823–1831 (2015)
11. Gurdal, O., Wright, A., Carpenter, C., Blackmore, M.: Investigation of the performance of different face milling inserts to improve rough milling of SA508 Grade 3 forgings. In: Proceedings of the ASME 2018 Pressure Vessels and Piping Conference, PVP 2018, Prague, pp. 1–9 (2018)
12. Wertheim, R., Satran, A., Ber, A.: Modifications of the cutting edge geometry and chip formation in milling. Ann. CIRP **43**(1), 63–68 (1994)
13. Li, X.P., Zheng, H.Q., Wong, Y.S., Nee, A.Y.C.: An approach to theoretical modeling and simulation of face milling forces. J. Manuf. Process. **2**(4), 225–240 (2000)
14. Rodrigues, A.R., Coelho, R.T.: Influence of the tool edge geometry on specific cutting at high speed cutting. J. Braz. Soc. Mech. Sci. Eng. **29**(3), 279–283 (2007)
15. Subramanian, M., Sakthivel, M., Sooryaprakash, K., Sudhakaran, R.: Optimization of cutting parameters for cutting force in shoulder milling of Al7075-T6 using response surface methodology and genetic algorithm. Procedia Eng. **64**, 690–700 (2013)
16. Mourão, A., Slătineanu, L., Gonçalves-Coelho, A.M.: The effect of wiper edge inserts on the specific cutting energy in face milling of aluminium alloys. Int. J. Mod. Manuf. Technol. **2** (2), 71–76 (2010)

Industry 4.0 Applications

A Conceptual Framework for Developing a Customized I 4.0 Education Scale: An Exploratory Research

Yunus Kaymaz(✉), İnanç Kabasakal, Ural Gökay Çiçekli, and Murat Kocamaz

Business Administration, Ege University, Bornova, Turkey
kaymazyunus@gmail.com, inanc.kabasakal@gmail.com, gokaycicekli@gmail.com, murat.kocamaz@gmail.com

Abstract. The fourth industrial revolution proposes digitization and networking by providing more productive, intelligent, controllable, and transparent factory environment. Although organizations have increasing attention to such a paradigm, changes in production and automation technologies led organizations to consider various areas. The necessity of individual qualifications and skills will eventually require more qualified managers. Furthermore, in a high technological environment which requires expertise on new materials, machines and information need more skilled labour. The increased complexity of workspaces eventually resulted in a need for a high level of education for the employees. The presented study contributes to research by providing a framework to structure an education scale in the context of Industry 4.0. Preliminary work for an I 4.0 education scale through an extensive literature review and academic reviews were provided.

Keywords: Industry 4.0 · Employment qualification · Production technologies

1 Introduction

As the introduction of the fourth industrial revolution took very first steps in Germany, many developed countries such as USA, France, and Japan have already been started nationwide programs for this industrial development. Although such initiatives mostly focus on the development of the current technology, various challenges have emerged. One of the crucial challenges of Industry 4.0 (I 4.0) is necessary of the skilled workforce for such industrial development. Reference [1] stated that in order to introduce I 4.0 to the business practices, employees with appropriate competences should be fulfilled. Reference [2] highlighted that there is a mismatch between the skill sets job applicants have and the skill sets they are expected to possess. Reference [3] proposed that due to the interconnectedness and the complexity of the processes, the requirements that employees need to have increased. Reference [4] highlighted the importance of the human resource in the context of I 4.0 and stated that there should be various qualification strategies for the current workforce because employees will be

N. M. Durakbasa and M. G. Gençyılmaz (Eds.): ISPR 2019, LNME, pp. 203–216, 2020.
https://doi.org/10.1007/978-3-030-31343-2_18

enabled to take on more strategic, creative and coordinating activities. Reference [5] indicated that there is a lack of knowledge about the future trends in job profiles and skills through the transformation of manufacturing to I 4.0 and companies should consider the skill gap between current ones and the trends. Reference [6] argued that due to the new technologies, employees would face a transformation of business models and work processes, which eventually lead to the necessity of high skilled labour for companies. Reference [7] introduced that different skills are applicable for specific I 4.0 biotypes, which implies that different workforce segments use a different subset of technologies. Thus, it is inevitable for companies to consider their workforces to evaluate by the requirements of I 4.0, specifically for the technological dimension of I 4.0. In this study, preliminary work for a customized education scale for the skilled workforce for companies have been structured. Through the proposed education scale presented in this study, the organizations which try to adopt I 4.0 in manufacturing and production technologies dimensions, may identify the gap between the actual level and the required level of employee skill and competence in selected dimensions. With regard to the literature reviews since the topic is relatively new, there is a little research on the subject. Although many of the available works focuses on social and knowledge skills of the employees with regard to I 4.0, this paper only focuses of the technological dimension. The originality of the paper focuses on the technological dimensions and sub-dimensions of I 4.0 with the context of the skill gap identification. The novelty of the study provides a customized structure of the education scale which aims to determine the gap between the actual level and the required level for each employee. Thus, it will be available for both managers and employees to define and determine the actual competency level and the required level for the job position.

2 Literature Review

This section of the study intensively analyses the dimensions as; Technology Enabled Quality Control, Virtual Product Development, Production Monitoring with Augmented Reality, Total Facility Maintenance and Internet of Things (IoT) and Manufacturing Execution Systems For Operational Excellence. All five dimensions has context which proposes the sub-dimensions of the main ones as seen in Table 1.

2.1 Technology Enabled Quality Control

The quality control systems are evolving on prevention before the error arises. This development, which extends from quality control to statistical process control, has taken the predictive quality control with I 4.0 applications. Such solutions are referred to as "Predictive Quality Management." The idea is automatically to predict quality issues before they arise and pinpoint the root cause. Predictions of the final tests are within specifications are based on historical data and past accuracy testing results. The results are expressed as a probability that a failure in the final test will occur [8].

Today, intelligent systems connect devices and sensors to a central place with collecting enormous amounts of data in seconds. This enables real-time analysis of predictive algorithms from structured and unstructured data and provides value-added information. With the help of intelligent systems, the predicted failures can be monitored in real time [9].

It has become possible to track many quality problems that may occur in production processes with nanotechnology. With the support of nanosensors and nanomaterials, possible quality failures in the production processes can be detected in real time. As an example, nanomaterials [10] and nanosensors [11] for innovative detection schemes of food-related compounds are already used.

Also, the usage of robotics for quality control and inspection are becoming more widespread, according to the developments in sensors. Robots can achieve more consistency in detecting quality problems and more reliable on inspections. Manual quality control techniques require fixtures and coordinate measuring devices. However, in robotics systems, the absence of a human may invalidate these techniques. The main benefit of robotic inspection is that products out of tolerance can be detected at an early stage in the manufacturing process.

2.2 Virtual Product Development

The global competition pressurizes businesses to adjust to quicker launch & delivery of products with reduced product lifetime [12]. To accomplish such constraints, businesses need to respond with flexible approaches [13] that benefit from digital technologies in developing and prototyping products. Computer-aided design (CAD) software is often used to design products. Accordingly, manufacturers require the organizational capability to utilize CAD tools. Besides, the assessment of organizational capability might be extended to meet the quality of VR (Virtual Reality) systems. Reference [14] noted that VR technology helps developing digital platforms to assess customer experience; and listed several criteria to assess the VR implementations; including visibility/viewability of users, ergonomics for ease of use, and aesthetic quality of the digital world is provided. Such criteria also can be taken into consideration to evaluate the organizational skills for VR in Virtual Product Development.

Rapid Prototyping (RP) corresponds to techniques that generate physical copies of product designs composed in CAD systems [15]. As an emerging technique in RP, 3D Printing has been significant progress that facilitates product development phase in manufacturing [16]. An important consideration in the use of 3D Printing technology is the accuracy of the models printed [17]. Besides, the complexity of geometry is also regarded as a success factor in the use of 3D Printing technology in Virtual Product Development [16].

2.3 Production Monitoring with Augmented Reality

Production monitoring category includes both IoT and AR (Augmented Reality) activities. IoT concept is available for both manufacturing and maintenance operations. Thus, the IoT concept can be seen under two categories. Intelligent manufacturing under IoT (1) comprises the upgrades of design, production, management and the integration of life cycle for a product which based on intelligent science and technology [18]. Reference [19] pointed out that for a product life cycle, data analytics, advanced materials and smart sensors can be used. Zhong et al. (2003) indicated that IoT-enabled manufacturing uses typical production resources by converting them into smart manufacturing objects which that can interact and interconnect automatically. Cloud manufacturing covers different production capacities which can be connected intelligently to the cloud with the help of IoT technologies such as RFID, barcodes, beacons in order to control and manage these resources remotely [18].

Reference [20] introduced the main components of the AR system as an information system, interaction system, and system for guided work [20]:

- Information system: Main aim of the information system is to deliver the necessary information for the respective situation. During the time at the production site, the planner takes all the necessary data.
- Interaction system: The task of the interaction system is the visualization of the virtual information onto the real environment.
- System for guided work: This contains various procedures which need to be followed to achieve a specific task.

Another issue in AR systems is maintenance support. Reference [21] stated that preventive and corrective maintenance activities are performed according to a series of pre-defined procedures which need to be performed by trained workers. In some cases, maintenance workers need help from supporting systems during site operations. In such a case, AR system will provide maintenance data management and allow remote collaboration. In their study, [22], provided a real case implication which uses remote maintenance support and according to the results, an increase over 30% for the awareness of machine tools' condition and a decrease over 25% for response time was achieved.

AR-based simulation systems will provide to combine both the physical experiences of human operators as well as database knowledge. AR technology provides this knowledge through rendering virtual information such as texts, images, sounds, and videos, into a real environment, thus providing a real-world enriched with information to the users [22]. In their work, [23] introduced the AR system as a tool for training for data collection. In their work, they compare the actual process and process with augmented reality, which the latter gave more feasible achieving times for activities.

Through the AR system, the augmented assembly provides the user to see an AR view as a step by step procedure in order to complete a specific real-world assembly [24]. Reference [22] stated that the augmented assembly benefited from both physical and virtual prototyping by analyzing the behavior and properties of planned products.

Considering robot programming and monitoring through AR system, [25] provided a user interface and the system architecture of an Internet-based telelaboratory, which allows researchers and students to remotely control and program two online educational robots. Furthermore, reference [26] provided some feedbacks on a tool which allow students and scientist to use the laboratory to program remotely several vision-based internet robots.

2.4 Total Facility Maintenance and IoT

Reference [27] provided architecture for IoT based smart factory, which includes the issues listed below:

- Smart machines: includes m2 m communication, communication of machines with devices and humans.
- Smart devices: includes mobile devices, field devices and operating devices in the factory.
- Smart manufacturing processes: includes dynamic, efficient, automated, and real-time process communication for the management and control of a highly dynamic manufacturing environment enabled by IoT.
- Smart engineering: includes product design and development, product engineering, production, and after-sales service. This could require the use of collected data from the manufacturing process in the planning process, and optimize machines
- Smart logistics: include smart logistics tools and processes.

The consumption of energy in facilities and the management of resources in factories pose a critical saving area for organizations. In this case, many researches focus on the integration of IoT and energy consumption and management. Reference [28] provided a system which connects lighting devices used as data collection platform use energy more efficiently in facilities and cities.

Reference [29] introduced a the use of web-based platforms like node.js and protocols such as OPC UA which can automate transferring information about anomalies, root causes and nominal data directly between cloud-based services and condition monitoring systems on the shop floor.

In the category of total facility maintenance and IoT, unlike the IoT (1) in the previous category, the IoT (2) provides information and data about predictive maintenance and employee productivity. Many researchers discuss predictive maintenance by using data generated in facilities [30, 59, 65]. Reference [30] introduced a platform analysing sound data and finds only anomaly data in cloud and orders maintenance automatically.

For the employee productivity context, [31] planned to collect data from the motion of the knowledge worker and correlates with productivity.

Reference [32] classified data mining tasks as below:

- Classification: proposes that data miner need to build a model which route incoming records into the correctly labelled areas.
- Estimation: includes the task of filling missing values in a special field of incoming records and often the regression techniques used for estimation.
- Segmentation: in a great amount of data, sometimes it is necessary to divide the population into smaller sub-populations with similar behaviour.
- Description/Prediction: descriptive data mining tries to find human interpretable patterns, associations, or correlations which describe the behaviour of data. On the other hand, predictive data mining performs inference on available data sets and attempts to predict the new data set behaviour.

Reference [33] provided a prediction algorithm which forecasts future values based on existing records, in order to estimate the possibility of a machine breakdown and therefore to support maintenance teams in planning appropriate maintenance interventions.

Reference [34] investigated the causes of occupational accidents that are frequently hidden in the operations and proposed hazardous work management guidelines to serve as references for safety management guidelines in the petrochemical industry. Furthermore, in order to understand the factors of job accidents, descriptive statistics and classification regression tree was used.

2.5 Manufacturing Execution Systems for Operational Excellence

Big data are generated from online transactions, emails, videos, audios, images, click streams, logs, posts, search queries, social networking interactions, science data, sensors and mobile phone applications [35, 36]. [37] defined big data as "big data technologies describe a new generation of technologies and architectures, designed to economically extract value from huge volumes of a wide variety of data, by enabling the high-velocity capture, discovery, and analysis". Big data has four component, such as volume, variety, velocity, and value.

MapReduce is a simple but powerful programming model for large-scale computing using a large number of clusters of commercial PCs to achieve automatic parallel processing and distribution [38, 39]. MapReduce has two functions as Map and Reduce [40].

Hadoop is widely used in big data applications for spam filtering, network searching, clickstream analysis, and social recommendation. Although technologies represented by Hadoop have achieved great success, such technologies are expected to fall behind and will be replaced, given the rapid development of big data [39].

Business intelligence (BI) is defined as the use of information and specialized analytical tools to enable informed decision making in a variety of organizational contexts [41–43]. BI can be the integration between the organizational and Information and Communication Technology (ICT) component by using a management model and a concept of integrated systems. Firms mainly use BI to understand the capabilities available. BI integrates data from internal and external sources with an automated data collection process for providing a useful information platform to decision makers [44]. BI requires analysts to deal with both structured and unstructured data to provide the intelligence needed in decision making [41]. Key characteristics of BI are data warehousing, predictive modelling, OLAP, Interactive visualization, Data mining, Information retrieval & extraction, spatial-temporal analysis, Person-centered analysis, location-aware analysis, context-relevant analysis, mobile visualization, real-time decision [45].

Artificial intelligence (AI) is a term that in its broadest sense would mean the ability of a machine or artifact to perform similar kinds of functions that characterize human thought [46, 47]. The main AI techniques are expert systems, fuzzy systems, artificial neural networks, and genetic algorithms.

Table 1. Main dimensions and the sub-dimensions of the technological context of I 4.0

	Category	Context	References
Technology enabled quality control	Data management	Predictive quality management	[8]
		Intelligent systems	[9]
	Nano-technology	Nanomaterials	[10]
		Nanosensors	[11]
	Robotics in inspection	Robotics	[50]
Virtual product development	Virtual reality	Visibility/viewability	[14]
		Ergonomics	[14]
		Aesthetic quality	[14]
	3-D printing	Complexity of geometry	[16]
		3D Printing Accuracy	[17, 49]
Production monitoring with augmented reality	IoT(1)	Intelligent manufacturing	[48, 57, 58]
		IoT enabled manufacturing	[60–62]
		Cloud manufacturing	[63, 64]
	Augmented reality	Main components: Information system, interaction system, system for guided work	[20]
		Remote maintenance support	[21] [22]
		AR-based simulation	[21, 23]
		Augmented assembly	[21, 24]
		AR assistance on robot programming and monitoring	[25, 26]
Total facility maintenance and IoT	Smart factory	Smart machines, smart devices, smart manufacturing processes, smart engineering, and smart logistics	[27]
		Energy management	[28]
		Condition monitoring systems	[29]
	IoT(2)	Predictive maintenance	[30, 59, 65]
		Employee productivity	[31]
	Data mining	Main categories: classification, estimation, segmentation, description/prediction	[32]
		Failure prediction	[33]
		Employee safety	[34]
Manufacturing execution systems for operational excellence	Big data	Main component: volume, variety, velocity, value	[37, 39, 54–56]
		MapReduce	[36, 39]
		Hadoop	[36, 39]
		High-Performance Computing Cluster (HPCC)	[36, 53]
	Business intelligence	Structured, semi-structured an unstructured data	[41, 45, 66]
		Real-time data warehousing	[41, 44, 45, 66]
		Automated anomaly and exception detection	[41, 45, 66]

(*continued*)

Table 1. (*continued*)

	Category	Context	References
		Data mining	[45, 66]
		Data visualization	[41, 44, 45, 66]
		Geographic information systems	[41, 45, 66]
	Artificial intelligence	Expert systems	[51, 52]
		Artificial neural networks	[49, 51]
		Fuzzy logic	[49, 51]
		Genetic algorithms	[49, 51]
		Hybrid system	[49, 51]

3 Industrial Application

As listed in the Table 1, main dimensions and sub-dimensions will provide extensive information about the required skill for employees. Figure 2, depicts the overall skill requirement and gap between the actual and prospected competence for employees. Each dimension has its own questionnaire provided from literature, industrial professionals and academicians (see Appendix). Through the related questions, the educational gap for each sub-dimension will be determined. As an example, Fig. 1 shows the overall skill gap for one employee which need training and education for his/her professional career. The employee in Fig. 1 shows proficiency in Total Facility Maintenance and Manufacturing Execution Systems for Operational Excellence and Technology Enabled Quality Control respectively but need to be trained for Virtual Product Development and Production monitoring with AR for his/her future skill competency for specific organizational position. Figure 2 which depict the Level 1 skill gaps, provides an example for Technology enabled quality control and Manufacturing Execution Systems for Operational Excellence dimensions. Each main dimension has various sub-dimensions and provides the skill gaps between actual and required ones. For instance, Manufacturing Execution Systems for Operational Excellence dimension has big data, AI and BI sub-dimensions and in this specific case, although the employee shows a sufficient performance in big data and BI sub-dimensions, he/she need to be perform better results for AI. Figure 3, which provides the Level 2 skill gaps for AI sub-dimensions, introduces that for a target position in the organization, the employee need to be educated most of the sub-dimensions of AI.

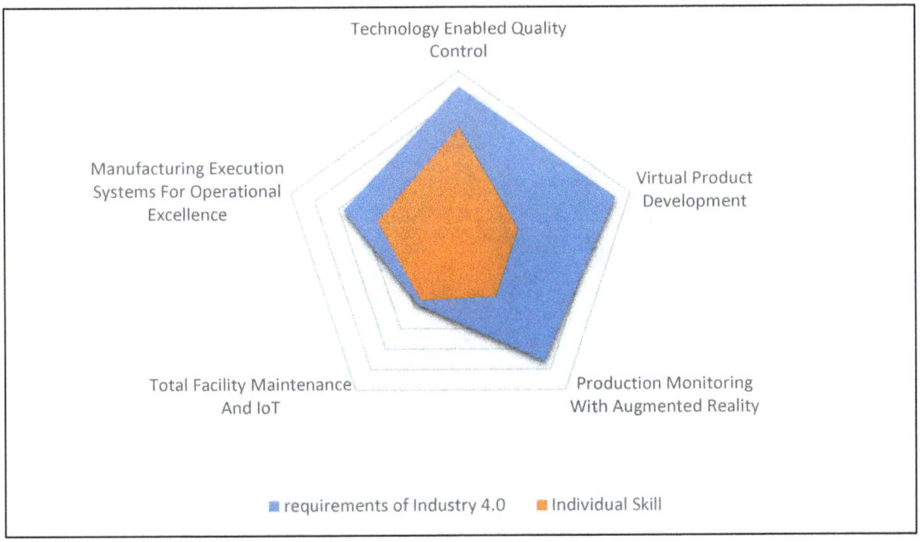

Fig. 1. I 4.0 overall skill gaps

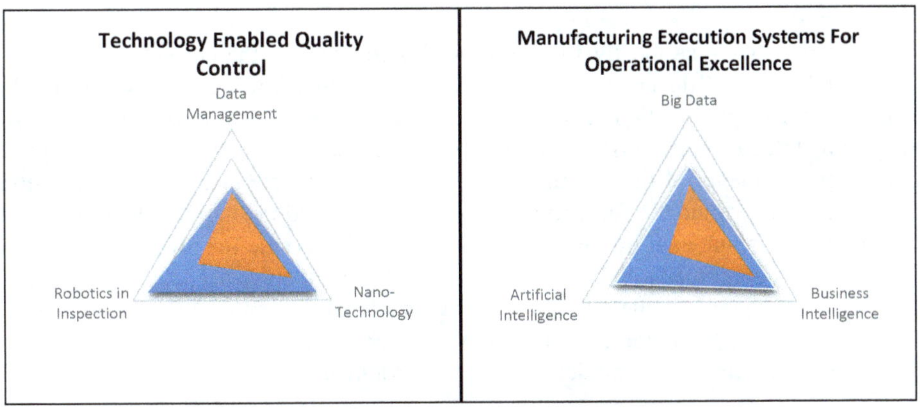

Fig. 2. I 4.0 level 1 skill gaps (technology enabled quality control, manufacturing execution systems for operational excellence)

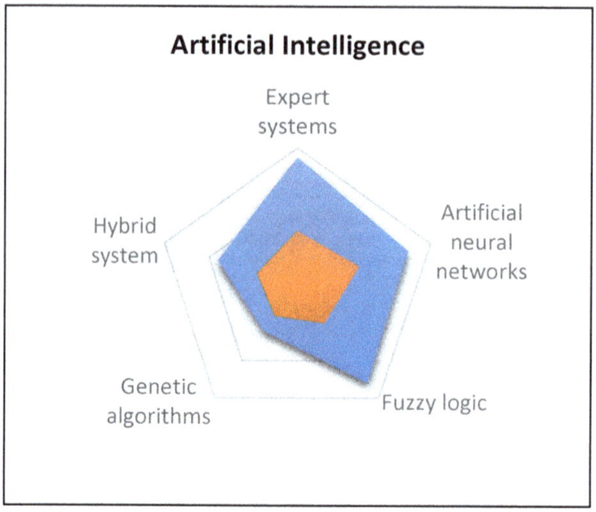

Fig. 3. I 4.0 level 2 skill gaps (artificial intelligence)

4 Conclusions

Although organizations mostly focus on the development of the current technologies towards I 4.0, one of the challenging area is the competency of the employee skills. In many cases, specific positions need necessary requirements with regard to I 4.0. This study aims to determine the skill gap between the actual competency of the employee and the requirements of the job position. The customized structure of the education scale aims to determine the gap between the actual level and the required level for each employee. For managerial implications this study will decrease the mismatch between job-employee assignments in the context of I 4.0. For future researches, some semi-structured interviews and detailed quantitative analyses in terms of skill competency may provide non-technological dimensions which need to be considered for the organizations in the transformation of I 4.0 environment.

Appendix

I 4.0 Level 2 Skill Gap Identification Form for Genetic Algorithms (GA) Sub-Dimension

	Question	1 (not sufficient)	2	3	4	5 (sufficient)
1	Knowledge about general terminology about the GA					
2	Ability to list the coding methods					
3	To provide the solution procedure of GA					
4	Knowledge about the crossover rate in genetic algorithm?					
5	Knowledge about GA selection operator					
6	Knowledge about mutation methods and procedures					

References

1. Grzelczak, A., Kosacka, M., Werner-Lewandowska, K.: Employees competences for Industry 4.0 in Poland–preliminary research results. DEStech Transactions on Engineering and Technology Research, (icpr) (2017)
2. Aulbur, W., Bigghe, R.: Skill development for Industry 4.0: BRICS skill development working group. Roland Berger GMBH (2016)
3. Gebhardt, J., Grimm, A., Neugebauer, L.M.: Developments 4.0 prospects on future requirements and impacts on work and vocational education. J. Techn. Educ. **3**(2), 117–133 (2015)
4. Hecklau, F., Galeitzke, M., Flachs, S., Kohl, H.: Holistic approach for human resource management in Industry 4.0. Procedia Cirp **54**, 1–6 (2016)
5. Pinzone, M., Fantini, P., Perini, S., Garavaglia, S., Taisch, M., Miragliotta, G.: Jobs and skills in Industry 4.0: an exploratory research. In: IFIP International Conference on Advances in Production Management Systems, pp. 282–288. Springer, Cham, September 2017
6. Prifti, L., Knigge, M., Kienegger, H., Krcmar, H.: A Competency Model for "Industrie 4.0". Employees (2017)
7. Fitsilis, P., Tsoutsa, P., Gerogiannis, V.: Industry 4.0: required personnel competences. Industrey 4.0 **3**(3), 130–133 (2018)
8. Gittler, T., Relea, E., Corti, D., Corani, G., Weiss, L., Cannizzaro, D., Wegener, K.: Towards predictive quality management in assembly systems with low quality low quantity data - a methodological approach. Procedia CIRP **79**, 125–130 (2019)
9. http://sapevents.be/AnalyticsBriefings/presentations/SAP_PQM.pdf
10. Mannino, S., Scampicchio, M.: Nanatechnology and food quality control. Vet. Res. Commun. **31**(1), 149–151 (2007)

11. Kuswandi, B., Futra, D., Heng, L.Y.: Nanosensors for the Detection of Food Contaminants, In: A. Grumezescu, A. Oprea (eds.) Nanotechnology Applications in Food. Academic Press (2017)
12. Manić, M., Miltenović, V., Stojković, M., Banić, M.: Feature models in virtual product development. Strojniski Vestnik/J. Mech. Eng. **56**(3) (2010)
13. Ale Ebrahim, N., Ahmed, S., Taha, Z.: Modified stage-gate: A conceptual model of virtual product development process. Afr. J. Mark. Manag. **1**(9), 211–219 (2009)
14. Berg, L.P., Vance, J.M.: Industry use of virtual reality in product design and manufacturing: a survey. Virtual Reality **21**(1), 1–17 (2017)
15. da Silva, G.C., Kaminski, P.C.: Selection of virtual and physical prototypes in the product development process. Int. J. Adv. Manuf. Technol. **84**(5–8), 1513–1530 (2016)
16. Lemu, H.G.: Study of capabilities and limitations of 3D printing technology. In: AIP Conference Proceedings, vol. 1431, No. 1, pp. 857–865. AIP, April 2012
17. Dimitrov, D., Van Wijck, W., Schreve, K., De Beer, N.: Investigating the achievable accuracy of three dimensional printing. Rapid Prototyping Journal **12**(1), 42–52 (2006)
18. Zhong, R.Y., Xu, X., Klotz, E., Newman, S.T.: Intelligent manufacturing in the context of industry 4.0: a review. Engineering **3**(5), 616–630 (2017)
19. Li, X., Lu, R., Liang, X., Shen, X., Chen, J., Lin, X.: Smart community: an Internet of Things application. IEEE Commun. Mag. **49**(11), 68–75 (2011)
20. Doil, F., Schreiber, W., Alt, T., Patron, C.: Augmented reality for manufacturing planning. In: Proceedings of the Workshop on Virtual Environments 2003, pp. 71–76. ACM, May 2003
21. Nee, A.Y.C., Ong, S.K., Chryssolouris, G., Mourtzis, D.: Augmented reality applications in design and manufacturing. CIRP Ann. **61**(2), 657–679 (2012)
22. Mourtzis, D., Vlachou, E., Zogopoulos, V., Fotini, X.: Integrated production and maintenance scheduling through machine monitoring and augmented reality: an industry 4.0 approach. In: IFIP Advances in Information and Communication Technology, pp. 354–362 (2017). https://doi.org/10.1007/978-3-319-66923-6_42
23. Mendoza, M., Mendoza, M., Mendoza, E., González, E.: Augmented reality as a tool of training for data collection on torque auditing. Procedia Comput. Sci. **75**, 5–11 (2015)
24. Hakkarainen, M., Woodward, C., Billinghurst, M.: Augmented assembly using a mobile phone. In: 2008 7th IEEE/ACM International Symposium on Mixed and Augmented Reality (2008)
25. Wirz, R., Marin, R. (n.d.). Remote programming of an Internet tele-lab for learning visual servoing techniques: a case study. In: 2004 IEEE International Conference on Systems, Man and Cybernetics
26. Marin, R., Sanz, P.J., Nebot, P., Wirz, R.: A multimodal interface to control a robot arm via the web: a case study on remote programming. IEEE Trans. Industr. Electron. **52**(6), 1506–1520 (2005)
27. Shrouf, F., Ordieres, J., Miragliotta, G.: Smart factories in industry 4.0: a review of the concept and of energy management approached in production based on the Internet of Things paradigm. In: 2014 IEEE International Conference on Industrial Engineering and Engineering Management (2014)
28. Primiceri, P., Visconti, P.: Solar-powered LED-based lighting facilities: an overview on recent technologies and embedded IoT devices to obtain wireless control, energy savings and quick maintenance. J. Eng. Appl. Sci. ARPN **12**(1), 140–150 (2017)
29. Fleischmann, H., Kohl, J., Franke, J., Reidt, A., Duchon, M., Krcmar, H.: Improving maintenance processes with distributed monitoring systems. In: 2016 IEEE 14th International Conference on Industrial Informatics (INDIN) (2016)

30. Yamato, Y., Fukumoto, Y., Kumazaki, H.: Predictive maintenance platform with sound stream analysis in edges. J. Inf. Process. **25**, 317–320 (2017)
31. Yano, K., Akitomi, T., Ara, K., Watanabe, J., Tsuji, S., Sato, N., Moriwaki, N.: Profiting from IoT: The key is very-large-scale happiness integration. In: 2015 Symposium on VLSI Circuits (VLSI Circuits) (2015)
32. Zangl, G., Oberwinkler, C.P.: Predictive data mining techniques for production optimization. In: Proceedings of SPE Annual Technical Conference and Exhibition (2004)
33. Bastos, P., Lopes, I., Pires, L.C.M.: Application of data mining in a maintenance system for failure prediction. In: Safety, Reliability and Risk Analysis: Beyond the Horizon: 22nd European Safety and Reliability, vol. 1, pp. 933–940 (2014)
34. Cheng, C.-W., Yao, H.-Q., Wu, T.-C.: Applying data mining techniques to analyze the causes of major occupational accidents in the petrochemical industry. J. Loss Prev. Process Ind. **26**(6), 1269–1278 (2013)
35. Eaton, C., Deroos, D., Deutsch, T., Lapis, G., Zikopoulos, P.C.: Understanding Big Data: Analytics for Enterprise Class Hadoop and Streaming Data. Mc Graw-Hill Companies (2012). 978-0-07-179053-6
36. Sagiroglu, S., Sinanc, D.: Big data: a review. In: 2013 International Conference on Collaboration Technologies and Systems (CTS), pp. 42–47. IEEE, May 2013
37. Gantz, J., Reinsel, D.: Extracting value from chaos. IDC iView, pp. 1–12 (2011)
38. Dean, J., Ghemawat, S.: Mapreduce: simplified data processing on large clusters. Commun. ACM **51**(1), 107–113 (2008)
39. Chen, M., Mao, S., Liu, Y.: Big data: A survey. Mobile networks and applications **19**(2), 171–209 (2014)
40. Bakshi, K.: Considerations for Big Data: Architecture and Approach. In: Aerospace Conference IEEE, Big Sky Montana, March 2012
41. Negash, S.: Business intelligence. Commun. Assoc. Inf. Syst. **13**, 177–195 (2004)
42. Rohloff, R.: Health-care BI: a tool for meaningful analysis. Healthc. Financ. Manag. **65**(5), 100–108 (2011)
43. Foshay, N., Kuziemsky, C.: Towards an implementation framework for business intelligence in healthcare. Int. J. Inf. Manage. **34**(1), 20–27 (2014)
44. Mettler, T., Vimarlund, V.: Understanding business intelligence in the context of healthcare. Health Inf. J. **15**(3), 254–264 (2009)
45. Chen, H., Chiang, R.H., Storey, V.C.: Business intelligence and analytics: From big data to big impact. MIS Q. **36**(4) (2012)
46. Kalogirou, S.A.: Artificial intelligence for the modeling and control of combustion processes: a review. Prog. Energy Combust. Sci. **29**, 515–566 (2003)
47. Mellit, A., Kalogirou, S.A., Hontoria, L., Shaari, S.: Artificial intelligence techniques for sizing photovoltaic systems: a review. Renew. Sustain. Energy Rev. **13**(2), 406–419 (2009)
48. Helu, M., Hedberg Jr., T.: Enabling smart manufacturing research and development using a product lifecycle test bed. Procedia Manuf **1**, 86–97 (2015)
49. Stopp, S., Wolff, T., Irlinger, F., Lueth, T.: A new method for printer calibration and contour accuracy manufacturing with 3D-print technology. Rapid Prototyping J. **14**(3), 167–172 (2008)
50. Dev-Anand, M., Selveraj, T., Kumanan, S., Ajith-Bosco-Raj, T.: Robotics in online inspection and quality control using moment algorithm. Adv. Prod. Eng. Manag. **7**, 27–38 (2012)
51. Mellit, A., Kalogirou, S.A.: Artificial intelligence techniques for photovoltaic applications: A review. Prog. Energy Combust. Sci. **34**(5), 574–632 (2008)
52. Kusiak, A.Y., Sunderesh, D., Heragu, S.: Expert systems and optimization. IEEE Trans. Software Eng. **15**(8), 1017–1020 (1989)

53. Kodi, A.K., Louri, A.: Energy-efficient and bandwidth reconfigurable photonic networks for high-performance computing (hpc) systems. IEEE J. Sel. Top. Quantum Electron. **17**(2), 384–395 (2011)
54. Gerhardt, B., Griffin, K., Klemann, R.: Unlocking Value in the Fragmented World of Big Data Analytics. Cisco Internet Business Solutions Group, June 2012. http://www.cisco.com/web/about/ac79/docs/sp/InformationInfomediaries.pdf
55. Russom, P.: Big data analytics. TDWI Best Pract. Report, Fourth Q. **19**(4), 1–34 (2011)
56. Singh, S., Singh, N.: Big data analytics. In: 2012 International Conference on Communication, Information & Computing Technology Mumbai India. IEEE, October 2011
57. Shen, W.M., Hao, Q., Yoon, H.J., Norrie, D.H.: Applications of agent-based systems in intelligent manufacturing: an updated review. Adv. Eng. Inform. **20**(4), 415–431 (2006)
58. Zijm, W.H.M.: Towards intelligent manufacturing planning and control systems. OR-Spectrum **22**(3), 313–345 (2000)
59. Kanawaday, A., Sane, A.: Machine learning for predictive maintenance of industrial machines using IoT sensor data. In: 2017 8th IEEE International Conference on Software Engineering and Service Science (ICSESS), pp. 87–90. IEEE, November 2017
60. Poon, T.C., Choy, K.L., Chow, H.K.H., Lau, H.C.W., Chan, F.T.S., Ho, K.C.: A RFID case-based logistics resource management system for managing order-picking operations in warehouses. Expert Syst. Appl. **36**(4), 8277–8301 (2009)
61. Huang, G.Q., Zhang, Y.F., Chen, X., Newman, S.T.: RFID-enabled real-time wireless manufacturing for adaptive assembly planning and control. J. Intell. Manuf. **19**(6), 701–713 (2008)
62. Zhang, Y.F., Jiang, P., Huang, G.: RFID-based smart Kanbans for Just-In-Time manufacturing. Int J Mater. Prod. Tech. **33**(1–2), 170–184 (2008)
63. Zhong, R.Y., Lan, S., Xu, C., Dai, Q., Huang, G.Q.: Visualization of RFID-enabled shopfloor logistics Big Data in Cloud Manufacturing. Int. J. Adv. Manuf. Tech. **84**(1–4), 5–16 (2016)
64. Qu, T., Lei, S.P., Wang, Z.Z., Nie, D.X., Chen, X., Huang, G.Q.: IoT-based real-time production logistics synchronization system under smart cloud manufacturing. Int. J. Adv. Manuf. Tech. **84**(1–4), 147–164 (2016)
65. Selcuk, S.: Predictive maintenance, its implementation and latest trends. Proc. Inst. Mech. Eng. Part B: J. Eng. Manuf. **231**(9), 1670–1679 (2016)
66. Sallam, R.L., Richardson, J., Hagerty, J., Hostmann, B.: Magic Quadrant for Business Intelligence Platforms. Gartner Group, Stamford (2011)

A Multi-criteria Decision-Making Model for Digital Transformation in Manufacturing: A Case Study from Automotive Supplier Industry

Hüseyin Furkan Beyaz(✉) and Nihan Yıldırım

Management Engineering, Istanbul Technical University, Istanbul, Turkey
{beyazh, yildirimni}@itu.edu.tr

Abstract. In digital transformation era, technology selection stands as a critical strategic technology management process for manufacturing companies in developing countries who struggle to adapt themselves to technological change despite their limited financial, technological and organizational resources and capabilities. In literature background, various scholars discussed that, regardless of scale, companies which had been able to apply lean production principles and techniques in their manufacturing processes, have the advantage of utilizing their process management (including measurement and control) systems as a leverage in effective needs and benefit analysis. However, determining both the various implications and costs together with the potential benefits of technology transfer requires a multidimensional analysis which includes technological, organizational, operational, economic, legal aspects in a common framework. Even if the company is competent in technology assessment, they may still face the challenge of combining multiple criteria that can enable contextual decision making for "appropriate" technology acquisition. In this framework, by utilizing a case study from a major automotive parts supplier company in Turkey, our study aims to explore the criteria and methods/techniques that can be used within a structured feasibility analysis (which may serve as a) decision making model for identifying appropriate technologies for effective digital transformation in accordance with the process improvement needs of manufacturing companies in developing country contexts. For this, after a detailed review of literature and similar practices in the selected industry, we defined the multiple dimensions of technology selection decision models. We also analysed current technology transfer needs of the selected company by process analysis techniques from lean production theory (Value Stream mapping, SPC) and also collected opinions of the practitioner experts. As the selected company is a supplier of a Japanese automotive manufacturer and has been applying lean production techniques for several decades, we could be able to use the information/data which the company provided. As well, to define the potential technologies that can respond to the defined needs, we applied technology auditing for available Industry 4.0 technologies with content analysis on literature and with interviews with the experts from the field. Together with the financial feasibility with NPV, ROI, Technology Acquisition Cost Analysis and Impact on labour productivity measures (where the horizon value is adapted for income/cost saving analysis), we designed a multi criteria decision making

N. M. Durakbasa and M. G. Gençyılmaz (Eds.): ISPR 2019, LNME, pp. 217–232, 2020.
https://doi.org/10.1007/978-3-030-31343-2_19

model which includes organizational, economic, technical and regulative factors and their sub-criteria, we proposed the feasibility analysis model. The proposed model is applied for the potential Industry 4.0 technologies which may respond to the process improvement needs of the studied company. To rank the feasibility results, we used TOPSIS multiple criteria decision-making method in the final phase.

Keywords: Multiple criteria decision-making · Technology selection · Digital transformation · Ready-use acquisition map · Feasibility assessment

1 Introduction

In the era of digital transformation, the practical implications of the technological change have the utmost priority for manufacturing companies who are in need of applying digitalization tools and technologies for higher levels of productivity and effectiveness. However, for deciding on the appropriate technology transfer, they face the challenge of applying multi-dimensioned decision-making tools which cover economic, technological, organizational and operational factors that can represent and evaluate their unique needs and potential solutions and insufficient.

In the current literature, it is known that companies which apply lean production techniques are more likely to better understand the linkage between their process needs and technological solutions as they have established systems for process performance measurement. As well, lean oriented companies have a cultural setting that enable them to adapt analytical tools in feasibility studies for technology selection, by being able to provide insights about their manufacturing process performances.

On the other hand, digital transformation and Industry 4.0 tools are diverse and offer numbers of possible process improvement potentials to manufacturing companies. However, these offered benefits are contingent with the needs and resources/ capabilities of the companies which will adapt these tools.

Hence finding the most optimized digital transformation tool for specified process problems stands as a practical problem, but also it provides a field of multidisciplinary research on proposing technology selection and feasibility frameworks and methodologies.

In this context, in our study, we reviewed current literature on technology selection and technology investment feasibility approaches to identify available methods and techniques which can be used in aligning process needs of the manufacturing companies with technological solutions. By using data provided from the studied company and the expert views, in the light of the literature we proposed a framework which includes the process analysis and multi-dimensioned feasibility framework that can be utilized in selecting digital transformation technologies for process and effectiveness improvement.

As this paper places the stress on how we prepared the technology selection mapping, the first part of the project dealing with the current system analyses and literature review is intended to kept at minimum so as to give enough room for explaining the main purpose.

2 Literature Review

Fourth Industrial Revolution which is called as Industry 4.0 and, in some settings, referred as digital transformation in manufacturing is expected to contribute to a significant increase in production quality and reduction in costs by eliminating wastes (mudas) and improving efficiency.

Manufacturing companies which struggle to optimize their operations, have to priorly understand the linkages and interaction between industry 4.0 and lean production techniques [1]. In recent years, researches on various operational excellence practices revealed that the lean production and Industry 4.0 must not be applied independently and sequentially but in an integrated harmony to create higher values [2].

In fact, there exist many common bases of industry 4.0 and lean production integration which can enable higher degrees of positive change and shorter information cycles. Hence it is required to analyse digital transformation strategies and roadmaps together with the lean production systems.

On the other hand, there is still need for research on the case studies and use cases of Industry 4.0 in manufacturing industry, especially from developing countries' perspectives. Use scenarios can be brought by detailed case studies on the best combination of production tools and digital technologies [2]. This is the basis of our research which we aimed to contribute to by adding business related financial issues. There are various studies on technology selection [3–6].

3 Methodology

Providing a sample for manufacturing companies, as a case study we conducted our analyses on a major automotive parts supplier company in Turkey, which have been applying lean production techniques till their establishment.

After we had reviewed the report [7] published by Boston Consulting Group and TUSİAD (Turkish Industry and Business Association), we chose to make our technology selection mapping approach, which focuses on digital transformation aspects, in the two sectors out of six, which were chosen by TUSİAD as pioneer ones in Turkey with regards to their contribution to Turkish economy. These two sectors were automotive and home appliances. To conduct out research, we communicated with two different companies running business in these sectors. Our aim was to enable information between these contributor sectors to flow as a means of increased efficiency, which in turn yields the whole economy. Upon the acceptance of the automotive supplier company, our study focused on this company as the case.

To collect data for analysing their current processes and identifying digital transformation opportunities, we conducted 4 factory visits, on-line interviews and face-to-face in-depth interviews with experts (Production Manager and Process Management Specialists) and observation. As well, we made content analysis on the process data of the company. Besides the studied company, to identify the alignment of digital transformation tools with most common processes of manufacturing industry, we also interviewed the manufacturing and Digital transformation experts of a major home appliances company from durable goods industry (which is the second leading industry according to TUSIAD and BCG [7]). Also, to rank these process-technology matches, we interviewed with a Smart Factory Technology Provider Company's IT Consultant as an expert and provided the scoring of the process-technology combinations.

Analysis phase of our study includes 5 basic steps. These are (1) literature review and content analysis on digital transformation tools and their usage areas, (2) analysis of current processes with the data provided from the company's systems and expert interviews (3) identification of digital transformation tools that can be utilized for process problems (4) proposing a feasibility framework for selecting the most appropriate technologies (5) Decision making for the technology by applying the scores/evaluations provided from experts and the financial analysis results in the proposed feasibility framework.

To give the reader a short map of the project, it is better to summarize our steps as follows: literature reviews on the topics which included digital transformation, Industry 4.0 tools and techniques, general problems faced in manufacturing practices, process improvement methods (Just-in-Time, TPS, Lean Manufacturing, VSM) [2, 8–13], current state analyses in the automotive supplier manufacturer, specifying the requirements in the company, observations in the home appliances manufacturer, compiling both sides so as to create a meaningful comparison and matching between the two companies, preparing the need-technology sets and tables, having the tables scored and matched by the company specialists and a technology consulting expert outside of the companies, feasibility analyses for selected need-technology matches, assessment of feasibility analyses and conclusion.

4 Analysis and Findings

In the first phase, process analysis techniques like Value Stream Mapping, Waste Analysis and value-added labour calculations are conducted with data provided from the company, from observations and the interviews with the experts and operators. In this phase the problems of the processes are identified.

As we have deduced from our current state analyses which included value stream mapping approach [12], value-added labour calculations and observations and surveys with the operators, we came up with a set of problems that are to be dealt. These existing problems are listed under the name of needs analysis below (Tables 1 and 2).

Table 1. Needs analysis outcomes

Where	Problem
Door framing lines	The cycle time which is highly close to that of the takt time
	The cycle time of the which are clearly higher than the others
Outside of the workcells	Not to be able to calculate the efficiency of operators working outside of the cells
Press-lines	Prolonged change-over time
Quality-control processes	Excess amount of work missed out
Dyeing processes	High rate of scrap, waste of raw material
Kanban	Problems arising due to the obsolete system of kanban

On the other hand, the Industry 4.0 tools and techniques that are now being deployed in other parts of the world for similar purposes were listed after the literature review [14–23] to create solutions to the problems founded in the company. This need-technology table is shown below; the table was shaped after having the cells, which means each and every need-technology match possibility, matched by specialists from the companies and scored by an expert outside of the companies. This matching meant the way forward for us to continue with the matched, selected ones in feasibility phase. Although we searched for and found lean manufacturing methods and approaches in the literature review, because the automotive supplier company is currently using these techniques and methods in its processes to minimize waste, we were forced to implement other technology solutions which accounted for digital transformation tools and techniques mentioned below.

Table 2. Need-technology matching and scoring

	Door framing lines	Outside of WorkCells	Press lines	Quality control process	Dyeing process	Kanban	Expert (Digital Transformation IT Consultant) Scoring (1–10)
Laser welding machine	AH						8
Simulation							
System integration							
Autonomous robots							
Internet of Things							
AGV		AH					9
Cloud							
Additive manufacturing							
Augmented reality				AH			8
RTLS		AH					10
Cyber security							
E-kanban						AH	10
Big data							
GPS		AH					8
RFID		AH					7
Smart camera				AH			8
3D laser scanning robot				AH			10

A: Expert Opinion from the automotive company matches, H: Expert opinion from the home appliances company matches

In feasibility framework design, there are two phases identified: (1) Feasibility of Highly Significant Dimension with Quantitative Approaches (2) Multi-Dimensional Evaluation with Qualitative Approach.

4.1 Technology Classification

As were found in the above tables, the technologies matched and scored are now going to be assessed on the basis of different feasibility steps. The technologies are grouped in four different categories with regards to their general aim. Tracking systems, for example, are selected to track desired objects, which in our paper are the objects carried by and with outside operators such as pallets (Fig. 1).

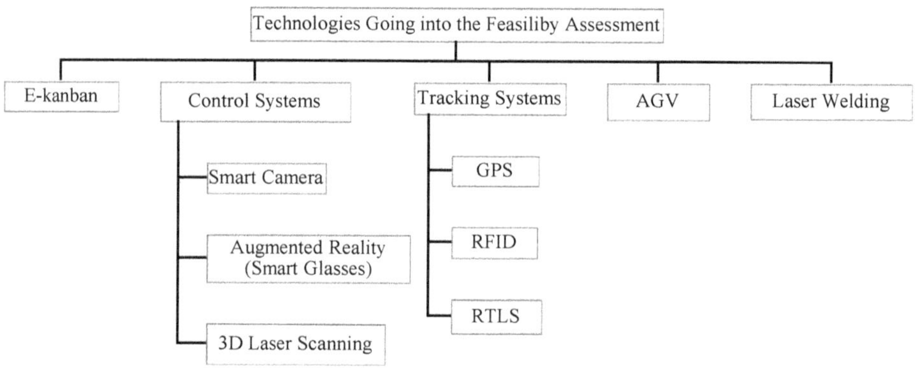

Fig. 1. Digital transformation technologies classification

We divided up the feasibility phase into four different steps, namely financial, organizational, technological and legal feasibilities.

For financial, technological and organizational feasibility, we prepared the below Table 3 thanks to the information we gathered from the literature review in general and The Toyota Way book for the most part [28]. The main reason to add sub-factors to each feasibility step was to consolidate the appropriateness of the assessment on the basis of real needs and details, after which a better comprehension of acquisition might be managed. By doing this, we also aimed at preparing a draft acquisition-map for the company for future investments. For this very reason, we had this table/map scored by managers and specialists from the company. Significance levels below were given according to the sub-factors' importance to the company. The reason why there exists some similar or the same parts is to put more emphasize on them than on others. For example, double use of labour and energy efficiency in technological and financial feasibility was to increase the significance level, which is also desirable for the company to assess the feasibility.

After having the significance levels given, we added them up to reach a total sum, which then be normalized into 1 in order for us to use the ratios in the last step of our project, TOPSIS, as weights of corresponding step-feasibilities.

At the far right of the Table 3, there is technology scoring, which helps us determine the effect of technology in terms of each sub-factor next to it. In AGV acquisition, for instance, how well the technology will increase the efficiencies or how big a layout change will be needed after the acquisition will be assessed by this scoring. The end of this scoring (qualitative and quantitative, explained later) appears at Appendix 1 where each sub-factors of each step of feasibility (technological, organizational, financial) for each technology is added up to reach a sum which will be used in the multi-criteria decision-making step as scores/points of technology acquisition. By following this techniques, we will end up with a decision-making model where each technology acquisition (which also means use-case scenarios) has its scores coming from both literature review and the views of managers in the company on the three feasibility steps, and these steps have their own weights coming from the scoring attained from the managers and specialists of the company.

Table 3. Feasibility scoring template

Significance level (1–10)	Technological feasibility	Technology scoring (1–10)
9	To be able to respond to the need	
7,5	Increase in labour efficiency	
7	Increase in energy efficiency	
8	Control and maintenance cost	
8	Impact on capacity utilization rate	
Significance level (1–10)	Organizational feasibility	Technology scoring (1–10)
7,5	Impact on factory design and layout	
9	Impact on production integration	
6,5	Time needed for adaptation	
7,5	To be able to used in the long run	
6,5	Impact on physical crowdedness	
Significance level (1–10)	Financial feasibility	Technology scoring (1–10)
9	Acquisition cost	
8,5	Orientation cost	
8	Increase in labour efficiency	
6,5	Increase in energy efficiency	
8,5	Rate of return	
7,5	Net present value	

As can be seen from Table 4, Total and Normalized values of significance scores by feasibility dimensions are given. In this respect, Financial Feasibility stands as the most significantly important feasibility dimension. In accordance with this finding, quantification of this dimension is prioritized and practiced in the following section. As well, normalized values will be used as weights in TOPSIS Decision Matrix in Table 10.

Table 4. Total and normalized values of significance scores by feasibility dimensions

Feasibility dimension	Total significance level score	Normalized values
Technological	39,5	0,32
Organizational	37	0,3
Financial	48	0,39
Total sum	124,5	1

4.2 Financial Feasibility by Quantification

In financial feasibility part, we mainly used Net Present Value method [24] for tangible assets where the period T is available for the technology in question. For technologies whose investment period T cannot be measured or estimated on the basis of NPV, namely intangible, software-based ones, we went for using Horizon Value method.

To calculate horizon value for the technologies, we needed "g", which means sectoral growth rate. After reviewing a report by PwC [25], by using numbers given, we calculated the annual average growth rate for 10 years, which is our period T for tangible investments (Table 5). Stock number below in the table represents the number of vehicles in Europe, which greatly differs in 2019 from in 2030.

Table 5.

Stock in 2019	Stock in 2030	Difference in stock	Growth rate
280	200	80	−3%

Figures attained from the report

$$280 * (1+g)^{11} = 200$$
$$0,715 = (1+g)^{11}$$
$$1+g = 0,95\, g = -3\%$$

By following simple math, we found the growth rate as above, which means, in the upcoming years, conventional automotive industry will face a decline in demand which will reflect a decline in supply as well.

For interest rate "r", we took two-year period bond yield rate which accounted for 20% [26]. We also presupposed for the company to increase the wages on average 14% each year in order for its labourers not to face a dramatic decline in purchasing power. Turkish Lira to Dollar and Euro rates were also based on the rate of the day of calculation [27]. Then we used these data in the feasibility assessment.

As a limitation of the data that can be provided from the company and the technology providers, only the sub-factors of "Acquisition Cost", "Increase in Labour Efficiency", "Rate of Return" and "Net Present Value" could be quantified and hence be used in the analyses.

Before using Table 3 as a guide for our decision-making step, we needed to clarify the points in the financial feasibility part. Having followed the financial techniques

(NPV, Horizon Value and Rate of Return), we prepared the Table 6 in order to be able to make a comparison between the technology acquisitions. However, these figures (NPV, HV) and ratios (Rate of Return) needed to be turned into the same 1–10 scale so that we can put them in the Table 3 to make sound comparison and add them up to reach a sum of scores for each feasibility part.

Table 6. Comparison of technology acquisition feasibility sub-factors cost/benefit values by technologies derived from financial feasibility

	E-KANBAN	RTLS	AGV	3-D LASER SCANNING
ACQUISITION COST	₺181.200,00	₺873.840,00	₺302.000,00	₺422.800,00
INCREASE IN LABOUR EFFICIENCY	₺2.482.664,61	₺5.590.968,66	₺7.165.785,17	₺454.234,71
RATE OF RETURN	12,7	5,4	22,73	0,074
NET PRESENT VALUE	₺2.431.527,87	₺5.010.031,53	₺6.863.785,17	₺31.434,71

To create correlations between the numbers below, we proposed to score the best in each row, which will appear on the general assessment table, 10, while giving 1 to the worst or the least efficient technology acquisition possibility. To clarify this point, the rating table for acquisition cost sub-factor is shown in the Table 7.

Table 7. Example financial rating for E-kanban technology (applied to other technologies)

EXAMPLE FINANCIAL RATING FOR E-KANBAN TECHNOLOGY (APPLIED TO OTHER TECHNOLOGIES)

	TECHNOLOGY	ACQUISITION COST	RATING	INDEX INTERVALS (THOUSAND)
1			₺1.812.000,00	1812+
2			₺1.630.800,00	1631-1812
3			₺1.449.600,00	1450-1631
4			₺1.268.400,00	1269-1450
5			₺1.087.200,00	906-1269
6	**RTLS**	₺873.840,00	₺906.000,00	725-906
7			₺724.800,00	544-725
8	**3-D LAZER**	₺422.800,00	₺543.600,00	363-544
9	**AGV**	₺302.000,00	₺362.400,00	182-363
10	**E-KANBAN**	₺181.200,00	₺181.200,00	0-181

After following the same rating method for increase in labour efficiency, net present value and rate of return sub-factors, we ended up with the Table 8, which represents the numbers that can now be used to make meaningful comparison with the rest of the first part of the Table 3. The combination of all these quantitative calculations were added in Appendix 1 which indeed is an extended version of Table 3, thereby applying to every technology acquisition.

As can be seen from this table, benefits of the Automated Guided Vehicles (AGV) had been the most advantageous technology to be invested in terms of the quantitative evaluation.

Table 8. Final scores of technologies by four sub factors of feasibility

	E-kanban	RTLS	AGV	3-D laser scanning
Acquisition cost	10	6	9	8
Increase in labour efficiency	4	8	10	1
Rate of return	6	3	10	1
Net present value	4	8	10	1

4.3 Multi-dimensional Evaluation by Qualitative Approach

For other dimensions of the technology evaluation, quantitative data and information are not available as the studied company and industry lack the digital transformation technology selection practices and so outcome measurements. Hence it is needed to utilize qualitative evaluations where the scores and ratings can be provided from expert opinions and literature review. Expert views are collected from the company's Production Manager and TPS (Toyota Production System) Engineers, the Smart Factory Company's Consultant, and also from the literature review at some parts.

This qualitative evaluation was prepared as shown in Appendix 1 where both qualitative and quantitative evaluations were combined. We then attained the Table 9 where every technology acquisition possibility (use-case scenarios) has its own total sum for each feasibility part. (Total Scores are derived from Appendix 1: Total Scores of Technologies by Feasibility Sub-Factors).

Table 9. Total scores of technologies by evaluation dimensions

Dimensions/criteria	E-kanban	RTLS	AGV	3-D laser scanning
Technological	274	297	265	154,5
Organizational	353,5	287	266	262,5
Financial	320,5	287	374,5	188

In the Table 10, we followed TOPSIS method (Technique for Order Preference by Similarity to Ideal Solution) that was introduced by Yoon ve Hwang (1980), that is based on the principle of closeness of decision points to ideal solution.

In the first part, it can be seen how our sums in the Table 9 were transmitted to the Table 10 as points (Total Scores are derived from Appendix 1: Total Scores of Technologies by Feasibility Sub-Factors). The weights were also transmitted from the Table 4 as the normalized values of significance levels given by the company specialists and managers.

Table 10. TOPSIS decision matrix with sum of squares and root of sum of squares (Standardized Decision Matrix R)

Technology solutions (use case scenarios)	Feasibility steps/their weights		
	Technological	Organizational	Financial
	0,32	0,3	0,39
E-kanban	274	353,5	368,5
AGV	273	255,5	369,5
RTLS	297	287	250,5
3D laser scanning robot	154,5	262,5	161
Sum of squares	**261684,3**	**341517,8**	**360993,8**
Root of sum of squares	**511,6**	**584,4**	**600,8**

Then we normalized the scores by Root of Total Sum of Squares in Table 11, by using the following formula.

$$r_{ij} = \frac{a_{ij}}{\sqrt{\sum_{k=1}^{m} a_{kj}^{2}}}$$

Table 11. TOPSIS standardized normalized decision matrix (r matrix)

Technology solutions (use case scenarios)	Feasibility steps/their weights		
	Technological	Organizational	Financial
	0,32	0,3	0,39
E-kanban	0,54	0,60	0,61
AGV	0,53	0,44	0,61
RTLS	0,58	0,49	0,42
3D laser scanning robot	0,30	0,45	0,27

As we proceed, we created the weighted normalized DM by multiplying each column of the matrix rj with its associated weighting factor, wj, as can be seen form Table 12. In the last rows of this matrix, also Positive * (max values in each dimensions) and Negative (−) Ideal Values (min values in each dimension) are calculated by the given formulas.

$$A^{*} = \left\{ (\max_{i} v_{ij} \big| j \in J), (\min_{i} v_{ij} \big| j \in J^{'} \right\}; A^{-} = \left\{ (\min_{i} v_{ij} \big| j \in J), (\max_{i} v_{ij} \big| j \in J^{'} \right\}$$

Table 12. TOPSIS weighted normalized decision matrix (V matrix)

Technology solutions (use case scenarios)	Feasibility steps/their weights		
	Technological	Organizational	Financial
	0,32	0,3	0,39
E-kanban	0,17	0,18	0,24
AGV	0,17	0,13	0,24
RTLS	0,19	0,15	0,16
3D laser scanning robot	0,10	0,13	0,10
Positive ideal A*	**0,19**	**0,18**	**0,24**
Negative ideal A-	**0,10**	**0,13**	**0,10**

To reach the optimum solution, we need to find the variation of each decision point (Alternative) of each criterion from ideal and negative ideal solution set Euclidian Distance Approach is used. Variations from the decision points which are called as İdeal separation (*) and Negative Ideal Separation (−) measures, were calculated by the given formulas and presented in Table 13.

$$S_i^* = \sqrt{\sum_{j=1}^{n} (v_{ij} - v_j^*)^2};\ S_i^- = \sqrt{\sum_{j=1}^{n} (v_{ij} - v_j^-)^2}$$

Table 13. TOPSIS positive and negative ideal matrices

Technology solutions (use case scenarios)	Feasibility steps/ their weights			Sum	Root of sum
	Technological	Organizational	Financial		
	0,32	0,3	0,39		
Positive ideal matrix					
E-kanban	0,0002	0,0000	0,0000	0,00	0,01
AGV	0,0002	0,0025	0,0000	0,00	0,05
RTLS	0,0000	0,0012	0,0060	0,01	0,08
3D laser scanning robot	0,0079	0,0022	0,0183	0,03	0,17
Negative ideal matrix					
E-kanban	0,006	0,0025	0,0181	0,03	0,16
AGV	0,005	0,0000	0,0183	0,02	0,15
RTLS	0,008	0,0003	0,0034	0,01	0,11
3D laser scanning robot	0,000	0,0000	0,0000	0,00	0,00

As the final phase of multicriteria decision making model, we applied "Relative Closeness to ideal solution for 3 decision points" with the below given formula and the results are presented in Table 14

$$C_i^* = \frac{S_i^-}{S_i^- + S_i^*}$$

Table 14. Technologies' relative closeness to ideal solution – final ranking for best use case

Technology solutions (use case scenarios)	S*	S^-	C*	Ranking
E-kanban	0,014	0,16	0,918	1
AGV	0,053	0,15	0,746	2
RTLS	0,084	0,11	0,560	3
3D laser scanning robot	0,169	0,00	0,021	4

At the final part, E-kanban technology was ranked first, which means that our best use case scenario is E-kanban technology solution in order for the company to increase efficiency in its manufacturing processes best. This finding is aligned with the results of financial feasibility sub-factor "Acquisition cost" but differed from other subfactors' evaluations like NPV, Rate of Return and increase in labour efficiency, as they concluded on AGV as the financially. most appropriate technology to be acquired by the studied company.

5 Conclusions

In era of digital transformation and Industry 4.0 adoption, appropriate technological tool selection stands as a challenge for manufacturing companies who are in need of increasing efficiency of their processes for a higher competitive performance. This challenge requires multi-dimensional/criteria, and also multi-disciplinary approaches to be applied to decision making models. Quantification of data to be used in these models also stands as another hard task in the lack of similar or equivalent practices, which is the actual situation today in developing countries like Turkey, who are stepping the first phases of digital transformation.

To provide a practical framework to practitioners and a methodology for researchers who work on technology acquisition decisions making models, we proposed a multi-dimensional, combined feasibility methodology in this paper. In this framework, we used both process measurement data of a lean manufacturing company and expert opinions for evaluating the appropriateness of digital transformation tools for process improvement needs. By using VSM and technology audit, we could list the process problems and potential Industry 4.0 technologies that can offer improvement. By creating a three dimensioned (financial, operational, technological) scoring model for feasibility, we evaluated the technologies. Financial feasibility included NPV, rate of return and acquisition cost sub-factors, which we could be able to calculate by using a hybrid approach including horizon value, index creation and estimations based on some assumptions. However, for technological and organizational sub-factors, we had to utilize expert opinions.

To combine the independent scorings TOPSIS method served as a useful tool, and concluded on the closest technology selection to ideal solution. Though the results from financial feasibility and multidimensional evaluation differed, two technologies "AGV" and "E-kanban" could be classified as the most appropriate ones. Findings make us think that, financial orientation and focus on return on investment may not be sufficient in decision making for disruptive technologies.

The findings also revealed that appropriate technology may differ by the company's resource base and process needs. Also, the technical expertise level and existing process management systems of the companies have impact on effective application of quantitative technology selection decision making models. Moreover, in such decision-making models, it is needed to define and customize multiple criteria with a break down structure and weighting, which will enable a relative comparison. As well, scoring various factors must depend on quantitative measures.

Study is expected to serve both to strategy makers in industry and also to the knowledge-base on multi-disciplinary technology selection approaches and usage of multi-criteria decision-making models for digital transformation. In further research, we aim to quantify the technological and organizational feasibility dimensions by using the outputs from digital transformation practices.

Appendix 1

Total Scores of Technologies by Feasibility Sub-Factors

SIGNIFICANCE LEVEL (1-10)	FEASIBILITY DIMENSION AND SUB FACTORS	TECHNOLOGY SCORING (1-10)							
		E-Kanban		AGV		RTLS		3-D Laser Scanner Robot	
	TECHNOLOGICAL FEASIBLITY	**Weighted Total**	**274**	**Weighted Total**	**265**	**Weighted Total**	**297**	**Weighted Total**	**154,5**
9	TO BE ALBE TO RESPOND TO THE NEED	10	90	9	81	10	90	6	54
7,5	INCREASE IN LABOUR EFFICIENCY	6	45	10	75	8	60	1	7,5
7	INCREASE IN ENERGY EFFICIECNY	5	35	3	21	5	35	3	21
8	CONTROL AND MAINTENANCE COST	9	72	4	32	8	64	6	48
8	IMPACT ON CAPACITY UTILIZATION RATE	4	32	7	56	6	48	3	24
	ORGANIZATIONAL FEASIBILITY	**Weighted Total**	**353,5**	**Weighted Total**	**266**	**Weighted Total**	**287**	**Weighted Total**	**262,5**
7,5	IMPACT ON FACTORY DESIGN AND LAYOUT	10	75	3	22,5	7	52,5	7	52,5
9	IMPACT ON PRODUCTION INTEGRATION	9	81	8	72	10	90	5	45
6,5	TIME NEEDED FOR ADAPTATION	10	65	8	52	8	52	9	58,5
7,5	TO BE ABLE TO USED IN THE LONG RUN	9	67,5	9	67,5	8	60	9	67,5
6,5	IMPACT ON PHYSICAL CROWDEDNESS	10	65	8	52	5	32,5	6	39
	FINANCIAL FEASIBILITY	**Weighted Total**	**320,5**	**Weighted Total**	**374,5**	**Weighted Total**	**287**	**Weighted Total**	**188**
9	ACQUISITION COST	10	90	9	81	6	54	8	72
8,5	ORIENTATION COST	10	85	4	34	6	51	7	59,5
8	INCREASE IN LABOUR EFFICIENCY	4	32	10	80	8	64	1	8
6,5	INCREASE IN ENERGY EFFICIECNY	5	32,5	3	19,5	5	32,5	5	32,5
8,5	RATE OF RETURN	6	51	10	85	3	25,5	1	8,5
7,5	NET PRESENT VALUE	4	30	10	75	8	60	1	7,5

References

1. Sanders, A., Elangeswaran, C., Wulfsberg, J.: Industry 4.0 implies lean manufacturing: research activities in Industry 4.0 function as enablers for lean manufacturing. J. Ind. Eng. Manag. **9**(3), 811–833 (2016)

2. Küpper, D., Heidemann, A., Ströhle, J., Spindelndreier, D., Knizek, C.: When Lean Meets Industry 4.0: The Next Level of Operational Excellence. Boston Consulting Group, Boston (2017)
3. Garbie, I.: Sustainability in Manufacturing Enterprises: Concepts, Analyses and Assessments for Industry 4.0. Springer International (2016)
4. Deloitte Development LLC: Advanced Technologies Initiative Manufacturing & Innovation. McKinsey & Company (2015). Industry 4.0: How to navigate digitization of the manufacturing sector (2015)
5. Nilsen, S., Nyberg, E.: The adoption of Industry 4.0-technologies in manufacturing – a multiple case study. (Master thesis). KTH Industrial Engineering and Management, Stockholm, Oesterreich (2016)
6. Frost & Sullivan: From concept to production: A 5-step approach towards successful Industry 4.0 projects, A Frost & Sullivan white paper with contributions from Bosch Software Innovations (2015)
7. TÜSİAD & BCG: Industry 4.0 in Turkey as an Imperative for Global Competitiveness, an Emerging Market Perspective (Yayın No: T/2016 03/576), İstanbul (2016)
8. Küpper, D., Kuhlmann, K., Köcher, S., Dauner, T., Burggräf, P.: The Factory of the Future. Boston Consulting Group, Boston (2016)
9. Monden, Y.: Toyota Production System. Institute of Industrial Engineers Press, Norcross (1983)
10. Ohno, T.: Toyota Production System: Beyond Large-Scale Production. CRC Press, New York (1988)
11. Roos, D., Womack, J., Jones, D.: The Machine That Changed the World: The Story of Lean Production. Google Scholar, New York (1991)
12. Rother, M., Shook, J.: Learning to See: Value-Stream Mapping to Create Value and Eliminate MUDA; A Lean Tool Kit Method and Workbook. The Lean Enterprise Institute, Brookline (1999). ISBN 978-0966784305
13. Womack, J.P., Jones, D.T.: Lean Thinking, revised edn. (2003). Presentation-Smart-Manufacturing-Sept-13-2017_Vienna.pdf
14. Bosch, Nexeed Track and Trace Starter Kit. https://www.bosch-connected-industry.com/en/connected-logistics/nexeedtrack-and-trace/nexeed-track-and-trace-starter-kit/. Accessed 20 Nov 2018
15. Bosch, Trackmytools. https://www.bosch.com/stories/trackmytools/. Accessed 20 Nov 2018
16. Bosch Software Innovations: Industry 4.0 Product Catalog (2017). https://www.boschpresse.de/pressportal/de/media/migrated_download/Bosch_Industrie_4.0_EN.pdf. Accessed 15 Nov 2018
17. Chalifoux, C.: Smart Cameras in a Manufacturing Environment: Today and in the Future. Quality, 6-7 (2018)
18. Finkenzeller, K.: RFID Handbook: Fundamentals and Applications in Contactless Smart Cards, Radio Frequency Identification and Near-Field Communication. Wiley, London (2010)
19. Houti, M., Abbadi, L.E., Abouabdellah, A.: E-Kanban the New Generation of Traditional Kanban System, and the Impact of Its Implementation in the Enterprise, s. 1–11 (2017)
20. Hofmann-Wellenhof, B., Lichtenegger, H., Collins, J.: Global Positioning System: Theory and Practice. Springer, New York (2012)
21. Jarupathirun, S., Ciganek, A.P., Chotiwankaewmanee, T., Kerdpitak, C.: Supply chain efficiencies through e-Kanban: a case study. Int. J. Comput. Internet Manag. **17**(1), 55-1 (2009)
22. Kuka: KUKA Mobil Platform 1500 (2018). https://www.kuka.com/tr-tr/%C3%BCr%C3%BCnler-hizmetler/mobilite/mobil-platformlar/kmp-1500#. Accessed 10 Dec 2018

23. Jetter, J., Eimecke, J., Rese, A.: Augmented reality tools for industrial applications: what are potential key performance indicators and who benefits? Comput. Hum. Behav. **87**, 18–33 (2018)
24. Brick, I.E.: Lecture Notes in Introduction to Corporate Finance (2017)
25. PWC: Five trends transforming the Automotive Industry (2018). https://eu-smartcities.eu/sites/default/files/2018–03/pwc-five-trendstransforming-the-automotive-industry.compressed.pdf. Accessed 27 December 2018
26. Bloomberg (2018). https://www.bloomberght.com/tahvil/faiz. Accessed 27 December 2018
27. Bloomberg (2018). https://www.bloomberght.com/doviz/. Accessed 27 December 2018
28. Liker, J.K., Meier, D.: The Toyota Way Fieldbook: A Practical Guide for Implementing Toyotas 4Ps. McGraw-Hill, New York (2006). ISBN: 0-07-144893-4

Aspects of Monitoring of Machining in the Context of Industry 4.0

Wojciech Zębala(✉) and Grzegorz Struzikiewicz

Production Engineering Institute,
Cracow University of Technology, Kraków, Poland
{zebala, struzikiewicz}@mech.pk.edu.pl

Abstract. The paper analyses the concept of a series of operations on data obtained from monitoring and supervising the machining processes in the aspect of the Industry 4.0 idea. The architecture of distracted data processing, i.e. *Edge Computing*, is compared with the architecture of centralized structure of data processing in the *Cloud.* An analysis is presented for a stand of physical phenomena monitoring in the cutting zone using vision systems. The analysis results aimed at defining the limitations of data processing systems with different architecture are presented.

Keywords: Industry 4.0 · Edge Computing · Monitoring · Machining · High-speed camera · Chip

1 Introduction

With development of the Industry 4.0 concept, the data processing which was carried out in the cloud system has been centralized again, and the function of terminal equipment has been reduced to providing and receiving processed information. Currently, the calculations are systematically transfer towards the cloud or remote data centers. It is anticipated that in a short time the functions implemented by the computer will be kept to a minimum. The computer will act as a terminal with access to data and the computing power in the cloud. However, as part of the Industry 4.0 concept, there are various solutions to supplement or expand this computing environment. It increases flexibility and gives users a wide range of solutions adapted to the requirements of different production or industrial systems. One of the solution is the *Edge Computing* concept with a distracted data processing architecture. In this approach, data are stored, processed and even analyzed directly in end devices, controllers or micro-data centers located in their immediate vicinity. Thanks to this, they often enable faster data transfer and analysis as well as quick response to registered irregularities. These advantages can be very important for the functioning of real systems operating in the manufacturing industry [1, 2].

The concept appeared a few years ago and is mainly related to the *internet of things (IoTs)*. It is generally based on the idea of processing and storage of data by end devices, controllers or micro-data centers operating near them, i.e. at the edge of the network. Locally processed data are then sent to a central database operating in the cloud or company data center. The definition of "*network edge*" depends on specific

N. M. Durakbasa and M. G. Gençyılmaz (Eds.): ISPR 2019, LNME, pp. 233–242, 2020.
https://doi.org/10.1007/978-3-030-31343-2_20

applications. For example, in telecommunications systems, this term may refer to telephone, automotive - car, manufacturing industry - a single machine or production line [2, 3]. Devices operating on the edge of the network can perform various functions, e.g. can be equipped with resources and software that allow the processing of data in contrast to devices that only provide transmission and reception of information from external systems or the cloud [4, 5].

Figure 1 shows a schematic comparison of architecture with a central and dispersed structure.

In a system with a central architecture, all information is stored in a central database and analyzed by applications operating in the central system. It follows that along with the popularization of the applications of the *IoTs*, the demand for transmission network capacity, storage capacity and the time needed to carry out calculations and analyzes is becoming more and more important [4, 5].

Processing the data near the place where they are created instead of sending all information to the cloud or company data centers that support the system, allows a substantial reduction in bandwidth requirements and allows the analysis of important data almost in real time. Instead of providing all information to central systems operating in the cloud, only relevant, statistical, valuable data are transferred to the *Cloud* for further analysis or regarding important situations requiring a specific response from the central system. This is particularly important in applications used in industry but also in telecommunications, healthcare and financial systems [4, 6, 7].

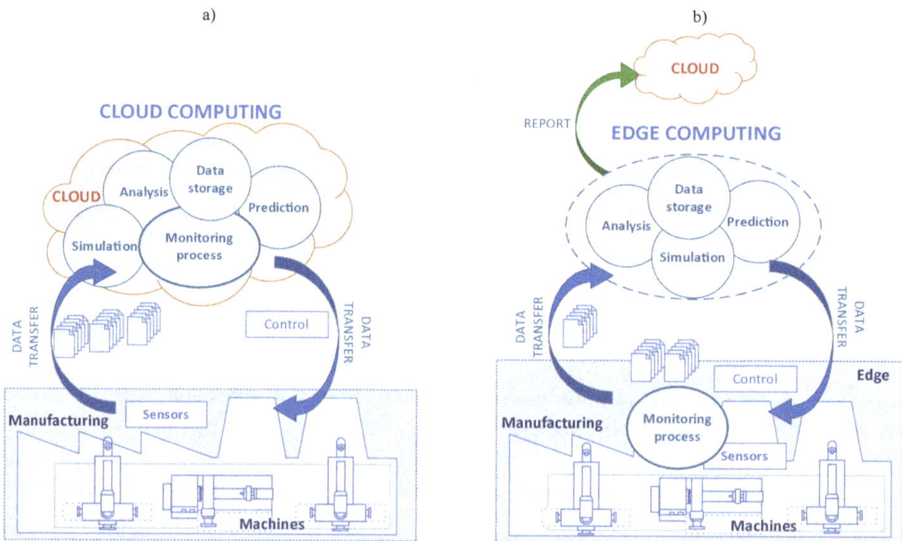

Fig. 1. The architecture of a data processing: (a) central system in the cloud (*Cloud Computing*), (b) dispersed system (*Edge Computing*)

With regard to the manufacturing industry, this application may also be a safer solution, regarding data protection and transfer stability. In the edge computing architecture, most of the data do not have to be transmitted over wide area networks and they remain close to the source, and as a result less data are transferred to the cloud or central system. It is easier to provide adequate access control, encryption and resistance to attacks on end devices. In many cases, ensuring stable connectivity with *IoTs* is difficult or maintaining a continuous data exchange with the central system is not an effective solution [5, 7, 8]. High requirements for efficient operation as well as small delays in time are set in particular by industrial production systems. For example, in systems for supervising or monitoring machining processes (widely understood as machining), delays calculated in milliseconds or a long time of data analysis are difficult to accept [9–13]. An inefficient supervision system generates gaps and increases manufacturing costs [5, 8, 13]. The production of machine parts on numerically controlled machine tools is particularly sensitive to these types of hazards. In the construction of machine tools, sensor sets are used that enable effective monitoring of the manufacturing process. The dimensional and shape accuracy as well as the quality of the workpiece depend on many factors occurring during cutting, like vibrations, high temperature in the cutting zone or the form the chips [14–16]. The basic qualification of chips is shown in Fig. 2.

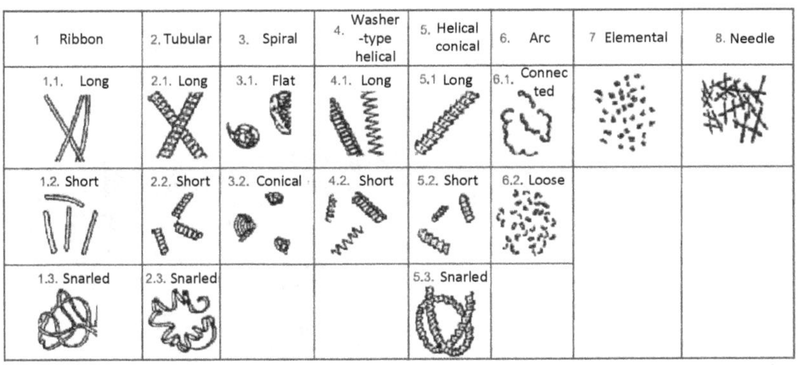

Fig. 2. Chip classification according to ISO3685

Obtaining the optimal machining conditions consistent with the Industry 4.0 idea requires, however, determining the limitations of the measurement techniques. The authors analyzed the system for monitoring the formation and flow of the chips in the turning process, using selected measuring techniques. The time and speed of transmission as well as the analysis of the results of measurements obtained with the use of a high-speed camera were analyzed. The analysis of the issue was carried out taking into account the *Edge Computing* concept and the *central cloud* system. The aim of the research was to determine the limitations of the presented measurement techniques.

2 Case Study

Figure 3 presents a diagram and a photograph of a measuring track for recording and analyzing the forms of chips obtained in the turning process. The research stand was built on the basis of the Knuth Masterturn 400 lathe. The measurement path included Phantom V5.2 high-speed camera from Vision Research, DEDOCOOL cold lighting set with 600 W power supply, Nikkor 200 mm macro lenses, necessary instrumentation (e.g. fastening systems, tripods) and a computer with software for recording and analysis of signals and film sequences. Figure 4 shows photographs of the used apparatus. A high-speed camera was directed to the cutting zone. The camera has been connected to the computer with a FireWire connector, used to transfer data from the camera directly to the computer's memory. The registrations of the chip forming and breaking process were carried out at recording speeds from 1000 to 3000 frames/s at a resolution of 512 × 512 up to 1152 × 1152 pixels. The high-speed camera had a maximum recording speed of 148,000 frames/s. The maximum resolution of 1152 × 896 pixels was achieved at 1000 frames/s. The camera allowed for recording in continuous mode and the maximum recording time of the film sequence was 4 s. The analyzes showed that the adopted registration speeds were sufficient to observe the chip formation and breaking process. The test stand made it possible to change the observation angle of the cutting zone, e.g. perpendicular to the rake surface of the cutting insert or to its clearance surface.

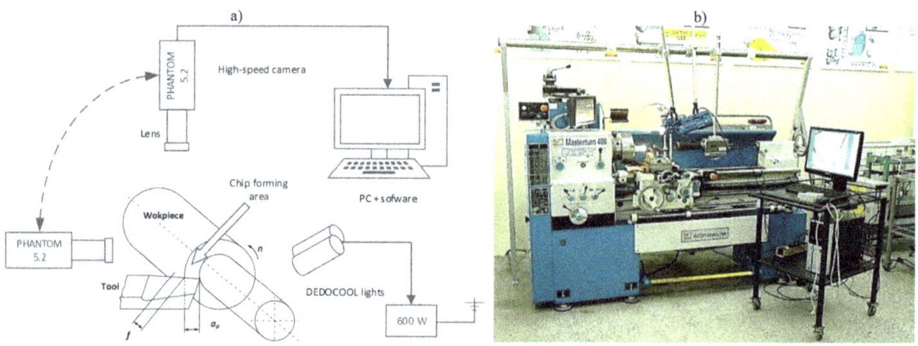

Fig. 3. Research stand: (a) block diagram of the measuring path, (b) photography of the test stand together with the measuring apparatus

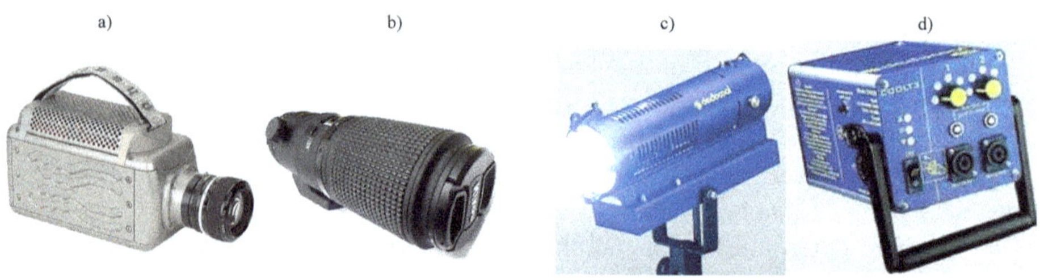

Fig. 4. Measuring apparatus: (a) high-speed camera, (b) lens, (c) cold spot illuminator (d) 600 W amplifier

Figure 5 shows examples of the forms of chips obtained in the turning process of various materials. Figure 5a shows the long washer-type helical form of chips obtained during longitudinal turning of Inconel 625 alloy. Figure 5b shows the long snarled chip form obtained during longitudinal turning of Ti6Al4V titanium alloy. A short and arc shape of chips formed during Inconel 718 turning are shown in Fig. 5c. In turn, Fig. 5d shows a long and spiral chips achieved during turning aluminum alloy.

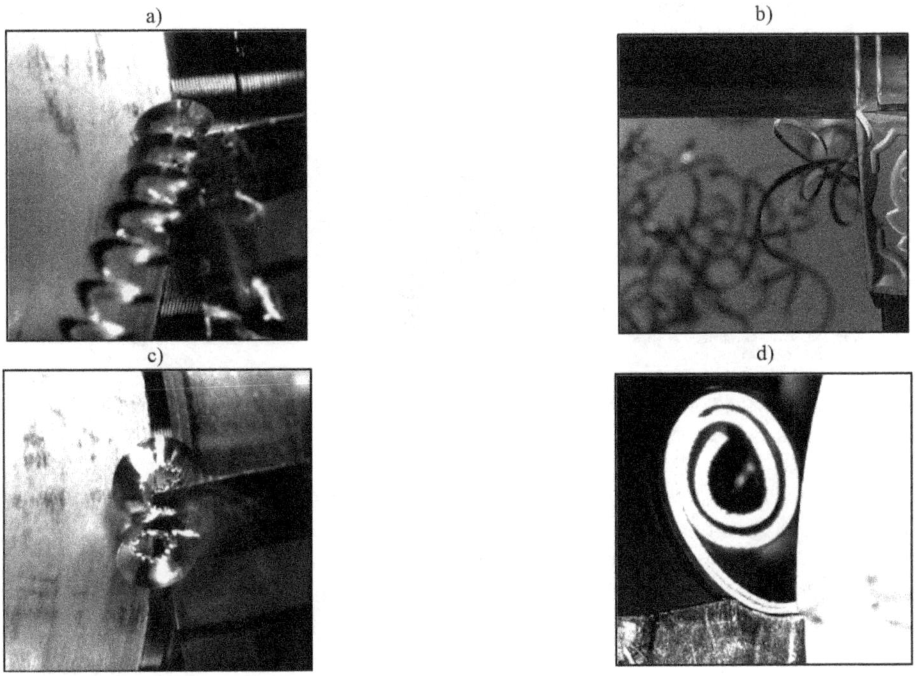

Fig. 5. Exemples frames from film sequences showing the form of chips: (a) long washer-type helical (b) long snarled, (c) short arc (d) long spiral - grooving

Figure 6 shows selected film frames from the registered sequence, showing the change of the form from short-arc to long washer-type helical chips for the case of longitudinal turning of the Inconel 625 alloy.

Fig. 6. The sequence of changing the form of chips from short arc to long. Time between frames is T_{ch} = 16875 μs, Inconel 625, v_c = 60 m/min, a_p = 1.5 mm, f = 0.25 mm/rev

Figure 5a and b show the correct chip breaking process. On the basis of the analysis of the stored fast-changing sequences, the chips breaking time was determined, T_{chip_break} = 16875 μs. Figure 6a–f show the individual film frames obtained in a constant time interval equal to T_{ch} = 16875 μs. The number of frames per accepted time interval with the recording frequency of 1600 frames/s was $i = 27$ frames. Figure 6c–f show longer and longer chips of a helical form. The presented analysis shows that the time needed to recognize the change in the form of chips and the response time through the supervisory system is *Treaction* = 50525 μs and requires registration and analysis of 81 film frames. The data volume of the film file recorded by the high-speed camera with the registered change in the form of the chip (shown in Fig. 6) is 153 MB (for a resolution of 768 × 768 pixels and a frequency of 1,600 frames/s).

Based on the analysis, an attempt was made to estimate the minimum transmission speed of measurement data through a monitoring system using a high-speed camera for supervise the change of the chip form.

The following assumptions have been adopted:

Tr – time of measurement/recording (it was assumed that the measurement time was *Tr* = 0.084275 s).

Wf – the size of the file in MB obtained for the time of measurement *Tr* (*Wf* = 153 MB).

Tt_i – transmission time of measurement data for a specific medium (transmission speed range *VTt* was from 1 MB/s to 10000 MB/s):

$$Tt_i = \frac{Wf}{Vt_t} \text{ (s)} \tag{1}$$

Ta – time of results analyzing of the measurement data, *Ta* = *Treaction* = 0.050525 s.

f – number of film files (sequences), necessary for supervising the whole machining process.

T_{totalC} – total time for *Cloud Computing* (2):

$$Ttotal_C = (Tr + Ta + Tt_i) \cdot f \text{ (s)} \tag{2}$$

T_{totalE} – total time for *Edge Computing* (3):

$$Ttotal_E = (Tr + Ta) \cdot f + Tt_{iR} \text{ (s)} \tag{3}$$

where Tt_{iR} – transmission time of the report (assuming the size of the report file 1 MB).

In the calculations, the coding, archiving, buffering and delay times of data transfer or differences in the time of reading data transmitted or lost data packets are omitted.

Table 1 presents the results of calculation of the file transfer time in the size of *Wf* = 153 MB for different transmission speeds and number of film sequences.

Table 1. The values of the transmission times of the high-speed camera sequence file for different transmission speeds and number of film sequences.

VTt (MB/s)	*Wf* (MB)	Tt_i (s)	*Ta* (s)	*Tr* (s)	*f*					Tt_{iR} (s)	$Ttotal_C$ (s)				
1	153	153	0.050525	0.084275	1	5	10	50	100	1	153.13	765.67	1531.35	7656.74	15313.48
10	153	15.3	0.050525	0.084275	1	5	10	50	100	0.1	15.43	77.17	154.35	771.74	1543.48
100	153	1.53	0.050525	0.084275	1	5	10	50	100	0.01	1.66	8.32	16.65	83.24	166.48
1000	153	0.153	0.050525	0.084275	1	5	10	50	100	0.001	0.29	1.44	2.88	14.39	28.78
10000	153	0.0153	0.050525	0.084275	1	5	10	50	100	0.0001	0.15	0.75	1.50	7.51	15.01
VTt (MB/s)	*Wf* (MB)	Tt_i (s)	*Ta* (s)	*Tr* (s)	*f*					Tt_{iR} (s)	$Ttotal_E$ (s)				
1	153	153	0.050525	0.084275	1	5	10	50	100	1	1.13	1.67	2.35	7.74	14.48
10	153	15.3	0.050525	0.084275	1	5	10	50	100	0.1	0.23	0.77	1.45	6.84	13.58
100	153	1.53	0.050525	0.084275	1	5	10	50	100	0.01	0.14	0.68	1.36	6.75	13.49
1000	153	0.153	0.050525	0.084275	1	5	10	50	100	0.001	0.14	0.68	1.35	6.74	13.48
10000	153	0.0153	0.050525	0.084275	1	5	10	50	100	0.0001	0.13	0.67	1.35	6.74	13.48

Figure 7 shows the dependence of the total time for the assumed transfer rates and the number of film sequences in the central (Fig. 7a) and dispersed (Fig. 7b) systems. For comparison, in Fig. 8a, the total time obtained according to *Cloud Computing* and *Edge Computing* for 100 and 1000 MB transfer are summarized. Figure 8b shows the impact of transfer speed on the total time for the central and dispersed systems.

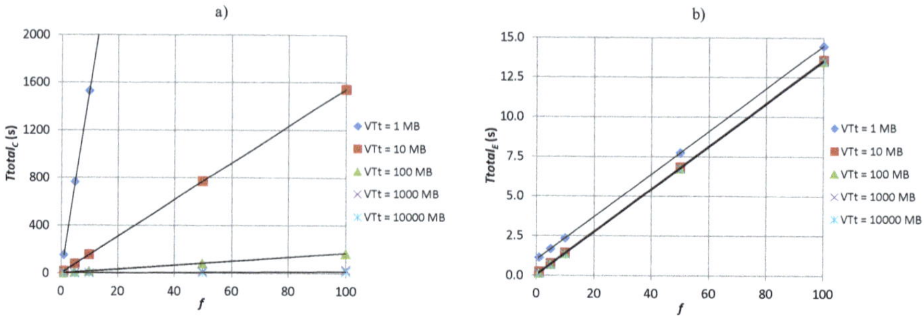

Fig. 7. The dependence between the total time at the assumed transfer rates and the number of movie sequences

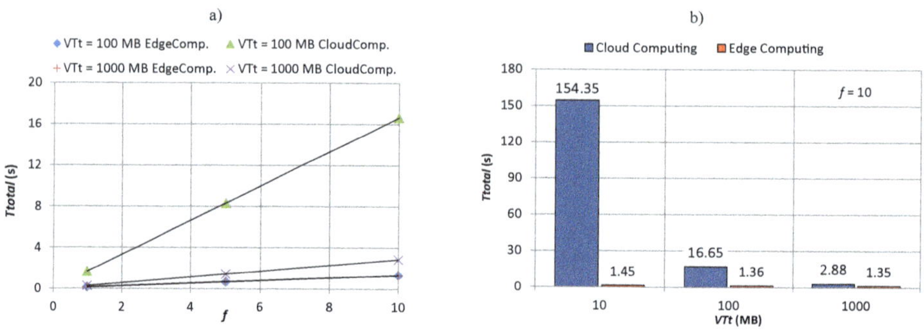

Fig. 8. Comparison of the total time for the central and dispersed systems: (a) depending on the number of sequences *f*, (b) depending on the transmission speed *VTt*

The graph in Fig. 9 shows the dependence of data transmission time as a function of data transmission speed. The presented dependence shows that the minimum data transfer speed necessary to perform the functions of the chip monitoring system using a high-speed camera is VTt_{min} = 3000 MB/s.

Analyzing the presented results, one can conclude that due to the size of files obtained during the registration of fast-changing sequences, and thus the time of their transmission and analysis, it is important to choose the place where they should be carried out. File transfer time can be shortened if data analysis is carried out directly in the machine's area (*Edge Computing*) and only the results of the analysis are sent to the cloud. In this approach, the coding time and archiving of transmitted data are also shortened. The times of data transfer delay may also be smaller, and the data may be stable and error-free.

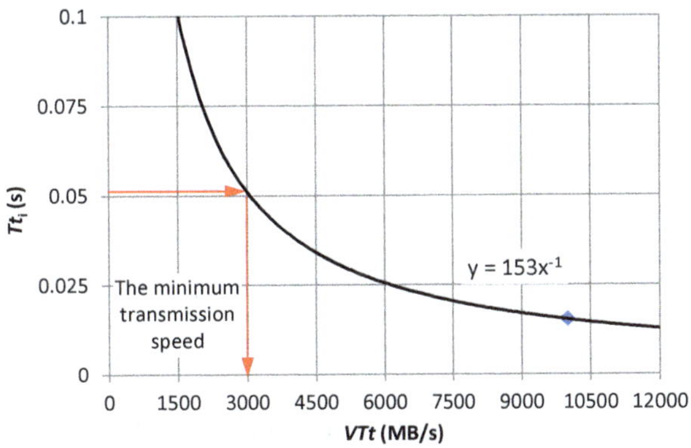

Fig. 9. The dependence between the data transmission time and file transmission speed

3 Conclusions

The following conclusions can be made:

- the use of a high-speed camera to supervise changes in the form of chips in the turning process requires registration, analysis and transmission of data of various sizes and formats. Therefore, the actual values of data transmission times and their analysis can be treated as restrictions on the application of selected measurement techniques in the Industry 4.0 concept.
- due to the total time of data transmission, an important point is the choice of architecture for the analysis of the results of measurement data, i.e. central architecture (*Cloud Computing*) or distracted (*Edge Computing*). This is particularly important in the case of large measurement data (files) such as the sequence of high-speed shots. Performing analysis of results directly in the machine tool area (*Edge Computing* architecture) and then sending only a ready report to the superior company management zones can significantly reduce the total time of data transfer. In addition, such a solution does not require large data volume encoding.

References

1. Zheng, P., Wang, H., Sang, Z., Zhong, R.Y., Liu, Y., Liu, C., Mubarok, K., Yu, S., Xu, X.: Smart manufacturing systems for Industry 4.0: conceptual framework, scenarios, and future perspectives. Front. Mech. Eng. **13**(2), 137–150 (2018)
2. Wang, Y., Ma, H.S., Yang, J.H., Wang, K.S.: Industry 4.0: a way from mass customization to mass personalization production. Adv. Manuf. **5**(4), 311–320 (2017)
3. Zhang, L., Luo, Y., Tao, F.: Cloud manufacturing: a new manufacturing paradigm. Enterp. Inf. Syst. **8**(2), 167–187 (2014)
4. Khan, W.Z., Ahmed, E., Hakak, S., Yaqoob, I., Ahmed, A.: Edge computing: a survey. Future Gener. Comput. Syst. **97**, 219–235 (2019)

5. Sittón-Candanedo, I., Alonso, R.S., Corchado, J.M., Rodríguez-González, S., Casado-Vara, R.: A review of edge computing reference architectures and a new global edge proposal. Future Gener. Comput. Syst. **99**, 278–294 (2019)
6. Shi, W., Cao, J., Zhang, Q., Li, Y., Xu, L.: Edge computing: vision and challenge. Internet Things J. **3**(5), 637–647 (2016)
7. Brecher, C., Ozdemir, D., Brockmann, M.: Introduction to integrative production technology. Prod. Eng.-Res. Dev. **11**(2), 93–95 (2017)
8. Cwikla, G.: Methods of manufacturing data acquisition for production management - a review. In: Modern Technologies in Industrial Engineering, vol. 837, pp. 618–623, March 2014
9. Struzikiewicz, G., Zębala, W., Rumian, K.: Application of Taguchi method to optimization of cutting force and temperature during turning of difficult to cut materials. In: Precision Machining VIII Book Series: Key Engineering Materials, Conference Proceedings, vol. 686, pp. 114–118, February 2016
10. Zębala, W., Kowalczyk, R., Matras, A.: Analysis and optimization of sintered carbides turning with PCD tools. Procedia Eng. **100**, 283–290 (2015)
11. Struzikiewicz, G., Zębala, W., Słodki, B.: Cutting parameters selection for sintered alloy AlSi10 Mg longitudinal turning. Measurement **138**, 39–53 (2019)
12. Słodki, B., Zębala, W., Struzikiewicz, G.: Turning titanium alloy, grade 5 ELI, with the implementation of high pressure coolant. Materials **12**, 768 (2019)
13. Zębala, W., Struzikiewicz, G., Franczyk, E.: Monitoring of machining in the context of Industry 4.0 – case study. In: ISPR 2018, Proceedings of the International Symposium for Production Research, pp. 660–673, August 2018
14. Boothroyd, G., Knight, W.A.: Fundamentals of Machining and Machine Tools. Marcel Dekker, New York (1989)
15. Van Luttervelt, C.A.: Neue entwicklungen in der spanuberwachung. VDI Berichte **988**, 431–477 (1993)
16. Davies, M.A., Chou, Y., Evans, C.J.: On chip morphology, tool wear and cutting mechanics in finish hard turning. CIRP Ann. Manuf. Technol. **45**(1), 77–82 (1996)

Assessment of Organizational Capability for Data Utilization – A Readiness Model in the Context of Industry 4.0

Mathias Nausch[1(✉)], Andreas Schumacher[1], and Wilfried Sihn[1,2]

[1] Fraunhofer Austria Research GmbH, Vienna, Austria
{mathias.nausch, andreas.schumacher}@fraunhofer.at
[2] University of Technology Vienna, Vienna, Austria
wilfried.sihn@tuwien.ac.at

Abstract. Increased utilization of digital elements and computer assistance in industrial enterprises leads to the acquisition of data encompassing all company levels and areas. Subsequently, manufacturing companies face large amounts of collected data from different sources and in various formats. Today this data is beginning to be viewed as a resource, comparable to material or machines. Thus, the ability to effectively collect and utilize internal and external data, is becoming an increasingly important factor in determining their economic success.

In this paper we develop a readiness model which can assess the Capability to Utilize Data in Industrial Enterprises (CUDIE). This model enables industrial decision makers to determine their organization's current status, to define strategies to increase data usage and to identify individual fields for improvement. Our model is based on various frameworks for readiness and maturity model development, as well as an Industry 4.0 maturity model that has already been applied in industrial settings. To assess readiness in real industry environments we transformed the theoretical CUDIE-model into a practically applicable tool that combines self-assessment with an external audit-based assessment.

Finally, in order to validate the model, and to collect feedback concerning its applicability and suitability, we tested our model on an industry use-case. There we could conclude that our method can be used to help companies to identify their current capability for data utilization, to derive areas of improvement and to define their future strategic steps.

Keywords: Industry 4.0 · Readiness model · Data utilization · Manufacturing industry · Assessment method

1 Introduction

In recent years, successive technological developments have fundamentally changed the industrial landscape. Increased utilization of digital elements and computer assistance in industrial enterprises, combined with higher complexity of production processes, has led to the acquisition of data on all company levels and in various company areas [1]. Today, this data is treated as a company resource or asset comparable to natural resources, machines or human resources. While industrial enterprises have

N. M. Durakbasa and M. G. Gençyılmaz (Eds.): ISPR 2019, LNME, pp. 243–252, 2020.
https://doi.org/10.1007/978-3-030-31343-2_21

already begun to see the necessity of implementing new data-utilization-technologies in order to stay competitive, they are facing the challenge of identifying the most suitable areas for implementation [2]. Moreover, the lack of organizational and technological readiness in companies to effectively collect and utilize internal and external data, is increasingly becoming a determining factor in their economic success. In particular, the biggest barriers towards successful data utilization (DU) are a lack of awareness of employees and decision makers as well as missing organizational structures and day-to-day processes to support DU [3].

As a result, when a company lacks the organizational capability to make use of their data they are often faced with the inability to define starting points and take action in any data related activities [4].

To face these challenges, we develop a readiness model, which can assess the Capability to Utilize Data in Industrial Enterprises (short: CUDIE). With this model, we aim to provide manufacturing enterprises insight into their organizational and structural readiness for DU by assessing the readiness levels of 23 dimensions in four organization layers.

The paper is structured as follows. In the next section, we describe the background and motivation of our research effort. Then we discuss our research methodology as well as the model-development. Finally, we present the model's application and validation in a use case and draw conclusions.

2 Readiness and Maturity Models in Data Utilization

In our application, readiness models are utilized for capturing and analysing an organization's development-starting point, before being able to progress towards higher levels of maturity [5]. Applying readiness models offers the advantage of avoiding conflicts of interest and views within the organisation by systematically following objective methods and tools. This allows for the identification of development gaps, which can then be closed by subsequent improvement actions [6]. Not being able to effectively close these gaps is often a direct result of missing knowledge regarding the current development status [7]. In regards to DU, according to Carolis et al. [8], we distinguish between "a readiness model for assessment of data utilisation" as a model assessing the capability of an organization to deploy data and "a maturity model of data utilisation" as a model representing how well an organization deploys data already.

In literature, various maturity and readiness models in the area of digitalization and Industry 4.0 already exist, and are taken into consideration in the development of our model. However, we could not identify a comprehensive model to assess the capability of DU, nor procedure methods of how to do so in theory. Regarding structure and design, existing publications suggest well tested designs for maturity and readiness models that we follow. One such example is the Business Intelligence Maturity Model (BIMM) from Näslund et al. [9], which consists of eight subcategories of BI with each of them assessed in five maturity stages. Comuzzi and Patel [10] developed a Big Data Maturity Model (BDMM), adopting a quality empirical approach, which is structured in 5 domains with 8 subdomains and has an assessment scale ranging from 0 to 5. In the area of knowledge management, Serna [11] presented a maturity model (MMKM) operationalized in 5 levels and 25 features, suggesting that knowledge management depends on human experience and social practices. Dinter [12] developed a Maturity Model (biMM) focusing on 3 relevant BI dimensions: functionality, technology and organization. This model consists of 11 categories and 22 objects and the assessment method consists of a questionnaire to support companies in assigning the degree of maturity ranging from 1 to 5 for each object. Davis [13] presents an Information Evolution Model (IEM) focusing on four key dimensions and providing a road map for evolving to higher maturity levels. Building on these models and the models, we reviewed during the development of our first Industry 4.0 maturity model in 2016 [5], we derive three specific development requirements:

- The model must operationalize the possibility of data usage in real industrial environments
- The model must offer an objective measurement-metric for readiness assessment
- The model must be transformed into a systematic measurement-method for practical application

Based on these requirements we define our research methodology.

3 Research Methodology and Model Development

For our model development, we followed a multidimensional approach consisting of expert interviews, literature research and survey research methods. Our literature review showed that the development methodology offered by Becker et al. [14] seems highly suitable for the development of the CUDIE-model. Therefore, our development will be based on Becker's step-by-step process for the development of maturity models which itself follows the design science requirements by Hevner et al. [15].

Following Becker's procedure, the development of CUDIE comprises of five steps (). In a first phase, we carried out a systematic literature review that resulted in 1000+ relevant works, from which we extracted 38 highly relevant publications around industrial DU. From these 38 publications, we isolated 15 in the field of industrial DU to develop our dimensions and attributes. The second phase focuses on defining the model's basic structure regarding dimensions, layers and assessment-items. Once the structure of the model is fixed, phase 3 focuses on filling the model content (model population) and phase 4 on the testing and validation. Following phase 4, we conduct a reassessment of the model structure and population based on first practical application feedback. Phase 5 consists of the development of suitable modes of representing and visualising the readiness levels (Fig. 1).

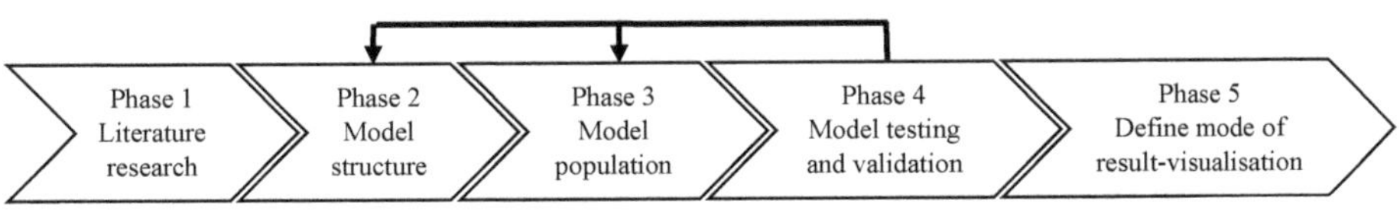

Fig. 1. Five phase readiness model development approach

To conduct our literature review, we developed a systematic approach based on other review guidelines [16–18] including search protocols, inclusion criteria and the definition of systematic search-term maps. Overall, we screened 1000+ publications and reviewed 200+ publications in depth. To define our model's structure, we used existing maturity and readiness models that were already successfully tested in practice [5]. To find suitable models, we screened publications for quality based on five criteria: relevance, added model content, systematic or standardized method of research, reliability and validity of data and the fact that the literature has been peer reviewed. We considered only literature fulfilling more than three of these five criteria. The outcome of this analysis resulted in a selection of 15 highly relevant publications that contributed to our model's design, as shown in Table 1.

From the most relevant literature [6, 7, 9, 10, 21, 29–32], we derived 23 dimensions of DU that reoccurred through the publications and we structured our model in four vertical layers:

- **Strategy:** The extent to which the strategic alignment and allocation of resources within the organisation reflects the intent of achieving a certain capability.
- **Governance:** A system of ensuring compliance to established rules and regulations.
- **Operations:** Based on a total quality management approach of analysing processes, we rate the degree to which they are defined, managed, measured, controlled, and effective.
- **Object:** The assessment of objects like products, reports or tools.

As shown in Table 2, we then assigned the 23 defined assessment dimensions to these four layers.

To operationalize the 23 DU-dimensions, we defined 72 attributes to describe and rate the individual degrees of readiness of each dimension. We described readiness levels for each item separately allowing for individual assessment e.g. of the current technological state-of-the-art and therefore increased measurement accuracy.

Table 3 shows an exemplary DU-dimension named "Information Quality (IQ) assurance" as well as the related attributes:

Each of the five levels of readiness describes different states of the three readiness attributes: IQ issue identification, IQ issue documentation and IQ issue prevention.

Following our five-phase development approach, the model's practical validation phase is conducted within a mid-sized manufacturing company using a developed validation form. The form's questions concern the completeness, the relevance, the comprehensiveness as well as applicability of the model. We defined the final content and structure of the CUDIE-model after iterating this process several times within different departments of the company.

During this iteration process, we also defined the model's final target group, which is made up of companies that are at least mid-sized, have a formalized management system and that work with data in mainly digital formats.

Table 1. Overview of derived dimensions from relevant publications

Data utilization dimensions	Relevant models														
	DWP-M [19]	EBIM [20]	PMS [21]	DQMM [22]	DM-CMM [23]	BPMM [24]	BIMM [9]	BDMM [10]	MMKM [11]	MD3M [25]	K3M [26]	KMM [27]	P-CCMM [28]	biMM [12]	IEM [13]
Access to information							x			x					x
Analytical abilities		x			x						x		x		x
Analytical tools & services					x		x	x						x	x
Data collection & integration			x	x										x	x
Data management		x		x			x	x	x	x				x	x
Data warehousing architecture	x	x	x	x			x	x						x	x
Data utilisation applications							x		x	x	x				
Data utilisation awareness (management)							x	x	x					x	x
Data utilisation awareness (staff)							x	x	x					x	x
Data utilisation policy		x	x				x	x	x	x	x		x	x	x
Data utilisation scope formalisation			x			x	x								
Data utilisation tool management	x							x							x
Data utilisation workflow formalisation			x		x		x		x			x		x	x
Data utilisation workflow integration			x		x		x		x					x	x
Information management							x	x	x	x					
Information Quality awareness		x		x			x			x					
Knowledge asset Sharing									x		x	x	x		x
Measurements			x		x	x									x

Table 2. CUDIE overview

Domain	Layer	Dimensions					
Utilisation of data in industrial enterprises	Strategy	Information Quality (IQ) awareness	Metadata awareness	DU awareness (Management)	DU policy	DU scope formalisation	DU workflow integration
	Governance	DU task Ownership formalisation	Data management	Information management	Access to information	DU tool management	DU workflow formalisation
	Operations	DU awareness (staff)	DU engagement	Analytical abilities	DU workflow synergy	Information Quality (IQ) assurance	Knowledge asset sharing
	Objects	Compatibility of the DU System-chain	Data collection & integration	Data warehousing architecture	Metrics Capability	Analytical tools & services	

Table 3. Attribute description based on the example of information quality assurance

Dimension	Level 1	Level 2	Level 3	Level 4	Level 5
Information Quality (IQ) assurance	• No sources for potential IQ issues have been **identified**	• Some potential sources for IQ issues have been **identified**	• All feasible potential sources for IQ issues have been **identified**	• All feasible potential sources for IQ issues have been **identified**	• All feasible potential sources for IQ issues have been **identified**
	• Occurred or potential IQ-issue-sources are not **documented** or evaluated and	• Occurred and potential IQ-issue-sources are **documented** and may be inconsistently evaluated	• Occurred and potential IQ-issue-sources are **documented** and are inconsistently evaluated	• Occurred and potential IQ-issue-sources are **documented** and are periodically evaluated	• Occurred and potential IQ-issue-sources are **documented** and periodically evaluated
	• IQ improvements are short-term solutions to overcome occurring issues. **Preventive** IQ improvements and long-term solutions are the exception	• IQ improvements are both **preventive** and short-term solutions to overcome occurring issues, long-term solutions are the exception	• IQ improvements are both **preventive** and long-term solutions, short term-solutions are the exception	• Most IQ improvements are **preventive** long-term solutions the occurrence of unidentified IQ issues is the exception and regular IQ improvement initiatives are supported and initiated throughout the entity	• Most IQ improvements are **preventive** and long-term solutions, the occurrence of unidentified IQ issues is the exception. Regular IQ improvement initiatives are supported and initiated throughout the entity, the potential cost of bad IQ has been calculated and sets the focus for IQ improvement initiatives

4 Model Application: Case-Study in Manufacturing Enterprises

In this section, the developed model and the assessment tool is applied use-case based for testing and validation in a manufacturing company in the field of steel-production, with around 200 employees. Thereby, we conducted the following steps:

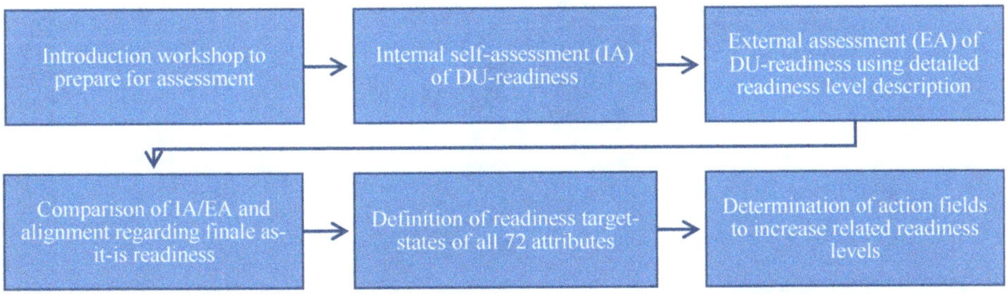

In order to create additional insights into the company's DU-readiness, we carried out an internal assessment (IA), where the company makes a self-assessment of their readiness, and external auditors make an external assessment (EA) of the same organization. We carried out the assessment in a workshop setup following our standardized assessment questionnaire for internal assessment. First, the company self-assessed readiness of all dimensions from 1 (lowest readiness) to 5 (highest readiness), based on their experience and without knowledge regarding detailed level description. In a second step, the company's IA-score of a dimension e.g. level 2 is compared to the respective level 2-description of the dimension (Table 3). If the description from the self-assessed readiness level also describes the company's actual readiness level from the external auditor's point of view, the internal assessment (IA) and the external assessment (EA) are fitting. In the case of differing internal and external assessments, a deeper analysis and discussion are required to align the company's self-reflection and external views. An example of an internal and external assessment is demonstrated for the layer "Operations" in Fig. 2.

The main information resulting from the evaluation, after the actual readiness levels, is the deviation of the IA from the EA, which indicates either a lack of knowledge or a lack of awareness in the self-assessment, or a lack of insight on the part of the external auditor. In both cases, discussion and deeper analysis of existing inconsistencies can create important insights for the company and decision makers about their current readiness level. Moreover, focusing on dimensions with low readiness and the definition of target-states for each attribute and dimension allows for the derivation of concrete and practical action fields and strategic development steps.

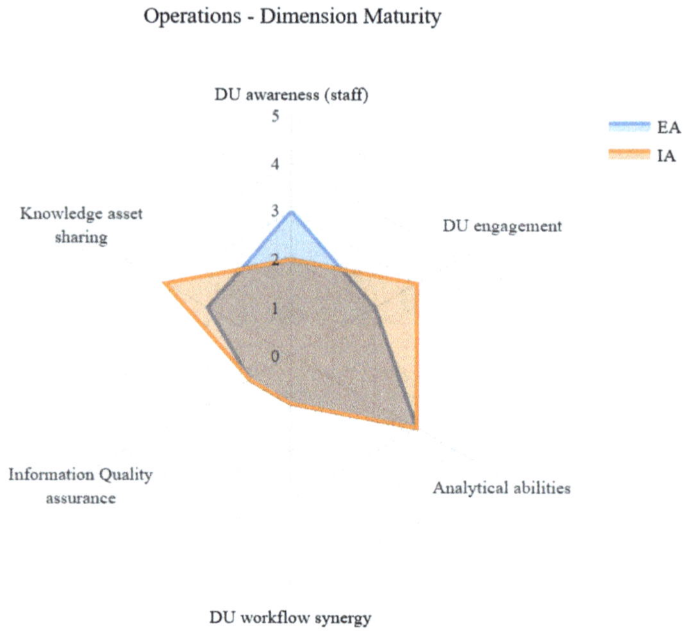

Fig. 2. Radar chart visualising the 6 dimensions of the operation layer

5 Conclusion

During this research effort, we aimed to develop a model to evaluate the readiness level of an organisation's capability to utilize data in the context of Industry 4.0. First practical applications of our model show that the assessment method, consisting of both internal and external assessments, serves as a valuable tool and offers guidance to assess readiness in DU. The CUDIE-model was explicitly developed for the application in a real-life scenario and, therefore, its practicality and transparency was a key focus throughout the entire development process. Feedback from the company where we tested the model confirmed that the assessment procedure is easy to understand and the two-step approach consisting of an internal self-assessment and afterwards detailed external assessment creates deeper awareness of how to reflect an organization's DU capability.

We find limitations of the CUDIE-model in its generality regarding the qualitative maturity level descriptions, which allow for assessment bias. Therefore, external auditors are essential to lead companies through the evaluation procedure including the determination of suitable action fields.

Our next research steps are the definition of a procedure model for systematic determination of target readiness levels as well as the model's implementation into a software tool.

References

1. Choudhary, A.K., Harding, J.A., Tiwari, M.K.: Data mining in manufacturing: a review based on the kind of knowledge. J. Intell. Manuf. **20**, 501–521 (2009)
2. Banerjee, A., Bandyopadhyay, T., Acharya, P.: Data analytics: hyped up aspirations or true potential? Vikalpa 1–12 (2013)
3. Pereira, A.C., Romero, F.: A review of the meanings and the implications of the Industry 4.0 concept. Procedia Manuf. **13**, 1206–1214 (2017)
4. Moeuf, A., Pellerin, R., Lamouri, S., Tamayo-Giraldo, S., Barbaray, R.: The industrial management of SMEs in the era of Industry 4.0. Int. J. Prod. Res. **56**, 1118–1136 (2018)
5. Schumacher, A., Erol, S., Sihn, W.: A maturity model for assessing Industry 4.0 readiness and maturity of manufacturing enterprises. Procedia CIRP **52**, 161–166 (2016)
6. de Bruin, T., Freeze, R., Kulkarni, U., Rosemann, M.: Understanding the main phases of developing a maturity assessment model. In: Australasian Conference on Information Systems (2005)
7. Mettler, T.: A design science research perspective on maturity models in information systems (2009). http://www.alexandria.unisg.ch/Publikationen/67707
8. de Carolis, A., Macchi, M., Kulvatunyou, B., Brundage, M.P., Terzi, S.: Maturity models and tools for enabling smart manufacturing systems: comparison and reflections for future developments. In: Ríos, J., Bernard, A., Bouras, A., Foufou, S. (eds.) Product Lifecycle Management and the Industry of the Future, pp. 23–35. Springer, Cham (2017)
9. Näslund, D., Sikander, E., Öberg, S.: Business Intelligence - a Maturity Model Covering Common Challenges (2014)
10. Comuzzi, M., Patel, A.: How organisations leverage Big Data: a maturity model. Ind. Manag. Data Syst. **116**, 1468–1492 (2016)
11. Serna, E.: Maturity model of knowledge management in the interpretativist perspective. Int. J. Inf. Manage. **32**, 365–371 (2012)
12. Dinter, B.: The maturing of a business intelligence maturity model. In: Proceedings of AMCIS 2012, pp. 1–10 (2012)
13. Davis, J.: Information Revolution: Using the Information Evolution Model to Grow Your Business. Wiley, Hoboken (2006)
14. Becker, J., Knackstedt, R., Pöppelbuß, J.: Developing maturity models for IT management. Bus. Inf. Syst. Eng. **1**, 213–222 (2009)
15. Hevner, A.R., March, S.T., Park, J., Ram, S.: Design science in information systems research. MIS Q. **28**, 75–105 (2004)
16. Okoli, C., Schabram, K.: A guide to conducting a systematic literature review of information systems research. Sprouts: Working Papers on Information Systems, vol. 10/26 (2010)
17. Fink, A.: Conducting Research Literature Reviews: From the Internet to Paper. Sage, Los Angeles (2014)
18. Levy, Y., Ellis, T.J.: A systems approach to conduct an effective literature review in support of information systems research. Inf. Sci. J. **9**, 181–212 (2006)
19. Sen, A., Ramamurthy, K., Sinha, A.P.: A model of data warehousing process maturity. IEEE Trans. Softw. Eng. **38**, 336–353 (2012)
20. Tan, C.-S., Sim, Y.-W., Yeoh, W.: A maturity model of enterprise business intelligence. In: CIBIMA, pp. 1–9 (2011)
21. Wettstein, T., Kueng, P.: A maturity model for performance measurement systems. Manag. Inf. Syst. **4**, 113–122 (2002)
22. Ryu, K.S., Park, J.S., Park, J.H.: A data quality management maturity model. ETRI J. **28**, 191–204 (2006)

23. Maya, K., Reuven, K.: A capability maturity model for knowledge-based decisionmaking. Inf. Knowl. Syst. Manag. 225–252 (2004)
24. Lee, J., Lee, D., Kang, S.: An Overview of the Business Process Maturity Model (BPMM). LNCS, pp. 384–395 (2007)
25. Spruit, M., Pietzka, K.: MD3M: the master data management maturity model. Comput. Hum. Behav. **51**, 1068–1076 (2015)
26. Liebowitz, J., Beckman, T.: Moving toward a knowledge management maturity model (K3M) for developing knowledge management strategy and implementation plans. In: Knowledge Management: An Evolutionary View, pp. 163–179 (2008)
27. Kulkarni, U., St. Louis, R.: Organizational self assessment of knowledge management maturity. In: AMCIS Proceedings, pp. 2542–2551 (2003)
28. Curtis, B., Hefley, B., Miller, S.: People Capability Maturity Model (P-CMM) Version 2.0, 2nd edn. Software Engineering Institute, Technical report (2009)
29. Maier, A.M., Moultrie, J., Clarkson, P.J.: Assessing organizational capabilities: reviewing and guiding the development of maturity grids. IEEE Trans. Eng. Manage. **59**, 138–159 (2012)
30. Andersen, E.S., Jessen, S.A.: Project maturity in organisations. Int. J. Project Manage. **21**, 457–461 (2003)
31. Hammer, M.: The process audit. Harvard Bus. Rev. **85**, 111-9, 122-3, 142 (2007)
32. Rud, O.P.: Business Intelligence Success Factors: Tools for Aligning Your Business in the Global Economy, vol. 18. Wiley, New York (2009)

Case Study of Digitization of the Production Cell

Michal Holub(✉), Zdenek Tuma, Jiri Kroupa, Jiri Kovar, and Petr Blecha

Department of Production Machines, Systems and Robotics, Faculty of Mechanical Engineering, Brno University of Technology, Brno, Czech Republic
Michal.Holub@vutbr.cz, {tuma,kroupa,kovar, blecha}@fme.vutbr.cz

Abstract. This paper focuses on the introduction of digitization in the production process. When deploying Industry-Oriented 4.0 components in the plant, emphasis is placed on HMI and workplace visualization. Designing a suitable way to visualize data obtained from the production process can have a significant impact on the workplace response. Timely and properly conducted responses to potential changes in the production process have a positive impact on the resulting quality of the production process. Continuous development of the elements of virtual and augmented reality also increases their usability in the field of data visualization from the production process. These technologies make it possible to meet the high demands on the clarity of a great deal of information.

This paper introduces the creation of a production cell into virtual and augmented reality. Particular emphasis is placed on the way of data visualization, including the environment, geometric accuracy of the CNC machine tool, and information from the safety parts. For efficient information handling, access to their display in virtual and augmented reality has been chosen.

The first part of the publication introduces methods for creating a virtual model based on photogrammetry. In the second part of the publication, procedures are presented for collecting and visualizing information on the geometric accuracy of the machine. Finally, procedures related to the risk analysis and functional safety of CNC machine tools are presented. In conclusion, the advantages, disadvantages, and recommendations of the presented solution, the critical places and the difficulty with the realization of the virtual workplace are referred to.

Keywords: Production cell · Digitalization · Industry 4.0 · Virtual reality

1 Introduction

The ever-increasing demands for increased productivity while preserving or further improving production quality lead to the necessary implementation of new technologies in the production process [1]. Other requirements that are constantly increasing include workplace safety and the associated implementation of new safety components for human-machine collaboration. Production productivity is primarily influenced by a

N. M. Durakbasa and M. G. Gençyılmaz (Eds.): ISPR 2019, LNME, pp. 253–262, 2020.
https://doi.org/10.1007/978-3-030-31343-2_22

degree of automation, and can be further enhanced by reducing the time of in-process control, unplanned shutdowns caused by machine failures, or an appropriately set philosophy of human-machine collaboration. By continuous ensuring the competence of the machine and the production process, the machine productivity is increased by limiting the production of nonconforming parts.

These basic aspects apply both to newly emerging production plants and to the modernization of workplaces already in operation. The basic pillar to increase the productivity is a suitably set degree of automation and monitoring of machine condition.

This publication presents proposals for monitoring the selected machine parameters, subsequent visualization of the information based on virtual and augmented reality and safety of the production cell. Continuous monitoring of machines and the production process is, above all, enabled by the dynamic development of IT industrial technologies for data acquisition and storage. A large number of current plants can never be modernized using a new machinery equipped with technologies in the context of Industry 4.0. Even in such plants, there are processes that have a high added value, and with increasing requirements for productivity or scrap rate, continuous operational monitoring is one of the solutions of how to meet growing customer demands. This solution is possible thanks to the universal HW for the collection of general signals, which have recently been on the rise.

One of the requirements of contemporary production is to have a sufficient amount of quality information on the machine operation. These are intended to ensure timely awareness of, for example, the impending change in the condition of the machine, which may result in the production of a non-conforming component or shutdown of the machine. This is a large amount of data that needs to be understood by the operator of the machine. One of the ways is to visualize data in a virtual environment or augmented reality. This publication presents the creation of a virtual workplace model for various types of objects, including the way of data visualization. The most common ways of creating virtual reality models arise from the available documentation based on 3D models or 2D drawing documentation [2–4].

To ensure a long-term competence of the production process, one of the possible monitoring parameters is the geometric and volumetric accuracy of CNC machine tool. There are a number of methodological approaches to increase the geometric accuracy. Very often, instruments working on the principle of circular interpolation are deployed for quick diagnostics [5–8]. The method based on the FEM model of the workpiece is presented as suitable for series production because of the laboriousness of the process setting [9], methods based on gravity [10], on the machine vision system [11], touch probe methods [12, 13], R-test-based methods [14–16].

Following the productivity of the production cell, we can divide the control of parameters describing the geometric accuracy of production machines into fully automatic, semi-automatic, and manual. The fully automatic control includes Touch Probe, R-Tests, use of Vision System and Multiline Technology (company ETALON-AG). Among the semi-automatic control systems, a Ballbar system can be sorted, e.g. with palletizing machines. For manual control, we can include traditional methods of machine tool control, such as single-axis, multi-axis and self-guiding laser interferometers (laser interferometer XL-80, XM60, LaserTRACER or Laser Tracker).

The information obtained on the basis of long-term regular measurements is further used to create a knowledge base describing the machine behaviour in the production process. This will make it possible to obtain the information needed to implement those measures that eliminate behavioural fluctuations. It is also possible to use the knowledge base to predict the dimensional and shape accuracy of the future workpiece [17]. Machine tools with automatic control must have a wide range of sensor systems to monitor the machine condition [18] and at the same time the requirements are posted as for machine - environment - process interactions [1].

The submitted publication presents a list of production cell requirements from both the legislative and user point of view for higher production accuracy and the ability to visualize the information from the working process. Creating a virtual workplace is based on the principle of photogrammetry. Furthermore, on the created model, data on geometric accuracy of the machine, which affects the production accuracy, are presented. The processing of the measured data is based on a data-driven approach where the input-output relationship is specified using machine learning methods throughout the entire use of the model.

2 Case Study of Digitization of the Production Cell

The case study is focused on a modernized production cell in the context of Industry 4.0 with a view to increasing production accuracy and data visualization quality. The following part of the publication is divided into the description of the production cell, the creation of the workplace virtual model, the structure of the production cell digitization, monitoring of geometric accuracy of machine tools and workplace safety.

2.1 Demonstrator - Production Cell

The case study is described on a modernized production cell, which is constantly being extended according to the Industry 4.0 concept. This testbed is designed to run various testing regimes in the areas of control, metrology, production planning, safety, data management, predictive maintenance, self-calibration of machines. These areas are oriented on the research of zero-defect manufacturing, safety, precision manufacturing, digital twinn and robotics. The original production cell consists of two CNC machine tools (three-axis vertical CNC milling centre, turning centre) of the industrial robot on the track motion.

Based on the requirements, the production cell was extended according to today's standards with safety elements and the elements supporting a full workplace automation, upgrade of control systems for mutual communication via a standardized OPC UA interface. The following figure shows the condition of the production cell after being extended with Industry 4.0 elements (Safety systems from ABB, Siemens Mindsphere, Siemens Sinumerik Inegrate, external device for temperature monitoring on the machine and its environment) (Fig. 1).

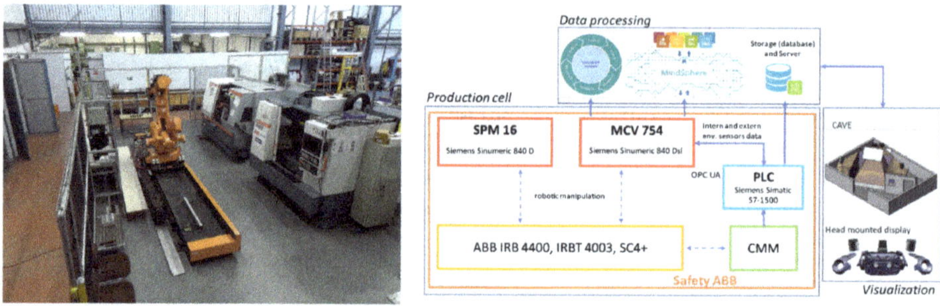

Fig. 1. Production cell, Production cell and communication layout

2.2 Creating a Virtual Model

Virtual reality visualization is supposed to provide maximum workplace credibility, including production process information. The resulting model is intended to immerse the production cell operator as much as possible in the production process without the necessity of the operator to be close to the workplace. In such set-up processes, a highly qualified person can operate multiple production sites from one location. This results in the requirements for creation of virtual reality models for production sites.

In our case, we create an object that is static and placed in the production hall under relatively constant conditions. The model will be used as a basis for creating a virtual twin and visualizing the real operation. We require a realistic model, i.e. including surface textures. To create a digital model, a machine shutdown is also needed and no intervention in the machine structure is possible.

The most accessible but very laborious method deals with shape reconstruction, measurement of dimensions and positioning of objects that are displayed in photos. The use of this technique is constantly on the rise, as the results of the acquired models reach high-end devices. However, unfortunately, it puts quite high demands on the acquired photos of the respective object. Also, the processing of acquired photos requires know-how.

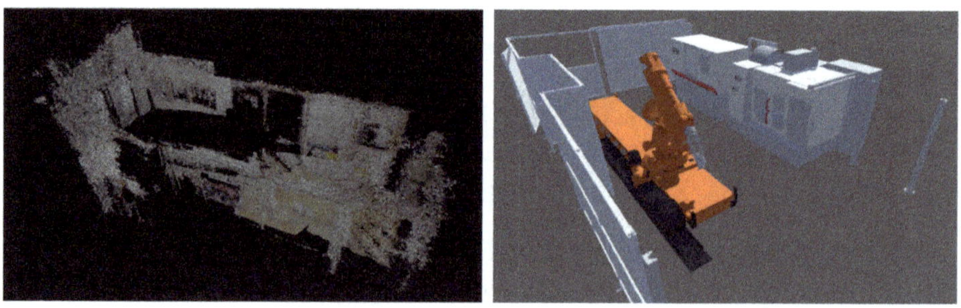

Fig. 2. Production cell model w/o correction, final production cell model

The process of collecting and evaluating the photos in testing of the method of photogrammetry. For this, a Canon 7D camera with fixed lenses with focal lengths of 50 mm and 20 mm was used. Unfortunately, even with a sufficient number of photos, it was not possible to create an overall model, as the object contains transparent and reflective surfaces that introduced a number of errors to the calculation in Autodesk Recap.

As shown in Fig. 2, it was necessary to correct the acquired model. To work with the model, MeshMixer software was used to seal holes and remove excessive objects. Then, the overall model of machine was imported into Autodesk Inventor, completed and supplemented with the remaining technologies. Subsequently, the model was imported into Unity 3D.

Reducing the time required to create a workplace model is possible in combination with, for example, optical scanning technology.

2.3 Structure of Digitalization

Digitizing a production cell is a term that, by its meaning, expresses, in general terms, the method of creating and operating a digital representation of a machine in machine-processable form. This representation, which, in the simplest form, is referred to as a model, must represent the essential elements (entities) and links between them. The definition of the problem situation determines which entities and links will be represented in a digital form. In this type of task, which is described above, it is typically the influence of ambient temperature and the temperature of individual parts of the machine on its production accuracy. Thus, the entities whose parameters are to be monitored are defined. The links between them are, in this case, represented by numerical dependence; therefore, they are given by their structure in the real part of the machine. A typical link is, for example, the measurable rise of temperature of the air under the roof of the hall, which has a delayed effect on the temperature of the machine part, and this, in turn, on the machine production accuracy.

The existing structure, entities, and links can then be digitally represented in several ways. An effective approach is a relatively widespread use of the approach called Physics based modelling. It is a convenient use of the possibilities offered by different simulation software solutions to create the most physically accurate digital representation of a machine (its important parts). The model is considered to be the more accurate, the more accurate and consistent are the calculations of the outputs compared to the measured ones based on the equal inputs and conditions of the machine for the equal time instant. However, even this model has its limitations. The most important ones are the limited possibilities of the used software solution to represent reality (e.g. transition states) and also the computational complexity of the model used, or the stability of the calculation.

An additional approach to create a suitable digital representation is the use of so-called data-driven methods; their aim is not primarily to use the appropriate mathematical apparatus to model e.g. hidden machine states, such as in the previous approach, but mainly to create the most accurate representation of the input-output relationship.

In this case, a combination of both approaches was used to benefit from their main advantages. The “stability” of digital representation of a real machine is obtained using a suitable model in Matlab/Simulink, “adaptability”, in turn, using a data-driven approach, where the input-output relationship is refined using machine learning methods over the entire use of the model. The software structure is described in Fig. 3. Measured data from the machine and its ambient environment are stored in a database solution after standard processing. These data are then used as model inputs in Matlab/Simulink and for machine learning algorithms. Other data from the state similar in terms of time and temperature are used to validate a learned algorithm. The result is a fairly accurate algorithm that is able to predict the temperatures of individual parts over the time frame from15 min to one hour with an accuracy of approximately one degree Celsius. In the long run, data can be used for analyses based on local changes in the system, e.g. by searching for extremes [19].

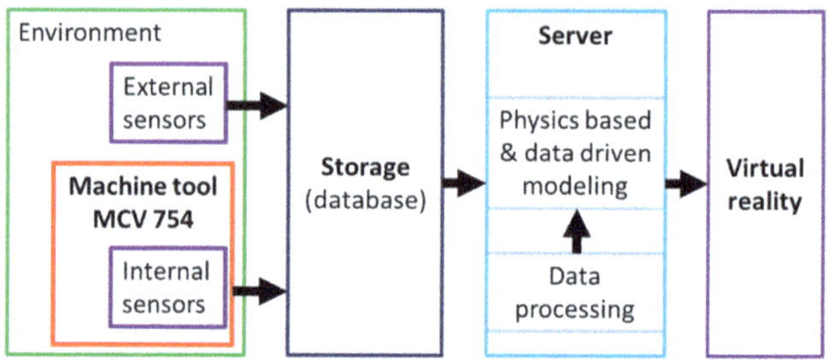

Fig. 3. Structure of digitalization

2.4 Geometric Accuracy of Machine Tool

Production accuracy monitoring is developed as part of the operation of the MCV three-axis milling centre. There are monitored and evaluated, for example, the dependence of machine geometric errors, volumetric errors and their mutual combination on the resulting dimensional and shape accuracy of the workpiece. Instruments such as Laser interferometer, Ballbar, LaserTRACER, Spindle error analyser and digital inclinometer are tested for monitoring and implementing of compensations. In terms of time demands and purchasing and operating costs, Ballbar is the most suitable for regular monitoring [5]. Measurements can be pre-defined in the form of a CNC program and fixtures. The measurement cycle itself can be performed on a single set-up with time demands depending on the set feed rate.

The results are evaluated on a PC, which can now be replaced by a tablet being a part of the machine. What is essential for monitoring of machine tool condition is to always have a measurement of geometric accuracy synchronized with the ambient condition measurement. These are mainly ambient temperatures and the temperatures at selected points of machine nodes.

The basic prerequisite for detecting as many error conditions caused by ambient fluctuations is a clear orientation in the data. The correct interpretation will then be determined by the quality of measured and processed data and the operator's experience. An appropriately set interpretation can accelerate the growth of operator 's experience. This is also the goal of deploying the augmented reality in the area of production data monitoring. Figure 4 shows a virtual environment view of a production cell workplace with a detailed view of a three-axis milling machine. Basic information related to the geometric accuracy of the machine is pre-set for the selected machine. Figure 4 on the right shows a view of the selected temperature sensor with relevant information.

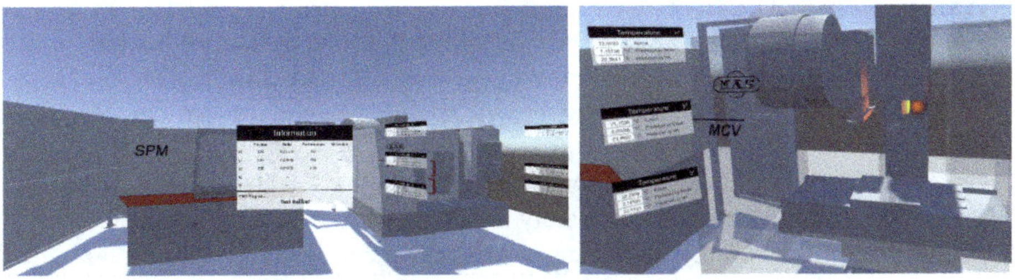

Fig. 4. Detail of workplace in the regime of milling machine monitoring, detailed view of axial temperature source

Figure 5 on the left shows a panel with selected information, which is part of a detail view shown on the machine. In Fig. 5 on the right, the temperature curve is then displayed in the selected time interval for the selected sensor on the machine. These values are linked to the assessment of the error obtained from the Ballbar measurement, i.e. from the circular interpolation test. It is then possible to process a detailed analysis of the results from a remote workplace and to effectively propose the measures directly with the machine operator or maintenance staff.

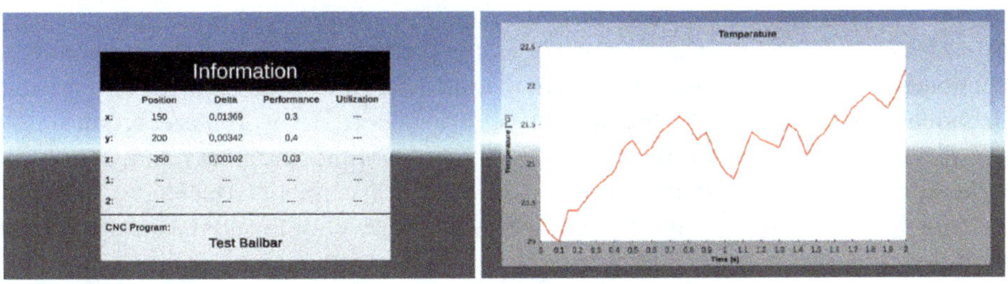

Fig. 5. Detail on information panel, detailed temperature curve in the selected time interval

2.5 Risk and Safety Analysis, Functional Safety Assessment

The safety of automated production cells enabling human-machine collaboration needs to be seen both from the point of view of fulfilling the requirements for new machinery by their manufacturer and from the point of view of ensuring health and safety at work and the possibility of maintaining the level of safety by the user. The basic requirements for machine safety and health and safety at work are given primarily in the following European Union legislative documents:

- Council Directive of 30 November 1989 concerning the minimum requirements for safety and health at work (first individual Directive within the meaning of Article 16 (1) of Directive 89/391 EEC) (89/654/EEC)
- Directive 2009/104/EC of the European Parliament and of the Council of 16 September 2009 on minimum health and safety requirements for the use of work equipment by operators at work (second individual Directive within the meaning of Article 16 (1) of Directive 89/391/EEC)
- Council Directive of 29 May 1990 on the minimum safety and health requirements for the manual handling of loads involving risks, in particular, spine damage, to workers (fourth individual Directive within the meaning of Article 16 (1) of Directive 89/391/EEC)
- Directive 2013/35/EU of the European Parliament and of the Council of 26 June 2013 on the minimum health and safety requirements for the exposure of workers to the risks arising from physical agents (electromagnetic fields) (twentieth individual directive within the meaning of Article 16 (1) of the Directive 89/391/EEC)
- Council Directive 92/58/EEC of 24 June 1992 on the minimum requirements for safety or health signs at the workplace (ninth individual Directive within the meaning of Article 16 (1) of Directive 89/391/EEC)
- Council Directive of 29 May 1990 on the minimum safety and health requirements for work with display screen equipment (fifth individual Directive within the meaning of Article 16 (1) of Directive 89/391/EEC)
- Decision No 768/2008/EC of the European Parliament and of the Council of 9 July 2008 on a common framework for the marketing of products.

It is the manufacturer's responsibility to specify EU legislation that applies to the respective product and to meet the EU requirements before the product is placed on the market or put into service. If necessary, the above- mentioned list shall be supplemented by other documents, e.g. machine directive 2006/42/EC, electrical directive 2014/35/EU etc. depending on the nature of the production cell. Based on the above-mentioned list of regulations, it is evident that the analysis and assessment of risks and functional safety relating to the design of a collaborative workplace requires the identification of risks related both to the construction of the machinery and, above all, to the operator's work with emphasis on the human-machine interface. The safety of such a workplace will then significantly influence the level of functional safety of both hardware and software. For hardware, the standard procedures of design and verification according to EN ISO 13849 can be used. In software verification, a digitized production cell model, allowing virtual simulation and visualization of all workflows carried out by a collaborative robot and operator, will find its significant application. Increased attention must be paid to the end-effector of the robot and the objects to be manipulated, since the consequences of the collision of the operator with these objects cannot be effectively treated by the construction of the robot itself. It is on the basis of this simulation that the trajectory of dangerous movements and their speed can be adjusted by intervention in the robot program so that the risk of a possible collision is reduced to an acceptable level.

3 Conclusions

Requirements for continuous increase in production efficiency lead to the implementation of new technologies and procedures. The Industry 4.0 concept targets the data collection and their processing; the subsequent analyses should improve the quality and productivity in industrial applications. One of the factors of correct data interpretation is their clarity and quick response. Virtual and augmented reality have been identified as possible tools for a credible interpretation of production operations. The photogrammetry technology was chosen for the creation of the virtual model; this technology was evaluated as sufficient in terms of the credibility of the workplace model, but as very time-consuming. It is not possible to generalize the identification of the most suitable method to create a virtual model. The method will always be chosen according to the available data and their quality.

The influence of machine geometric accuracy on production accuracy is unambiguous. For this reason, geometric errors of the machine are monitored and evaluated for a long time. Based on this, it is possible to predict the future condition of the product. However, it will be different to assess the shape, dimensions, or roughness of the workpiece surface. Therefore, it is desirable to be able to display to the workplace operators only those data that affect the predicted parameter.

These data are processed in the structure presented under the workplace digitization. The data-driven method is applied here. Last but not least, it is essential that the production cell shall meet all legislative requirements that are also related to workplace safety.

The presented solution offers an option of very effective communication in solving undesirable conditions in production with highly qualified staff. These employees can simultaneously operate multiple workplaces from a remote workplace, always with the online data necessary to evaluate the respective situation. It is precisely the credibility of the operation and the availability of the information that form the basis for the proposal of an appropriate measure to eliminate undesirable conditions during production process.

Acknowledgment. These results were obtained with financial support of the Faculty of Mechanical Engineering, Brno University of Technology (grant no. FSI-S-17-4477).

References

1. Blecha, P., Durakbasa, N., Holub, M.: Digitized production – its potentials and hazards. In: Proceedings of the International Symposium Production Research 2018, pp. 402–411. Springer, Cham (2019). https://doi.org/10.1007/978-3-319-92267-6_35
2. Ruzarovsky, R., Holubek, R., Sobrino, D.R.D., Janicek, M.: The simulation of conveyor control system using the virtual commissioning and virtual reality. Adv. Sci. Technol. J. **12**, 164–171 (2018). https://doi.org/10.12913/22998624/100349
3. Kroupa, J., Tuma, Z., Kovar, J., Singule, V.: Virtual laboratory for study of construction of machine tools. MM Sci. J. **2018**, 2503–2506 (2018). https://doi.org/10.17973/MMSJ.2018_11_2017100

4. Kovar, J., Mouralova, K., Ksica, F., Kroupa, J., Andrs, O., Hadas, Z.: Virtual reality in context of Industry 4.0 proposed projects at Brno University of Technology. In: Proceedings of 2016 17th International Conference Mechatronics - Mechatronika, ME 2016 (2017)
5. Holub, M., Bradac, F., Pokorny, Z., Jelinek, A.: Application of a ballbar fordiagnostics of cnc machine tools. MM Sci. J. **12**, 2601–2605 (2018). https://doi.org/10.17973/MMSJ.2018_12_2018032
6. Archenti, A.: Prediction of machined part accuracy from machining system capability. CIRP Ann. **63**, 505–508 (2014). https://doi.org/10.1016/J.CIRP.2014.03.040
7. Szipka, K., Laspas, T., Archenti, A.: Measurement and analysis of machine tool errors under quasi-static and loaded conditions. Precis. Eng. **51**, 59–67 (2018). https://doi.org/10.1016/J.PRECISIONENG.2017.07.011
8. Vichare, P., Nassehi, A., Flynn, J.M., Newman, S.T.: Through life machine tool capability modelling. Procedia Manuf. **16**, 171–178 (2018). https://doi.org/10.1016/J.PROMFG.2018.10.163
9. Wąsik, M., Kolka, A.: Machining accuracy improvement by compensation of machine and workpiece deformation. Procedia Manuf. **11**, 2187–2194 (2017). https://doi.org/10.1016/J.PROMFG.2017.07.365
10. Belforte, G., Bona, B., Canuto, E., Donati, F., Ferraris, F., Gorini, I., Morei, S., Peisino, M., Sartori, S., Levi, R.: Coordinate measuring machines and machine tools self calibration and error correction. CIRP Ann. **36**, 359–364 (1987). https://doi.org/10.1016/S0007-8506(07)62622-5
11. Mendikute, A., Leizea, I., Yagüe-Fabra, J.A., Zatarain, M.: Self-calibration technique for on-machine spindle-mounted vision systems. Measurement **113**, 71–81 (2018). https://doi.org/10.1016/J.MEASUREMENT.2017.08.029
12. Givi, M., Mayer, J.R.R.: Validation of volumetric error compensation for a five-axis machine using surface mismatch producing tests and on-machine touch probing. Int. J. Mach. Tools Manuf **87**, 89–95 (2014). https://doi.org/10.1016/J.IJMACHTOOLS.2014.08.001
13. Mutilba, U., Gomez-Acedo, E., Sandá, A., Vega, I., Yagüe-Fabra, J.A.: Uncertainty assessment for on-machine tool measurement: an alternative approach to the ISO 15530-3 technical specification. Precis. Eng. (2019). https://doi.org/10.1016/J.PRECISIONENG.2019.03.005
14. Florussen, G.H.J., Spaan, H.A.M., Spaan-Burke, T.M.: Verifying the accuracy of five-axis machine tool focused on kinematic ISO tests using a torus-shaped test work piece. Procedia Manuf. **14**, 58–65 (2017). https://doi.org/10.1016/J.PROMFG.2017.11.007
15. Florussen, G.H.J., Spaan, H.A.M.: Dynamic R-test for rotary tables on 5-axes machine tools. Procedia CIRP **1**, 536–539 (2012). https://doi.org/10.1016/J.PROCIR.2012.04.095
16. Brecher, C., Behrens, J., Klatte, M., Lee, T.H., Tzanetos, F.: Measurement and analysis of thermo-elastic deviation of five-axis machine tool using dynamic R-test. Procedia CIRP **77**, 521–524 (2018). https://doi.org/10.1016/J.PROCIR.2018.08.244
17. Theissen, N., Laspas, T., Szipka, K., Archenti, A.: Virtual machining system simulator: analysis of machine tool accuracy. Procedia Manuf. **25**, 338–343 (2018). https://doi.org/10.1016/J.PROMFG.2018.06.101
18. Fujishima, M., Ohno, K., Nishikawa, S., Nishimura, K., Sakamoto, M., Kawai, K.: Study of sensing technologies for machine tools. CIRP J. Manuf. Sci. Technol. **14**, 71–75 (2016). https://doi.org/10.1016/j.cirpj.2016.05.005
19. Augste, J., Holub, M., Knoflíček, R., Novotny, T., Vyroubal, J.: Monitoring of energy flows in the production machines. In: Mechatronics 2013, pp. 1–7. Springer, Berlin (2014). https://doi.org/10.1007/978-3-319-02294-9-1. Recent Technol. Sci. Adv

Comparative Analysis of the Most Industrialized Cities in Turkey from the Perspective of Industry 4.0

Zeynep Gergin[1(✉)], Doğan Aybars İlhan[1], Fadime Üney-Yüksektepe[1], Mehmet Güneş Gençyılmaz[2], Uğurcan Dündar[1], and Ali İhsan Çavdarlı[3]

[1] Department of Industrial Engineering, İstanbul Kültür University, İstanbul, Turkey
{z.gergin, a.ilhan, f.yuksektepe, u.dundar}@iku.edu.tr
[2] Department of Industrial Engineering, İstanbul Aydın University, İstanbul, Turkey
gunesgencyilmaz@aydin.edu.tr
[3] KOSGEB Small and Medium Sized Enterprises Development Organization, Ankara, Turkey
ali.cavdarli@kosgeb.gov.tr

Abstract. It can be declared that the adaptation of companies to the new industrial revolution, so called Industry 4.0, is not a choice but an issue of survival. The structure of business processes are changing with the newly introduced, radical and highly value adding technological drivers, such as internet of things, cloud computing, autonomous robots, augmented reality, cyber security system, etc. In the globalized trade markets none of the companies can close their eyes to this new era and ignore the required change in their operations. Small and Medium Sized Enterprises (SMEs) have 99.8% enterprise share of Turkey, and consequently have obvious contribution not only to the local economy but also to the global competition strength of the country. In that sense, catching up the fourth industrial revolution has to be the prior agenda of the SMEs, and also of the government institutions and the universities from the perspective of giving the required support. This study is the continuum of the broad research that is initiated in March 2018. The purpose is to identify the current situation of SMEs in Turkey with respect to their various implementations for Industry 4.0 transformation. A specifically designed questionnaire is applied to SME's with the support of KOSGEB (Small and Medium Sized Enterprises Development Organization), and in this stage of the research, the companies located in four major cities, namely İstanbul, Ankara, İzmir and Bursa, are analyzed and compared. The study aims to find out the level of awareness and readiness of SMEs in these cities and measure their Industry 4.0 scores. The final target is to support the Industry 4.0 implementations by defining the current situation and proposing suitable actions for adaptation based on varying results.

Keywords: Industry 4.0 · Turkish economy · Turkish Industry · Industry 4.0 scorecard · Industrial revolution

N. M. Durakbasa and M. G. Gençyılmaz (Eds.): ISPR 2019, LNME, pp. 263–277, 2020.
https://doi.org/10.1007/978-3-030-31343-2_23

1 Introduction

In recent years, Industry 4.0 caught researchers' attention and consequently studies on the topic has been increasing rapidly. A review displays two main groups of studies can be seen in the literature; First group focuses mainly on the definitions and aspects of Industry 4.0, whereas the second group focuses on the evaluation of enterprises according to Industry 4.0 requirements and on enablers, barriers, and other factors that have effect on adopting Industry 4.0 practices. As the aspects of Industry 4.0 adaptations vary in a wide range, these second group of studies commonly make evaluations considering not all the technology drivers of Industry 4.0, but generally some selected part of these. Literature shows that, most of the studies are focusing on defining Industry 4.0 and its constituent parts. Some also mention the core ideas of Industry 4.0 and requirements for a success [1]. The Platform of Industry 4.0 defines the concept as "the intelligent networking of machines and processes for industry with the help of information and communication technologies" [2]. In another resource, the definition introduces the revolution and declared that "Industry 4.0 defines The Fourth Industrial Revolution as being a new level in the organization and management of value chain in the products and production systems lifecycles covering all the chain from development and production order to distribution of a product to the end user and recycling starting from idea stage." [3]. Readers are referred to the related studies for further explanations of Industry 4.0 [4–8].

PWC published a report in 2016 that investigates the readiness of companies in more than 26 countries, 9 major industrial sectors and with more than 2000 respondents [5]. They declared that Industry 4.0 is not a future trend anymore, it is currently in companies' agendas. Furthermore, findings show that core businesses are attuning themselves to the trend much faster than others. Also, attendees of the survey believe that digitization will be doubled until 2020. Based on a survey applied for Turkish companies by Turkish Industry and Business Association (TÜSİAD), the output report highlights that the Turkish Industry is aware of what Industry 4.0 will bring with pros and cons, and is willing to attune itself to Industry 4.0 [6]. The question is whether the companies have the required knowledge and infrastructure for this, and do all of them have to follow the same roadmap as they may be practicing different readiness.

Literature also reveals some studies on assessing the readiness and factors affecting the adaptations. In one of the studies, readiness for Industry 4.0 in manufacturing for EU countries is assessed with respect to two main categories, namely Industry 4.0 infrastructure and big data maturity [9]. In another study, authors evaluate the factors affecting the implementation of Industry 4.0 on SMEs in Peru [10]. Another one specifies the drivers and barriers for adopting Industry 4.0 based on the Small and Medium Sized Enterprises (SMEs) in Romania [11]. Authors develop a model for evaluating Industry 4.0 readiness for enterprises [12]. Another study analyzes the difficulties of SMEs in Industry 4.0 applications by using Analytical Hierarch Process (AHP) and Analytical Network Process (ANP) methods [13]. A conceptual future readiness model is developed to evaluate Industry 4.0 readiness [14], and a conference paper provides useful steps to develop Industry 4.0 scenarios for companies. [2]. The proceeding study of this current one also propose Industry 4.0 scorecard for Turkish

SMEs by using a survey assessing the Industry 4.0 implementations [15]. For further reading, readers are referred to the detailed analysis article [16] and the systematic literature review [17] covering last five years.

SMEs are defined as the enterprises that has less than 250 employees and less than 125 million Turkish Liras (TL) endorsement or balance [18]. According to Turkish Statistical Institute (TÜİK), SMEs have the 99.8% of the enterprises share of Turkey [19]. Hence, they have an obvious contribution not only to the local economy but also to the global competition strength of the country. In that sense, catching up the fourth industrial revolution has to be the prior agenda of the SMEs, and also of the government institutions and the universities from the perspective of giving the required support.

The stages of Industry 4.0 adaptation can be grouped under four proceeding categories;

1. Defining the current situation of the companies and/or sectors
2. Informing the stakeholders about the current situation
3. Developing plans and actions based on the current situation
4. Adopting the Industry 4.0 implementations

This study is the continuum of the broad research that was initiated in March 2018. The purpose is to identify the current situation of SMEs in Turkey with respect to their various implementations for Industry 4.0 transformation. A questionnaire is prepared and applied to SME's with the support of KOSGEB (Small and Medium Sized Enterprises Development Organization). First output of the research was delivered in August 2018 which has focused on analyzing the current situation of SMEs operating in Marmara region [15]. In this proceeding study, four major cities, İstanbul, Ankara, İzmir and Bursa, are analyzed and compared. Similar to the first study outputs, this study aims to find out the level of awareness and readiness of SMEs in these cities and measure the Industry 4.0 scores. As being the second step of the first pioneering study, the target is to support the Industry 4.0 implementations by defining the current situation and proposing suitable actions for adaptation based on varying results.

2 Methodology

The questionnaire designed for the previous research [15] is used for this current study. There are 28 questions prepared by academicians considering the relevant literature including papers, technical reports, and comments of KOSGEB and SME representatives. It aims to collect information on SMEs' profile, current industrialization level, and questions on awareness and readiness of SME's on Industry 4.0 are also included.

First section consists of five questions for collecting general information about the sector, location and number of employees regarding the company, and also the role of the respondent in the company who is answering the questions. This section is used for comparative analysis between the sectors and different sizes of companies.

Second section is the main body of the questionnaire including 12 questions. All the questions are designed to measure the industrialization score of the company with a scale of 1.0 to 4.0. Industry 4.0 adoptions on various processes such as, order taking, product design/development, work order management, on-line traceability of operations, purchasing, production technology and hardware usage, software/automation and software usage, quality-control, inventory and warehouse management, maintenance management, and mobile application usage. All questions have differing options to be chosen by the respondent, and for each answer a value is designed for measuring the industrialization score between 1.0 and 4.0. The definitions and requirements of all industrial revolutions are used for this purpose. In this perspective, the following concepts are taken into consideration;

Industry 1.0 Level: The operations are realized manually. There is no electrical system introduced to the processes.

Industry 2.0 Level: Some systematic implementations are introduced to the system. Electricity is used instead of manpower. Personal computers, simple software and Internet are present in the company, but not integrated with each other and not for managing the processes.

Industry 3.0 Level: All operations are realized in a systematic manner. Automation and sensors are used in the operations, but they are not in collaboration with each other. There are systems for data collection, but the data are not used for online decisions. The infrastructure of the company is ready for the full transformation.

Industry 4.0 Level: The company is in the adaptation phase. The internal operations and those with the stakeholders are realized in an integrated manner via digitalization. Most of the technological drivers such as big data, augmented reality, horizontal and vertical integration etc. are implemented in the system.

A sample question from this part showing the given scores based on the defined levels is given in Table 1;

Table 1. A sample question for scoring industrialization level [19]

Which one(s) is/are most suitable for work order system in your production environment?	Score
Verbal communication is used to give an order and to get feedback about the order	1.0
Written communication is used to give an order without a prepared form	1.3
Order is sent to corresponding units via PC using a prepared form	2.8
A work order management system/software is used	3.3
An e-system/software, consisting of e-tickets attached to the products, is used	4.0

Third section of the questionnaire has ten questions to collect data on awareness and readiness of the company for the new revolution. There are questions asking the challenges that the SMEs are facing and the enablers for SMEs to invest in or adopt Industry 4.0 technologies. The responses for these questions are used for comparing the differences between sectors and the four cities. Moreover, the results can help to develop appropriate strategies for different sectors and/or cities.

The validity and reliability of the questionnaire was tested in the first study [15] hence there is not any test implemented for this second stage. The data collected with the same questionnaire since the beginning of the research in May 2018 until April 2019 is used for this current study.

3 Implementation and Results

This study examines the industrialization levels of the SMEs in four major cities of Turkey, namely Istanbul, Ankara, Izmir and Bursa. The responses to the questionnaire on an e-survey platform have been continued since May 2018 until April 2019. Totally, 623 SMEs from the selected four major cities participated in the survey. As shown in Table 2, almost all sectors are represented for each city. No company has responded from beverage and tobacco production, hence, these industries are left out of the scope.

Table 2. Sector information on the samples

Sectors	Ankara	Bursa	İstanbul	İzmir
Manufacture of food products	3	13	4	4
Manufacture of textile products	3	29	22	6
Outfit manufacturing	1	3	14	5
Manufacturing leather and its bi-products	–	1	9	3
Manufacture of wood and cork products (except furniture)	3	1	2	1
Manufacture of paper and paper products	–	1	10	3
Printing and copying recorded media materials	1	–	1	1
Manufacturing chemicals and chemical products	1	4	11	4
Manufacturing basic pharmaceutics and pharmaceutics materials	3	–	4	1
Manufacturing rubber and plastic materials	6	19	34	3
Manufacturing non-metallic minerals	–	1	3	1
Base metal industry	5	21	15	6
Manufacture of fabricated metal products. Except machinery and equipment	7	19	23	7
Manufacturing computers. Electronics and optic materials	5	3	9	1
Manufacturing electronic equipment	10	2	19	3
Manufacture of machinery and equipment not elsewhere classified	7	18	22	12
Manufacture of motor vehicles, trailers and semi-trailers	3	7	4	2
Manufacture of other transport vehicles	2	1	–	1
Furniture manufacturing	5	17	5	1
Other productions	21	18	45	13
Installation and repair of machinery and equipment	4	2	15	4
Sample size	**90**	**180**	**271**	**82**
Total number of companies	**26553**	**23086**	**129892**	**27756**
Confidence level	**95%**	**95%**	**95%**	**95%**
Confidence interval	**10%**	**7.3%**	**6%**	**11%**

The Table 2 also displays the total number of companies in every city based on the Turkish Statistical Institute (TUIK) 2016 records. Regarding the sample sizes, the confidence level is set to 95%, and the minimum confidence interval is calculated as 11%. Consequently, it can be mentioned that the results based on the samples have high reliability.

The sample also represents the population regarding the size of the companies. Table 3 displays the distribution SMEs contributed to the survey which mentioned their number of employees. 588 out of 623 companies responded to this question, and based on the answers, considering 4 given groups (1–9; 10–49; 50–249 and 250+), most of the companies responded have 10–49 number of employees.

Table 3. Size information on the samples

Number of employees	1–9	10–49	50–249	250+	Total
Ankara	13	45	17	75	**150**
Bursa	12	79	53	3	**147**
İstanbul	46	111	67	2	**226**
İzmir	11	34	18	2	**65**
	82	**269**	**155**	**82**	**588**

The findings from the 12 questions in the second section for measuring the industrialization levels are summarized below, in Table 4. The results considering all processes show that, average industrialization score of the 623 companies of the focused cities is 2.47. The individual scores of the cities İstanbul, Ankara, İzmir and Bursa are 2.46, 2.45, 2.41 and 2.53 respectively. It is seen that, the average scores of the cities are very closed to each other. Moreover, process based scores are also similar for each city. In other words there is not any significant difference observed between these four major cities. All the industrialization scores are closed to the general average of 2.47. High scores coming from the mobile applications is an expected result as the major cities provide high infrastructure. On-line traceability of operations and software usage during operations are the major contributors for increasing the average. On the other hand, it is seen that order receiving and warehouse management process are still behind Industry 2.0 level, and require improvement in the first hand.

Table 4. Overall results for industrialization level

Process	All	İstanbul	Ankara	İzmir	Bursa
Order receiving	1.90	1.85	2.15	1.77	1.90
R&D	2.63	2.64	2.64	2.52	2.66
Work order	2.48	2.46	2.42	2.28	2.62
On-line traceability	3.61	3.61	3.51	3.60	3.66
Purchasing	2.62	2.57	2.67	2.67	2.64
Production technology	3.11	3.11	3.06	3.12	3.13

(continued)

Table 4. (*continued*)

Process	All	İstanbul	Ankara	İzmir	Bursa
Software usage for operations	2.94	2.94	2.93	2.88	2.96
Quality control	2.57	2.58	2.55	2.56	2.59
Inventory management	2.51	2.52	2.41	2.45	2.57
Warehouse management	1.58	1.53	1.36	1.53	1.77
Maintenance management	2.10	2.11	2.07	1.98	2.15
Mobile applications	3.58	3.58	3.59	3.59	3.58
Average	**2.47**	**2.46**	**2.45**	**2.41**	**2.53**

When the company scores are examined companies are categorized under three main levels. To obtain that evaluation, the industrialization score measured for each company is rounded to the nearest integer. For instance, if the score is between 1.50 to 2.49, the company's score is assumed as at 2.0 level. As a result of this analysis, 206 companies are categorized under the group of 2.0 level. Majority of the companies exist in the group of 3.0 level with their scores between 2.5 to 3.0. Only 3 of the companies that are located in Bursa have scores close to 4.0. Consequently, it can be said that the overall score 2.47 is a representative overall score for SMEs. Figure 1 shows the graphical representation of overall scores of the companies in differing locations.

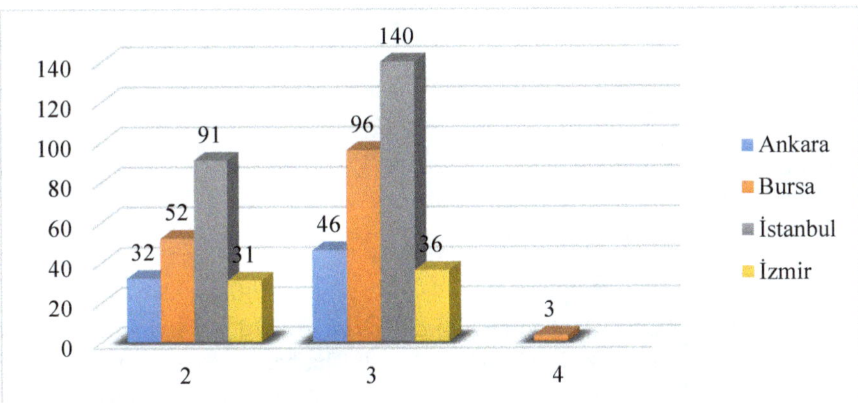

Fig. 1. Overall industrialization scores

When industries in different cities are analyzed separately, it is observed that they also have close average scores. As displayed in Table 5, the average industrialization score for "Printing and copying recorded media materials" industry has the maximum with a value of 3.15 in İstanbul, "Manufacturing basic pharmaceutics and pharmaceutics materials" with a value of 2.95 in Ankara, "Manufacturing non-metallic minerals" with a value of 3.37 in İzmir and "Manufacturing computers, electronics and optic materials" with a value of 3.06 in Bursa. There are very few sectors that passed the 3.0 level, however most of them are above 2.5.

Table 5. Average scores of industries in four cities

Sector	İstanbul	Ankara	İzmir	Bursa
Manufacture of wood and cork products (except furniture)	2.63	2.45	1.92	2.98
Base metal industry	2.63	2.70	2.66	2.76
Manufacture of machinery and equipment not elsewhere classified	2.63	2.50	2.61	2.54
Manufacturing computers, electronics and optic materials	2.90	2.72	2.70	3.06
Manufacturing leather and its bi-products	2.64	#N/A	2.28	2.27
Other productions	2.58	2.68	2.61	2.61
Manufacturing non-metallic minerals	2.58	#N/A	3.37	2.54
Manufacturing electronic equipment	2.68	2.60	2.79	3.01
Manufacture of food products	2.76	2.12	2.58	2.50
Outfit manufacturing	2.89	2.39	2.42	2.63
Manufacture of paper and paper products	2.65	#N/A	2.45	2.46
Manufacturing rubber and plastic materials	2.59	2.60	2.55	2.91
Printing and copying recorded media materials	3.15	2.51	2.40	#N/A
Manufacturing chemicals and chemical products	2.50	2.37	2.58	2.77
Installation and repair of machinery and equipment	2.57	2.85	2.34	2.45
Manufacture of fabricated metal products, except machinery and equipment	2.62	2.47	2.57	2.69
Furniture manufacturing	2.44	2.43	2.07	2.56
Manufacture of motor vehicles, trailers and semi-trailers	2.63	2.70	2.51	2.77
Manufacture of textile products	2.60	2.41	2.43	2.70
Manufacturing basic pharmaceutics and pharmaceutics materials	2.74	2.95	3.13	#N/A
Manufacture of other transport vehicles	#N/A	3.20	#N/A	2.30

Regarding the awareness, readiness and interest of SMEs for Industry 4.0, the analysis results are summarized in Tables 6 and 7. The percentages in these tables say that, most of the SMEs under consideration are neither currently using high technologies nor planning to invest and use them.

"Preemptive maintenance" implementations are the most preferred Industry 4.0 adaptation currently, with a percentage of 38.34%. All the cities display close values except Ankara with 20.63% regarding preemptive maintenance. Other adaptations are "automation" implementations and "sensor system" usage with 33.12% and 31.16% values respectively. Bursa has the higher value of 41.43% for automation usage. When compared with other major cities, Ankara has the lowest values with respect to these technologies. "Virtual reality" and "augmented reality" implementations are very rare in all cities with 1.76% and 2.31% average values counted. Istanbul companies show the highest value of 3D Printing usage, but it is still at a rate of 18.91%. On the other hand, this is almost three times higher than İzmir's value which is 6.25%. On the

Table 6. Awareness of technologies (currently in use)

	All	İstanbul	Ankara	İzmir	Bursa
3D printing	13.92%	**18.91%**	10.14%	6.25%	12.03%
Sensor systems	**31.16%**	35.64%	12.33%	31.15%	34.53%
Virtual reality	**1.76%**	3.08%	1.41%	1.69%	0.00%
Cloud technologies	21.67%	23.65%	14.08%	**32.76%**	17.91%
Internet of things	5.53%	8.06%	3.08%	3.45%	4.00%
Additive (3D) manufacturing	6.98%	7.41%	5.80%	8.77%	6.20%
Improved robot technologies	9.13%	10.82%	1.45%	5.17%	12.50%
Automation	**33.12%**	32.34%	21.43%	30.00%	**41.43%**
Modeling and simulation	26.77%	30.00%	13.04%	31.03%	27.41%
Online education	12.13%	15.71%	5.88%	15.25%	8.66%
Electronic performance reports	17.73%	17.55%	13.43%	17.24%	15.27%
Big data	6.38%	8.02%	6.06%	3.45%	5.47%
Augmented reality	**2.31%**	2.17%	1.54%	3.51%	2.33%
Cyber security	24.89%	24.47%	18.46%	**33.90%**	24.62%
Integrated product development	15.86%	14.59%	10.94%	15.79%	20.16%
Predictive maintenance	15.72%	14.81%	7.81%	12.28%	22.48%
Preemptive maintenance	**38.34%**	44.04%	20.63%	37.29%	38.93%
Smart warehouses	2.71%	2.66%	0.00%	3.39%	3.88%
Mobile systems	17.26%	**24.34%**	5.97%	17.86%	12.69%

Table 7. Awareness of technologies (planned to invest and use)

	All	İstanbul	Ankara	İzmir	Bursa
3D printing	6.21%	5.97%	5.80%	6.25%	6.77%
Sensor systems	8.00%	8.42%	9.59%	9.84%	5.76%
Virtual reality	4.62%	5.64%	2.82%	3.39%	4.62%
Cloud technologies	6.44%	7.88%	4.23%	1.72%	7.46%
Internet of things	5.99%	6.45%	3.08%	3.45%	8.00%
Additive (3D) manufacturing	3.60%	3.17%	2.90%	3.51%	4.65%
Improved robot technologies	7.57%	7.73%	10.14%	12.07%	3.91%
Automation	**12.95%**	13.93%	15.71%	16.67%	8.57%
Modeling and simulation	7.74%	9.47%	10.14%	10.34%	2.96%
Online education	4.94%	4.71%	2.94%	6.78%	5.51%
Electronic performance reports	8.03%	7.45%	10.45%	12.07%	7.63%
Big data	4.33%	4.81%	3.03%	6.90%	3.13%
Augmented reality	3.93%	4.35%	1.54%	5.26%	4.65%
Cyber security	7.24%	6.38%	7.69%	8.47%	7.69%
Integrated product development	10.11%	8.65%	14.06%	14.04%	8.53%
Predictive maintenance	5.47%	4.23%	14.06%	7.02%	2.33%
Preemptive maintenance	7.17%	6.22%	14.29%	11.86%	3.05%
Smart warehouses	**10.61%**	10.11%	13.43%	20.34%	5.43%
Mobile systems	**10.99%**	11.11%	14.93%	16.07%	6.72%

contrary, İzmir companies display considerably higher usage of "cloud technologies" and "cyber security" implementations compared with other cities. "Mobile systems" are seen more in Istanbul, whereas "integrated product developments" is observed more in İzmir companies. Overall picture displays that companies in Ankara are behind the others regarding 17 technologies out of total 19.

When the future plans of the companies are analyzed (Table 7), it is observed that the companies require motivation and support, as we see that none of the technologies got high preference of plans for investment. The highest score is just 12.95% for "automation". The followers are "mobile systems", "smart warehouses" and "integrated product development" with the scores of 10.99%, 10.61% and 10.11% respectively. Comparative results show that companies in İzmir is more willing in investment than the other cities, and Ankara display the lowest percentages.

Benefits of Industry 4.0 from the perspective of SMEs are also analyzed and the results for four cities are shown below, in Fig. 2. All companies think that "increase in productivity" is the highest benefit of Industry 4.0 implementation, while "decrease in costs" stands in the second position. "Improve demand forecast" and 'increase in demand' also have considerably high occurrence. There is no considerable difference observed between the four cities related to this analysis.

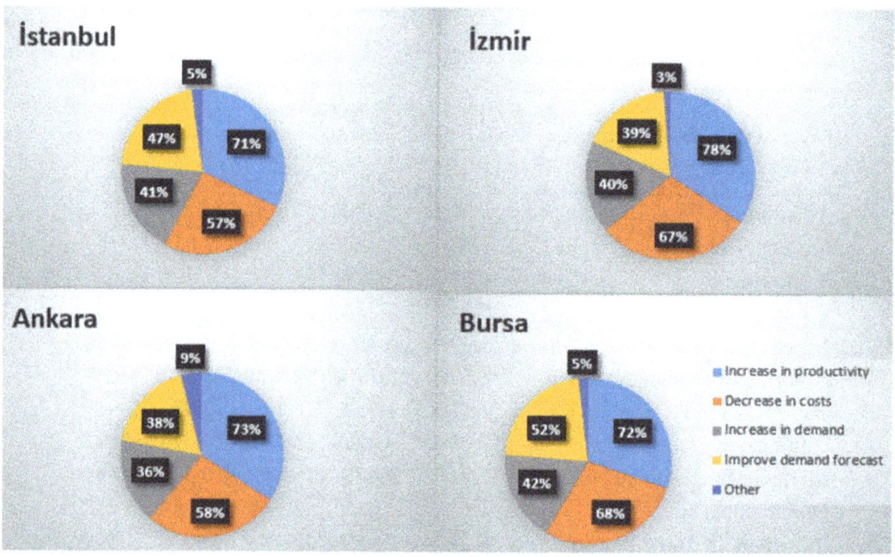

Fig. 2. Benefits of Industry 4.0

Table 8 summarizes the comments of SMEs on the barriers of Industry 4.0. Majority of the respondents think that, "high cost of new technologies", "financial resource shortage", "high cost or lack of consultancy" and "lack of support from universities or government" are among the important barriers. Responses to these options are more than 50%. "Uncertainty of return on investment" is the follower with a value of 49.81%. Differences in answers between the four cities are observed only for three factors; "Financial resource shortage" is perceived as the highest barrier for all cities except İzmir. İzmir companies think "high cost or lack of consultancy" is the biggest constraint, where as "resistance of employees for change" is the greatest barrier for companies in Bursa. Companies in Ankara think "complexity of new technologies" is the lowest barrier faced, and also answers from Ankara display "weak technical support from suppliers" is considerably a lower barrier compared to other cities.

Table 8. Barriers of Industry 4.0

Barrier	All	İstanbul	Ankara	İzmir	Bursa
Current hardware is not suitable	42.81%	46.67%	36.36%	41.46%	40.40%
Current software is not suitable	44.11%	45.77%	45.65%	45.45%	40.40%
Financial resource shortage	**59.39%**	**65.48%**	57.69%	45.45%	56.92%
High cost of the new technologies	**58.80%**	64.44%	56.25%	53.85%	54.41%
Weak technical support from suppliers	41.85%	43.79%	**25.49%**	43.48%	46.23%
Complexity of new technologies	22.96%	21.69%	19.64%	13.04%	29.84%
Mismatch of work flow in the facility	29.73%	28.10%	31.37%	28.26%	31.67%
Lack of knowledge on new technologies	37.13%	34.25%	34.78%	39.53%	41.41%
Lack of education of workers in production	44.79%	42.52%	45.24%	51.61%	45.45%
Technical personnel's insufficient knowledge/capability	42.07%	40.30%	42.50%	40.54%	44.90%
Resistance of employees to change	33.80%	30.07%	34.69%	28.57%	**40.54%**
Resistance of top management to change	17.95%	**13.69%**	23.21%	24.49%	18.80%
Lack of support from universities/government	**50.59%**	54.29%	45.24%	48.39%	49.35%
High cost or lack of consultancy	**56.00%**	57.43%	51.43%	56.76%	55.84%
Low demand	37.76%	42.47%	36.73%	37.50%	31.73%
Lack of support of the big clients	35.26%	37.41%	38.30%	39.02%	29.41%
Uncertainty of return on investment	**49.81%**	47.27%	48.72%	55.56%	51.19%
Hesitation on data security and privacy	28.99%	30.14%	34.69%	30.43%	24.04%

When the results on "information sources on Industry 4.0" are analyzed (Table 9), although all of the listed sources are considered very important, "websites" are observed as the major source with a rank of 73.36%. Companies in Ankara prefer web usage slightly more than the other cities. "Website" sources are the prior answer for every city except İzmir which mentions "trade fairs" as the first runner of information sources. Rest of the overall scores display that "participating to trade fairs" and "articles in search engines" are the proceeding information sources after the web sites. Although articles are preferred, they are not from printed journals. All the city answers show that "papers from journals" and "direct mails" are the least preferred information sources. SMEs also declared their interest in government foundations, and working with consultants, as well as practicing academic collaborations like attending congresses or working with universities.

Table 9. Information sources for Industry 4.0

Information source	All	İstanbul	Ankara	İzmir	Bursa
Websites	**73.36%**	71.88%	80.95%	75.00%	71.58%
Articles on search engines	**64.65%**	63.57%	70.45%	69.23%	62.24%
Trade fairs	**63.02%**	56.30%	69.05%	81.08%	62.89%
Direct mails	44.48%	46.10%	46.00%	43.59%	41.58%
Papers in journals	44.09%	43.83%	34.00%	53.49%	45.30%
Trade unions	55.66%	50.00%	55.32%	62.50%	61.05%
Presentations in congress	57.32%	51.77%	60.00%	71.43%	59.14%
Consumers	52.13%	53.13%	53.66%	55.56%	49.00%
Holding company	56.15%	55.07%	53.19%	66.67%	54.84%
Equipment vendors	54.34%	50.99%	48.98%	66.67%	56.73%
Consultants	56.12%	53.38%	57.45%	62.79%	56.70%
Universities	55.22%	48.87%	60.47%	65.71%	58.14%
Government foundations	57.09%	54.17%	59.46%	67.74%	56.16%

4 Discussion of the Results and Conclusions

In the globalized trade markets, none of the companies can ignore the required change in their operations brought by the new and highly value adding technological drivers, such as internet of things, cloud computing, autonomous robots, augmented reality, cyber security system, etc. Due to their great enterprise share of 99.8%, SMEs have an obvious contribution not only to the local economy but also to the global competition strength of the country. Hence, catching up the fourth industrial revolution has to be the prior agenda of the SMEs, and also of the government institutions and the universities from the perspective of giving the required support. With that motivation, this is the continuum of the broad research that is initiated in March 2018, with a purpose to identify the current situation of SMEs in Turkey evaluating their various implementations for Industry 4.0 transformation. A questionnaire is prepared and applied to SME's with the support of KOSGEB (Small and Medium Sized Enterprises Development Organization), and the responses to the questionnaire are collected via an e-survey platform together with sent e-mails. In this current stage of the ongoing study, four major cities, İstanbul, Ankara, İzmir and Bursa, are analyzed and compared.

This study measures the Industry 4.0 scores, and finds out the level of awareness and readiness of SMEs in these cities with a target to support the Industry 4.0 implementations by defining the current situation and proposing suitable actions for adaptation based on varying results.

When the process implementations are observed, average industrialization score of the 623 companies of the focused cities is 2.47. The individual scores of the cities, İstanbul, Ankara, İzmir and Bursa are 2.46, 2.45, 2.41 and 2.53 respectively. It is found that, the average scores of the cities are very closed to each other. Moreover, process based scores are also close to each other for each city. In other words, there is not any significant difference observed between the implementations of these four major cities. When the companies are grouped under 3 main levels of industrialization, majority of the companies exist in the group of 3.0 level with their scores between 2.5 to 3.0. Only 3 of the companies that are located in Bursa have scores close to 4.0.

Differences between sectors are also analyzed; The leading sectors of Industry 4.0 are "Printing and copying recorded media materials" industry in İstanbul with a score of 3.15, "Manufacturing basic pharmaceutics and pharmaceutics materials" in Ankara with a score of 2.95, "Manufacturing non-metallic minerals" in İzmir with a score of 3.37, and "Manufacturing computers, electronics and optic materials" in Bursa with a score of 3.06. There are very few sectors that passed the 3.0 score, however most of them are above 2.5.

"Preemptive maintenance" implementations are the most preferred Industry 4.0 adaptation currently, with a percentage of 38.34%. "Virtual reality" and "augmented reality" implementations are very rare in all cities with 1.76% and 2.31% average values counted. Istanbul companies show the highest value of 3D Printing usage, but it is still at a rate of 18.91%. İzmir companies display considerably higher usage of "cloud technologies" and "cyber security" implementations compared with other cities. "Mobile systems" are seen more in Istanbul, whereas "integrated product developments" is observed more in İzmir companies. Overall picture displays that companies in Ankara are behind the others regarding 17 technologies out of total 19.

When the future plans of the companies are analyzed, it is observed that the companies require motivation and support, as we see that none of the technologies got high preference of plans for investment. The highest score is just 12.95% showing the plans on investing for "automation". The followers are "mobile systems", "smart warehouses" and "integrated product development" with the values of 10.99%, 10.61% and 10.11% respectively. Comparative results show that companies in İzmir are more willing in investment than the other cities, and Ankara display the lowest percentages.

Regarding the benefits of Industry 4.0, all companies think that "increase in productivity" is the highest benefit of Industry 4.0 implementation, while "decrease in costs" stands in the second position. "Improve demand forecast" and 'increase in demand' also have high occurrence. There is no considerable difference observed between the four cities related to this analysis.

Majority of the respondents think that, "high cost of new technologies", "financial resource shortage", "high cost or lack of consultancy" and "lack of support from universities or government" are among the important barriers, and "uncertainty of return on investment" is the follower. Differences in answers between the four cities are observed only for three factors; "Financial resource shortage" is perceived as the highest barrier for all cities except İzmir. İzmir companies think "high cost or lack of consultancy" is the biggest constraint, where as "resistance of employees for change" is the greatest barrier for companies in Bursa. Companies in Ankara think "complexity of new technologies" is the lowest barrier faced, and also answers from Ankara display "weak technical support from suppliers" is considerably a lower barrier compared to other cities.

Analysis on "information sources on Industry 4.0" display that, although all of the listed sources are considered very important, "websites" are the major source with an average rank of 73.36%. "Website" sources are the prior answer from every city except İzmir, which mentions "trade fairs" as the first runner of information sources. Rest of the overall scores display that "participating to trade fairs" and "articles in search engines" are the proceeding information sources after the web sites. SMEs also declare their interest in government foundations, and working with consultants, as well as practicing academic collaborations like attending congresses or working with universities.

Based on the findings of this research it can be concluded that SMEs are willing to cooperate with the government institutions and universities, and they need financial support as well as information on Industry 4.0 adoption. Based on the comparative tables, the companies in Ankara have the prior need for support. Since it is known that there are various support programs developed by the government, it can be proposed that, government has to be more active in announcement of these programs, collaborations of the companies and the academicians have to be encouraged. Different programs can be designed for different cities, as the barriers are varying, such as resistance of employees is higher in Bursa, while high costs have to be considered for İstanbul. The future studies are the continuing phases of this research in order to analyze all regions of Turkey, and to propose and develop actions to enhance the Industry 4.0 adaptation of Turkish SMEs.

Acknowledgment. This research is made possible with the support of KOSGEB. The researchers express their gratitude to KOSGEB Management for their support on this study and for the ongoing research to cover all regions of Turkey.

This study is supported by İstanbul Kültür University Technology and Projects Support Unit (TPDB) under the Support Program for National and International Activities (ULEP). Authors would like to thank Istanbul Kültür University [ULEP-2019-U0708/53/15] for funding conference attendance.

References

1. Drath, R., Horch, A.: Industrie 4.0: Hit or Hype? IEEE Ind. Electron. Mag. **8**(2), 56–58 (2014)
2. Hermann, M., Pentek, T., Otto, B.: Design principles for Industrie 4.0. In: 49th Hawaii International Conference on System Sciences (2016)
3. Siemens. "Endüstri 4.0 Yolunda" (2016)
4. Impuls. "Industrie 4.0-Readiness", Aachen, Köln (2015)
5. PriceWaterhouseCoopers (PWC). Industry 4.0: building the digital enterprise. https://www.pwc.com/gx/en/industries/industries-4.0/landing-page/industry-4.0-building-your-digital-enterprise-april-2016.pdf

6. Turkish Industry & Business Association (TÜSİAD). Industry 4.0 in Turkey as an Imperative for Global Competitiveness: An Emerging Market Perspective, TÜSİAD-T/2016-03/576
7. McKinsey. Industry 4.0: how to navigate digitization of the manufacture sector. https://www.mckinsey.com/business-functions/operations/our-insights/industry-four-point-o-how-to-navigae-the-digitization-of-the-manufacturing-sector
8. Kagermann, H., Lukas, W., Wahlster, W.: Industrie 4.0: Mit dem Internet der Dinge auf dem Weg zur 4. industriellen Revolution. VDI nachrichten, vol. 13 (2011)
9. Castelo-Branco, I., Cruz-Jesus, F., Oliveira, T.: Assessing Industry 4.0 readiness in manufacturing: evidence for the european union. Comput. Ind. **107**, 22–32 (2019)
10. Huang, C.J., Chicoma, E.D.T., Huang, Y.H.: Evaluating the factors that are affecting the implementation of Industry 4.0 technologies in manufacturing MSMEs, the case of Peru. Processes **7**, 161 (2019)
11. Türkes, MCa, Oncioiu, I., Aslam, H.D., Pantelescu, A.M., Topor, D.I., Capusneanu, S.: Drivers and barriers in using Industry 4.0: a perspective of SMEs in Romania. Processes **7**, 153 (2019)
12. Basl, J., Doucek, P.: A metamodel for evaluating enterprise readiness in the context of Industry 4.0. Information **10**, 89 (2019)
13. Sevinç, A., Gür, S., Eren, T.: Analysis of the difficulties of SMEs in Industry 4.0 applications by analytical hierarchy process and analytical network process. Processes **6**, 264 (2018)
14. Botha, A.P.: Rapidly arriving futures: future readiness for Industry 4.0. S. Afr. J. Ind. Eng. **29**(3), 148–160 (2018). Special edition
15. Gergin, Z., Yüksektepe, F., Gençyılmaz, M.G., Aktin, A.T., Gülen, K.G., İlhan, D.A., Dündar, U., Cebeci, Ö., Çavdarlı, A.İ.: Industry 4.0 scorecard of Turkish SMEs. In: Proceedings of the International Symposium for Production Research, pp. 426–437 (2018)
16. Muhuri, P.K., Shukla, A.K., Abraham, A.: Industry 4.0: a bibliometric analysis and detailed overview. Eng. Appl. Artif. Intell. **78**, 218–235 (2019)
17. Piccarozzi, M., Aquilani, B., Gatti, C.: Industry 4.0 in management studies: a systematic literature review. Sustainability **10**, 3821 (2018)
18. Ministry of Science, Industry and Technology, Kosgeb, Kobi Stratejisi ve Eylem Planı (KSEP) 2015–2018 (2016)
19. Turkish Statistical Institute (TÜİK), Small and Medium Sized Enterprises Statistics, 2016, Press release no. 21540, November 2017

From Chaos to Calm: Industry 4.0 Practices of Turkish White Goods Companies

Nihan Yıldırım and Kübra Şimşek Demirbağ(✉)

Management Engineering, Istanbul Technical University, Istanbul, Turkey
{yildirimni, simsekkub}@itu.edu.tr

Abstract. Just as in the previous industrial revolutions, the rules of the game were rewritten and a new paradigm that is known as "Industry 4.0" was born. However, it is seen that everyone does not understand the concept of Industry 4.0 in a similar manner. For instance, Germany, where the concept was first introduced, is one of the technologically advanced and oldest countries, while labor is abundant and cheap in China, which is the most populous country in the world. On the other hand, Turkey is a developing country that is technologically dependent on center countries and has an import-intense manufacturing structure. Due to the potential changes in productivity level and flexibility capability of companies that Industry 4.0 technologies can enable in manufacturing industries, strategies followed by companies from different countries and industrial contexts should be different and be customized to local conditions, but these differences were not considered in many recent studies in the related literature. As it is understood, there is a need for scientific studies based on real practices of Industry 4.0. In this context, this study aims to explore the current practices of two Turkish white goods manufacturing companies through a case study. The first company offers products and services around the world and the second company is the supplier of the first one.

Keywords: Industry 4.0 · Fourth industrial revolution · White goods · Durable goods · Case study

1 Introduction

At the end of the 18th century, the first industrial revolution or mechanization took place thanks to steam engine, power loom and iron processing [1]. With the development of railways, mechanization was spread from Britain, where the revolution began, to Europe. At the dawn of the 20th century, the introduction of mass production based on the division of labor powered by electrical energy caused the second industrial revolution to begin [2, 3]. Additionally, the birth of the third industrial revolution, also known as the digital revolution, corresponds to the release of the first programmable logic controller, Modicon 084, in 1969 [4]. Thanks to these controllers, significant breakthroughs have been made in the control and automation of production technologies. Due to the industrial revolutions, especially in Western Europe and the USA, surplus production has emerged. In this way, developed countries started to transfer their financial resources and knowledge to the less developed countries, which

N. M. Durakbasa and M. G. Gençyılmaz (Eds.): ISPR 2019, LNME, pp. 278–287, 2020.
https://doi.org/10.1007/978-3-030-31343-2_24

had low-cost resources such as labor, raw material, and semi-finished products. In this period; countries like China, India, Indonesia, Brazil, Taiwan, Thailand, Turkey, and Argentina had become critical markets and sources of labor for the developed countries [5]. However, developed countries have lost their know-how to developing countries. Particularly Asian countries like China are very successful in reverse engineering and they have achieved advanced levels of digitization. New competitors from Asia are now a threat to manufacturing companies in developed countries, which have to follow a new path to overcome this threat. There is now a new trend that is called "reshoring" for center economies that once shifted their manufacturing to developing or underdeveloped countries. Reshoring means in broad terms as the process of returning the manufacturing to the company's original country [6]. Concordantly, several projects around the world have begun to be created by countries to protect their manufacturing such as "High Tech Strategy 2020" of Germany and "Industrial Internet Consortium" of the USA [7]. As it is understood, the concept of Industry 4.0 was originated to maintain manufacturing superiority against developing countries. Kagermann [8], president of the German National Academy of Science and Engineering (acatech), conveys that "INDUSTRIE 4.0 is the German strategic initiative to take up a pioneering role in industrial IT which is currently revolutionizing the manufacturing engineering sector". For the first time, the German Federal Government announced Industry 4.0 (as Industrie 4.0) as one of the strategic initiatives under the "High-Tech Strategy 2020" action plan in November 2011. Primary implementation recommendations of the project were formed by the "Industrie 4.0 Working Group" under the coordination of acatech and presented as a report to the German government [9].

Industry 4.0 represents the fourth industrial revolution. According to their findings from the related literature, Hermann et al. [10] define Industry 4.0 as "a collective term for technologies and concepts of value chain organization" and there are four components of Industry 4.0: Internet of things (IoT), Internet of Services (IoS), Cyber-Physical Systems (CPS), and Smart Factory. Thanks to the IoT, cyber-physical systems communicate in real time with humans and each other and take decentralized decisions through following the processes in smart factories. The participants of the value chain use internal and cross-organizational services provided via the IoS [10]. Taking into account all of these, Industry 4.0 presented as a German Government Strategy has triggered a new competition across all over the world. Policy initiatives and strategies were created by many other governments, as well. For instance, "Made in China 2025" of China [11], "Catapult" of the United Kingdom [12], "Society 5.0" of Japan [13] are some other examples of national policy initiatives.

Turkish manufacturing companies face the challenge of losing their competitive advantage in the global value chain which has utilized labor cost advantage for decades. Due to operating in a context of technology follower country, which lacks the capital accumulation needed for effective technology transfer in industrial transformations, manufacturing companies in developing countries like Turkey are in need of customized approaches and models of effective technology transfer and capacity building, which are derived from detailed and comparative case studies. However, although there are high numbers of publications on research and practice for Industry 4.0 adaptation in literature, studies that point out contextual differences and propose unique models to manufacturing companies in developing countries are very immature.

Realistic data are generally unavailable, and knowledge based on detailed use-case scenarios concerning Industry 4.0 tools from developing countries is very limited. As it is understood, there is a need for scientific studies based on real practices. In this context, this paper aims to explore the current practices of two Turkish white goods manufacturing companies through a case study. In the case study, we investigated two manufacturing companies from different tiers of the value chain to provide differences between supplier and customer perspectives. The first company offers products and services around the world and the second company is the supplier of the first one.

The next two sections of the paper describe the methodology and present the findings, respectively. Finally, in the last section, the results of the study, suggestions for future studies and the limitations of the study were presented.

2 Methodology

Qualitative research techniques enable researchers to understand a social phenomenon through the perspectives of relevant individuals and to present the current structure and processes. In this study, the case analysis method, which is one of the qualitative research techniques, was used. Case studies are preferred when the form of research questions consists of "how" and "why" questions, the researcher has little or no control over behavioral events, and the focus is on contemporary events [15]. Also, while the boundary between phenomenon and context is unclear, case studies have a high potential to elicit causality in situations that are complex for statistical and experimental methods. Case studies organize rich and diverse data sources such as interviews, archival records, documents, observations, and artifacts, and collecting data from multiple sources is a strong aspect of case analysis [15]. Especially, interviews are one of the most efficient ways to collect rich data if the phenomenon is rare and consists of a series of separate events [16]. Therefore, in this study, primary data is collected through structured expert interviews and analyzed as a comparative case study. Additionally, Industry 4.0 maturity levels were measured through "The Digital Maturity Model" [14] to determine the digital transformation levels of the companies. After addressing the questions of the model consisting of four dimensions (culture, organization, technology, and customer insights) and 28 items to companies in 1–5 Likert Scale (1: Completely disagree, 5: Completely agree), the digital maturity levels of the companies were calculated in percentages as seen in the Fig. 1.

In order to ensure consistency, the authors observed the production facilities of the companies and compared the information on the company websites and Internet news about the companies with interview reports and survey results. Figure 1 illustrates the research process from data collection to analysis and Table 1 shows the questions of the interview conducted with the companies. As it was aimed to determine the current status of the companies regarding Industry 4.0, in the light of the information in related literature, the research subjects were chosen as "knowledge level about Industry 4.0", "Industry 4.0 readiness and maturity level", "Industry 4.0 investments", "Industry 4.0 tools and technologies", "motives and advantages", and "barriers".

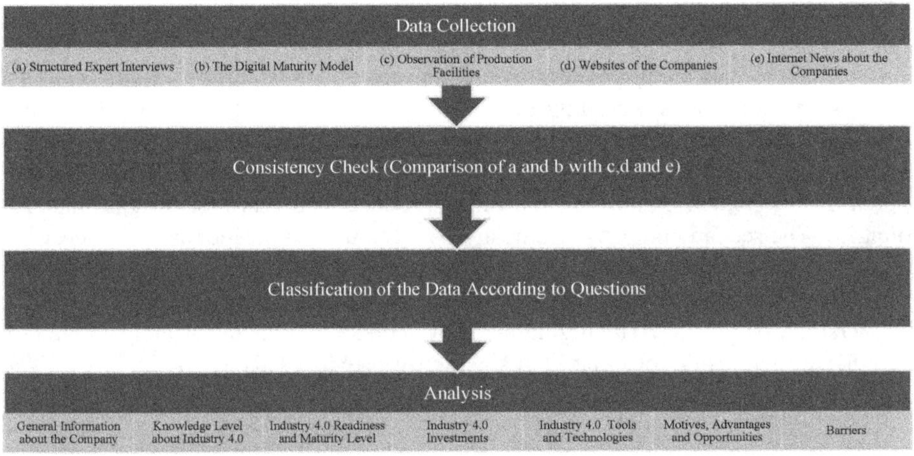

Fig. 1. The research process

Table 1. Interview questions

Research issue	Interview questions
Knowledge level about Industry 4.0	Have you encountered the concept of Industry 4.0 before? What does Industry 4.0 mean to you?
Industry 4.0 readiness and maturity level	What do you think is your company's digital transformation level?
	How do you manage technology? What are the actions you take regarding this issue?
	Do you perform statistical process control? What is your level of lean production maturity?
Industry 4.0 investment	What investments did you make for Industry 4.0? Have you achieved a positive return on investment?
Industry 4.0 tools and technologies	Which technology and tools do you use in your production processes? What is the policy of your company on these technological developments?
Motives and advantages	Do you think your company is affected positively by digital transformation? What will be the advantages?
	What are the motives of your company to transition to Industry 4.0? Do you think the transition is optional or is it a necessity? Why is that?
Barriers	What do you think are the challenges and barriers in the transition to Industry 4.0? Why is that?

3 Findings

This section presents the responses of the companies to the interview questions and the scores of the companies according to "The Digital Maturity Model" [14]. Interviews were conducted with two different Turkish companies that operate in the durable consumer goods industry. The first company (Company A) operates in 7 countries with approximately 30,000 employees and offers products and services to more than 100 countries. The second company (Company B) has a unique position as a producer and supplier of cooker burner systems with 160 employees and serves to main industries, especially to the white goods industry. At the same time, Company B is one of the suppliers of Company A. The digital maturity scores of Company A and Company B according to the results of "The Digital Maturity Model" [14] are 64.28% and 74%, respectively.

3.1 Company A

In the interview with Company A, "Production Technologies and Industry 4.0 Manager" and one of "Production Technologies Senior Specialists" were interviewed.

(1) *Knowledge Level about Industry 4.0:* Industry 4.0 is a marketing strategy and brand created by the German government. The aim is to ensure whether the European community is positioned correctly to increase its income and gain a place in the world market. Thus, third-world countries and developing countries such as Turkey also had the opportunity to question where they are in the market. In this way, developing countries are examining their competencies to produce value-added products and not to be dependent on center countries in technology. In Turkey, "100% domestic and national initiatives" are consolidating and are being integrated into socioeconomic, political and economic structures. Namely, Industry 4.0 is not a concept limited to autonomous vehicles, augmented reality or robots. It is a holistic structure and development strategy.

(2) *Industry 4.0 Readiness and Maturity Level:* We have an advisory board that consists of academicians from different universities around the world. Every year, members of the board come together for 3 days and give suggestions on technology management. We have also built a model for lean manufacturing for two years because leanness is a KPI for our businesses. There is a methodology and a road map of this. I think we follow the right strategy to focus on the leanness and digital transformation maturity index. We are twice as good in terms of leanness compared to last year. We use leanness and maturity measurements to make benchmarks. Both businesses can compare themselves with others, and we can monitor their progress. We use lean production as a prerequisite. Our systems have to be fundamentally suitable for lean production. If the systems are lean, digital technologies can support them. In brief, you can't be digital unless you're lean.

(3) *Industry 4.0 Investments:* We set the first investment budget for Industry 4.0 in 2017. The investments we made in the "Industry 4.0" budget item are approximately 2% of all our investments, but in fact, all our investments are indirectly for digital transformation and Industry 4.0. We mainly invested in a lab that enables

investigating the technology enablers because only theoretical training is not enough. It is necessary to touch technology enablers and understand how they work. We already made investments in robot and automation before Industry 4.0 wave, but now we are planning everything, making assessments and investing in the relevant technology if it improves our leanness and flexibility. Since we are still in the transition period, it is too early to expect positive returns from investments.

(4) *Industry 4.0 Tools and Technologies:* Table 2 presents the Industry 4.0 technologies used in the manufacturing processes of Company A.

Table 2. Industry 4.0 tools and technologies used in the manufacturing processes of Company A

	Internal production	Montage	Notes
Robot/Automation	x	x	
Sensor	x	x	
Big data/Data analytics	Pilot Project (plastic injection)	x (function test)	
Simulation	x	x	
Additive manufacturing	Pilot Project (moulding)		R&D uses it in prototyping
AR/VR			Pilot projects have been made but not suitable for use
Artificial Intelligence			They are far behind in AI
Cloud	x	x	They have their cloud systems. The use of Amazon and Azure cloud services is quite limited

(5) *Motive and Advantages:* Of course, cost reduction and profit improvement are the greatest advantages of digital transformation. The most critical issue is the increase in the value we create for the customer. As our company evolves digitally, it will be more visible and accessible to the customer. Needless to say, the main goal is to improve profitability, but the only way to increase profit goes through capturing the voice of the customer. Company A is a global company, which is one of the main brands of Turkey and the market leader in many other countries. Therefore, as we grow, we have no alternative but to make our cumbersome structure dynamic and digitalize. Another source of motivation in digital transformation is our competitors, especially China. Hence, the fear of being disrupted is the greatest motive that drives us to survive. Another one is globalization. If you are a global player, you have no choice but to digitalize.

(6) *Barriers:* Firstly, competence and vision are very critical issues. When people read or learn something about Industry 4.0, they get excited. They think we'll be successful if we buy robots or other enablers. Unfortunately, you don't digitalize these big organizations when you buy a few robots. Also, the strategy is essential

when the roles are appointed because of the fact that authority confusion may cause chaos. However, we have overcome these obstacles. Agility is quite significant. After taking a decision, the decision can completely change while we are performing it. Therefore, it is necessary to work in gray environments, because in real life there is not only black and white. Besides, the status quo management style does not work here, because we have encountered with the technologies of the younger generation. In short, baby boomers have no place in the management. One who cannot use new technologies cannot be a top manager. Furthermore, the traditional IT structure needs to be updated very quickly. Finally, why should engineers work in production instead of working in banks? Though, as we need competent engineers, we must encourage engineers and create a proper environment and ambiance. I mean, we have to make sure our factories are the right places for competent engineers to work.

3.2 Company B

"The Founder and Managing Director" of Company B was interviewed.

(1) *Knowledge Level about Industry 4.0:* We have been dealing with Industry 4.0 more intensively for the last two years. There are several definitions accepted by everyone, but for me, the meaning of Industry 4.0 is that the communication of technologies we currently use or will use with each other and monitoring the communication continuously. Technological advances of Industry 4.0 such as 3D printers, AR/VR, robots, cybersecurity, cloud computing are known for a long time. Thus, the difference of Industry 4.0 is the communication between these technologies.

(2) Industry 4.0 Readiness and Maturity Level: We do not have a department dedicated to technology management. However, we use different methods like benchmarking. We especially participate in fairs in Turkey and abroad. For example, we never miss the Hannover, Automatica, and WIN Fairs and Robot Investments Forum and Exhibition. We've been doing statistical process control for a long time and have a project that started two years ago about lean production. Due to lean production, there is an improvement in buffer stocks, and the workload of the department of Production Planning department has decreased. When we tried to measure our digital maturity level, we found that our automation systems were not visible. Therefore, we focused on digital twin, MES and ERP. If we look at the circumstances of our own, in my opinion, our digital readiness is around 6–7 out of 10.

(3) *Industry 4.0 Investments:* We made investments in 2017 and 2018, and we plan to continue in 2019. We have probably allocated money from nearly 10% of our budget to Industry 4.0 investments. Previously, we did not intend to make these investments very fast, because, we did not prefer to use a quick method as in many other investment decisions. First, we tried to understand and learn about the developments of Industry 4.0. The two-year period in our sector is a reasonable period in terms of return on investment. We divided the investments into two parts like those, which will return in the short term and long term. We have already had

the positive returns of our short-term investments. Providing a decrease in poor-quality costs and productivity growth was the first benefit of our short-term investments.

(4) *Industry 4.0 Tools and Technologies:* Industry 4.0 technologies used in the manufacturing processes of Company B are shown in Table 3.

Table 3. Industry 4.0 tools and technologies used in the manufacturing processes of Company B

	Moulding	Machining	Sandblasting	Montage	Control	Packaging	Notes
Robot/Automation		x		x	x		
Sensor	x	x		x	x		
Big data/Data analytics	x	x	x	x	x	x	
Simulation	x	x	x	x	x	x	
Additive manufacturing							There's no plan because there's no need. Maybe it could be planned after 2020
AR/VR							Pilot project
Artificial Intelligence							They are far behind in AI
Cloud	x	x	x	x	x	x	

(5) *Motive and Advantages:* Digital transformation is not a preference but a necessity. In my opinion, our company will be positively affected by digital transformation. Thanks to Industry 4.0, we started following other companies in the sector, especially our foreign competitors, and we moved away from daily concerns. Even if our costs are sometimes high, our customers prefer us owing to our digital transformation investments. They think that working with our company will be more profitable in the long run. In short, being preferred is the most significant advantage of Industry 4.0.

(6) *Barriers:* The biggest obstacle we face is the lack of competent employees and suppliers. Apart from that, there is a need for financial resources for the investments in medium- and long-term. Unfortunately, both equity of subsidiary industry companies and government investment supports are insufficient.

4 Discussion and Conclusion

Considering observations and the answers to the questions of interviews and the maturity model, Company A and B are compared with each other in this paper. Firstly, both companies are knowledgeable about Industry 4.0. Company A perceives Industry

4.0 as the development strategy of developed countries, while Company B focuses on the technical definition of the concept. Both companies work to be ready for Industry 4.0. Therefore, they perform lean manufacturing practices and statistical process controls. Additionally, both companies have measured the Industry 4.0 readiness levels to overcome the deficiencies. For example, Company A conducts leanness measurements, has a specific leanness strategy and roadmap, and brings together academicians from different countries each year to determine the level of digital maturity and to address their shortcomings. On the other hand, Company B uses methods such as participation to fairs and benchmarking to understand the level of digital maturity and to build a right digitalization strategy. Companies have made Industry 4.0 investments, especially for the last two years. Robot/automation, sensors, simulation, big data/data analytics are the tools and technologies used in the manufacturing processes of both companies. Company B has a pilot project to evaluate the feasibility of AR/VR technologies for its manufacturing processes while Company A completed the pilot project regarding AR/VR and decided that this technology is not suitable for use. Company A uses 3D printing only for prototyping limitedly. Actually, according to companies, additive manufacturing is not a necessity for manufacturing processes. Finally, firms think they are far behind in artificial intelligence. In that case, companies use Industry 4.0 tools and technologies in their manufacturing processes in a similar way. According to both companies, digital transformation is not a preference but of the necessity for companies to survive. This paper also explores the motives and advantages of Industry 4.0 and digital transformation for white goods industry. The most important motives are globalization, customers, competitors, and surviving for Company A, which is a large and global company. On the other hand, Company B, which is a medium-sized supplier company, thinks the most significant motive to digital transformation is competitors and its first advantage on the way of digitalization is preferability by customers. Moreover, for both companies, one of the barriers of the transition to Industry 4.0 is the lack of competent employees. Company B believes that non-competent suppliers also make the transition to Industry 4.0 difficult. Also, lack of financial resources and government supports are among the most critical problems for SMEs due to the need for investments. However, large-scaled Company A does not have any problems in accessing financial resources. Company A asserts that the cumbersome structure caused by being a large-scale company beclouds digital transformation. Traditional and status quo approaches, lack of vision, authority confusion and traditional IT structure are also obstacles to digital transformation.

Based on the findings, these companies, which are active in the white goods sector, are highly aware of Industry 4.0. Especially, considering the challenges faced by companies and the motives and advantages of transition to Industry 4.0, company size is very significant. For example, while the most critical barrier for Company B, a medium-sized supplier, is the lack of financial resources, the problems stemming from the size of the Company A, a global player, are the barriers to transition to Industry 4.0. Then, the size of the companies should be taken into consideration in further research. The current paper makes a comparative case analysis considering two white goods companies. In future studies, cross-sectoral comparisons can also be made.

References

1. Jänicke, M., Jacob, K.: A Third Industrial Revolution? Solutions to the crisis of resource-intensive growth. Solutions to the Crisis of Resource-Intensive Growth, FFU Report (2009)
2. Abramovici, M., Göbel, J.C., Neges, M.: Smart engineering as enabler for the 4th industrial revolution. In: Fathi, M. (ed.) Integrated Systems: Innovations and Applications, pp. 163–170. Springer, Cham (2015)
3. Lukač, D.: The fourth ICT-based industrial revolution "Industry 4.0"—HMI and the case of CAE/CAD innovation with EPLAN P8. In: 23rd Telecommunications Forum Telfor (TELFOR), pp. 835–838. IEEE (2015)
4. Bonciu, F.: Evaluation of the impact of the 4th industrial revolution on the labor market. Rom. Econ. Bus. Rev. **12**(2), 7–16 (2016)
5. Siemens: Siemens: Endüstri 4.0 Yolunda (2016). https://www.slideshare.net/mkugu/endustri-40-siemens
6. Ellram, L.M.: Offshoring, reshoring and the manufacturing location decision. J. Supply Chain. Manag. **49**(2), 3–5 (2013)
7. Kiel, D., Arnold, C., Collisi, M., Voigt, K.I.: The impact of the industrial Internet of Things on established business models. In: Proceedings of the 25th International Association for Management of Technology (IAMOT) Conference, pp. 673–695 (2016)
8. MacDougall, W.: Industrie 4.0: smart manufacturing for the future. Germany Trade & Invest (2014)
9. Kagermann, H.: Recommendations for implementing the strategic initiative Industrie 4.0: final report of the Industrie 4.0 working group. In: Wahlster, W., Helbig, J. (eds.) Frankfurt (2013). https://www.acatech.de/wp-content/uploads/2018/03/Final_report__Industrie_4.0_accessible.pdf
10. Hermann, M., Pentek, T., Otto, B.: Design principles for Industrie 4.0 scenarios. In: 2016 49th Hawaii International Conference on System Sciences (HICSS), pp. 3928–3937. IEEE (2016)
11. The Made in China Website (2019). http://english.gov.cn/2016special/madeinchina2025/
12. Catapult Website (2019). https://catapult.org.uk/
13. Society 5.0 Website (2019). https://www8.cao.go.jp/cstp/english/society5_0/index.html
14. Gill, M., VanBoskirk, S.: The Digital Maturity Model 4.0. Benchmarks: Digital Business Transformation Playbook. Forrester Research Inc. (2016). https://forrester.nitro-digital.com/pdf/Forrester-s%20Digital%20Maturity%20Model%204.0.pdf
15. Yin, R.K.: Case study Research: Design and Methods, 5th edn. SAGE Publications, Thousand Oaks (2014)
16. Eisenhardt, K.M., Graebner, M.E.: Theory building from cases: opportunities and challenges. Acad. Manag. J. **50**(1), 25–32 (2007)

How Science Communication Delivers Meaningful Frameworks About Industry 4.0 to the Public Sphere

Murat Can Becerir and Lutz Peschke(✉)

Department of Communication and Design,
Bilkent University Ankara, Ankara, Turkey
{can.becerir, lutz.peschke}@bilkent.edu.tr

Abstract. With the emergence of new media and interactive media technologies, the mediatized transformative process began. Specifically within the science communication context, the way we perceive issues transformed drastically. With the mediatization of forms of understanding and literacy, the public perception towards many topics, including environmental sciences has gained recognition. This research paper will be focusing on the connection between Industry 4.0 and science communication. By making a valuable contribution to the communication process through constructing a bridge between the industry and the general public, science communication provides a smooth and effective transition of mutually beneficial relationships. This paper concentrates upon the notion of sustainability and its relevance to the industrial revolution, which is also known as Industry 4.0. The paper will be exploring the ways in which how humanity can stay within the safe operating limits with the existing technology and resources. By taking into consideration the notion of sustainability, the paper will attempt to discover possible ways that how people can construct their discourse on the big framework where everything is interconnected to one another. By meaning, Industry 4.0 indicates the connection between automation, computers, smart technologies and the cloud system in which the flow of information has increased in unprecedented levels. Since Industry 4.0 is already happening and has an immense impact on our lives, science communication plays a big part in generating the accessible and relatable message of the implications regarding the age of Industry 4.0. From the automobile industry to the electronic products that we use every day, the majority of the markets use renewable energy in order to lower the costs of production and increase their profits. A disruption in the sustainable development of any given society may cause irreparable damages which can result in colossal economic and political instability. Hence, science communication plays a vital role in terms of providing an understanding of possible pros and cons regarding the practices of the industrialists and the business world. Since it can be quite hard to comprehend the full scale of the consequences of technical occupations and practices, science communication will be the vanguard in terms of constructing a meaningful framework for the general society. This paper will explore what might happen in the lack of understanding of sustainability in Industry 4.0 through renewable energy consumption. Particularly, how a situation looks like when a country experiences a huge setback for its economy, massive loss of life and property in an unsettled global economy. What kind of a

N. M. Durakbasa and M. G. Gençyılmaz (Eds.): ISPR 2019, LNME, pp. 288–300, 2020.
https://doi.org/10.1007/978-3-030-31343-2_25

scenario would people witness with a disruption in one part of the world, in a world of interconnected production of supply chains that stretch across the world? Therefore, this paper will explore the possible implications of Industry 4.0 through renewable energy consumption on a grand scale.

Keywords: Sustainable development goals · Renewable energy · Industry 4.0 · Science communication

1 Introduction

The world in which we are living is unique in relation to its era. In the status quo, it is possible to argue that we live in a global village where everything is interconnected with one another. The uniqueness of this interconnectivity is primarily derived from the recent technological advancements. This era of interconnectivity is also known as "Industry 4.0". Equally significant, the topic of renewable energy for sustainable societies is fiercely debated not only amongst the academic realm but also in the governmental and business world. For this reason, this paper will attempt to establish a relationship between Industry 4.0 and renewable energy through the medium of science communication. The research will be investigating what kinds of knowledge can be delivered to the public sphere through science communication in relation to the notions of Industry 4.0 and sustainability? This paper asserts that science communication is a pivotal phenomenon in terms of informing and delivering meaningful insights regarding Industry 4.0 to public sphere. It is vital to recognize the role that science communication plays when it comes to conveying particular messages to a targeted audience. Science communication is the essential medium in which experts of the relative fields desires to educate masses on particular subjects. This paper will construct a framework where the readers/audience will become familiar and even maximize the utility of the notions such as sustainability, renewable energy, Industry 4.0 and the Public Sphere.

2 Implications of Industry 4.0 Through the Lens of Experts

In order to delineate how Industry 4.0 can have an immense impact on the welfare of the society, the Boston Consulting Group in Germany conducted research. They found four different ways in which Industry 4.0 will be highly influential in the industrial expansion and improvement for Germany. The first stage they discovered was Productivity. BCG experts concluded that there is an increasing number of companies that are embracing Industry 4.0 applications. According to Rüßmann et al. [1] a growth of up to 150 billion dollars amongst the German manufacturing. This growth will be possible by maximizing the utility of sustainable systems. Secondly, revenue growth proved to be a significant factor which will affect the economy in Germany. BCG proposes that revenue growth will be encouraged through the Industry 4.0 applications. Within the foreseeable future, customer's are being expected to demand more stylized and customized products for themselves in which will require more enhanced equipment for the manufacturers to produce such complex goods. As a result, an estimated additional 30 billion euro's will be generated which is considered as a substantial amount for the revenue growth. In fact, this increase also represents 1% of Germany's GDP.

The third way in which the effect of Industry 4.0 will be observed will be in is employment. BCG experts propose that in alignment with the revenue growth, the employment rate will be increasing approximately 6% within the next decade. Not only the rate will increase 6% in general, but also particularly in the technical/engineering fields, the rates will even go further. In accordance with the new technologies which is emerging with Industry 4.0, the on-demand skills such as IT are also becoming more popular. Thus, more employees will be needed to serve in this area and employers will be creating new fields to meet the needs of the newly emerging business sector. Employing more people even by theory, simulates economic prosperity within a society since it means more people will be able to afford purchase goods and manufacturers will be sustaining their businesses. Finally, as the fourth-factor investment plays a key role. BCG experts assert that German manufacturers will be investing an estimated amount of 250 within the next decade.

2.1 Production and Sustainability

It's worthwhile to note that the growth and stimuli of the private sector in relation to Industry 4.0 will be affecting every stage of the production. From the design phase to the after sales/marketing, the shift from the conventional way of conducting business to a virtually endless digital network will be vital to observe. With the agency delivered by enhanced IT systems, the production processes will be fully optimized through the integration of them to each other. As a result, in the new system, the production systems will be fully automated in which means a much more effective and sustainable production culture. Moreover, one will be able to preserve the energy surplus and use the exact unit of energy needed for a particular type of production.

Industry 4.0 will be providing humanity with a new set of tools which will revolutionize the way we conduct our capital affairs consumption habits and consumer culture. Notions such as quality, speed, flexibility, and productivity gain a whole new level with the new age which is upon us. Improving the effectiveness of these notions leads to a more productive future, lucrative business ventures and customers who can afford to purchase the goods since their cost will be cheaper. Industry 4.0 will be constructing the base for the upcoming new business models, production types, diverse jobs, and various different innovations.

The sections aforementioned in relation to Industry 4.0 proves to be a vital point both for ordinary people and the private sector. For this reason, science communication has to be incorporated into these phenomena in such an effective way that people in the public sphere can be aware of the innovations that will change their lives. Raising awareness of the issue of Industry 4.0 which will also be connected to sustainability enables us to enhance our comprehension towards our future selves. Hence, highlighting the crucial importance of science communication in terms of reconstructing our mentality towards technology, production, and their effects will be beneficial for people who can live their lives in the best possible way. Science communication can serve as a tool for informing the public sphere on imperative topics which will have a considerable amount of impact on our lives.

3 Theoretical Aspect "Public Sphere"

Before focusing on concrete events and case studies, it is important to be familiar with the concept of the Public Sphere. Habermas asserts that "a realm of our social life in which something approaching public opinion can be formed" [2]. Today, the public sphere largely consists of media which includes the internet, tv and print materials. In the earlier ages, when people had no knowledge regarding the internet or the tv, the medium of the sphere was different. In the 21st century, the active public sphere consists of the world wide web, TV and screen technologies in general. With the advancement of technology and innovations, our public sphere expanded to a much larger extent. By connecting to the internet, one gains access to the global village in which we're living in.

Media should not be merely associated with TV, but also linked with the internet which revolutionized the concept of media itself. Anyone who has access to the internet also gains access to a platform in which they can connect with a global community and participate in a new form of the public sphere. Since Industry 4.0 is significantly related to internet technology, Internet of Things (IoT) to be more specific, it is crucial to recognize the significance of the relationship between the internet and Industry 4.0. Hence, this paper will be referring to matters in the grand scale of media considering the public sphere.

4 Sustainability as a Concept

Moreover, sustainable development is one of the core themes in which this paper will be focusing on. In order to provide a general understanding of sustainability, it would be useful to overview the sustainable development goals (SDGs) of the United Nations. The total 17 goals encapsulate the Essentials for the transformation of the world including zero hunger, no poverty, gender equality, affordable and clean energy, sustainable cities and communities and so on [3]. In a general sense, these goals exist with the purpose of creating a sustainable and highly functioning environment. and leaving it as a heritage to the new generations to come. Jeffrey Sachs [4] defines the age we're living in as "We have entered a new era. Global society is interconnected as never before. […] I suggest that we have arrived in the Age of Sustainable Development". in his book the Age of Sustainability.

In order to illustrate the state of public awareness to sustainability, this paper will be focusing on the intersection of public engagement and Industry 4.0. To accomplish this, we will be focusing on science festivals within the scope of public engagement with particular themes and issues. Concrete case studies are beneficial for the sake of delineating how the public engages with certain topics in the public sphere.

5 Science Communication as a Concept

In order to initiate a discussion and provide an enhanced form of understanding towards Industry 4.0 through science communication, one should focus on science and technology studies. Equally significant with science communication, science and technology studies (STS) puts forth worthwhile notions which support our comprehension towards innovation, technology, research, and responsibility. According to Stilgoe and Guston, STS had the primary function of providing an understanding of the course of the results and process of science and technology management. RRI, which stands for Responsible Research and Innovation sets a couple of goals in order to provoke a meaningful discussion. "First, on the environmental implications and applications of nanotechnology and, second, on integrating social sciences, as a way of divining and making sense of societal concerns, into the nano-scale science and engineering research itself" [5]. The goal is to establish a bridge where nanotechnology, social sciences, engineering, and societal concerns can be discussed and made sense of on the common ground. Stilgoe and Guston assert that the notion of innovation is being interpreted as the offsprings of conventional presumptions concerning scientific autonomy and authority.

"Revealing the normative models of publics that are being enacted in science and policy, understanding and challenging "expert" presumptions about relevant public issues, describing the diversity of other cultures in which understandings might differ and be able to contribute, and, finally, putting this in the context of historical and philosophical perspectives on science as public knowledge" [5]. By comprehending and challenging general knowledge about public issues, STS aims to deliver new, meaningful and creative ways of rethinking pressing matters at hand. Concentrating on the different cultural contexts can be considered as a pivotal aspect since meaning can be derived from many different cultural aspects of a given society. Hence, STS underscores the presence and the functionality of responsible innovation which is both an adversary and an opportunity where the conventional way of conduct and innovative way of practicing matters can be closely studied.

5.1 Science Communication in Germany

The relation of science and public was subject to a big change in the course of the last century. This was mainly associated with the temporal development of science. In the 17th and 18th century, neutral scientist started to work on a view on the world which differed from the religious dogmatized ideology of that time. But these activities were only possible with the help of patronages which could only be provided by the aristocracy. For instance, Galileo served at the court of the Medicis which gave him the status of an independent researcher. In principle, he was the first researcher in the modern understanding [6]. Until that time, the society did not know the differentiation of different scientific subjects neither the separation of scientists and non-scientists. Hochadel called the 18th century the "golden age of lay science" [7]. The central goal was to present experiments with big surprise effect to a broad and amazed audience. With increasing differentiation of science, scientific work moved increasingly to enclosed laboratories which were specially built for those issues. The public participation and attestation of results were complicated by the aid of precise measurements which become increasingly an important component of scientific research. This shift of science to enclosed laboratories changed the science communication crucially. On the one hand, results were primarily shared and discussed with peers of the collaborative laboratory. On the other hand, spectacular results and discoveries were presented to the public in a kind of secondary communication [8].

The displacement of scientific productions to an enclosed location, where only selected people and peers had access to was conventionally called professionalization of science (ibid.). In a way of countermovement, there arose a movement of popularisation of natural science in the 19th century.

Associations were founded which were supported and funded by well-educated and interested citizens. Popular scientific presentations and lectures were organized and institutions were established, where lay people could execute their scientific work. One example is the Urania in Berlin where Alexander von Humboldt presented his studies in his famous lecture about the cosmos to a diverse audience in 1827/28. Among the audience was King Friedrich Wilhelm III as well as master bricklayers [9].

The term popularisation is grounded in the meaning of popularity which is understood in term of intelligibility, for the people, etc. [10]. However, from the very beginning, this concept stuck to the image of the shallow and the trivial. Kretschmann worked out that Liebig repudiated all popularising forms of representations in is Chemische Briefe (Chemical Letters). He stated that the diverse connotations of the term popularisation complicate the modeling significantly. According to him, older approaches regard popularisation as knowledge transfer in a hierarchical way from a small circle of experts to a non-specific public. It means the imparting action remained oriented towards academic science production. The firstly academically generated knowledge will be later communicated to the lay people in a simplified way. However, the audience was involved neither in the production nor in the distribution of knowledge (ibid 21). Daum talks about a "diffusionistic model" in this context [11]. In reply to this model, Shinn and Whitley offered an interactionistic view, where scientists, popularisers, and public are the actors of interactive communication between producers and recipients [12]. Popularisers are almost media-based actors. In 1851 the first world exhibition was executed in London. It was a new format of popularisation of science and technology. Within the scope of the technological progress, it should be demonstrated to the broad western public that the technological progress enhanced the living conditions in the world and that the still continuing progress effects the conditions in an ongoing positive way.

But at the end of the 20th century, there was a big change in the societal acceptance and interest in science and technology. Schmidt compared the relation of (not only) the young generation to science with the meeting of Jim Knopf and Lukas the locomotive driver with the "pseudo-giant" Mr. Tur Tur in Michael Ende's same-named novel [13]. The main feature of the pseudo-giant was the phenomenon that he did not become bigger the closer Jim and Lukas came to Mr. Tur Tur but smaller [14]. This described phenomenon is completely unrealistic from the point of view of a natural scientist since it contradicts all observations in nature. But with a look from outside to the system natural science, the science itself corresponds with the feature of Mr. Tur Tur. Young people often do not decide to study science and technology because they are too afraid of the complex discipline. The reason is almost the fear of the distance to the subject. But the more people are engaged with science the more they get the feeling of their own potential to manage the complex problems of science and the fear of dealing with disappears [15].

According to the model of Shinn and Whitley, science communication in the sense of science popularisation means that science and society have to be involved in the process of popularisation in the same way. For a long time, the opinion of science and especially of scientists was that the public has to approach to science. Otherwise, the risk of trivialization of science and the decrease of the educational level would be significant. This understanding was very common in the institution of science. It enabled an undisturbed study for the scientists. But at the same time, a kind of ivory tower arose around the scientists which they left rarely. It was not necessary to leave the ivory tower. Economical enterprises will fail on the market if they would develop and design a product which is not equivalent to the need of the consumers. But as opposed to economic enterprises, scientists are spared the failure, since scientists proceed their research always under the condition not to know the result of their activities. Even a negative result is understood as positive recognition. Accordingly, there was no direct need to get in contact with the public. The communication and discourse happened in a closed shop which developed an own "language". The internal audience, the peer was regarded as the relevance audience [15].

Within the scope of the dramatic drop in the numbers of student beginners in science in Germany at the end of the 20th century, the scientific community has to rethink its concept of science communication. The drop was an indicator that the voluntary engagement with science disappeared. This caused a fatal cognitive discrepancy of the society between the products of science which are regarded as highly systemically important, e.g. smart technology, and the interested in the scientific disciplines which are the basis of the products. To counteract this discrepancy, leading research institutions and associations, universities as well economy which was represented by the Stifterverband für die Deutsche Wissenschaft and the Federal Ministry for Education and Research negotiated a memorandum in 1999. The signatories agreed and committed oneself to enable the dialogue of science with society. Systems of incentives should be established which encourages scientists to get in dialogue with the society without losing their scientific reputation [16].

The process of the described public understanding of science and humanities (PUSH) was methodically introduced by a process of fuzzification of science [17]. It is understood as activities which transform precise descriptions of the scientific world into communicable hermeneutic forms to enable the connection of scientific phenomena with experiences of the life world and to support the process of reception (ibid., 78). This kind of scientific communication was a completely new and unfamiliar experience for the very most of German scientist. They were used to present their research topics to a peer audience. According to Neidhardt, science in Germany was used to loneliness and freedom [18]. But now they stood in pedestrian areas and shopping malls and presented their research activities to a shopping public which primarily did not come to this location to participate in science. When the Federal Ministry of Education and Research announced the "Year of Physics" in 2000 which was executed by iserund-schmidt – agency for science communication, the resentments among the scientists were very high to present their scientific research in the way described above. But in the course of the year of physics, they realized that this process of science communication for the public changed additionally the communication process between the peers. Scientist felt satisfaction during their presentations to a non-scientific public [19]. Schmidt stated that within the scope of estimation of the initiative "Science in Dialogue" a modernisation of the scientific process was started [13]. The consequence of the turn was a sustainable redevelopment of the communication policy: Scientific institutions turned from an object of public discourses to an active player (ebd.: 38). The feedback of the turn in Germany in the public was significant and it was newsworthy in the science magazine nature [20].

But the objective to place science in media and to make science accessible for a broad public was discussed in a controversial way. While Schmidt understands the steady media presence as a requirement to guarantee the return of science to the society, Weingart sees an abyss between professional science and the communicators in the media. He states that medical criteria for news values contradict the media criteria for scientific integrity. According to him, the increasing mediatisation of societal communication generates an increasing need for entertainment among the non-professional audience which is served by the conditions of media productions. It proceeds in a way that the media follow the prevalent dramaturgic formats considering the narrative, temporal and visual design a format [8, 21]. But the science edutainment concept provoked criticism and defensive attitudes. According to Weingart, Public Understanding of Science is sales campaigns which are directed to the mass media attention for event and entertainment, because they assume an unstructured and unspecific audience [8]. Science exhibitions follow just the principle of edutainment since the claim is not only education but also entertainment. He takes for granted that education is not the only value in the mediatised society. The media are forced to generate special kicks to reach new clients (ebd. 25). As a consequence, the specificity of science disappears because of the indistinguishability of conventional media formats. He brings up the question, which consequences regarding the legitimacy of science will arise. According to Weingart and Neidhardt, the safeguarding of scientific talents will not be reached with help of PR campaigns, but the setting the focus on peers which are resulted in the context of research and education in universities. The most important duty of scientists for the public will proceed in the universities and students are the most important audience which is mainly mistreated by the educational system [8, 18].

In contrast, Kretschmann stated that the ambitions of popularisation in Germany are more criticised the more entertainment the format consists of. This problem does not exist in the Anglo-Saxon countries [10]. He points to the fact that research, politics, and economy demand knowledge reception of the public which only can be proceeded voluntarily. But the broad public will bring this performance only when the media offer a certain degree of entertainment. Hence, the basic need of divertissement, relaxation, and dissipation has to be considered in the concept of science popularisation. Newest studies show that communication processes are more creative and effective than instructions which are based on pure information (ibid.).

5.2 IdeenExpo in Hanover

The IdeenExpo is a science festival which occurs every two years on the fair trade area in Hanover/Germany. It was initiated in 2007 by the Minister-President of Lower-Saxony at that time Christian Wulff. He was inspired by the IdeenPark, which was executed 2006 on the Expo Plaza close to the fair trade area. The organizing company was ThyssenKrupp. Member of the supervisory board of IdeenExpo is the employers' association NiedersachsenMetall. Sponsors of the science festival are enterprises, like Volkswagen or Sennheiser electronics. The first IdeenExpo 2007 was designed as a big technique show with 250 exhibits, workshops, experimental presentation, and talk events. The center of IdeenExpo was from the beginning until today an open-air science show on nine days at noon time where the TV moderator presented exciting and fascinating highlights of science. The IdeenExpo started with 160,000 visitors in 2007 and increased every year to 360,000 visitors in 2017. The aim of the science festival is to attract young people with science and technology. The target group is not only potential student beginners but also apprentices. Accordingly, the exhibition and interactive area dominantly consist of crafting and hands-on experiences. Besides, show acts are designed to introduce in jobs based on formal training which are widely unknown as a technical profession, e.g. mechatronic technician or skyscraper windows cleaner. Consequently, the science festival does not limit its focus on potential academicians but includes also students who will finish school after middle school graduation.

In the context of public understanding of science in Germany, with the establishment of IdeenPark and IdeenExpo, it was the first time that an economic entity realized a science festival with the goal to attract young people for science. It can be understood as an executive step to fulfill the self-commitment of the economy to invest in public understanding of science formates which politics, economy, and academia signed within the scope of the symposium "Science and Society" in Bonn in 1999 [16]. After 2007, the successful penetration of dialogic exhibits in the meaning of public understanding of science could be observed. While 2007 the event was predominantly designed as a technique show, in 2017 the hands-on experiments in the context of joining in formates increased. Additionally, the IdeenExpo contributes to compensating the significant mistakes of the 1970 s and 1980 s that the educational and governmental system created a system to increase the chance for children of families with low and middle income to study since university education was regarded as the chance to overcome educational and job discrimination. As a consequence, the majority of children were forced to study at universities and after 10 until 20 years the lack of technicians, plumbers, electricians and other workers based on formal trainings became significantly high. With the science festival IdeenExpo, young students from middle and high schools had the chance to learn about job perspectives beyond academic education.

6 Industry 4.0 as a Phenomenon

Industry 4.0 marks our age as a unique era in which undergoes a significant transformation through an unprecedented increase in the level of interconnectivity. What humanity already had was the connection between people and machines. With the emergence of Industry 4.0, the World witnessed much broader connectivity encapsulating both human-machine and machine to machine connectivity. "Full automation and digitization processes, and the use of electronics and information technologies (IT) in manufacturing and services in a private environment" [22]. Through being fully automated and digitized, our age has been revolutionized. When one investigates the recent changes that took place in people's daily lives, it becomes quite visible to identify the changes and revolutions. Considering that people are able to order food directly from their computer or smartphone to their fridge and being able to order various different kinds of goods from any location on the globe to their doorstep will indeed have a significant impact regarding the reconstruction of small and medium scale corporations.

For the sake of demonstrating how Industry 4.0 leads to the future of factories and smart factories which is connected to efficiency as well, its worthwhile to concentrate on Roblek's input on factories. "The smart factory will be more intelligent, flexible, and dynamic. Manufacturing will be equipped with sensors, actors, and autonomous systems. Machines and equipment will have the ability to improve processes through self-optimization and autonomous decision making" [22]. With these improvements, it's possible to have a more efficient and sustainable production system which would benefit both the customer (affordability) and the manufacturer (cost efficient).

"Industry 4.0 will be characterized by greater flexibility in the production process. It will be possible to create products for-fitted to the needs of a single client at a relatively low marginal cost" [22]. Hence, the corporations and manufacturers will be benefiting from the low marginal cost which will be projected on the consumers as well. With this technological revolution, the goal of achieving sustainable economic growth within society seems to be more feasible. Furthermore, Roblek puts forth that smart cities will be benefiting significantly from these improvements. The existence and enhancement of smart cities are considered to be the offspring of IT technology and data-based systems. "Universality in industrial applications: Internet of Things, Internet of Services, Internet of Media, big data, communications inter-machines and cyber-physical systems using interoperability, decentralization and full virtualization certainly will affect different course of many phenomena than is apparent from past experience" [22]. As stated by Magruk, today and in the foreseeable future the way we communicate with each other and with our environment will be drastically affected [23, 24].

6.1 Industry 4.0 and Renewable Energy

Renewable energy is being discussed in a wide range of bodies including governmental institutions, the private sector and in the academic circle. In fact, this demonstrates the severity of the situation globally. Increasingly, more and more funding is being allocated by governments and companies. In academia, Universities and academic conferences grant a separate funding/scholarship to encourage research on renewable energy and how it can be more efficient and accessible worldwide. Since this situation is being dealt within the era of Industry 4.0, it becomes even more significant to identify which entities are benefiting from renewable energy and in what ways can it contribute to the welfare of the society. In general, smart goods manufactured by smart factories use smart grids as a source of energy.

"A smart grid dynamically matches the energy generation of suppliers using renewable energies with the energy demand of consumers, e.g. smart factories or smart homes, by using short-term energy storages for buffering. Within a smart grid, energy consumers and suppliers can be the same" [24]. In the status quo, we have the tendency to identify multiple different entities as smart. The reason for this is that with the current technological capabilities and advancements lead both the consumers and companies to name today's goods as smart to demonstrate their various skills and efficiency. By utilizing the short term energy storages, it seems that the smart entities are maximizing their efficiency. Stock and Selinger asserts that by being involved in self-sufficient supply as well as taking advantage of the supply brought by the external smart grid, the factory will be existing both as a consumer and as a supplier. This will be all made possible through the factories increasing usage of renewable energy resources.

6.2 Economic, Social and Environmental Aspects of Industry 4.0

By allocating resources effectively such as materials, products, and energy, indeed economic benefits are unavoidable. Stock and Seliger argue that in social terms, it is possible to enhance the efficiency and productivity of the workers by incorporating the latest ICT technological with the virtual reality headsets. "Sustainable business models significantly create positive or reduce negative impacts for the environment or society or they can even fundamentally contribute to solving an environmental or social problem" [24].

Thus, sustainability has such an impact that it can radically alter the shape of the business itself which also means that it can regulate the flow of it too. In other words, sustainability has the agency to control economic growth as well. By allocating the right amount and type of material to the right place, sustainability also affects the environment as well. In essence, sustainable systems do not require any external resources since they can generate the necessary energy/power within themselves. Therefore, achieving sustainable development is vital for the existence and evolution of Industry 4.0.

7 Conclusion

As the paper deals with various notions such as sustainability, renewable energy and Industry 4.0, we have witnessed how crucially significant they are for us today and for the foreseeable future. Hence, science communication needs to be used as effectively as possible to communicate with people about relevant content. The communication has to be so effective that an ordinary person in the public sphere should be able to comprehend the basic concepts alongside with its pros and cons. Thus, it would be possible to pave the way for enhanced comprehension regarding our surroundings. Possessing knowledge about Industry 4.0, sustainability, renewable energy and its implications for humanity will enable us to live our lives more wisely and efficiently.

Acknowledgment. This research paper was generated within the scope of the project "POWER: Empowerment of Youth on Renewable Energy for Sustainable Societies", which is funded by the Erasmus+ Program of the European Union 2017-3-TR01-KA205-048402.

References

1. Rüßmann, M., Lorenz, M., Gerbert, P., Waldner, M., Justus, J., Engel, P., Harnisch, M.: Industry 4.0: the future of productivity and growth in manufacturing industries. Boston Consult. Group **9**(1), 54–89 (2015)
2. Habermas, J., Lennox, S., Lennox, F.: The public sphere: an encyclopedia article (1964). New German Critique **3**, 49–55 (1974)
3. United Nations Development Programme Sustainable Development Goals 2019. https://www.undp.org/content/undp/en/home/sustainable-development-goals.html. Accessed 14 May 2019
4. Sachs, J.D.: The End of Poverty: Economic Possibilities for our Time. Penguin, London (2006)
5. Felt, U., Fouché, R., Miller, C.A., Smith-Doerr, L. (eds.): The Handbook of Science and Technology Studies. MIT Press, Cambridge (2017)
6. Biagioli, M.: Galilei, der Höfling. Entdeckung und Etikette: Vom Aufstieg der neuen Wissenschaft. S. Fischer, Frankfurt am Main (1999)
7. Hochadel, O.: Öffentliche Wissenschaft. Elektrizität in der deutschen Aufklärung. Wallstein, Göttingen (2003)
8. Weingart, P.: Die Wissenschaft der Öffentlichkeit Essays zum Verhältnis von Wissenschaft Medien und Öffentlichkeit. Velbrück Wissenschaft, Weilerswist, Weilerswist (2005)
9. Orland, B.: Reisen zum Mittelpunkt der Erde. Anfänge und Aspekte einer Geschichte der Populärwissenschaft. Kultur/Technik, vol. 3, p. 48 (1996)
10. Kretschmann, C.: Wissenspopularisierung. Verfahren und Beschreibungsmodelle – ein Aufriss. In Boden, P., Müller, D. (eds.) Populäres Wissen im medialen Wandel seit 1850. Berlin: Kulturverlag Kadmos, Berlin, p. 20 (2009)
11. Daum, A.: Wissenschaftspopularisierung im 19. Jahrhundert. R. Oldenbourg, München (1998)
12. Whitley, R.: Knowledge producers and knowledge acquirers. Popularisation as a relation between scientific fields and their publics. In: Shinn, T., Whitley, R. (eds.) Expository Science: Forms and Function of Popularisation. Springer Science + Business Media, Luxemburg/Berlin (1985)
13. Schmidt, M.: Raus aus dem Scheinriesen-Dilemma. In: Mager, B., Hamacher, H. (eds.) Marketing und Kommunikation von Forschung. Research Goes Public. Köln International School of Design, Köln (2003)
14. Ende, M.: Jim Knopf und Lukas der Lokomotivführer. Thienemann, Stuttgart (2004)
15. Peschke, L.: Wissen sichtbar machen – von der Verbindung des Nützlichen mit dem Schönen. In: Seuser, K. (ed.) Zukunft im Technikjournalismus. Hochschule Bonn Rhein-Sieg, Sankt Augustin (2014)

16. Stifterverband für die Deutsche Wissenschaft. Dialog Wissenschaft und Gesellschaft. Symposium "Public Understanding of Science and Humanities – International and German Perspectives". Stifterverband für die Deutsche Wissenschaft, Essen (1999)
17. Peschke, L.: Infografiken. Visualität und Wissensaneignung in der mediatisierten Welt. Springer VS, Wiesbaden (2019)
18. Neidhardt, F.: Wissenschaft als öffentliche Angelegenheit. WZB-Vorlesungen 3. Berlin: Wissenschaftszentrum Berlin, Berlin (2002)
19. Lahr-Nilles, E., Wandelt, K.: Eine Stadt aus Licht und Laser/Wissenschaft im Kaufhaus - eine gelungene Premiere/"Stein der Weisen"—Physik trifft Kunst. Physikalische Blätter **56** (9), 18–21 (2000)
20. Nature: Germans engage with physics, Nature, vol. 406, p. 6798 (24 8 2000)
21. Donges, P., Imhoff, K.: Öffentlichkeit im Wandel. In: Jarren, O., Bonfadelli, H. (eds.) Einführung in die Publizistikwissenschaft. Paul Haupt, Bern (2001)
22. Roblek, V., Meško, M., Krapež, A.: A complex view of industry 4.0. Sage Open **6**(2), 2158244016653987 (2016)
23. Magruk, A.: Uncertainty in the sphere of the industry 4.0–potential areas to research. Bus. Manag. Educ. **14**(2), 275–291 (2016)
24. Stock, T., Seliger, G.: Opportunities of sustainable manufacturing in industry 4.0. Procedia Cirp **40**, 536–541 (2016)

Industry 4.0 Operationalization Based on an Integrated Framework of Industrial Digitalization and Automation

Andreas Schumacher[1,2(✉)], Christian Schumacher[3], and Wilfried Sihn[1,2]

[1] University of Technology Vienna, Vienna, Austria
wilfried.sihn@tuwien.ac.at
[2] Fraunhofer Austria Research GmbH, Vienna, Austria
andreas.schumacher@fraunhofer.at
[3] University of Economics Vienna, Vienna, Austria
christian.schumacher@wu.ac.at

Abstract. The utilization of Digitalization and Automation (DA) is currently one of the determining factors of increasing prosperity, productivity, and efficiency. Although widely promoted and incentivized by policy- and decision makers, industrial enterprises seem reluctant to actively push DA-development in their organizations. From a scientific point of view, we find missing operationalization resulting in a lack of analysis and understanding of basic DA-elements in real production environments as the main barrier towards Industry 4.0- and DA-implementation. Thus, in this paper we introduce a novel approach to transfer abstract Industry 4.0-concepts into industrial environments through the utilization of basic concepts of industrial digitalization and automation. Based on this framework we develop a model to assess the degree of digitalization and automation of value creation factors and organizational factors (DAVO). Knowledge about the company's DAVO-degree enables their decision makers to derive operational Industry 4.0 target-states, define strategies thus decide targeted investments into organizational and technological developments. We applied the developed and operationalized measurement-metric and model in an empirical study in the Austrian industry assessing the DAVO-degree of 200 industrial companies. Besides insights into their maturity status, we find clear evidence in our study for the need of operationalized Industry 4.0-concepts. Furthermore, the DAVO-approach seems to encourage practitioners to reflect their organization's DA-status with more accuracy thus enables them to derive more targeted and sustainable strategic steps.

Keywords: Industry 4.0 · Digitalization · Automation · Value creation processes · Empirical studies

N. M. Durakbasa and M. G. Gençyılmaz (Eds.): ISPR 2019, LNME, pp. 301–310, 2020.
https://doi.org/10.1007/978-3-030-31343-2_26

1 Introduction

Currently, the utilization of Digitalization and Automation (DA) in integrated manners is one of the main determining factors of increasing prosperity, productivity, and efficiency in both - society and industry. Although widely promoted and incentivized by policy- and decision makers, industrial enterprises seem reluctant to actively push DA-development in their organization. Practitioners name high abstraction and complexity of DA-concepts - i.e. through Industry 4.0-applications - as well as missing practical methods and tools to assess and increase their company's DA-grade as the main barrier towards realization [1]. From a scientific point of view, we identified the underlying problem as missing practical methods to assess DA that itself bases on the lack of analysis and operationalization of Industry 4.0 in real production environments [2].

To tackle these real-world problems, we aim for the development of an integrated concept of DA to reduce the complexity of Industry 4.0-concepts as well as a measurement-metric, -method as well as tool-support to evaluate and improve the DAVO-degree in industrial companies (see Fig. 1).

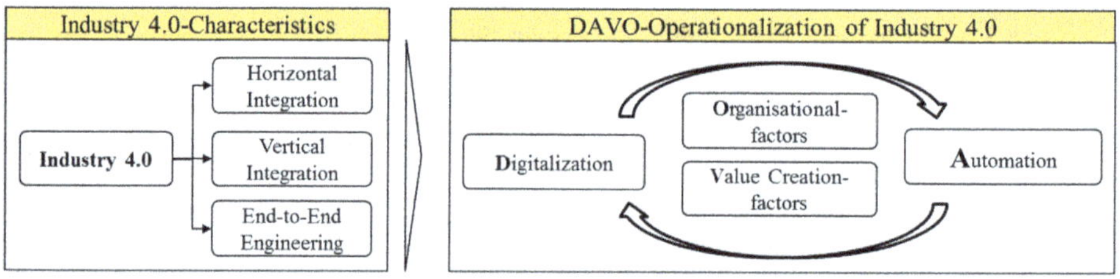

Fig. 1. Operationalization of industry 4.0 through four basic DAVO-elements

Knowledge about the company's DAVO-degree enables decision makers to derive DAVO-target-states, define strategies and investments into organizational and technological developments. To realize this research-project, we firstly define an integrated concept of industrial digitalization and automation. Next, we develop the DAVO-measurement metric and method through to relation of DA-elements with value creation factors and organizational factors. To allow for the model's testing, validation and re-work in early development phases, and to generate first insights into the industry's DA-development status, we evaluate the degree of digitalization and automation in 200 industrial companies in Austria.

2 Measuring of Digitalization and Automation – State of the Art

At the Industry 4.0-initiative's start, the German platform for Industry 4.0 defines Industry 4.0 as "[…] the intelligent networking of machines and processes in industry with the help of information and communication technology" [3]. However over time, scientist as well as practitioners increasingly use the term Industry 4.0 as a synonym or collective term for the implementation of various DA approaches - such as smart products, more flexible work arrangements, or the digital integration of customers. To consider both understandings, we focus on Industry 4.0's basic utilization for efficient production and include relevant organizational and external aspects such as products, customers or strategic elements.

To capture the state of the art, we conducted an extensive literature research on published research in the field of Industry 4.0 and industrial DA, as well as models and methods to allow their assessment. In Fig. 2 we summarized the search results after applying a criteria-based two-step filter-procedure on all findings resulting in 600 + relevant findings and around 25 publications that present the state of the art regarding specific Industry 4.0 and DA assessment models and methods.

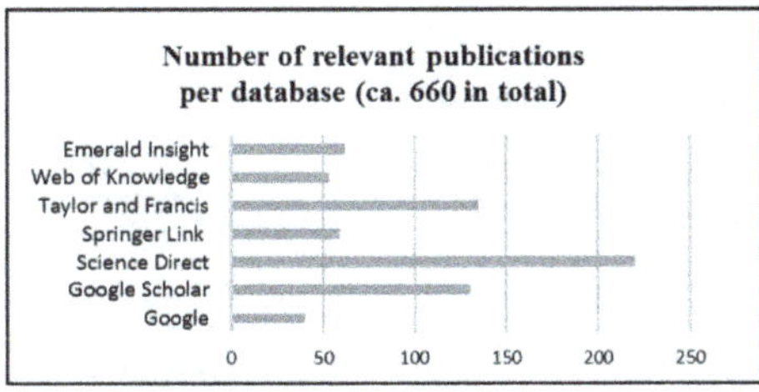

Assessment focus	Authors (extract)
General degree of digitalization	[4], [5], [6], [7], [8], [9], [10], [11], [12]
Degree of readiness or maturity in Industry 4.0	[13], [14], [15], [16], [17], [18], [19], [20], [21], [22]

Fig. 2. Overview of literature capturing most relevant assessment models

Several approaches exist for the overall assessment of an organization's degree of digitalization, which, however, do not specifically focus on industrial companies and do not show transparency with regard to their assessment logic. Scientifically more sound approaches evaluate DA without a comprehensive focus on value creation, consider the DA-degree in isolated areas, or utilize generic measurement-metrics and scales. A large group of assessment approaches focus on DA in the form of Industry 4.0 from various ankles – as our own Industry 4.0 maturity model developed in 2015 [2]. Overall, the approaches we reviewed show at least one of the following characteristics:

- Missing holistic approach to measure and assess the overall degree of DA of value creation and organizational factors - resulting in isolated assessments
- Measurement-metrics using abstract concepts such as integration, interconnection or intelligence of objects, systems or processes - resulting in difficulties to assess concepts accurately due to missing relevance for practitioners
- Mostly qualitative scales for DA-measurement - resulting in difficulties to assess concepts objectively
- Missing practical tools to assess and evaluate DA-concepts - resulting in missing real-world relevance

To our best knowledge, we did not find publications that present approaches with scientific rigour to measure the DAVO-degree in industrial enterprises. We therefore aim to close this research gap and define our two research goals targeted in this paper.

1. Firstly, the development of an integration DAVO-concept and -model capturing the core-characteristics of industrial digitalization and automation as well as relevant value creation and organizational factors.
2. Second, the transformation of the abstract DAVO-model into measurable items using quantitative scales allowing for objective and accurate measurement.

3 Operationalization of Industry 4.0 – the DAVO-Approach

To transform abstract Industry 4.0-concepts into practically observable DA-elements, initial definitions of all relevant concepts are required. We therefore distinguish the ideas and terms of digitization, digitalization, automation and Industry 4.0 and state our working definitions in Table 1:

Table 1. Working definitions for concepts

Digitization	Digitalization	Automation	Autonomy	Industry 4.0
Describes the technical conversion of continuous analog, noisy and smoothly varying information into clear bits of 1 s and 0 s	Describes the social implications of increased computer-assistance, new media and communication platforms for economy, society and culture and working environments	Describes the implementation of technology, software and programs to accomplish a procedural outcome with little or no human interference	Describes self-governing systems, processes or objects that operate and interoperate in conditions to reach defined outcomes and goals in automated manners	Describes the utilization of modern information- und communication technology to interconnect machines and processes in industrial environments

These four concepts of digitization, digitalization, automation and autonomy are distinct, yet merging concepts. We consider digitalization thereby as the most generic concept including social, organizational and employee-focused aspects besides technology. The concept of digitization focuses on technical aspects within the area of digitalization and we therefore relate to both - when referring to digitalization. Automation describes processes with little human interference - not necessarily by application of digital elements. However, to operationalize Industry 4.0 in industrial environments, we implicate automation by state-of-the-art technology, which requires high digitization. Finally, autonomy captures systems, processes that run in automated, decentral, and self-adapting manners, and are therefore considered smart or intelligent. It becomes clear that one concept cannot exist without the other, as any kind of automated and autonomous processes nowadays require digital elements to work without human interference and any kind of digitization requires elements to automatically handle and display digital data. An integrated application of these approaches on value creation factors and relevant organizational factors finally leads to the integrated concept DAVO - Degree of Digitization and Automation of Value creation factors and Organizational factors (in industrial enterprises) - see Fig. 3.

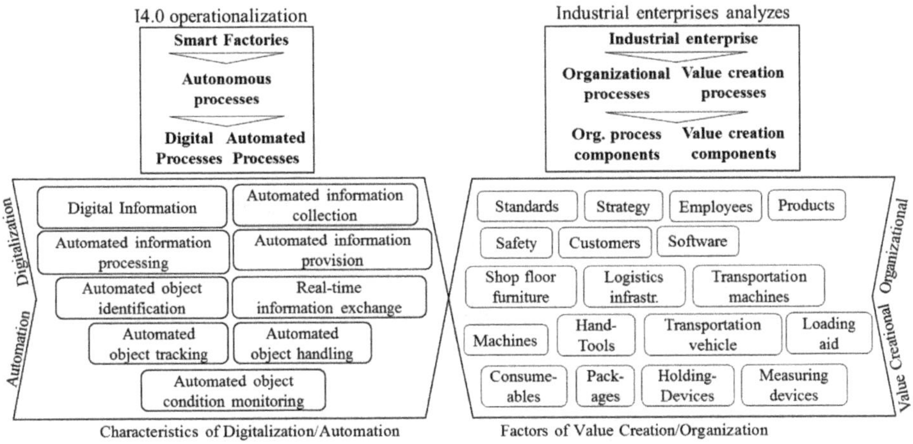

Fig. 3. Operationalization of Industry 4.0 through DAVO-approach

To operationalize Industry 4.0, we follow a multi-step approach: firstly, we depict Industry 4.0 as connected and integrated Smart Factories. Next, taking a process view, we divide the smart factory in smart organizational processes and smart value creation processes. We then argue that the concept of smart processes can be operationalized further by applying the concept of autonomous processes that operate in flexible and optimized manners to reach defined organizational outcomes and goals. Autonomous processes themselves can be composed with a combination of digital and automated processes on a physical and information level. Finally, we operationalize these two types of processes further by utilizing basic characteristics of digitalization and automation applied on all elements along them.

In a next step, we derive nine basic characteristics of digital and automated processes and match them with seven organizational and 11 value creation factors. Experts provided ratings for the relation's relevance and non-relevant relations were eliminated. Finally, we carried out additional literature research to define measurement-items for each relation that can be observed and assessed in real production environments. With the development of the DAVO-relation model shown in Fig. 3, we reached our first defined research goal.

4 Empirical DAVO-Measurement

Our second goal requires testing, if our level of Industry 4.0 operationalization allows for objective and accurate assessment. Therefore, we transformed the model into a preliminary online-assessment tool and defined the following survey set-up:

- Assessment of 200 Austrian industrial companies from the area of machine- and steel-related production
- Utilization of a questionnaire for online-self-assessment using 30 assessment-items
- Definition of 4 DAVO-levels to categorize all companies based on their progressiveness
- Definition of 6 preliminary DAVO-dimensions to allow for more detailed evaluation
- Half of all assessment-questions use qualitative Likert-Scales and half quantitative scales for further analysis of DAVO-measurement-metric

We define four distinct and progressive DAVO-levels (see Table 2) to categorize the assessed companies based on the average over all 30 measurement-items:

Table 2. Definition of DAVO-levels

DAVO-Levels (1-Lowest until 4-Highest)			
Level 1	Level 2	Level 3	Level 4
• No concrete software solutions planned/No modern ICT for employees • No or little basic data collection from machines • No autonomous internal transports/warehouses • No data collection for customers or data-based services • Digital customer integration not planned • No concrete digital training courses	• Concrete software solutions planned/selectively modern ICT for employees • Selective collection of extended machine data • selectively autonomous transports/warehouses • Data collection planned for customers or selectively data-based services • Digital customer integration planned • Concrete digital training planned	• Partially concrete software solutions and modern shop floor ICT • Selective collection of special machine data • Partially autonomous internal transports/warehouses • Partial data collection for customers or data-based services • Partial digital customer integration • Partially already digital trainings	• Comprehensive software solutions and modern shop floor ICT for MA • Comprehensive collection of special machine data • Largely autonomous internal transports/warehouses • Comprehensive data collection for customers or data-based services • Comprehensive digital customer integration • Comprehensive digital training courses underway

To assign a company to a DAVO-level, we relate the scale-ranges of all 30 questions to the four levels, depending on the current state-of-the-art in the respective field. Besides four DAVO-Levels, we group all assessment-items into six DAVO-dimensions to create additional insights during data analysis.

Overall, we collected around 7000 observations from around 200 companies through a survey-tool over the assessment-period of 4 weeks including one reminder that resulted in around 8% participation increase. Using the software-packages Tableau™ and Stata™ for data-analysis, we created a detailed assessment-report. The overall DAVO-assessment results for all 200 companies are shown in Fig. 4.

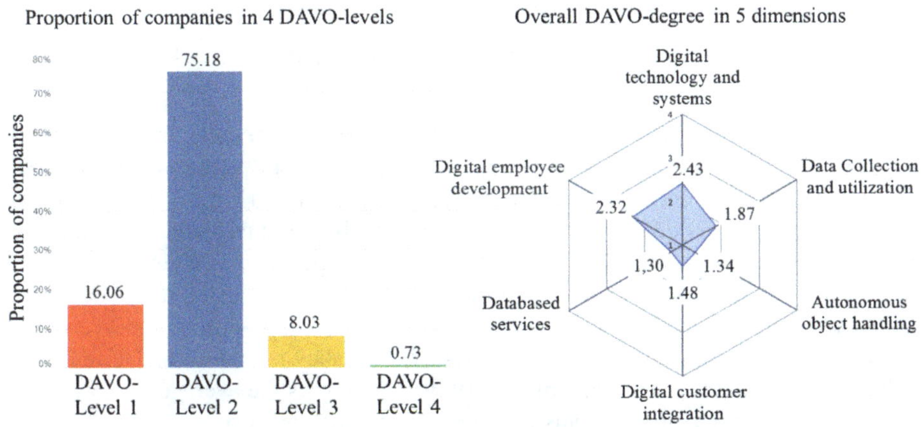

Fig. 4. Summary – assessment results DAVO-levels and DAVO-dimensions

Whilst around 75% of all assessed companies show DAVO-level 2, only one company rated itself with the highest possible overall DAVO-level 4. The dimension digital technology and systems (DAVO 2.43) and digital employee development (DAVO 2.32) are the ones most developed. On the contrary, the dimensions databased services (DAVO 1.30) and autonomous object handling (DAVO 1.34) are least developed. On a more detailed level, questions aiming to measure data collection at customers and digital customer integration revealed areas with low development. Around 75% off all assessed companies do not plan to collect data at customers (cross-value over three related questions), 75% do not plan to digitally integrate customers into the company (cross-value over five related questions) and around 66% of the companies currently do not offer any internal trainings regarding databased business models.

We find from high participation rate, as well as collected feebdack through the survey, that the operationalization of DA-elements seems to enable and encourage practitioners evaluate their own organizations, as no specific knowledge regarding digitalization, automation or Industry 4.0 is required. In Fig. 5 we show an exemplary question that operationalizes the Industry 4.0-concept vertical system integration by rating the proportion of automated exchange of data between machines and IT-Systems (e.i. MES or ERP).

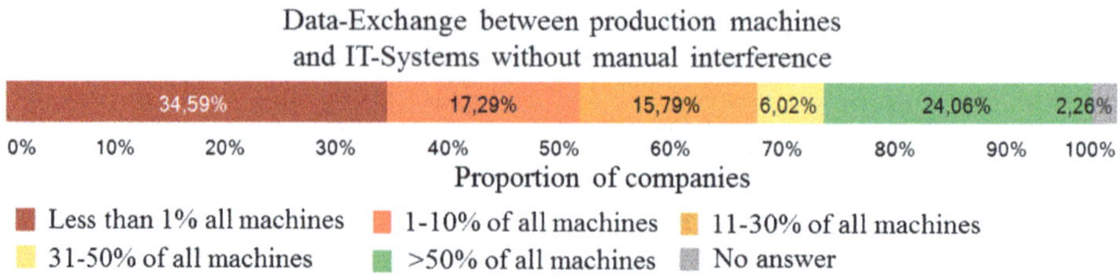

Fig. 5. Exemplary assessment of questions using quantitative scale

By operationalization of Industry 4.0-concepts into observable DA-elements in daily work-processes of pracitioners, no knowledge regarding Industry 4.0 is required. Moreover, instead of using Likert-scales reaching from "no machine" until "all machines", we defined quantiative assessment-scales that require participants to assess their as-it-is-status without subjective interpretations in percentage-scales.

To examine this research goal of objective DAVO-measurement in more detail, we defined two groups of questions in our survey, 15 questions with qualitative Liker-scales and 15 questions with quantitative scales as shown in Fig. 6. During data-analyzis, we created two new variables, whereby one shows the average maturity level over all qualitatively assessed items and one the maturity level over all quantitatively assessed items - see Fig. 6 on the right-side.

```
. sum Average_Qant_Attr

    Variable |        Obs        Mean    Std. Dev.       Min        Max
-------------+---------------------------------------------------------
Average_Qa~r |        204    .4244014    .1461467         .2   .8666667

. sum Average_QualAttribute

    Variable |        Obs        Mean    Std. Dev.       Min        Max
-------------+---------------------------------------------------------
Average_Qu~e |        215    .6067304    .1304711   .3428572          1
```

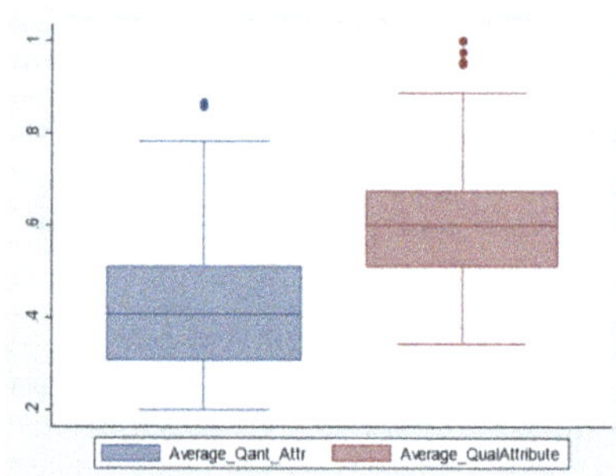

Fig. 6. Comparison of self-assessment using qualitative and quantitative assessment-scales

By comparing the maturity level of the two new variables (left: Average_Quant-Attr and right: Average_QualAttribute) we can show, that participants self-assessed their organization's maturity 20% higher if qualitative scales are used (compare mean-values). As the two different scales are equally occuring in all 6 DAVO-dimensions and all 30 questions, different technological readiness levels cannot explain the significant assessment-deviation.

We therefore conclude that quantiative scales encourage participants to reflect their organization's as-it-is-status in more detailed and with less subjective assessment-bias. We therefore reached our second research goal presented in this paper, which was to transform the abstract DAVO-model into measurable items using quantitative scales allowing for objective and accurate measurement. We could furthermore show the clear need for objectivication of measurement-metrics in Industry 4.0 and industrial DA.

5 Conclusions and Outlook

In this paper, we aimed for the development of a novel DAVO-model and approach to operationalize the concepts of Industry 4.0 and to allow for objective and accurate assessment through defined DAVO-elements. We thereby presented our first development steps and carried out a survey to test and validate our method and to collect practitioner's feedback. Regarding our general DAVO-development, we conclude positively, that our encompassing approach can serve practitioners for strategic developments as well as scientists to base their future research on a scientifically sound and practically relevant model of industrial digitalization and automation.

Our next research step is the finalization of all DAVO-elements based on the relation model. Next, we will develop a systematic assessment and evaluation method and finally define a procedure model that guides companies from their first assessment until the definition of target-states and action fields.

Acknowledgment. Research and results presented in this paper were fully enabled by funding from the Austria Research Promotion Agency (FFG)/grand no. 872637.

References

1. Schumacher, A., Nemeth, T., Sihn, W.: Roadmapping towards industrial digitalization based on an Industry 4.0 maturity model for manufacturing enterprises. In: Proceedings 12th CIRP Conference on Intelligent Computation in Manufacturing Engineering, CIRP ICME, pp. 409–414. Gulf of Naples (2018)
2. Schumacher, A., Erol, S., Sihn, W.: A maturity model for assessing Industry 4.0 readiness and maturity of manufacturing enterprises. In: Proceedings 6th CARV International Conference on Changeable, Agile, Reconfigurable and Virtual Production, pp. 161–166. Bath (2016)
3. Plattform Industrie 4.0 (2019) Plattform Industrie 4.0 homepage. https://www.plattform-i40.de/PI40/Navigation/EN/Industrie40/WhatIsIndustrie40/what-is-industrie40.html
4. Berghaus, S., Back, A.: Gestaltungsbereiche der digitalen transformation von unternehmen: entwicklung eines reifegrademodells. Die Unternehmung **70**, 98–123 (2016)
5. Åsa, F., Stahre, J., Dencker, K.: Level of automation analysis in manufacturing systems. In: Karwowski, W., Salvendy, G. (eds.) Advances in Human Factors, Ergonomics, and Safety in Manufacturing and Service Industries, pp. 233–242. CRC Press (2016)
6. Kotarba, M.: Measuring digitalization – key metrics. Found. Manag. **9**, 123–138 (2015)
7. Booz & Company (2011) Stragedyand homepage. https://www.strategyand.pwc.com/media/file/Strategyand-Measuring-Industry-Digitization-Leaders-Laggards-Digital-Economy.pdf
8. Deloitte (2017) Deloitte homepage. https://www2.deloitte.com/content/dam/Deloitte/global/Documents
9. Boston Consulting Group (2018) Boston Consulting Group homepage. https://www.bcg.com/capabilities/technology-digital/digital-acceleration-index.aspx

10. PricewaterhouseCoopers (2018) Pricewaterhouse Coopers homepage. https://digital.pwc.com/en/products/digital-fitness.html
11. Accenture website (2016). https://www.accenture.com
12. Fraunhofer ISI (2018) Fraunhofer ISI homepage. https://www.isi.fraunhofer.de/de/themen
13. Schumacher, A., Erol, S., Sihn, W.: Automation, digitization and digitalization and their implications for manufacturing process. In: Proceedings International Scientific Conference "Innovation and Sustainability", Bucharest (2016)
14. Ganzarain, J., Nekane, E.: Three stage maturity model in SME's toward industry 4.0. J. Ind. Eng. Manag. **9**, 1119–1128 (2016)
15. Jodlbauer, H., Schagerl, M.: Reifegradmodell industrie 4.0 - ein vorgehensmodell zur identifikation von industrie 4.0 potentialen. In: Proceedings Informatik, pp. 1473–1487. Bonn (2016)
16. Klötzer, C., Pflaum, A.: Toward the development of a maturity model for digitalization within the manufacturing industry's supply chain. In: Proceedings 50th HICSS Hawaii International Conference on System Sciences, Waikoloa Village (2017)
17. de Carolis, A., Macchi, M., Negri, E., Terzi, S.: A maturity model for assessing the digital readiness of manufacturing companies. In: Lödding, H., Riedel, R., Thoben, K.D. (eds.) Advances in Production Management Systems. The Path to Intelligent, Collaborative and Sustainable Manufacturing, pp. 13–20. Springer Verlag, Heidelberg (2017). https://doi.org/10.1007/978-3-319-66923-6_2
18. Rößler, M.R., Haschemi, M.: Smart factory assessment (SFA): eine methodik zur integralen reifegradbewertung von produktion und logistik hinsichtlich lean und industrie 4.0. ZWF Zeitschrift für wirtschaftlichen Fabrikbetrieb **112**, 699–703 (2017)
19. Leyh, et al.: Assessing the IT and software landscapes of industry 4.0-enterprises: the maturity model SIMMI 4.0. Bus. Inf. Process. **277**, 103–119 (2017)
20. Schott, P., et al.: A maturity model to organize the multidimensionality of digitalization in smart factories. In: Yilmaz, Ö.F., Tüfekçí, S. (eds.) Applied Optimization Methodologies in Manufacturing Systems, pp. 20–42. IGI Global, Nürnberg (2017)
21. Issa, A., Hatiboglu, B., Bildstein, A., Bauernhansl, T.: Industrie 4.0 roadmap: framework for digital transformation based on the concepts of capability maturity and alignment. In: Proceedings 51th CIRP Conference on Manufacturing Systems, pp. 973–978, Stockholm (2018)
22. Rajnai, Z., Kocsis, I.: Assessing industry 4.0 readiness of enterprises. In: Proceedings 16th IEEE World Symposium on Applied Machine Intelligence and Informatics (SAMI), pp. 225–230, Kosice (2018)

The New Paradigm in the Task of Direction-Coordination of Innovative Technological Projects in the Environment of INDUSTRY 4.0

M. Numan Durakbasa[1], Jorge M. Bauer[1,2,3(✉)], Esteban Capuano[2], J. Martin Acosta[3], and J. Carlos Diaz[3]

[1] AuM, TU-Wien, Vienna, Austria
durakbasa@ift.tuwien.ac.at, doktorjbauer@yahoo.com.ar
[2] Robotic, UTN-FRBA Argentina, Depto Industrial, Buenos Aires, Argentina
[3] Robotic, University Palermo Argentina, Buenos Aires, Argentina
dazjuancarlos@gmail.com

Abstract. The Internet and the associated data highway have radically changed lifestyles and forms of production/services in a transcendent manner. Many authors characterize this qualitative change as a new industrial revolution. Exchange emails/information in electronic format and video conferences is already part of daily uses applied by vast sectors of the world population for the most varied purposes. Computers, tablets and cell phones are the access and interface nodes to this process. For several years now, Telepresence has allowed us to operate remote equipment more easily and monitor with online vision in order to control processes of different types that occur in almost any point of the global village from almost anywhere in the global village. INDUSTRY 4.0 has as its central objective the use of all these tools in an integrated manner and to reinforce a path of continuous improvement, increasing the efficiency and effectiveness of organizations.

The present work reflects the experience in two new aspects, opportunities and difficulties that arise with the new paradigms of the organization of teams for creative and participative workers. On the one hand, the difference between implementing Telepresence and not just telecontrol is analyzed. On the other hand, we present a new small software modules complementary to those already developed/implemented in our technological research team, that underpin a "Virtual Engineering Office" with its similarities and differences to the "Virtual Classroom".

Keywords: Virtual Engineering Office · Telepresence · Home Office · Concurrent Engineering · Remote work

1 Introduction

We live a permanent acceleration of the evolution of the organizational forms of production and fundamentally the forms that human work is organized. In relatively few decades, a very short time in the history of humanity, we can observe enormous changes, stages that experts model as industrial revolutions. Starting with the steam

N. M. Durakbasa and M. G. Gençyılmaz (Eds.): ISPR 2019, LNME, pp. 311–318, 2020.
https://doi.org/10.1007/978-3-030-31343-2_27

engine and the ingenious invention for dynamic control of this machine made by Watt, the period that began is known as the first industrial revolution (18th century). The social changes that this stage generates have been widely studied and analyzed by experts from the most varied disciplines of knowledge. The experts define the second industrial revolution, linked to the use of electricity (19th century), the third industrial revolution with micro-electronics (1950), the fourth industrial revolution with computerized data processing (1970), and the fifth industrial revolution with the widespread use of the Internet and the data highway.

Regardless of the name with which we characterize each stage, we will all agree that the enormous changes of each stage, radically modified the life of large sectors of the world population. In the case we analyze, in each stage the forms of the organization of work and the role of the different actors in the new forms of organization were modified substantially. Observing that the radical transformations in the organization of work forms are developed in less and less time we can appreciate that one or two generations today lived and participated in more than one of the stages that we've mentioned before. With this situation, people not only must learn permanently "what's new", but must also leave old incorporated paradigms to be an active part of the new possibilities. Generating tools to take advantage of the experience of many years of worker-activity and re-empower it in the new times, is a challenge to which we seek to contribute. In this direction, we present a new module that we incorporate into our Teleoperation – Telepresence System, as tools that we apply in the model of global cooperation in technological research at the university worker-teams distributed in the global village. Just like the experiences of CIM (Computer Integrated Manufacturing), IMS (Intelligent Manufacturing Systems), MFIF (Multifunction Inteligent Factories) and today the well-known experiences of all kinds in INDUSTRY 4.0, we can see that tools that were developed and experimented in university fields. quickly were projected and adapted to organizations of all kinds. The increasingly close integration of knowledge, technology and interaction of new forms of work, will surely continue to be the distinctive feature of this process.

2 Experience in Teleoperation/Telepresence at Work of University Groups

The rapid development of internet technology has enable additional approaches in University's Laboratory's by applying telepresence [1]. Effectiveness of teleoperation and telepresence can be maintained by development of necessary expertise on the role and value of the laboratory work [2]. Step by step we develop or adapt different applications in the environment of the "Austria-Argentina" university cooperation. In order to reach the goal of remote laboratory concept an important stage was to build up a telepresence system in the high precision and nanotechnologie laboratory (AuM-TU-Wien Karlsplatz,), so that remote experimental work is not only conducted in the framework of measuring instrument software but also supported by video and audio instrument room [3, 4, 5]. The general model of this construction with its main parts can be seen in Figs. 1 and 2.

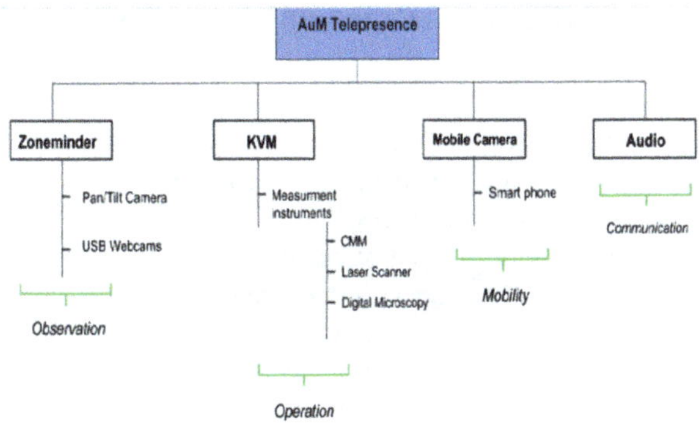

Fig. 1. Main components of the telepresence system

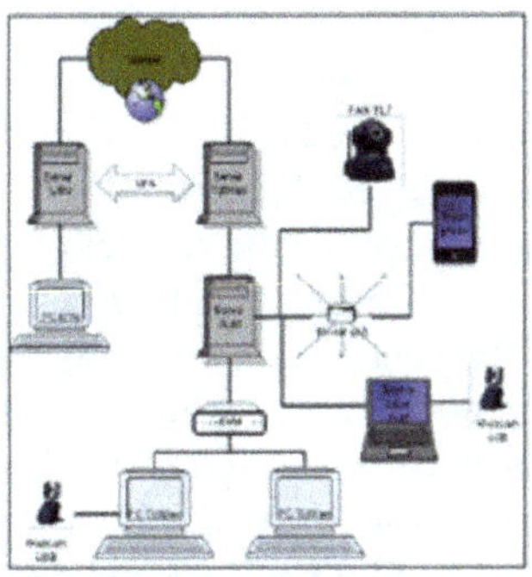

Fig. 2. Telepresence working environment at the AuM high precision nanometrology laboratory

This model already includes, from its first modules (2003), four different aspects of the Telepresence and Teleoperation System, namely observation, operation, mobility and audio. The specific equipment of micrometrology and nanometrology or the robotic units used in our laboratories, initially were not conceived by their manufacturers for teleoperation tasks. Our tasks for one interface to link its to the Internet-Network without interfering the specific software of the instrumental-specific equipment, were the first and main challenge. Today we can observe more and more instruments and equipment prepared from its origin to connect to the network of networks. The suppliers saw the opportunities that this also provides for their own organization, facilitating support tasks, maintenance, and even updating of part of the software. The real thing that is observed is that special equipment of laboratories every time has built in factory greater possibilities to work in a shared and cooperative way with teleoperation/telepresence. Based on this aspect, we understand that what we call the bottleneck were transferred to other aspects in the implementation of one integrated Telepresence/Teleoperation System. Sometimes it would seem that technological equipment can change in key aspects faster than many human beings, particulary some human habits and forms to do parts of his work.

3 Teleoperation and Telepresence as Part of a Working Group

Incorporating telepresence/teleoperation schemes into a real teamwork, is much more than install the software structure that sustains his work. Not only does it need an active acceptance of all the team members. According to some experiences, it is necessary to clarify doubts that many collaborators express implicitly or explicitly. Video cameras are every day part of our life in the most diverse areas. Nevertheless, video cameras are often perceived as control units, as "telecontrol", invasive of the privacy live. In this sense, video and cameras act consciously or unconsciously against the comfort environment that we must generate in each collaborator so that he can freely and creatively work towards the assigned objectives. Generating tools that aim to differentiate "Telepresence" from "Telecontrol" is important in our conception of work.

In this way we have developed a software module in which all the cameras in a room are linked to a computer. At the moment one person remotely accesses for a "telepresencial" section, his image and the specific images that he is remotely receiving are displayed, preferably in a largesized monitor at the laboratory. Remote camera images are not transmitted and received directly, but only the complete image formed on the computer monitor by the orders of the remote operator, that is the same image that the person working in the observed laboratory can see. The sought effect is the same as if the remote observer had entered the room, as his presence and who he is, are instantly perceived and transparently known what he is looking. In this sense, what is sought is that nobody feels his spied on. Even automatically if any of those present in the laboratory or who accesses it remotely believe it necessary, a trigger conversation via audio start. So we can complement the images and increase the remote collaborative participation as if physically the remote colleague had entered through the laboratory door. The system is the same if there are no people working at the moment at the laboratory. In that case nobody is perceived spied and simply in a telepresence way we operate and observe equipment and robot at a distance.

4 The "Working Up" Module, as Part of the-Virtual Engineering Office

The module "Working UP" was developed in Argentina, specifically by collaborators of the robotic group of the University of Palermo for its own use and shared in the international network. Its general objective is to incorporate, step by step, tools that promote synergy between team-collaborators to work in different locations and with a flexible schedule, generating a "Virtual Engineering Office". As we previously mentioned, the difference between "Telepresence" and "Telecontrol" should be perceived instantaneously as a substantial aspect in each module of the system. As we already enunciated, our experience tells us that many collaborators who perceive a Tele-CONTROL do not work comfortably to make an autonomous creative work with team criteria based only on the aspects of direction by objectives.

On the other hand, the coordination of the working groups and human resources that work with Telepresence, Home Office and Concurrent Engineering for Research & Development, also needs elements that in classical organizational structures, "a Office of Classical Engineering", were implied by the made of a simultaneous work of team-members in one building/office and in a common time band. The leader, (the coordinator) usually sees from his desk the members of his team or by taking a few steps he can approach each one if he considers it necessary. Without losing the capacity to apply management methodologies by objectives, experience tells us that the leader of a group also needs tools that allow him to evaluate the performance of each team-member, and to know when to contact immediately to analyze together different aspects - interface of the common project and/or coordinate other aspects, in spite of working in different locations and having only small time overlays.

Experience also indicates us that it is sometimes easier for young people (the millennials) to insert themselves in a scheme of "Virtual Engineering Office" as for many of those who today make tasks of coordination and project leadership. These people are usually professionals with more work experience, which in most cases implies more age, so they did not have similar experiences to work with their bosses in a "Virtual Engineering Office", their experience are to interact with his leader at an organization as a "Classical Engineering Office". They must "unlearn" certain forms of leadership, and learn others. They conserve conscious or unconscious old paradigms of organizational direction with all its positive effects and some negative or obsolete parts. Building bridges, to improve the leadership/coordination capacity of groups for work-people distributed at the global village, is the function of the second element proposed as software-module and organizational procedures to use at Hierarchical Organization Structures or at Work Matrix Structure, is the objective of the module "Working UP-Robotic"

With its similarities and enormous differences to the famous "Virtual Classroom", for the "Virtual Engineering Office ", we need to rethink and make experience with dynamic coordination tools feedback in fast and fluid bonds. The "Virtual Engineering Office " should allow us to re-empowering a project leader with years of experience, formed in a more traditional organizational scheme, basically allowing some aspects which are usual in a large number of organizations that operate at common buildings and unified schedule.

The zero version of the "Working UP" module is currently being used and perfectionated. It works as an APP in the cellular environment with Android operating system & a server in the cloud, and allows:

- Do "Check In", (Fig. 3) and also registering in optional way the work that is "thought to do", by the team-worker, his plan for his work at "VEO" (not more than 200 characters) Fig. 4
- Do "Check Out", and also registering in optional form the work that he "has done", the solved work or work-problems in its passage through the "VEO" (not more than 200 characters)
- See the members of the different groups associated with the leadership or members of his working-team that are "ACTIVE" in the "Virtual Engineering Office" Fig. 5

- Making a simple click on the name of an "ACTIVE" person in the "Virtual Engineering Office", a call is automatically established that links in audio and if necessary image, the team-leader with team-member, two team-members or a member with his team-leader, to interact in a dynamic telepresence way to clarify doubts, coordination task or other aspects that in a classic engineering office were solved with a small face-to-face interaction.

Fig. 3. Check In

Fig. 4. Text box

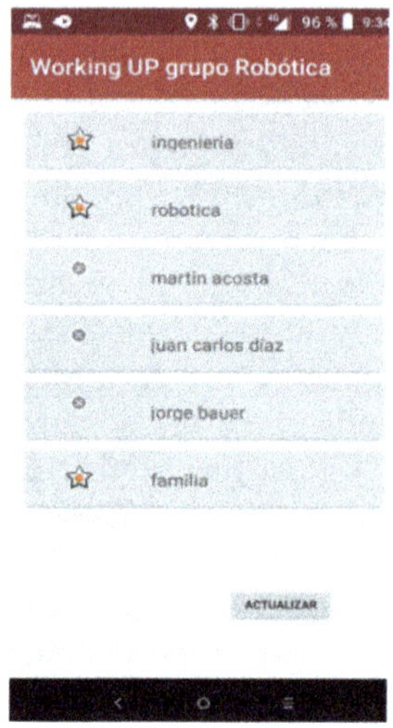

Fig. 5. Active team worker

5 "Working Up", also as a Way to Care for the Worker

A no less important aspect that we seek to incorporate as a procedure and habit, is also a form to take care of the worker at its various levels and the modus "Working UP" favors this philosophy.

The intensive use of cellular resources in the company gives initially a perception of freedom to the worker which in many cases is transformed in the opposite in a short time. It is observed that at the end, people many times work 24 h a day without having personal time for relaxation or for personal concerns.

The "Working UP" module lets you know which team- member is active in the VEO, triggering only a communication with the team workers who show up in that situation. Of course there are the urgent queries, which if they exist must be made through the classic cellular communication channel. The "Working UP" tool helps to make this situation the great exception and not the rule. Go back at that form to the way in which the team-worker of a project has assigned a series of hours in which he is in "Work-Activities" mode, hours that he fulfills not only with flexible schedule and also at the point of the global village that is more comfortable or useful for its work at project in the VEO.

Automatic records in the application server, allow to do both for the task of headship or for information of each member a series of reports, for example:

- Reports of the time in the VEO in a determined period for each team-worker.
- Place/places in which team-worker make his activity. This is based on the GPS information of the cell phone at the time that Check-In or Check-Out was done. Its is useful to sub-discriminate the time that each one is affected to lab or other tasks.
- Summaries or keywords of the activity that make one team-worker which enter at "VEO".
- This information also allows to be contrasted indirectly with the management scheme by objectives, having a perception if the assigned effort, quantified in assigned "engineering hours", was of the same order of the effort that the worker really has to do or it is only a part. One minimum historical register is always a material that is valued when completing the classic reports that every engineering project has.

6 Informatic Security

In times when IT security is a growing concern, the global TelePresence System must include these mechanisms, so that each worker can freely and creatively make the assigned objectives, enhancing a dynamic interaction between expensive laboratory equipment and also his human partners, to perform a true transdisciplinary work. The Telepresence System created, enables with classic passwords, each one in the use of its different modules. The own norms of computer security are fulfilled in the point-to-point links, on the basis of the basic software used for said process. In general, more and more university units are protected his computer links coming from abroad with "2 factor authentication", one technology that was originally implemented in banking transactions, where the security system must be strict, since the motivation to violate it is generally high. Data used in research/development are also subject to confidentiality, which means that the computer security of the system in general is not an issue that should be neglected. In any case, according to our point of view, the important thing here is to allow to do by "Telepresence" the same set off activities as if a team-worker enters the "building - Classic Engineering Office".

7 Conclusion and Future Work

In this global project, all modules are added to increase Teleoperation/Telepresence systematically from the first unit implemented in AuM-TU-Wien Micro and Nano-Metrologie Labor in 2003. The objective remains the same, to enhance the positive and useful aspects of Telepresence with simple, low-cost tools that enhance action in the global inter-university network for collaborative projects of creative participation with computer tools that empower and order this way of working. Future actions will continue this line of action, perfecting the benefits of each module, generating new applications and also transferring knowledge and experience to students in the first instance and to industrial companies in second place.

References

1. Krause, C., Strunz, U.: Vision system for teleoperating mobile robots, telepräsenz bei mobilen robotern. Forsch. Ingenieurwes. **63**(1), 18–26 (1997). https://doi.org/10.1007/PL00010747
2. Ma, J., Nickerson, J.V.: Hands-on simulated and remote laboratories: a comparative literature review. ACM Comput. Surv. **38**(3), 7 (2006)
3. Durakbasa, N., Bauer, J.M., Bas, G., Riep, D.: Micro and nanometrology laboratory applications in industry and education by telepresence and teleoperation remotely in the global village. In: Vortrag: 11th International Conference on Remote Engineering and Virtual Instrumentation REV2014, Porto, Portugal, 26.02.2014–28.02.2014
4. Bauer, M., Durakbasa, M.N., Bas, G., Güclü, E., Kopacek, P.: Telepresence in education. Elsevier ScienceDirect IFAC-PapersOnLine **48**(24), 178–182 (2015)
5. Durakbasa, M.N., Bas, G., Bauer, J.M., Riepl, D.: A model of global cooperation to enable intelligent metrology applications by means of an integrated teleoperation system. Int. J. Online Eng. (iJOE) **11**(3), 14–19 (2015)

The Turkish Automotive Industry in the Era of Digital Technologies and Autonomous Cars

Ahmet Tezcan Tekin(✉), Lerzan Özkale, and Derya Gültekin-Karakaş

Istanbul Technical University, Sariyer, Turkey
{tekina, ozkale, dkaraka}@itu.edu.tr

Abstract. Technological developments accelerating in the 21st century are making great changes in the automotive sector. Due to the reliance on technologies such as artificial intelligence, big data, robotics, cloud computing and internet of objects, self-driving cars have an essential place in Industry 4.0. However, it seems that these technologies will change the global market share of developing and less developed countries' in the automotive industry. The rising trend of reshoring activities of developed countries are expected to affect the global automotive industry. In this paper, we aim to investigate the impact of Industry 4.0 and the digital transformation on the automotive industry. Also, we aim to put a brief description of the social impacts of reshoring and how developing and less developed countries will keep or increase their market share in the global automotive industry.

Keywords: Automotive industry · Turkish automotive industry · Digital transformation · Autonomous cars · Reshoring

1 Introduction

The automotive industry has a vital role in the development of modern economies with its large domain and contribution to other industries. The automotive industry provides motor vehicles that are needed by many other sectors such as tourism, construction, agriculture and transportation. Thus, rapidly occurring changes in the automotive sector in terms of production, employment, exportation and importation affect related sectors considerably. [1].

Automotive is one of the largest sectors in developed countries. Global production of passenger cars hit 73.4 million units while total production of motor vehicles reached 97.3 million units in 2017 [2]. China, US and Japan are the key motor vehicle producers realizing more than half of the world production while Germany is the lead player in the European Region with a global share of 6% only [2]. The sector is characterized by high competition, especially in developed countries as the industry has reached saturation. Therefore, the importance of technology is increasing in this competitive environment.

The automotive industry has been one of the most rapidly growing sectors in Turkey [3]. The sector plays an important role in the Turkish economy considering relatively higher added value compared to other sectors as well as its positive effect on employment and transfer of advanced technologies. The automotive production

N. M. Durakbasa and M. G. Gençyılmaz (Eds.): ISPR 2019, LNME, pp. 319–327, 2020.
https://doi.org/10.1007/978-3-030-31343-2_28

reached nearly 1.7 million units, recording an increase of 14% in 2017. Turkey ranked 1st in the Central and Eastern European region, 5th in Europe and 14th in global automotive production in 2017. The rising trend in vehicle production in Turkey is shown in Fig. 1.

The global automotive industry has been going through a restructuring process under the joint effects of Industry 4.0, digital transformation, autonomous vehicle technologies and reshoring. In this study, we aim to examine how developing countries in general and Turkey in particular will take a market position in the global automotive industry which digital transformation affects crucially. Accordingly, on the basis of a survey of the relevant literature and analysis of secondary data, we discuss whether it is possible for the Turkish automotive industry to catch up with digital transformation and maintain its position in vehicle manufacturing. Then, we investigate if the country has any action plan regarding the adaptation of Industry 4.0 and the use of autonomous cars technologies.

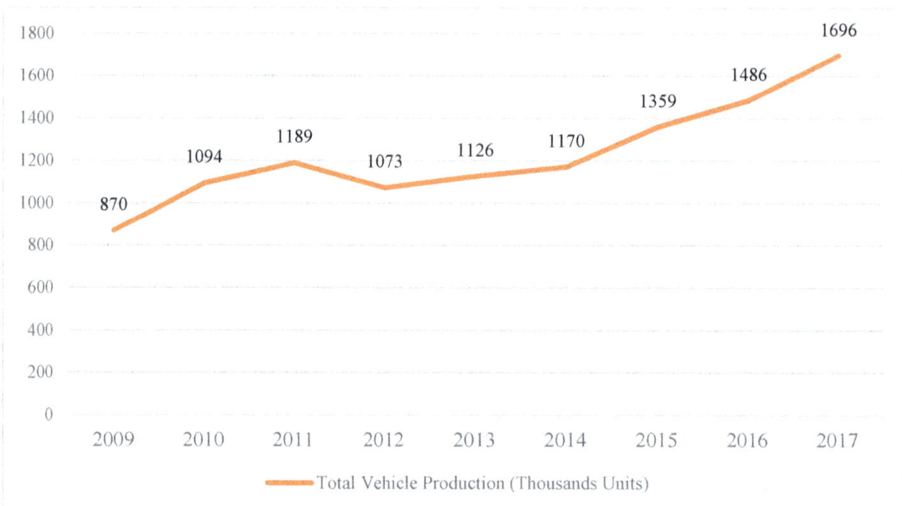

Fig. 1. Passenger and commercial vehicle production in Turkey (Thousands of units), 2009–2017 [3].

2 Industry 4.0 Impact on Automotive Industry

Industry 4.0 has already led to immense transformations in many sectors. One of the most highly affected industries is automotive. Accordingly, it is expected that there will be changes in cost, employment, productivity, speed, investment and income issues in the automotive industry. Automotive firms have been under the huge impact of digital transformation, given competitive market conditions and expectations of their customers [5, 6]. The automotive sector has been one of the fastest adapting sectors to Industry 4.0, owing to the increase of data exchanges and automation in production technologies in recent years [4].

The development and commercial release of Connected and Autonomous Vehicles is largely carried out by the aim of quicker, more reliable and safer vehicles producing more robust and resilient infrastructure of transportation [4]. Autonomous cars are also expected to contribute to the vision of zero accidents and zero emission. Zero emission plans of developed countries affect directly the industry's future. UK, Germany, Norway, and France are among the countries that have announced plans to forbid gasoline and diesel vehicles. As well, China, with the world's largest automobile market, has announced its zero-emission plan. Besides, it is expected that in 2025, 450 million vehicles will be connected to each other in the USA, the EU and China while fully autonomous cars will become widespread starting from 2028 [7].

The literature indicates that R&D investments and the strength of economies are related to each other. Therefore, developed countries spend more on R&D investments than developing countries [8]. The shares of R & D expenditures in GDP were analysed for selected developed countries and Turkey for the period of 2005-2017 (See Fig. 2).

This study [9] indicates that developed countries which are more ready than developing ones for autonomous car technologies spend more than the others. As a matter of fact, the actual levels of Industry 4.0 and R&D expenditures are correlated. Turkey has reached only one per cent of R&D expenditures in GDP and approaches 50% of the average of 28 European countries and 40% of the average of OECD countries. These results explain Turkey's lag in Industry 4.0.

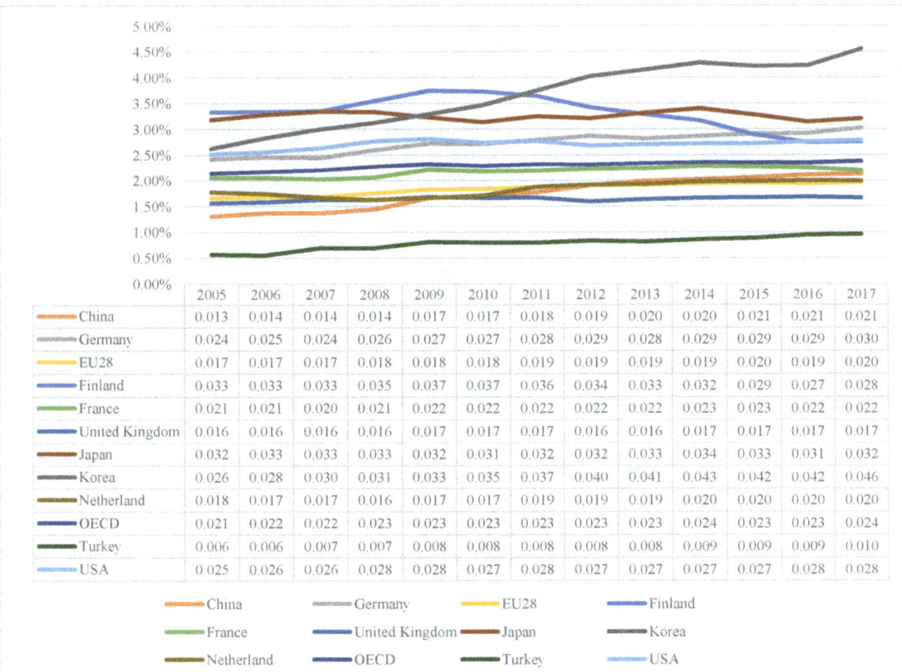

	2005	2006	2007	2008	2009	2010	2011	2012	2013	2014	2015	2016	2017
China	0.013	0.014	0.014	0.014	0.017	0.017	0.018	0.019	0.020	0.020	0.021	0.021	0.021
Germany	0.024	0.025	0.024	0.026	0.027	0.027	0.028	0.029	0.028	0.029	0.029	0.029	0.030
EU28	0.017	0.017	0.017	0.018	0.018	0.018	0.019	0.019	0.019	0.019	0.020	0.019	0.020
Finland	0.033	0.033	0.033	0.035	0.037	0.037	0.036	0.034	0.033	0.032	0.029	0.027	0.028
France	0.021	0.021	0.020	0.021	0.022	0.022	0.022	0.022	0.022	0.023	0.023	0.022	0.022
United Kingdom	0.016	0.016	0.016	0.016	0.017	0.017	0.017	0.016	0.016	0.017	0.017	0.017	0.017
Japan	0.032	0.033	0.033	0.033	0.032	0.031	0.032	0.032	0.033	0.034	0.033	0.031	0.032
Korea	0.026	0.028	0.030	0.031	0.033	0.035	0.037	0.040	0.041	0.043	0.042	0.042	0.046
Netherland	0.018	0.017	0.017	0.016	0.017	0.017	0.019	0.019	0.019	0.020	0.020	0.020	0.020
OECD	0.021	0.022	0.022	0.023	0.023	0.023	0.023	0.023	0.023	0.024	0.023	0.023	0.024
Turkey	0.006	0.006	0.007	0.007	0.008	0.008	0.008	0.008	0.008	0.009	0.009	0.009	0.010
USA	0.025	0.026	0.026	0.028	0.028	0.027	0.028	0.027	0.027	0.027	0.027	0.028	0.028

Fig. 2. Shares of R&D expenditures in GDP for selected countries, 2009–2017 [8].

Patents are another indicator of field related researches. They are also related to R&D expenditures. In addition to the unconventional car companies, technology giants such as Samsung, Google and LG also operate in the driverless cars field. Over the 2010–2015 period, there were in excess of 5,700 driverless car technologies patents registered [10]. This number reached nearly 13,000 in 2017 and almost 4,000 of them in 2017 alone [11]. Hence, the exponential growth in researches and patents indicate that technology giants and automotive companies have to collaborate for the success in this field.

3 Digital Transformation in Automotive Industry

New technologies that pave the way for digital transformation are truly groundbreaking. All actors in the automotive value chain must review and prioritize their strategic priorities and invest accordingly in order to stay competitive. According to the IDC World Digital Transformation Expenditure Guide [14], the money spent on digital conversion in the world is expected to grow by $ 2.1 trillion by 2021 and by an annual compound growth rate of 17.9% between 2016 and 2021. Information is at the heart of the new digital ecosystem and therefore needs to be treated equally with all other entities. In order to maximize the value in terms of organization, the necessary governance and investments should be exhibited [12].

The digital transformation period in the automotive industry focuses on the new technologies increasing the importance of Industry 4.0 such as robotics, internet of things, big data, cloud solutions and 3D printers [13]. Driverless car technologies cover almost all of these technologies. The greatest impact of digital transformation has been the emergence of the data economy. Data-driven new business models have become widespread. Many organizations are trying to figure out what data advantage they may have against their competitors, how they can collect and organize them, and how they can use it to improve business operations and customer service [12].

Automotive production is one of the most data-producing industries worldwide, and many institutions in this sector have begun to understand the importance of digital transformation and data economy [14]. The potential of data analytics to create more certainty in customer relationship goals and superior efficiency in business operations have transformed these new business models into a gold mine for companies in the automotive industry. The question is where and how automotive actors will not continue to invest in data.

Also, another observation is that the western countries leading the digital transformation race in the industry have made significant progress in this journey. Although digital transformation and production are considered to shift back to western countries, China has been one of the leading countries in the world [15]. Considering the high-technology product exports and the ratio of R & D investments to GDP, it is evident that there is a long way to go in order to reach the pioneer countries in the digital transformation race.

Measuring the level of competence of companies that use digital technology transformation in Turkey in order to realize the digital transformation of the industry will help separating areas of improvement and points to focus. [15]. Comprehensive research evaluating the competence levels of the technology user companies was applied to technology companies in Turkey. The findings of this research show that while the majority of companies state that they have high levels of knowledge and interest in digital transformation in the industry, the proportion of companies considering to be ready for transformation is relatively low. Moreover, companies in Turkey state that the biggest obstacles to digital transformation are high investment costs and uncertainty about return on investment [15].

4 Reshoring in the Global Automotive Industry

Another trend in the sector was that, starting from the mid-2000 s onwards, transnational companies started to reshore manufacturing operations to their home countries or closer regions because of political, economic and performance-related concerns. Especially automotive companies have brought many jobs back to their home countries. The rising trend in reshoring has been also related to the desire of global companies to keep high-tech R&D activities in their home countries accompanying the developments in digital technologies.

An interesting work by Baldwin [16] is worth mentioning about the possible radical transformation that digitalisation could bring in the globalisation trend in the 21st century. Baldwin argues that this will only happen if the cost of moving people falls in the future as much as the cost of moving ideas has in the recent past. Stating that labour is made mobile by robots enabling workers to offer their services remotely, Baldwin sees it a major factor to reduce labour cost differences in the world, thus the name of his book Great Convergence. If Baldwin's expectation is realized, such a transformation of the labour market would certainly impact the reshoring effects.

On the other hand, developing digital technologies such as big data, artificial intelligence and robotics which are at the heart of Industry 4.0 and digital transformation is changing the classical manufacturing methods. According to De Backer [17], the growing use of these new digital technologies will make labour costs relatively less important for competitive advantage in a number of manufacturing industries. Also, De Backer [17] states that increasing automation technologies which are important for Industry 4.0 such as using robotics may erode the labour cost advantage of developing countries as labour costs will represent a smaller share of total costs. The developing digitalisation is relied upon progressively taking into account lower-cost and top notch generation in created economies, therefore disheartening offshoring from these nations and favouring reshoring. Some starter aftereffects of expected OECD work propose a negative connection between robot use and offshoring crosswise over nations and ventures [18].

Reshoring has been observed in a broad of range of manufacturing industries from low tech industries such as clothing and furniture to high tech industries such as automotive and electronic appliances in Europe and the US [19]. However, the reshoring level is changing according to industries' characteristics. Companies from the automotive industry which needs a high level of specialization tend to return earlier than other companies which operate in low tech industries. One of the main reasons of this differentiation is that high tech industries need strong supplier relationship which requires a high level of technical knowledge and R&D capability [19]. Because driverless car technology needs this knowledge and R&D capability, in the short term it is expected to effect reshoring activities in developed countries.

In the literature it is hard to find reshoring rates in the global automotive industry specifically. Since the automotive industry is one of the key industries in the manufacturing sector, we can build a relationship between reshoring in manufacturing and reshoring in the automotive industry. But still, for example in Germany which is one of the main players of the automotive industry, about 3% of the German manufacturing companies reshored parts of their foreign production capacities back to Germany from 2013 to 2015 [19]. Subsequently, there is as of now one reshoring organization against three offshoring organizations. An offer of around 3% of reshoring organizations is surely not a major pattern.

Nonetheless, it is an applicable marvel. At the point when extrapolated to the whole German assembling division, outright numbers account really for around 500 to 550 German organizations performing reshoring exercises every year. Further time series examination of board information demonstrates that each fourth to sixth offshoring movement is countered by a reshoring action inside two to five years [20].

5 Social Impacts of Reshoring

As mentioned above, reshoring is a part of Industry 4.0 period. In this perspective, implications of Industry 4.0 on human labour and work organization will impact significantly job redundancies for the blue-collar workforce and a necessary shift towards more white-collar workforce which requires a highly skilled complex and more intense focus on continuous learning and education [21]. This situation is also explained by Martin Ford [22] as the conventional way to financial improvement has been to manufacture production lines which utilize expansive quantities of incompetent specialists. As artificial intelligence and mechanical technology advance there will be less and less requirement for such work escalated processing plants of this sort. Quite a bit of this generation will finish up being "reshored" to created nations where it will be delivered using highly automated facilities.

In light of these developments, we can claim that the market share of countries in high-skill based industries such as automotive and electronics will change with the Industry 4.0 period. Therefore, if developing countries including Turkey does not want to lose their global market share, they need to retake position by providing high-skilled labour in these fields. Accordingly, it is necessary to adjust the education system and infrastructure while industry 4.0 technologies are adapted for developing countries. Such a change should focus on the adaptation of suitable educational programmes and support entrepreneurship that can increase the skills of information technologies in the country [23].

On the other hand, with the offshoring activities in high-tech industries, corporate companies originating from developed countries provide job opportunities for developing countries' high-skilled workforce. Also, those companies could contribute to technological development in developing countries by transferring knowledge.

However, in addition to the possible transformation led by globalisation as Baldwin sees it [15], whether offshoring and participation in global value chains could open the door to social upgrading, in other words to an improvement in work conditions and rights of the working class in developing countries is an issue open to debate [24–27]. It is quite certain that if developing countries cannot provide the necessary skills for the manufacturing of high technology products like autonomous cars, developed countries will reshore their manufacturing activities from developing countries, causing job redundancies in developing countries.

6 Conclusion

The automotive industry is one of the largest industries in the world. Also, the production amount of vehicles is increasing every year. With the help of Industry 4.0, the focus of the automotive industry shifted to hype technologies such as big data and artificial intelligence which are crucial for future vehicle technologies. These technologies are also correlated with the digital transformation period. Also, it is seen that the R&D expenditures of countries are critical for the level of development. However, there are some findings in the literature which point out that developing countries are not ready for digital transformation literally.

In additional, Industry 4.0 has led to big transformations for many sectors. Automotive is one of the fields that is affected by this transformation with the high-tech domain. Autonomous cars can be considered as an output of interaction between Automotive Sector and Industry 4.0. Countries should adopt this transformation in the production process as well as their education system to succeed in the highly competitive market. Therefore, investments made in remarkable value-added sectors such as automotive will help countries to develop.

One of the most important things to remember is that R&D related technologies speed up the reshoring process. Therefore, the market share of developing and less developed countries' in the global automotive industry will fall down. That is because these technologies are not dependent on low-level workforce which is the strongest side of the developing and less developed countries. Therefore, if developing countries including Turkey do not want to stay out of the race, they need to increase the IT-related skills and innovation abilities of the workforce. The way for it is Industry 4.0 supported school curricula, training, and university programs.

Despite a fierce international competition in the automotive sector and general economic crises impacting it both at national and international arena, Turkish automotive industry continues to enlarge and use strategic management tools at micro and macro level. One of the strongest evidence for this strategic behaviour is the establishment of the series of International Automotive Engineering Conference in 2016, by the sector's leading association OSD (Automotive Manufacturers' Association). One of the main conference topics has been autonomous car technologies since its second gathering in 2017, encouraging researchers to continue working in this field in the near future.

References

1. Otomotiv Sektör Raporu. http://www.osd.org.tr/sites/1/upload/files/2016_Degerlendirme_Raporu-2250.pdf
2. International Organization of Motor Vehicle Manufacturers. http://www.oica.net/category/production-statistics/2017statistics/
3. Görener, Ö., Görener, A.: Türk otomotiv sektörünün ülke ekonomisine katkilari ve geleceğe yönelik sektörel beklentiler. J. Yaşar Univ. **3**(10), 1213–1232 (2008)
4. Automotive Manufacturing Association. http://www.osd.org.tr/sites/1/upload/files/Yillik_2017-2224.pdf
5. Lucas, H.C., Goh, J.M.: Disruptive technology: how kodak missed the digital photography revolution. J. Strateg. Inf. Syst. **18**(1), 46–55 (2009)
6. Porter, M.E., Heppelmann, J.E.: How smart, connected products are transforming competition. Harvard Bus. Rev. **92**(11), 11–64 (2014)
7. Parkinson, S., et al.: Cyber threats facing autonomous and connected vehicles: future challenges. IEEE Trans. Intell. Transp. Syst. **18**(11), 2898–2915 (2017)
8. Tekin, A.T., Özkale, N.L., Ayhan, G.: The importance of R & D investments in information and communication technologies and tax policies applied through information and communication technologies. In: International R&D, Innovation and Technology Management Congress, In IRDITECH 2018 Uluslararası Ar-Ge İnovasyon ve Teknoloji Yönetimi Kongresi Bildiriler Kitabi, pp. 155–166 (2018)
9. OECD, Gross domestic spending on R&D (indicator) (2019). https://doi.org/10.1787/d8b068b4-en. Accessed on 06 May 2019
10. Liuima, J.: Autonomous Vehicles Impact – Motor Vehicles Industry (2016). https://blog.euromonitor.com/2016/08/autonomous-vehicles-impact-motor-vehicles-industry.html
11. European Patent Office. Patents and self driving vehicles, the inventions behind automated driving (2018). http://documents.epo.org/projects/babylon/eponet.nsf/0/65910DF6D3F02057C125833C004DB1E6/$File/self_driving_vehicles_study_en.pdf
12. IDC, Geleceğe Hazırlanan Türkiye'deki Otomotiv Şirketleri Dijital Dönüşümün Potansiyelini Nasıl Kullanabilir? (2018). https://otomotiv.kocsistem.com.tr/download.html?token=eyJ0aW1lc3RhbXAiOjE1NTczNTQxOTksImxhbmciOiJ0ciIsImhhc2giOiI3ZTBjNzNmZmE5NDI0MTE4MjUyMjYzZWJhODI2OWE3YyJ9
13. Tekin, A.T., Özkale, N.L., Öztayşi, B.: Big data concept in small and medium enterprises: how big data effects productivity. In: Industrial Engineering in the Big Data Era. Selected Papers from the Global Joint Conference on Industrial Engineering and its Application Areas, GJCIE 2018, June 21–22, 2018, Nevsehir, Turkey, pp. 363–376 (2018)
14. IDC, Worldwide Semiannual Digital Transformation Spending Guide (2018). http://www.idc.com/getdoc.jsp?containerId=IDC_P32575
15. TUSIAD. Türkiye'nin Sanayide Dijital Dönüşüm Yetkinliği, Türkiye'nin 4. Sanayi Devrimi Yayın No: TÜSİAD-T/2017,12 – 589 (2017). https://cdnendustri40.4flyy.com/file/997b726707494962941dc353aae22fa1/turkiye-nin-sanayide-dijital-donusum-yetkinligi.pdf
16. Baldwin, R.: The Great Convergence: Information Technology and the New Globalization. Harvard University Press, Cambridge (2016)
17. De Backer, K., et al.: Reshoring: Myth or Reality?. OECD Science, Technology and Industry Policy Papers, No. 27, OECD Publishing, Paris (2016). http://dx.doi.org/10.1787/5jm56frbm38s-en
18. OECD. Robots in Manufacturing: A First Look at the Evidence, forthcoming, OECD Publishing, Paris (2015)

19. Makers. Measuring reshoring trends in the EU and in the US. Ref Ares (2018). 64134-05/01/2018
20. Kinkel, S.: Future and impact of backshoring – some conclusions from 15 years of research on German practices. J. Purchasing Supply Manag. **20**(1), 63–65 (2014)
21. Bonekamp, L., Sure, M.: Consequences of industry 4.0 on human labour and work organisation. J. Bus. Media Psychol., 6(1), 33–40 (2015). www.journal-bmp.de
22. Ford, M.: Why Developing Countries Will Be Left Behind By Automation. Two problems developing countries face in an AI future (2019). https://thriveglobal.com/stories/why-developing-countries-will-be-left-behind-by-automation/
23. Boston Consulting Group, Industry 4.0: The Future of Productivity and Growth in Manufacturing Industries (2015). https://www.bcg.com/publications/2015/engineered_products_project_business_industry_4_future_productivity_growth_manufacturing_industries.aspx
24. Dicken, P.: Lobal Shift-Mapping the Changing Contours of the World Economy, (7th Edition). Sage, London (2015)
25. Gereffi, G., Fernandez-Stark, K.: Global Value Chain Analysis: A Primer Center on Globalization Governance and Competitiveness. Duke University, Durham (2011)
26. Barrientos, S., Gereffi, G., Rossi, A.: Economic and social upgrading in global production networks: a new paradigm for a changing world. Int. Lab. Rev. **150**(3–4), 319–340 (2011)
27. Kurtulmuş, M.M., Tanyilmaz, K.: Uluslararası Üretim Zincirleri ve Emek: Türkiye Ototomotiv Sektörü Deneyimi, in Finansallaşma Kiskacında Türkiye'de Devlet, Sermaye Birikimi ve Emek, Bedirhanoğulları, P., Çelik, Ö., Mihçi, H. (Eds.) Nota Bene, pp. 185–220 (2017)

Tools of Industry 4.0 on Die Casting Production Systems

Ali Serdar Vanli[1(✉)], Anil Akdogan[1], and M. Numan Durakbasa[2]

[1] Department of Mechanical Engineering, Yildiz Technical University, Istanbul, Turkey
{svanli,nomak}@yildiz.edu.tr

[2] Research Group of Industrial Metrology and Adaptronic Systems, Institute for Production Engineering and Photonic Technology, TU Wien, Vienna, Austria
durakbasa@tuwien.ac.at

Abstract. In Industrial Revolution 4.0 applications the production is increasingly digitized. Moreover, component-driven production is supported. Machines used in production systems exchange very big amounts of process data with one another. The most impressive part is the big amounts of data which can be analyzed for production improvements. Of course, die casting industry has faced these challenges very close. Because each component of an integrated manufacturing system of die casting process requires managing the big data. According to the Industry 4.0, it is possible to adapt various tools and concepts to the manufacturing system with the help of modern information technology and the machine flexibilization. If users can access the process data easily the material flow can be analyzed and optimization along the process can be done. Although there are some challenges in applications, practical potentials of the tools of Industrial Revolution 4.0 especially in die casting production systems offers many advantages especially in management of big data for process improvements.

Keywords: Industry 4.0 · Die casting · Integrated manufacturing systems

1 Introduction

Industry is the branch of activity in bulk production by changing the nature or form of a substance by means of machinery and the related equipment and tools. The first industrial revolution was based on the usage of water and steam power machines in industry. After that, during the second industrial revolution period, is more known as the mass production period, the most evolving industry has emerged from the need for standardized products. The first mass production line was to manufacture automobiles with serial production lines. This is referred to today as Industry 2.0 and is still the main building block of production. With the third industrial revolution, manufacturing automation has progressed. With computer controlled production, more precise, efficient and better standards have been produced. Of course, the main characteristics of the Industrial Revolutions were technological, socioeconomic, and cultural.

As the last, Industry 4.0 focus of the research policy of the German government in Industry. Their aim was to define scientific targets which shows the current and future

N. M. Durakbasa and M. G. Gençyılmaz (Eds.): ISPR 2019, LNME, pp. 328–334, 2020.
https://doi.org/10.1007/978-3-030-31343-2_29

challenges of society and industry. The main features of this new era, in which the interconnected processes are in communication, the objects communicating over the Internet, collecting data and completely changing the production process, emphasize the interaction of the machines with people. The industrial revolution, manifests itself in the subject of many manufacturing. Today, especially integrated mass production systems benefit from the incomes of the new industrial revolution. The high pressure die casting industry also takes its share from these developments. Because the high pressure die casting industry is one of the best examples of integrated mass production systems with many components [1].

2 Smart Factories

Mass production, which was prevailing over the last decades, is increasingly displaced by the production of product variants and of individually customized single parts. A second aspect that comes into play is the lack of utilization of modern information technologies in the manufacturing industry. A "Smart Factory" is intended to be the solution for the challenges of the manufacturing industry. Smart factories contain highly flexible production systems, which are capable of producing single individual parts with the required quality and economic efficiency [2]. A "Smart Factory" is offers a highly digitized and integrated production facilities that relies on smart manufacturing which connects the plant to other entities in the digital supply network, enabling more effective supply chain management. In the course of Industry 4.0 there are well known set of systems and concepts such as "The Internet of Things", "Cyber-Physical Systems" and "The Big Data Analytics" were defined. This paper mentions the common systems and concepts of Industry 4.0 at the frame of die casting production systems. But less known, in general almost unmentioned, are "Flexible Production Systems" and "The Component-Driven Production". Figure 1 shows an Industry 4.0 adapted high pressure die casting smart foundry.

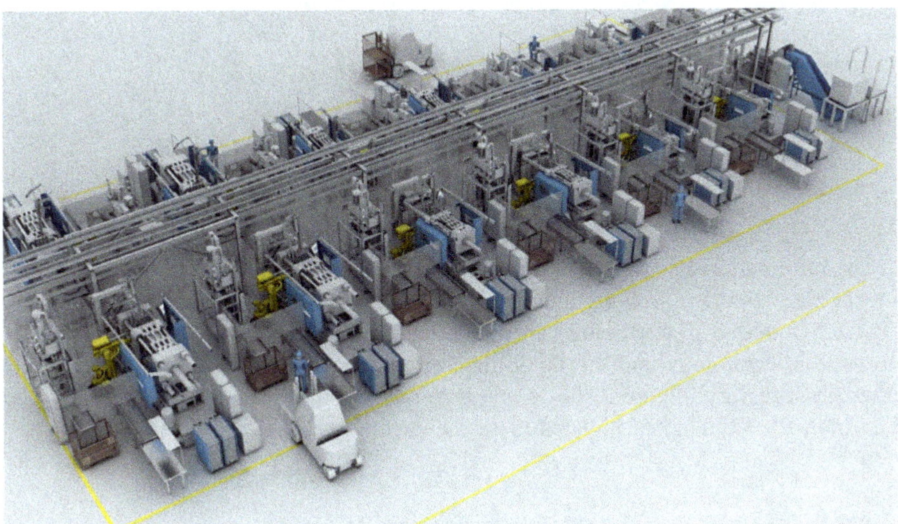

Fig. 1. Industry 4.0 adapted smart foundry

3 Die Casting Production Systems

High pressure die casting is a unique casting method for manufacturing fully automatic, high efficiency and high quality parts with a large scale of weight. An integrated mass production system can be established with a die casting machine, a melting furnace, a molten metal transfer system and a mold heating/cooling system. Such an integrated system allows continuous high quality and efficient production by a well-designed operating conditions. An Industry 4.0 adaptable pressure die casting system offers very important and economic advantages that other casting methods cannot provide. The crucial manufacturing parameters like gate velocity and intensification pressure for better quality parts can be collected as the big data about manufacturing and be analyzed as computer aided quality parameters which were used for continuous improvement of the process. The obtained big data from manufacturing process could be used to increase the product quality expectations like high density and satisfactory mechanical specifications [3].

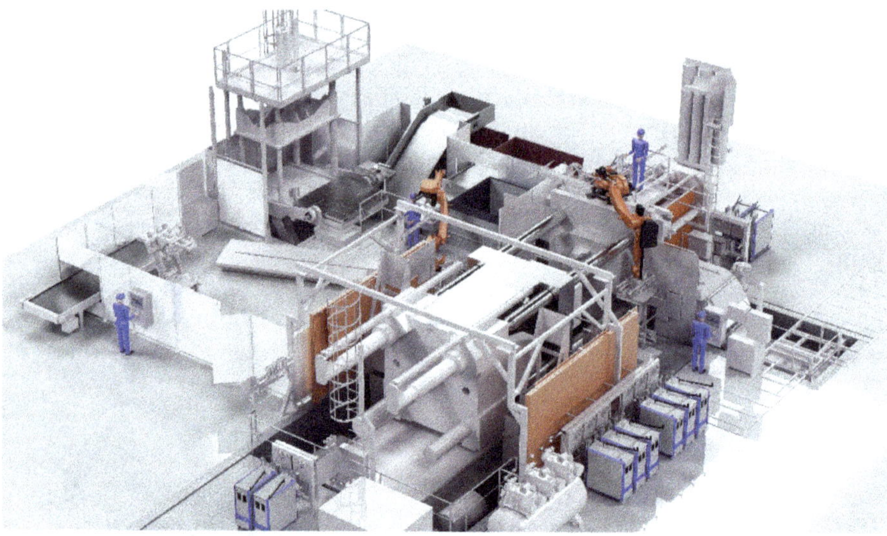

Fig. 2. Production system of an industry 4.0 adapted high pressure die casting cell

The information technology that is used for production machines should be modern IT technology to drive such big amount of data. One of the core aspects of Industry 4.0 is the use modern IT technology in manufacturing systems [4]. The use of modern information technology and the flexibilization of foundry machines and foundry processes are necessary. Figure 2 shows a very good example of a mass production system of an Industry 4.0 adapted high pressure die casting cell.

4 Tools of Industry 4.0 on Die Casting Production Systems

In this paper, the well-known set of tools used for die casting production systems in industry 4.0 basis such as "The Internet of Things" important especially in data collection, "The Big Data Analytics" important especially in data analysis and "Cyber-Physical Systems" were respectively defined (Fig. 3).

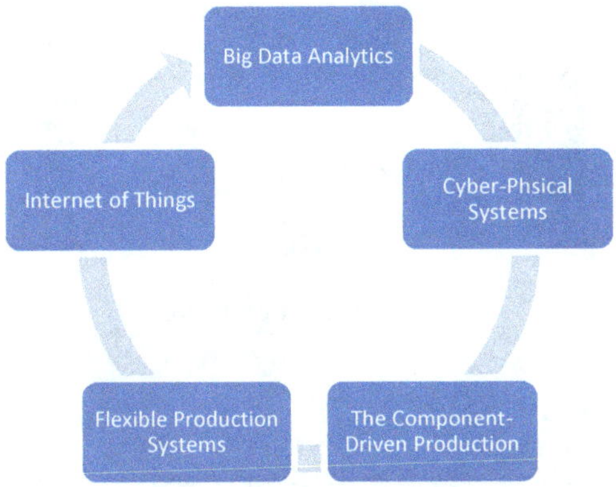

Fig. 3. Tools of industry 4.0

4.1 Internet of Things

Over the past years we have heart of several project and many applications and cases related to the Internet of Things. They were maybe about a series of technologies or devices, maybe were about platforms and applications that are connected to a network or devices. Data collection and analyses is the subject of the big data concept. However, the collection of this data even it is a big one, is directly related to the internet of things. If we think about the die casting process, there are hundreds of thousands process data come from each shot belongs to several process parameters like gate velocity and intensification pressure. Of course, the users receive value from the ability to collect data from connected devices (Fig. 4). Information about each processing step can be transferred internally and externally between the single processes of the process chain allowing to individually adjusting machines and processes to achieve quality and economic efficiency.

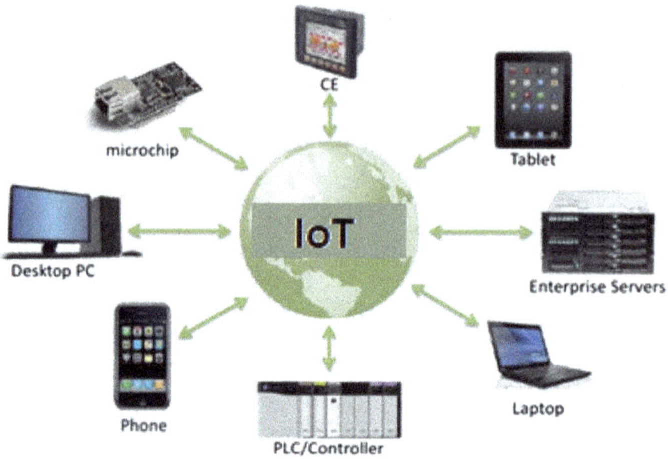

Fig. 4. Internet of Things

4.2 Big Data Analytics

In integrated high pressure die systems mostly produce a significantly greater number of data sources and higher information density which will lead to an enormous increase in significance of data driven analysis and control systems. Smart foundries uses analyzing algorithm and software tools and control tasks for many years like other manufacturing industries. The problem in data acquisition is requiring high effort and cannot be automated. The other problem is the mystery in the history of the part. They need to be marked with unique codes for identification. Because of the economic concerns of this application, knowledge-based approach will be necessary to include data-driven systems into the control of casting processes [2].

4.3 Cyber Physical Systems

The virtual or digital representation of an object or a process can be supplied by a cyber physical system. For instance, in die casting industry many manufacturers uses simulation programs in different variations of process parameters already. Differently, a cyber physical system represents digital interaction of the part with the virtual representation. Although many modern machines have Programmable Logic Controllers and have digital interfaces for automation and data transfer. The data is stored in the machines or section controllers but is not present for a comprehensive analysis. However, an integration of computation with physical processes need to be defined by both cyber and physical parts of the system. So that, optimize the production and adjusts machines, processes and planning towards a multi-criteria would be possible (Fig. 5).

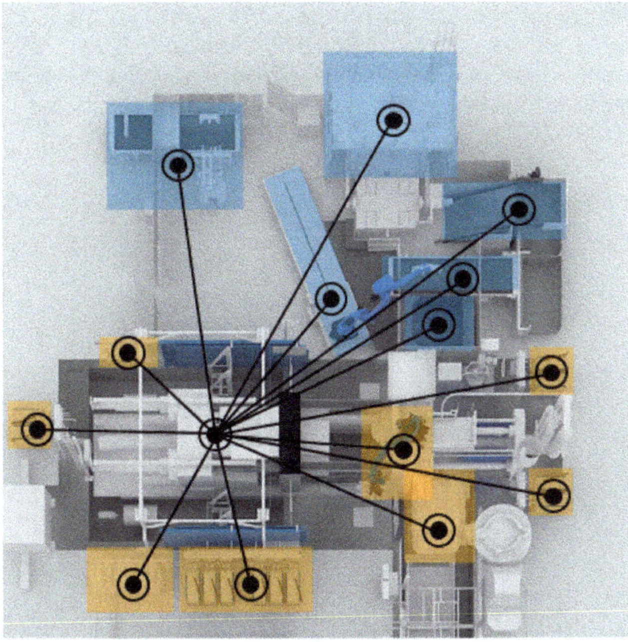

Fig. 5. Cyber and physical parts of a die-casting system

5 Conclusions

The major task of Industry 4.0 is to realize a highly flexible production system that is capable to produce casting parts even with small and/or specialized lot sizes. The Internet of Things, Big Data Analytics and Cyber-Physical Systems are the main supporting systems to archive this goal. The use of modern information technology and the flexibilization of machines and/or processes are necessary to be succeeded. If machine and process data are accessible without barriers, there are lots of possibilities to link these and to carry out comprehensive optimization along the process chain of die casting. The ability to interact with data and to expand the capabilities of the physical world through computation and communication are the key renders for future technology developments.

References

1. Vanli, A.S., Akdoğan, A.: Manufacturing automation for magnesium die casting. In: Proceedings of the 12th International Conference on Measurement and Quality Control – Cyber Physical Issue, vol. 26, pp. 122–130. May (2019). https://doi.org/10.1007/978-3-030-18177-2_13
2. Vanli, A.S., Akdoğan, A., Kerber, K., Özbek, Ş., Durakbaşa, M.N.: Smart die casting foundry according to industrial revolution 40. Acta Techn. Napocensis Ser.: Appl. Math. Mech. Eng. **61**(4), 787–792 (2018)

3. Vanli, A.S.: Optimization of process parameters for product quality in high pressure die casting of magnesium alloys. Ph.D. thesis, Yildiz Technical University, Istanbul (2013)
4. Durakbasa, N.M., Bauer, J., Poszvek, G.: Advanced metrology and intelligent quality automation for industry 4.0-based precision manufacturing systems. Solid State Phenoma **261**, 432–439 (2017)

Usage of Industry 4.0 Technologies Among ISO 500 Manufacturing Firms in Turkey

Alper Kaan Erkut[1(✉)], Seçkin Polat[2], and Umut Asan[2]

[1] Faculty of Business and Economics, Humboldt University of Berlin, Berlin, Germany
kaanerkut@gmail.com
[2] Industrial Engineering Department, Istanbul Technical University, İstanbul, Turkey
{polatsec,asanu}@itu.edu.tr

Abstract. This paper presents the results of a research on the usage of Industry 4.0 (I4.0) technologies among ISO 500 manufacturing firms in Turkey and the factors affecting the usage rate. For this purpose, secondary data has been collected and examined. News, reports and other relevant information obtained from the internet about the I4.0 technologies used by ISO 500 manufacturing firms were used in the research. The descriptive results indicate that the usage rates of the I4.0 technologies vary among the ISO 500 manufacturing firms. A difference in usage is also observed across the categories of firms. I4.0 technologies are mostly used in discrete manufacturing firms. According to the results of the regression analysis, the rate of usage is most affected by the type of manufacturing technology and the number of employees. Finally, the usage rates were compared with the ones of the study carried out by Boston Consulting Group (BCG) for TUSIAD (Turkish Industry and Business Association).

Keywords: Industry 4.0 · Technologies of Industry 4.0 · ISO 500 firms · Manufacturing firms · Regression analysis

1 Introduction

This paper presents the results of a research on the usage of Industry 4.0 (I4.0) technologies among ISO 500 manufacturing firms in Turkey and the factors affecting the usage rates. The First Industrial Revolution, which was based on early developments and started with the invention of the steam engine in 1776, has made a splash on the Fourth Industrial Revolution with various technological inventions. Just as the First Industrial Revolution has led to radical changes, the Fourth Industrial Revolution will also lead to radical changes. The change requires to be ready, and being ready requires to know what the present and tomorrow will be.

There are various studies in the literature about the current state of I4.0 technologies [1–6]. Some of these studies were conducted from an academic perspective, while others were conducted from a practitioner's perspective. Moreover, the studies have different concerns on the status of I4.0 technologies. Some studies consider the current state of a particular I4.0 technology [2, 3], some studies cover the state of overall I4.0

N. M. Durakbasa and M. G. Gençyılmaz (Eds.): ISPR 2019, LNME, pp. 335–347, 2020.
https://doi.org/10.1007/978-3-030-31343-2_30

technologies [1], while some studies consider the use of I4.0 technologies in a particular country [5]. The research conducted in this paper presents the overall usage of I4.0 technologies in Turkey from an academic perspective and compares the findings with those reported by TUSIAD (2017) [5].

This study focuses on presenting a picture of today. For this purpose, secondary data has been collected and examined. News, reports and other relevant information obtained from the internet about the I4.0 technologies used by ISO 500 manufacturing firms were used in the research. Prepared by Istanbul Industry Chamber, ISO 500 (ISO 2017) [7] is a list of the largest industrial firms in Turkey ranked according to their sales turnover. In this study, only manufacturing firms in ISO 500 firms were analysed. The ranking belongs to the year of 2017. The manufacturing firms are divided into four categories according to the manufacturing technology they use and the product type they produce. Based on the use of I4.0 technologies obtained from the research on the internet, technology usage rates were calculated overall and based on firm categories. Then, a regression analysis was performed to determine the factors that are effective on the rate of usage.

The paper consists of four main sections. In the first section, the technologies that constitute Industry 4.0 are briefly explained. The second part is devoted to research on how much I4.0 technologies are used among ISO 500 manufacturing firms in Turkey. The following section presents the results of the regression analysis. Finally, the last section covers the results.

2 Industry 4.0 Technologies

Industry 4.0 is a combination of technologies. Different researchers suggest different technology combinations for I4.0. For example, Smart City Applications are considered by some researchers within the set of I4.0 technologies and some are not included in this group. The I4.0 technologies that are used among ISO 500 firms are briefly explained below.

Decision Making by Artificial Intelligence: Artificial intelligence, which is a set of algorithms that overcomes the problems caused by human capacity limitation, can find the best solutions in big data environment. Artifical intelligence applications such as Netflix and Amazon have long been put into our lives. In the future, an important part of the work done today will be realized through artificial intelligence [8].

The Internet of Things: The Internet of objects, which enable objects and machines to communicate with each other with the help of RFID, sensors, smart phones and mobilizing mechanisms, makes the connected objects and devices stronger and creates very important values for the user [9, 10]. The fact that the machines interact with each other on the internet enables the future errors to be predicted by the data and the system to adapt to the changes [11].

Decision Making with Big Data: Today, large volumes and variety of data has become a valuable asset for organizations [12]. With big data, faster and better decisions can be made in a wide range of industries and applications [8].

Robots: Robots that have been developed to help people in complex or dangerous jobs can now communicate with each other and become more interoperable and work faster and more efficiently with zero error in routine jobs [11]. Robots affect many jobs in many sectors, from manufacturing to agriculture, from retail to service [8].

3D Printers: 3D printers, which are the means of additive or layered production, are able to produce three-dimensional physical products by printing in layers on the layer by moving from the digital 3D drawings or templates, thus creating the possibility to produce customer-specific products with a complex and difficult design [8, 11].

Blockchain: The situations where transactions between individuals and organizations are carried out by a central third element, which cannot be controlled by anyone, are changed by blockchain technology [13]. Blockchain eliminates the third institutions with a centrifugal structure. With the platform created by the blockchain, everyone can see what is real by looking at the data obtained. A global electronic spreadsheet program, global electronic bookkeeping analogies are made to the blockchain [14].

Cloud Technology: Due to big data, the importance of data storage areas increased. Cloud technology with both large storage areas of the data occurred and big data was computed and processed. In addition, cloud technology has eliminated the necessity of organizations to use high-cost servers. With cloud computing, data, photos and documents exchanged between different devices accelerated; getting electronic tickets for airplane and bus trips, making hotel or cinema reservations made easy [9]. Systems that monitor and control production processes can also be cloud-based [11].

Sensors: Sensors are one of the indispensable elements of Industry 4.0 and are small devices that enable the physical world to turn into a digital world. Through them, the machines detect the physical environment surrounding them [15]. In this way, the devices are in communication with each other, they carry out common-purpose operations. In addition, it is possible to store and analyze the data from the devices and products with the help of sensors. In time, the sensors have diminished in size, have easy integration and have been increased technologically. With intelligent sensors, malfunctions can be seen and intervened immediately.

RFID: For the realization of Industry 4.0, not only machines but also raw materials and semi-finished products must be communicated with each other [16]. RFID plays a special role in this communication. Objects can be connected to the Internet by attaching small tags to objects; in this way, the supply chain can be managed more effectively and costs can be reduced. According to Ashton, RFID is an object that frees computers from dependence on human data entry [17].

Augmented Reality: Augmented reality is the addition of virtual objects to real objects [18]. While objects are virtual in virtual reality, virtual objects are combined with real objects in augmented reality. Augmented reality is also called replicated reality. Augmented reality applications can be used with mobile devices such as glasses or mobile phones developed for this purpose. Areas of use can range from medicine to education, training and marketing. In the future, augmented reality will be used more in providing information that will facilitate an employee's decision-making [11].

Simulation: Since the Monte Carlo simulation developed during the construction of the atomic bomb in Los Alamos during World War II, the simulation is used in the I4.0 context to find answers to what if type questions in all environments. Simulation in end-to-end digitalization is an important tool and advanced simulation software plays a key role in developing virtual prototypes [16]. In Germany, a machine tool supplier has developed a virtual machine that simulates the processing of parts using real data on physical machine tools; thus, the preparation time of actual machines can be reduced by up to 80% [11].

Data Analysis/ Modeling: Large volume data encourages predictive analytics [19]. Technologies covered in big data can process large volumes of data and analyze them in real time. New technologies require different data models and data conversion formats [20].

3 Research

Among ISO 500 firms, it was investigated which I4.0 technologies were used by those who made manufacturing. This research is based on data collected from the internet. The names of the manufacturing firms in ISO 500 were searched in Google and the firms using I4.0 technologies were determined. Then, data was collected by reading the related documents. This approach does not provide in-depth information on the use of such technologies, but rather general information on their usage. Even if a company uses I4.0 technologies but there is no internet registration on this issue, the company was omitted from the research. Therefore, this study does not claim that everything about the use of I4.0 technologies in manufacturing firms in ISO 500 is covered. The findings should be viewed in light of these limitations. In this study, two basic usage rates were calculated. How these calculations are made is illustrated by the following four companies and three technologies (see Fig. 1).

		I4.0 Technologies			**Usage rate of I4.0 Technologies**	**Average Usage rate**
Firm	**Firm Category**	**T1**	**T2**	**T3**		
F1	C1		1*	1	2/3	
F2	C1	1	1		2/3	
F3	C2	1	1	1	3/3	
F4	C2	1			1/3	
Usage rate of I4.0 Technologies for All Firms		3/4	¾	2/4		8/12
Usage rate of I4.0 Technologies for Category 1(C1)		1/2	2/2	1/2		4/6
Usage rate of I4.0 Technologies for Category 1(C2)		2/2	½	1/2		4/6

*: "1" indicates that firm uses the I4.0 technoloy.

Fig. 1. Example for the method to calculating usage rate of I4.0 technologies

It was determined that 252 firms among ISO 500 manufacturing firms used I4.0 technologies and the results of the usage rates calculated in accordance with the method described in Fig. 1 are given in Table 1. These usage rates belong to manufacturing companies using I4.0 technologies in ISO 500; therefore the results cannot be generalized to all companies in the Turkish industry. According to the results of the research, I4.0 technologies used in in the manufacturing industry are: 3D printers, robots, cloud technology, RFID, sensors, simulation, artificial intelligence, big data, data analysis,

Table 1. Usage rates of I4.0 technologies among iso 500 manufacturing firms

Firm categories	Usage rate of I4.0 technologies												Total usage rate of I4.0 technologies
	3D printers	Cloud	Robot	RFID	Sensors	Simulation	Augmented reality	Big data	Data analysis	Artificial intelligent	Cyber security	Internet of Things	
Manufacturing firms	0.02	0.75	0.24	0.31	0.76	0.55	0.01	0.07	1.00	0.03	0.11	0.07	0.33
Discrete manufacturing firms	0.07	0.78	0.42	0.36	0.85	0.73	0.03	0.15	1.00	0.07	0.22	0.12	0.40
Continous manufacturing firms	0.01	0.73	0.17	0.29	0.85	0.48	0.00	0.04	0.99	0.02	0.07	0.05	0.31
Consumer product firms	0.00	0.78	0.17	0.29	0.76	0.43	0.00	0.08	1.00	0.03	0.08	0.06	0.31
Industrial product firms	0.03	0.74	0.26	0.32	0.89	0.59	0.01	0.07	0.99	0.03	0.12	0.07	0.34

augmented reality, the Internet of objects and cyber security. According to this finding, it can be deduced that wearable technologies and blockchain, which are considered as I4.0 technologies, are not in the application area of ISO 500 manufacturing firms or that the usage area is narrow.

3.1 Use of I4.0 Technologies According to Company Categories

In this study, only manufacturing firms were taken into consideration. Manufacturing companies are divided into two categories as discrete manufacturing and continuous manufacturing in terms of technologies. Likewise, firms are divided into two categories according to whether they produce consumers or industrial products. In case of doubt, firms are not assigned to any category when they are assigned to these categories. This should be taken into account in the interpretation of the rates of use of technologies by category. The fact that firms are categorized into discrete manufacturing and continuous manufacturing also includes a subjective evaluation. This issue should also be considered when interpreting rates.

According to usage rates statistics, technologies can be divided into 4 main groups. These main groups can be named as those with high rate of use, moderate use rate, low use rate and very low degree of use. The use rates of 70% and above were assigned to the high group, the usage rates between 60% and 25% were assigned to the middle group, the usage rates between 20% and 10% were assigned to the low group and the usage rates to 9% and below were assigned to the very low group.

Considering all manufacturing companies in the study, technologies with very high usage rates are data analysis, cloud technology and sensors. Technologies with high usage rates are simulation and RFID. Robots can also join this group with a 24% usage rate. Technologies with a low usage rate are cyber security, big data and the Internet of things. Technologies with very low usage are artificial intelligence, 3D printers and augmented reality. In the manufacturing companies, data analysis is the technology with the highest usage of 99%, and the augmented reality technology is the technology with lowest usage of 0.7%. In general, the usage rate of I4.0 technologies of manufacturing firms is 33%. Differences occur when manufacturing firms are categorized as discrete and continuous manufacturing.

The usage rate of I4.0 technologies in discrete manufacturing companies is 40%. This ratio indicates that I4.0 technologies are mostly used in discrete manufacturing companies. This may be due to the fact that the manufacturing technologies of the continuous manufacturing firms already contain a large amount of automation. In discrete manufacturing firms, data analysis, sensors, simulassion and cloud are the technologies most commonly used. The use of robot and RFID in discrete manufacturing companies is moderate. Technologies with the lowest usage rate are 3D printers, augmented reality and artificial intelligence.

The usage rate of I4.0 technologies in continuous manufacturing companies is 31%. This ratio indicates that I4.0 technologies are used less than discrete manufacturing companies. The reason for this, as mentioned above, is that the manufacturing technologies of the continuous manufacturing firms already have a large degree of automation. Data analysis in continuous manufacturing companies, sensors and cloud are the technologies most widely used. In continuous manufacturing firms, the rate of use of the simulation and RFID is moderate. The technologies with the lowest usage rate are 3D printers, augmented reality, artificial intelligence, big lot, cyber security and the internet of objects.

The biggest differences between discrete manufacturing and continuous manufacturing are in robots, simulation, big data and cyber security and objects on the Internet. All of these technologies are more common in discrete manufacturing firms. Even though the absolute usage rate is very low, 3D printers are used much more in batch production than in continuous production.

The rate of usage of I4.0 technologies in industrial product manufacturers is 35%. Data analysis, sensors and cloud are the most widely used technologies in industrial product manufacturers. In terms of these results, industrial products companies are more similar to the continous manufacturing companies. Usage rate of simulation, robot and RFID in industrial production companies is moderate. Technologies with the lowest usage rate are 3D printers, augmented reality, artificial intelligence, big data and the Internet of things. Industrial product companies differ in terms of robot from continuous manufacturing companies. In terms of technologies with the lowest usage rate, the biggest difference between the companies producing industrial products and continuous manufacturing firms is seen in cyber security. Cyber security is used more in companies producing industrial products than in continuous manufacturing companies.

The usage rate of I4.0 technologies of companies producing consumer products is 31%. In terms of this ratio, it is very similar to the production companies. Data analysis, sensors and cloud are the technologies which are used in consumer products companies. In terms of these results, consumer products companies are more similar to the continuous manufacturing companies. The rate of use of robot and RFID in consumer goods producing companies is moderate. Technologies with the lowest usage rate are 3D printers, augmented reality, artificial intelligence, big data, cyber security and the Internet of things. The companies that produce consumer products differ from the continuous manufacturing firms in terms of robot. In terms of technologies with the lowest usage rate, the companies that produce consumer products are similar to the continuous manufacturing firms. In general, consumer products producing companies are more similar to continous manufacturing companies in terms of usage rates of the technologies.

In terms of technology usage rates, the difference between discrete manufacturing firms and continuous manufacturing firms is greater than the difference between consumer products and industrial products. The similarity of the companies producing consumer products to the continuous manufacturing companies in terms of the use of I4.0 technologies is higher than the similarity of the firms producing industrial products to the discrete manufacturing companies.

3.2 Usage Rates of I4.0 Technologies

When a study is done on the usage of the technologies, the following results appear. Technologies with the lowest usage rate are 3D printers, artificial intelligence, augmented reality, big data and internet of objects. Wearable technologies and blockchains, which have never been used in the companies covered by this research, can be added to these technologies. The most commonly used technologies are cloud technology, data analysis and sensors. As mentioned earlier, these ratios were calculated among the firms using I4.0 technologies in ISO 500 firms.

The overall usage rate of 3D printers is 2%. The most common use is in the discrete manufacturing companies with a rate of 7%. Discrete manufacturing companies use 3D technology more than ten times that of continuous manufacturing firms. 3D printers are used more in industrial products companies than in consumer products.

Data analysis is the most widely used technology with the highest rate of 99% overall use. Usage rates of 3D printers in all company categories are very close to this rate.

Sensors are the second most commonly used I4.0 technology with 76% overall usage rate. With the usage rate of 89%, it is mostly used in industrial products producing companies. The lowest usage rate is 76% for consumer products.

The overall usage rate of cloud technology is 75%. This is the second most commonly used I4.0 technology with sensors after data analysis. The highest usage rate belongs to discrete manufacturing companies. Cloud technology is used at lowest level in industrial companies and continuous manufacturing companies.

The overall usage rate of robotics technology is 24%. The highest usage rate belongs to discrete manufacturing companies with 42%. The lowest usage rates are about 17% for the production of consumer products. The overall usage rate of RIFD technology is 31%. The highest usage rate belongs to the discrete manufacturing companies with 37%. The lowest usage rate is continuous manufacturing companies and consumer products companies with 29%. The overall rate of use of the simulation is 55%. In discrete manufacturing, simulation is used at a rate of 73%. Simulation is used in consumer products with 43% minimum.

Augmented reality is one of the least used technologies with a 0.7% overall usage rate. It is used in discrete manufacturing companies with a rate of 2%. The overall rate of use of big data is 7%. Big data is used at the most in discrete manufacturing companies with a rate of 15%. The lowest rate is 4% for continuous manufacturing companies.

Artificial intelligence is one of the least used I4.0 technologies and its overall usage rate is 3%. It is used at least in continuous manufacturing companies with a rate of 2%. The highest usage companies discrete manufacturing companies. Consumer products companies and industrial products companies use it in similar rates.

The overall average use of cyber security technologies is 11%. The highest usage rate is in discrete manufacturing firms. The rate of use in these firms is 22%. The lowest usage is in continuous manufacturing companies and consumer products companies. The overall use of Internet of Things is 7%. The highest usage rate is 12% for discrete production. The rates of use of other categories are close to 6%.

3.3 Statistical Analysis

Regression analysis was performed in order to determine factors that have effects on the usage rates of I4.0 technologies of manufacturing firms. The effect of two main factors on the rate of usage is considered. These main factors are divided into external and internal factors. The firm size and firm experience for internal factors are defined as two variables. For external factors, type of product and type of manufacturing technologies are defined as two variables. It was thought that external factors would have an impact on what I4.0 technologies should be used by a company and internal factors would have an impact on how companies could implement I4.0 technologies. The technology usage rates in Table 1 above provide clues that the type of manufacturing technology can be effective. The table also provides clues on the effect of the product type. The variable of firm size, which can be considered as one of the internal factors, is that the company can meet the resources needed to implement both I4.0 technologies; and may affect the resilience to cash flow uncertainty when I4.0 technology is introduced. In this study, the size was measured by the number of employees. The number of employees is not only an indicator of the resource size of a firm but also has an effect on the necessity of using I4.0 technologies. The reason for this can be explained as follows: I4.0 technologies will greatly reduce the number of employees; this can be perceived as an attractive opportunity for companies with a high number of employees and may increase interest in I4.0 technologies. Another variable that is one of the internal factors is firm experience. The effect of company experience can be in two ways. If a firm's performance is good and its experience is high, this company may not want to switch to a new technology. On the other hand, if the performance is not good and the experience is low, then this company may think that moving to the new technological environment is a better strategy. Firm experience was measured by firm age in this study.

As stated above, 252 of ISO 500 manufacturing firms use I4.0 technologies. But in the regression analysis, 212 of the 252 firms were used. This is due to the fact that all data is available for only 212 firms. Information on the number of employees of some 252 firms was not suitable for use in the regression analysis and therefore such firms were excluded from the regression analysis. For example, some firms have expressed the number of employees as "over 1000". These firms were not included in the regression analysis. Similarly, firms which manufacturing technology or product category could not be determined by the authors were not included in the regression analysis. The values of the indicators used in the regression analysis are given in the Table 2.

Table 2. Regression model indicators

Indicators	Values
Number of firms	212
Number of discrete manufacturing firms	66
Number of continuous manufacturing firms	146
Number of industrial product firms	157
Number of consumer product firms	55
Average number of employees	1414
Standart deviation of number of employees	1582
Average firm ages (years)	42
Standart deviation of firm ages (years)	19

Of the 212 manufacturing firms, 66 are discrete manufacturing firms and 146 are continuous production firms. 157 of 212 firms produce industrial products while 55 of them produce consumer products. The average number of employees of the firms is 1414 person and the standard deviation is 1512 person. The average age of the firms is 42 years. Standart deviation of firms' ages is 19 years.

The results of the regression model are given below. According to these results, the model is significant at level 0.000 (see Table 3) and R^2 has a value of 0.184 (see Table 4).

Table 3. Analysis of variance

Model		Sum of squares	df	Mean square	F	Sig.
1	Regression	1.005	4	.251	11.644	.000
	Residual	4.466	207	.022		
	Total	5.470	211			

Table 4. Model summary

Model	R	R square	Adjusted R square	Std. error of the estimate
1	.429	.184	.168	.14688

All variables are significant at values below 0.10 (see Table 5). According to the standardized Beta coefficients, the most effective factors are the type of manufacturing technology and the number of employees. The way that the number of employees to influence the I4.0 usage rate was explained above. It can be said that the reason why this variable has the highest effect is this influence way. The lowest effect variable is firm age. In parallel with this finding, the lowest level of significance is the firm age. The effect of manufacturing technology type is more than the effect of product type. This finding is in line with the results of the rate of usage in the categories of firms in relation to the technology usage rates described above. When firms' manufacturing technologies are discrete, firms use I4.0 technologies. This finding resulting from the regression analysis is consistent with the I4.0 technology usage rate statistics described above. According to the results, the highest usage rate belongs to discrete manufacturing companies. According to the standardized Beta values, companies producing industrial products use more I4.0 technology.

The findings of this study were compared with the findings of the study conducted by BCG consulting firm for TUSIAD [5]. For this, firstly the research characteristics of both studies were compared (See Table 6).

The data on I4.0 usage presented in the research by BCG carried out for TUSIAD, were collected through questionnaires from firms in manufacturing industry in Turkey operating in 10 different sectors [5]. All firms in TUSIAD study are users of I4.0 technologies. There is no clarity in the TUSIAD report about how these firms were selected. The size of the firms is measured by annual turnover. In the report, there is no information about the usage rates according to the firm's manufacturing technologies and

Table 5. Coefficients

Model		Unstandardized coefficients		Standardized coefficients	t	Sig.
		B	Std. Error	Beta		
1	(Constant)	.198	.032		6.251	.000
	N of employee	2.904E-5	.000	.285	4.437	.000
	Firm age	.001	.001	.106	1.644	.102
	Manufacturing technology type	.086	.022	.249	3.929	.000
	Product type	.052	.023	.142	2.253	.025

Table 6. Research characteristics of studies

Research characteristics	This research	TUSIAD (2017)
Country	Turkey	Turkey
Firm type	Manufacturing	Manufacturing
Number of firms	252	108
Firm source	ISO 500	No information
Relation to I4.0 technology	User of I4.0 technology	User of I4.0 technology
Data source type	Secondary	Primary
Data collection method	Internet	Survey with firms at question
Measurement for firm size	Number of employee	Annual sales
Statistical analysis	Yes	No
Number of I4.0 technologies	12	12
Types of I4.0 technologies	Most of technology are same in both researches	

product types. Only information on the use of automation technologies according to firm size is available. In BCG‘s research, the use of technology was determined in 5 levels. These levels are implementation, dissemination, successful implementation in pilot application, pilot application and not use. As seen Table 6, both studies have similar and different aspects in terms of research characteristics. Both studies are focused on firms using I4.0 technologies in Turkey. The technologies used in researches are largely the same. I4.0 technology usage rates in this study and usage rates in BCG's research are shown in Table 7. Numbers for TUSIAD in Table 7 shows rates of implementation that corresponds to the first level of use of technologies mentioned above.

The most commonly used I4.0 technologies in the TUSIAD study are cyber security (53%), sensors (35%) and robots and automation (24%). The least used 3 technologies are additive production (4%), augmented reality (5%) and artificial intelligence (2%). Usage rates of 3D printers, robots, augmented reality, artificial intelligence and the Internet of objects are very close to each other in both studies. Contrast to this similarity, there are big differences in the usage rate of some technologies between this study and TUSIAD study. These technologies are cloud, RFID, sensors, simulation and cyber security. The difference in the usage rate of technology may be due to characteristics of the studies. Table 6 indicates that there are some differences in data source type, data collection method, firm source, and I4.0 technologies between the studies.

In the TUSIAD study, it was emphasized that automation technology utilization rates were related to firm size, which is similar to finding of the study of ISO 500 Manufacturing Firms presented in this paper. This finding of TUSIAD study is not a result of hypothesis testing but rather of descriptive statistics, but finding presented in in this paper is a result of a regression analysis.

Table 7. Comparision of usage rates

Studies	Usage Rates of I4.0 Technologies											
	3D printers	Cloud	Robot	RFID	Sensors	Simulation	Augmented reality	Big data	Data analysis	Artificial intelligent	Cyber security	Internet of Things
This study (ISO 500 Manufacturing Firms)	0.02	0.75	0.24	0.31	0.76	0.55	0.01	0.07	1.00	0.03	0.11	0.07
TUSIAD study	0.04	0.10	0.24	0.03 0.09	0.35	0.11	0.05	0.14	NA	0.02	0.53	0.07

4 Conclusion

In this paper, it is tried to present the current state of the usage of Industry 4.0 technologies among ISO 500 manufacturing firms in Turkey. Care should be taken to generalize this picture to the whole Turkish manufacturing industry. Furthermore, attention should be paid to the interpretation of the findings because of the above-mentioned reasons (categorization, based on the interpretation of the data on the Internet, etc.). The results of the research indicate that the usage rates of the I4.0 technologies vary among the ISO 500 manufacturing firms. There is also differences among the categories of companies. I4.0 technologies are mostly used in discrete manufacturing companies. According to the results of the regression analysis, the rate of usage is most affected by the type of manufacturing technology. As the number of employees increases, usage rate of I4.0 technologies by firms is increasing. Discrete manufacturing technology, one of the manufacturing technologies, increases the usage rate of Industry 4.0 technologies.

References

1. Kang, H.S., Lee, J.Y., Choi, S., Kim, H., Park, J.H., Son, J.Y., Kim, B.H., Do Noh, S.: Smart manufacturing: past research, present findings, and future directions. Int. J. Precision Eng. Manuf.-Green Technol. **3**(1), 111–128 (2016)
2. Chen, T., Tsai, H.R.: Ubiquitous manufacturing: current practices, challenges, and opportunities. Robot. Comput.-Integrated Manuf. **45**, 126–132 (2017)
3. Wan, J., Tang, S., Yan, H., Li, D., Wang, S., Vasilakos, A.V.: Cloud robotics: current status and open issues. IEEE Access **4**, 2797–2807 (2016)

4. Chen, Y., Dong, F.: Robot machining: recent development and future research issues. Int. J. Adv. Manuf. Technol. **66**(9–12), 1489–1497 (2013)
5. TUSIAD and Boston Consulting Group (BCG). Turkey's digital transformation competence in industry, Publication Number: TÜSİAD-T/2017,12 – 589, 2017. (In Turkish)
6. Lu, Y.: Industry 4.0: a survey on technologies, applications and open research issues. J. Ind. Inf. Integr. **6**, 1–10 (2017)
7. ISO (İstanbul Sanayi Odası) The 500 largest industrial companies in Turkey (in Turkish) (2017). http://www.iso500.org.tr/
8. Schwab, K.: The Fourth Industrail Revolution. Penguin, Great Britian (2017)
9. Greengard, S.: Nesnelerin İnterneti. Optimist Yayım Dağıtım, İstanbul (2017)
10. Mario, H., Tobias, P., Otto, B.: Design principles for Industrie 4.0 scenarios. In: 49th Hawaii International Conference on System Sciences (HICSS), pp. 3928–3937. IEEE, January (2016)
11. Rüssmann, M., Lorenz, M., Gerbert, P., Waldner, M., Justus, J., Engel, P., Harnisch, M.: Industry 4.0: the future of productivity and growth in manufacturing industries, Boston Consulting Group, **9**(1), 54–89 (2015)
12. Zyskind, G., Oz, N.: Decentralizing privacy: using blockchain to protect personal data. In: IEEE Security and Privacy Workshops, pp. 180–184. IEEE, May (2015)
13. Yli-Huumo, J., Ko, D., Choi, S., Park, S., Smolander, K.: Where is current research on blockchain technology?—a systematic review. PLoS ONE **11**(10), e0163477 (2016)
14. Tapscott, D., Tapscott, A.: Blockchain Revolution: How the Technology Behind Bitcoin is Changing Money, Business, and the World. Penguin, London (2016)
15. Lee, J., Kao, H., Yang, S.: Service innovation and smart analytics for Industry 4.0 and big data environment. Procedia CIRP **16**, 3–8 (2014)
16. Brettel, M., Friederichsen, N., Keller, M., Rosenberg, M.: How virtualization, decentralization and network building change the manufacturing landscape: an industry 4.0 Perspective. Int. J. Inf. Commun. Eng. **8**(1), 37–44 (2014)
17. Ashton, K.: That 'internet of things' thing. RFID Journal **22**, 97–114 (2009)
18. Taskan, L.: What is augmented reality? What are the areas of use? (In Turkish) (2018). http://www.teknolo.com/augmented-reality-nedir-hangi-alanlarda-kullanilabilir/
19. Schmidt, R., Möhring, M., Härting, R.-C., Reichstein, C., Neumaier, P., Jozinović, P.: Industry 4.0 - potentials for creating smart products: empirical research results. In: Abramowicz, W. (ed.) BIS 2015. LNBIP, vol. 208, pp. 16–27. Springer, Cham (2015). https://doi.org/10.1007/978-3-319-19027-3_2
20. Lasi, H., Fettke, P., Kemper, H.G., Feld, T., Hoffmann, M.: Industry 4.0. Bus. Inf. Syst. Eng. **6**(4), 239–242 (2014)

IoT and Production Systems, Lean Production

A Prediction Model for Fault Detection in Molding Process Based on Logistic Regression Technique

İnanç Kabasakal(✉), Fatma Demircan Keskin, Aydin Koçak, and Haluk Soyuer

Department of Business Administration, Ege University, İzmir, Turkey
{inanc.kabasakal, fatma.demircan.keskin, aydin.kocak, haluk.soyuer}@ege.edu.tr

Abstract. Industry 4.0 is often described as a wave of transformation that enforces the digitalization of processes to create integrated and autonomous systems. In this regard, the collection of process data is a necessity to analyze data with advanced techniques for various purposes. Statistical techniques in machine learning might provide solutions for fault detection and other tasks in manufacturing processes. In our study, a learning model is proposed for a fault prediction task with the use of Logistic Regression. The data used for the analysis involve measurements from sequential processes carried out in a large-scale wheel rim manufacturer. The pre-processing and analysis of process data was introduced along with a case study. Moreover, findings of the model were presented and the potential use of the model will be discussed.

Keywords: Machine learning · Logistic regression · Binary classification · Fault prediction · Industry 4.0

1 Introduction

The rapid progress in Information Technology (IT) is a major driver for change in many industries. Such transformation is often cited as the Industry 4.0 or Industrial Internet, and this phenomenon has attracted many researchers within the last decade. Internet of Things (IoT), Robotics, Artificial Intelligence (AI), 3D Printing, Virtual/Augmented Reality, Big Data, and Cloud Computing are often mentioned among the enabling technologies for Industry 4.0.

The use of sensors, IoT devices, and networking of machinery enable a flow of data in manufacturing plants. Capturing business value with leveraging such data streams is an essential issue for organizations [1]. Reference [2] explored the big-data approach by a major airliner and mentioned the use of records and real-time operational data for predictive maintenance to save errors, increase the lifetime of the equipment and improve safety. Connectivity is a prerequisite to create value for businesses [3], especially for manufacturers that head towards the integration of digital technologies into monitoring & controlling of operations in real-time. Beyond the collection of data

N. M. Durakbasa and M. G. Gençyılmaz (Eds.): ISPR 2019, LNME, pp. 351–360, 2020.
https://doi.org/10.1007/978-3-030-31343-2_31

through a digital framework (IoT, sensors, etc.), machine learning tasks help to obtain useful results from sensor data.

In this study, a real world problem of a wheel rim manufacturer is addressed by using actual measurements taken during the molding process, and a binary logistic regression model aiming to predict failures is introduced. After pre-processing the initial data set, 29 variables are included in the model. The model can predict the actual fault category with a 96.8% success rate.

2 Industry 4.0 and Smart Factory

In the Industry 4.0 vision, a focal area of transformation is the manufacturing landscape. In the last decades, computer-aided design and manufacturing, expert systems, robotics had already been integrated into manufacturing. Furthermore, with the prevalence of digitalization, an extensive level of automation might be anticipated in manufacturing processes.

In intelligent manufacturing; IoT, Robotics, Artificial Intelligence, CPS, Cloud Computing, Big Data Analytics, and other digital technologies are regarded as key technologies [4]. Moreover, the effective use of digital technologies provides various opportunities for real-time monitoring and control over the processes in manufacturing.

As an essential step in the digitalization of the manufacturing environment, data collection is required to monitor processes in manufacturing. Each successive subtask within a process might be monitored by tracking relevant data. For such purpose; sensors, or more generally IoT systems, provide great potential in collecting and sending contextual data.

A major implication of logging process data might be noted as the potential use of data in descriptive and predictive analysis. For instance; the use of signal processing models that rely on pattern recognition, neural networks, and other advanced data-driven techniques, help to sense potential failures for predictive maintenance [5].

Another use of data analysis is reducing faults, finding root causes, and improving quality. In general, Information Technology has been a strategic tool to improve quality and increase business performance [6]. IoT is used to obtain real-time operational data from the machinery and processes, and provide input for advanced analytics to keep track of quality [7]. To track performance in manufacturing; monitoring of the processes and applying advanced data analytics techniques are crucial [8].

In prior research, data analysis techniques have been employed to extract valuable information from the process data. In this way, models can be developed for the prediction of outputs or relevant data. In Machine Learning, supervised methods such as Support Vector Machines (SVM's), Random Forests, Logistic Regression Models, and Neural Networks [9] are often used to analyze data for classification. As an example; [10] proposed a Machine Learning model to predict the performance of air conditioning systems, and utilized Logistics Regression and Random Forest methods for prediction. In another study for prediction of health risks [11], Logistic Regression and Support Vector Machine (SVM) based techniques were utilized to classify streaming sensor data.

3 Case Study

Our case study demonstrates a prediction model over a process dataset obtained from a wheel rim manufacturer in Manisa Industrial Zone, Turkey. In particular, the data consists of a spreadsheet that involves measurements within various steps of the molding process. As listed on the official website [12], the manufacturer is a large-scale industry and has a production capacity of 5 million rims in a year. Besides, a 75% share in the OEM passenger car steel wheel market indicates the supremacy of the manufacturer in the Turkish market.

The data obtained for our case involves various measurements during the molding process that involve several steps of the entire wheel rim manufacturing process. The molding process takes place in several steps. Initially, the aluminum-alloy is poured into a tank in melted form. The tank is kept at a constant temperature during the molding process. Afterward, the tank is pressurized with air. Then, the melted alloy is blown out and poured into a mold. The aluminum alloy in the mold is subsequently cooled in successive stages.

The dataset for the case study involved 137 columns for 594 products manufactured between 26th and 30th of November in 2018. All records were for a single mold processed in the same casting machine, and there were 41 units (6.9%) with faults. A batch with such considerably high fault rate was intentionally chosen for the study to train the learning model with a smaller dataset.

In the original dataset, one of the columns holds for the product order during the molding process. 8 of the columns indicate the type of fault as well as details if any. The fault details in those 8 columns were marked manually for our study. The other columns include various measurements recorded during the entire molding process. Additionally, the data involve columns for target temperatures measured in heating and cooling processes.

The detailed descriptions of the columns are listed in Table 1.

Table 1. Detailed descriptions for the columns in dataset

No	Variable name	Range
1	Shift (sh)	[1–3]
2	Hour (hr)	[0–23]
3	Humidity (hmd)	[21.8–34.2]
4	Temperature (tmp)	[29.4–52.2]
5	metal_temperature_difference (mtd)	[−11–4]
6	aluminium_transfer (at)	[0–5]
7	maximum_pressure_value (mpv)	[5.8–7.2]
8	cooling_temperature (wt)	[21–30]
9	cycle_time (ct)	[295–37009]
10	thermocouple_1_diff (t1d)	[−113–55]
11	thermocouple_2_diff (t2d)	[−192–29]

(*continued*)

Table 1. (*continued*)

No	Variable name	Range
12	phase_1_pressure (ph1p)	[131.3–149.5]
13	phase_2_time (ph2t)	[25–40]
14	phase_2_pressure (ph2p)	[301.3–516.9]
15	phase_4_time (ph4t)	[23–210]
16	h1_flow_average_value (h1f)	[0–1205.2]
17	h2_flow_average_value (h2f)	[0–602.4]
18	h3_flow_average_value (h3f)	[0–624.8]
19	h7_flow_average_value (h7f)	[1200.6–1204.8]
20	h8_flow_average_value (h8f)	[785.2–798.4]
21	h9_flow_average_value (h9f)	[795.9–800.8]
22	h10_flow_average_value (h10f)	[795.8–800.6]
23	h11_flow_average_value (h11f)	[0–770.6]
24	h12_flow_average_value (h12f)	[0–604.8]
25	s1.1_side_core_cooling_flow_difference (s1.1sd)	[−16–0.4]
26	s1.2_side_core_cooling_flow_difference (s1.2sd)	[−16–0.1]
27	s1.3_side_core_cooling_flow_average_value (s1.3sa)	[0–15.4]
28	s1.4_side_core_cooling_flow_average_value (s1.4sa)	[0–16.3]
29	s2_bottom_core_cooling_flow_average_value (s2ba)	[0–17]

3.1 Data Pre-processing

Due to the nature of the molding technique being used, the process is prone to potential faults in manufacturing. The faults are detected with measurements and visual inspections conducted by operators. In particular, three individual forms of scrap can be listed as in Table 2.

Table 2. Types of scrap in the entire molding process

Fault code	Description
1	Start-up Scrap
2	Shrinkage
3	Visual Scrap

The model proposed in our study performs binary prediction and classifies a sample as 'Fault' and 'Normal'. As the next step of data pre-processing phase, the records that refer to the faults were marked with a binary identifier (1) in a new column.

The manufacturing plant operates continuously in three shifts, and dataset involved molding time recorded for each wheel rim. The measurements were recorded from a single casting machine and rim model. Accordingly, a numeric column was added into the original dataset to identify the shift. The molding time was reduced to an hourly basis.

An important change in the dataset was the reduction of target and actual values measured in several sub processes. Most of the columns in the analyzed data involved measurements obtained in consecutive steps of cooling. The cooling temperature is recorded for each phase. The data obtained involves both of the actual and ideal temperatures for the recordings during cooling. Rather than the raw values for ideal and the actual temperatures (or pressures), the deviation from the ideal might be attributed as a trigger for faults. For this reason, the difference of the target and the actual values were calculated into new columns. In the analysis, the difference is taken into account in the model rather than the actual and ideal values.

After the data pre-processing phase, a dataset of 30 columns and 594 rows was analyzed with a binary logistic regression model. In the next subsection, the model will be detailed along with the findings.

The model proposed in our study performs binary prediction and classifies a sample as ‘Fault’ and ‘Normal’. As the next step of data pre-processing phase, the records that refer to the faults were marked with a binary identifier (1) in a new column.

The manufacturing plant operates continuously in three shifts, and dataset involved molding time recorded for each wheel rim. The measurements were recorded from a single casting machine and rim model. Accordingly, a numeric column was added into the original dataset to identify the shift. The molding time was reduced to an hourly basis.

An important change in the dataset was the reduction of target and actual values measured in several sub-processes. Most of the columns in the analyzed data involved measurements obtained in consecutive steps of cooling. The data obtained involves both of the actual and ideal temperatures for the recordings during cooling. Rather than the raw values for ideal and the actual temperatures (or pressures), the deviation from the ideal might be attributed as a trigger for faults. For this reason, the difference of the target and the actual values were calculated into new columns. In the analysis, the difference is taken into account in the model rather than the actual and ideal values.

After the data pre-processing phase, a dataset of 30 columns and 594 rows was analyzed with a binary logistic regression model. In the next subsection, the model will be detailed along with the findings.

4 Binary Logistic Regression Model

Logistic regression is an approach that has a wide field of application allowing a categorical dependent variable to be predicted using categorical or continuous independent variables [13].

In this study, a real problem of a wheel rim manufacturer firm’s molding process is addressed. During the molding process, quality failures may occur depending on various reasons, and rims are divided into two categories as “Fault” and “Normal.”

The most probable fault causes were determined based on the firm’s experts’ opinions, and after pre-processing the data set, 29 explanatory variables were included in the analysis. One of the explanatory variables is categorical, and the others are continuously scaled. Since the dependent variable is binary, the independent variable set includes both categorical and continuous variables; binary logistic regression is selected to be used to model the probability of classifying wheel rims as fault and normal. Yi implies the classification of the ith wheel rims' classification. Yi = 0 if the wheel rim i is classified as normal, Yi = 1 otherwise. Let $\pi(x)$ represents the conditional mean of Y given x, the logistic regression model is ([14, 15]):

$$\pi(x) = \frac{e^{\beta_0 + \beta_1 x_1 + \ldots + \beta_p x_p}}{1 + \beta_0 + \beta_1 x_1 + \ldots + \beta_p x_p} \tag{1}$$

The logistic transformation of the model is (de Menezes et al. 2017:63):

$$logit = \ln\left(\frac{\pi(x)}{1 - \pi(x)}\right) = \beta_0 + \beta_1 x_1 + \ldots + \beta_p x_p \tag{2}$$

where $\beta_{0,}\beta_1, \ldots, \beta_p$ are the regression coefficients.

The binary logistic regression model parameters were estimated using the maximum likelihood method in SPSS 15.0. The goodness of fit of the obtained model was tested using the Hosmer-Lemeshow test, and the explanatory power of the model was evaluated based on the Nagelkerke R Square value. Results of the Hosmer-Lemeshow test are given in Table 3.

Table 3. Hosmer and lemeshow test results

χ^2	df	p-value
2.942	8	0.938

The p-value of the Hosmer–Lemeshow test is 0.938 (> 0.05) implies that actual values of the dependent variable are not significantly different from the values predicted by the model. Nagelkerke R-square value of the model, including all considered explanatory variables, is obtained as 0.751.

A summary of the model results that includes coefficient, odd ratio and p-value of each variable are presented in Table 4.

The logit model is:

$$\begin{aligned} logit = 929.071 &- .672\,sh + .343\,hr - .028\,hmd + .034\,tmp - .012\,mtd + 1.635\,at \\ &+ .030\,mpv + 1.553\,wt - .001\,ct + .220\,t1d - .137\,t2d + .009\,ph1p \\ &+ .765\,ph2t + .058\,ph2p - .631\,ph4t + .088\,h1f - .228\,h2f + .390\,h3f - .166\,h7f \\ &+ .295\,h8f + .547\,h9f - 1.587\,h10f + .139\,h11f - .429\,h12f + 10.754\,s1.1sd \\ &+ 2.484\,s1.2sd - 4.721\,s1.3sa - 5.185\,s1.4sa + .803\,s2ba \end{aligned}$$

The p-values indicate that, among the independent variables considered in the model, aluminium_transfer (at), thermocouple_1_diff (t1d), phase_2_pressure (ph2p), h10_flow_average_value (h10f) and s1.1_side_core_cooling_flow_difference (s1.1sd) can significantly predict the fault classification at.%5 significance level, thermocouple_2_diff (t2d), and h11_flow_average_value (h11f) at the level of 10% significance level. The most significant predictors are s1.1_side_core_cooling_flow_difference (s1.1sd) and phase_2_pressure (ph2p), respectively.

Table 4. Model results summary

Variable	β	p-value	Odd ratio	Variable	β	p-value	Odd ratio
Intercept	929.071	.319	–	ph4t	−.631	.247	.532
sh	−.672	.745	.511	h1f	.088	.351	1.092
hr	.343	.243	1.410	h2f	−.228	.526	.796
hmd	−.028	.962	.973	h3f	.390	.136	1.477
tmp	.034	.710	1.035	h7f	−.166	.833	.847
mtd	−.012	.970	.989	h8f	.295	.309	1.343
at	1.635	.046	5.127	h9f	.547	.626	1.729
mpv	.030	.989	1.030	h10f	−1.587	.043	.204
wt	1.553	.139	4.725	h11f	.139	.079	1.149
ct	−.001	.390	.999	h12f	−.429	.142	.651
t1d	.220	.032	1.246	s1.1sd	10.754	.013	46806.611
t2d	−.137	.065	.872	s1.2sd	2.484	.555	11.992
ph1p	.009	.934	1.009	s1.3sa	−4.721	.254	.009
ph2t	.765	.232	2.149	s1.4sa	−5.185	.332	.006
ph2p	.058	.022	1.060	s2ba	.803	.403	2.232

The model provides a binary category based on the measurements recorded during the molding process. In this manner, the model can be described as a classification model in machine learning. The class values of 0 and 1 represent the normal products versus the faults, respectively.

To evaluate the performance of the classification model, 2/3 of the data was randomly picked for training. Then, the predictions for the remaining part (1/3) were noted for evaluation. Specifically, the size of training data set and test data set were 405 and 189. For classification models, a confusion matrix contains the count of correct and incorrect predictions for each class [16]. Accordingly, the confusion matrix for the Binary Binary Logistic Regression (BLR) Model is presented in Table 5.

Table 5. Confusion matrix for the BLR model

	Predicted category		Percentage correct
Actual category	0	1	
0	164	12	93.20%
1	6	7	53.80%
Overall			90.50%

As the confusion matrix demonstrates, the model correctly predicts 164 of the 176 observations in Category 0 with a success rate of 93.20%. On the other hand, 7 out of 13 observations were predicted correctly for Category 1. The count total errors in predictions was found 18. In particular, 12 normal samples were predicted as faults, and the 6 faults were not detected by the model.

To assess the regression model for our case further, a 10-fold cross validation procedure was conducted. As noted in [17], multi-fold (k-fold) cross validation with a high number (k) reduces bias in predictions, but increases the variance as well as computational cost. Accordingly, a moderate value for k = 10 is chosen, and data was randomly split into 10 equal-sized groups. Next, one group was chosen for testing, while the remaining 9 were used in training. With such approach, all records in the dataset have been used interchangeably for training and testing through 10-fold cross validation. The correct and incorrect predictions have been demonstrated as a confusion matrix in Table 6.

Table 6. Confusion matrix for the BLR model with 10-fold cross-validation

	Predicted category		Percentage correct
Actual category	0	1	
0	544	9	98.37%
1	22	19	46.34%
Overall			94.78%

As the results suggest, the overall percentage of correct predictions 94.78% with 10-fold cross validation. It could be noted that the rate of success for predicting unfaulty products has been considerably high. On the other hand, correct predictions for faults have been found dramatically lower.

5 Conclusions

Industry 4.0, the industrial vision of the 2010 s, has transformative effects on the manufacturing systems. Real-time and accurate data collection, making inferences and creating value with advanced data analytics, and driving the manufacturing operations with the aid of digital assets, can be mentioned among the critical steps on the road of such vision. An essential use of the collected process data is to predict the possible faults in the investigated processes and ensure quality by taking the necessary cautions.

In this study, a wheel rim manufacturer's real problem that occurs during the molding process is addressed with actual collected data by the firm. The objective of the study is to build a model to predict whether a process will result in a fault. The data consists of sensor data that involve various measurements taken through the molding process. Specifically, the dataset obtained for the study involves data recorded for 594 parts continuously processed in four days.

The molding process occasionally results in start-up scraps, shrinkage, and visual scraps. The faults were assigned with fault codes of 1, 2 and 3. In addition, the normal products were also marked with a code of 0. Due to the lack of faulty samples for categories 2 & 3, the functionality of the logistic regression model was restricted to binary classification. The model was initially trained with 2/3 of the dataset, and tested with the remaining 1/3. In this scenario, 90.50% of predictions has been found correct. Subsequently, the dataset was split and a 10-fold cross validation procedure was conducted to reduce the bias. In this setting, the overall percentage of correct predictions has been recorded as 94.78%. A limitation of the model for both cases was the relatively lower success rates in predicting the faulty samples.

The BLR model proposed in this study might be tested on larger datasets with more observations and faults in further studies. Besides, alternative prediction and classification models such as Multinomial Logistic Regression and Artificial Neural Networks might be applied along with a comparison of techniques in terms of prediction performance.

Acknowledgments. We would like to thank Maxion İnci Wheel Group and Logo Business Solutions for their support and contribution in this study.

References

1. Anand, A., Coltman, T., Sharma, R.: Four steps to realizing business value from digital data streams. MIS Q. Executive **15**(4), 259–277 (2016)
2. Chen, H.M., Schütz, R., Kazman, R., Matthes, F.: How lufthansa capitalized on big data for business model renovation. MIS Q. Executive **16**(1), 19–34 (2017)
3. Ives, B., Palese, B., Rodriguez, J.A.: Enhancing customer service through the internet of things and digital data streams. MIS Q. Executive **15**(4), 279–297 (2016)
4. Zhong, R.Y., Xu, X., Klotz, E., Newman, S.T.: Intelligent manufacturing in the context of industry 4.0: a review. Engineering **3**(5), 616–630 (2017)
5. Hashemian, H.M.: State-of-the-art predictive maintenance techniques. IEEE Trans. Instrum. Meas. **60**(1), 226–236 (2010)
6. O'Brien, J.A.: Introduction to Information Systems. Irwin McGraw-Hill, USA (2000)
7. Aberdeen Group: IOT and analytics - Better manufacturing decisions in the era of Industry 4.0 (2017). https://www.ibm.com/downloads/cas/ZPB2PN2G
8. Almada-Lobo, F.: The Industry 4.0 revolution and the future of manufacturing execution systems (MES). J. Innov. Manag. **3**(4), 16–21 (2016)
9. Caruana, R., Niculescu-Mizil, A.: An empirical comparison of supervised learning algorithms. In: Proceedings of the 23rd International Conference on Machine learning, pp. 161–168. ACM (2006)
10. Candanedo, I.S., Nieves, E.H., González, S.R., Martín, M.T.S., Briones, A.G.: Machine learning predictive model for industry 4.0. In: Uden, L., Hadzima, B., Ting, I.-H. (eds.) KMO 2018. CCIS, vol. 877, pp. 501–510. Springer, Cham (2018). https://doi.org/10.1007/978-3-319-95204-8_42
11. Pandey, P.S.: Machine learning and IoT for prediction and detection of stress. In: Proceedings of 17th International Conference on Computational Science and its Applications (ICCSA), pp. 1–5. IEEE (2017)
12. Maxion İnci: Companies and Brands (2019). http://www.inciholding.com/en/companies-and-brands/production/maxion-inci
13. King, J.E.: Binary logistic regression. In: Osborne, J.W. (ed.) Best Practices in Quantitative Methods, pp. 358–384. Sage Publications, USA (2008)
14. Hosmer, D.W., Lemeshow, S.: Applied Logistic Regression. John Wiley and Sons, New York (2000)

15. de Menezes, F.S., Liska, G.R., Cirillo, M.A., Vivanco, M.J.: Data classification with binary response through the Boosting algorithm and logistic regression. Expert Syst. Appl. **69**, 62–73 (2017)
16. Giudici, P.: Data mining model comparison. In: Maimon, O., Rokach, L. (eds.) Data Mining and Knowledge Discovery Handbook, pp. 641–654. Springer, Boston (2009)
17. Hastie, T., Tibshirani, R., Friedman, J.: The Elements of Statistical Learning. Springer Press, Second Edition, USA (2017)

Automated Quality Assurance Applications in the Rise of IoT

Gamze Ugur Tuncer[1,2](✉), Can Davutoğlu[1], and M. Numan Durakbasa[2]

[1] Industrial Metrology and Adaptronic Systems, Institute of Production Engineering and Photonic Technology TU Wien, Vienna, Austria
{gamze.ugurtuncer, can.davutoglu}@qitasc.com
[2] QiTASC GmbH, Vienna, Austria
numan.durakbasa@tuwien.ac.at

Abstract. Together with the rise of industry 4.0 and emergence of 5G possibility, swift technological developments take place in the IoT field. The increasing customer attention on smart home applications and systems containing IoT devices, necessitates diminishing the time to market and also maintain the required quality while sustaining the profit.

Quality in IoT systems can only be ensured if all the involved systems and components (the devices and their sensors, gateways, apps and websites, backend systems and transmission protocols) are viewed and checked together.

With intaQt® (INtelligent Test Automation & Quality-assurance Tool) it is possible to perform the tests of these different elements automatically under one context and collect the evidences in milliseconds basis. This increases the efficiency of IoT testing and decreasing the time to market dramatically.

Keywords: Quality assurance · Test automation · Smart home · IoT testing · End to end testing

1 Introduction: Automated Quality Assurance Concept

Quality Assurance (QA) and testing are evolving in the direction to maximizing business outcomes and end-user satisfaction. Increasing concentration to end-user satisfaction shapes the QA objectives and IT trends towards digitalization of businesses and adoption of agile, DevOps and Cloud systems. The emphasis on end-user satisfaction underlines the importance of end-to-end testing. Furthermore, together with swift technological, developments the systems to be tested and assured becomes more and more complicated.

These phenomena highlight the importance of automation of testing of end-to-end processes. On the other hand, as the way of smart, intelligent and perceptive QA, test automation currently is not fully exploited and incorporated by the companies. According to World Quality Report 2017–18 [1], the average level of automation in QA is around 16 percent. Lack of experienced and skilled resources is the main reason for this. Therefore, it has utmost importance to explore the ways to increase the level of test automation in QA.

N. M. Durakbasa and M. G. Gençyılmaz (Eds.): ISPR 2019, LNME, pp. 361–368, 2020.
https://doi.org/10.1007/978-3-030-31343-2_32

Driving forces for automation in the ICT industry:

- Agile Projects require automation
- Complex, mashed applications can no longer be manually tested. (Various technologies, less time, high complexity)
- Projects ask for transparency (Automation provides an immediate snapshot of a project at anytime)
- Innovations and new approaches facilitate automation
- Increasing cost pressure in software development

Test automation is not necessarily the replication of existing manual test processes. Automation enters to picture in various steps of QA. These might be counted as planning and modelling the best test scenario, provision of all elements and systems, execution of test, data/evidence collection, verification (logs, Call Detail Records, etc.) and reporting.

The companies which adopt intelligent automation approaches elevate their ability to overcome time to market pressure together with the optimum level of quality to secure their business outcomes. With test automation the companies can make their projects transparent, so that they can take the right decisions, fulfil their projects faster and cheaper, reduce timelines due to parallel tasks and continuously growing repositories, verify features and services any time and in required accuracy which means reduced risk.

Automation helps the QA team to automate routine tasks and execute test cases many times which is not possible for a human to perform. For instance, one can test a mobile app's functionality on multiple devices without manually testing all the features recurrently. This accelerates time to market for the apps [2].

Benefits of automation testing can be summarized as:

1. Automated testing saves time by eliminating the need to write extensive test scripts.
2. The test scripts that need to be written using the automation tools can be reused.
3. It can also be used alongside manual testing procedures to crosscheck the test results.
4. Bugs can be detected during the development phase itself, which leads to shorter product development lifecycles.
5. The tests can be run 24/7 without the need for manual input thus ensuring thorough quality checks.
6. The tests scripts can be executed on the entire network or multiple devices simultaneously.
7. The reports that are generated are extensive and give the software tester details such as scripts executed, scheduled, bugs found and fixed, etc.
8. Automated testing also includes testing of web applications on all the browsers.
9. Automated testing is perfect for user environment simulation because graphical user interface (GUI) testing is very time consuming and consists of repetitive actions for the software tester. With automation tools the tester can run the script and make the test fast and accurate.

2 Test Automation with IntaQt®

Test software intaQt® is an INtelligent Test Automation & Quality-assurance Tool. It is used to create, manage and execute end-to-end automated tests that reflect real-life scenarios, customer experiences and complex interactions with backend systems, different domains and technologies in a single test case, as stated in detail Intact Handbook [3].

It is a unique solution for an important need of performing full-stack automation where rule-based validation capability (such as Call Detail Record Verification/XML-Verification/Trace Verification) and reporting is possible within the same framework.

Some possibilities that are a part of intaQt® test automation are as follows:

- Web-UI Automation
- Protocol & Network Simulation. CAP, Diameter, SIP, etc.
- APP iPhone Automation
- Mobile Device Automation
- PC-Client/ MacOS/ Windows Automation
- APP-Tablet Automation
- APP-Android Automation
- Trace Integration & Automation
- Attenuator Automation
- VoIP-Phone Automation
- Remote Access Automation
- Roaming Automation
- IoT Automation
- Backend, IF, Automation/ Emulation, SOAP, REST, Ldap, SQL, ssh, sftp, etc.
- Automated Data Correlation & Verification
- Automated Reporting (HP-ALM, Jira, etc.)

Automation is possible in various phases of test process as interpreted in Fig. 1. End-to-end concept incorporates planning and modelling the test framework, configuring all elements and systems which will take part, test automation, data collection and rule-based verification and finally reporting [3].

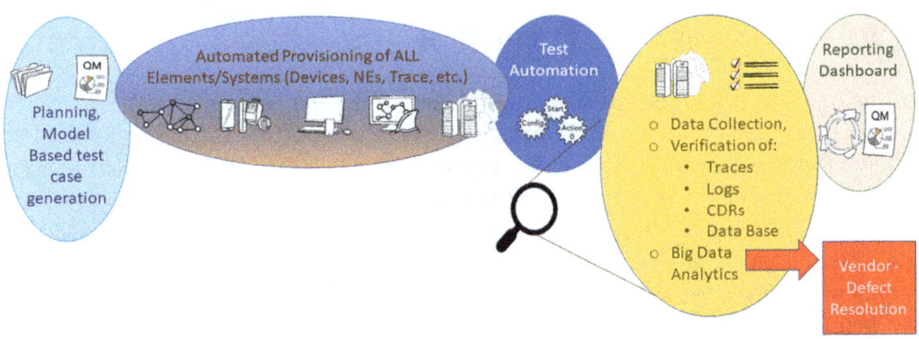

Fig. 1. Interpretation of test automation framework with IntaQt®

Figure 2 shows some open sources integrated to intaQt® that enable automation of various interfaces [4]:

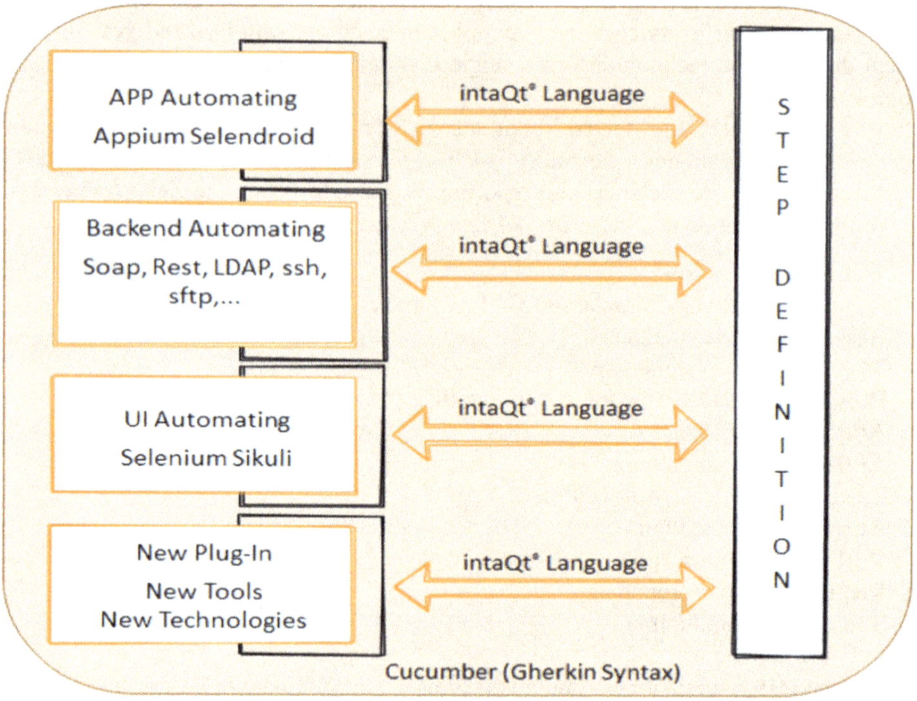

Fig. 2. Integration and IntaQt® language

- Appium, Selendroid for App automation
- SOAP, REST, LDAP, ssh, sftp, etc. for Backend automation
- Selenium, sikuli for UI Test Automation

3 IoT Test Automation Framework

IoT and smart home products are lately taking more and more place in the market which is very sensitive to easy operability. The IoT devices are also expected to communicate with each other. Therefore, a new product should be integrated into the existing smart home in a user-friendly way and should function with ease. It is also fundamental that the data logged in to these devices/applications and/or produced via usage is well protected [7].

Interconnection of products, services and data is realised via open interfaces (APIs) and the cloud. The interaction of apps or control panels, gateways and cloud-based services enables the smart services. Moreover, these software components are being updated continuously [4].

Hence the quality in IoT systems can only be ensured if all the involved systems and components (the devices and their sensors, gateways, apps and websites, backend systems and transmission protocols) are viewed and checked together. Another goal of IoT testing is to ensure that smart devices are both working properly and communicating with their app correctly.

The steadily increasing complexity of these systems, with increasingly interconnected devices and interfaces, creates challenges for the manufacturers. They need to integrate continuous testing and smart tests into their processes to ensure the quality of their smart home systems throughout the product's life cycle.

"Smart testing" means continuously and automatedly performing user actions and evaluating many different test scenarios for connected devices and systems integrating and verifying all necessary devices, protocols, interfaces, systems and user interfaces [5].

For example, the following components should be operated and the results should be verified automatically:

- User interfaces such as apps, websites or voice services
- Smart devices by pressing keys, settings via controller, checking by reading device displays
- Collecting and evaluating information from gateways, communication between the participating systems and systems connected via interfaces.

It is only possible to test the billion combinations of changing device firmware, advanced software, new app versions, and integrated cloud services with automated and configurable workflows [5].

4 Smart Home Applications

Some examples of Smart Home applications include controlling a room's thermostat over the Internet, having an alarm system that notifies the user about events, verifying that a smoke alarm detects particle density changes and controlling both optical and magnetic door/window contacts. Since many of these devices involve physical movement, such as a camera turning on/off or a window opening/closing, automated IoT testing is uniquely challenging.

intaQt® is able to operate IoT tests via test environment based on 3D print outs to hold and manipulate IoT devices, actors and sensors and control them. intaQt® automates and tests characteristics of interconnected devices that are used for Smart Home automation and verifies the end-to-end behaviour of these scenarios. Actions are automatically triggered via actors, such as operating small motors. Then sensors notify intaQt® about the changes of state which are aimed to be tested. At the same time, expected result is also checked whether it is displayed on the customer's app [6].

intaQt® includes the following functionality that enables IoT test case creation:

- Support for both Android and iOS devices when writing IoT tests
- Use intaQt®'s UI Steps language to write "Views", which define app elements and the actions to perform on them

- Create test cases consisting of scenarios that define the steps that will be executed to automate the Smart Home app
- Develop custom models and functions to customize behaviours
- Configure local and remote devices, network interfaces, reporting services, context objects, etc.

5 IoT Test Automation with IntaQt®

Together with the rise of industry 4.0 and emergence of 5G possibility, smart home applications with IoT devices are also taking more place in the market [8]. On the other hand, testing is getting more complicated which stresses the importance of automation. intaQt® has an automated framework for IoT testing, consisting of electronic and mechanical devices which operate the IoT devices and their App communication.

The environment is simulated through embedding the IoT devices to the housings which are designed and printed out from 3D printers. The devices are operated/triggered automatically as if a human or an environmental condition is generating, via tools like electronic motors, magnets, heaters etc. and their functionality are tracked via sensors, cameras, thermometers etc. and the automatic orchestration and result evaluation of the test is realised through intaQt®.

The physical connections of all the devices are realised through Raspberry Pi control unit and virtual connections to backend server and smart home App are realised through qivicon and network. Machine to machine area network is defined as the "device and gateway domain" of test framework. Back end server and App environment is defined as "network domain" of test framework. The test automation framework is operated and verified step by step via intaQt® as shown in Fig. 3.

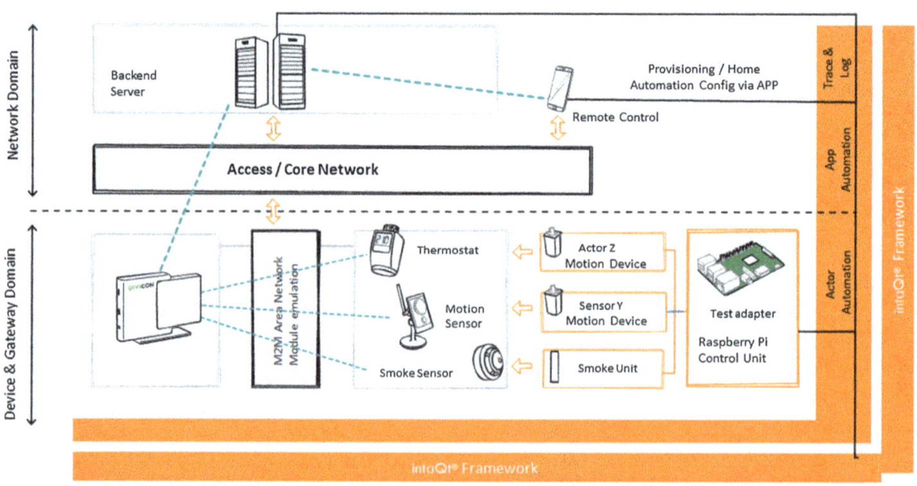

Fig. 3. IntaQt® automation setUp for IoT - 3 devices

As a concrete automated use case of a smoke detector test by intaQt®, the smart home App is loaded to the telephone and started, then the smoke in a room is produced automatically via a small machine which burns oil. The smoke detector operated and sent an alarm message to the App informing about the fire at the room. After the verification of message delivery, the smoke is ventilated via fan, the alarm stopped and a message about tranquillity is sent to the App which is also verified.

IntaQt® realises the whole operation and result verification in milliseconds automatically. Therefore, it is very convenient to swiftly and accurately detect where and when the bugs have occurred, enabling the engineers to fix quicker and the system to go live faster.

6 Conclusions

As industry 4.0 evolves towards more sophistication and 5G possibility is ready, the market for smart home applications and IoT devices expands rapidly. End-users are in need of smart solutions which work easily and quickly, also are simply compatible with the existing smart systems at home, work efficiently with their apps and also are keeping their personal information protected. The quick evolution, complex environments and need to perform end-to-end testing requires automation. Therefore, test automation should be considered at the centre of IoT concept.

IntaQt® provides a unique solution for IoT end-to-end test automation, where testers are able to perform various environments' testing without changing any context within one test case scenario automatically. This enables a swift testing possibility with accurate test results, evidence collection and evaluation which dramatically eases the defect correction and reduces the time to market.

References

1. World Quality Report 2018–19 Tenth Edition. https://www.capgemini.com/resources/world-quality-report-2018-19-australia-new-zealand/
2. Softweb Solutions. Automated Testing. https://www.softwebsolutions.com/resources/qa-automation-consulting-services.html
3. Davutoglu, C., urakbasa, N.M.D., Bas, G., Guclu, E.: Improved quality and reliability in telecommunication network testing by means of a remote online test system. iJOE **10**(4), 66 (2014)
4. Zehender, M., Davutoglu, C.: IintaQt® (Intelligent Test Automation and Quality-assurance Tool) Handbook, Austria 2012–2019
5. TÜV Rheinland. Smart Home Services Brauchen Smart Testing. Gebäude Digital. https://www.gebaeudedigital.de/haus-und-gebaeudeautomation/qualitaet-dauerhaft-sicherstellen/
6. Use Cases/ Internet of Things Testing. Qitasc web site (2019). https://www.qitasc.com/usecases/iot

7. TÜV Rheinland. Smart-Home, Hohe Qualität von Smart-Home-Services mit automatisierten Tests. Haus und Elektronik 2/2019. https://www.beam-verlag.de/fachzeitschriften/haus-und-elektronik/fachartikel/
8. Bauer, A., Bas, G., Durakbasa, N.M., Kräuter, L., Ugur–Tuncer, G.: Measurement technology & quality & justicia in industry 4.0. In: ISPR 2018: Proceedings of the International Symposium for Production Research 2018, pp 438–450, August (2018)

Miscellaneous Topics

A Stochastic Approach for Effective Workforce Planning: A Case of a Call Centre in Turkey

Ece Üstün, Buket Baritugan, Aziz Kemal Konyalioğlu(✉), and Ferhan Çebi

Department of Management Engineering, Istanbul Technical University, Istanbul, Turkey
eceustn@gmail.com, buketbaritugan@gmail.com, {konyalioglua, cebife}@itu.edu.tr

Abstract. In this study, customer service activities of the company, which is actively serving in the e-commerce sector in Turkey, provided through the call center were reviewed and a solution was proposed to improve the existing problems. In this respect, the literature research has been conducted on e-commerce sector and call centres, increasing the customer demands with the campaigns implemented by the companies serving in the e-commerce sector in certain periods, and thus the effect of the intensity experienced on customer services in call centres has been examined from different perspectives. Later on, with a company that provides international services, including Turkey it has reached to more detailed information and data relating to jointly carrying out a study subject. Surveys were carried out with the planning team and team leaders of the company, moreover, current problems were investigated and suggestions for solutions were evaluated. The data obtained, the previous year's statistics of the company were examined, and the needs were determined. In order to develop solutions, different concepts have been created and evaluated along with one of the concepts created in line with these results has been selected. The design of the study was prepared according to the results obtained from these stages. Finally, risk analyses were made, the points that should be prevented by the company were reached, and ideas were presented for the future stages.

Keywords: Erlang distribution · Call centre · Workforce planning · Segmentation planning · E-commerce

1 Introduction

E-commerce sector is developing in our country as in other countries recently. Many e-commerce companies continue to serve in our country and their numbers are increasing. In this case, it increases the competition between firms and makes it difficult to come forward in the sector. For this reason, as well as giving importance to the issues such as the quality and diversity of their products, the companies have begun to care about their communication with the customers. At this point, call centres come to the forefront as the channel where the companies communicate directly with the customer. Therefore, the improvement of the service provided by the call centre is very important among the companies. However, some problems can be experiences.

N. M. Durakbasa and M. G. Gençyılmaz (Eds.): ISPR 2019, LNME, pp. 371–388, 2020.
https://doi.org/10.1007/978-3-030-31343-2_33

To illustrate, the quality of the service provided to customers mostly decreases due to lack of planning in call centers. In addition, customers need to wait even longer periods. Such problems reduce customer satisfaction and lead customers to shop from other e-commerce sites. Customers give importance to factors such as product quality, security, etc. Additionally, they started to consider the customer service quality before and after shopping while evaluating the brand and its products. The rapidly expanding e-commerce sector has also increased competition, and customers are quickly starting to shop from competitors if they are not satisfied. In this case, e-commerce sites are affected negatively in terms of revenue, brand image, etc. Therefore, the call centers which work with these e-commerce sites should make their plans such as the number of employees and employee competencies very carefully. However, it is still seen that this issue is not given enough importance and workforce planning can be made for shorter periods rather than long term.

Due to the rapid development of the e-commerce sector and increasing competition, the companies serving in this sector apply various campaigns depending on the seasons or on special days to attract customers. Increasing customer demand in these periods causes more problems in existing processes. Therefore, call traffic increases in call centers working with these companies. The company works with the Webhelp company as a call center. The increase in calls made to the call center during these periods leads to a decrease in the quality of service provided to the customer and increase in the waiting time due to the shortcomings in the planning of the number of staffs.

Basically, this study aims to improve the process of determining the number of call center agents that can meet customer demand increase during special days and campaign periods. Furthermore, the study aims to put forward an analyse about the data of the past years to determine the periodic increases in the number of calls during year, determine the changes in service quality and waiting time during high call volume periods through customer feedbacks, define current workforce planning method deficiencies, determine demand increase in special days and promotion periods through demand forecasting methods, determine the desired service quality and waiting time objectives and plan the number of agents using one of the appropriate methods. The study includes five main sections which are an introduction section, a literature review section, methodology and application section and finally a conclusion section.

2 Literature Review

Nowadays, rapid developments in information and communication technologies have increased the use of mobile devices in every aspect of our daily life and the concepts of information society and e-commerce have gained a strategic importance in terms of economic development, competitive advantage and social welfare. According to the World Trade Organization (WTO), electronic commerce is the production, advertisement, sales and distribution of goods and services through telecommunication networks (Canpolat, 2001) [1]. Expectations from e-commerce can be summarized as benefiting from the globalization process which is an irresistible fact, providing competitive advantage, getting more share from international trade and contributing to the sustainability of social and economic development of our country.

Multi-channel and pure online are the types of retail. They also include subcategories according to the type of product they sell such as Electronics, Travel, etc. as seen in the Fig. 1.

Fig. 1. Scope of estimation model

The company is in the Multi-Channel/Clothing & Footwear category since it serves both on the internet site and two stores in Istanbul.

As it explained in TÜSİAD The Driving Force of the Economy in the Digitalized World: E-Commerce Report, the classical trade, which started its life with a single channel, continues with multiple and cross channel structures. The main goal is to reach the structure called omni-channel with a shopping experience which is complementary of different channels.

Single channel (Traditional): The customer interacts with the brand over a single channel. The brand also interacts with the customer over a single type of channel.

Multi-Channel (Real): The customer can reach the brand through different channels, but these channels act independently. Brand interacts with customers who do not see the customer singularly and are not integrated in operational terms.

Cross Channel (Interim Goal): The customer interacts with the same brand over different channels. In the brand, the user can perceive as a single user, but interacts in an operational way through independently moving channels.

Full Channel (Final Goal): It sees a brand not just a brand channel but a brand. The brand also sees the user as a single customer both strategically and operationally.

Full Channel service not only enhances customer satisfaction, but also contributes to the efficiency of many operations, from customer service to customer retention, if it is set up correctly (Kantarci, Özalp, et al., 2017) [2].

2.1 Strong Growth Potential of E-Commerce Sector in Turkey

Due to the fact that electronic commerce has a dynamic structure and is constantly affected by the developments in information and communication sectors, the growth potential is quite high. As stated in the TÜBISAD E-Commerce in Turkey 2017 Market Size Report, there is a strong growth potential of e-commerce sector in Turkey (2018) [4].

Fig. 2. E-commerce estimation model

As can be seen from the Fig. 2 above, to determine the e-commerce volume, forecast model is generated using the number of visits, the rate of return to shopping and the average basket size information.

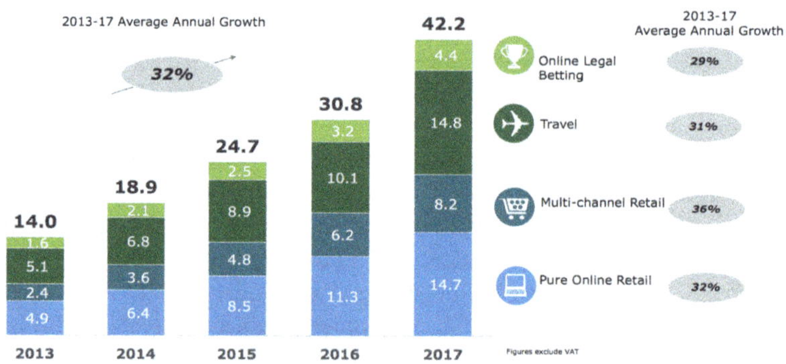

Fig. 3. E-commerce market size in Turkey

While calculating the change in the size of the market by years, four different categories were examined. These are Online Legal Betting, Travel, Multi-Channel Retail and Pure Online Retail. Between 2013 and 2017, the market size of e-commerce in Turkey increased by 32% and reached 42.2 billion TL from 14 billion TL. Moreover, multichannel retail volume grew by 36% and reached 8.2 billion TL from 2.4 billion TL during these years. The highest annual growth rate is also in the multi-channel retail (TÜBISAD E-Commerce in Turkey 2017 Market Size Report, 2018) [3] as seen in Fig. 3.

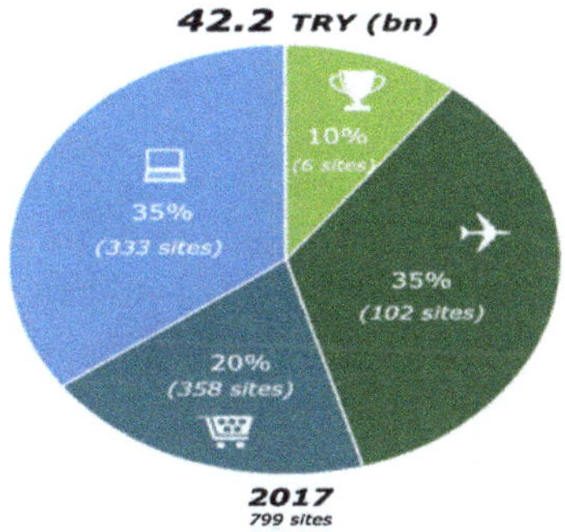

Fig. 4. Category mix and number of sites

Additionally, in 2017, the multi-channel retail category constitutes 20% of the total market volume with 358 web-sites given in the Fig. 4. The categories with the largest share in total market volume are travel and pure- online retail with 35% volume. Online legal betting has at least market volume with 10% and 6 web-sites. According to The Driving Force of the Economy in the Digitalized World: E-Commerce, Both demand and supply side dynamics support e-commerce growth expectations. The growth potential of the internet penetration in the less developed countries, the growing young population and the inclination of the young population to the technology show that customer demand in the sector has increased. With the developing technologies, the expectations of the customers are increasing. Speed and convenience are the most important expectations. On the supply side, in the classical retail sector, costs have increased and profit margins have started to decrease with the increase in competition. Despite these negativities in classic retail, the cost advantages of online sales and opportunities of the potential market make the internet more interesting for businesses (Kantarci, Özalp, et al. 2017) [2].

There are several queue model studies on call centers in the literature. For instance, according to Ma et al., a new performance evaluation methodology has been proposed for the service provided to people in the call center service cycle at the level of interaction with customer relations. A customer group-based queue model has been created to enable performance analysis at the interaction level using a temporary performance criterion within certain time intervals and in this study aimed to enable individual evaluation of team performance. Evaluated performance criteria are average response time, average talk time, queue waiting time, hourly call and service level. In addition, the proposed system has been tested by applying different scenarios in real life (2011) [4]. Moreover, Ertogral and Bamuqabel stated, a shift plan has been created for the call center employees of a telecommunication company serving in two different languages and queue model approach and the simulation model was used. In addition, the minimum number of operators, the most cost-effective shift system and the service level of the call center were determined in this study (2008) [5]. Van Buuren et al., presented and compared three discrete event simulation models for emergency service call centers. These queue models provide a new perspective for emergency call centers and can be used in capacity and workforce planning (2016) [6]. Legros et al., have developed a markov chain method for evaluating the performance of the system for a call center model that has a callback option that allows for an outgoing call to be converted to an outgoing call (2017) [7]. Lin et al., have developed multi-channel queuing models to estimate the border crossing delay with the help of the predictive passenger information system (2014) [8]. Aktekin, sought to resolve the inadequacy of studies on the uncertainty of input distributions and their impact on call center management. For this purpose, Bayesian parametric and semi-parametric mixed models have been analysed (2014) [9].

3 Current System of the Company

One of the important issues in call centers is that there are special terms used in the company. Furthermore, all employees are expected to dominate these terms.

ACHT Duration (Average Call Handling Time): The number of times that customer agents serve for the answered calls is the average value per answered call.

ACW (After-Call Work ACW): Processes that must be performed immediately after a transaction due to an incoming transaction type. This can often be done by data entry, filling in forms, or making external crests. The agent cannot answer another incoming call while in this state.

AHT (Agent Handling Time): It is the time a representative spends with the caller during the interview.

Abandon Call: Also called lost call. The caller hangs up the phone before reaching a representative.

SL (Service Level): This is the ratio of the incoming calls to the total number of calls received within the target period. The higher the SL value, the higher the number of calls answered in the first targeted second.

AR (Answer Rate): It is the ratio of the calls received by the customer representatives to the total number of incoming calls.

3.1 Current System Analysis

This section examines the current system of call answering systems used in the company. In the current system, while The company is a supra system, the subsystems of the Webhelp call center are identified as customers, telephone and agents. The main modules that make up the Webhelp call center are as follows.

Customers: People who are shopping from the company site and calling the call center for any problems or questions.

Phone: It is used to ensure communication between customers and call center employees.

Customer Representatives: Employees who receive incoming calls to the call center by customers.

3.2 Process Flowchart of the Existing System

The customer calls the call center and the call appears in the system and then the customer starts waiting in the queue. If the customer does not wait until the call is answered, he/she leaves the queue and waits to solve the problem if the call is answered. If the agent who answered the customer's problem does not solve the problem, he/she transfers it to the back office and the problem is solved. If problems cannot be solved in back office, the problem is saved.

3.3 Current System Limitations

There are many limiting factors in call centers. The most familiar of these is the turnover rating. The reason why this ratio is quite high is the difficulty of the customer representatives in working conditions. Therefore, there is a need for new employees and these employees have to undergo certain training. The high turnover rate also increases the time spent on training the operators. This may cause problems in the quality of service provided to the company and the customer. In addition, there may be situations in which previously unscheduled campaigns are put into practice. Therefore, it may be necessary to re-arrange the number of employees or working hours. Therefore, it is important that the planning system is flexible and easy to implement. Another system limit is to plan the break times of the operators. At certain times of the day, the need of the employees increases, but if they are not planned correctly, there may be an unlikely number of employees to reach a sufficient number of employees when the incoming calls are intense. When planning, it should be considered in such cases. Main limitations of the company can be followed as;

- Turnover Rate
- Training time for customer representatives
- Unscheduled campaigns
- Customer representatives' break times
- Shift (8 h) working system
- Attendance status of customer representatives

4 Methodology and Application

4.1 Segmentation of Customer Representatives

Customer contact time with clients may vary for many reasons; for example, the problem that the customer has experienced may be a problem that has not been encountered before, or the customer may be asking questions other than his or her own problem. Therefore, although the call duration of the customer representative is not a criterion for the performance of each call, the situation is different when we look at the average of the duration of all calls. The following table lists the AHT (Average Handling Time) of customer representatives grouped according to their experience. As the experience of customer representatives increases, the decrease in AHTs is evident as seen in Table 1.

Table 1. AHT of customer representatives

Experience	Average AHT
0–2 Month	315 s
2–4 Month	293 s
4–6 Month	288 s
6–8 Month	241 s
8 + Month	233 s

According to Manap, the learning curve that determines the relationships between the number of repeats of a work element and the performance increase plays an important role in determining the optimum processing time (2013) [10]. The fact that customer representatives complete the call in a shorter period of time indicates the existence of the learning curve.

Considering the AHTs to divide customer representatives into three classes in terms of experience; It is seen that there is little difference between the AHTs of 6 to 8 months experienced and 8 and more months experienced customer representatives. In addition, a similar situation is valid for 2–4 months experienced and 4–6 months experienced customer representatives. For this reason, a class of 8 and more experienced customer representatives has been formed. A further class was created for 2–4 months experienced and 4–6 months experienced customer representatives. Finally, a group of 0–2 months experienced customer representatives have been created.

The group consisting of 6–8 months experienced and 8 and more months experienced customer representatives is named as experienced agent. The other group consisting of 2–4 months experienced and 4–6 months experienced customer representatives was given the name of medium experienced agent. Lastly, the group consisting of 0–2 months experienced customer representatives was named as less experienced agent.

4.2 Customer Segmentation

Customers who currently have their data in company data warehouse; it was decided to classify 10% as upper segment, 30% as middle segment and 60% as lower segment. Segments are named by considering that each segment customer can feel special these names are the most important to the less important, respectively, platinum, and gold, silver. Using the calculated weights of the criteria, all customers are ranked from the best score to the worst score; the first 10% group platinum, the next 30% group gold, the last 60% group silver customers.

(1) *Determination of Customer Segmentation Criteria:* Segmentation of the customers according to the strategic importance of the company is very important. Accordingly, the company decides what kind of service to be provided to the customer segment. Before this stage, the point to be noted is whether the customer needs to be given this service. In other words, the company will spend more effort for the customer will increase the satisfaction of the customers should be known. In short, to make the segmentation successful, it is necessary to make sure that the

customer is sensitive to this service. With this survey, it is tried to learn what kind of customers waiting time or talking with experienced agent is an important criterion. In line with this questionnaire, waiting time and talking with experienced agent are important for customers in frequent shopper, large amount shopper categories.

(2) *AHP Method:* The importance of these criteria was compared with the pairwise comparison matrix by the senior management of the company. AHP method was used in this comparison. As a result of the calculations, the weights of the criteria were determined. Their weights are from highest to lowest, respectively total shopping amount, frequency of shopping, frequency of problem with service received and high level AHT (Average Handling Time). Senior management checked the consistency of weights according to first ratings and final weights accepted respectively given as in Tables 2, 3 and 4.

Table 2. Pairwise comparison

Pairwise comparison	Frequency of shopping	Frequency of problem with service received	High level AHT (Average Handling Time)	Total shopping amount
Frequency of shopping	1	3	5	1/3
Frequency of problem with service received	1/3	1	3	1/5
High level AHT (Average Handling Time)	1/5	1/3	1	1/7
Total shopping amount	3	5	7	1

Table 3. Pairwise comparison calculations

Pairwise comparison	Frequency of shopping	Frequency of problem with service received	High level AHT (Average Handling Time)	Total shopping amount	Weights
Frequency of shopping	0,2206*	0,3214	0,3125	0,1989	0,263
Frequency of problem with service received	0,0735	0,1071	0,1875	0,1193	0,122
High Level AHT (Average Handling Time)	0,0441	0,0357	0,0625	0,0852	0,057
Total shopping amount	0,6618	0,5357	0,4375	0,5966	0,558

Table 4. AHP rating

Pairwise Comparison	Weights	Final weights
Frequency of shopping	0,263	26%
Frequency of problem with service received	0,122	12%
High level AHT (Average Handling Time)	0,057	6%
Total shopping amount	0,558	56%

$$*Calculation\ of\ AHP: \frac{1}{1+\frac{1}{3}+\frac{1}{5}+3} = 0,2206$$

4.3 Erlang Calculation

As stated Robbins. Medeiros, and Harrison, Erlang formulation is accepted (2010) [11]. Sample data which is retrieved from Call Center Helper website for a call were used to illustrate the steps of the Erlang formulation.

Step 1: In the first stage, the inputs for the Erlang formulation are determined given in the Table 5.

Table 5. Inputs

Inputs	Values
Number of calls	100
In a period of minutes	30
Average handling time (seconds)	180
Required service level	80%
Target answer time (seconds)	20
Maximum occupancy	85%
Shrinkage	30%

Step 2: At this stage, the hourly incoming call is calculated by using number of calls and in a period of minutes as follows.

$$(100 \div 30\,\text{min}) * 60\,\text{min} = 200 \text{ calls per hour} \tag{1}$$

Step 3: At this stage traffic intensity is calculated by using average handling time and calls per hour as follows. This value is shown as A in the formulation.

$$(180\text{ sec} * 200) \div 3600\,\text{sec} = 10 \tag{2}$$

Step 4: At this stage, the raw number of agents is estimated and accepted by adding 1 an A value. This value is shown as N in the formulation.

Step 5: At this stage probability a call waits is calculated by using following erlang formulation.

$$P_W = \frac{\frac{A^N}{N!}\frac{N}{N-A}}{\left(\sum_{\mathrm{i}=0}^{N-1}\frac{A^i}{i!}\right) + \frac{A^N}{N!}\frac{N}{N-A}} \tag{3}$$

A: Traffic Intensity N: Raw Number of Agents

The 'ACT' formula in Excel is used for the factorial parts in this formulation while the 'POWER' formula in Excel is used for exponential parts.

Step 6: When we break the formula into pieces and simplify; the following part of the formula is equal to X.

$$\frac{A^N}{N!}\frac{N}{N-A} = X \tag{4}$$

The following part of the formula is equal to Y.

$$\left(\sum_{\mathrm{i}=0}^{N-1}\frac{A^i}{i!}\right) = Y \tag{5}$$

Step 7: The X value is the top raw of the formula is calculated as follows.

$$(\mathrm{AN}\ /\ \mathrm{N!}) * (\mathrm{N}/(\mathrm{N}-\mathrm{A})) = (1011/11!) * (11/(11-10)) = 27557,319$$

Step 8: The Y value is the down raw of the formula is calculated as follows.

$$\left(\sum_{\mathrm{i}=0}^{N-1}\frac{A^i}{i!}\right) \tag{6}$$

N = 11, N-1 = 10 so calculation i = 0 until i = 10 is needed. Table 6 shows the calculations of Erlang Distributions.

Table 6. Calculations

i	i!	A^i	$A^i/i!$	$\Sigma\ A^i/i!$
0	1	1	1	1
1	1	10	10	11
2	2	100	50	61
3	6	1000	166,7	227,7
4	24	10000	416,7	644,3
5	120	100000	833,3	1477,7
6	720	1000000	1388,9	2866,6
7	5040	10000000	1984,1	4850,7
8	40320	100000000	2480,2	7330,8
9	362880	1000000000	2755,7	10086,6
10	3628800	10000000000	2755,7	12842,3

Step 9: The X and Y values calculated in step 10 and step 11 in the formulation are replaced and the probability a call waits is calculated as follows.

$$P_W = \frac{\frac{A^N}{N!}\frac{N}{N-A}}{\left(\sum_{i=0}^{N-1}\frac{A^i}{i!}\right) + \frac{A^N}{N!}\frac{N}{N-A}} \tag{7}$$

$$P_W = X/(Y + X) \tag{8}$$

$$P_W = 27557/(12842 + 27557) = 0.6821 = 68\% \tag{9}$$

Step 10: The service level in this section is calculated using the following service level formula.

$$Service\,Level = 1 - \left(P_W * e - (N - A) * \left(\frac{Target\,Answer\,Time}{AHT}\right)\right) \tag{10}$$

The 'EXP' formula in Excel is used for the Euler number.

$$SL = 1 - (0.6821 * EXP\left(-(11 - 10) * \left(\frac{20}{180}\right)\right) \tag{11}$$

$$SL = 1 - (0.6821 * EXP(-0.11111)) = 0.389613812 = 39\% \tag{11}$$

Step 11: Since the service level calculated in this section is less than 80%, the number of agents must be increased and rechecked. When the number of agents is 12, the service level is %64 after the previous calculations are made. Since the calculated service level is still below 80%, the number of agents is increased until it reaches 80%. The number of agents is 14; therefore, the service level is 88,8% and is greater than 80%, in this case the number of raw agents required is determined.

Step 12: In this section, average speed of answer is calculated as follows.

$$\text{Average Speed of Answer (ASA)} = P_W * \frac{AHT}{N - A} \tag{12}$$

$$\text{Average Speed of Answer (ASA)} = 0.1741 * \frac{180}{14 - 10} = 7.8sec \tag{13}$$

Step 13: In this section, occupancy is calculated as follows.

$$Occupancy = Traffic\,Intensity \div Raw\,Agents \tag{14}$$

$$Occupancy = 10 \div 14 = 71,4\% \tag{15}$$

Step 14: In this section, the effect of shrinkage on the number of agents is calculated as follows.

$$\textit{Number of Agents Required} = \textit{Raw Agents}/\left(1 - \left(\frac{\textit{Shrinkage}\%}{100}\right)\right) \tag{16}$$

$$\textit{Number of Agents Required} = 14 \div 0.7 = 20 \tag{17}$$

Step 15: Lastly, the data obtained in this section are summarized in the Table 7.

Table 7. KPI summary

KPI	Values
Total number of agents required	20
Service level	0,89
Probability a call waits	0,17413
Average speed of answer	43684

All these steps are formulated in excel and used to calculate the number of agents required.

4.4 Determining the Number of Customer Representatives Required

The following table shows the total number of agents, which is calculated with Erlang formulation, needed in 30-minute periods over the day of November 1 as an example as seen in the Table 8.

Table 8. Required agents

Time interval	Incoming calls	Agents required
08:00	52	13
08:30	96	21
09:00	203	41
09:30	251	49
10:00	279	54
10:30	307	59
11:00	327	63
11:30	308	60
12:00	320	62
12:30	296	58
13:00	320	62
13:30	343	66
14:00	361	69
14:30	374	71

(continued)

Table 8. (*continued*)

Time interval	Incoming calls	Agents required
15:00	378	72
15:30	364	70
16:00	338	65
16:30	345	66
17:00	311	60
17:30	264	70
18:00	227	45
18:30	180	37
19:00	186	38
19:30	176	36
20:00	189	38
20:30	181	37
21:00	198	40
21:30	172	35
22:00	160	33
22:30	141	29
23:00	121	26
23:30	110	24
00:00	2	1
00:30	2	1
01:00	2	1
01:30	2	1
02:00	2	1
02:30	2	1
03:00	0	0
03:30	0	0
04:00	0	0
04:30	0	0
05:00	0	0
05:30	1	1
06:00	1	1
06:30	1	1
07:00	2	2
07:30	4	2

4.5 Determining the Segmented Number of Customer Representatives Required

Customer representative segmentation is calculated with 10%, 30%, 60% of the total number of customer representatives required; experienced agent, medium experienced agent, less experienced agent numbers were determined. The important point here is that the number of experienced agents and middle-experienced agent numbers that are calculated as decimal numbers are rounded down to the lower levels, whereas the number of less experienced agents is rounded up to the upper level. In this case, if the total agent need is not equal to the total requirement for these three agent groups, 1 is added to the number of less experienced agents. In this way, the required capacity has been reached as seen in the Table 9.

Table 9. Segmented required agents

Time interval	Incoming calls	Agents required	Experienced	Medium experienced	Less experienced	Control	Adjusted less experienced
08:00	52	13	1	3	8	1	9
08:30	96	21	2	6	13	0	13
09:00	203	41	4	12	25	0	25
09:30	251	49	4	14	30	1	31
10:00	279	54	5	16	33	0	33
10:30	307	59	5	17	36	1	37
11:00	327	63	6	18	38	1	39
11:30	308	60	6	18	36	0	36
12:00	320	62	6	18	38	0	38
12:30	296	58	5	17	35	1	36
13:00	320	62	6	18	38	0	38
13:30	343	66	6	19	40	1	41
14:00	361	69	6	20	42	1	43
14:30	374	71	7	21	43	0	43
15:00	378	72	7	21	44	0	44
15:30	364	70	7	21	42	0	42
16:00	338	65	6	19	39	1	40
16:30	345	66	6	19	40	1	41
17:00	311	60	6	18	36	0	36
17:30	264	70	5	15	32	0	32
18:00	227	45	4	13	27	1	28
18:30	180	37	3	11	23	0	23
19:00	186	38	3	11	23	1	24
19:30	176	36	3	10	22	1	23
20:00	189	38	3	11	23	1	24
20:30	181	37	3	11	23	0	23
21:00	198	40	4	12	24	0	24
21:30	172	35	3	10	21	1	22
22:00	160	33	3	9	20	1	21
22:30	141	29	2	8	18	1	19
23:00	121	26	2	7	16	1	17
23:30	110	24	2	7	15	0	15
00:00	2	2	0	0	2	0	2
00:30	2	1	0	0	1	0	1
01:00	2	1	0	0	1	0	1

(continued)

Table 9. (*continued*)

Time interval	Incoming calls	Agents required	Experienced	Medium experienced	Less experienced	Control	Adjusted less experienced
01:30	2	1	0	0	1	0	1
02:00	2	1	0	0	1	0	1
02:30	2	1	0	0	1	0	1
03:00	0	0	0	0	0	0	0
03:30	0	0	0	0	0	0	0
04:00	0	0	0	0	0	0	0
04:30	0	0	0	0	0	0	0
05:00	0	0	0	0	0	0	0
05:30	1	1	0	0	1	0	1
06:00	1	1	0	0	1	0	1
06:30	1	1	0	0	1	0	1
07:00	2	2	0	0	2	0	2
07:30	4	2	0	0	2	0	2

4.6 Workforce Schedule

Employees in the call center work 8 h a day and have one-hour breaks. Customer representatives were also placed on the shift schedule with this constraint. Three different categories of customer representative needs were met for every half hour interval. Due to the flexible decision-making process during the check-in and check-out times, the customers were prevented from being idle at the call center during the hours without call density. In this way, optimum the number of customer representatives needed was decided and the inefficiency was tried to be prevented.

As an application, risk management plan has been constructed. Changing the process of workforce planning, which is one of the most critical processes of call centers, is an important decision for the company. For this reason, the company must determine its risks and evaluate the severity of these risks. With the brainstorming technique, the risks that may occur during this process change have been identified and the responses to these risks are determined. As shown in the table below; the probabilities and effects of the identified risks were scored from 1 to 5 and the effects and probability are multiplied to find the value of each risk. As a result, it was determined that the prevention plan for risks which have high values should be done as seen in the Table 10.

Table 10. Risk analysis

Risk	Potential response	Risk action	Root causes	Probability	Impact	Probability *impact
More than enough customer representatives can be recruited	Transfer of employees to other Webhelp companies needed	Mitigate	Inadequate planning	0,7	0,7	0,49
There may be a large number of customer representatives who quit work	Transfer from other Webhelp companies to this company as the training process will not be necessary	Accept	Lack of motivation, not getting salary, natural causes...	0,3	0,9	0,27
Technical malfunction may affect incoming calls in the call center	Ensuring regular technical maintenance and repairs	Accept	Insufficient maintenance and control	0,3	0,9	0,27
The breaks of customer representatives may intensify at certain hours	Additions to restrictions on the use of breaks	Mitigate	Intensive working hours, unfavorable working conditions	0,9	0,7	0,63
Workers in planning may not be able to adapt to the system	Giving necessary training about new planning method	Mitigate	Lack of training	0,3	0,5	0,15
Employees trained to move to the new system can leave the job	Consideration of these issues before the necessary trainings are given	Accept	Lack of motivation, not getting salary, natural causes...	0,3	0,9	0,27
Excessive interaction of employees can lead to lower productivity	Making rotations between employees' seating or shifts	Accept	Lack of control	0,5	0,5	0,25

5 Conclusion

To conclude, this study was aimed to improve the process of determining the number of call center agents that can meet customer demand increase during special days and campaign periods. The start point of the study to conduct the literature research to understand the dynamics of the e-commerce sector and the call center sector and to determine their impact on each other. It is seen that the quality of service provided by call centers to customers is an important factor that affects the relationship of e-commerce companies with the customer and thus the success of the sector. After the literature search, surveys were conducted with customers, call center planning team and customer representative team leaders. Additionally, the opinions of the experts who work in the company were consulted. Based on the analysis of the results of the surveys and company data, the main problems were determined. Inadequate workforce planning processes for the periods that are critical for e-commerce and call center sectors such as campaign periods have been identified as the prominent problem. Considering the constraints on the planning of workforce, the concepts that were designed to minimize the effects of the problems identified in the previous stage were developed and evaluated. After the evaluation of the concepts, it was concluded that the most effective solution was to segment and match the customers and customer representatives and to plan accordingly. Firstly, the criteria to be used for segmentation and their significance were determined. In the light of this information, customers are divided into three main categories: platinum, gold and silver. Moreover, customer representatives were categorized as experienced, medium level experienced and less experienced. In this way, it is aimed to provide high quality customer service to the strategically most important customer group for e-commerce Company. For further development, total shopping amount, frequency of shopping, frequency of problem with service received and high level AHT (Average Handling Time). As a result of the implementation of this method, which is planned for workforce planning, the change in customer satisfaction level and cost analysis are determined as the future study.

References

1. Canpolat, Ö.: E-Ticaret ve Türkiye'deki Gelişmeler, Hukuk Müşavirliği. T.C Sanayi Ve Ticaret Başkanlığı, Ankara (2001)
2. Kantarcı, Ö, Özalp, M., Ezginsoy, C., Özaşkınlı, O., Cavlak, C.: Digitalleşen Dünyada Ekonominin İtici Gücü: E-Ticaret. TÜSİAD (2017)
3. E-Commerce in Turkey, 2017 Market Size Report, TÜBİSAD (2018)
4. Ma, J., Kim, N., Rothrock, L.: Performance assessment in an interactive call center workforce simulation. Simul. Model. Pract. Theor. **19**(1), 227–238 (2011). https://doi.org/10.1016/j.simpat.2010.06.008
5. Ertogral, K., Bamuqabel, B.: Developing staff schedules for a bilingual telecommunication call center with flexible workers. Comput. Ind. Eng. **54**(1), 118–127 (2008). https://doi.org/10.1016/j.cie.2007.06.040
6. Buuren, M.V., Kommer, G.J., Mei, R.V., Bhulai, S.: EMS call center models with and without function differentiation: a comparison. Oper. Res. Health Care **12**, 16–28 (2017). https://doi.org/10.1016/j.orhc.2016.12.001
7. Legros, B., Jouini, O., Koole, G.: Optimal scheduling in call centers with a callback option. Performance Eval. **95**, 1–40 (2016). https://doi.org/10.1016/j.peva.2015.09.002
8. Lin, L., Wang, Q., Sadek, A.W.: Border crossing delay prediction using transient multi-server queueing models. Transp. Res. Part A: Policy Pract. **64**, 65–91 (2014). https://doi.org/10.1016/j.tra.2014.03.013
9. Aktekin, T.: Call center service process analysis: Bayesian parametric and semi-parametric mixture modeling. Eur. J. Oper. Res. **234**(3), 709–719 (2014). https://doi.org/10.1016/j.ejor.2013.10.064
10. Manap, G.: Paranın zaman değerive öğrenme etkisi altindaki kesikli zaman-değişken talepli parti büyüklüğü modelleri (unpublished master's thesis). Süleyman demirel üniversitesi sosyal bilimler enstitüsü işletme anabilim dali (2003)
11. Robbins, R., Medeiros, D.J., Harrison, T.P.: Does the Erlang C Model Fit in Real Call Centers? (2010)

An Analytical Approach for Analysing the Impact of Risks on Production Planning: Case of Öztiryakiler

Eda Telatar, Talha Bekeç, Aylin Başaran, Nihan Balıkçı, Berk Bilgin, Esra İlbay, and Tülin Aktin(✉)

Department of Industrial Engineering,
Istanbul Kültür University, Istanbul, Turkey
edatelatar@hotmail.com, talhabekec@hotmail.com,
esr_ilbay@hotmail.com, aylinbsrnn@gmail.com,
balikcinihann@gmail.com, berkbilgin01@gmail.com,
t.aktin@iku.edu.tr

Abstract. Accurate and applicable production plans are a must for manufacturing companies. Although companies tend to prepare ideal production plans, some exogenous factors can affect their validity. These risks, which occur unexpectedly, will have a negative influence on the plan. This study aims to determine the exogenous factors affecting the success of production planning of square and rectangular food containers manufactured by Öztiryakiler, and analyse their impacts on the plans. The risk factors are evaluated using Failure Mode and Effects Analysis, and their risk priority numbers are calculated. A mixed-integer linear programming model with the objective of total cost minimisation is developed to obtain the production plan of containers. Initially, an ideal data set is used as input; hence, this model's output displays a risk-free plan. Similarly, for each of the risk factor scenarios, mathematical models are solved with risk-related data. GAMS software and CPLEX solver is utilised in the solution of all models. Finally, for each of the selected high risk alternative, the expected total costs are calculated. This is achieved by multiplying the normalized risk priority number obtained from the Failure Mode and Effects Analysis with the corresponding optimal total cost of the risky plan. This analysis highlights the most critical risks, and comparison with the risk-free plan helps in proposing system improvements.

Keywords: Production planning · Production risks · Failure Mode and Effects Analysis · Mathematical modelling · Kitchen equipment manufacturing

1 Introduction

In order to cope up with the fierce competition in the markets, it is essential for companies to identify risks and exogenous factors that affect their planning processes. These risks should be taken into account when preparing a production plan. Consideration of the most commonly encountered and critical risks during production will minimise the effects of potential failures.

N. M. Durakbasa and M. G. Gençyılmaz (Eds.): ISPR 2019, LNME, pp. 389–402, 2020.
https://doi.org/10.1007/978-3-030-31343-2_34

Integrating risks into production planning has been studied by several researchers. Reference [1] proposes an approach that incorporates operational risks into the planning system, so as to reduce the risk level. The approach's feasibility is demonstrated through an exemplary scenario. Production planning models that consider uncertainty in an explicit manner are advantageous over those models, which ignore this aspect. With the aim of introducing the reader to these studies, Reference [2] provides a literature review consisting of 87 citations on production planning under uncertainty. Reference [3] presents a model and a methodology for production planning in a supply chain. The approach being stochastic, thus fits to the case of a highly variable make-to-order environment, having multiple resources of uncertainty.

The aim of this study is to identify the risks affecting the production plans of the square and rectangular stainless steel containers manufactured by Öztiryakiler, and to analyse their impacts on the plan.

Established in 1958, Öztiryakiler is one of the first kitchen equipment manufacturers in Turkey. It has 1300 employees and operates in a closed production area of 149,500 m^2. The company produces more than 4750 product types in the industrial kitchen sector, and provides planning, production and service support for all process from project planning to setting up of kitchens. Öztiryakiler is the leader in domestic markets, and with its distribution channels and authorized services, the company also serves foreign markets in 5 continents and over 100 countries. It is among the first 15 companies in its sector throughout the world [4].

The proposed methodology starts with the analysis of the current production planning system, and exogenous factors which have a direct effect on the planning process are taken into account. Failure Mode and Effects Analysis (FMEA) is used to evaluate these risk factors, and their risk priority numbers (RPN) are calculated. Economic crisis, raw materials stuck in customs, fluctuations in the exchange rate, cancellation of orders, arrival of an unplanned high priority order, increase in mould changing time, machine breakdowns, health issues and resignation of workers are among the 21 factors under consideration.

The production plans are determined with the help of a mixed-integer linear programming (MILP) model. The objective function of the developed model is to minimise the total cost, which consists of raw material and production costs. In the second phase of the study, by using an ideal data set, a risk-free production plan is obtained. The risk factors, which are studied in FMEA, are analysed as scenarios. The input data of these scenarios consist of risk-affected data that display various deviations from the ideal situation. All models are solved using GAMS software and CPLEX solver.

In the third phase, the outputs from the first two phases are combined to calculate the expected total cost (ETC) of each risk scenario. For this purpose, the critical risk factors with an RPN above a threshold value are selected for further analysis. The ETC value is found by multiplying the normalized RPN with the total cost of the corresponding production plan of each scenario. By comparing the cost of the risk-free production plan with the cost of each risk scenario, the impacts of these factors are evaluated and improvement opportunities are proposed.

The study is organized as follows: The proposed methodology is presented in Sect. 2. The implementation together with the numerical results are summarized under Sect. 3. Finally, the study is concluded in Sect. 4.

2 Methodology

The flowchart of the methodology is given in Fig. 1. It starts by analysing the current production system, and collecting the relevant data. This phase has been completed with the support of the production planning manager of the container group at Öztiryakiler. The methods employed in the study are summarized in the following subsections.

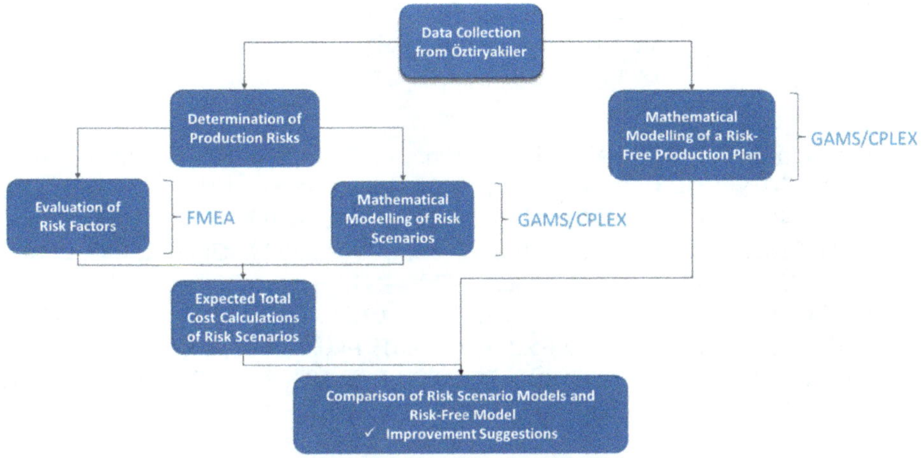

Fig. 1. Flowchart of the methodology

2.1 Failure Mode and Effects Analysis

FMEA is a widely used tool to evaluate and prioritise risks. In order to overcome the disadvantages in RPN calculations, Reference [5] proposes a new risk prioritisation model, and tests the model performance by the help of some examples. On the other hand, researchers have also developed hybrid models that integrate FMEA with other methods, and used these approaches in risk analysis [6–8].

After examining the current production system of Öztiryakiler, FMEA is selected to evaluate the various risk factors that the company faces. This constitutes the first step in examining the effects of these risks on the production plan. Initially, potential failure modes are specified and grouped under four process functions as; estimation-based, production-based, facility-based and sales-based. With respect to each potential failure mode, potential effect(s) are identified, and the affecting exogenous factors are written as potential cause(s) of failure. Severity (S) and occurrence (O) values of these effects are assigned. Then, the current process controls regarding each factor is indicated after discussions with the various departments involved in these processes. Detection (D) values are given by considering the extent of which the failures are detected by performing current process controls. Eventually, RPN is calculated as the multiplication of the respective S, O, and D values. All values are obtained by utilising a 10-point scale.

The risk factors that have an RPN greater than 200 are considered as critical, and are selected for investigating their impact on the total cost of the production plan generated.

2.2 Mathematical Modelling

An MILP model with the aim of total cost minimisation is developed to determine the optimum production plans. This model is run using different parameters that represent both the risk-free situation and the cases with risk factors. Each risk scenario employs specific risk-related data that deviates from the ideal situation.

Indices, parameters and decision variables of the proposed model are defined below:

Indices:

i product index (11020, 11040, 11055, 11065, 11100, 11150, 11200, 12020, 12040, 12055, 12065, 12100, 12150, 12200, 13020, 13040, 13065, 13100, 13150, 13200, 14020, 14040, 14065, 14100, 14150, 14200, 16065, 16100, 16150, 16200, 19065, 19100, 21020, 21040, 21065, 21100, 21150, 21200, 23020, 23040, 23065, 23100, 23150, 23200, 24020, 24040, 24065, 24100, 24150)

k machine index (01.01, 01.02, 01.03, 01.05, 01.06, 01.07, 01.14, 110.0, 02.33, 05.01, 05.02, 05.03, 05.04, 05.05, 05.08, 14.01, 14.03, 14.04, 14.15, 02.01, 02.02, 02.03, 02.04, 02.06, 02.09, 02.13, 02.16, 02.17, 02.18, 02.19, 02.20, 02.23, 02.37, 02.38, 03.01, 03.02, 03.04, 03.07, 03.08, 03.16, 03.20, 03.26, 03.27)

Parameters:

dpm number of working days in the month (January 2018)
hpd number of working hours per day
sph number of seconds per hour
sr scrap rate of products (percentage)
d_i demand of product i in the given month (units)
$inv0_i$ initial inventory of product i in the given month (units)
cs_i cost of stainless steel used in one unit of product i (TL per kg)
rm_s_i stainless steel used in one unit of product i (kg per unit)
c_i unit cost of product i excluding cost of stainless steel used (TL)
pt_{ik} unit processing time of product i at machine k (seconds)
st_{ik} setup time of product i required at machine k (seconds)

Decision Variables:

X_{ik} amount of product i passing from machine k (units)
MS stainless steel requirement of the given order (kg)
TS_k number of seconds used at machine k
ATS_k additional number of seconds required at workstation k

The developed MILP model is as follows:
Objective function:

$$minz = \sum_i (c_i + (cs_i \cdot rm\_s_i)).X_{i,01.01} \tag{0}$$

Subject to:

$$(1 - sr) \cdot X_{ik} \geq d_i - inv0_i; \qquad \forall i, k \text{ and } pt_{ik} > 0 \tag{1}$$

$$\sum_i (pt_{ik} \cdot X_{ik}) + st_{ik} \leq TS_k + ATS_k; \qquad \forall k \tag{2}$$

$$TS_k \leq dpm \cdot hpd \cdot sph; \qquad \forall k \tag{3}$$

$$\sum_i rm\_s_i \cdot X_{i,01.01} = MS \tag{4}$$

$$X_{ik} \geq 0 \text{ and integer}; \qquad \forall i, k \tag{5}$$

$$MS, TS_k, ATS_k \geq 0; \qquad \forall k \tag{6}$$

Objective function (0) minimises the total production and material costs. In order to satisfy the demand requirements, Constraint set (1) determines the production amount of each product at each machine on its route by taking into account the scrap rate, and the initial inventory. Constraint set (2) guarantees that the time (in seconds) required to manufacture all products at a certain machine cannot exceed its capacity, and determines the extra time required, if any. The upper bound on each machine (in seconds) is set in Constraint (3). The stainless steel requirement of all products is calculated in Constraint (4). Constraints (5) and (6) define the integer and continuous decision variables, respectively.

2.3 Expected Total Cost Analysis of Scenarios

As explained in Sect. 2.1, the risk factors that have an impact on the production plan are assessed with FMEA, and their RPN are calculated. Among the 21 factors examined, risks having an RPN of 200 and above are defined as critical and selected for further analysis. In order to use these RPN's in the expected total cost analysis, their values are normalized. The normalized RPN's constitute the first component of this analysis.

The second component comes from Sect. 2.2, where the production plans corresponding to the selected risk factors are obtained via mathematical modelling. The total cost of these scenarios are input to the analysis.

The expected total cost calculations are performed using Eq. (7) below:

$$ETC_j = NRPN_j \cdot TC_j; \qquad \forall j \tag{7}$$

where:

ETC_j	expected total production cost of risk scenario j (TL)
$NRPN_j$	normalized risk priority number (RPN) of risk scenario j
TC_j	total production cost of risk scenario j (TL)

This analysis highlights the most critical risks and assists in recommending system improvements.

3 Implementation and Results

Among the various industrial kitchen equipment models manufactured by Öztiryakiler, the planning of square and rectangular stainless steel containers are taken into account for implementing the proposed methodology. These models are selected by the manager in charge of their production, because of their popularity. As it can be seen from the indices presented in Sect. 2.2, the mathematical model includes 49 products and 43 machines, which are involved in the products' manufacturing processes.

The subsequent sections introduce the input data, and summarize the results of the different phases of the proposed methodology.

3.1 Input Data

The data used in the mathematical model is obtained from the company, and belongs to January 2018. There are 22 working days in the month. The plant operates 8.5 hours per day.

The scrap rate of products is given as 8%.

Due to space limitation, only a part of product and machine codes are given in Tables 1 and 2, respectively. The last three digits of the product codes in Table 1 indicate the depth of the container. Table 2 displays some of the different machines engaged in manufacturing the products.

Table 1. Definition of product codes

Product	Definition
11020	Container (530 mm × 325 mm, 20 mm depth)
11040	Container (530 mm × 325 mm, 40 mm depth)
…	…
14200	Container (265 mm × 162 mm, 200 mm depth)
16065	Container (176 mm × 162 mm, 65 mm depth)
…	…
24100	Container (530 mm × 162 mm, 100 mm depth)
24150	Container (530 mm × 162 mm, 150 mm depth)

Table 2. Definition of machine codes

Machine	Definition
01.01	Roll Cutting
01.02	Hydraulic Scissors
01.03	Eccentric Scissors
...	...
110.0	Hand Polish
...	...
14.01	Press Line
14.04	Washing Tunnel
...	...
03.27	Hydraulic Press 160 tons

The containers follow a multi-stage manufacturing process at Öztiryakiler. All products start from the roll cutting department (01.01), and leave the system from one of the washing tunnels (05.05, 14.04, 14.15). In between, they visit different cutting, polishing, press departments. Three sample routes belonging to products 11020, 13040, and 16065 are listed in Table 3.

Table 3. Routes of three products

11020	13040	16065
01.01	01.01	01.01
01.02	01.02	01.02
01.03	01.03	01.03
02.38	03.27	01.14
02.20	03.26	03.16
110.0	03.04	02.17
14.04	14.03	03.04
	14.15	14.03
		14.15

In order to give an idea about the processing and setup times, the data of product 11020 are provided in Table 4. Since the times provided by the company correspond to a lot, they are converted into a unit basis before inputting to the model.

Table 4. Processing time and setup time of product 11020 on machines

Machine	Processing time (seconds/unit)	Setup time (seconds/unit)
01.01	2.030	0.312
01.02	4.320	0.156
01.03	6.489	0.226
02.38	38.880	1.405
02.20	33.259	4.580
110.0	38.880	0.000
14.04	10.809	0.000

3.2 Failure Mode and Effects Analysis Results

As a result of the analysis of the current system and meetings with the production manager, 21 potential causes of failure are identified. These are grouped under four process functions: estimation-based, production-based, facility-based, and sales-based. Each cause of failure is scored in terms of severity, occurrence and detection, and its RPN is calculated by multiplying these three points. The RPNs highlight the types of failures together with their causes that should be solved by applying corrective actions. The threshold value for RPN is specified by the company as 200. Consequently, four critical factors with an PRN greater than 200 are selected as factors which impose a risk on the production planning process. These are explained in Table 5.

Table 5. First four high RPN values

Potential failure mode	Potential causes of failure	RPN
Loss of machine time	Production planning failures associated with increased mould changing time	540
Problems in sales forecasting	Arrival of an unplanned high priority order	480
Waste of time	Irregular arrangement of machines as a consequence of the multi-floor facility	360
Wrong inventory levels	Recording of damaged products in the system as non-defective products (Machine cannot identify the damaged product)	200

The RPN scores are recalculated after the improvement proposals. Predictions and suggestions are discussed with the staff responsible from each activity. The scores after the second calculation are expected to be below the initially accepted threshold. This analysis is provided in Sect. 3.5 below.

3.3 Mathematical Modelling Results

In order to obtain the risk-free production plan, the MILP model presented in Sect. 2.2 is first solved by using an ideal data set. Then, the outcome of the FMEA is observed and the factors having the highest RPNs, which are listed in Table 5, are analysed as risk scenarios. The input data of these scenarios consist of risk-affected data that display various deviations from the ideal situation. All models are solved optimally using GAMS software and CPLEX solver [9], and their solution time is in seconds.

(1) *Results of the Risk-Free Model:* The objective function value of the ideal model gives a total cost of 2,680,531.82 TL. All machines are fully utilised (673,200.00 s) during the month. It is observed that, four of the machines, listed in Table 6, require additional time to satisfy the orders. This can be achieved using overtime.

Table 6. Additional time required at workstation *k* in risk-free plan

Machine	Additional number of seconds required
05.01 (Electro Polishing Machine)	660,956.432
05.02 (Electro Polishing Machine)	949,684.707
14.03 (Automatic Burr Cleaning)	306,512.202
03.16 (Hydraulic Press 160 tons)	519,978.229

Stainless steel is the basic raw material of all containers. A total of 89,602.779 kg of stainless steel is used to support the production plan.

(2) *Results of the Risk Scenarios:* In accordance with the outcome of the FMEA, the four potential causes of failure provided in Table 5 are further studied as risk scenarios. The results are discussed below:

Scenario 1 (Increase in mould changing time): The highest RPN corresponds to production planning failures associated with increased mould changing time, which is measured as 15 minutes on the average. On the other hand, the average production time of a product is 17 seconds. Hence, in this 15-minutes loss of machine time, an average of 53 products could have been produced, which would contribute to the profit of the company.

Mould changes are only performed on press machines, and there are 22 press machines in the production line. Since this change will affect the setup times and setup cost, in order to observe its impacts on the production plan, the setup times are modified, and a setup cost is added to the objective function:

$$minz_1 = \sum_i (c_i + (cs_i \cdot rm\_s_i) + asc) \cdot X_{i,01.01} \tag{8}$$

where:

asc: additional setup cost per unit of product due to mould changing at hydraulic presses (TL)

The total cost of Scenario 1 is calculated as 2,862,477.63 TL., displaying an increase of 6.8% in the cost of the risk-free plan.

Scenario 2 (Unexpected high priority customer): The second highest RPN is related to the arrival of an unplanned high priority order, which makes the sales forecasting obsolete. As the company prioritizes some customers, it is observed that there is an accumulation of orders and a delay in the planned production due to this unexpected arrival. This case is reflected as a 3% increase in the demand of products.

As a result of this increase in demand, the total cost is found as 2,812,345.68 TL., which is 4.91% higher than the ideal plan's cost. Moreover, the additional time required to satisfy the orders given in Table 6 are further increased.

Scenario 3 (Inappropriate layout): The third highest RPN results from the irregular arrangement of machines as a consequence of the multi-floor facility, which leads to waste of time during the production process. In order to satisfy the orders, the company has to extend its working hours. With the aim of analysing the impacts of this situation, c_i, which is defined as the unit cost of product i excluding cost of stainless steel used, is doubled. Moreover, the daily working hours is increased to 17, corresponding to a day shift and a night shift, each consisting of 8.5 hours.

The total cost of the scenario is determined as 4,002,101.50 TL. Compared with the risk-free model, an additional cost of 1,321,569.68 TL is incurred by Öztiryakiler. All machines are working 1,346,400.00 seconds, and only machine 05.02 requires 276,484.707 seconds of additional time.

Scenario 4 (False recording of damaged products): The fourth highest RPN comes from recording of damaged products in the system as non-defective products. Since machines cannot identify the damaged products, this situation leads to wrong inventory levels.

In the press department, two types of product counting are performed; counting by workers and by machines. The foreman then double checks to identify the defective products. However, it takes a long time for these counts to reach the office so that they can be input to the system. Meanwhile, if defective products are produced at the shop floor, since the machines cannot detect them during the process, they are considered as non-defective until the double checking results arrive at the office. This causes unnecessary use of raw material and labour.

Assuming that some undetected defective products will exist, in order to satisfy the customer orders, this case is reflected by a 6% increase in demand, which results in a total cost of 2,944,371.28 TL. The company faces an additional cost of 263,839.46 TL. Hence, the return of every 6% of the resources spent due to this false recording leads to a cost increase of 9.82%.

With this scenario, a total of 98,546.477 kg of stainless steel is used to support the production plan. Not only the additional time required by the machines reported in Table 6 increase, but machine 02.19 (Hydraulic Press 100 tons) also needs an extra time of 32,567.155 seconds.

3.4 Expected Total Cost Analysis Results of Scenarios

As explained in Sect. 3.2, out of 21 potential causes of failure identified in the planning system of the company, the ones having an PRN of 200 and above are taken as critical risk factors, and their impacts on the cost of the production plan are inspected through mathematical modelling.

In order to further examine these four cases, their RPNs are normalized, and the corresponding expected total costs are calculated using Eq. (7). The results of this analysis are presented in Table 7.

Table 7. Expected total cost analysis of critical risks

Risk Scenario, j		RPN_j	$NRPN_j$	TC_j (TL)	ETC_j (TL)
1	Production planning failures associated with increased mould changing time	540	0.3417	2,862,477.63	978,108.61
2	Arrival of an unplanned high priority order	480	0.3037	2,812,345.68	854,109.38
3	Irregular arrangement of machines as a consequence of the multi-floor facility	360	0.2278	4,002,101.50	911,678.72
4	Recording of damaged products in the system as non-defective products (Machine cannot identify the damaged product)	200	0.1265	2,944,371.28	372,462.97

The expected total cost values in the table indicate that, the importance ranking of the risks are changed as 1-3-2-4. It is obvious that, even though it has a low normalized RPN, its high total cost has positioned the significance of Scenario 3 to the second place.

3.5 Improvement Suggestions

As the final step of the methodology, some recommendations are offered to overcome these risks. The suggestions are mutually developed with the manager of the production line. Recalculated RPNs are shown in Table 8. Although the detection and occurrence values are changed, the first two RPNs are still above the threshold value.

The use of rollers or conveyor systems to move the moulds to the machines can help in decreasing the mould changing time, which will lead to the minimisation of lost machine hours. Another approach is to change the mould design. Innovative designs can provide a faster transition by taking the shape of each product with the aid of internal gears.

It should be noted that, the existence of unexpected high priority customers is quite unavoidable in real life. To minimise the adverse effect of this situation on the production plans, a new inventory policy can be adopted to cope with the extra demand. Different customer service levels and safety stock levels can be evaluated before selecting the best alternative for the company.

The current multi-floor facility often triggers bottleneck situations and capacity shortages, which again results in a waste of time. As a suggestion, a new location can

Table 8. Recalculated RPNs

Potential cause of failure	RPN (initial)	Current process control	Recommended action	*S*	*O* (initial)	*O* (final)	*D* (initial)	*D* (final)	RPN (final)
Production planning failures associated with mould changing time	540	None	Rollers or conveyor systems can be used to move the moulds to the machines	9	6	4	10	7	252
Arrival of an unplanned high priority order	480	None	An optimal inventory policy can be adopted to satisfy the extra demand	8	6	6	10	5	240
Irregular arrangement of machines as a consequence of the multi-floor facility	360	Layout analysis and some evaluation techniques	A new location can be considered for a simpler production system	9	10	4	4	4	144
Recording of damaged products in the system as non-defective products (Machine cannot identify the damaged product)	200	Foreman double checks to identify the defective products	A barcode or laser system can be installed at the exit points of the machines to provide online data flow	8	5	2	5	2	32

be considered to establish a simpler and lean manufacturing system. The layout of this new location can be designed using analytical techniques. Although the initial investment cost might be high, it will affect the future profit margin positively. If the company does not want to modify its existing facility area or move to another location, in order to deliver the orders on time, they have to consider a change in their shift system, and start working in 2-shifts, rather than in 1-shift.

Products can be counted incorrectly due to machine-based or human-based errors. Human-based errors are resulting from the carelessness of employees. This system is controlled by the foreman's double check, but it is often overlooked when it originates from a machine. A barcode or laser system can be installed at the exit points of the machines to provide online data flow. The installation of such a system will also constitute the first step in creating the company's big data.

Within its current facility, Öztiryakiler has allocated an area to produce a limited volume of one of its container models, where a few machines are connected through a conveyor belt. There are no workers in this system, and only some of the machines involved in the manufacturing process are available. This implementation can be perceived as a starting point for a larger fully automatic system that can benefit from Industry 4.0 principles. The first, third, and fourth risk scenarios presented in Table 7 are consequences of inefficient manual operations and layout structure. It is believed that, integration of Industry 4.0 into the present manufacturing system will help to minimize these risk factors and maximize the productivity.

As it can be observed from Table 8, most of the risks can be avoided with additional investments. Hence, incorporation of these suggestions in the company's future strategic plans becomes a managerial decision.

4 Conclusion

This study aims to analyse the risks that have an impact on the production planning process. It is crucial for companies to identify these critical factors, and to act proactively against their consequences. The proposed analytical approach is implemented at Öztiryakiler's square and rectangular food containers' manufacturing line.

Initially, after examining the current system and collecting the relevant data, Failure Mode and Effects Analysis is used to evaluate the risk factors. The factors with an RPN of 200 and above are defined as critical, and consequently, four out of 21 factors are selected for the risk scenarios. These are: (1) Production planning failures associated with increased mould changing time, (2) Arrival of an unplanned high priority order, (3) Irregular arrangement of machines as a consequence of the multi-floor facility, and (4) Recording of damaged products in the system as non-defective products.

The second phase determines the total cost of the production plans through a mixed-integer linear programming model. In this phase, both risk-free and risk scenario models are solved optimally using GAMS/CPLEX solver. The difference between these models lie in the input data, and each scenario tends to represent the risk factor by variations in the data set employed.

In the third phase, by combining the outputs of the first two phases, the expected total cost of each risk scenario is calculated. This analysis changes the risk ranking of the factors; such that, the factors at the second and third position exchange places. By comparing the risk-free production plan with each risk scenario, the impacts of these factors are evaluated and recommendations to improve the system are proposed.

The results of the study indicate that, various factors affect the production plans of the company, and unexpected risks not only can cause extra costs during production, but also may lead to delays in manufacturing the products. Therefore, any factor that is not considered at the beginning of the production period can interrupt and change the plan. Hence, it is a wise decision to take preventive actions against these risks at the beginning of the planning period, since risks encountered during production with no initial actions taken, will reduce the possibility of overcoming the undesirable outcomes. Most of the risks can be prevented by making additional investments, and as always, this decision is to be left to the company management.

Acknowledgement. We would like to thank Mr. Murat ÜNAL and Mr. Tahsin ÖZTİRYAKİ from Öztiryakiler for their continuous support in the realisation of this study, and providing great convenience during the data collection and implementation phases.

References

1. Klöber-Koch, J., Braunreuther, S., Reinhart, G.: Predictive production planning considering the operative risk in a manufacturing system. Procedia CIRP **63**, 360–365 (2017)
2. Mula, J., Poler, R., García-Sabater, J.P., Lario, F.C.: Models for production planning under uncertainty: a review. Int. J. Prod. Econ. **103**, 271–285 (2006)
3. Kütner, R., Majak, J.: Multistage optimisation strategy for solving production planning problems. IFAC Proc. Volumes **42**, 332–337 (2009)
4. Öztiryakiler website (2019). https://oztiryakiler.com.tr/
5. Anes, V., Henriques, E., Freitas, M., Reis, L.: A new risk prioritization model for failure mode and effects analysis. Qual. Reliab. Eng. Int. **34**, 516–528 (2018)
6. Fattahi, R., Khalilzadeh, M.: Risk evaluation using a novel hybrid method based on FMEA, extended MULTIMOORA, and AHP methods under fuzzy environment. Safety Sci. **102**, 290–300 (2018)
7. Mizrak Özfirat, P.: A new risk analysis methodology integrating fuzzy prioritization method and failure modes and effects analysis. J. Fac. Eng. Archit. Gazi Univ. **29**(4), 755–768 (2014)
8. Wang, W., Liu, X., Qin, Y.: A risk evaluation and prioritization method for FMEA with prospect theory and choquet integral. Safety Sci. **110**, 152–163 (2018)
9. IBM ILOG CPLEX website (2012). www.ibm.com/software/integration/optimization/cplex-optimization-studio/

Cold Chain Logistics Firm Selection by Using AHP-VIKOR Integrated Method and a Case Study in Food Industry

Elif Caloglu Buyukselcuk(✉)

Department of Industrial Engineering, Fenerbahce University, Istanbul, Turkey
elif.buyukselcuk@fbu.edu.tr

Abstract. With the Industry 4.0, a new trend has started for all sectors and enterprises have entered into a period of intense working and investment to turn into intelligent systems. Cold chain logistics also has the same trend, and companies have pushed the button to be involved in this process. Due to changing lifestyles and generally decreasing prices, the world market for perishable products such as processed foods and cooled products is growing. Cold chain transportation is an important issue in basic food and health product logistics to prevent the growth of harmful microorganisms and to ensure that the products remain intact. In this study, it is aimed to determine the right cold chain logistics firm for company EDB that is a small and medium-sized enterprise (SME) operating in the food industry. Firstly, the selection criteria are defined according to the literature survey and expert views. AHP approach has been used to calculate the weight of them. After the determination of weight of criteria, VIKOR approach has been used to identify the best cold chain logistics firm for company EDB. The results were obtained by using Expert Choice 11.0 and Excel calculator finally the best one has been determined.

Keywords: Cold chain · Industry 4.0 · AHP · VIKOR · Decision making

1 Introduction

With the increasing of the globalization and consumption in the world in an environment where trade is concentrated, supply chain should be managed good in order to be permanent. Cold logistics has more specific features than other logistics applications. Damage and deterioration rates and initial investment and operating costs are higher than other logistics branches. Therefore, it requires more intensive expertise and technology. According to another opinion, cold chain logistics is not just a logistics process; beyond that, it is activated by the output of the first production area and it contains all processes about the storage, keeping, and transporting up to end consumer [1].

Cold chain is a temperature-controlled supply chain that starts from the production points to the end consumer. There must be a perfect flow every step of this chain to prevent the product from degrading. In cold chain logistics, especially besides adjusting the temperature, control is of great importance. However, many significant subjects are critical to enable the freshness, viability and usability of the perishable products in cold

N. M. Durakbasa and M. G. Gençyılmaz (Eds.): ISPR 2019, LNME, pp. 403–415, 2020.
https://doi.org/10.1007/978-3-030-31343-2_35

chain transport. Perishable products can be divided into two categories: living products such as vegetables, fruits, flowers, live seafood and non-living products such as dairy products, meat, medicines, blood, processed food products. If these products are not given the importance, it is inevitable that there will be some problems to endanger the human health. According to the results of a study conducted in the USA until 2012, 3000 people died and 120,000 people became ill as a result of not giving these products importance. In perishables products sector, there is a direct correlation between the cold chain performance and the quality delivered to the customer. Therefore, firms must work with the right cold chain transportation firms [2].

Also in Turkey, a number of regulations is made about the packaging, storage and transportation of foods in 2018. According to this regulation, they will be transported by refrigerated vehicles suitable for pallet use after being pre-cooled. Besides, transportation should be done by closed or tilted transport vehicles. During the transportation of different goods in the same vehicles, measures should be taken to prevent the goods from damaging each other and the goods that can cause harm to each other should not be carried in the same vehicle. In addition, the maintenance and inspection of the vehicles to be used during transportation is also covered by this legislation [3].

The aim of this study is to determine the most suitable cold chain transporter for company EDB that is a small and medium-sized firm operates in the food industry in Istanbul. Literature review has been given in the second section, methodology of study has been explained in the third section, results have been explained in the fourth section and conclusion has been given in the fifth section, respectively.

2 Literature Review

The determination of the best cold chain logistics firm is a multi-criteria decision making (MCDM) problem and it is a complex issue. MCDM is a combination methodology of comparing, ranking and choosing of multiple alternatives that have multiple alternatives. It depends on a matrix like that decision matrix, evaluation matrix or payoff matrix [4]. To solve this MCDM problem, firstly the selection criteria and weights are determined and then the best alternative is selected among the other alternatives.

2.1 Selection Criteria for Cold Chain Logistics

Because of the strategic significance of the selection process, the selection criteria are determined correctly. Selection criteria for cold chain logistics which frequently come out in the literature are classified into three main groups that are named Conventional Criteria, Vehicle Fleet Criteria and Firm Criteria in connection with this study.

(1) *Conventional Criteria:* Transportation cost, payment terms and flexibility, delivery time and quality, return policies and warranties, reject rate, packaging quality are considered as criteria for conventional criteria group [5]. The transportation cost should be at minimum level. The firm should be flexible and helpful in paying. The delivery is made in time and correctly. Returns and agreements

about the faulty products are the important criteria. Reject rate is at what percentage the products are rejected by the quality control. The correct packaging for the products are significant.

(2) *Vehicle Fleet Criteria:* Adequacy and capacity of vehicles, vehicle age and maintenance, in-vehicle loading, vehicle traceability, in-vehicle technological competence are considered as criteria for vehicle fleet criteria group. A cold chain logistics firm should have vehicles with new, technically equipped and convenient storage and transportation [6].

(3) *Firm General Criteria:* Firm reliability, knowledge and experience of personnel, sector image, level of IT systems and technical resources, cooperation with partners are considered as criteria for firm general criteria group. In today's conditions, past performance of the firm, skilled workforce, technological competence, reputation in the market, open communication with partners are the vital factors [7].

All these criteria, sub-criteria, and their codes have been represented in Table 1.

Table 1. Criteria and their codes

Main criteria	Sub-criteria	Code	Main criteria	Sub-criteria	Code	Main criteria	Sub-criteria	Code
Conventional Criteria	Transportation cost	CC1	Vehicle Fleet Criteria	Adequacy and capacity of vehicles	VFC1	Firm General Criteria	Firm reliability	FGC1
	Payment terms and flexibility	CC2		Vehicle age and maintenance	VFC2		Knowledge and experience of personnel	FGC2
	Delivery time and quality	CC3		In-vehicle loading	VFC3		Sector image	FGC3
	Packaging quality	CC4		Vehicle traceability	VFC4		Level of IT systems and technical resources	FGC4
	Return policies and warranties	CC5		In-vehicle technological competence	VFC5		Cooperation with partners	FGC5
	Reject rate	CC6						

2.2 Analytic Hierarchy Process (AHP)

This approach was developed by Thomas L. Saaty in 1970s to solve the complicated decision making problems. In this model, the problem that is the subject of the decision problem is divided into components and arranged in a hierarchical structure. There is a purpose at the top of the hierarchy. Under this objective, the hierarchy is completed with criteria, sub-criteria, and alternatives, respectively. The system approach theory is at the basis of AHP. Unlike conventional decision-making techniques, qualitative (quantitative) values are taken into account not only in quantitative terms. A large number of criteria is determined by a pairwise comparison in a hierarchy of criteria. If these comparisons are concluded to be consistent, the comparison results are moved from the bottom step to the top of the hierarchy. The degree to which our options meet our needs is determined [8].

This approach is easy to use and it is user friendly method. It can be adjusted in many sized problems and it does not require intensive data. On the other hand, there can be many problems between criteria and alternatives due to the interdependence. It causes inconsistency among judgment. Application areas area planning and supply chain management, resource management and strategic decisions [9].

In the literature, many academicians have been applied AHP approach to solve their decision problems easily. A location selection problem for a military airport in Turkey has been solved by AHP integrated PROMETHEE and VIKOR methods to select the best alternative [10]. Awasthi and her colleagues used fuzzy AHP approach for the determination of the criteria weights in multi-tier sustainable global supplier selection problem and then, fuzzy VIKOR approach has been applied to rate supplier performances against the evaluation criteria [11]. Garg also studied AHP approach for evaluation of criteria and then applied fuzzy technique for order performance by similarity to ideal solution (FTOPSIS) for the selection of strategic alliance partner in Indian airline industry [12]. AHP approach steps follow as:

Step 1. The decision problem is formed as a hierarchy that goal at the top level, criteria with sub-criteria at the middle level and alternatives at the bottom level. Figure 1 shows a generic hierarchic structure.

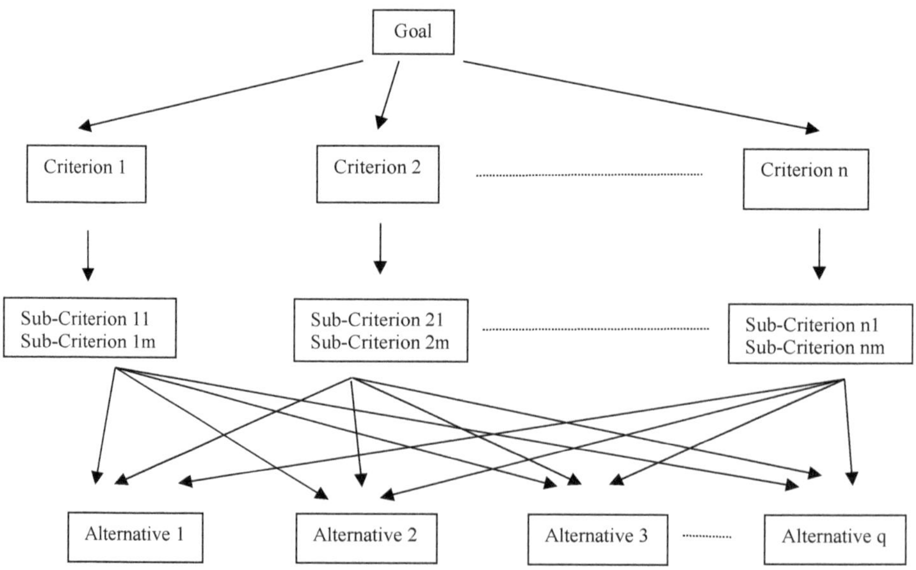

Fig. 1. Generic hierarchic structure for AHP

Step 2. Data are collected from decision makers or experts and they can rate the comparison as equal, high importance, strong importance, very strong importance and extreme importance.
Step 3. The square matrix from the pair wise comparisons of various criteria is determined. Comparison of alternatives is made according to the Table 2.

Table 2. Importance value(s) for situations in comparison process [13]

Situations	Value(s)
Equal importance of both factors	1
Moderate importance of one factor over another	3
Strong importance of one factor over another	5
Very strong importance of one factor over another	7
Extreme importance of one factor over another	9
Intermediate values between two adjacent judgements	2,4,6,8

Step 4. In this step, normalized pair wise comparison matrices are prepared by dividing each column value to column summary. After that, priority vector is determined for each row by calculating rows' average.
Step 5. In this step, the consistency of the matrix is evaluated. The consistency index CI, is calculated below:

$$\mathrm{CI} = (\lambda_{max} - \mathrm{n})/(\mathrm{n} - 1)$$

where λ_{max} is the maximum eigenvalue of the judgement matrix. This CI can be compared with that of a random matrix, RI. The term of consistency ratio is derived from CI/RI and CR value should be less than 0.1 [13].

Step 6. In this step, the degree of each alternative is multiplied with the weights of the sub-criteria and aggregation is made to obtain local ratings according to all criteria. After that they are multiplied by the weights of the criteria and aggregation is made to obtain global ratings.

2.3 VIKOR (Vise Kriterijumska Optimizacija I Kompromisno Resenje)

This approach was developed by Serafim Opricovic to handle decision problems with contradictory and different unit criteria in 1998 [14]. It has also been defined as a multi criteria ranking index that aims to determine the decision that is the closest to ideal solution [15]. VIKOR method aims to the majority group of maximum benefits and the minimum of individual remorse of competitors. Calculations are quite simple and clear.

Many researchers have used VIKOR approach to solve their MCDM problems in the literature. Štirbanović and their colleagues have used to TOPSIS and VIKOR approaches for the evaluation of flotation machines [16]. Also VIKOR approach has been used to select the suitable improvement alternative goals with the corresponding weights that are determined by the decision-making trial and evaluation laboratory method with the concepts of the analytic network process (DANP) [17]. To determine the best repair material for concrete structures, a procedure coupled with the VIKOR approach has been applied [18].

The steps of VIKOR approach follow as:

Step 1. The best f_i^* and the worst f_i^- values of all criterion functions are determined where i = 1, 2,…, n is the index number of the criterion function. If i[th] function represents a benefit or a maximization than $f_i^* = \max_j f_{ij}, f_i^- = \min_j f_{ij}$, where j = 1, 2, …, m is the index number of alternatives. If i[th] function represents a loss or a minimization than $f_i^- = \max_j f_{ij}, f_i^* = \min_j f_{ij}$.

Step 2. Sj indicates the maximum group utility and the minimum individual regret of the opponent Rj are calculated by using the weighted and normalized Manhattan distance formula and the weighted and normalized Chebyshev distance formula, respectively.

$$S_j = \sum_{i=1}^{n} \frac{w_i\left(f_i^* - f_{ij}\right)}{\left(f_i^* - f_i^-\right)} \qquad R_j = \max_i \left[\frac{w_i\left(f_i^* - f_{ij}\right)}{\left(f_i^* - f_i^-\right)}\right]$$

Step 3. The Qj values are calculated to rank the alternatives as follow:

$$Q_j = \frac{v\left(S_j - S^-\right)}{S^* - S^-} + \frac{(1 - v)\left(R_j - R^-\right)}{R^* - R^-}$$

where;

$$S^* = \max_j S_j, S^- = \min_j S_j$$

$R^* = \max_j R_j, R^- = \min_j R_j$ for benefit criteria, and v is the voting rule and it should be between [0,1].

Step 4. The decreasing sorting order of the alternatives is made according to the Q, R, and S values. As a result, there are three different ranking lists.

Step 5. The results satisfy the two following conditions:

C_1: "Acceptable Advantage"

$$Q(a'') - Q(a') \geq DQ$$

where a' and a'' are the first alternative and the second alternative in the rank list according to the Q value, respectively. DQ is defined like that:

$$DQ = \frac{1}{j - 1}$$

where j is the number of all alternatives.

This condition indicates that the best alternative is obviously the most advantageous alternative according to the other alternatives.

C_2: "Acceptable Stability in Decision Making"

Alternative (a') must also be the best alternative in the rank lists due to the S and R values. This compromise solution will provide a stability in the decision making process which may be ballot by majority rule (if v > 0.5), by consensus (v = 0.5) or with vote (if v < 0.5).

If none of the conditions are met, a compromise solution set is recommended like that in below:

- Alternative a' and alternative a'' if only condition 2 (C_2) is not met,
- Alternative $a', a'', a''', \ldots, a^m$, if condition 1 ($C_1$) is not satisfied, where a^m is determined by the formula

$$Q(a^m) - Q(a') < DQ$$

where m is the maximum number that satisfies the condition. With this inequality, a compromise set will be revealed. In this evaluation result, an increasing rank is made according to Q values and the alternative that has the lowest Q value will be chosen.

3 Research Methodology

Company EDB is a small and medium-sized enterprise operating in the food industry in Istanbul. In this study, it is aimed to determine the preferable cold chain logistics firm and flow chart is represented in Fig. 2.

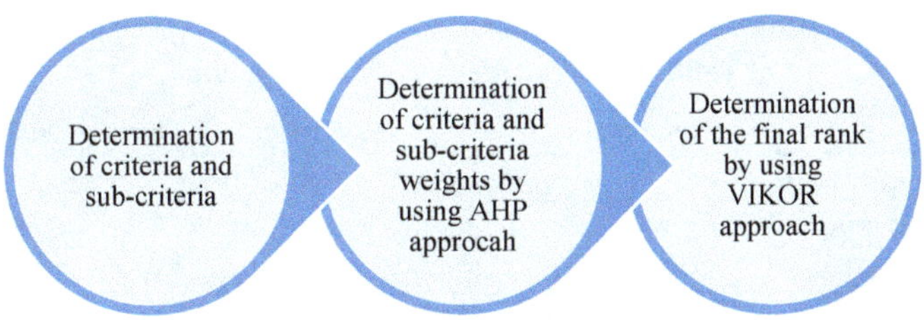

Fig. 2. Flow chart for the cold chain logistics firm selection process

Academicians and experts decided the selection criteria according to the results of the literature research. The hierarchical structure of problem is illustrated in Fig. 3.

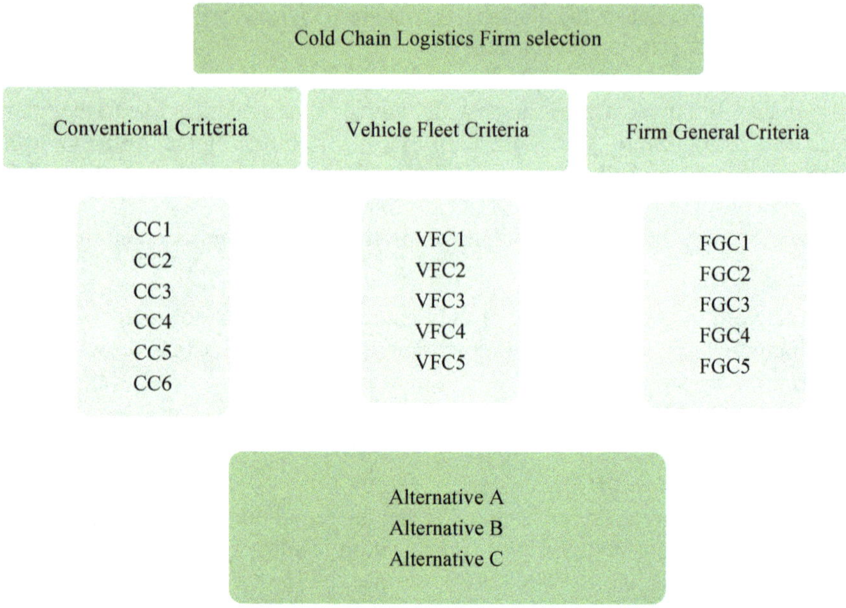

Fig. 3. Hierarchical structure to determine cold chain logistics firm

The decision-maker team established by authorized persons in company EDB evaluated criteria, sub-criteria and alternatives by using linguistic terms that were given in Table 2. Bilateral comparison matrix of main criteria groups and sub-criteria groups that formed by the team and they are shown in the following Tables 3 and 4.

Table 3. Main criteria evaluation results

Criteria	Conventional criteria	Vehicle fleet criteria	Firm general criteria
Conventional criteria	1	1/3	2
Vehicle fleet criteria		1	3
Firm general criteria			1

Table 4. Sub-criteria evaluation results for conventional criteria

Conventional Criteria	CC1	CC2	CC3	CC4	CC5	CC6
CC1	1	1	1	3	2	2
CC2		1	1/3	1/2	1/2	1
CC3			1	3	3	2
CC4				1	1/3	2
CC5					1	1
CC6						1

These comparison matrices have been introduced to Expert Choice AHP solver and then results are shown in the following section (Tables 5 and 6).

Table 5. Sub-criteria evaluation results for vehicle fleet criteria

Vehicle Fleet Criteria	VFC1	VFC2	VFC3	VFC4	VFC5
VFC1	1	3	1/5	1	1
VFC2		1	1/5	1/3	1/3
VFC3			1	5	3
VFC4				1	1/3
VFC5					1

Table 6. Sub-criteria evaluation results for firm general criteria

Firm General Criteria	FGC1	FGC2	FGC3	FGC4	FGC5
FGC1	1	5	1	2	2
FGC2		1	1/5	3	1/2
FGC3			1	5	3
FGC4				1	1/3
FGC5					1

4 Results

In this section, the results obtained from the evaluations made will be examined. Firstly, the results acquired from the AHP approach will be evaluated and the results acquired from the VIKOR approach will be evaluated.

4.1 Selection of Cold Chain Logistics Firm by Using AHP

In this section, results for the determination of the best cold chain logistics firm are represented. First of all, the main criteria results from the AHP solver can be seen in Fig. 4. As seen figure, the inconsistency ratio for all the pairwise matrices is less than 0.1. Main criteria has been examined, the weights of CC, VFC and FGC are 0.249, 0.594, and 0.157, respectively. The most important main criteria group is determined as vehicle fleet criteria group. For conventional criteria, CC3 (0.286), CC1 (0.230) and CC5 (0.158) are the significant criteria for AHP. For vehicle fleet criteria, VFC3 (0.5), VFC5 (0.193) and VFC1 (0.137) are the significant criteria for AHP. For firm general criteria, FGC3 (0.37), FGC1 (0.302) and FGC5 (0.157) are the important criteria for AHP.

Alternative 2 has been determined the best cold chain logistic firm by using AHP approach and it has been shown in Fig. 5. Alternative 1 is the second and Alternative 3 is the third one.

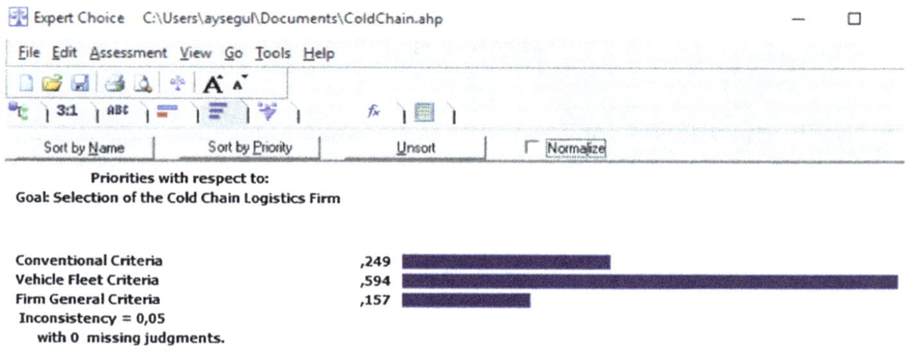

Fig. 4. Main criteria results

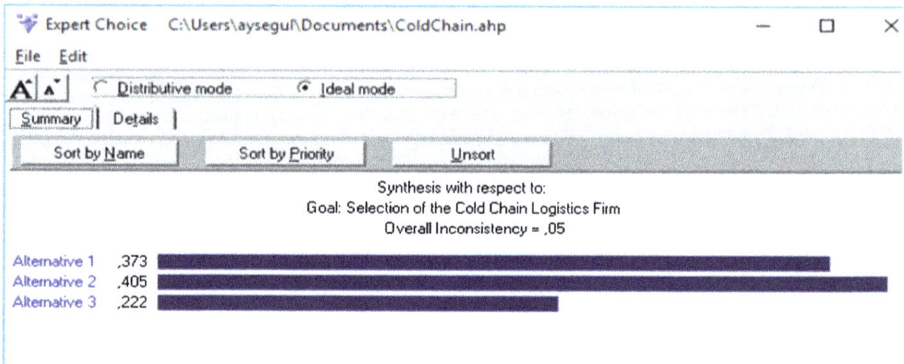

Fig. 5. Ranking of alternatives

4.2 Selection of Cold Chain Logistics Firm by Using VIKOR

In this part of the study, the weights of sub-criteria found by AHP approach, VIKOR approach is used to solve the decision making problem. Decision makers evaluated each alternative in the framework of sub-criteria. The decision making team in company EDB evaluated each sub criteria according to the scale given in Table 7. The values of CC1, CC6 and VFC3 criteria has been obtained from the decision making team.

Table 7. Scale used in evaluation

Linguistic Term	Value
Very bad	1
Bad	2
Medium	3
Good	4
Very Good	5

S, R, and Q values for each alternative have been determined by using VIKOR approach in Excel calculator. These results have been showed in Table 8. In Table 9, the ranking of alternatives has been represented.

Table 8. S,R and Q values for each alternative

			Q				
	S	R	v = 0	v = 0.25	v = 0.50	v = 0.75	v = 1.00
Alternative 1	1.18	0.25	0.19	1.32	2.46	3.60	4.74
Alternative 2	0.68	0.19	0.00	0.88	1.75	2.63	3.50
Alternative 3	1.89	0.50	1.00	1.70	2.39	3.09	3.78

Table 9. Ranking due to S, R and Q values

			Ranking for Q value				
	Ranking for S value	Ranking for R value	v = 0	v = 0.25	v = 0.50	v = 0.75	v = 1.00
Alternative 1	2	2	2	2	3	3	3
Alternative 2	1	1	1	1	1	1	1
Alternative 3	3	3	3	3	2	2	2

Condition 1: The DQ value is assumed that 0.25 because the number of alternatives is smaller than 4. For the condition of v = 0, 0.19–0 < 0.25, advantage requirement is not satisfied. For the condition of v = 0.25, 1.32–0.88 > 0.25, advantage requirement is satisfied and it can be said that Alternative 2 has advantage. For the condition of v = 0.5, 2.39–1.75 > 0.25, advantage requirement is satisfied and it can be said that Alternative 2 has advantage. For the condition of v = 0.75, 3.09–2.63 > 0.25, advantage requirement is satisfied and it can be said that Alternative 2 has advantage. For the condition of v = 1.00, 3.78–3.50 > 0.25, advantage requirement is satisfied and it can be said that Alternative 2 has advantage. Under this conditions, Alternative 2 is the best one for selection of cold chain logistics.

Condition 2: Alternative 2 is the best one due to the Q values. Besides the same alternative also ranked in the first place according to the S values and it can be said that the decision is stable.

5 Conclusions

In this study, the cold chain logistics process of company EDB operating in the food industry is discussed and it is aimed to determine the criteria to be considered in selecting the best logistics firm for this company. AHP approach was used to calculate the weight of decision criteria and both AHP and VIKOR approaches were used in the evaluation of alternatives. The AHP approach was preferred due to its effective and easy use, the fact that the problem consisted of many criteria and was based on the decision-making team idea. Also, in this approach, if the 'acceptable advantage' condition cannot be achieved, compromise solutions can be easily determined and it can be possible to determine the same order of importance without making pairwise comparisons for alternatives. In addition, it can be easily adapted to multi-criteria decision-making problems.

According to the results, vehicle fleet criteria are the most significant main criteria. It is usual that obtain these results like that because this study has been applied to the food industry, where cold chain logistics is important. More specific criteria such as in-vehicle cooling system, in-vehicle stowage, storage and vehicle traceability have become prominent according to the traditional criteria such as cost and quality for this study. The weights of in-vehicle loading, in-vehicle technological competence and adequacy and capacity of vehicles are 0.5, 0.193 and 0.137, respectively. On the other hand, the reliability of the company, market share and image and cooperation with partners are among the important criteria. In addition to these criteria, transportation costs, delivery time and quality, and return and guarantee conditions can be included in the evaluation criteria. Undoubtedly, as a result of these evaluations, alternative 2 was found to be the best alternative according to the evaluations made by the company where the application was carried out. Alternative 1 may be preferred when alternative 2 is not sufficient. Considering the vehicle fleet characteristics as the selection criteria which are an important factors during the transportation of the perishable products in the food industry will be an important guide for the enterprises operating in this sector.

This study was conducted in a small and medium-sized enterprise operating in the food industry in Istanbul. There may be a number of limitations for companies in other regions of Turkey and larger. Besides it can also create constraints for other sectors operating in cold chain logistics.

In addition to the criteria subject to evaluation, a number of new criteria in the scope of Industry 4.0 for cold chain logistics can be considered. Companies operating in the cold chain logistics sector other than the food industry may also be considered. In addition, these results can be compared with new results that obtained from other multi-criteria decision-making techniques.

Acknowledgment. I am grateful to my dear teacher Prof. Semra Birgun for his support during the realization of this work. Also I would like to thank company EDB personnel for their support, comments and help.

References

1. Sen, A.: Tedarik zinciri yönetiminde soğuk lojistik uygulamalarinin etkinliğinin arttirilmasina yönelik bir çalişma. MSc Thesis, Dokuz Eylul University, Izmir, Turkey (2008)
2. Onursal, F.S., Birgun, S., Yazici, S.: Using theory of constraints thinking processes to provide solutions for rejection problems in cold chain logistics. In: International Conference on Eurasian Economies, pp. 332–340. Uzbekistan (2018)
3. The Ministry of Trade regulation homepage (2019). http://kms.kaysis.gov.tr/Home/Goster135770

4. Whaiduzzaman, M.D., Gani, A., Anuar, N.B., Shiraz, M., Haque, M.N., Haque, I.T.: Cloud service selection using multicriteria decision analysis. Sci. World J. **2014**, 1–10 (2014)
5. Joshi, R., Banwet, D.K., Shankar, R.: A delphi-AHP-TOPSIS based benchmarking framework performance improvement of a cold chain. Expert Syst. Appl. **38**(8), 10170–10182 (2011)
6. Saravanan, S., Anubama, B.: Selection of cold chaing logistics service providers in pharmaceutical industry with reference to India. Int. J. Manag. Value Sup. Chains (IJMVSC) **8**(2), 1–12 (2017)
7. Korucuk, S.: Soğuk zincir taşımacılığı yapan işletmelerde 3PL firma seçimi: İstanbul örneği. Igdir Univ. J. Soc. Sci. **16**, 341–366 (2018)
8. Caloglu Buyukselcuk, E., Uner, O., Ozkan, C.: Kobilerde Çalışma Koşulları-Çalışan Memnuniyeti İlişkisi. In: V.Ulusal Uretim Araştırmaları Sempozyumu, pp. 105–110, Turkey (2005)
9. Velasquez, M., Hester, P.T.: An analysis of multi-criteria decision making methods. Int. J. Oper. Res. **10**(2), 56–66 (2013)
10. Sennaroglu, B., Varlik Celebi, G.: A military airport location selection by AHP integrated PROMETHEE and VIKOR methods. Transp. Res. Part D: Transp. Environ. **59**, 160–173 (2018)
11. Awasthi, A., Govindan, K., Gold, S.: Multi-tier sustainable global supplier selection using a fuzzy AHP-VIKOR based approach. Int. J. Prod. Econ. **195**, 106–117 (2018)
12. Garg, C.P.: A robust hybrid decision model for evaluation and selection of the strategic alliance partner in the airline industry. J. Air Transp. Manag. **52**, 55–66 (2016)
13. Saaty, T.L.: Decision making with the analytic hierarchy process. Int. J. Serv. Sci. **1**, 83–98 (2008)
14. Opricovic, S.: Multi-criteria optimization of civil engineering systems. Faculty of Civil Engineering (1998)
15. Caloglu Buyukselcuk, E.: Green and innovative supplier selection model via MCDM techniques and a case study in automotive industry. PhD Thesis, Marmara University, Istanbul, Turkey (2017)
16. Štirbanović, Z., Stanujkić, D., Miljanović, I., Milanović, D.: Application of MCDM methods for flotation machine selection. Min. Eng. **137**, 140–146 (2019)
17. Pineda, P.J.G., Liou, J.J.H., Hsu, C.C., Chuang, Y.C.: An integrated MCDM model for improving airline operational and financial performance. J. Air Transp. Manag. **68**, 103–117 (2018)
18. Kiani, B., Liang, R.Y., Gross, J.: Material selection for repair of structural concrete using VIKOR method. Case Stud. Constr. Mat. **8**, 489–497 (2018)

Determination of Field Temperature for Composite Materials Using Empirical Methods

Gabriel Nicodim Ciusca[1(✉)], Teodor Potra[2], Vasile Ceclan[1], and Sorin Dumitru Grozav[1]

[1] Department of Manufacturing Engineering, Technical University of ClujNapoca, Cluj-Napoca City, Cluj County, Romania
{Gabriel.Ciusca,Vasile.Ceclan, Sorin.Grozav}@tcm.utcluj.ro
[2] Department of Mathematics, Technical University of ClujNapoca, Cluj-Napoca City, Cluj County, Romania
Potra.Teodor@meth.utcluj.ro

Abstract. In the present research is granted a mathematical model in order to obtain the field temperature for composite materials. This model has its origin in Fourier- Kirchhoff equation. This mathematical model is useful in analysis finite elements. Other areas of use of them are also presented such as their structure. In the present work is conferred the importance of finite elements using analysis in composite materials area, too.

Keywords: Composite material · Mathematical model · Field temperature · Finite elements · Phenomenological

1 Introduction

Composite materials are used in industry more and more in the last decade in multiple ways. We can find them in aeronautics, in machine building industry and in consumer goods industry. These materials have been developed lately. In aeronautics industry there are airplanes which content composite materials up to 50%, for example Airbus 380 and Boeing 787. These materials are also used in Formula One cars [1], so the first designed body was made of carbon fibers in 2002. Nowadays other important companies use them to different body elements and not only one type at once [2]. Composite materials are made of two or more components whose properties completing each other resulting a material with superior qualities than each material by itself [3]. By technical point of view composite materials artificially created represent a combination of at least two materials particularly from chemical point of view and have properties which any other component can't have separately [4]. By the nature of material point of view we can divide them into natural/ biological materials (wood, moss, straw, etc.) and synthetic (metals, ceramics, polimers, etc.).There is also represented the mathematical model of phenomenological breaking type of composite materials using biaxially balanced fabrics utilized for research with finite elements [5].

N. M. Durakbasa and M. G. Gençyılmaz (Eds.): ISPR 2019, LNME, pp. 416–421, 2020.
https://doi.org/10.1007/978-3-030-31343-2_36

At other researches there was presented the manufacturing method to obtain bent shape tubes with variable cross-section from fibre reinforced composites with application in high performance handlebars was used for mountain-bicycles [6]. Other research has been achieved about recycling carbon – fiber reinforced materials [7]. Further researches pointed at using a software analysis with finite elements for determination of mechanical characteristics for unidirectional composite material [8].

Other researches pointed to determination of adhesion resistance, for instance composite materials adherence to different temperature (45 °C and 65 °C), after thermoforming with experimental methods and modeling methods as well. The empyric model assisted to identifying the contribution of the adhesive to transfer of demand [9]. Multi – line modeling has been studied in order to characterise nonlinear behavior of anisotropic material through numerical tehnics based on Kalman filter, by which they identify statistically multi-linear coefficients which build up the multi-linear model through probability modeling of previous knowledge and empyrical knowledge [10].

2 Field Temperature Calculation in Composite Materials

An uneven body M is considered as a reunion of N parallelepipeds $M_n, M_1 = M_2 = \ldots = M_N$.

$$M = \bigcup_{n-1}^{N} M_n, \overset{o}{M}_i \cap \overset{o}{M}_j = \phi, i \neq j, \tag{1}$$

M is a convex and closed set of numbers.

Every M_N parallelepiped is a composite formed out of L materials, structured in parallelepipeds too, M_k, $(k = 1 \div L)$ considered to be homogenous and isotropic.

$$M_n = \bigcup_{k=1}^{L} M_{k,}(n = 1 \div N) \tag{2}$$

In these suppositions the Fourier-Kirchhoff equation has the following form in the stationary instance.

$$-\left[\frac{\partial}{\partial x}\left(\lambda_x(x,y,z)\frac{\partial\theta}{\partial x}\right) + \frac{\partial}{\partial y}\left(\lambda_y(x,y,z)\frac{\partial\theta}{\partial y}\right) + \frac{\partial}{\partial z}\left(\lambda_z(x,y,z)\frac{\partial\theta}{\partial z}\right)\right] = S_q(x,y,z), \tag{3}$$

where: λ_x, λ_y, λ_z- are continuous functions defined on the M set and periods belonging to the M_n period.

$\theta = \theta(x, y, z)$- represents the temperature field.

S_q (x, y, z) - the heat source.

The reason we use substantial derivative instead of simple derivative if time is that we must take into consideration the pressure, the specific volume, the enthalpy and the speed of the place function. The quantity of heat conduction and heat transfer depends on incoming and outgoing heat.

In case of certain conditions posed at a pinch correctly, it will become a Dirichlet or Neumann problem, or a mix problem with a unique solution. The homogenous method leads to the calculation of constant coefficients $\overline{\lambda_x}$, $\overline{\lambda_y}$, $\overline{\lambda_z}$ from the variable coefficients, $\lambda_x(x,y,z), \lambda_y(x,y,z)$ and $\lambda_z(x,y,z)$.

Constant coefficient prediction will be obtained by using the asymptotic expansion method, or the finite element method.

Supposing that a θ function admits a asymptote expansion in multiple steps in the form of:

$$\theta^{\varepsilon}(x,y,z) = u^{o}(x,y,z) + \varepsilon u^{1}(x,y,z,u,v,w) + \varepsilon^{2}u^{2}(x,y,z,u,v,w) + \ldots \tag{4}$$

Where

$$u^{i}(x,y,z,u,v,w), i = 1,2\ldots,$$

are periods from the εMnperiod in u,v,w variables given by:

$$u = \frac{x}{\varepsilon}, v = \frac{y}{\varepsilon}, w = \frac{z}{\varepsilon} \tag{5}$$

And ε is a low positive parameter, ε > 0.

If the following $\theta^{\varepsilon}(x,y,z)$ shall be inserted in the (3) equation and it is identified by ε it's coefficients ε^{-1} and ε^{o} add up to:

$$\frac{\partial p_x^o}{\partial u} + \frac{\partial p_y^o}{\partial v} + \frac{\partial p_z^o}{\partial w} = 0 \tag{6}$$

and

$$-\left[\frac{\partial p_x^o}{\partial x} + \frac{\partial p_y^o}{\partial y} + \frac{\partial p_z^o}{\partial z} + \frac{\partial p_x^1}{\partial u} + \frac{\partial p_y^1}{\partial v} + \frac{\partial p_z^1}{\partial w}\right] = S_q(x,y,z) \tag{7}$$

where:

$$\begin{aligned}
p_x^o &= \lambda_x(u,v,w)\left(\frac{\partial u^o}{\partial x} + \frac{\partial u^1}{\partial u}\right) \\
p_y^o &= \lambda_y(u,v,w)\left(\frac{\partial u^o}{\partial y} + \frac{\partial u^1}{\partial v}\right) \\
p_z^o &= \lambda_z(u,v,w)\left(\frac{\partial u^o}{\partial z} + \frac{\partial u^1}{\partial w}\right)
\end{aligned} \tag{8}$$

$$p_x^1 = \lambda_x(u,v,w)\left(\frac{\partial u^1}{\partial x} + \frac{\partial u^2}{\partial u}\right)$$
$$p_y^1 = \lambda_y(u,v,w)\left(\frac{\partial u^1}{\partial y} + \frac{\partial u^2}{\partial v}\right) \quad (9)$$
$$p_z^1 = \lambda_z(u,v,w)\left(\frac{\partial u^1}{\partial z} + \frac{\partial u^2}{\partial w}\right)$$

For the determination of the homogenous coefficients equation we use a mediation operator:

$$\langle\theta\rangle = \frac{1}{|V_n|}\int_{V_n} \theta(x,y,z)dM, \quad (10)$$

Where $|V_n|$ represents the parallelepiped's volume V_{n_1}, $(n = 1 \div N)$

If is taken in account the fact that, p_x^1, p_y^1, p_z^1 are periods of εV_n category [11], applying the Gauss-Ostrogradski formula in (7) and the mediation operator (10) is then obtained:

$$-\frac{\partial}{\partial x}\langle p_x^o\rangle - \frac{\partial}{\partial y}\left\langle p_y^o\right\rangle - \frac{\partial}{\partial z}\langle p_z^o\rangle = S_q(x,y,z) \quad (11)$$

The result is the Fourier-Kirchhoff homogenous Eq. (11).

Its coefficients are calculated with the formula:

$$\bar{\lambda}_x = \langle\lambda_x\rangle + \left\langle \lambda_x \frac{\partial T_x}{\partial u}\right\rangle$$
$$\bar{\lambda}_y = \langle\lambda_y\rangle + \left\langle \lambda_y \frac{\partial T_y}{\partial u}\right\rangle \quad (12)$$
$$\bar{\lambda}_z = \langle\lambda_z\rangle + \left\langle \lambda_z \frac{\partial T_z}{\partial u}\right\rangle$$

where:

T_x, T_y, T_z, are the solutions of the next limit variation problems:

$$\int_{V_n}\left(\lambda_x \frac{\partial T_x}{\partial u}\cdot\frac{\partial B}{\partial u} + \lambda_y \frac{\partial T_x}{\partial v}\cdot\frac{\partial B}{\partial v} + \lambda_z \frac{\partial T_x}{\partial w}\cdot\frac{\partial B}{\partial w}\right)dV = \int_{V_n}\frac{\partial \lambda_x}{\partial u} BdM \forall B \in H^1(V_n) \quad (13)$$

$$\int_{V_n}\left(\lambda_x \frac{\partial T_y}{\partial u}\cdot\frac{\partial B}{\partial u} + \lambda_y \frac{\partial T_y}{\partial v}\cdot\frac{\partial B}{\partial v} + \lambda_z \frac{\partial T_y}{\partial w}\cdot\frac{\partial B}{\partial w}\right)dV = \int_{V_n}\frac{\partial \lambda_y}{\partial u} BdM \forall B \in H^1(V_n) \quad (14)$$

$$\int_{V_n} \left(\lambda_x \frac{\partial T_z}{\partial u} \cdot \frac{\partial B}{\partial u} + \lambda_y \frac{\partial T_z}{\partial v} \cdot \frac{\partial B}{\partial v} + \lambda_z \frac{\partial T_z}{\partial w} \cdot \frac{\partial B}{\partial w} \right) dV = \int_{V_n} \frac{\partial \lambda_z}{\partial u} BdM \forall B \in H^1(V_n) \quad (15)$$

$$H^1(V_n) = \left\{ B : M \rightarrow \prod / B \varepsilon V_n aclass C^1 Subolevperiod, \langle B \rangle = 0 \right\} \quad (16)$$

Solution to the (13), (14), (15) problems is found in $H^1(V_n)$ and it can be shown that [12] the variational problems (13), (14) and (15) formulated from the Subolev $H^1(V_n)$ space perspective have a unique solution.

To determine, T_x, T_y, T_z, a finite element type of process is applied with a local basis of functions $\{\phi_k\}_{k=1}$, having this form:

$$\phi_k(x, y, z) = (A_k z + B_k) \cdot (C_k x + D_k y + E_k) \quad (17)$$

attached to $\overset{(e)}{V}$ prism type elements having six nodes.

The coefficients, $\lambda_x, \lambda_y, \lambda_z$, appear in the form of:

$$\begin{aligned} \lambda_x &= \sum_{k=1}^{6} \lambda_{x_k} \cdot \phi_k(x, y, z) \\ \lambda_y &= \sum_{k=1}^{6} \lambda_{y_k} \cdot \phi_k(x, y, z) \\ \lambda_z &= \sum_{k=1}^{6} \lambda_{z_k} \cdot \phi_k(x, y, z) \end{aligned} \quad (18)$$

Where

$$\lambda_{x_k}, \lambda_{y_k}, \lambda_{z_k} (k = 1 \div 6)$$

are constant data in the nodes belonging to $\overset{(e)}{V}$.

This way the homogenous Eq. (11) becomes an equation containing constant coefficients, given by (12).

3 Conclusions

The general software analysis with finite elements in which a complete process analysis of composite materials will be simulated, will be developed for the entire manufacturing process of these materials. Using this mathematical model has succeeded the transformation of homogeneous equation from field temperature of composite material to equations with constant coefficients. The process of realization of composite materials is easier to determine with finite element methods. Data obtained at experimental attempt are compared with analytical results.

Contrasting with obtained results of finite element methods it can be noted the fact that the present composite material behaviour analysed with mathematical model is more precise for an anisotropic material than for an isotropic one.

The mathematic model has a keen role at finite element methods development because this is approximate, and for a better approximation is needed to utilize a big number of elements in order to reach one precise result ever possible.

One essential disadvantage of analytical calculation methods it represents the imposing of a big number of calculation hypothesis, fact that leads to distancing from the real problem.

References

1. Bere, P., Neamtu, C.: Methodology for evaluate the form deviations for formula one nose car. Central Eur. J. Eng. **4**(2), 148–154 (2014)
2. Bere, P., Gutiu, M.: Fabricareamaterialeor composite. EdituraTechnica-UTM, Chisinau (2018)
3. Somotecam, S.: Compozite. Calcul de rezistenta, Editura U.T. Pres,ClujNapoca (2000)
4. Sabau, E.: Comportareamecanica a materialelorcompozite, EdituraMS LOGO, Chisinau (2019)
5. Bere, P., Berceand, P., Nemes, O.: Phenomenological fracture model for biaxial fibre reinforced composites. Comp. Part B: Eng. **43**(5), 2237–2243 (2012)
6. Bere, P., Dudescu, M.C., Balc, N., Berce, P., Iurianand, A.M., Nemes, O.: Design and analysis of carbon/epoxy composite bicycle handlebar. Mat. Plast. **51**(2), 145–149 (2014)
7. Ceclan, V.A., Bere, P., Borzan, M., Grozav, S., Borzan, C.: Development of environmental technology for carbon fibre reinforced materials recycling. Mat. Plast. **50**(2), 79–83 (2013)
8. Sabau, E., Popescu, A., Vilau, C.: Mechanical behavior of composite materials using the finite element analysis. MATEC Web Conf. **137**, 08006 (2017)
9. Wu, S., Xu, W., Balamurugan, G.P., Thompson, M.R., Nielsen, K.E., Brandys, F.A.: Recovery behaviour of shape memory polyurethane based laminates after thermoforming. Smart Mat. Struct. **26**(11), 115002 (2017)
10. Furukawa, T., Pan, J.W., Cheng, J.Q., Michopoulos, J.G., Wada, Y.: Multi-linear modeling for characterization of nonlinear behavior of anisotropic materials. In: ASME International Design Engineering Technical Conferences/Computers Information in Engineering Conference, vol. 2 (2012). http://orcid.org/0000-0001-7004-6838
11. Potra, T., Mădărăşan, T., Dreve, M.: Ometodă de omogenizareprin element finitaplicată la câmpului de temperaturăînmateriale composite. BuletinŞtiinţific al InstitutuluiPolitehnic, Cluj-Napoca (1989)
12. Bensousan, A., Lions, J.L., Papanicolaous, G.: Asymtotic Analysis for Periodic Structures. North-Holland, Amsterdam (1978)

Effect of the Voxel Size on the Final Scan Accuracy in Computed Tomography

Michaela Kritikos(✉)

Slovak University of Technology,
Faculty of Materials Science and Technology in Trnava, Trnava, Slovakia
michaela.samardziova@stuba.sk

Abstract. Production of different kind of parts has increasing character. They are different according to materials, shapes, production technology and accuracy. Possibilities of their quality and accuracy control have to taken into consideration. In addition to optical (non – contact method) and contact (CMM) methods there are applied progressive technologies which bring new opportunities into the metrology processes. One of them is application of X - ray beam in measurement process. Computed tomography finds application in the field of different parts quality evaluation. This new method has many unexamined areas which are necessary to research. One of them is effect of voxel size on quality and accuracy of achieved model.

Keywords: Measurement · Computed tomography · X – ray · Voxel · Models accuracy

1 Introduction

Today X-ray computed tomography (CT) finds applications in three major fields.

1. CT scanners for medical imaging go back to the early 1970s [1, 2]: 1st scanner built by Nobel Prize winner Hounsfield in 1969 and 1st patient brain scan performed at the Atkinson Morley Hospital, Wimbledon, UK, in October 1971.
2. Since 1980, CT became popular for material analysis and non destructive testing (NDT) [3]: e.g. for observing the inner structure of materials (e.g. fiber-reinforced plastics) and detecting material defects. The cabinets of material testing CT devices are often small and aimed at hosting samples cut out from larger objects.
3. More recently, CT technology entered the application field of dimensional metrology, as an alternative to tactile or optical 3D coordinate measuring machines, measuring arms, fringe projection systems, etc. The first attempts to perform dimensional measurement using existing CT scanners appeared around 1991 [1, 5] but accuracy was not better than about 0.1 mm. An important breakthrough came in 2005, when the 1st dedicated dimensional CT machine [4] was exhibited at the Control Fair in Germany. From then on, several other machine vendors started offering larger and more powerful industrial CT devices able to host full workpieces and radiate through their large volume and/or mass [e.g. 4, 6].

N. M. Durakbasa and M. G. Gençyılmaz (Eds.): ISPR 2019, LNME, pp. 422–431, 2020.
https://doi.org/10.1007/978-3-030-31343-2_37

So far, dimensional CT metrology is the only technology able to measure as well the inner as the outer geometry of a component without need to cut it through and destroy it. As such, it is the only technology for industrial quality control of workpieces having nonaccessible internal features (e.g. components produced by additive manufacturing) or multi-material components (e.g. 2K injection molded plastic parts or plastic parts with metallic inserts). The advent of those production methods favoring the trend for more part feature integration and yielding parts with complex internal geometries or multi-material components, is a major incentive that boosts the demand for dimensional CT metrology: industry can no longer accept that intricate components produced by additive manufacturing or multi-material injection molding escape any geometrical and tolerance quality control for the only reason that there is no non-destructive method to measure the inner or internal geometry. Dimensional CT metrology is also interesting for quality control of assemblies in assembled states, since the geometry and dimensions of components may differ in unassembled and assembled state. An assembly may fail geometrically, even if all individual elements of the assembly meet the tolerances when unassembled. CT being a non-contact measuring technique, it also might become a competitor to other novel noncontact quality inspection methods, like fringe projection, laser scanners, etc. [1].

1.1 Basic Principle

X-ray tomography uses the ability of X-ray radiation to penetrate objects. On the way through an object, part of the impinging radiation is absorbed. The longer the radiographic length of the object, the less radiation escapes from the opposite side. The absorption also depends on the material. An X-ray detector (sensor) captures the escaping X-ray radiation as a two-dimensional radiographic image. At detector sizes of approximately 50 mm to 400 mm, a large portion of the measured object can be captured in a single image.

In order to use tomography on an object, several hundred two-dimensional radiographic images are made in sequence, with the measured object in various rotated positions (Fig. 1). The object is located on a rotating table for this purpose, which is gradually rotated step by step. The three-dimensional information about the measured object contained in this series of images is extracted using a suitable mathematical process and is made available as a "voxel image". Each voxel (volume pixel) embodies the X-ray absorption by the measured object for a defined location in the measured volume. Similar to two-dimensional image processing, the actual measured points are calculated from the voxel data using a suitable threshold process.

The sensors currently used capture up to 4 million image points. Typically, several hundred thousand to a few million measurement points are derived in the measured volume. These points are distributed evenly across the surface of the part being measured. Geometries in the interior of the measured object, such as hollow cavities or undercuts, are also captured. The measurement points can be evaluated using the familiar methods of coordinate measuring technology.

Similar to measurement using image processing, it is possible to change the magnification using tomography (Fig. 1b) in order to capture small parts with higher magnification, or larger parts completely with lower magnification. To do this, either the measured object is positioned in the radiation path or the X-ray components (X-ray source and detector) are moved in an axial direction relative to the measured object.

In some cases, the size of the sensor or the number of pixels available is still not enough to meet the requirements for the measuring task. In such cases, several images are stitched together (Fig. 1c) by moving the rotary table with the measured object relative to the X-ray components. Reconstruction of the voxel volume image is then accomplished on the basis of the stitched 2-D radiographic images [7].

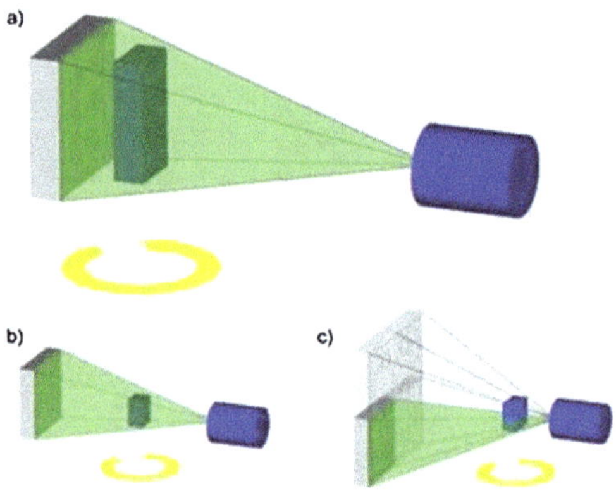

Fig. 1. X – ray tomography: (A) Basic principle: similar to a pinhole camera, the radiation emitted by a point X – ray source travels through the measured object to a flat sensor – images are taken for various rotated positions, (B) Magnification adjustment: movement of the measured object relative to the sensor and to the radiation source, (C) Raster tomography: images from various positions are stitched [7]

The measurement target (sample) is positioned between the X-ray generator and the X-ray detector, as shown in the figure below. X-ray fluoroscopic data is collected from every angle by rotating the sample 360 degrees, and tomographic images (CT images) are calculated [8] (Fig. 2).

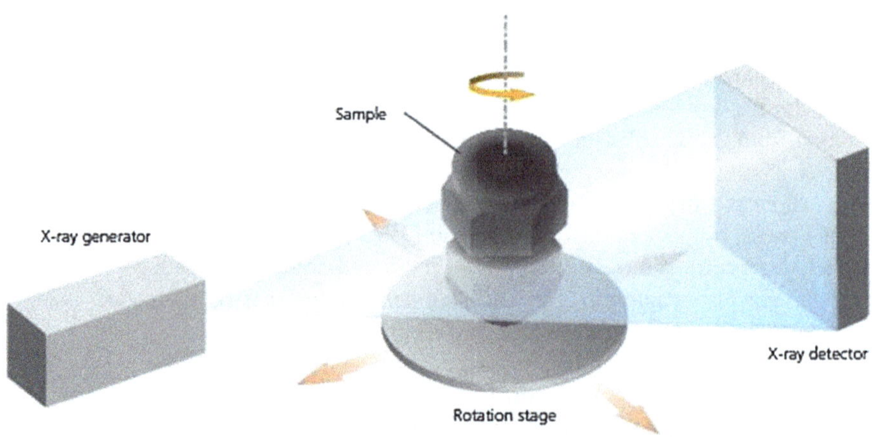

Fig. 2. Schematic representation of industrial tomograph [8]

There is still not enough information or experimental results about how big effect has a voxel size on the final accuracy of the scan. Some papers are dealing with this topic as [9] (Fig. 3).

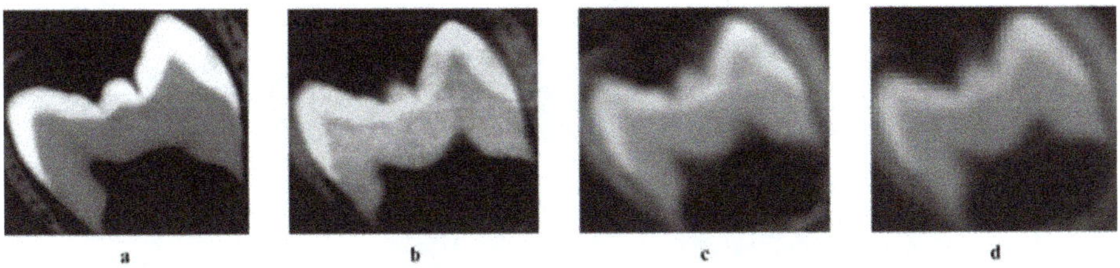

Fig. 3. CT tooth scans with different voxel sizes. (a) voxel size 41 μm (b) voxel size 76 μm (c) voxel size 200 μm (d) voxel size 300 μm [9]

2 Experimental Investigation

The aim of the experiment was in comparison of the accuracy of the scans with different voxel size.

METROTOM 1500 (Fig. 4) – X-ray from Zeiss company was used for getting data for this investigation.

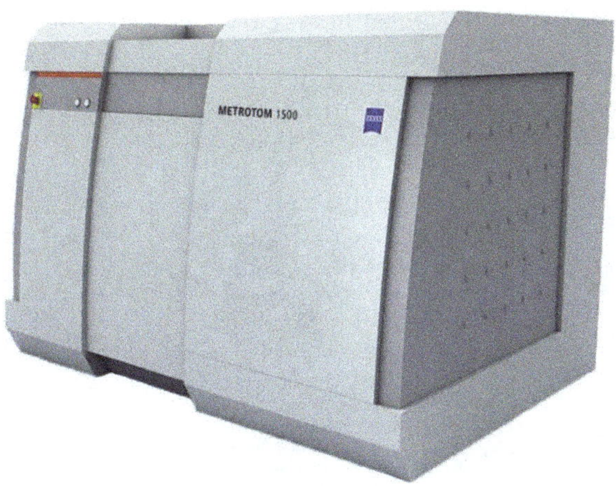

Fig. 4. METROTOM 1500 Zeiss [10]

It is equipment for nondestructive testing of parts made of plastic, ceramics or composite materials, as well as magnesium, aluminum and steel with the resolution of 1024 × 1024 (half) and 2048 × 2048 pixels (full resolution). Scanning by full resolution is leading to half voxel size in comparison to scanning by half resolution.

Software used for scanning of part was METROTOM OS (Fig. 5). Number of images was 1150.

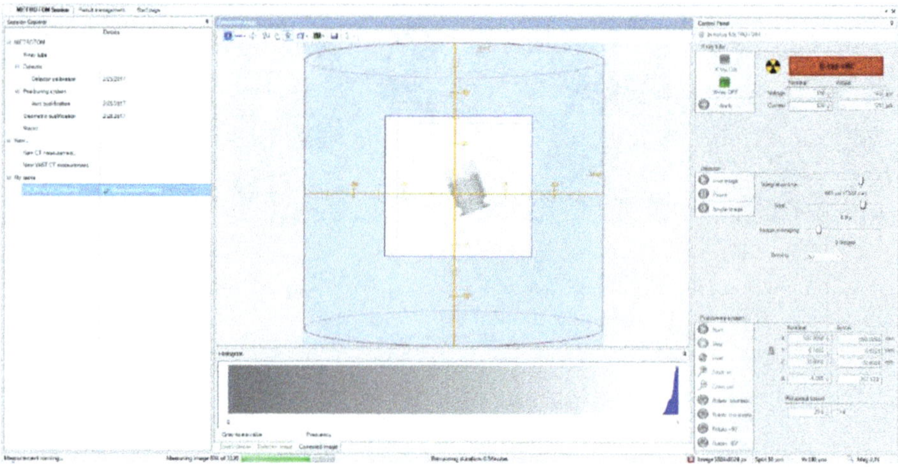

Fig. 5. METROTOM OS graphical user interface

Coordinate measuring machine CENTERMAX (Fig. 6) from Zeiss company with tactile sensor VAST XTR gold was used for getting data, which were used as references for comparison with data gotten from X-ray.

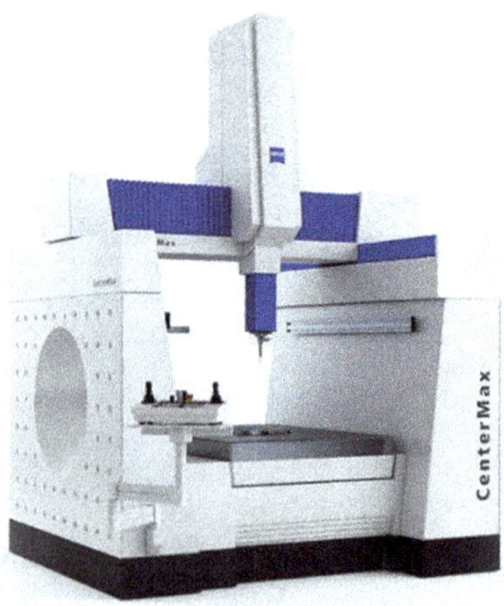

Fig. 6. CenterMax Zeiss [11]

Both machines were calibrated before realization of experiment.

CALYPSO software (Fig. 7) was used for data evaluation in both cases. Values were obtained using the same measuring plan with programmed strategies for elements measurement.

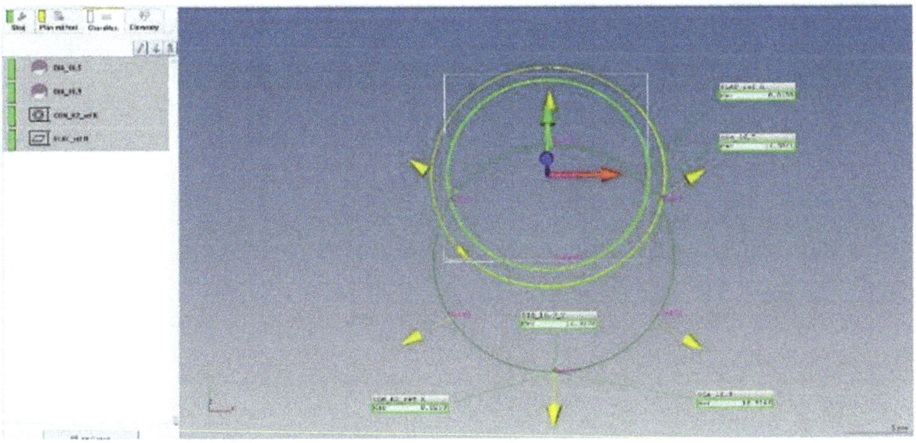

Fig. 7. Calypso software interface

Measured part was made of plastic material (Fig. 8) with the biggest diameter of 17 mm. Density of the scanned material is very important to consider when using X – ray tomography. It is more difficult to scan materials with higher density. It depends on the performance of the machine used for getting data.

Fig. 8. Measured part

Process of experiment:

1. Measuring plan programming in Calypso software.
2. Measurement of the part on CenterMax to get reference values.
3. Part's scanning with different voxel sizes (half and full resolution) in METROTOM 1500.
4. Evaluation of the achieved scans.

The voxel size depends on the resolution of detector and the distance of the scanned object from X – ray lamp.

Part was scanning in 5 different positions from X – ray lamp with different voxel sizes with half and full resolution of detector.

Positions: 150 mm, 350 mm, 550 mm, 750 mm, 950 mm.

Resolution: 2048 × 2048 (full), 1024 × 1024 (half).

Evaluated characteristics were 2 × DIAMETER, FLATNESS and CONCENTRICITY (Table 1).

Table 1. Reference values achieved on CenterMax

Measured characteristics	Reference (measured) value
Diameter	16,5061 mm
Diameter	16,9046 mm
Concentricity	0,0293 mm
Flatness	0,0188 mm

Scans with different voxel sizes were compared (Fig. 9).

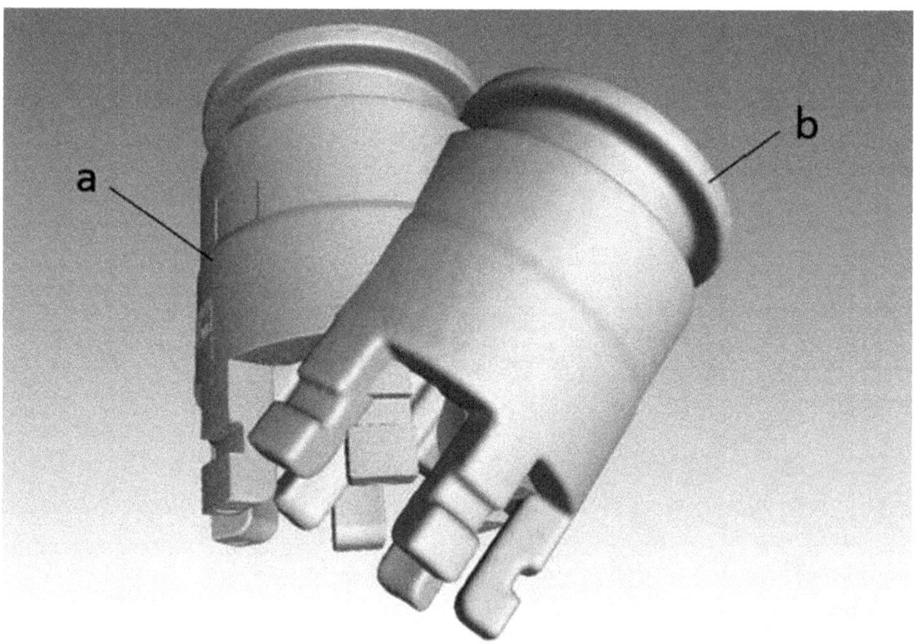

Fig. 9. Comparison of scans with the smallest and the highest voxel size. (a) full resolution, Vx: 20 μm, position: 150 mm, (b) half resolution, Vx: 251 μm, position: 950 mm

It is obvious that voxel size has a big effect on the surface accuracy of the scan as it is seen from Fig. XXX. There is a scan of the part with the smallest and the highest voxel size. Full resolution enables to get more details of the surface with its roughness and the edges are sharper. Half resolution scan with voxel size of 251 μm is smoother and the edges are rounded.

3 Evaluation of Scans with Different Voxel Size

The evaluation process of different characteristics (2 × diameter, flatness and concentricity) was done after the measurement of the part with the tactile CMM and scanning of the part on CT. Programmed measuring plan was created by Calypso software how it was mentioned above. The option metrotomography has been set for evaluation of the cloud of points from CT scanning. Every scan with different voxel size was evaluated 3 times and it was used the average value of each characteristic.

Evaluated values are shown in the Table 2 for half resolution of detector and in the Table 3 for full resolution in different positions from X – ray lamp (with corresponding voxel sizes).

Table 2. Full resolution

Voxel size/position	20 μm, 150 mm	47 μm, 350 mm	73 μm, 550 mm	99 μm, 750 mm	126 μm, 950 mm	Reference values
Diameter 1 [mm]	16,4841	16,4929	***16,5090***	16,5138	16,5119	16,5061
Diameter 2 [mm]	16,9019	16,9165	16,9150	16,9194	***16,9064***	16,9046
Concentricity (mm)	***0,0277***	0,0388	0,0305	0,0356	0,0356	0,0293
Flatness (mm)	0,0100	0,0130	0,0132	0,0131	***0,0139***	0,0188

Table 3. Half resolution

Voxel size/position	40 μm, 150 mm	93 μm, 350 mm	164 μm, 550 mm	199 μm, 750 mm	251 μm, 950 mm	Reference values
Diameter 1 [mm]	16,4907	16,4994	***16,5079***	16,5240	16,5017	16,5061
Diameter 2 [mm]	16,9128	16,9170	16,9076	***16,9061***	16,9070	16,9046
Concentricity (mm)	0,0266	0,0319	***0,0289***	0,0417	0,0620	0,0293
Flatness (mm)	0,0103	***0,0178***	0,0220	0,0150	0,0219	0,0188

Values with the smallest difference from reference values were obtained by using half detector resolution in all cases. It is obvious the most accurate position for part's scanning for evaluation of diameter, concentricity and flatness is between 350 and 750 mm from X – ray lamp. How it is seen the smallest voxel sizes does not lead to the most accurate results.

The smallest difference from reference value (0,0004 mm) was achieved in the case of evaluation of concentricity with voxel size 164 μm. On the other hand the highest difference from reference value was achieved in evaluation of concentricity in the position of 950 mm from X – Ray lamp with appropriate voxel value of 251 μm.

When dealing with stability of the process, scanning by the full resolution gives more stable results with not high differences in achieved values between the smallest and the highest voxel sizes. This experimental investigation is dealing with very important topic in the field of metrology at present. The Computed tomography is method for geometrical products specifications evaluation. There is possibility to get data with different voxel sizes which present their accuracy. Scanning using full resolution gives half voxel size models but it is more time consuming than scanning using half resolution (the time of scanning is double). These results can lead to higher productivity by application of half resolution in the most measuring tasks.

- it has been proven that the smallest voxel size doesn't lead to the highest accuracy achievement,
- it was evaluated 4 different variables – 2 times diameter, concentricity and flatness. The most accurate values were achieved with voxel size of 164 μm and half resolution of the detector,
- there is visible difference in the scans with the smallest and the highest voxel size, the full resolution enables to get more details of the surface with its roughness and the edges are sharper. Half resolution's scans are smoother and the edges are rounded. But it has not been proven that this knowledge affects the accuracy in positive way according to evaluation of above mentioned GPS characteristics.

4 Conclusions

This experimental investigation is dealing with very important topic in the field of metrology at present. The Computed tomography is method for geometrical products specifications evaluation. There is possibility to get data with different voxel sizes which present their accuracy. Scanning using full resolution gives half voxel size models but it is more time consuming than scanning using half resolution (the time of scanning is double). These results can lead to higher productivity by application of half resolution in the most measuring tasks.

- It has been proven that the smallest voxel size doesn't lead to the highest accuracy achievement.
- It was evaluated 4 different variables – 2 times diameter, concentricity and flatness. The most accurate values were achieved with voxel size of 164 μm and half resolution of the detector.
- There is visible difference in the scans with the smallest and the highest voxel size, the full resolution enables to get more details of the surface with its roughness and the edges are sharper. Half resolution's scans are smoother and the edges are rounded. But it has not been proven that this knowledge affects the accuracy in positive way according to evaluation of above mentioned GPS characteristics.

Acknowledgment. This work was supported by the Slovak Research and Development Agency under the Contract no. APVV-18-0418.

References

1. Kruth, J.P., Bartscher, M., Carmignto, S., Schmitt, R., De Chiffre, L., Weckenmann, A.: Computed tomography for dimensional metrology. CIRP Ann. Manuf. Technol. **60**, 821–842 (2011)
2. Ambrose, J., Hounsfield, G.: Computerized Transverse Axial Tomography. Br. J. Radiol. **46**, 1023–1047 (1973)
3. Reimers, P., Goebbels, J.: New possibilities of non-destructive evaluation by X-ray computed tomography. Mater. Eval. **41**, 732–737 (1983)
4. Werth Messtechnik (2011). www.werth.de/de/unser-angebot/
5. Mitchell, K.W.: A generalized approach to wall thickness measurements in CT images. In: Topical Proceedings of Industrial Computerized Tomography, pp. 120–124. ASNT (1989). ISBN 0–931403-89-8
6. GE/phoenix Prufbericht, 3D Prazisionsprufung CT, Prufbericht zur Bestimmung der Messgenauigkeit von CT-Systemen, pp. 1–8 (2009)
7. http://werthinc.com/computer-tomography-principle-x-ray-tomography/
8. https://www.shimadzu.com/an/ndi/ct/smx90ct.html
9. Kato, A., Ohno, N.: Construction of three-dimensional tooth model by micro-computed tomography and application for data sharing (2009). https://www.ncbi.nlm.nih.gov/pubmed/18386082
10. ZEISS 2015 METROTOM 1500: Návod k obsluze. ISBN 61212-7090112
11. https://static.industryarena.com/files/i/143/505/61366e2dadeace5fe0242a67d606edcb_MW_1200_.png

Industry 5.0: Is the Manufacturing Industry on the Cusp of a New Revolution?

Mary Doyle-Kent[1(✉)] and Peter Kopacek[2]

[1] Department of Engineering Technology, Waterford Institute of Technology, Waterford, Ireland
MDKENT@WIT.IE

[2] Department of Mechanical Engineering and Management, TU Wien, Vienna, Austria
KOPACEK@IHRT.TUWIEN.AC.AT

Abstract. It is widely accepted that the advent of robotics and automation in previous industrial revolutions brought about paradigm shifts in the manufacturing industry worldwide. It is therefore conceivable that the fifth industrial revolution will bring about a similar shift in norms and make fundamental changes in our approach to industry and manufacturing. From the first industrial revolution to the fourth, science and technological innovations, training and capital, were the key elements in evolving how products are conceived, designed and manufactured. Looking forward to the fifth industrial revolution it is likely that there will be a paradigm change in how industry will evolve given the tools of industry 4.0; the internet of things, digitization, blockchain, advanced materials, additive manufacture, artificial intelligence and robotics, drones, energy technology, biotechnology, neurotechnology and virtual and augmented reality.

Keywords: Technology · Industrial revolution · Robotics · Social impact · Industry 4.0 · Industry 5.0 · Engineering ethics

1 Introduction

The fifth industrial revolution will bring challenges for humanity. The destruction of our environment, as well as increasing income inequality, will become a worldwide crisis in the future. The few will possess most of the world's wealth and control. Automation will increasingly change the dynamics of the workforce. Work will be classed into either 'highly exposed to automation', or 'less exposed to automation', with a view to introducing robotics wherever possible. The new type of work created in an automated environment will require a highly developed set of skills which will not be within the reach of most. Current online security techniques will become outdated. International laws, regulation and governance will need to be reinvented to be agile and relevant to the emerging challenges. The fifth industrial revolution has yet to materialise and is currently an exercise in predicting what will come about, based on future technologies and trends. From the literature there are different understandings and

N. M. Durakbasa and M. G. Gençyılmaz (Eds.): ISPR 2019, LNME, pp. 432–441, 2020.
https://doi.org/10.1007/978-3-030-31343-2_38

visions of post-industry 4.0. In this paper an outline of different theories is put forward. These opinions and theories have been gathered from very different viewpoints and sources.

2 Looking Back

To understand the effect of radical changes on industry let us look back at previous industrial revolutions that have taken place throughout the world. Figure 1 shows, in chronological order, the sequence of industrial revolutions that took place from the invention of steam powered machines to the present day. It projects forward to the next industrial revolution, industry 5.0. It has been recorded that from the 1700's, when steam power was introduced, that the dynamics of manufacturing radically changed; how work was undertaken, how factories were laid out, how human job descriptions changed and how this influenced the quality of life of the workers.

In the eighteen century the first industrial revolution changed manufacturing processes across Europe by the introduction of mechanisation. This meant that previous hand production methods were replaced by machines powered by steam and water. One of the impacts of the first industrial revolution was an unprecedented rise in population growth. The standard of living began to improve, and life changed dramatically for the working classes. It is widely accepted that the United Kingdom (UK) lead the way. What were the influencing factors in the UK that optimised conditions to facilitate these changes? According to Allen in 2011 "the high real wage (also) stimulated product innovation since it meant that Britain had a broader mass

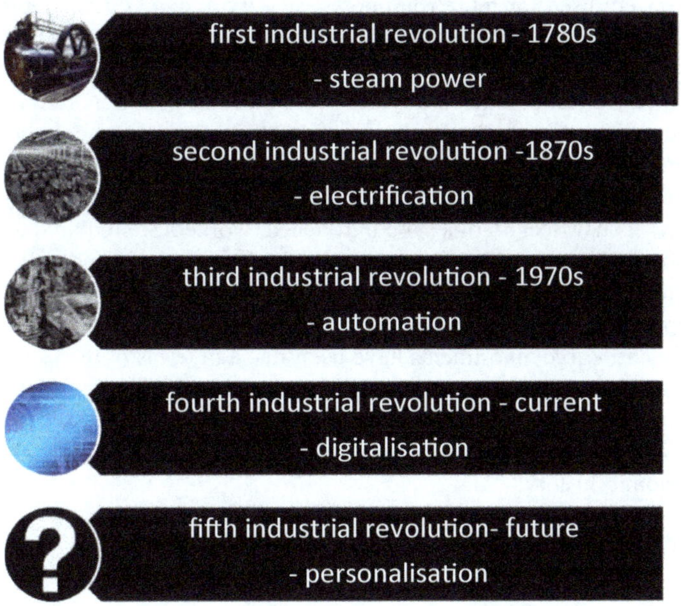

Fig. 1. The sequence of industrial revolutions throughout the ages.

market for 'luxury' consumer goods including imports from east Asia. The supply of technology was also augmented by the high real wage. It meant that the population at large was better placed to buy education and training than their counterparts elsewhere in the world. The resulting high rates of literacy and numeracy contributed to invention and innovation as seen in the Fig. 1 [1].

In the 1870's the introduction of electricity and the roll-out of electrification in the 1880's led to the second industrial revolution. Ian McNeil in his book An Encyclopaedia of the History of Technology shares the different stages of technological advancement by citing the developments in technology over the centuries. McNeil states that the development of the incandescent light bulb both by TA Edison in the USA and JW Swan in England made public lighting a reality. Ferranti's Deptford power station went into operation in 1889 and he says that electricity was born at this time. This invention has been the biggest transformation of our daily lives, bringing convenience, comfort, transport and wellbeing [2].

Manufacturing plants evolved from being powered by steam and water to using electric power at this time. Electric powered motors were much more efficient than the previously used steam, water and human power. Electric motors from this era were around 30% efficient. Hydroelectric and chemical power were also invented at this stage including chemical fuel from petroleum.

The introduction of automatic equipment into manufacturing companies became popular in the 1970 and 1980's. The first automated equipment was used for tasks that were challenging to humans such as welding, materials handling and product assembly. In heavy manufacturing industries, such as steel and automobile manufacturing, robotics played an important role as the working environment proved to be unfit for humans. In the 1980's robots became more sophisticated and less expensive. The result was that they were used for more routine types of work across diverse processes.

In 1997 Jeremy Greenwood wrote an insightful book entitled The Third Industrial Revolution. Technology, Productivity and Income Inequality. Here he looks at the effects of innovative technologies introduced into industry over hundreds of years, and, more specifically, what were the knock-on effects of this new technology both on the workforce and on the economy. He states, "this thesis (here) connects the rate of technological progress to the level of income inequality and productivity growth". He goes on to say "as new technologies are introduced, workers are required to upskill and become proficient in the use of these technologies, or, new more highly skilled employees are hired to do this work. As a result, these more highly skilled employees will earn more money and often the productivity will fall initially as the workers become proficient. Large investments have been made in terms of the technologies and the up skilling of the workers and there with be a time lapse before full potential is reached. Greenwood states on page 2 that "the coincidence of rapid technological change, widening inequality, and slowing productivity growth is not without precedence in economic history" [3].

Professor Peter Kopacek, in 2018 describes industry 4.0 in his paper titled Development trends in Cost Oriented Production Automation. He states that "The industry 4.0 vision is not limited to automation of a single production facility. It incorporates integration across core functions, from production, materials sourcing, supply chain, and warehousing all the way to sale of the final product. This high level of integration and visibility across business processes will enable greater operational efficiency, responsive manufacturing, and improved product design" [4].

He continues to explain how workers will be affected by the automation and technological advances by explaining that work will change dramatically for the workers. He says that "real-time oriented control will transform work content, work processes and the working environment. Industry 4.0 yields to dramatic changes like:

- New skills and structures in the workforce.
- More collaboration required across all functions within the company, with its customers, and with its suppliers (especially tool vendors and IT).
- Need for advanced cyber security increasing exponentially.
- Investment requirements increasing."

Klaus Schwab in his book The Fourth Industrial Revolution explains that the fourth industrial revolution extends and transforms digital technologies. They are connected to one another and this connection allows the digital technologies and capabilities to expand. Information storage, processing and communication are the vital elements. The fourth industrial revolution technologies disrupt creating new value sources, and the new digital technologies will be the 'usual' going forward. (P20–21). The digital networks facilitate the movement and manufacture of physical products by knowledge transfer. Digital products can be reproduced at very low costs. In the third industrial revolution physical systems disappeared into the digital, for example digital recording of music, and, in the fourth the digital is remerging into the physical, for example, 3d printing [5].

In summary industry 4.0 already greatly impacts factory workers, they will need to upskill to 'futureproof' themselves to remain useful for their employers.

3 Looking Forward

Automation and robotics excel in the manufacture of standardised products using standard manufacturing processes in high volumes to an excellent quality standard. When creativity or customisation is expected the human being is key. The solution is the collaboration of robots and the humans. Traditional robots cannot work side by side with humans but collaborative robots or "cobots" are designed to work in sync with human employees. The robot prepares the product in a rigorous manner, meeting specification requirements to a high standard, and the human then takes over to add the finishing touches to the product. Kopacek states that "in order to perform collaborative tasks cobots are designed to be safe around humans (using sensors, force limiting and rounder geometries than traditional robots), to be lightweight (for them to be moved from task to task) and to be easy to implement and use without skills in programming.

Above all, a collaborative robot is not a replacement robot; it assists workers rather than replaces them" [4].

In 2018 Deloitte wrote a piece entitled 'The skills gap in US manufacturing 2015–2025 outlook' and stressed that humans need to be put back into the loop but to do this, new and differently skilled employees, are required to fill new job specifications. In summary they noted that the skills gap is outlined in the following themes;

- "Putting humans in the loop
- Expanding digital and "soft" skills
- Leveraging the digital toolbox" [6].

According to Accenture, industry 5.0 is where each actual product to be manufactured will be based on the individual customer needs. The final product will be designed for the individual based on their personal requirements, otherwise known as 'personalisation'. The product will be manufactured in an agile manner, taking the digital requirements and using radical new manufacturing techniques and materials. These products, in some cases can be an extension of the human mind or body which will require new radical ethical protocols and governance to be put in place. It is predicted that the products manufactured by fifth industrial revolution will bring new technologies to the mass population. This means manufacturing will not be just agile and lean, but automated, digital and data driven. Products will be of extremely high quality and available at more affordable prices. These products must be the best available and the supply chain optimised to support manufacturing as seen in the Fig. 2.

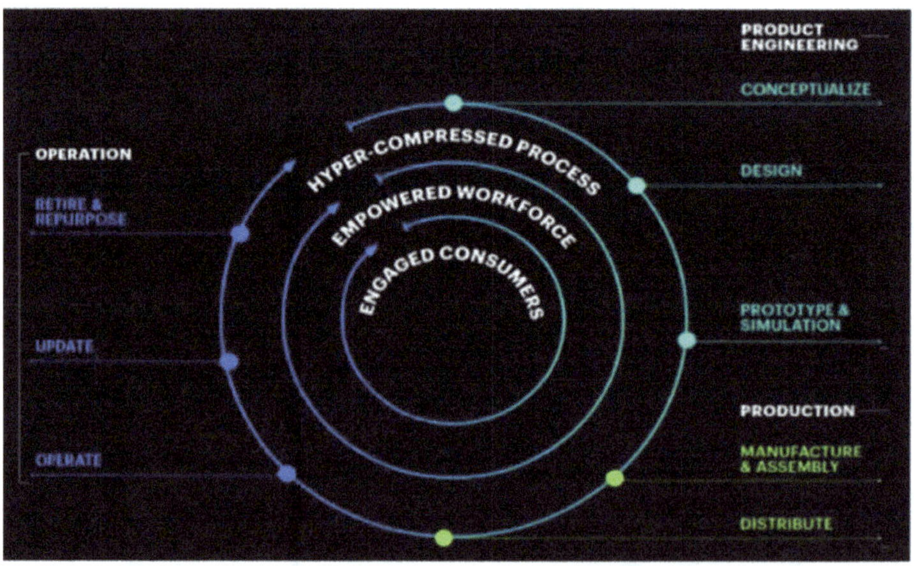

Fig. 2. Product manufacturing reinvented [7].

Phill Cartwright, executive chairman of the "Centre for Modelling & Simulation", looks forward to the fifth industrial revolution. He predicts the roles humans will play in an automated manufacturing environment in an online article from Raconteur.net. Moving from industry 4.0 to industry 5.0, will lead to the creation of higher-value jobs than ever before. The reason for this is that the freedom of design and the associated responsibility is handed back to the human designer/engineer. Cartwright consistent with a recent study from Meggitt explains that "the workspace doesn't become smaller in terms of a manufacturing cell around the human being; it actually becomes bigger. The human being has more responsibility and you end up with a bigger, lighter environment that's safer than the previous environment. The manufacturing operative within the manufacturing cell starts to become more involved in the design process rather than the manufacturing process, which is more, or less, automated. It allows freedom of design to work with you and it enables products that are more bespoke and personal" [7]. This theory is illustrated in Fig. 2. It shows the strong relationship between the engaged consumer, the empowered workforce and the hyper compressed manufacturing processes.

Korcomptenz are a technology company stating that they provide a competitive advantage for their customers with innovative technology solutions by addressing business challenges and driving growth. They explain that industry is "currently still in the midst of industry 4.0, where manufacturing has become "smart" through the development of AI, the Cloud, the Internet of Things, and other such systems. The basic principle behind the fourth industrial revolution was that through linking machines and other intelligent devices, manufacturers could create smart networks throughout the value chain (from materials to production) that could control each other" [8].

According to Korcomptenz industry 5.0 is rapidly approaching and it will bring with it an increased human touch back to manufacturing. It is thought that "where industry 4.0 put smart technology at the forefront of manufacturing, 5.0 will encourage increased collaboration between humans and smart systems." Bringing the high speed, optimised automation together in a meaningful way with the cognitive, critical thinking of the human being in industry 5.0 will mean human and machine will work together in a revolutionary manner to solve problems of the future and produce personalised products of high value.

There are also questions raised about how humans will fare with the technological advances. Martin Ford, in a publication called "The rise of the Robots. Technology and the Threat of Mass Unemployment", portrays a more negative viewpoint of the impact of industry 5.0.

Ford explains that in the 1980s and 1990s the evolution of work landscape required the workforce to become more information technology (IT) literate. The 1990s saw the tech bubble and creation of millions of new jobs, especially in IT sector. Wages increased with this new type of work but fell short of productivity gains. Recession was followed by jobless recovery as workers, many from the mid-salary range, failed to find employment.

In 2000-plus IT continued its acceleration and productivity rose; businesses took full advantage of all new innovations. Many good jobs disappeared as corporations offshored, automated and outsourced using cloud computing. Throughout the developed world computers and machines were increasingly replacing workers rather than making them more valuable. The share of national income going into labour declined. The jobs market continued to polarize. Joblessness became the norm [9].

It was noted that factors contributing to the "disruptive economics" include; IT, globalization, politics and growth of financial sector.

Looking into the future Ford predicts that IT will continue to evolve at an alarming rate. Coupled with the IT changes many more jobs will be pulled into automation and robotics. Low wage jobs will be replaced with self-service automated technologies and increasingly intelligent algorithms will threaten higher skilled work.

Ford cited researchers Frey & Osbourne from the University of Oxford who predict that over the next 20 years approximately 47% of US jobs will be vulnerable to automation. This, in turn, will have a knock-on effect of reducing customer spending which could, in turn, result in a secondary wave of job losses affecting all occupations.

Referencing Fig. 3 below, which was part of the "World Bank Group Flagship Report" released in 2019 titled "The changing face of work", it is noted that the citizens of the European Union believe that technology is a positive element in the workplace. Two thirds believe that technology will benefit society and quality of life.

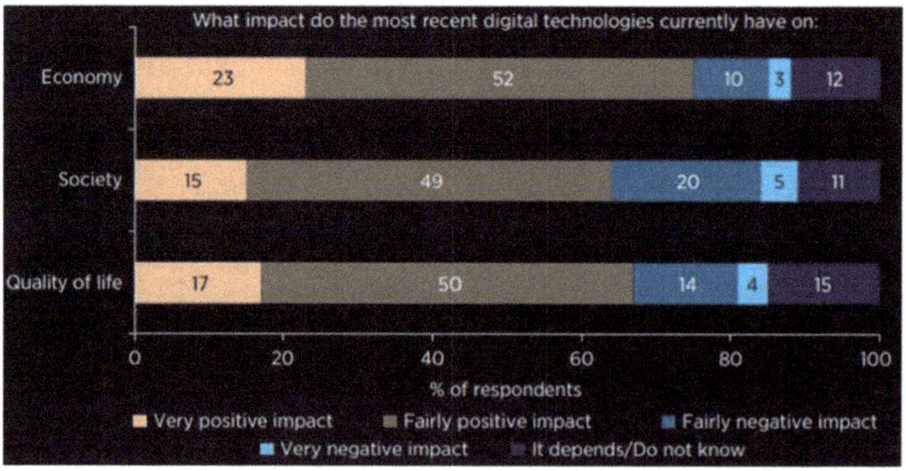

Fig. 3. A survey indicating that technology is improving the European Economy, society and quality of life. The source is WDR 2019 team, based on special Eurobarmeter 460 "Attitudes Towards the Impact of Digitization and Automation on Daily Life" Question 1, European Commission, 2017 [10].

"Despite this optimism, concerns about the future remain. People living in advanced economies are anxious about the sweeping impact of technology on employment. They hold a view that rising inequality, compounded by the advent of the gig economy (organizations contract with independent workers for short-term engagements), is encouraging a race to the bottom in working conditions." So, what factors can help improve working conditions? "Investing in human capital is the priority to make the most of this evolving economic opportunity. Three types of skills are increasingly important in labour markets: ***advanced cognitive skills such as complex problem-solving, socio-behavioural skills such as teamwork, and skill combinations that are predictive of adaptability such as reasoning and self-efficacy***. Building these skills requires strong human capital foundations and lifelong learning." The report continues "at the economywide level, human capital is positively correlated with the overall level of adoption of advanced technologies. Firms with a higher share of educated workers do better at innovating. Individuals with stronger human capital reap higher economic returns from new technologies. By contrast, when technological disruptions are met with inadequate human capital, the existing social order may be undermined" [10].

Karl Schwab in his book concludes that human wellbeing and development can continue on a positive upward trend. Previous industrial revolutions have enriched and progressed the human race and that we need to solve the negative aspects of technological evolution for the common good. These include environmental damage and growing wage inequality. He states "involving all relevant stake holder groups will help us overcome the core challenges ahead – distributing the benefits of technological disruptions, containing the inevitable externalities and ensuring that emerging technologies empower, rather than determine, all of us human beings. Finding solutions for governance challenges of the fourth (and fifth) industrial revolution will require governments, businesses and individuals to make the right strategic decisions about how to develop and deploy new technologies" [5].

4 Questions Raised

The literature review in the previous section has outlined how the next industrial revolution will fundamentally evolve the manufacturing industry worldwide given the tools of industry 4.0 and future requirements of the customer, worker and the environment. At this moment in time there is no consensus on what this new industry will look like but there is consensus that there will be a change to the current norms, a paradigm shift driven by digital. It also raises many questions and the following Fig. 4 summarises seven key areas that will need to be addressed. If these are not addressed, then the result may be a negative force for humanity going forward.

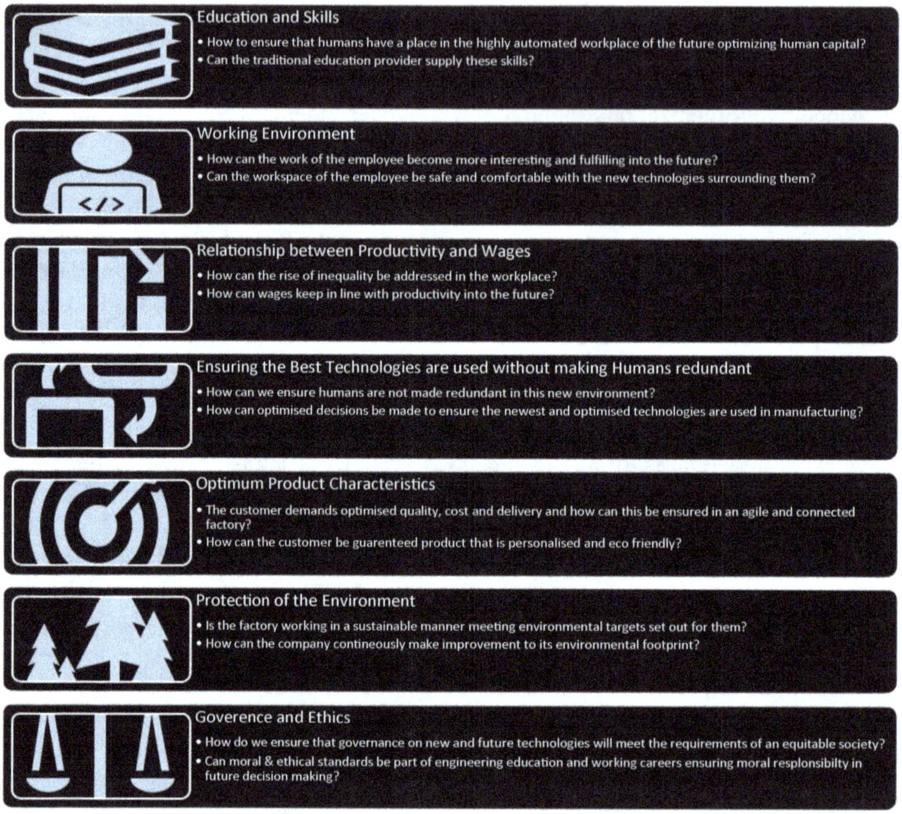

Fig. 4. Questions raised from the literature reviewed

5 Conclusions

This paper has looks at the history of manufacturing worldwide from the invention of steam and water powered machines to the current digital economy. The fifth industrial revolution will bring about a similar shift in norms and require fundamental changes in our approach to industry and manufacturing, as well as to the environment which is declining at an ever-increasing rate.

There is no definite vision now as to what the next industrial revolution will bring in terms of technology, but, it is envisaged that the workforce of today will be very different to the workforce of the future generations.

Governments will have to invest in human capital to make the workers of the future relevant and useful in the industries of the future. Governance and ethics will be increasingly important so that the correct decisions for mankind will be made.

A number of key areas and questions were highlighted in the previous section. These can be the start of a 'think tank' for future technologists, scientists, industrialists, engineers and politicians, all of whom will have to work together if the world is to have a sustainable and healthy future.

References

1. Allen, R.C.: Why the industrial revolution was British: commerce, induced invention, and the scientific revolution 1. Econ. Hist. Rev. **64**(2), 357–384 (2011)
2. McNeil, I. (ed.): An Encyclopaedia of the History of Technology. Routledge, London (2002)
3. Greenwood, J.: The Third Industrial Revolution: Technology, Productivity, and Income Inequality, No. 435. American Enterprise Institute (1997)
4. Kopacek, P.: Development trends in cost oriented production automation. IFAC-PapersOnLine **51**(30), 39–43 (2018)
5. Schwab, K., Davis, N.: Shaping the Future of the Fourth Industrial Revolution. Currency, New York (2018)
6. Deloitte: The skills gap in US manufacturing 2015–2025 outlook (2018). http://www2.deloitte.com/us/en/pages/manufacturing/articles/boiling-point-the-skills-gap-in-us-manufacturing.html. Accessed 25 June 2019
7. Accenture: Accenture-Industry-X0-POV-RGB (2019). https://www.accenture.com/_acnmedia/PDF-63/Accenture-Industry-X0-POV-RGB.pdf. Accessed 20 June 2019
8. Korcomptenz: Korcomptenz.com (2019). https://www.korcomptenz.com/blog/industry-5-0. Accessed 20 June 2019
9. Ford, M.: Rise of the Robots: Technology and the Threat of a Jobless Future. Basic Books, New York (2015)
10. Worldbank: The Changing Nature of Work (2019). http://documents.worldbank.org/curated/en/816281518818814423/pdf/2019-WDR-Report.pdf. Accessed 03 July 2019

Upscaling of Collaborative Disassembly Lines for Mobile Phones – Economic and Environmental Considerations

Rainer Pamminger[1(✉)], Sebastian Glaser[1], Roland Ambrosch[2], and Mario Genner[2]

[1] Institute of Engineering Design and Product Development, Vienna University of Technology, Vienna, Austria
{rainer.pamminger, sebastian.glaser}@tuwien.ac.at
[2] CDP Center for Digital Production GmbH, Vienna, Austria
{roland.ambrosch, mario.genner}@acdp.at

Abstract. To enable advanced mobile phone recycling and circular economy strategies like reuse of mobile phone components, a modular collaborative disassembly machinery demonstrator has been developed. The demonstrator consists of flexible usable dismantling modules, a collaborative robot and a human workspace. The remaining challenge is to scale the demonstrator from the prototype level up to line variants in series applications. This is achieved through simulation of different scenarios where cycle time, workload, economic - as well as environmental aspects are determined. The objective of the study is to define a virtualise disassembly line with blocks and algorithms in order to perform a simulation of different line variants and disassembly scenarios. With this quick evaluation next to the consideration of technical aspects, economic and environmental implications are simulated.

Keywords: Upscaling · Simulation · Modular · Collaborative · Economic · Circular environmental considerations

1 Introduction

The current end-of-life strategy of electronics is focused mainly on recycling. The existing recycling technologies for electronics scrap focus on shredding devices into smaller fractions for recovery of prevailing valuable materials from integrated printed circuit boards. In this recycling and its metallurgic processes, low volume materials such as rare earths metals and still functioning valuable integrated circuits (processors, memories, controllers, etc.) meaning a lot of resources are lost.

In order to realise a more efficient use of resources the EC will push the shift from a linear to a circular economy. Within a Circular Economy products and its components should be used as long as possible through strategies like reuse, remanufacturing or repair and if this is not feasible any more the materials should be recycled in an efficient way [1, 2]. In order to support the concept of a circular economy an economic feasible and environmental friendly disassembly of electronic products is key as it enables

N. M. Durakbasa and M. G. Gençyılmaz (Eds.): ISPR 2019, LNME, pp. 442–451, 2020.
https://doi.org/10.1007/978-3-030-31343-2_39

strategies like reuse or remanufacturing. Also additional legal requirements can be expected. The removal of the battery of smart phones is already mandatory [3] and more directives are on its way. First standards on have already been published on the European Commission [4].

The approach developed in the sustainablySMART project [5], implies a modular collaborative disassembly machinery to gain batteries and printed circuit boards with the integrated circuits from different types of mobile phones for further advanced end-of-life strategies (Fig. 1).

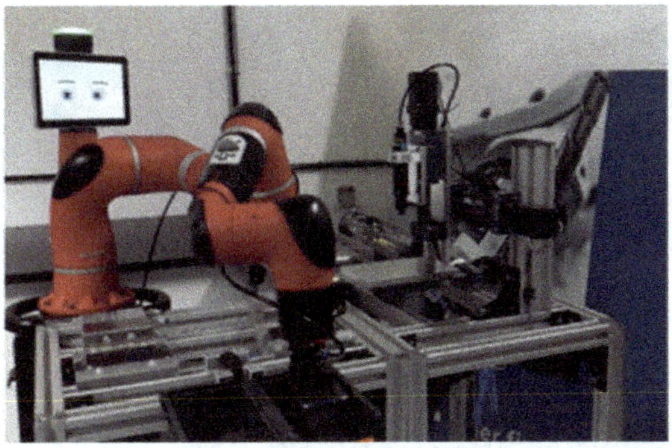

Fig. 1. Disassembly machinery demonstrator with Sawyer-Robot. Source: sustainablySMART

A collaborative design, the interaction between robots and humans, makes it possible to combine disassembly steps in an effective and flexible manner. It is preferable when there are rapid changes in mobile phone models and the disassembly line have to be configured quickly, or when certain dismantling steps are too expensive to fully automate [6].

Today different mechanical designs of information technology devices like mobile phones or tablets exists on the market using different joining techniques like screws, clips and adhesives. Schischke et al. revealed very large differences in the complexity of the devices, quantified by a significant increase in the number of work steps required for dismantling [7]. The (analysed) performance of a whole disassembly line is essentially influenced by these mechanical designs. Every mobile phone variation and component has its own dismantling process which could be realised either manually or automatically.

When dismantling a mobile phone, the following steps need to be performed: Cover removal: opening the mobile phone either mechanically by loosening snap-fit fasteners (if it is a non-glued cover), or by detaching glued covers or screens via heating; Battery extraction: mechanically or via heat utilization; extraction of internal covers and PCBs: unscrewing and removal of electrical parts like vibration motors, speakers, microphones and cameras.

To fulfil these disassembly steps listed above, three modules were developed within the sustainablySMART project. A clamping module for mechanical extraction steps, a heating module to dissolve glue, and a unscrew module for the extraction of the PCBs.

The technical viability has been demonstrated by designing and realizing flexible dismantling modules being able to dismantle the different mobile phones.

Figure 2 illustrates the three basic dismantling modules to fulfil these disassembly steps listed above. A clamping module for mechanical extraction steps, a heating module to dissolve glue, and a unscrew module for the extraction of the PCBs. The disassembly line can consist of various arrangements of these modules beside collaborative robots and workspace for humans.

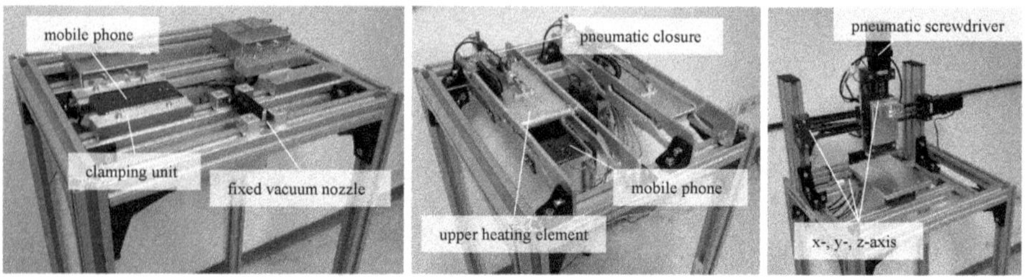

Fig. 2. Dismantling modules for a flexible mobile phone disassembly process: clamping module/heating module/unscrew module. Source: ProAutomation

(1) *Clamping module:* The clamping module offers two clamping units. Such a clamping unit is similar to a bench vice, and holds a smartphone while for example the handling robot removes the back cover and extracts the battery. To increase the versatility, it also offers a fixed vacuum nozzle, which can grip phones that does not fit into the clamping unit.
(2) *Heating module:* To detach glued covers, screens, or batteries, the heating module offers two heating units. A phone is pushed beside two elastic heating elements, and pneumatic closure ensures a good heat transfer.
(3) *Unscrew module:* The unscrew module has a pneumatic screwdriver mounted on a linear x- and z-axis. The clamped phone moves via a third axis in the y-direction. With these setup internal covers and PCBs can be detached mechanically via unscrewing.

The next task is to scale the demonstrator from the prototype level up to series applications. This will be achieved through digital simulation.

2 Objectives

The objective of the study is to define a virtualise disassembly line with blocks and algorithms in order to perform a simulation of different line variants and disassembly scenarios. With this evaluation technical aspects as well as economic and environmental implications are simulated.

The focus for the modelling is on the finding of appropriate line setups for a collaborative disassembly process. The line consists of flexibly usable modules, collaborative robots and human workspaces. Depending on the arrangement of the separate modules, typical economic and environmental aspects like cycle time, workload, and process flow and energy consumption differ and are therefore simulated.

Questions to be answered are: How does the line setup look like to keep the costs low from an economic perspective, respectively how can such a line setup be found? How does a line setup look like regarding minimal environmental impacts? And how does the line look like if both economic and environmental aspects are considered?

3 Methods

This paper shows a method for an upscaling process of the collaborative disassembly line for mobile phones through simulation in respect to a line-view. While simulating operation periods using different market scenarios a lot of data are gained and analysed regarding peaks and surges of production efficiency.

Simulating industrial systems like production lines is a common method for testing scenarios and planning industrial systems. A virtual model representing as close as possible the real process which consist out of underlying algorithms is created and different experiments are carried out. Different simulation granularities enable to focus on different aspects. Since a real prototype of the dismantling modules was built, and their cycle times as well as their energy consumptions were measured in detail, the following work simulates disassembly line variants on a higher abstraction level. The modelled dismantling modules (e.g. unscrewing unit) are blocks, having inputs and outputs and internal algorithms which will be pre-defined. By using these blocks, serial, parallel and combined line arrangements will be simulated and compared. Upcoming bottlenecks and appropriate countermeasures (such as additional parallel dismantling modules) can be weighted by the throughput rate and the energy consumption.

The methodical approach for the environmental evaluation is built upon the ISO 14955 standard [8] regarding the energy measurements of machine tools. First these measurements are used as input for the line simulation to gain the "environmental results" for each simulated scenario, and additionally they are used to identify the most energy intensive process steps and the major consumers. Once these environmental weak points are detected, this can be used as a starting point for further improvement of the dismantling modules and the dismantling process. For example reducing the energy consumption by using more efficient dismantling technologies or changing the dismantling process.

3.1 Upscaling Through Simulation

To encounter pros and contras of various collaborative disassembly line layouts a simulation is necessary. To achieve that, simulation tools like "FlexSIM/Visual Components" are used. The dismantling process is virtualised in three areas, so that an analysis of different line layouts can be realised flexible and statements about cycle time, workload and environmental impact can be made. These three sections are shown in Fig. 3 below.

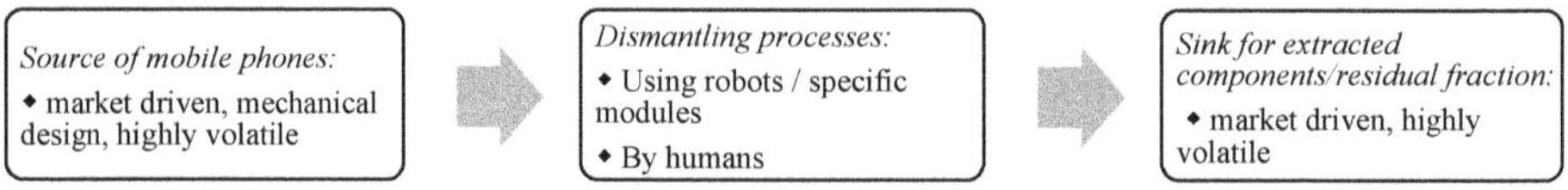

Fig. 3. The three basic areas for simulation

One of the main uncertainties in the dismantling of phones is the market driven source of old, disposed mobile phones. Depending on the phone model and its mechanical design, different dismantling steps are needed and as a result different dismantling equipment (modules) is needed. Following the idea that smartphones are exchanged at least once every three years [9] the sales numbers of those years within Europe are used as an input.

This means the flexible line layout (robots, humans and changing module arrangement) is realised in dependence of the supplied mobile phones. Additionally, the workload of the human worker is analysed and possible cycle time optimizing redundancies of certain dismantling modules could be tested in dependence of the mechanical design of supplied phones. Furthermore, the cycle times of every station in contrast to the whole cycle time is analysed.

The relevant outputs of the dismantling process are the battery (pollutant), the printed circuit board and the residual fraction including displays and other small parts such as vibration units or speakers. Within the simulation two different scenarios are analysed: in the first scenario only the extraction of the pollutant (battery) is analysed and in the second one the dismantling of the phone. This approach is followed since legislation does require the separation of batteries before shredding the rest of the smartphone. The second scenario is analysed to enable reuse or advanced recycling to fulfil possible future requirement from an upcoming enlargement of the ecodesign directive (resource efficiency) [3]. Another advantage of this simulation is a better understanding of the availability of high quality fractions of different materials for the recyclers and its economic implications and the possibilities of new resale business cases.

Following this strategy and abstracting the dismantling line to blocks means that the simulation can be faster done and a longer duration can be evaluated. But the abstraction also means blurring the exact results (like for the cycle time). Nevertheless, the abstraction is done with respect to the measurements of the built-up prototypes (cycle times and energy consumptions), and knowing that the difference in cycle time and energy consumption is minor between different phone models.

3.2 Environmental Assessment

When assessing the environmental performance of a product, usually a so called Life Cycle Assessment acc. to the ISO 14040-44 framework is carried out [10]. A product generates emissions and consumes resources through its entire life. As a result, such an assessment must cover the product's full life cycle (raw materials, manufacturing, distribution, product use and product's end of life [11]. When identifying the environmental weak point of machine tools and also automated manufacturing lines, in most cases the energy consumption of the use phase will stick out, their typical profile is "use-intensive".

Therefore the ISO Standard 14955 addresses just the energy efficiency of machine tools during the use stage, i.e. the working life of the machine tool. Although the dismantling modules are not a typical machine tools like a milling machine or a turning lathe, it seems useful to follow these standard. Beside the environmental evaluation, also improvement potentials can be found and the energy efficiency of the disassembly machinery can be maximised for the planned industrial implementation. More potential benefits acc. ISO 14955 are: a higher energy efficiency cause cost reduction in machine tools operations, attraction of environmentally conscious investors, an increased competitiveness and increased knowledge about the product.

Going one step further, these environmental results serve as an additional input for the upscaling simulation as already mentioned before. This way, not only the most efficient line variant from an economical and technical point of view, but also from an environmental point of view can be identified. To get the relevant energy consumption data for the simulation input, the main procedure using the ISO 14955 Standard is summed up in 4 tasks below: defining the system boundaries, defining the generalised machine tool functions, mapping those functions to the components and parts and performing the measurements.

(1) *System boundaries:* According to the ISO standard system boundaries and the relevant energy flows which will be measured have to be defined. In this study, two energy flows remain relevant, compressed air and electrical energy (heat- and air-exchange were neglected).
(2) *Generalised machine tool functions:* According to the ISO 14955 a functional approach for describing the machine tool is used. The ISO standard suggests 6 standardized functions: machine operation, process conditioning, workpiece handling, tool-handling/die change, recyclables/waste handling and machine cooling/heating. Because the disassembly machinery defers quite a bit compared to a typical machine tool, only two out of 6 generalized machine tool functions stays relevant: machine operation and workpiece handling.
(3) *Functional mapping:* Because the energy consumption per machine tool function is of interest, the components (and measurements) are assigned to these functions.
(4) *Energy measurements:* There are two different ways to access energy consumption data, performing measurements or using data from datasheets. Generally speaking, declared energy consumptions in datasheets are often max or average ratings and do not match the actual situation.

In this study, all measurements were performed on-site at ProAutomation in Vienna. If necessary, multiple measurements were performed and the mean value was calculated. In some cases the power demand of the single components were determined from the module's power consumption. In addition to gather data from measurements, assumptions based on datasheets were taken.

4 Results and Discussion

4.1 Selection of Simulation Abstraction

For the underlying process a prototype was built-up and measurements were taken regarding cycle time and energy consumption for two different phone types (Samsung S3 and S6). Those measurements showed that mobile phones with a different mechanical design have a slightly different cycle time since the battery was once pulled out by a robot and one pulled out by a human worker but in general the overall output was almost the same. The energy consumption of the robot and the dismantling stations is also not differing too much. The underlying reason is the built-up of phones and materials used.

Mobile phones are either structured from the display side or from the battery side, meaning that the head of screws are either orientated towards the back or towards the front of the phone. This is a big difference from the mechatronic point of view for the automation system but no difference in cycle time since the equipment for both variants is almost equal.

Regarding the different materials used mainly the glue or screws, which are holding the battery in place, are making a difference. On one hand batteries in three to four-year-old phones are either screwed to the phone or glued to the phone. In case the batteries are screwed an unscrewing station is used with a shorter cycle time and higher investment for one station and in case the batteries are glued a heating station is used with a longer cycle time and lower investments. Investing in two heating stations (which are running parallel for a higher throughput) results in the same costs as investing in one unscrewing station which leads to the fact that economical values are equal. Looking at the environmental impact it is clear that the heating stations needs significant more energy than an unscrewing station.

4.2 Basic Scenario for Point of Reference

The energy consumption needed to disassemble one mobile phone was calculated in a basic scenario. In this scenario the disassembly line layout is similar to the disassembly machinery demonstrator. The outcome serves as point of reference for the environmental simulation results.

In the basic scenario (Fig. 4) a mobile phone similar to Samsung S6 is disassembled. The robot takes the phone and pushes it inside the heating unit. The heating module heats up the phone and the adhesive dissolves. Then the phone is clamped in the clamping module and the robot pulls off the back cover and extract the battery. In the last step all screws fixing the PCB get loosened, afterwards the phone is moved further to a worker who separates the PCB among other electronic parts. As it can be seen in Fig. 2, the clamping- and heating module are designed for two phones each (two clamping and two heating mechanism) to achieve a higher through put. In the assessed basic scenario only one of each of these "stations" is used.

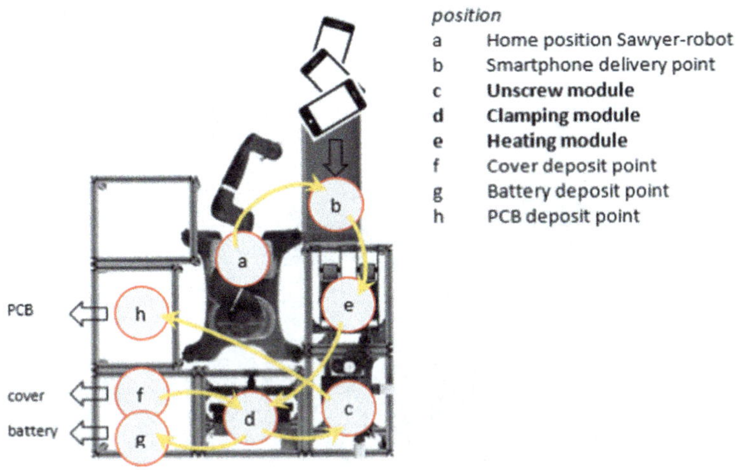

Fig. 4. Basic scenario, necessary movements for mobile phone disassembly scenario

Figure 5 shows the average power demand of every operating state (OFF/Standby/Warm-up/Processing). It is obvious that the standby state compared to the processing state needs quite some power (standby > 50% of processing). In the assumed shift regime scenario with only 2 h standby and 16 h processing it does not affect the result per phone as it would in a scenario with a longer standby state.

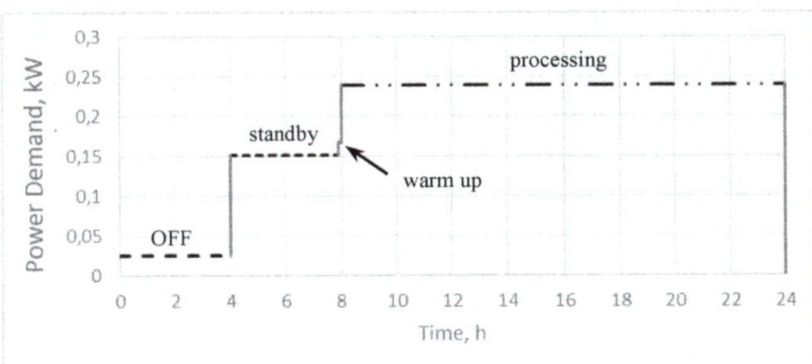

Fig. 5. Shift regime scenario, (average) power demand of operating states

(1) *Results - per type of energy:* The main type of energy used by the disassembly machinery is electricity as it can be seen in Fig. 6, on the right side. But the unscrew module needs for almost half of its energy demand compressed air (48% compressed air in total, rest 52% electricity). This is caused by the demand of the pneumatic screwdriver, which needs about 86% of the total compressed air demand of the unscrew module. In the chosen shift regime scenario, about 400 phones become disassembled per 24 h (145 s per phone, cycle time disassembly machinery). According to the shift regime including the energy consumption

during the operating states OFF, Standby, and Warm-up the overall energy consumption to disassemble one phone adds up to 11,4 Wh (without any attention to the data accuracy). With attention to the data accuracy and assumptions, the upper limit of energy demand per phone disassembly is in a range between 10 Wh and 17 Wh.

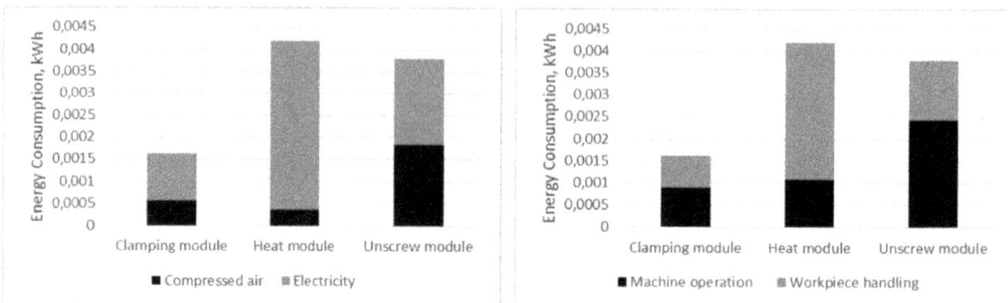

Fig. 6. Energy consumption of modules processing one phone. Left: Types of energy/Right: Machine tool functions

(2) *Results - per generalized machine tool functions acc. ISO 14955:* As suggested in the ISO 14955 Standard the energy consumption is identified per (generalized) machine tool function. The energy consumption of the two remaining generalised machine tool functions (workpiece handling, machine operation) are showed per module in Fig. 6, on the left side. The heat module needs about 74% energy for workpiece handling, caused by complex movement required when clamping or removing a smartphone.

5 Conclusions

Within this paper the topic of upscaling disassembly lines and it's surrounding is described, leading to the necessity of simulations. Simulating different layouts and different market situations means simulating a high variety of inputs. Within this paper it is shown that different mobile phone types have to be evaluated and the huge variety of phone types have to be considered. To get an economical effective line layout the flexibility for future phones has to be guaranteed due to the fact that the marked situation is rapidly changing. Valuable inputs for the simulation are selling numbers of previous years since those numbers give indications of the upcoming market situation. With this knowledge the existing machine layouts and combinations of modules could be extended years before the recycling of the mobile phones or adapted to the changed circumstances. By using simulations, that developing stage is improved. Set-up times of automation equipment play a big role in the final simulation results and also have to be minimized in collaborative approaches.

Future work will have to focus on possible assessments that allow to evaluate results of simulations regarding economic and environmental results. For example, it is possible that one line layout delivers high throughput and high efficiency but low scores on environmental aspects like energy consumption coming from a higher use of electrical energy for the generation of heat or compressed air needed for the mobile phone disassembly steps.

As a next step, the simulation will get carried out in the upcoming moths. The templates for the blocks will be programmed, and the various scenarios will be build and analysed to gain in-depth knowledge about the economic and environmental aspects of disassembly line variants in real life series applications.

Acknowledgement. The research leading to these results has received funding from the European Union's Horizon 2020 research and innovation program under grant agreement No 680604.

References

1. European Commission: Communication from the Commission to the European Parliament, the Council, the European Economic and Social Committee and the Committee of the Regions: Closing the loop - An EU action plan for the Circular Economy, Brussels (2015)
2. Ellen MacArthur Foundation, SUN, McKinsey Center for Business and Environment (Hg.): Growth Within: A Circular Economy Vision for a competitive Europe (2015)
3. Directive 2012/19/EU of the European Parliament and of the Council of 4 July 2012 on waste electrical and electronic equipment (WEEE)
4. CEN and CENELEC: New standards on material efficiency aspects for ecodesign (2019). https://www.cen.eu/news/brief-news/Pages/News-2019-012.aspx
5. Sustainable Smart, Mobile Devices Lifecycles through Advanced Re-design, Reliability, and Re-use and Remanufacturing Technologies, Project ID: 680604 (2019). https://www.sustainably-smart.eu
6. Ambrosch, R., Ambrosch, E., Pamminger, R., et al.: Automation of smartphone disassembly: collaborative approach. In: International Conference on Going Green CARE INNOVATION 2018, Vienna (2018)
7. Schischke, K., Nissen, N.F., Stobbe, L., et al.: Ansätze zur stofflichen Verwertung von Tablets aus Sicht des Produktdesigns. In: Thome-Kozmiensky, K.J. (ed.) Recycling und Rohstoffe, Bd.7, pp. 217–230. TK Verlag (2014)
8. DIN EN ISO 14955-1, Machine tools – Environmental evaluation of machine tools – Part1: Design methodology for energy-efficient machine tools (2012)
9. Statista: Bitkom Research (2019). https://de.statista.com/statistik/daten/studie/514801/umfrage/umfrage-zum-alter-der-genutzten-smartphones-in-deutschland/
10. DIN EN ISO 14040-44: Environmental management – Life cycle assessment – Principles and framework (2006)
11. Wimmer, W., Züst, R., Lee, K.M.: Ecodesign Implementation: A Systematic Guidance on Integrating Environmental Considerations into Product Development. Springer, Dordrecht (2004)

Using Process Mining Approach for Machining Operations

Zeynep Altan[1(✉)] and Semra Birgün[2]

[1] Software Engineering Department, Beykent University, Istanbul, Turkey
zeynepaltan@beykent.edu.tr
[2] Industrial Engineering Department, Fenerbahçe University, Istanbul, Turkey
semra.birgun@fbu.edu.tr

Abstract. In the Industry 4.0 world, both service and manufacturing companies should review their systems and processes, remove any application that causes waste, ensure lean flow and change business models if necessary, in order to fulfill the requirements of this trend. Introducing Industry 4.0 on a problematic system or process might harm it enough to cause the company disappear instead of benefiting it. For applications correctly decided to be built upon a correct system, data flow must be accurate and timely. And at this stage, data amount that increases with process mining and complexity of the big data will be solved and more information will be obtained about real production processes and data. In this study, a prototype is developed using the data of a previously studied manufacturing research. This prototype handles only one phase of the manufacturing process and extracts all the initial possible pathways of this phase through process mining.

Keywords: Batch production · Event logs · Manufacturing process analysis · Process improvement · Process mining

1 Introduction

The documentations of the processes are substantial for many developers while developing any product in the software engineering discipline. Even though the working software instead of documentation is determined as the second rule of agile manifesto [1], today's software sector demands the documentation in solving complex problems [2]. Logs that are kept in the development stages of any product are a part of the management system and they also explain the roles of developers in the process. Logs about different activities carried out in the process in classic systems are not included. Systems developed over time have started to feature the activity models in the processes [3]. This architectural model has entered the world of software as Model Driven Software Engineering. From here, the transition to process models in which activity logs in the management system documentation processes took place.

N. M. Durakbasa and M. G. Gençyılmaz (Eds.): ISPR 2019, LNME, pp. 452–464, 2020.
https://doi.org/10.1007/978-3-030-31343-2_40

Process mining is a discipline in which business processes based on event logs are analyzed. The discipline has made a non-erasable entrance to the sector with a manifesto. Process mining manifesto was written by members and supporters of IEEE Task Force [4]. Its objective is the discovery, control and development of real processes from easily accessible logs from the current information systems. According to the manifesto, detailed information can be obtained from data sets with process mining methods and development of processes can be supported in such an ever-changing competitive business environment.

In process modeling, instead of assumed process values, recorded real event logs and performance related and compliance related analysis are made. The purposes of analyses may vary according to the business processes. Business processes can be generalized as manufacturing, distribution, logistics, supply chain, accounts payable, service management - IT etc. The purpose of the analysis, in other words the methods in this context may be the comparison of the process realized to the target process. Another main purpose of analysis with process mining methods can be to observe the current situation before deciding on the design of a new application. This requires the analysis and understanding of the current structure. Determining the most common paths and main insights about the process will illuminate the decision makers. Big projects that extract knowledge from event logs and their development tools [5] are partners of international organizations on works about various industrial subjects [6].

In his study, Van der Aalst [7] explained the modeling of processes in detail by using event data. Here, the relations between logs and models was explained with conceptualization and visualization. Van der Aalst's classification as log abstraction and model abstraction targets combining different types of behavior of processes instead of examining the different techniques of the processes. In this study, Petri Nets [8] and Structured Process Models [9] from activity sets are given as the examples of different process models.

Suriadi et al. [10] carried out a study on the process behavior of a big insurance company and in this case study, they employed the life circle of process mining model developed by Van der Aalst called planning, extraction, analysis, interpret, improvement. Techniques used for data analysis transform the problem that can be modeled in a complex structure into a simple process model that is easy to read.

Varbeek et al. [11] worked on process discovery, process conformance and process improvement in order to develop a tool named Divide and Conquer. They generally automated the process mining approach with working applications containing many activities.

Lu et al. [12] refined the event logs and found and removed the tasks to be duplicated in determining the related processes. The study was then applied in the hospital project. Knoll, Reinhart and Prüglmeier [13], Son et al. [14], Sani, Zelst and Van der Aalst [15], Turner et al. [16], Tax, Genga and Zannone [17] and Gupta [18] are some of the authors that work in the process mining field.

In the next parts of the study, after giving a brief information about the methodology aiming at the effective application of process mining to different projects, an earlier research done on manufacturing sector has been restructured using its data.

2 Applied Process Mining Methodology

There is not one single approach used in the process mining applications. However, what is aimed at the end of all different approaches is to increase the performance of the process and check the compliance of the process with the rules. One of the process mining approaches in project management is the examination in three steps of initialization, analysis and iterations and improvement [19]. The purpose of this methodology called Process Mining Project Management - PM2 is to make sure the processes of the systems developed are in compliance with each other. Thus, the performance of the system processes will increase. Applications in which this approach is used can be both purpose related problems and data related problems. Life Circle of Process Mining Projects (PMPL) [20] is the development of product as iteratively in the scale of the life circle process.

In the PM^2 methodology, during the analysis phase, the data in the processes of the current situation are used. For example, the conformance of these processes to the ideal behavior can be checked or done discovery analysis or done performance analysis can be carried out. The results of these analyses will naturally ensure the improvement of the process. The second important task of the approach in the planning phase is the study of the knowledge extraction stage. The most crucial part of the process mining is the determination of event data.

It is known that the analysis of the information that consists of the business processes in the process mining is different from the data mining techniques. Although large data sets, i.e. big data are processed in both approaches, data tables and the relations between these tables are analyzed in data mining. However, in process mining, the relations among data patterns are not extracted. Instead, process relationships in the data sets are analyzed. Another important feature of process mining is that working processes eliminate the data collection condition of the existing data set. Therefore, process mining is carried out with the performance analyses of the process model that strengthens the relationship between data set and business processes.

The extraction stage of the PM2 methodology is important in process mining. Clearing of this phase is impossible by deciding the granularity of the event data. Because data set that is compared as the event log will vary according to the characteristics of each application. The elements of the set may be quality, time, resource or cost according to the business process. Key attributes can be defined as caseID, activities and timestamps event logs. In various studies, it is possible to determine event logs with different techniques. For example, interview techniques such as asking questions in surveys are one of the preferred approaches in this process. So, the events can be decided with the question driven approach.

The purpose of the data processing phase is to obtain different forms of current event data of event logs. In other words, as event data is used as input in the data process model, event logs are used in mining and analysis phases. Mining/analysis stage that aims to increase the performance investigates the product's verification/validation (V/V) analyses for the improvement possibilities of processes. PM2 methodology that consists of main phases of initialization, analysis and iterations and process improvement will be taken as the fundamental approach. In this study, a similar flow will be observed and the related extractions will be decided through planning on the initialization phase.

3 An Application from Manufacturing Industry

The study takes the Ph.D. thesis [21] that analyses the manufacturing of parts in the manufacturing industry as resource. The study referenced is completed through different processes including raw material machining, heat treatment and grinding. Only the first phase of this flow referenced as remodeling is decided to be renewed with process mining. This flow is the generalized modeling of the transfer of raw material from the storage to the heat treatment operations in different machines. That modeling is a prototype. The reason for the review of the existing project with a process mining approach is to increase the performance of related business process and check its conformance to the given rules. In this context, the business process of the study will consist of the planning phase's inputs as an as-is system. This data is the current state data during manufacturing. Due to the structure of the followed methodology, the existing state's processes are used in the analysis phase.

In this prototype study, process mining is carried out with the aim of improving the machining process of the manufacturing company mentioned above. The company operates in the automotive industry and manufactures propeller shafts, pump drive shaft, power take-off shafts and their components and spare parts. Despite 3 shifts per day, the company needs 900 min/day overtime. The examination of the past manufacturing data shows that high level of variability in demand and manufacturing time cause imbalances in production and overtime is needed in order to rush production as a result of extension of lead time. As a result of this, the transition to lean production is initialized to improve the system, planning the manufacturing processes with more realistic data is decided and process mining is employed with the aim of collecting the correct data at the correct time.

Based on the company's past demands, the manufacturing process of the propeller shaft (which is the main product) has been selected as the pilot area for lean production process. Universal joint kits (UJKs) product family determined among 3 families after the cellular production studies on the process in which products are produced in discrete batches are the subject of this study. The production flow of the UJKs brought from the warehouse are carried out in three stages of machining, grinding and heat treatment and the produced UJKs are assembled to propeller shafts or packaged as the spare parts kit. The description of machines used in the application about machining processes is given in Table 1.

Table 1. Machinery to be used in process

Machine ID	Machine definition
M15	Semi-automatic lathe
M50	Drilling and tapping
M51	Special milling
M52	Special milling
M58	Manual press

The raw material to be examined for its transfer from the warehouse to the heat treatment are UJKs which are always ready for use. It is accepted that there is no limit for these UJKs and they are always ready at the warehouse. The batch size is 100 units. The first raw material batch from the warehouse will enter the queue of the first machine (M15) to be processed. After machining operations finish, the batch from the last operation will be first to wait to enter the furnace. As the furnace charge is 1000 units, the furnace entrance takes place after 5 batches are completed. Accumulating of the work-in-process (WIP) in front of the furnace makes the solution of the problem complex. It will be important to reveal and display both the WIPs in front of the furnace and the empty capacity waiting and/or overloading in the stages before coming to the furnace. The fact that these analyses will be determined for each machine will make the process mining important.

Following definitions are given that were taken from the system-as-is, in other words from the current system, in order to complete the information about the event logs (Fig. 1).

(i) Six different resources, i.e. routes P05, P12, P20, P24, P31, P41 are obtained from the existent system.
(ii) The system will complete an iteration when a batch with 1000 units is processed from storage to furnace. Each batch at the beginning has 100 pieces. They enter machining one by one and are carried to the other machine after getting their operation completed. In other words, parts (in batch of 100 units) that are transmitted to enter machining in the M15 machine from the storage come with one of these batch numbers.
(iii) P05, P12, P20, P24, P31, P41 have different paths and these also demonstrate the probable routes of parts in transition from Storage to the Furnace. For example, P12, P24, P31, P05, P24 and P41 routes are individually transferred from the Storage for machining in M15 machine. The transfer time from the Storage to the first machine-M15 is neglected. Then, the related parts will follow their own routes to reach the Furnace.
(iv) The following activities will take place in each communication between machines.
    (a) Transport - T1
    (b) Processing
    (c) Transport to Furnace – T3 (This activity only takes place during the process of moving the parts to the furnace).

In this study, an existent manufacturing application [21] is restructured using Disco tool [22] handling the action times of the production at each machine. Action times are taken as event logs. The results extracted from the probable event time of the current model are analysed. Resource annotations are given about the expression used by Disco in Table 2. Resources are the all probable paths from the beginning (Storage) to the target (Furnace). The flow of the routes (resources as used in Disco) can be followed from Fig. 1. Each route on manufacturing processes include all processing flows of different UJK:

(i) 50% of the UJKs that completed their turning operations in M15 machine are directly sent to heat treatment and accumulated before heat treatment.
(ii) 20% of the UJKs that are turned from the M15 machine are sent to the M51 special milling machine queue.

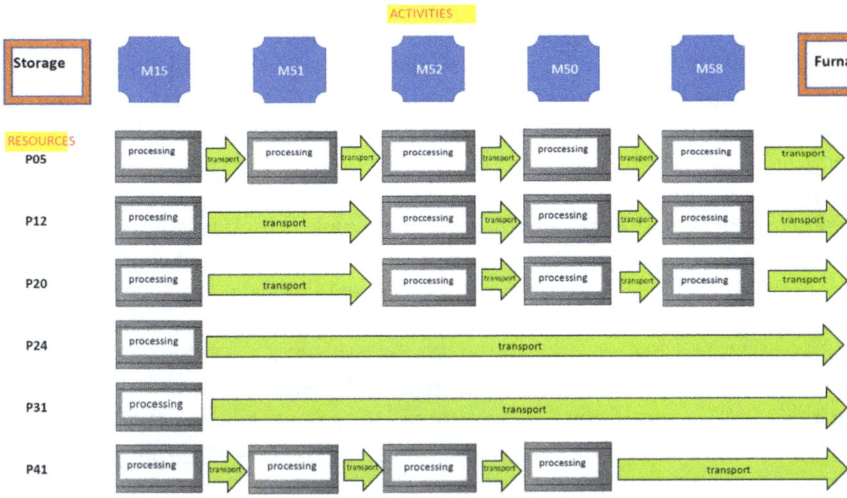

Fig. 1. Flow of resources from storage to furnace for each batch

(iii) Parts machined in M51 machine enter the queue of M52 machine which is the special milling machine.
(iv) Also 30% of the turned UJKs from M15 machine enter the M52 queue.
(v) Transactions in the queue which consist of 50% of the total are rotated to the M50 drilling and tapping queue.
(vi) Machining activity is carried out in M50.
(vii) 25% of the UJKs from the M50 machine are sent to the M58 manual press and 75% are sent to the heat treatment queue.
(viii) UJKs pressed on the M58 machine are sent to the heat treatment queue.

Table 2. Case IDs and their UJKs corresponding to resources for the first iteration

Resources	Case IDs	UJK (100 Cluster)
P05	000-006	1 2
P12	002-003-008	1 2 3
P20	013-014-009	2 1 3
P24	004-005-010-011-012	2 3 4 1 5
P31	015-016-017-018-019	3 2 1 4 5
P41	001-007	1 2

The probable distributions of different routes are explained as Case IDs. 20 different caseIDs are specified changing according to various UJKs. More clearly, with the processing of the existing data values, the number of probable different cases where the 1000 pieces initiated to be processed in batches of 100 in the M15 machine were obtained as 20. For example, for the P24 route 200 items are processes by caseID 004, 300 items are processed by caseID 005, 400 items are processes by caseID 010, 100 items are processes by Case ID 011 and 500 items are processes by case ID 012. P31 route complements the different items corresponding to caseIDs of P24 route. Each paths of P3 and P24 routes include 500 items in total. In this sequence, caseID 012 in P24 route and caseID 019 in P31 route are processed independently.

As explained in the previous section, the system in which process mining will take place will use the real data of the current system. Since the event logs are determined by the average processing times of the current study, there is no need to utilize any approach of PM2 methodology to determine the event logs as the initialization step. Table 3 is organized from the existing study to include average process times of machines of per item for different routes. Event logs for each route are obtained by processing these average times from an accepted start time. Processing time (mins/unit) has been distributed randomly for each route in each machine.

Table 3. Processing times for each resource (route)

Route	Mach.	Processing times (min/unit)														
P05	M15	5,17	1,67	0,95	2,37	**2,10**	1,59	4,05	3,16	3,00	2,56	1,69	4,29	0,95	2,58	4,5
P12		1,76	**2,27**	2,77	2,32	2,16	1,71	2,22	1,84	1,74	4,24	2,03	1,76	2,76	3,0	2,84
P20		1,48	2,3	2,44	1,69	3,99	**0,87**	2,20	2,76	2,7	3,91	0,92	2,16	3,91	1,48	2,3
P24		1,71	2,35	2,57	2,85	2,61	0,84	1,76	4,89	1,79	2,26	3,18	2,35	2,4	2,43	3,02
P31		1,81	0,89	4,17	0,85	2,22	4,84	2,20	2,39	2,65	2,92	4,09	1,69	2,21	1,79	1,77
P41		2,61	2,7	2,57	3,21	3,88										
P05	M50	1,55	1,48	1,88	1,71	1,80	1,50	2,0	2,25	1,50	1,86	1,43	1,64	1,50	1,68	1,42
P12		1,48	2,19	1,73	2,75	1,59	1,81	2,89	1,61	2,4	1,78	1,55	1,94	1,50	1,61	1,33
P20		1,68	1,67	1,6	1,67	1,92	1,35	1,80	1,72	1,62	1,42	1,70	2,22	2,25	1,64	1,68
P41		1,96	1,51	1,69	1,50	1,96	1,93	1,62	1,58	2,3	1,70					
P05	M51	2,75	1,42	1,5	1,59	1,68	1,58	1,68	2,25	1,74	1,9	2,7	1,5	1,5	1,7	2,0
P41		1,87	1,96	2,25	2,22	1,59	1,78	2,73	3,75	3,22	1,96	1,87	1,6	1,61	2,4	1,81
		3,4	2,3	2,3	1,93	1,95	1,81	1,62								
P05	M52	1,35	2,65	2,79	1,8	1,5	1,58	2,25	1,96	1,67	1,54	1,8	1,5	1,46	1,5	1,81
P12		2,3	1,61	1,67	1,54	1,18	1,98	1,42	1,59	1,86	1,93	2,4	1,67	1,42	1,5	1,72
P20		1,33	1,73	1,68	1,62	1,96	1,73	1,67	1,67	1,61	2,75	1,68	1,96	1,6	1,81	2,19
P41		1,35	1,55	2,20	1,5	2,0	2,25	1,59	1,61	1,8						
P05	M58	0,5	0,9	0,7	0,6	0,41	0,36	0,66	0,72	0,8	0,55	1,2	0,49	0,35	0,39	0,48
P12		0,32	0,67	0,53	0,38	0,59	0,58	0,33	0,21	0,4	0,64	0,21	0,98	0,96	0,8	0,42
P20		1,17														

The analysis results and performance evaluations of the model from the Storage to the Furnace for all possible events were obtained by using Disco tool. Visual solution of the processed data can be seen in Fig. 2. Figure 2-a shows the result of the three entities (processing, transporting and waiting), which are defined as activities according to absolute frequencies and mean processing times. Moreover, frequency values between the entities are also given. In Fig. 2b, the information taken from each machine and their entities are analyzed separately. It is also possible to obtain total, maximum and minimum duration times for each action and machine. These values will help to decide the best probable route after many iterations.

Table 4 and Fig. 3 gives detailed information about the statistics. Table 4 summarizes the results of Disco. An attribute called variant has been automatically created in Disco. This attribute shows how many actions are completed in each event. For example, Disco groups the case 010 and the case 018 with variant number 5. Variant 5 shows that the routes include 5 events from the Storage to the Furnace. While probable path P24 with case 010 continues 1 day 8 h, the other probable path P31 lasts 20 h 27 min with case 018. The similarity of both route is that they complete their paths with 5 different events.

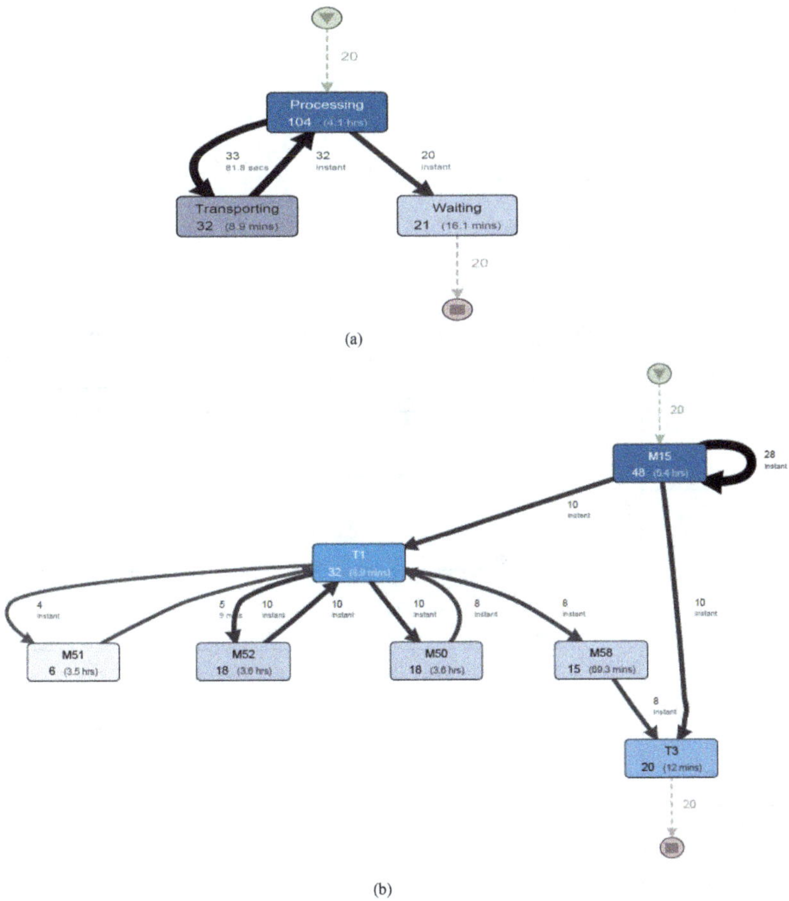

Fig. 2. Map of the activities between absolute frequency and mean duration (A) According to total actions (B) According to machines

Activity details of two different cases (case 003 and case 014) for one iteration is presented in Fig. 3. Processing of case 003 in the machine M15 finishes at 10.27 and the processing of case 014 starts 10.27 in the same machine. Because the route P12 and the route P20 complements each other as 300 items. Since the route P12 processes 100 items, all activities in each machine will be completed with one event. For example the lead time in machine M15 continues 87 min (0,87 min per item –Table 3). On the other hand, the route P20 processes 200 items, and all activities in each machine will be completed with two consequtive events. The lead time in machine M15 continues 210 and 227 min sequentially (2,1 and 2.27 min per item –Table 3). The total lead time of M15 machine with P20 route lasts 7 h 17 min. Transporting time is accepted as 5 min for 100 items, and 10 min for 200 items totally. Since the duration times mins/unit has been taken randomly, it is precision to take more then one iterations to decide the best route to process at machine M15.

Table 4. Distribution of events in terms of cases and their treatments

Variant Number	Event Amount	Case IDs	Resources (No. of Batches)	Processing Lead Time
1	8	001 003 013	P41 P12 P20	16 HRS 24 MINS 9 HRS 4MINS 14 HRS 47 MINS
2	12	002 014	P12 P20	18HRS 05 MINS 1 DAY 5 HRS
3	3	004 015	P24 P31	16HRS 17 MINS 10HRS 26 MINS
4	4	005 016	P24 P31	1DAY 1 HR 14HRS 44 MINS
5	5	010 018	P24 P31	1DAY 8 HRS 20HRS 27 MINS
6	6	012 019	P24 P31	1DAY 16 HRS 1DAY 1HR
7	2	011 017	P24 P31	7HRS 35 MINS 5HRS 23 MINS
8	10	000	P05	13HRS 47 MINS
9	15	006	P05	20HRS 21 MINS
10	16	008	P12	1 DAY 3 HRS
11	16	009	P20	1 DAY 20 HRS

Total processing times complementing each other with different routes (P05–P41 and P24–P31) for all other case pairs can be given similar to the Fig. 3, and it is also possible for other case pairs of P12 –P20 routes. Figure 4 summarizes the all statistics of the model.

Another example is given in Table 5 for a single iteration process of the 1000 parts product in which the processes are carried out. Parties following the P12 and P20 routes account for 30% of all processing batches and are processed in the M15 machine as given in Table 5.

Fig. 3. Total duration times of case 003 and case 014 for the routes of P12 and P20

Since 1000 pieces were required for the furnace charge, it has been stated that the raw material output from the warehouse was accepted as 1000 pieces. In the exemplary study, the batches were treated with three different probabilities as 500, 300 or 200 parts in M15 machine with P24 and P31, P12 and P20, P05 and P41 routine couples, respectively. As the analysis results of each route pair will be discussed in the model, the start time in the M15 machine is the same (02.01.2018, at 9: 00). For example, for the P12–P20 route pair, the first operations of the M15 machine start with P12 route. P12 starts the processing with 100 items and 200 items possibilities, respectively. In this case, P20 completes the activities of processing 30% (300 items) with 200 items and 100 items, respectively. In addition, two individual activities were started to process from the start time for 300 items by P12 and P20 routes, seperately. Thus, the model created 6 different case IDs for P12–P20 pairs: case 003, case 002, case 008 for P12 route and case 14, case 013, case 009 for P20 route.

Fig. 4. Global statistics of all cases

Table 5. Results of sample cases

Case	Total lead time
2 pieces P12 (Case 002) and 1 piece P20 (Case 013)	18 h 05 min (12 events - Case 002) + 14 h 47 min (8 events - Case 013)
1 piece P12 (Case 003) and 2 pieces P20 (Case 014)	9 h 4 min (8 events - Case 003) + 1 day 5 h (12 events - Case 014)
3 pieces P12 (Case 008)	1 day 3 h (16 events)
3 pieces P20 (Case 009)	1 day 20 h (16 events)

In the continuation of the study, all party groups will be considered at the same time and the flow which gives the shortest lead time will be calculated according to the results.

4 Conclusions

Process mining projects generally aim at the improvement of processes effectively. Therefore, the target of a process mining project is to achieve the time and cost reduction and to increase the performance of the system. The methodology accomplishes these criteria by using event data. Whenever event data is obtained as event logs, process model orders the activities in a process. Tool support of process mining methodology is also important to analyze the problem interactively, and to reduce the time-consuming.

This study gives a small process mining application analyzing probable paths of a manufacturing model. Different lead times of machine loads are evaluated. The data used in the model is taken on a variety of parts of an existent manufacturing system. This is a prototype application. It is aimed to determine the possible amounts of various routes in order to decide the transition times of different batches into the oven in an optimum time. The effective use of machines will be realized by analyzing the process times after the starting time of the case pairs according to different parameters.

In the system, which deals with the processing of raw material from warehouse to machining operations and up to the heat treatment, statistical results are obtained about the combination of batches of UJKs with different batch numbers in different possibilities. In later studies, detailed analyses of this prototype and the application of it to all processes in the production of raw material.

For big projects at different sectors process mining will soon become very important since it extracts knowledge for event logs recorded by the system. Nowadays, event logs are rarely used to analyze the underlying projects. Utilizing the process mining tool, it is possible both to control business processes and to discover their performance.

References

1. Manifesto for Agile Software Development (2001). https://agilemanifesto.org/
2. Flahiff, J.: Integrating agile in a waterfall world. Project Management Institute Global Congress 2011, Europe, the Middle East and Africa -EMEA, Ireland (2011)
3. Barrios, J., Nurcan, S.: Model driven architectures for enterprise information systems. In: International Conference on Advanced Information Systems Engineering CAiSE 2004. Lecture Notes in Computer Science, vol 3084. Springer (2004)
4. Process Mining Manifesto (2009). https://www.win.tue.nl/ieeetfpm/lib/exe/fetch.php?media=shared:process_mining_manifesto-small.pdf
5. vander Aalst, W.M.P.: Process Discovery from Event Data: Relating Models and Logs Through Abstractions (2018). https://onlinelibrary.wiley.com/doi/abs/10.1002/widm.1244
6. HSPI Consulting. Process Mining: A Database of Applications (2017)
7. van der Aalst, W.M.P.: Decomposing petri nets for process mining –a generic approach. Distrib. Parallel Databases **31**(4), 471–507 (2013)
8. Thamizharasan, R., Appavoo, K.: A review on software process mining using petri nets. Asian J. Appl. Sci. **9**(3), 131–142 (2016)
9. Leemans, S.J.J., Fahland, D., van der Aalst, W.M.P.: Discovering block-structured process models from incomplete event logs. In: Ciardo, G., Kindler, E. (eds.) Application and Theory of Petri Nets and Concurrency, PETRI NETS 2014. Lecture Notes in Computer Science, vol. 8489. Springer, Cham (2014)
10. Suriadi, S., Wynn, M.T., Ouyang, C., ter Hofstede, A.H.M., van Dijk, N.: Understanding process behaviours in a large insurance company in Austria: a case study. In: CAiSE 2013. LNCS, vol. 7908, pp. 449–464. Springer (2013)
11. Verbeek, H.M.W., van der Aalst, W.M.P., Munoz-Gama. J.: Divide and conquer: a tool framework for supporting decomposed discovery in process mining. Comput. J. **60**(11), 1649–1674 (2017)

12. Lu, X., Fahland, D., van den Biggelaar, F.J.H.M., van der Aalst, W.M.P.: Handling duplicated tasks in process discovery by refining event labels. In: International Conference on Business Process Management BPM 2016: Business Process Management, pp. 90–107 (2016)
13. Knoll, D., Reinhart, G., Prüglmeier, M.: Enabling value stream mapping for internal logistics using multidimensional process mining. Expert Syst. Appl. **124**, 30–142 (2019)
14. Son, S., Yahya, B.N., Song, M., Choi, S., Hyeon, J., Lee, B., Jang, Y., Sung, N. Process Minig for manufacturing analysis: a case study (2014). https://www.researchgate.net/publication/271910986
15. Sani, M.F., van Zelst, S.J., van der Aalst, W.M.P.: Repairing outlier behaviour in event logs (2018). https://www.researchgate.net/publication/325785319
16. Turner, C.J., Tiwari, A., Olaiya, R., Xu, Y.: Business process mining: from theory to practice. Bus. Process. Manag. J. **18**(3), 493–512 (2012)
17. Tax, N., Genga, L., Zannone, N.: On the use of Hiererchial Substrace mining for efficient local process model mining. https://www.researchgate.net/publication/321748426.2017
18. Gupta, E.: Process mining algorithms. Int. J. Adv. Res. Sci. Eng. IJARSE **3**(11) (2014). http://www.ijarse.com
19. van Eck, M.L., Lu, X., Leemans, S.J.J., van der Aalst, W.M.P.: PM^2: a process mining project methodology. In: Zdravkovic, J., Kirikova, M., Johannesson, P. (eds.) Advanced Information Systems Engineering, CAiSE 2015, pp. 297–313. Springer, Cham (2015)
20. van der Heijden, T.H.C.: Process mining project methodology: developing a general approach to apply process mining in practice. Master of Science in Operations Management and Logistics. School of Industrial Engineering, TUE, Eindhoven (2012)
21. Durmusoglu, S.: Tam zamanında imalat sisteminin simülasyon ile analizi ve uygulanabilirliğinin etüdü. PhD thesis, İstanbul Technical University, İstanbul, Turkey (1989)
22. Disco User Guide, Fluxicon (2018). http://processminingbook.com/reference.html

Quality Management

Increase the Success Rate of the Molds Due to the Measurement Mistakes in AKPA Aluminum

Dilaranur Çalışkan[1], Dilara Demir[1], Eda Gamze Aday[1], Kübra Türk[1], Bahadır Tunaboylu[1](✉), Selim Zaim[1], and Tuğçe Beldek[2]

[1] Department of Industrial Engineering, Istanbul Şehir University, Istanbul, Turkey
{dilanurcaliskan, dilarademir96, edaaday, kubraturk}@std.sehir.edu.tr, {btunaboylu, selimzaim}@sehir.edu.tr

[2] Department of Management Engineering, Istanbul Technical University, Istanbul, Turkey
beldek@itu.edu.tr

Abstract. The aim of the project was to increase the success rate of molds due to the measurement mistakes in AKPA Aluminum. Several problems of defects were identified by using 5S and six sigma methods and Pareto analysis, the quality function deployment (QFD ANALYSIS) with the help of Ishikawa Diagram. The project duration included the progress in four quarter periods using Tricolor that helped to collect all the data from the production line. Along with the data collection, they were analyzed for potential improvements. In addition, the results were provided on the tables with the grouping of the failure fields. Finally, the defects and reasons of the defective molds were analyzed and optimized using same methods.

Keywords: 5S · Six sigma · Ishikawa methods · Pareto analysis · Quality function deployment · Tricolor

1 Introduction

In the project, the reasons of mold failure is analyzed. The goal of the project is the decrease of the mold errors. Therefore, the production and success rate of the firm will be increased. Thus, the data is collected and analyzed with the methods and technics. 5S, six sigma, Ishıkawa methods, Pareto analysis and QFD analysis are provided to analyze the reasons of the defects and those methods are provided to how to solve and optimize the amount of the mold failures. In addition, Tricolor method is provided the process monitoring. In addition, the control chart, statistic, and grouping the field is researched and the improvement of the mold failure by looking data. The Tricolor program is helped to follow company needs in the testing part. Therefore, it is collected all system as a whole piece by following from order Entry to Loading, Dispatch List, and Invoicing.

N. M. Durakbasa and M. G. Gençyılmaz (Eds.): ISPR 2019, LNME, pp. 467–480, 2020.
https://doi.org/10.1007/978-3-030-31343-2_41

As a result, the manufacturing was influenced positively and costumer's satisfaction is increased according to results and improvement. In the end of the project, profit and success rate, customer satisfaction will be increased.

2 Literature Review

According to some data's from researches in AKPA AS, it is analyzed that failed products are mostly caused by mold failures. We optimized the mold failures with using some methods such as six sigma, 5S, Ishikawa methods that help us to finding reasons of production problems. Ishikawa diagram helps to find out main causes that leads with the main production problem [1]. The diagram shows both causes and sub-causes and sometimes it provides global risks [2, 4]. 5S is the beginning step of lean management and consists of five steps: Selection, Systematics, Cleaning, Standardization and Self Discipline [3]. 5S helps to organize the operation, where it is adapted, so that the process is improved in terms of time and quality. Six Sigma is another lean tool that improve customer satisfaction and business performance [5]. Lastly quality function deployment (QFD) is another method to decrease product or service design duration in order to meet customer needs [6].

In our last project, the defects of mold is fixed with using Tricolor program. The Tricolor is an ERP system that helps to follow company needs and it collects all system as a whole piece by following from order Entry to Loading, Dispatch List, and Invoicing and then provides necessary and useful information about the production line and the products to the authorities. Furthermore, about contract manufacturing like Eloxal, Dyeing, Overlaying and Mechanical Affairs, these can be followed up easily. Authorities can get the latest information about the products on the screen that chefs entered from the production line. Also daily capacity and balance of the workload can be determined and checked. Another point is while getting the reports from Tricolor, all modules and printouts has different formats and it can be converted to Excel, Word, Html, Jpg and PDF to send by email. The Program contains parts as listed; Customer Information, Die and Profile Descriptions, Billet Descriptions, Order, Warehouse and Process Waiting List Follow-up, Planning, Production Phase, Quality, Delivery/ Shipment. Thanks to Tricolor, project process can be analyze without needing anything else. Pareto analysis and QFD (Quality Function Deployment) modeling also used to produce more comprehensive, detailed and easier solutions to our problems.

3 Methodology

In the beginning of the project, it is observed all the production lines and all stages of the processes. In addition; it is investigated the problems that the company faces in the production line. It is peaked the problem in the mold room which caused by measurement issues. Different methods is used for which are Ishikawa Diagram, 5S and Six Sigma. After that, it is done calculations on the measurement part and searching where way to obstacle the issues. After that, it is worked with mold part on all presses from October of 2018 to now. AKPA A.Ş rate of deadline success is one of the important subject followed for customer satisfaction. The rate of mold success is a vital parameter for our project that affects deadline success. The aim of mold success rate in 2018 is 85%, even though the company below the goal. Making optimization on the parameters that affects mold success will be affect the increase of deadline success rate in a good way. The goal is set to reduce mold issues in all presses, caused by measurement mistakes from 2.3% to 0.7% in four months, which will make acquisition on incidental work rate. 12% of failure rate is caused by measurement mistakes.

Secondly, it is worked on mold room in Quality Control. By increasing efficiency, reaching the goal of customer order deadline rate, and rise the interrelationships between customer and company. In October 2018, the data of first six months, the mold success rate is 81%, failure rate 19%. 16% of failure rate is the lineated molds. (Total 3%, 0.4). 23.58% of lineated molds are collected on 'Orhan presses. (Total 0.716%). By making 70% optimization on 'Orhan press', we gain acquisition with 0.50% optimization.

In the last part of the project, it is focused to optimization part with implementation of the methods. In that part, Ishikawa, 5S, and 6 sigma methods provide to obtain reasons of the mold failures. In addition, the Tricolor program that helps us to follow company needs in the testing part. Therefore, it is collected all system as a whole piece by following from order Entry to Loading, Dispatch List, and Invoicing. After that, the program provides necessary and useful information about the production line and the products to the authorities. It can be followed up easily. It is obtained the latest information about all products on the screen that chefs entered from the production line. Moreover, it is determined and checked the daily capacity and balance of the workload step by step. According to program reports, the reasons of the optimization is obtained during to the testing part. Therefore, it can be analyzed the percentage of the improvement in the mold failures. In other words, it can be analyzed the percentage of the success rate of the failure production with aid of the Tricolor program. Using the Pareto and quality function deployment (QFD ANLYSIS) is provided to decide the most significant defects of mold. Therefore, it can be decreased the amount of the mold failures in the production. The Ishikawa Diagram of the project is shown by Fig. 1.

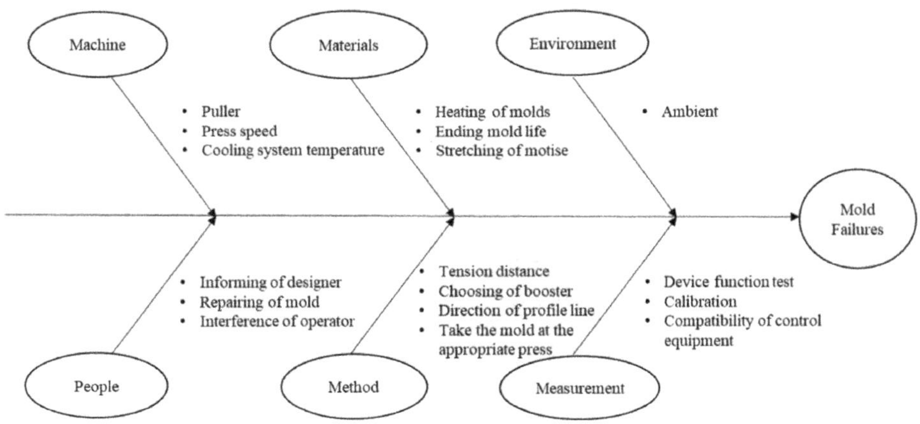

Fig. 1. Ishikawa diagram of the project

First of all, it is worked on reducing of lineated errors, which decrease the success rate of Orhan press's mold. According to the data in 2018, 16% of the molds in the production lineated from the line error and 23% of this error was collected at Orhan pres. As of October 2018, our aim of the project is the data success rate was 81% and the failure rate was 19%. 16% of these failures are striped patterns. 23% of striped descending molds were collected at Orhan Pres. By achieving a 70% improvement in Orhan Pres; it is achieved an improvement by achieving an overall improvement of 0.50%. Figure 2 shows types of the mold failures.

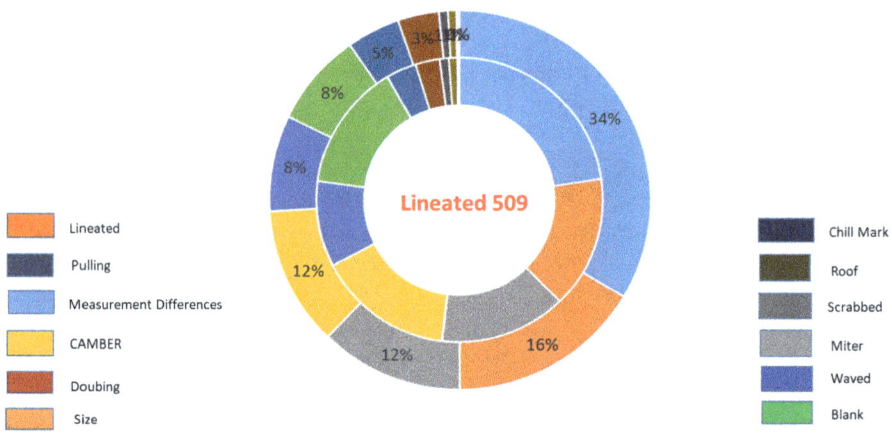

Fig. 2. Types of the mold failures

Mold failure types are represented in numbers in Table 1.

Table 1. Mold failures in numbers

Measurement differences	1038
Lineated	509
Size	379
Camber	368
Waved	255
Miter	246
Pulling	143
Daubing	108
Scrabbled	24
Roof/swelling	22
Chill mark	6
(blank)	1
Grand total	3099

Work Listed to be done for the problems that comes out from Ishikawa Diagram:

1- Print Speed
2- Cleaning of front surface
3- Infusion of dirt based on raw material
4- Temperature of billet
5- Temperature of mold
6- Mold did not clean properly
7- The fail of mold leveling
8- End of the "tenefer's" life span
9- Metals that sticks on over
10- More usage of graphite oil
11- Tenefer's surface fail cause by putting mold more on oven
12- Interval service cleaning
13- Committing wrong or more
14- Internal work length less than normal

In development part of the mold failure, it is focused on caused by defected product and defected products in which part.

3.1 Investigation in Material Part

(1) Mold temperature: While the same mold is looked at two different temperature, the effects of mold temperature on measure is analyzed. Table 2 shows size differences under different temperatures.

Table 2. Size differences under different temperatures

Mold temperature	Billet temperature	Size	Tolerance	Stated size
412	446	76.5	0.7	76.80–76.80
		34.3	0.5	34.14–34.16
		45.5	0.5	45.67–45.42
484	450	76.5	0.7	76.26–76.35
		34.3	0.5	34.68–34.41
		45.5	0.5	45.37–45.40
484	400	76.5	0.7	76.39–76.27
		34.3	0.5	34.38–34.33
		45.5	0.5	45.43–45.38

(2) Billet Temperature: Changes of billet temperature on measurement is investigated. Moreover, mold production with number 3962 in 24/11/2018, billet temperature is changed and examined the effects on measurement of mold. We determined the data as;

1. Sample: 450 °C 17.90 mm
2. Sample: 480 °C 17.46 mm
3. Sample: 410 °C 16.9–5 mm

(3) Feather Stretch: Generally, the production problems can be at production of front profile, so the reviews was done at front profile.

3.2 Investigation in Measurement Part

(1) Devices and Function Test: Errors of measurement that work profiles are measured and confirmed. In addition, tolerance of profile is checked. According to the standard of TS EN, miter tolerance is given as ±0, 40 mm and cap size for in the length 53,53 mm is given as ±0, 70 mm.

(2) Availability of Control Equipment and Control of Ended Production: Controls are done according to quality and control plans. This process is more critical to affect the production for profiles directly.

3.3 Investigation in Method Part

The investigation in method part was separated in five effects.

(1) Stretch Range: Effects of elongation distance' on measure changefulness were searched and analysed. In 19.11.2018 on Fatih Press, the mold with number 9670, stretch range is changed and measurement changes observed. When stretch range are 4.5–5.5–6.5, it is observed that there is no change. In 24.11.2018 on Fatih Press, the mold with number 10227 is taken and stretch range effect on measurement is observed again. 56.05 mm interval measuring ±1.4 mm tolerance current. After that, change observed in between 57.37 and 56.15.

(2) Mold Taken in Suitable Press: Examples have measure errors were searched from mold in inappropriate press by considering the like trunk diameter, parameters. Mold did not take suitable press and it is gotten the defected profile. Profile shaped as square due to production in wrong press machine.

(3) Choosing Booster: Effects of using the inappropriate booster on measure errors are analysed. When Mold with number 10245 go to the produce at Fatih Press, production started with number 26 although writing Booster with number 9 taking at mold card. 90° of the product was broken end of the press.

(4) Profile Flow Side: Walking surface of profiles are altered and the list is done. Then, poka-yoke was implemented at mold. There are no documents related to the molds produced with double pins or molds with walking direction changed. Double Pinned mold that we found in the achieve are given below that: 518, 4348, 5002, 8121, 1694, 748, and 3709. Molds having 518, 4348, 5002, 8121 are symmetrical molds. Therefore, walking on both surfaces can be carried out in a healthy way. Therefore, there is no need to close one of the pins in these molds.

(5) Graphite Use When Necessary: The values given in the table are molds which are produced by using graphite construction and these were determined from quality reports. Graphite molds were followed. Number of molds are given by days in Table 3.

Table 3. Number of molds by days

Date	Number of molds
14.11.2018	9
15.11.2018	8
17.11.2018	2
19.11.2018	2
20.11.2018	1
21.11.2018	3
22.11.2018	2
24.11.2018	5

3.4 Investigation in Machine Part

Press Speed of Operators Producing Profiles: According to the samples taken from the Kanuni and Fatih press machines, 431 profiles were taken from the product which was produced by the profile 2647 and their press speeds were determined. According to the samples taken from the Kanuni and Fatih press machines, 431 profiles were taken from the product which was produced by the profile 2647 and their press speeds were determined. Press speeds are given at Table 4.

Table 4. Press speeds

Operator	Press speed (M/Min)	Press
Yaşar Yazgan	9.4	Osman
Polat Şirin	8.5	Osman
Özgür Gürsel	6.7	Fatih
Ali Ocak	5.5	Fatih

3.5 Investigation in Human Factor Part

(1) Informing the Designer: It is very important that the designer is fully informed during the proposal stage. Critical measures, apparatus used, functional work should be clarified before design.
(2) Mold Repair: Since the molds which were revised were not followed systematically, it was determined that the first version of the design was used in the mold backups. The staff who made the mold repair will inform the mold card only as the process was done. The process is usually not written.
(3) Operator Intervention: The subjects that are in the initiative of the operator were identified and standardization work was started on these issues. (press speed, use of graffiti etc.)
(4) Proficiency of Operator: Trainings are organized for the personnel who perform measurement control and the deficiencies are completed.

4 Results and Discussion

Data of the failures in all production is gotten. The first and second processes show the failure type and number of mold. In first part of the project, it is analyzed the failure type of the mold and (1st and 2nd Process). In second part (3th Process) it is analyzed the different type of failure and it is examined improvement of manufacturing. (Ratio of improvements are on the table). These is improvement. Thanks to this improvement, production quality has increased. In addition, profit was provided in terms of cost. Accordingly, production has become more efficient. It is collected data of the failures in all production. The first and second processes show the failure type and number of mold. In first part of the project, it is analyzed the failure type of the mold and (1st and 2nd Process). In second part (3th Process) it is analyzed the different type of failure and it is examined improvement of manufacturing. (Ratio of improvements are on the table). These improvement. Thanks to this improvement, production quality has increased. In addition, profit was provided in terms of cost. Accordingly, production has become more efficient.

4.1 Pareto Analysis

Pareto analysis is used to differentiate the important reasons in a problem from the less important reasons. There is a rule known as 80/20 in the Pareto analysis. According to this rule, 80% of the results in most cases are caused by 20% of the causes. The Pareto principle is used not only to show a result, but also to help make decisions based on results. Table 5 shows the first part of the process, which is the analyze part and Fig. 3 shows the related Pareto chart.

Table 5. 1st part of the process: analyze part

First part of process		
Mold failure	Amount of mold	Percentage (%)
Height difference	1038	33
Stripe	509	16
Size	379	12
Curved	368	12
Wavy	255	8
Miter	246	8
Pulling	143	5
Daubing	108	3
Broken roof	22	1
Missing interwork	0	0
Tear	0	0

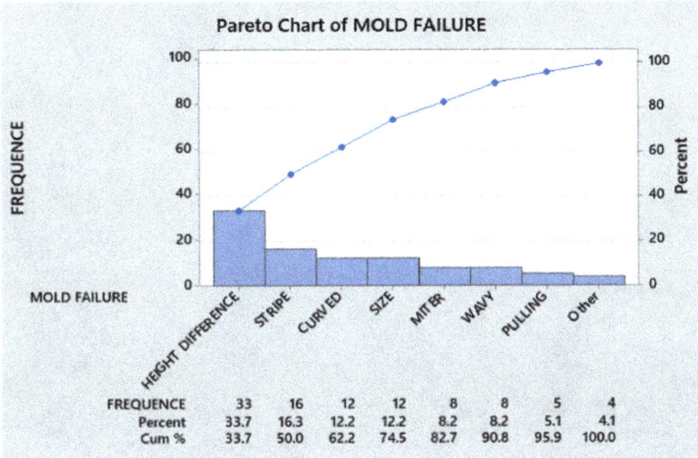

Fig. 3. First process' Pareto chart of mold failure

It is clearly shown in the graph that, first four problems are the most important issues of the process according to 80% rule. Since these are; height difference, stripe, curved. According to results, higher focus should be necessary on first four problems than others. These problems that caused to get failure mold and decreased the quality of the production, damaged profit of the firm. If the amount of failure production can be optimized; quality and profit of the firm will be increased. Thus, continuous improvement can be achieved. Table 6 shows the second part of the process, which is the test part and Fig. 4 shows the related Pareto chart.

Table 6. 2nd part of the process: test part

Second part of process		
Mold failure	Amount of mold	Percentage (%)
Height difference	649	24
Stripe	324	12
Size	291	11
Missing interwork	261	10
Wavy	202	7
Curved	185	7
Pulling	157	6
Miter	107	4
Daubing	77	3
Broken roof	76	3
Tear	74	3
Smash	47	2
Color	42	2

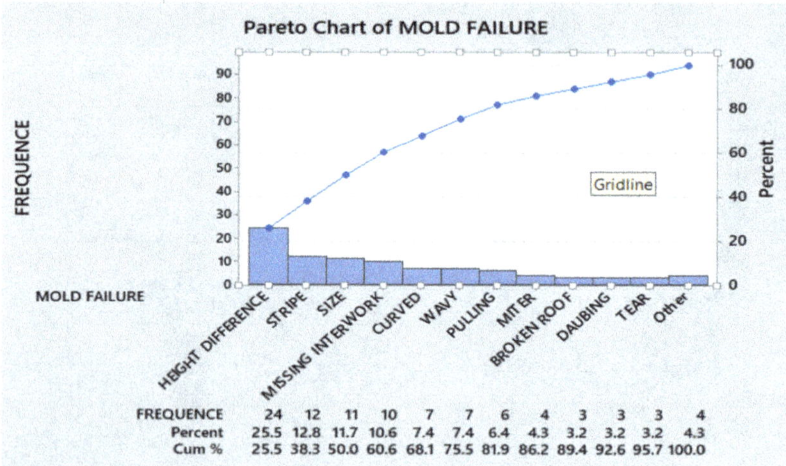

Fig. 4. Second process' Pareto chart of mold failure

The next part of the process is improved. When the two graphs are compared, it can be seen that the error rates are decreased. The frequency of height difference was decreased from 33 to 24, which was the best improvement in the process. Furthermore, there are improvement for other problems such as; stripe, size etc. According to the results of the second part is that our test part; more focus is given on how to optimized the failure product. Table 7 shows the third part of the process, which is after the test part and Fig. 5 shows the related Pareto chart.

Table 7. 3rd part of the process: after the test part

Third part of process		
Mold failure	Amount of mold	Percentage (%)
Height difference	649	21%
Stripe	324	11%
Size	291	10%
Missing interwork	261	8%
Wavy	202	3%
Curved	185	9%
Pulling	157	6%
Miter	107	5%
Daubing	77	4%
Broken roof	76	5%
Tear	74	5%
Smash	47	2%
Color	42	2%

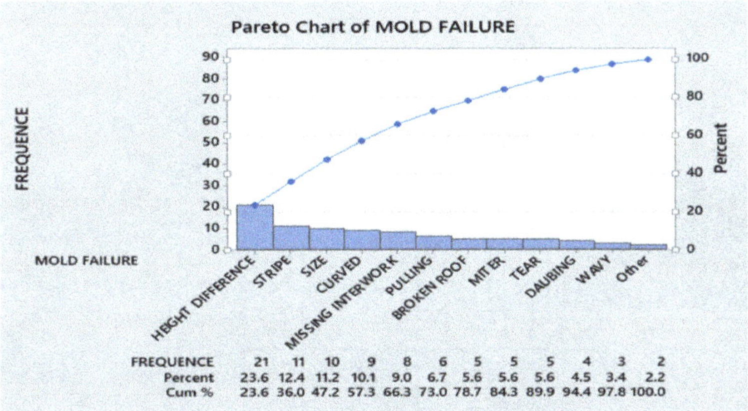

Fig. 5. Third process' Pareto chart of mold failure

In the last process, mold failures were improved at the best. The height difference has the highest percentage than others. The improvements were more applied for high difference. It can be seen from the test that the frequency of the mold failure had the lowest level in the three periods. It is observable from the graph that amount of the frequency of the height difference into failures was decreased from 33.7 in the first part to 23.6 in the last part. In the stripe failures was decreased from 16.4 to 11.2. When the graph size failures are analyzed it can be clearly seen that it was decreased from 12.2 in the first part to 11.7 in the second part then decreased to 11.2. It may be seen from that it can be optimized and decreased amount of the mold failures. It contributed to increased quality and profit of the production in the firm. If all analysis will continued to apply, success rate will increased. Also, in the end of the analysis, the rate of the mold failures can be equal to zero.

4.2 Quality Function Deployment (QFD Analysis)

QFD collects data of customer expectations and connect it to the customer voice with processes. It helps to translate us the customer requirements into specific offering specifications, furthermore it prioritizes possible offering specifications and make trade-off decisions based on weighted customer requirements and ranked competitive assessment. QFD has 4 deployment process. In this project, QFD model is used for quality control process. House of quality is conducted for QFD analysis and is given in Fig. 6.

In first part of process, mold failures were detected. As seen from the QFD model, mold errors and their causes are modeled. In order to find the relationship between the two, the importance of symbolizing their connections with each other is shown. Thus, the relationship between errors and causes has been analyzed and this analysis has made it easier for the factory to solve problems. Figure 7 shows the HoQ results of the first part of the process.

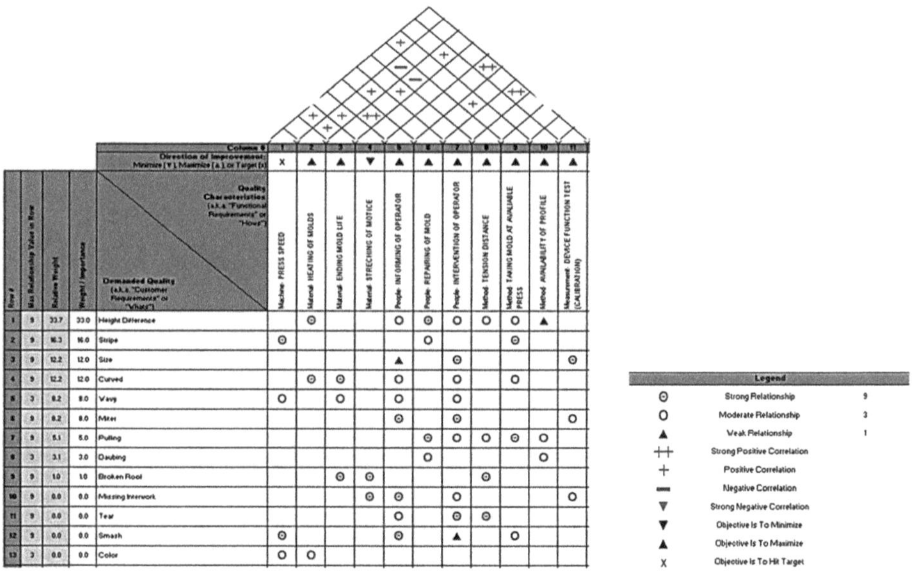

Fig. 6. House of quality (HoQ)

Max Relationship Value in Column	9	9	9	9	9	9	9	9	9	3	9
Weight / Importance	171.4	413.3	143.9	9.2	248.0	407.1	361.2	125.5	330.6	58.2	134.7
Relative Weight	7.1	17.2	6.0	0.4	10.3	16.9	15.0	5.2	13.8	2.4	5.6

Fig. 7. 1st part of the process

The data of the first process is located below the schema. The data of the second process prepared because of improvements are as follows.

Max Relationship Value in Column	9	9	9	9	9	9	9	9	9	3	9
Weight / Importance	162.8	303.2	118.1	124.5	295.7	335.1	346.8	153.2	277.7	54.3	150.0
Relative Weight	7.0	13.1	5.1	5.4	12.7	14.4	14.9	6.6	12.0	2.3	6.5

Fig. 8. 2nd part of the process

When the first improvements were made, it was observed that the data improved. Figure 8 shows the HoQ results of the second part of the process.

Max Relationship Value in Column	9	9	9	9	9	9	9	9	9	3	9
Weight / Importance	134.8	306.7	151.7	131.5	280.9	323.6	361.8	192.1	276.4	57.3	144.9
Relative Weight	5.7	13.0	6.4	5.6	11.9	13.7	15.3	8.1	11.7	2.4	6.1

Fig. 9. 3rd part of the process

Improvements have been made as seen in the last process. For example, in measurement-device function test importance was increased from 134.7 to 150.0 then decreased to 144.9. Furthermore its relative weight was changed as 5.6, 6.5, and then 6.1 respectively to their quarters. All these data shows that the modeling has been improved and solutions have been successfully achieved. Figure 9 shows the HoQ results of the third part of the process.

5 Conclusion

The aim of the project is to reduce the error rates of mold defects in AKPA Aluminium. In the first part of project, it analyzed that failed product are caused by mold failures. After that the required data were collected and analyzed with 5S, 6 Sigma, Ishikawa methods for identify reasons of mold failures in the production line. The goal is to find the factors that reflects failed product caused by measurement problem. In addition, making profit from money, time, material and layout. Therefore, customer satisfaction, material and money profit increases. In the implementation part, Tricolor program was used to help monitor the company needs in the test part. Therefore, it collects all system as a whole piece by following from order Entry to Loading, Dispatch List, and Invoicing. In addition, information can be easily retrieved from the program. The program provides all information necessary in the production line. It can get the latest information on all products on the screen.

In addition, the capacity and work load balance of products can be determined and controlled. According to the program reports, the reasons for optimization are being sought in the test part. Therefore, the success rate of unsuccessful production is analyzed by the Tricolor program. In the last stage of the process, the development of the process with Pareto was analyzed. Pareto helped to find the most important problems in the method. The first four problems continue to reduce the problem rate. The method is then useful for improvement steps. Moreover, since the most important problems were reduced and quality improved, the analysis was performed with Quality Function Deployment Analysis. The problem level is based on current information and expert. Therefore, customer satisfaction has increased and the success of the production race has increased.

References

1. Basic Tools for Process Improvement, 3 May 1995. Balanced Scorecard Institute. http://www.balancedscorecard.org/Portals/0/PDF/c-ediag.pdf. Accessed 20 Dec 2009
2. Ciocoiu, C.N.: Managementul riscului. Teorii, practici, metodologii. ASE, Bucharest (2008)
3. Falkowski, P., Kitowski, P.: The 5S methodology as a tool for improving organization of production. PhD Interdiscip. J. **4**(1), 127–133 (2013)
4. Ilie, G., Ciocoiu, C.N.: Application of fishbone diagram to determine the risk of an event with multiple causes. Manag. Res. Pract. **2**(1), 1–20 (2010)

5. Summers, D.C.: Six Sigma: Basic Tools and Techniques. Pearson/Prentice Hall, Upper Saddle Rive (2007)
6. Kwong, C.K., Bai, H.: A fuzzy AHP approach to the determination of importance weights of customer requirements in quality function deployment. J. Intell. Manuf. **13**(5), 367–377 (2002)

Industrial Computed Tomography for Nondestructive Inspection of Additive Manufactured Parts

Binnur Sagbas[1(✉)] and M. Numan Durakbasa[2]

[1] Department of Mechanical Engineering, Yildiz Technical University, Istanbul, Turkey
bsagbas@gmail.com

[2] Department of Interchangeable Manufacturing and Industrial Metrology, Institute for Production Engineering and Laser Technology, TU Wien, Vienna, Austria
numan.durakbasa@tuwien.ac.at

Abstract. Powder bed fusion technology is one of the additive manufacturing methods that build up near net-shape complex parts from powder form of the material. Direct Metal Laser Sintering (DMLS) and Selective Laser Sintering (SLS) are primary powder bed fusion technology methods for metal and polymer materials respectively. AlSi10Mg is one of the most widely used materials in DMLS applications while polyamide 2200 (PA) is frequently used in high quality functional parts production and various prototyping in SLS applications. By developing technology, AM has begun to provide solution for generating final products with complex geometries. Dimensional and structural inspection of the AM parts without any destruction has become important issue in parallel with the development of the manufacturing technology. Industrial X-Ray Computed Tomography (X-CT) is one of the promising methods that provide opportunity to nondestructive inspection of inner structure and porosity of AM manufactured part. In this study, nondestructive porosity measurement of DMLS manufactured AlSi10Mg test part and SLS manufactured PA 2200 functional part were applied by X-CT method. Density of manufactured parts was calculated by measured volume data. Moreover, density of AlSi10Mg part was defined by Archimedes method and the results were compared with each other. It can be concluded that X-CT was an effective technique to define porosity and pore distribution of the AM parts in comparison with Archimedes porosity measurements.

Keywords: Additive manufacturing · Industrial X-CT · Non-destructive inspection · Porosity measurement · Quality control

1 Introduction

Additive manufacturing (AM) is rapidly growing technology that provides opportunity to create complex geometries layer-by-layer. Being suitable to integrate Industry 4.0 smart manufacturing system, it is seen as the production method of the future [1, 2].

N. M. Durakbasa and M. G. Gençyılmaz (Eds.): ISPR 2019, LNME, pp. 481–490, 2020.
https://doi.org/10.1007/978-3-030-31343-2_42

Direct Metal Laser Sintering (DMLS) and Selective Laser Sintering (SLS) are powder bed fusion technology based additive manufacturing (AM) methods that build up near net-shape complex parts from powder form of the material. By having good corrosion resistance, mechanical strength and thermal properties AlSi10Mg is one of the most widely used materials in DMLS applications [3]. Similarly, with its high mechanical and thermal resistance, polyamide (PA) is frequently used in high quality functional parts production and various prototyping in SLS applications. Moreover, PA is biocompatible and widely used in medical applications. PA products can protect their forms for a long time [4].

With development of novel manufacturing technologies, the need for new measurement systems and procedures for inspection of complex parts is increasing. AM has wide range of application area such as from automotive and medical sector to defense and aerospace industries [5, 6]. Surface texture characterization, geometrical and dimensional properties, density, hardness and porosity measurements become critical factors that affect functionality and service life of the parts. Porosity is the void fraction which defines a measure of the void spaces in material structure. It is generally reported as a percentage between 0–100% or as a fraction between 0–1. Measurement of porosity would be a destructive method if conventional metallurgical measurement methods were used. Because of the nature of powder bed fusion technology, each manufactured part contains different fraction of porosity which depends on the used material and process parameters. So porosity inspection would be necessary for each functional part like as dimensional control. But with destructive methods 100% inspection is not possible. Industrial X-ray computed tomography (X-CT) is one of the promising methods that provide opportunity to nondestructive inspection of inner structure and porosity of AM manufactured part [7–9]. Although many researches have been done in recent years, standardization studies are still in draft stage for X-CT measurement methods. By continuing research and recent developments it is becoming an appealing alternative both for structural, geometrical and surface topography measurements [10–12].

In this study, nondestructive porosity measurement of DMLS manufactured AlSi10Mg test part and SLS manufactured PA 2200 functional part were applied by X-CT method. Density of manufactured parts was calculated by measured volume data and weight of the parts. Moreover, density of AlSi10Mg part was defined by Archimedes method and the results were compared with each other.

2 Materials and Methods

2.1 Additive Manufacturing of the Samples

In this study, density and porosity measurements of additively manufactured AlSi10Mg and PA materials samples were achieved by Archimedes method and XCT technique. AlSi10Mg alloy samples were manufactured by DMLS with EOS M290 which uses 400 W Yb-fibre laser. 3D geometry of the samples was directly produced from CAD data under argon atmosphere by melting AlSi10Mg powders. The powder, produced by gas atomization method, was provided by EOS. The medium size of the powder is about 30 μm. Chemical composition can be seen in Table 1 as stated by the producer in its data sheets [13]. Also, SEM image of the powder and manufactured sample can be seen in Fig. 1.

Table 1. Chemical composition of AlSi10Mg

Al (%)	Si (%)	Fe (%)	Cu (%)	Mn (%)	Mg (%)	Ni (%)	Zn (%)	Pb (%)	Sn (%)	Ti (%)
Balance	9.0–11.0	0.55	0.05	0.45	0.2–0.45	0.05	0.10	0.05	0.05	0.15

Fig. 1. SEM image of the AlSi10Mg powder (a), photograph of AlSi10Mg sample (b)

PA sample was manufactured by EOS Formiga P110 3D SLS system that uses CO2 Laser with 30 W. Medium diameter of the PA powder was about 56 μm. Manufactured part can be seen in Fig. 2.

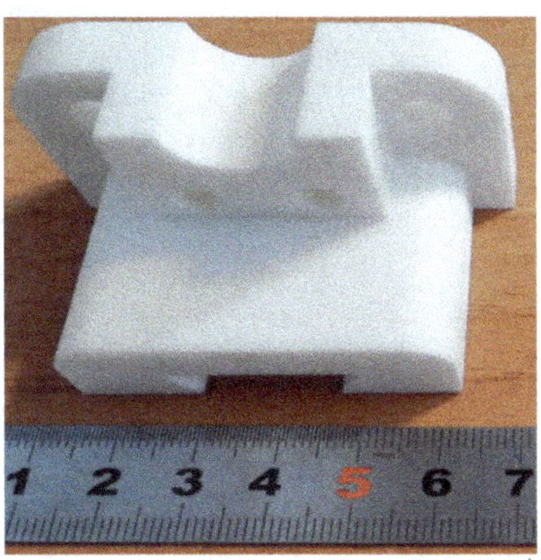

Fig. 2. PA 2200 manufactured part

2.2 Density Measurement

The Archimedes method was used for defining density of the metal sample. Principle of the method based on the difference of ascending force between the test sample's weight measured in air and submerged into a reference fluid [14]. In this study, density of the sample was defined by an electronic balance Precisa XB220A which works according to the Archimedes principle and give directly density value of the sample. Ethanol was used as the reference fluid. The measured density was compared to the sample material's nominal reference density for calculating the porosity values.

2.3 X-ray Computed Tomography Measurements

XCT Porosity measurements were taken by Werth Tomo Scope Technology XS in the High Precision Measurement and Nano Metrology Laboratory at the AuM-TUWien (The Department of Interchangeable Manufacturing and Industrial Metrology/ Production Metrology and Quality, Vienna University of Technology). Photographs of the XCT system and the sample, fixed on rotating table can be seen in Fig. 3.

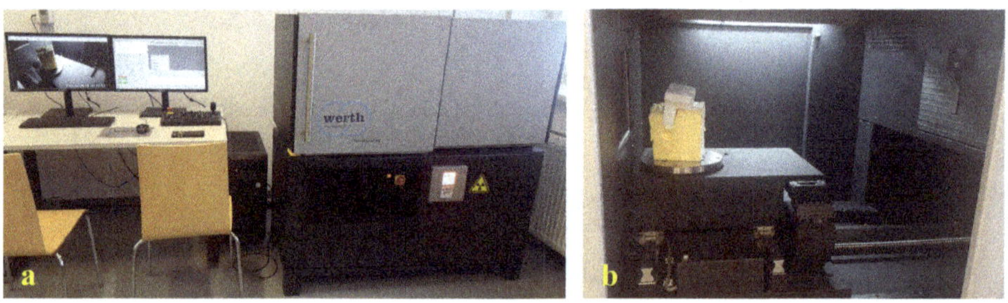

Fig. 3. Werth Tomo scope technology XS in the high precision measurement and nano metrology laboratory at the AuM-TUWien (a). The XCT system and the sample fixed on rotating table (b).

XCT provides three-dimensional representative geometry of a workpiece by taking number of X-ray images around the rotation axis of the workpiece. The system contains an X-ray source tube and a detector [15]. A 360° rotating table placed between X-ray source and detector. During measurement process, the workpiece is fixed on rotating table and radiographic images are recorded by the detector at different angular positions. At the end these images construct 3D model of the workpiece by using proper algorithms [9, 16]. Properties of Werth Tomo Scope Technology XS X-CT can be seen in Table 2.

Table 2. Properties of Werth Tomo scope technology XS X-CT

Machine type	Coordinate measuring machine with CT device
Probing systems	Compact X-ray sensor
Modes of operation	Linear path control
Measuring software	WinWerth®
Voltage measuring machine	230 V (115 V) ±10%
Frequency	50–60 Hz
X-Ray source	Transmission target: 130 kV or 160 kV
Resolution	0.1 μm
Detector	Surface area: from 65 × 57 mm^2 to 146 × 114 mm^2 Number of pixels: from (1024 × 1024) pixel to (2940 × 2304) pixel
Maximum Permissible Error	P: 7.5 μm E: (7.5 + L/50) μm (Where L = measuring length in mm comparable to ISO 10360 and VDI/VDE 2617)

Measurement parameters of the X-CT system have to be arranged according to dimensions and density of the workpiece. Resolution of the image decreases by increase in workpiece dimensions while it reduces magnification of scan. So voxels sizes of the scan increases and image quality decreases. Moreover, it is difficult to take X-CT measurements on high density material workpiece than low density one because of the poor X-ray penetration [9]. Therefore, different X-CT scan parameters were selected for measurement of AlSi10Mg and PA2200 parts. Measurement parameters can be seen in Table 3.

Table 3. X-CT scan parameters

Parameter	AlSi10Mg	PA2200
Voltage (kV)	148	146
Current (μA)	336	408
Filter material	Silver	Copper
Filter thickness (mm)	1	0.2
Voxel size (μm)	8.57	34.07

3 Results and Discussion

The application of X-CT for non-destructive inspection of density and porosity measurements of AM parts has been studied by different researchers. It is reported that X-CT was promising technique for nondestructive inspection of the AM parts in comparison with SEM surface analysis and Archimedes porosity measurements [14, 17]. Porosity and density measurements results for both AlSi10Mg and PA2200 samples were reported in the Table 4.

Table 4. Porosity and density measurement results

	AlSi10Mg		PA2200	
	Porosity, %	Density, g/cm^3	Porosity, %	Density, g/cm^3
Archimedes	0.075	2.67	–	–
X-CT	0.040	2.63	0.323	0.9784

Density of AlSi10Mg was recorded about 2.67 g/cm^3 by Archimedes method. Porosity of the sample was calculated by using theoretical bulk density as 2.68 g/cm^3 [18] and it is defined as 0.075%. In X-CT measurements, total volume of the AlSi10Mg was measured as 996.3742 mm^3 while the volume of the pores was recorded as 0,401 mm^3. Therefore, relative porosity of the part was calculated by the ratio of the total porosity volume to the total sample volume and it was found as 0.040% Density of AlSi10Mg was recorded about 2.67 g/cm^3 while porosity of the sample was calculated as 0.075% by Archimedes method [19]. In this study, for calculating the real density of the sample XCT result for volume without pores and weight of the sample in the air, recorded as 2.6174 g, were used and the density was determined as 2.63 g/cm^3. X-CT scan results can be seen in Fig. 4.

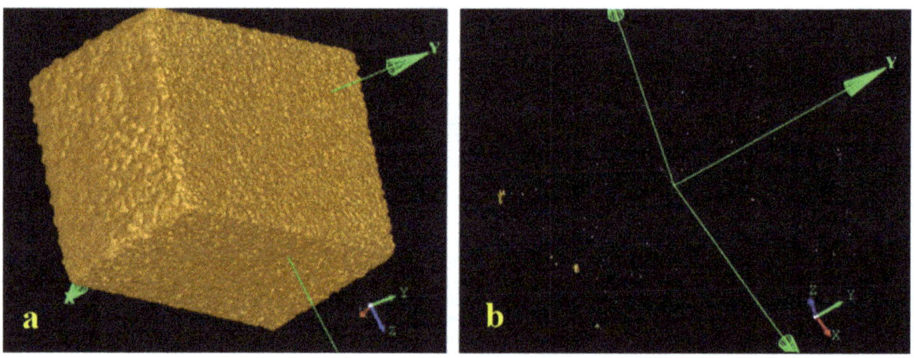

Fig. 4. X-CT scan results; total geometry of the AlSi10Mg part (a), distribution of the pores (b).

However lower density value was resulted by high porosity of the part, higher porosity and higher density values were measured by Archimedes method than X-CT. This may aroused by the precision of the used device and principle of measurement method. For absolute density measurement, Archimedes method may provide accurate results. But for calculating relative density and porosity percentage, nominal density of the material has to be used as reference which cannot be considered reliable for non-homogeneous parts as stated in the reference [19].

For PA2200 functional part, total volume was recorded as 41585.3176 mm^3 while the volume of the pores was recorded as 134.1885 mm^3 by XCT. Porosity of the PA2200 part was calculated as 0.3226%. So, the real density was calculated as

0.9784 g/cm^3 by using XCT result for the volume without pores and weight of the sample in the air, recorded as 40.5571 g. Because of the dimensions of the PA2200 functional part, it was not possible to define density by Archimedes method. X-CT scan results of PA 2200 can be seen in Fig. 5.

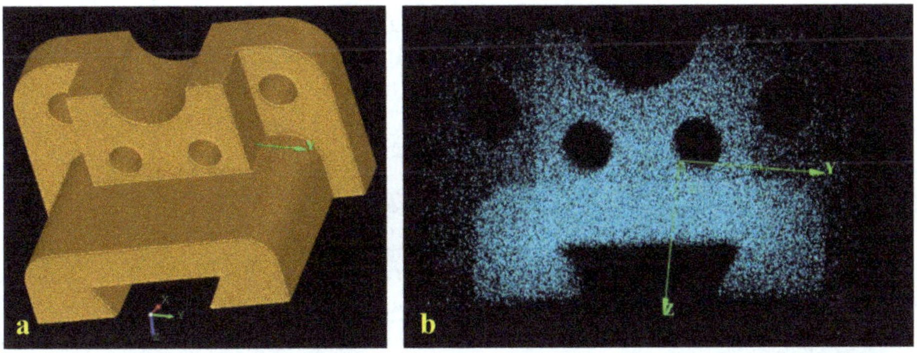

Fig. 5. X-CT scan results; total geometry of the PA2200 part (a), distribution of the pores (b).

However conventional methods give porosity volume of the sample, they don't provide enough data about localized porosity in the inner structure of the part. For functional parts, pore distribution is important factor that effects crack propagation and mechanical strength. Besides measurement of pore volume, X-CT provides opportunity to define pore distribution in different region of the part by taking sections from the data. Inspection of the functional part by sectioning measured geometry data can be seen in Fig. 6.

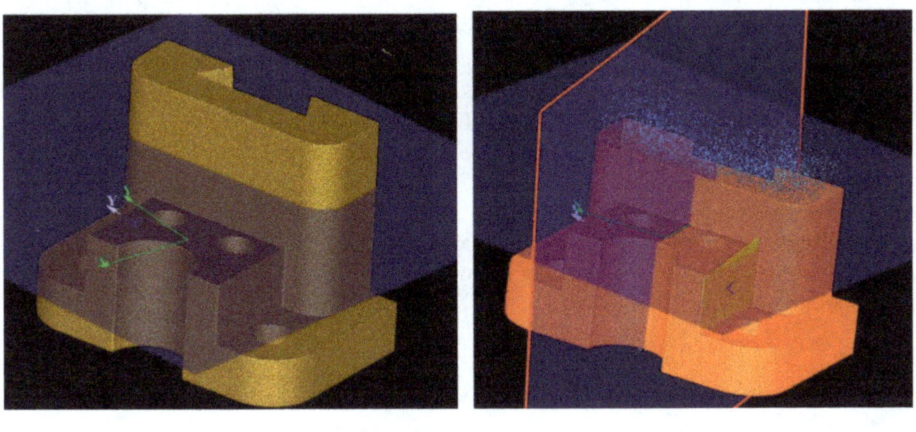

Fig. 6. Inspection of the functional part by sectioning measured geometry data

Archimedes method is relatively economical and quick method for nondestructive inspection of density and porosity of the parts. But it is difficult to assess pore distribution by this method. X-CT is an effective and promising method for defining size, shape and distribution of the pores beside their volume. The basic concern of the X-CT method may be the resolution of the system which affects the ability of the method to detect pores smaller than the minimum voxel size [9]. X-CT is rapidly developing system for providing solution to restriction of the method. In future studies, it would be better to focus on the optimization of the X-CT measurement parameters and uncertainty analysis on different materials to develop measurement procedures for additive manufacturing.

4 Conclusions and Future Perspectives

In this study, nondestructive porosity measurement of DMLS manufactured AlSi10Mg test sample and SLS manufactured PA 2200 functional part were applied by X-CT method. Density of manufactured parts was calculated by measured volume data. Moreover, density of AlSi10Mg part was defined by Archimedes method and the results were evaluated in terms of the abilities and restrictions of the systems. It can be concluded that;

- Archimedes method is relatively economical and quick method for nondestructive inspection of density and porosity of the parts however, it is difficult to assess pore distribution by this method,
- X-CT is promising technique for nondestructive inspection of the AM parts,
- X-CT provides opportunity for defining size and distribution of the pores in functional part,
- The precision of the method restricted by its resolution.

For further studies, it would be complementary to take repeated porosity measurements by X-CT method with different measurement parameters. Moreover uncertainty analysis would be helpful to develop measurement procedures for additive manufacturing. The studies are being continued at the High Precision Measurement and Nano Metrology Laboratory at the AuM-TUWien.

Acknowledgement. Direct Metal Laser Sintered AlSi10Mg and Selective Laser Sintered PA2200 parts were manufactured at Aluminum Test Training and Research Center, Fatih Sultan Mehmet Vakif University, Turkey. X-CT measurements were taken at High Precision Measurement and Nano Metrology Laboratory at the AuM-TUWien, Vienna, Austria. The rest of the measurements were applied at Yildiz Technical University laboratories.

References

1. Galba, M.J., Reischle, T.: Additive manufacturing of metals using powder-based technology. In: Bandyopadhyay, A., Bose, S. (eds.) Additive manufacturing, pp. 97–142. CRC Press/Taylor & Francis Group, Boca Raton (2016)

2. Sagbas, B.: Investigating the tribological performance of AlSi10Mg parts manufactured by direct metal laser sintering. In: 4th International Congress on 3D Printing (Additive Manufacturing) Technologies and Digital Industry, Antalya, Turkey, 11–14 April 2019 (2019)
3. Girelli, L., Tocci, M., Montesano, L., Gelfi, M., Pola, A.: Optimization of heat treatment parameters for additive manufacturing and gravity casting AlSi10Mg alloy. IOP Conf. Series: Mat. Sci. Eng. **264**, 2017 (2017). https://doi.org/10.1088/1757-899X/264/1/012016
4. PA2200 Material Data Sheet: EOS GmbH-Electro Optical Systems (2008)
5. Lorusso, M., Aversa, A., Manfredi, D., Calignano, F., Ambrosio, E.P., Ugues, D., Pavese, M.: Tribological Behavior of aluminum alloy AlSi10Mg-TiB2 composites produced by direct metal laser sintering (DMLS). J Mat. Eng. Perf. **25**(8) (2016). https://doi.org/10.1007/s11665-016-2190-5
6. Vinoth, K.S., Subramanian, R., Dharmalingam, S., Anandavel, B.: Optimization of dry sliding wear conditions for AlSi10Mg/SiCp composites using response surface: genetic algorithm approach. Ind. Lub. Trib. **66**(5), 593–600 (2014). https://doi.org/10.1108/ILT-02-2012-0016
7. Townsend, A., Racasan, R., Leach, R., Senin, N., Thompson, A., Ramsey, A., Bate, D., Woolliams, P., Brown, S., Blunt, L.: An interlaboratory comparison of X-ray computed tomography measurement for texture and dimensional characterization of additively manufactured parts. Addit. Manuf. **23**, 422–432 (2018)
8. Thompson, A., Körner, L., Senin, N., Lawes, S., Maskery, I., Leach, R.: Measurement of internal surfaces of additively manufactured parts by X-ray computed tomography. In: 7th Conference on Industrial Computed Tomography, Leuven, Belgium (2017)
9. Thompson, A., Maskery, I., Leach, R.K.: X-ray computed tomography for additive manufacturing: a review. Meas. Sci. Technol. **27**, 07200 (2016)
10. Sagbas, B., Boyacı, T.H., Durakbaşa, N.M.: Precision metrology for additive manufacturing. In: Proceedings of the International Symposium for Production Research 2018. Springer (2019). ISBN 978-3-319-92267-6
11. Durakbasa, N.M., Demircioglu, P., Bauer, J., Bogrekci, I., Bas, G., Bodur O., Poszvek G.: Additive miniaturized-manufactured gear parts validated by various measuring methods. In: Proceedings of the 12th International Conference on Measurement and Quality Control - Cyber Physical Issue. IMEKO TC 14 2019. Springer (2019). ISSN 2195–4364
12. Villarraga-Gómez, H., Seifi, M., Uchiyama, Y.: Assessing the structural ıntegrity of additive manufactured metal parts with X-ray CT. In: Summer Topical Meeting, vol. 64, pp. 151–155. American Society for Precision Engineering (2016)
13. EOS Aluminium AlSi10Mg Material Datasheet: EOS GmbH-Electro Optical Systems (2014)
14. Spierings, A., Schneider, M., Eggenberger, R.: Comparison of density measurement techniques for additive manufactured metallic parts. Rapid Prototyp. J. **17**, 380–386 (2011)
15. ISO 15708-2:2018-11 Non-destructive testing-Radiation methods for computed tomography- Part 2: Principles, equipment and samples
16. Villarraga-Gómez, H., Lee, C.B., Smith, S.T.: Dimensional metrology with X-ray CT: a comparison with CMM measurements on internal features and compliant structures. Precis. Eng. **51**, 291–307 (2018)
17. Van Bael, S., Kerckhofs, G., Moesen, M., Pyka, G., Schrooten, J., Kruth, J.P.: Micro-CT-based improvement of geometrical and mechanical controllability of selective laser melted Ti6Al4 V porous structures Mat. Sci. Eng. A **528**, 7423–7431 (2011)

18. Krishnan, M., Atzeni, E., Canali, R., Calignano, F., Manfredi, D., Ambrosio, E., Iuliano, L.: On the effect of process parameters on properties of AlSi10Mg parts produced by DMLS. Rapid Prototyp. J. **20**(6), 449–458 (2014)
19. Wits, W.W., Carmignato, S., Zanini, F., Vaneker, T.H.J.: Porosity testing methods for the quality assessment of selective laser melted parts. CIRP Ann. **65**, 201–204 (2016)

Managing Quality Perception Along the Customer Journey: A Behavioral Economics Approach

Florian Teleaba[1(✉)], Sorin Popescu[2], and Roxana Santa[1]

[1] T.U. Cluj-Napoca, Cluj-Napoca, Romania
florian.teleaba@gmail.com, roxanasantaa@yahoo.com
[2] Design Engineering and Robotics, T.U. Cluj-Napoca, Cluj-Napoca, Romania
sorin.popescu@muri.utcluj.ro

Abstract. Human beings are not completely and always rational and are influenced by several cognitive biases and heuristics in how they evaluate and perceive the quality of a product or service offered to them; moreover, their expectations and perceptions of quality varies along customer journey and product lifecycle. Therefore, professionals dealing with product development, marketing and selling cannot deliver results if they do not understand these two perspectives and how to manage effectively customer perception of product quality. The usual perspective of quality management on quality perception on the one hand, and behavioral economics – which deals with how cognitive biases and heuristics affect economic decisions of people in real-life – on the other hand, have been both researched extensively. Yet there are just few research publications proposing a structured knowledge at the intersection of these two domains. This paper proposes a conceptual framework on how to use behavioral economics concepts and understanding in quality management, bringing together these two domains in a synthesized and exhaustive manner. The paper focuses on the influence of cognitive biases and heuristics on how quality of products or services is perceived, and how this influence occurs across customer journey, and consolidates other factors which may influence expected and perceived quality. It proposes a structured approach to manage those perceptions. It provides readers from the quality management and new product development research areas with an exhaustive list of the behavioral (cognitive) biases and heuristics that influence how people (customers) perceive the quality of products and services, and it provides readers from the behavioral economics research area a new, deeper, perspective on quality perception and its role in economic decision making. The paper does not propose itself at this stage to test and demonstrate the theoretical contributions it makes, but rather to make professionals from these domains aware of the possible implications – hypotheses remain to be tested in future research.

Keywords: Behavioral economics · Cognitive biases · Heuristics · Quality management · Quality perception · Perceived quality · New product development · Customer journey · Decision-making

N. M. Durakbasa and M. G. Gençyılmaz (Eds.): ISPR 2019, LNME, pp. 491–507, 2020.
https://doi.org/10.1007/978-3-030-31343-2_43

1 Introduction

For the last two decades or so, the world has been discussing about Industry 4.0, expected to bring unprecedented transformation in the production area. With these trends of automation, data exchange, Internet of Things or cognitive computing, and even more together with the concept of Globalization 4.0, production is prone to shift to faster-to-market, more international, more adapted products to a wider and wider group of customers. Yet this comes with an increasing challenge: how will customers perceive the quality of products in this context?

Quality perception appears on the table of many quality management, product development and marketing professionals, and has been researched well in the academic literature. More importantly, several researches have indicated the difference between what adequate and desired quality for customers are, and between the expected quality and the perceived quality of products or services.

Yet how we as human beings make economics decisions, and in this matter form ourselves a perception about the quality of a product or service, is not always completely rational, logic, coherent, and decisions are not always optimal as neoclassical economics would predict. Behavioral economics research has taught us there are various cognitive biases and heuristics, errors, that affect our decision-making. These are triggered by various other factors like context, emotions, social influences.

Consider the following example [1]: you go to a store wanting to buy a new TV. An enormous variety is in front of you, depicting the typical impressive image of a flower slowly blooming to reveal the ultra-high definition characteristics and quality of image. You must make a choice though, and you narrowed down your options to two products: a fairly good one for 400$, or a 500$ deluxe option with impressive and highly attractive characteristics yet above your budget. Despite how irresistible second one is, you decide to not go over your budget and choose the first one. On the moment, you feel you made a mediocre choice and have to settle in for a mediocre life – sigh. Yet once you reach home, you realize the TV looks very fine, in fact, it looks amazing – amazing display, color, sound, amazing quality! – you don't even figure out why you considered the second option in the first place. So why is this happening? It's because of a term in behavioral economics field called 'distinction bias', a tendency to look too much on small quantitative differences when comparing options, or in more exact words, a distinction between how you perceive a product when being in joint-evaluation mode (two or more options comparing that is) versus in single-evaluation mode [2]. In fact, research shows there is even a preference reversal between joint and separate evaluation of alternatives [3], not only a difference in perceived quality or performance.

After providing a brief background with examples of research in the areas of quality perception and behavioral economics concepts touching (directly or mostly indirectly) on the notion of quality perception, this paper brings for the first-time these two domains together in a systematic, synthesized manner, covering most important behavioral economics concepts grouped in six specific dimensions, and explains the potential influence on the perceived quality of product or service. As we see in the above example, the way we perceive quality of products varies at different moments in our customer journey. Therefore the paper explains why and how these implications need to be considered at different moments in the customer journey, providing a comprehensive logical scheme of how to do it and factors to consider. The paper also proposes for this an approach consisting of 5 steps, named suggestively SENSE (Study the customer journey and decision-making process, Evaluate the role of quality in customer decision making process, Narrow down to desired/target quality role and objectives, Set relevant interventions to drive quality perception in the targeted direction, and Evaluate results and adapt continuously).

2 Background

Some of the most notable work on how quality is perceived by consumers is unarguably the work of Parasuraman, Berry and Zeithaml on the difference between expected and perceived quality [4], and the important separation they make later on between adequate and desired quality: adequate quality is the minimum level of quality that customer finds acceptable, while the desired quality is the level the customer hopes to receive (from either a new product, or service, or from customer service activities) [5]. Perceived quality thus can be different than the objective, expected value of the product or service, and needs to be understood relative to the concepts of adequate and desired levels of quality. An even more interesting separation was made by the same famous authors between quality of products and services: in case of products or goods, the consumer has many tangibles cues to evaluate quality, like technical characteristics, performance, materials, color, etc. In case of services, the tangible evidence is more limited, and consumers must depend on other 'soft' cues, and as such they defined a series of ten determinants of service quality later on grouped into five dimensions: reliability, tangibles, assurance, empathy, responsiveness – the famous SERVQUAL model [6]. Though it addressed the notion of perceived quality, quality management literature overall has not been concerned with what psychological, cognitive, behavioral factors influence that perception of quality.

Behavioral economics literature on the other hand deals a lot with cognitive biases and heuristics and how these influence our decision making in real life contexts. While one could argue that people would tend to maximize the economic expected utility (or quality being offered) from an economic transaction, product or service, according to Herbert Simon, people tend to make decisions by "satisficing", a combination of satisfying and sufficient, rather than optimizing or maximizing utility [7]. As such, people usually choose options or decide based on the basic criteria that is met, not based on analyzing and maximizing utility or finding the optimal option weighing all criteria and information. Several concepts from behavioral economics touch upon the idea of how quality is perceived by consumers and people, yet not always directly, and not systematically.

One of the most famous behavioral economists and recent winner of the Nobel prize for his contributions to the field, Richard Thaler, argued that people think of value in relative rather than absolute terms. He introduced the term "mental accounting" [8], showing people treat money differently depending on money origin or intended use. He also showed how people derive pleasure not only from the product quality or value itself, but also from the quality of the process or deal of getting that product – term called "transaction utility". According to his research, people derive or feel/perceive a loss if they give up to something for which they already incurred a cost, failing to consider the opportunity costs, term called "sunk-cost fallacy" [9], that also influences in turn the quality or value of the good. Similarly, once owning a product or service, the value assigned to it increases versus when it is now owned, regardless of the real market value, term called "endowment effect" [10]. This is observed even more for goods that are considered symbolic of experiential. The experience itself is an important factor. Hedonic adaptation occurs when you get used to changes in life experience, returning to a relatively stable base of happiness. According to research by Mochon, Norton and Ariely [11], repetition of smaller positive hedonic boosts, or positive experiences, has a more lasting effect on our wellbeing than major life events. Another interesting paper by Huang et al. [12], proposed 3 hypotheses, rooted on this exact idea, that incremental improvements in a product or customer service will make the customers perceive quality as desired, even if it is not the case in reality. They base their hypotheses on other behavioral economics principles, on Kahneman's two cognitive systems theory [13], loss aversion and reference dependence [14], and also on attribute substitution effect [15]. Similar theory has been developed by the authors of this paper relating quality perception to the frequency of product innovation and improvement, arguing that customer churn/deflection can increase, and perceived quality can be unfavourable relative to competition, even if still better objectively, if frequency of product innovation is not sustained due to an innate bias of people to try what is new – try-new bias [16].

Research exists also on how quality and perceived quality fit together with other factors of choice of customers, such as price or brand. For example, a direct positive relationship has been shown between the brand name and the perceived quality of consumer products from that brand by Rao and Monroe [17]. According to Maheswaran, Mackie and Chaiken, even using brand names alone as salient cues can trigger evaluations about the product quality [18]. Reliance on such cues is known as salience and it refers to information that stands out, seems relevant or new, and is more likely to affect our thinking, evaluation and actions about a product [19]. On price, research exists on how for example price is correlated with quality, as there is a tendency to think that higher price typically means higher quality. And in fact, research showed there is somehow a correlation explaining this direction – see Rao and Monrore [17], citing Tellis and Wernerfelt [20].

Next to affect and availability, representativeness heuristic is one of the most important heuristics researched in behavioral economics. Its influence on perceived quality has been shown for example in the rating of the quality of a local product from a generic store being higher it its packaging was designed to resemble a national brand [21].

Another major part of behavioral economics field has been focused on choice making. Choices are often made relative to the available options of the offer, and not in absolute terms. Distinction bias described in the introduction giving the TV example is one. Yet there are several other biases or heuristics that may influence perceived quality. Asymmetrically dominated choice effect for instance occurs when people's preferences for one option change when introducing a third option, similar to one of the previous two, but less attractive. As an example, people are likelier to choose a high-quality pen instead of cash if there is a third option in the form of a low-quality pen

[22]. As such, we see that choices can be presented in different ways, highlighting either positive or negative aspects, as in the riskiness behind (risky choice framing), or the attributes – attribute framing. On the latter, Levin, Schneider and Gaeth provide an example of beef meat described as 95% lean versus being described 5% fat, which rationally represents the same thing, yet influence how customers perceive it [23]. Other known concept researched in relation to choices and decision-making is "extremeness aversion" – that is the influence of extreme options making the middle one to seem satisfactory or desirable; people tend to avoid choosing the extremes. Here both background context, defined by prior existing options, and the local context, defined by the choice set, are important considerations [24].

Also, presence of choices triggers often a certain decision tree for customers. Elimination-by-aspects introduced by Tversky ever since 1972 refers to a heuristic where people gradually reduce the number of options they consider from a choice set starting with the aspects they see most significant, evaluating one cue at a time until fewer options remain [25]. Not only that such a funnel in decision-making can exist, but sequential decision-making is shown to facilitate certain comparisons at the different stages of the choice process [26] and breaking down the decision process in more stages can sometimes yield superior decision making [27]. We observe as such again the importance of perceived quality not only at the moment of purchase, but along the entire customer journey, from consideration to researching, comparing, and only after the final decision-making.

The more "fast and frugal" view of behavioral economics lead by Gigerenzer sees choice making rather as ecologically rational, versus irrational, based on heuristics such as "take the first" or "take the best" (as the names suggest, making decisions of choice based on the alternative or factor that comes first to mind in the former, and based on the one attribute that is deemed most important – discriminant factor – in the latter) [28].

Behavioral economics research concentrated also on the social influence. Social proof bias, or herd behavior for example, has been show several times to influence economic decision-making of people, and in fact has been discussed in psychology, behavioral finance (ex. collective irrationality of investors), politics, science and other fields – it is sometimes referred to as, or put in relation with, "information cascades" [29].

In all these cases, be those choice making related or social influence, we observe overall that perceived quality can be influenced by several cognitive biases and heuristics, yet this influence has not been researched sufficiently, with only a handful of examples existing – on the one hand, quality management or product development professionals or researchers did not focus too much on the impact of behavioral economics, on the other hand behavioral economists did not focus that much on how quality is perceived, rather on how the actual decision is made or what that is.

Lastly, despite not being classified as behavioral economics, it is worth mentioning Kano's model that separates between three types of needs or expectations of customers in terms of quality: must's, those attributes or properties of product that satisfy the basic needs or expectations; want's, those that customers are able to still specify loudly from their mind, and provide a higher level of satisfaction and differentiation of product from competition; and wow's, those that excite and delight customers beyond expectation, that are unexpected in fact [30]. Authors see this as yet another layer of complexity in how customers perceive the quality of the product, given the difference between known or stated characteristics that make up the perception, and surprising factors which cannot be stated upfront and form/influence the perception only ex-post – again, the importance of customer journey.

3 Proposed Conceptual Framework to Use Behavioral Economics in Influencing Perceived Quality of Product or Service

As we have seen in the previous pages, several notes on the influence of some behavioral economics concepts over the quality perception have been drawn in the research literature, where notable research was identified. Some research exists therefore on these implications, however not extensively and holistically, and gaps to our understanding of what drives perception of quality from a cognitive and behavioral point of view exist.

The authors leverage previous work performed on the influence of behavioral economics concepts over new product development [16], where more than 60, most important, behavioral economics were brought together, analyzed, and allocated for the first time to a conceptual framework of how to utilize them in different phases of the new product development process, structured in five main dimensions: utility and value perception, uncertainty and risk, probabilities and weights, temporal, social, and choice. That allocation has been done systematically and it is based on reviewing behavioral economics literature extensively – some of the most important papers are mentioned in the previous background section, while some other important ones addressing some of these concepts are considered [31–37, 39]. Further references and details on each concept can be found in the behavioral economics literature as each concept mentioned below is a consecrated term.

This time, authors leverage this clusterization and provide a new conceptual framework on how to consider these dimensions and their concepts in quality management, i.e. how these cognitive biases and heuristics may influence the perception of the quality of a (new) product or service. This analysis contains already a drill-down of only those behavioral economics concepts considered relevant for the understanding of quality perception – a short list that is. For more behavioral and cognitive concepts within each dimension see [16].

Table 1 below provides a synthesis of this new conceptual framework proposed, by selecting those appropriate and relevant behavioral economic concepts within each dimension and commenting on what is the potential influence on quality perception.

Table 1. Conceptual framework: potential influence of behavioral economics on perception of quality of products or services

Dimension	Behavioral and cognitive aspects	Potential influence/hypothesis
Utility and value perception in decision making	Experienced utility Procedural utility Transaction utility	The perceived quality may be influenced by these types of utility consumers derive, despite other core standard attributes that make up the quality of a product or service (like functional of technical specs)
	Remembered utility	Potential influence of past utility evaluations on present utility evaluation. See also influence of peak-end rule below
	Endowment effect	The perceived quality depending on whether the product is owned or not, and even if tested or not (utilized for a short period of time)
	Try-new bias	Potential impact on how quality is perceived simply because a product/service is new or has a new functionality/feature, regardless of its objective value-add *(To be seen in relation to loss aversion and reference dependence also)*
	Satisficing	Potential influence on overall quality perception. Expected to show that same level of perceived quality can be achieved through less actual quality, due to focus only on some factors through which consumers assess the product quality as sufficient and satisfactory *(To be seen also in relation to other concepts of decision making in dealing with choices, for ex. take the best/first heuristic, evaluation by aspects, etc.)*
	Anchoring Affect heuristic	Anchor information or salient features that trigger feelings, especially in presence of asymmetric or incomplete information, may trigger System 1 evaluations that may not be followed by more rational System 2 ones
	Scarcity	Potential influence on quality depending on the scarcity of the product or service
	Sunk-cost fallacy	Potential influence depending on how much money has been invested or paid for the product/service and depending on distance toward achieving the goal/job to be done through that product or service
	Hedonic adaptation	Frequency of occurrence of the benefit of the product/service may influence how quality is perceived, ex. more frequent experience of small benefits may lead to overall higher perceived quality, vs. rare experience of high benefits
	Pain-of-paying [38]	The different pain of paying between spendthrifts and tightwads may influence also a difference in quality perceived between the two groups
Decision making under uncertainty and risk	Loss aversion Reference point/dependence Prospect theory	Potential influence depending on reference point consumers use to compare with. In addition, the perceived 'loss' they could have in switching to a new product, and/or in comparison to the new benefits, may influence perceived quality standalone and in comparison, with competition
	Availability heuristic	Standing out information, cues, etc. may influence the way quality is perceived, most of the times rather unconsciously

(continued)

Table 1. (*continued*)

Dimension	Behavioral and cognitive aspects	Potential influence/hypothesis
	Representativeness heuristics Recognition heuristic	Unlike above anchoring and affect heuristic effects, it is argued that these perceptions of quality could actually hold longer in time, meaning they will not necessarily trigger a System 2 rational evaluation after some period of time (when factors and context or available information do not change)
	Confirmation bias	By tapping on to what consumers need/prefer (even what they say they need/prefer) through communication of benefits, they may have a confirmation bias that could be reflected in their quality perception
	Control premium	By offering more 'control' over how consumers interact, use, even construct the final product/service, perceived quality may increase (control premium)
	Risk as feelings	Emotions and feelings at the moment of decision making are treated in this model as an anticipatory factor. By understanding consumers' biases (or prediction power on the other hand) in how they will feel about the product/service in the future, quality perception could be influenced through smarter targeting or communication
	Information avoidance	Perceived quality may be influence in both directions. On one hand, too much information may influence consumers to avoid taking it into account, remain prey to other biases (ex. take the first, take the best heuristics, evaluation by aspects, etc.) and look only at certain benefits or factors in evaluating quality. On the other hand, offering insufficient information may produce in fact the same effect, also due to a 'satisficing' behavior, or could actually not be sufficient to move them from the 'status-quo'. In both cases, perceived quality may be below actual quality, and below adequate quality level in some cases
Decision making under uncertainty and risk – Probabilities and weights dimensions	Certainty and possibility effects Naïve allocation/1-N heuristic Category size bias	The way consumers think of probabilities either at extremes (going from a chance of 0 to some, or going from a 99% to 100% chance), how they think of probabilities between a wider set of choices (also triggered by other aspects of satisficing behavior), or how they evaluate chance (or direct quality of a product) based on the size of category where it comes from, all could lead to biased perceived quality
	Ratio bias	The right way (or wrong way) of communicating benefits for example, either in percentages or absolute numbers could influence positively (or negatively respectively) the quality perception
	Optimism bias Hindsight bias	Perceived quality may be influence, similarly as with confirmation bias, by tapping on some salient features/information/aspects of the product/service that occupy a major portion of consumers' attention or thoughts, leading to biased impressions about quality
	Ellsberg paradox (ambiguity aversion)	Failing to articulate clearly the benefits for example (or offering too much information – see above information avoidance) may trigger an ambiguity aversion of consumers and consequently biased perception of quality

(*continued*)

Table 1. *(continued)*

Dimension	Behavioral and cognitive aspects	Potential influence/hypothesis
Decision making when dealing with time aspects	Time discounting Hyperbolic discounting Present bias Planning fallacy Dual-self model (and Myopic loss aversion)	Understanding these biases of consumers of placing more importance on immediate gains, myopic planning and sight, and so on, could be leveraged both ways: avoid negative evaluation of quality (by focusing for example in communication only on the long term benefits), or avoiding overestimated perceived quality at the beginning without a real background – companies may be 'biased' in trying this, yet a backfire effect could exist if after a time, consumers realize actual quality is below their initial evaluation
	Diversification bias	The bias of preferring to keep options open for the future, have more features available, etc. may negatively influence perceived quality in cases too much focus is placed on simplicity for example. A balance of 'easy to use, easy to understand' with 'more features, options in case you will need them' should be found. On the other hand, offering too many 'diverse' alternatives or options may again trigger a negative evaluation of quality (connected also with choice overload and information overload potentially)
	Peak-end rule (evaluation by moments)	Direct influence expected over how consumers perceive the quality, by looking/remembering mostly the peak-end moments from interacting with the product (either before or after acquisition), as opposed to taking a rational evaluation approach and judge quality through the sum of all interactions. See also potential influence of remembered utility above
Decision making under social influence	Social norms Social proof Herd behavior	Leveraging social proof, norms or trying to influence decision making through herd behavior may influence quality perception. In most cases, it is expected to influence it positively, yet caution should exist as backfire effect may exist (if initial expectations or perceptions are not confirmed practically). In some cases a social proof through the wrong segments, with which consumers do not identify with, or that are subject to controversial aspects in society in that moment, could have a negative effect over perceived quality
	Cognitive dissonance	It is expected that any cognitive dissonance that is triggered in consumers mind (through communication or consumer behavior that occurs) can either have positive or negative influence over quality perception, depending on whether consumers are able eliminate that dissonance (in which case, a sentiment of frustration and desire of avoidance can appear). It would be recommended to avoid leveraging such cognitive dissonance, unless it is easy for consumers to eliminate it through product/service interaction (acquisition or usage)
	Commitment	By leveraging this and making consumers 'commit' to a certain behavior, along the progress or after the progress is achieved the perceived quality may deviate from the more rational evaluation

(continued)

Table 1. (*continued*)

Dimension	Behavioral and cognitive aspects	Potential influence/hypothesis
Decision making when dealing with choices	Salience	Information that stands out, seems relevant or new, emphasizing either positive or negative information and so on, is expected to have major influence over perceived quality
	Status-quo bias	The status-quo bias in consumers may negatively influence the perceived quality or produce underestimations, even though actual quality is at a certain level, simply because people will have a tendency to not make the acquisition or use the product/service. In addition, it can influence also the perception of reference point – see reference dependence and loss aversion above
	Choice architecture Default options Framing effect Extremeness aversion Asymmetrical dominated choice	The way product/service choices are shown/exist, the frame (positive or negative for ex.), the existence of extreme choices, asymmetrical choices – all factors are expected to influence the perceived quality of a product or service
	Choice overload	Choice overload is expected to influence the perceived quality of the choices at hand. In addition, even after the decision to select an alternative has been made, choice overload is expected to continue to influence the perceived quality of the final selected option. This effect could be in both ways, both positive and negative, and factors influencing the direction should be studied, and it is expected to come also through other cognitive and behavioral biases
	Less is better effect Distinction bias Evaluability hypothesis Elimination by aspects Take the best heuristic Take the first heuristic Decision staging/choice bracketing	The way consumers evaluate choices and their attributes, features, and so on, is expected to influence perceived quality. Again, this can happen in both directions, positive or negative influence over perception. Decision staging/choice bracketing is expected to influence the quality perception not only at the moment of after the decision is made, but also during the process of deciding. The perceived quality during the decision-making process, and the perceived quality at the moment or after the moment of deciding, are expected to be different – factors influencing this difference and direction of difference (which is lower/higher) should be studied

Authors suggest that even more, the treatment of customers and potential customers given the suggested ideas in Table 1 should not be done only homogeneously, and in fact there likely exist segments of customers that display different behavioral or cognitive typologies – research in the segmentation of customers depending on their behavioral or cognitive typology, i.e. cognitive biases or heuristics, is rather at the beginning, a virgin field.

As one of the questions readers of these paper may have is "How or when should I use this conceptual framework?", the paper proposes the following steps in understanding and influencing the quality perception of customers when developing new products or services or when marketing and selling them, presented in Table 2 below, and named suggestively the **SENSE approach**.

Table 2. Proposed approach: steps in managing properly how your customers will perceive the quality of your (new) product or service and improving sales and customer satisfaction

Step	Step description	Actions to take	Recommended methods
S	Study customer journey and understand the decision-making process	Map customer journey in following recommended steps: awareness, research, compare, test, decide-buy, fulfillment, usage, re-engagement, disposal. (Please note that this is a suggested departure point; actual steps of customer journey may vary) Understand main actions used by customers in each step and on which channels (ex. physical location/shop, online, mobile, social media, etc.) Understand and map the customers' decision-making process at each step of the journey (ex. how do they compare, how do they narrow down list of choices and decide, based on which factors) Identify and analyze on the decision-making process how customers: • define quality (ex. design, functionality, features, technical specifications, materials, finishing, look and feel, etc.) • perceive quality of product/service • how they evaluate quality, when and in what order they consider the factors that define 'quality' in their view • what other factors become important in their quality perception (context, biases, heuristics) – utilize Table 1 in this paper. *Note: how quality is defined needs to be done from the customer perspective, accounting both for 'hard' elements as well as for 'soft' elements like design, look and feel, which may matter as much or even more to customers* When/where relevant, perform customer segmentation to differentiate: • customer journey by different relevant segments • customer decision making process (ex. behavioral biases and heuristics) by different relevant segment ***Note: utilize for this step the logical scheme in*** Fig. 1 ***provided further after this table***	Quantitative research (survey) – in particular more important if required to perform customer segmentation Qualitative research: customer interviews (if relevant, also focus groups) Ethnographic observations (ex. observing real customers in real life context)

(*continued*)

Table 2. (*continued*)

Step	Step description	Actions to take	Recommended methods
E	Evaluate current quality role in customer decision making	Based on previous step, identify main factors of choice within relevant target segments. Recommended and most important (at minimum) to use are: • quality • price • promotion • brand (and marketing communication) • variety (of product offer, or within one product, of functionalities or features) • availability (of stock) *For all above factors, it is important to identify the exact sub-factors which are important and their influence, for ex. a simple message in communication that taps on social norm bias like "95% of people recognized this product as very good" or a simple "Most sold" tag may influence perceived quality beyond standard brand-quality-price relationship* Similarly, it is recommended to identify factors of: • customer satisfaction • customer recommendation (NPS – net promoter score) • customer loyalty (high share of wallet) Compute calculations to identify accurately the importance of each of these factors in consumer decision making process, at each relevant step in the customer journey *Note: while a low price and high quality may many times appear most important factors of choice when asking customers directly their preference, reality may differ. Thus, authors recommend relying on quantitative techniques and experimentation & testing to separate what customers "say" from what they actually "do" – see right*	Direct research through quantitative survey and based on discrete choice modelling, factor analysis and SEM (Structural and/or Simultaneous Equations Modelling) QFD in selected cases (products with heavy technical/technology focus) Indirect research based on historical purchase data if available (with similar techniques as above)

(*continued*)

Table 2. (*continued*)

Step	Step description	Actions to take	Recommended methods
N	Narrow down to desired/target quality role and objectives at specific moments in customer journey	Based on findings from first two steps: • synthesize or prioritize which should be the main business goals: improve acquiring of new customers, improve share of wallet in existing customers/loyalty, retain customers, win-back lost customers, improve customer satisfaction, recommendation, etc. • the target customer journey that the company needs to stimulate/implement • the desired role that quality should play in customer decision-making process, in the different relevant customer journey steps (role here covers also level of perceived quality) • the desired quality level that should be achieved, if case, that requires actual product specifications modification and a (new) product development process	For the second and third bullet point on the left, consider also the possibility of actual defining and testing preliminary ideas with customers in interviews or focus groups – either direct expression of interest/preference/opinion, or indirect observation of customer reactions to changes
S	Set relevant interventions to drive perceived quality in the targeted direction	Define and implement specific actions targeted at changing perception of quality of product or service. These will vary from case by case, from: • simple context or framing modifications, for instance in communication and marketing (ex. displaying 3 alternatives instead of 2 to make middle option seem of adequate or even desired quality) to • more complex modifications (ex. developing a new product with attention to the actual specifications that are considered by customer, and how & when, along the different steps of the customer journey: researching, comparing, deciding, buying, using the product, given biases and heuristics that affect their perception, for instance elimination by aspects or take the best or first heuristics)	Testing of actions recommended in real decision-making contexts (ex. real store with real merchandize for purchase), using test & learn methodology (ex. using one store for test and one as control group)
E	Evaluate results, incorporate in new product development (if case) and adapt continuously	This step refers to evaluating the results of the behavior-perception-change actions designed in first steps on a recurrent basis, and continuously adapt Authors recommend management of the perceived quality and influencing it in the right direction through these steps should be a continuous improvement exercise rather than just a "project"	

Authors propose to treat perceived quality of product or service by customers in the following logical scheme, described below in Fig. 1.

Note that it is important to understand that:

(a) there will be a difference between perceived quality and objective quality, perceived quality and expected quality, differences triggered by a series of factors,
(b) while objective quality likely varies only along product lifecycle (i.e. performance in utilization reduces after a number of years), expected quality may vary along both customer journey steps and product lifecycle (during usage until disposal/ replacement),
(c) perceived quality likely varies along the customer journey and later along product lifecycle as customers may change their perceptions along the steps, both until purchase decision and later while using the product or service and re-engaging with the company (either for a re-purchase or other reasons)
(d) the factors described as influencing expected and perceived quality may contribute with different relative powers in each of the customer journey steps (ex. factors important for forming a perception of quality in the research phase may be different than those important in the comparing or testing phase, or in purchasing)
(e) when referring to possibility to perform customer segmentation, in step S of the proposed **SENSE** methodology, segmentation of customers can be performed not only on standard moderating factors such as age or income, but in fact on the factors influencing expected and perceived quality. These can range from more intuitive or simple to separate, like customer needs or preferences, to differences in the importance awarded to the typical purchase factors (utility, brand, price, etc.), to more complex criteria to manage practically like consumer limbic type [40]. (These latter factors are considered more complex to manage in real life as for example allocating a new customer to one limbic type may require rich data for profiling or complex real-time profiling based on advanced analytical models and/or multiple questions).

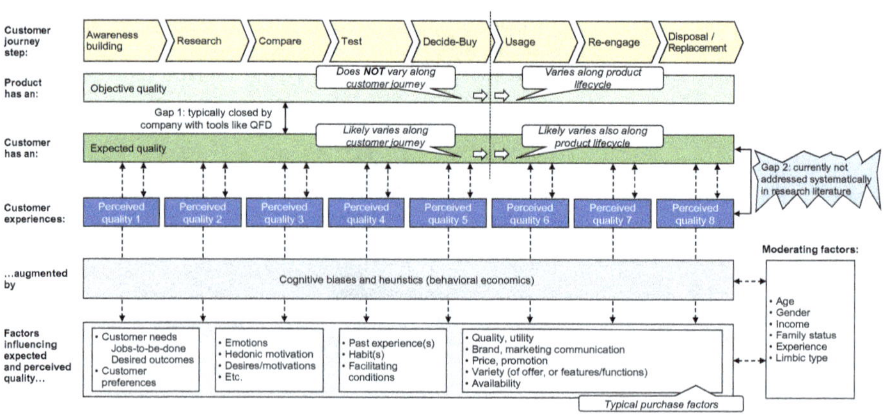

Fig. 1. Quality perception and factors across customer journey

4 Conclusions

Quality management has been extensively researched as a field of study, with a focus on translating customer needs into product specifications and further ensuring product is developed matching customer requirements. Furthermore, studies have shown the differences in types of quality, either at performance level, between adequate or desired quality, and at perception level, between expected quality and actual perceived quality.

Behavioral economics has also benefited of extended research along the last few decades, with popularity of this field growing tremendously in the last decade or so along the Nobel prize winnings of Daniel Kahneman and Richard Thaler, and with the growing utilization of behavioral economics in both public policy and business alike.

Yet there exists little to no extensive research at the intersection of these two fields of studies, research to address in a systematic and exhaustive matter how various cognitive biases and heuristics (studied by behavioral economists) influence the perception of quality of a product or service (studied by quality management or product development researchers and professionals). Furthermore, as this paper discusses, the way quality is perceived varies both along customer journey (from having an underlying need or becoming aware of a need to researching, comparing, deciding, buying a product that serves that need, using it, re-engaging with the company, disposing it) and along product lifecycle. This only adds complexity to the way quality perception needs to be managed and when, drawing attention that it is not sufficient anymore to close the gap between objective and expected quality (with tools like QFD), but also to close several other gaps along the customer journey between objective and expected quality on one hand and perceived quality on the other hand – closing gaps which in turn means from simple actions targeted at managing perception directly (without changing the product or service) to more complex interventions needed (changing the product or service along with managing perception directly).

The paper assumes as contribution and novelty offering a conceptual framework that explains the potential influences of cognitive biases and heuristics on perception of quality, covering most important behavioral economics concepts. It thus represents an exhaustive list that both researchers and professionals can use as a start, without fearing of missing other important elements or concepts from sight. It also provides a comprehensive set of guidelines, or steps, named suggestively the SENSE approach, on how to utilize this conceptual framework and further design actions and interventions aimed at managing the perception of quality. Finally, it provides a logical scheme that summarizes in a comprehensive and exhaustive manner the factors influencing expected and perceived quality of product or service along customer journey, including potential moderating factors. This logical scheme and complete set of factors is derived from multiple angles, from various research in quality management, product development, behavioral economics, consumer psychology, to experience of authors in their consulting careers.

The ideas and hypotheses described in the paper constitute a promising background for further research in the areas of quality perception and behavioral economics alike.

References

1. Eyal, N.: Nir & Far website/blog, example. https://www.nirandfar.com/distinction-bias/
2. Hsee, C.K., Zhang, J.: Distinction bias: misprediction and mischoice due to joint evaluation. J. Pers. Soc. Psychol. **86**(5), 680–695 (2004)
3. Hsee, C.K.: The evaluability hypothesis: an explanation for preference reversals between joint and separate evaluations of alternatives. J. Organ. Behav. Hum. Decis. Process. **67**(3), 247–257 (1996)
4. Parasuraman, A., Berry, L.L., Zeithaml, V.A.: A Conceptual Model of Service Quality and Its Implications for Future Research. J. Mark. **49**(4), 41–50 (1985)
5. Parasuraman, A., Berry, L.L., Zeithaml, V.A.: Understanding Customer Expectations of Service. Sloan Manag. Rev. **32**(3), 39–48 (1991)
6. Parasuraman, A., Zeithaml, V.A., et al.: SERVQUAL: a multiple-item scale for measuring customer perceptions of service quality. J. Retail. **64**(1), 12–40 (1988)
7. Simon, H.: Rational choice and the structure of the environment. Psychol. Rev. **63**(2), 129–138 (1956)
8. Thaler, R.H.: Mental accounting and consumer choice. Mark. Sci. **4**(3), 199–214 (1985)
9. Thaler, R.H.: Mental accounting matters. J. Behav. Decis. Mak. **12**, 183–206 (1999)
10. Kahneman, D., Knetsch, J., Thaler, R.: Anomalies: the endowment effect, loss aversion, and the status-quo bias. J. Econ. Perspect. **5**(1), 193–206 (1991)
11. Mochon, D., Norton, M.I., Ariely, D.: Getting off the hedonic treadmill, one step at a time: The impact of regular religious practice and exercise on wellbeing. J. Econ. Psychol. **29**, 632–642 (2008)
12. Huang, M.H., Pei-Hsian, T., Wang, L., Chun-Shun, L.: A behavioral economics approach to customer relationship management. In: PACIS Proceedings 58 (2005)
13. Kahneman, D.: Thinking, Fast and Slow. Allen Lane, London (2011)
14. Kahneman, D., Tversky, A.: Prospect theory: an analysis of decision under risk. Econometrica **47**, 263–291 (1979)
15. Kahneman, D., Frederick, S.: Representativeness revisited: attribute substitution in intuitive judgment. In: Gilovich, T., Griffin, D., Kahneman, D. (eds.) Heuristics of Intuitive Judgment: Extensions and Applications, pp. 49–81. Cambridge University Press, New York (2002)
16. Teleaba, F., Popescu, S.: A behavioral economics perspective over lean versus 10x improvement in new product development – brief review of existing literature. In: International Conference on Production Research – Africa, Europe and Middle East, 5th International Conference on Quality and Innovation in Engineering and Management, 2018, published in Journal ACTA TECHNICA NAPOCENSIS. Series: Applied Mathematics, Mechanics, and Engineering, Vol. 61, Issue IV, November 2018
17. Rao, A.R., Monroe, K.B.: The effect of price, brand name, and store name on buyers' perceptions of product quality: an integrative review. J. Mark. Res. **XXVI**, 351–357 (1989)
18. Maheswaran, D., Mackie, D.M., Chaiken, S.: Brand name as a heuristic cue: the effects of task importance and expectancy confirmation on consumer judgments. J. Consum. Psychol. **1**, 317–336 (1992)
19. Dolan, P., Hallsworth, M., Halpern, D., King, D., Vlaev, I.: MINDSPACE: Influencing Behavior Through Public Policy. Cabinet Office, London (2010)
20. Tellis, G.J., Wernerfelt, B.: Competitive price and quality under asymmetric information. Mark. Sci. **6**(Summer), 240–253 (1987)
21. Kardes, F.R., Posavac, S.S., Cronley, M.L.: Consumer inference: a review of processes, bases, and judgment contexts. J. Consum. Psychol. **14**(3), 230–256 (2004)

22. Bateman, I.J., Munro, A., Poe, G.L.: Decoy effects in choice experiments and contingent valuation: asymmetric dominance. Land Econ. **84**(1), 115–127 (2008)
23. Levin, I.P., Schneider, S.L., Gaeth, G.J.: All frames are not created equal: a typology and critical analysis of framing effects. Organ. Behav. Hum. Decis. Process. **76**, 149–188 (1998)
24. Tversky, A., Simonson, I.: Context-dependent preferences. Manag. Sci. **39**(10), 1179–1189 (1993)
25. Tversky, A.: Elimination by aspects: a theory of choice. Psychol. Rev. **79**, 281–299 (1972)
26. Johnson, E.J., Shu, S.B., Dellaert, B.G.C., Fox, C.R., Goldstein, D.G., Haubl, G., Larrick, R. P., Payne, J.W., Peters, E., Schkade, D., Wansink, B., Weber, E.U.: Beyond nudges: tools of a choice architecture. Mark. Lett. **23**, 487–504 (2012)
27. Read, D., Loewenstein, G., Rabin, M.: Choice bracketing. J. Risk Uncertain. **19**(1–3), 171–197 (1999)
28. Gigerenzer, G., Gaissmaier, W.: Heuristic decision making. Annu. Rev. Psychol. **62**(1), 451–482 (2011)
29. Bikhchandi, S., Hirschleifer, D., Welch, I.: A theory of fads, fashion, custom and cultural change as informational cascades. J. Polit. Econ. **100**, 992–1026 (1992)
30. https://www.kanomodel.com/ or on American Society for Quality website: https://asq.org/quality-resources/kano-model
31. Slovic, P., Finucane, M.L., Peters, E., MacGregor, D.G.: The affect heuristic. In: Gilovich, T., Griffin, D., Kahneman, D. (eds.) Heuristics and Biases: The Psychology of Intuitive Judgment, pp. 397–420. Cambridge University Press, New York (2002)
32. Tversky, A., Kahneman, D.: Judgment under uncertainty: heuristics and biases. Sci. (New Ser.) **185**, 1124–1131 (1974)
33. Wansink, B., Kent, R.J., Hoch, S.J.: An anchoring and adjustment model of purchase quantity decisions. J. Mark. Res. **35**(1), 71–81 (1998)
34. Read, D., Loewenstein, G.: Diversification bias: explaining the discrepancy in variety seeking between combined and separated choices. J. Exp. Psychol. Appl. **1**, 34–49 (1995)
35. Iyengar, S., Lepper, M.: When choice is demotivating: can one desire too much of a good thing? J. Pers. Soc. Psychol. **79**, 995–1006 (2000)
36. Kahneman, D., Tversky, A.: Evaluation by moments: past and future. In: Kahneman, D., Tversky, A. (eds.) Choices, Values and Frames, pp. 2–23. Cambridge University Press, New York (1999)
37. Thaler, R.H., Sunstein, C.: Nudge: Improving Decisions About Health, Wealth, and Happiness, New Haven. Yale University Press, CT (2008)
38. Rick, S.I., Cryder, C.E., Loewenstein, G.: Tightwads and spendthrifts. J. Consum. Res. **34**, 767–782 (2008)
39. Isaac, M., Brough, A.: Judging a part by the size of its whole: the category size bias in probability judgements. J. Consum. Res. **41**, 310–325 (2014)
40. See more details: https://www.trinityp3.com/2015/05/neuromarketing-insights-customer-journey/, http://www.haeusel.com/limbic/, and https://www.nymphenburg.de/limbic-map.html

Total Quality Management Through Defect Detection in Manufacturing Processes Using Machine Learning Algorithms

Almira S. Kahya(✉), Selin Şişmanoğlu, Zeynep Erçin, and Hatice Camgöz Akdağ

Management Engineering Department,
Istanbul Technical University, Istanbul, Turkey
{kahyaal, sismanoglu, ercinz, camgozakdag}@itu.edu.tr

Abstract. Total Quality Management is the new raising value of all industries. The more it is revealed that TQM is one of the key success factors for the companies, the more it is being absorbed by the industries. This study aims to analyze TQM approaches considering its history and development worldwide while observing manufacturing industry with machine learning applications in order to identify the defects in the process before completed which contributes continuous improvement to the system. Also, different descriptions of quality according to the customer satisfaction will be examined.

Keywords: Total Quality Management · Quality · Manufacturing · Customer satisfaction · Machine learning applications · Continuous improvement · Problem solving · Defect detection

1 Introduction

In todays' highly competitive markets, firms have been striving to ensure their production process without failure while satisfying customers' expectations with defect free product. Profit margins and the product quality are highly fragile elements and affected by defect ratios in manufacturing due to nonconforming internal failure costs. Rosyidi, Nugroho, Jauhari, Suhardi, and Hamada [1] state that the organization must invest in various resources which identify with achieving quality objectives to diminish the quality costs, so as to accomplish the desired quality improvement. In this paper, a manufacturing process of a product in detergents sector will be observed while using Machine Learning (ML) Algorithms to identify the root causes and prevent the system from possible unwanted cost and loss of time with comparing different computer-intensive methods that are implemented through R Studio Programme. Logistic regression, decision tree, and linear discriminant analysis, to name a few.

Improvements in technology, increasing competition environment made people to create new things. The number of entrepreneurs is highly augmented in recent years. Everybody is seeking to make new products, implement new processes. On the other hand, existing organizations should develop their products or service quality to not to scratch the competition in global market. Like the all products, services and processes;

N. M. Durakbasa and M. G. Gençyılmaz (Eds.): ISPR 2019, LNME, pp. 508–516, 2020.
https://doi.org/10.1007/978-3-030-31343-2_44

the new products or new process ideas are assessed in terms of their qualities. The feature, performance, reliability and durability are the main factors which form quality. According to Shewhart and Levitt, quality is defined as conformance to specifications; quality of conformance relates to the degree to which a product meets certain design standards [2]. However, quality has not a unique definition. Juran [3, 4] states that quality is fitness for use, Leffler [5] asserts that quality is product desirable attributes, Taguchi emphasizes that quality is loss avoidance and Ryall and Kruithof [6] indicate that quality is meeting customer expectations. All of the products or processes can have some deficiencies which can be decreased by effective quality management methods. The main purpose should be to have at least deficiency. Based on the official web site of the ASQ Quality wordbook, quality management is defined as a system which main goal is to accomplish customer desires during development in the processes with minimum cost, its components are quality planning, quality assurance (defect prevention), quality control (which includes product inspection and other elements, such as competence), and quality improvement [7]. These four parts are very important to determine errors and minimize it.

This paper begins with introduction part and then continue with literature review which is explained total quality management and applications in life. In the following parts, methodology & application in machine learning which will be used in quality to find deficiencies will be analyzed. The last part which is conclusion part, will give the results of machine learning to resolve the deficiencies.

2 Literature Review

2.1 Problem Solving Tools in Quality

In this competitive market conditions, quality improvement is an inevitable phase in manufacturing. According to Tsou and Chen [8], products that has unsatisfactory quality lead to decrease efficiency, enlarge cost of business operations, and lessen customer satisfaction. PDSA cycle and Six Sigma's DMAIC steps are common methods to identify and understand the current conditions of the system and how to approach an existing problem in a problem solving.

There are seven quality tools which are particularly Flow Chart, Control Chart, Check Sheet, Histogram, Pareto Diagram, Cause and Effect Diagram, Scatter Diagram to identify the nonconformance in the process and try to solve the problems in the quality management. The common goal of those tools is to identify the deficiencies in the process and improve the system eventually. The techniques and components of total quality management is used in many areas like Six Sigma, lean management departments. Thus, the relationship among the customer and company develops which results to reduce deficiencies in manufacturing process. Ertosun, Zehir and Müceldilli [9] assert that the main factor to have competitive advantage is changed from quality to novelty with changing world conditions. Thanks to novelty which enables to create new products, enter new markets for companies and permits to modify themselves rapidly. Moreover, WHY-WHY Diagram technique is a milestone for this case, since it looks for the root causes of the problem by asking "Why" five times.

2.2 Machine Learning Approaches

Statistical applications of quality such as Six Sigma that is a part of Statistical Process Control tries to prevent defects in the products or services by practicing statistical methods to control the processes. Like Six Sigma, Machine Learning is one of the arising approaches that use statistical applications. Therefore, Machine Learning is one of the new computer-based approaches, which can act and learn like humankind, can be used for finding deficiencies in the process to set an example. It is more accurate than traditional rules. Sebastiani [10] describes that, thanks to this methodology, interferences are not required from expert witnesses for the development of distributing an alternate arrangement of classifications which will be resulted to a significant reserve funds as far as master labor and accomplishment by human specialists. Moreover, machine learning applications can be used to determine the deficiencies in the process not only after the process completed but also it is possible to specify the root cause of the problems like WHY-WHY Diagram technique. There is a huge advantage of machine learning is a time saving approach and the computers do the procedure without faults unlike individuals who have to use analytical thinking to improve the overall system with minimum defects and maximum quality to ensure total quality management standards. Abdaziz [11] explains that, all association staff individuals during the time spent covering clients' desire through using critical thinking strategies to improve the nature of every hierarchical item and services are connected by total quality management. It is used in every area like automotive, manufacturing process such as detergent, cancer screening, sales analysis. The methods which are used in machine learning are differentiated according to dataset characteristics. Decision Trees, Logistic regression, Boosted Trees, Linear Discriminant Analysis etc. are classification methods which can be used in this case to classify whether a product is defective or non-defective. Since we have only 2 classes as defective and non-defective whose values are 1, 0 respectively. "As Escobar and Morales-Menendez [12] stated, A learning process (LP) and pattern recognition (PR) strategy for a knowledge-based (KB) ISCS is presented, aimed at detecting rare quality events from manufacturing systems." The defect detection is formulated as a binary classification problem, in which the l1-regularized logistic regression (LR) is used as the learning algorithm. The most relevant properties are shown by the product of ideas which is used for forecasting model. Data visualization is vital manner for both the owner of the facility and client to make the data more clear and easy to interpret. Therefore, the machine learning techniques should not include unnecessary variable that leads to prevent model to be fit and absence of full comprehension which will cause the inconsequential properties. Decision trees is also used to promote product and determine errors and reduce them in total quality management with acceptable visual support. Yussupova, Kovács, Boykol and Bogdanoval [13] describe that the fundamental things for the client which are the most critical item perspectives are distinguished and displayed from decision trees.

The four level of total quality management which are quality control, quality assurance, continuous quality improvement, quality award models will be completed and will reveal deficiencies and causes. By using several Machine Learning models, quality control will be completed easily and defects will be reduced before production is completed which improves the production process with a higher yield. However, it should be considered as Nakajima [14] stated, in ML-based systems any minor changes in training dataset, can lead to a major influence on the test results which has to be considered carefully in order not to end up accusing an important variable as useless. Therefore, this delicacy might rise a new objection to quality tools. In this study validation set approach is used to determine the samples precisely, since there is a huge observation set.

3 Method

3.1 Logistic Regression

Logistic Regression is a type of classification method and can use logistic function formula simply as follows:

$$\mathrm{P(X)} = \frac{e^{\beta_0 + X\beta_1}}{1 + e^{\beta_0 + X\beta_1}} \tag{1}$$

While using maximum likelihood method to fit the non-linear model. Here, the formula looks for the probability of an observation that comes from that class, not for the values.

Therefore, the goal of the formula given in the Eq. (2) is to find the maximum value of the beta that is the regression coefficient.

$$\mathrm{L}(\beta_0, \beta_1) = \prod\nolimits_{i:y_i=1} p(x_i) \prod\nolimits_{i':y_i} (1 - p(x_{i'})) \tag{2}$$

R will be doing this procedure behind it.

3.2 Decision Tree

After plotting the tree, we examined the confusion matrix to find out the error rate of classification with this formula as given in the Eq. (3):

$$E = 1 - argmax_c(\hat{\pi}_{mc}) \tag{3}$$

Those variables increase the explanatory power of the model, even though they are not directly affecting the class determination by the model.

3.3 Linear Discriminant Analysis

LDA is less direct approach in comparison with Logistic Regression Model, however; in some cases, it can be more reliable. Therefore, it uses the Bayes' Theorem to assign the variables to the classes with the formula given in the Eq. (4):

$$P_r(Y = k|X = x) = \frac{\pi_k f_k(x)}{\sum_{l=1}^{K} \pi_i f_i(x)} \tag{4}$$

4 Application

4.1 Dataset and Variables

In this study, there are mainly 9 variables. For example, date is represented as date in the fit model for dataset, on the other hand; experience variable has been shown in the model as Exp_1, Exp_2 and Exp_3 according to corresponding values in data cleaning process of the study. The model is fit according to 14 variables as indicated in the Table 1. The output is Defective referring to defect information of coloring. This variable will be the target variable in prediction stage. Additionally, Defective is a qualitative response and could have tried to clarify by those 9 variables as stated below table.

Table 1. Variables and dataset

Variable	Type of variable in dataset	Explanation of variables in the dataset
Date	Date	Data of measurement
Shift	Shift_1, Shift_2, Shift_3	Measurement shift period
Experience	Exp_1, Exp_2, Exp_3	Experience information of measurement (a metric related with the measurement, not related with operator)
Percentage of dye (%)	DyePercentage	Percentage of coloring
Flow rate	'Flow rate'	Flow rate of coloring
Teflon	Teflon	Teflon rate of part
Thickness	Thickness	Thickness rate of coloring
Part	Part_a, Part_b, Part_o	Which part of product is colored

4.2 Logistic Regression

To classify either the product is defective or non-defective we fit a logistic regression model with glm() function. This fitted model explains the deficiency through the variables of date, time and shift of the production,

condition of the environment with experience categorical variable, dye percentage, flow rate of coloring, teflon rate of part, thickness rate of coloring, colored part of the product.

After fitting binomial logistic regression model, it is examined that thickness variable has the highest explanatory power of the deficiency. As seen in the Variable Importance graph below which is created by varImp() function (Fig. 1).

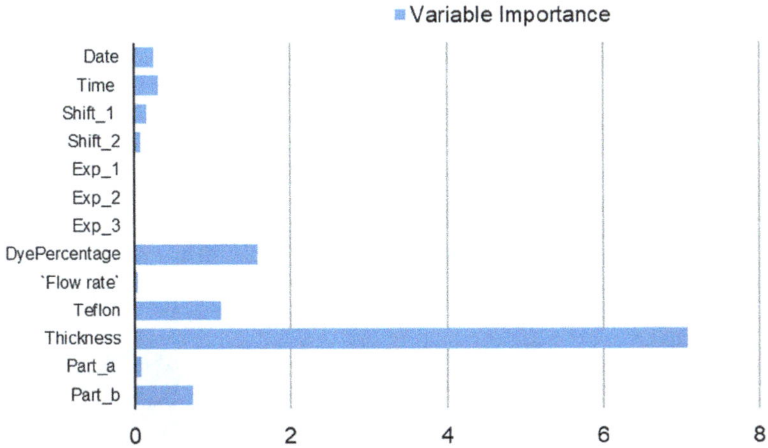

Fig. 1. Importance of the variables. Shift 3 and Part_o are not included in the chart because R found atomic importance in the model.

Although Teflon rate and Dye percentage seem to have a significant effect in the model, they do not tend to improve accuracy. We cannot say that other variables are useless, however to detect defection thickness should be examined firstly. With this machine learning algorithm, the defects are identified after the production which has a non-conformance cost for the firm (Fig. 2).

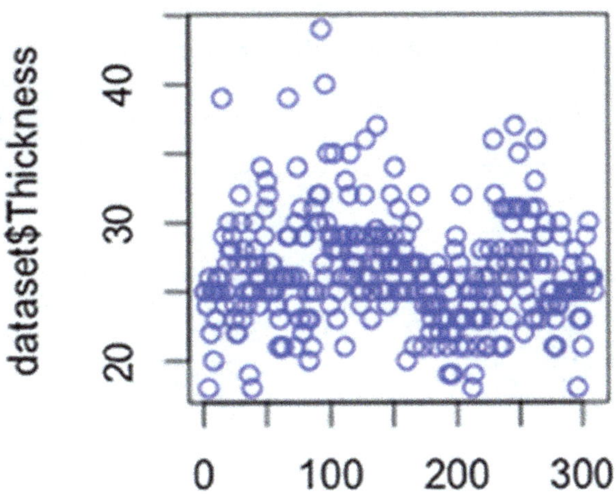

Fig. 2. Scatter plot of thickness variable

4.3 Decision Tree

We also plotted this data in a decision tree to examine if another possible solution or relations occur. But the results are the same as below (Fig. 3):

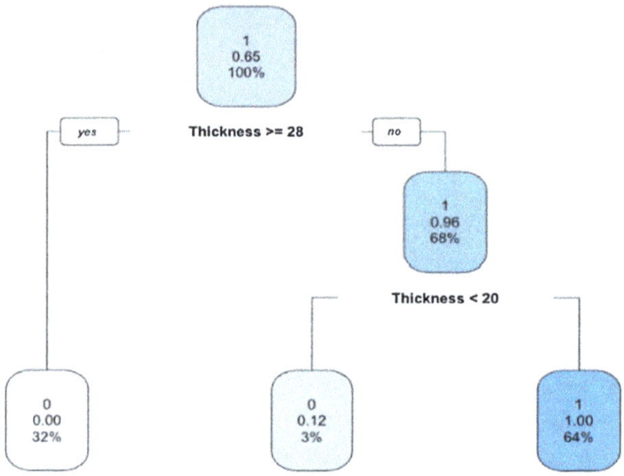

Fig. 3. Decision tree of thickness variable

Even though our number of observation is not that high to see the prediction accuracy of our model we split the dataset into test and training samples with the proportion of 80:20 and applied our model to the training data. Then, we test the rest to reveal whether to ensure the model can predict with low error rate which means model explains the data convincible. The prediction results and the actual states of the deficiencies are shown in the confusion matrixes with their test prediction accuracy rate (Table 2).

Table 2. Accuracy of alternative models

GLM with all variables: Accuracy : 87,09%

	Non-Defective	Defective
Non-Defective	42	2
Defective	1	17

GLM with Thickness variable: Accuracy : 88,70%

	Non-Defective	Defective
Non-Defective	43	7
Defective	0	12

Lda with all variables: Accuracy : 95,71%

	Non-Defective	Defective
Non-Defective	19	1
Defective	2	48

Lda with Thickness variable : Accuracy : 98,57%

	Non-Defective	Defective
Non-Defective	21	1
Defective	0	48

As seen in the accuracy tables linear discriminant analysis (lda) outperforms the other methods. James, Witten, Hastie and Tibshirani [15] stated if the number of observation is relatively smaller to number of predictors, and the predictors in each class has Gaussian distribution, linear discriminant analysis offer more reliable model than logistic regression. Also in lda.fit, model performs more accurate if the p = 1 and variable is thickness. Below lda.fit graphs are attached to reveal the distribution of defected (group 1) and non-defected (group 0) products (Fig. 4).

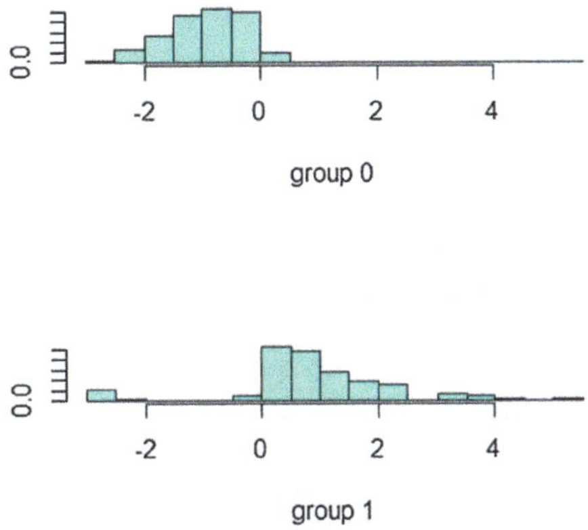

Fig. 4. Distribution of defected (group 1) and non-defected (group 0) products

5 Conclusion

In conclusion, TQM is related with all the phases of a production starting from planning to customer satisfaction. Since any failure that is not detected will reduce customer satisfaction and will bring non-conformance costs, it is crucial to capture failure possibilities in production processes. Amiri, Noghondarianb and Safaeic [16] explained that even though implementing such a system may lead to a rise in administrative costs, total cost of quality can be balanced with lower costs of failure. In order to reduce cost of failure, machine learning algorithms can be installed to the processes. In our example, before starting the manufacturing process, if the part of the production which some preventive action should be taken to control the products' color thickness. This will lead less failure at the end of the manufacturing, therefore an increase in the probability of customer satisfaction.

References

1. IEEE: Quality improvement by variance reduction of component using learning investment allocation model. In: 2016 IEEE International Conference on Industrial Engineering and Engineering Management (IEEM) Industrial Engineering and Engineering Management (IEEM), p. 39 (2016)
2. Elshaer, I.: What is the Meaning of Quality? MPRA, pp. 3–7 (2012). https://mpra.ub.uni-muenchen.de/57345/1/MPRA_paper_57345.pdf
3. Juran, J.M.: Quality Control Handbook. McGraw-Hill, New York (1974)
4. Kemp, S.: Quality Management Demystified. McGraw-Hill, New York (2006)
5. Leffler, K.B.: Ambiguous changes in product quality. Am. Econ. Rev. **72**(5), 956 (1982)
6. ISO 9000:2008: The Quality Management System. Fundamentals and vocabulary
7. Asq.org: Quality Assurance vs Quality Control: Definitions & Differences | ASQ (2019). https://asq.org/quality-resources/quality-assurance-vs-control. Accessed 24 Feb 2019
8. Tsou, J., Chen, J.: Case study: quality improvement model in a car seat assembly line. Prod. Plan. Control. **16**(7), 681 (2005). https://doi.org/10.1080/09537280500249223

9. Zehir, S., Zehir, C., Mücedilli, B., Ertosun, Ö.: Total quality management practices' effects on quality performance and innovative performance. Procedia Soc. Behav. Sci. **41**, 274 (2012). https://www.sciencedirect.com/science/article/pii/S1877042812009111
10. Sebastiani F.: Machine Learning in Automated Text Categorization (2001). https://arxiv.org/pdf/cs/0110053.pdf
11. Al-Qahtani, A., Abdaziz, N.: The impact of Total Quality Management on organizational performance. Eur. J. Bus. Manag. 1 (2015). ISSN 2222-1905 https://www.researchgate.net/publication/294886200_The_impact_of_Total_Quality_Management_on_organizational_performance
12. Escobar, C.A., Morales-Menendez, R.: Machine learning techniques for quality control in high conformance manufacturing environment (2018). https://doi.org/10.1177/1687814018755519
13. Yussupova, N., Kovács, G., Boyko, M., Bogdanova, D.: Models and methods for quality management based on artificial intelligence applications. Acta Polytechnica Hungarica **13** (2016). https://www.uni-obuda.hu/journal/Yussupova_Kovacs_Boyko_Bogdanova_67.pdf
14. Nakajima, S.: Quality assurance of machine learning software. In: 2018 IEEE 7th Global Conference on Consumer Electronics (GCCE), Nara, Japan, pp. 601–604. IEEE (2018)
15. James, G., Witten, D., Hastie, T., Tibshirani, R.: An introduction to Statistical Learning with Applications in R, pp. 138–139. Springer, New York (2013)
16. Amiri, F., Noghondarian, K., Safaei, A.: Evaluating the performance of variable scheme X-bar control chart: a Taguchi loss approach. Int. J. Prod. Res. **52**(18), 5385–5395 (2014). https://doi.org/10.1080/00207543.2014.906762

Simulation and Modelling

An Implementation of Cloud Based Simulation in Production

Kamil Öncü Şen[1(✉)], M. Numan Durakbasa[2], Gökçen Baş[2], Gizem Şen[1], and Osman Akçatepe[3]

[1] The Scientific and Technological Research Council of Turkey, Ankara, Turkey
{oncu.sen,gizem.sen}@tubitak.gov.tr
[2] Technische Universitat Wien, Vienna, Austria
{durakbasa,goekcen.bas}@tuwien.ac.at
[3] University of Hildesheim, Hildesheim, Germany
osman@ismll.uni-hildesheim.de

Abstract. Production systems of today are developing an ever-increasing dependency on information systems, as a result of which intelligent systems are being transformed into smart systems. A production facility can be called a 'smart factory' provided that all the processes are being managed, which can be done only if sensor based an intelligent measurement data derived from production systems are analyzed and processed. To increase the quality of production and decrease energy waste, cloud computing, edge computing, energy management systems (EMS) and continuously updating simulation system must be applied to facilities. All these cutting-edge ICT technologies help the designers make more effective planning, control the Total Quality Management (TQM), and build more efficient production systems whose components interact with each other in real time. To accomplish this, previously we have set up a prototype complete which includes with sensor-based intelligent measurement nodes with a semi-edge computing and cloud-computing system to run on the real production process. And with this study we have strengthened the designed prototype with cloud-based simulation for:

- Modeling the energy consumption of the air conditioning system, which ensures the control of environmental conditions,
- Modeling the production quality that under the influence of environmental conditions.

With this new model, the product quality can be standardized by controlling the environmental conditions of the production lines as well as avoiding unnecessary energy expenditure thus ensuring cost efficiency.

Keywords: Cloud computing · Edge computing · Internet of Things (IoT) · TQM · Simulation

N. M. Durakbasa and M. G. Gençyılmaz (Eds.): ISPR 2019, LNME, pp. 519–524, 2020.
https://doi.org/10.1007/978-3-030-31343-2_45

1 Introduction

Manufacturers are focusing on to meet continuously increasing quality demand and working for reducing energy consumption on production using with Smart Production (SP) and all these structures developed with paradigms of cloud computing, edge computing and statistical prediction models across Internet of Things (IoT) to become the industrial production more and more monitorable, manageable and analyzable [1–4]. Our project has been carried out firstly by developing an intelligent system for controlling the production's environmental conditions using with cloud computing and edge computing structures (even though for the allowing the system to provide non-stop service when the internet connection is interrupted) [5]. This structure focuses on the monitoring and the controlling the live system to keep the production in the standardization objective. The system can be automatized for keeping the system in a determined confidence interval. This confidence interval may cause unnecessary energy consumption during the keeping process. To prevent unnecessary energy consumption simulation structure implemented into the decision-making process. This simulation structure is using for the simulating the energy consumption for the air-conditioning the production facility and the production quality. Both these items are important cross-cutting issues for the manufacturers, and with this work we offer a solution for balancing the quality management and energy management.

2 Method

In this article, the environmental condition data was used. All the environmental condition data was gathered from the live system via Wireless Sensor Networks (WSN). The used WSN system gathered the large amounts of environmental data from the production line's environmental conditions for the decision making and controlling. In this article, only the temperature data was used. For the wireless communication, ZigBee protocol was chosen to gather the temperature data form the edge point of the production line [8]. Figure 1 includes an overview of the designed system:

For the data collection a measurement node was designed and implemented into the live production line that includes temperature sensor. It can also make communication (sending data to cloud and receiving command from cloud) with cloud. That measurement node called intelligent because it can not only collect the data but also can adjust the air condition system. That designed node can provide the flexible and functional measurement for the keeping the production quality under the determined specs and quality standards. The designed intelligent measurement process can be reviewed with the following steps:

- Automatization the measurement
- Automatization the evaluation of measured environmental temperature data
- Off-line or/and on-line adjustment
- Interactive communication between cloud
- Implementation the decision

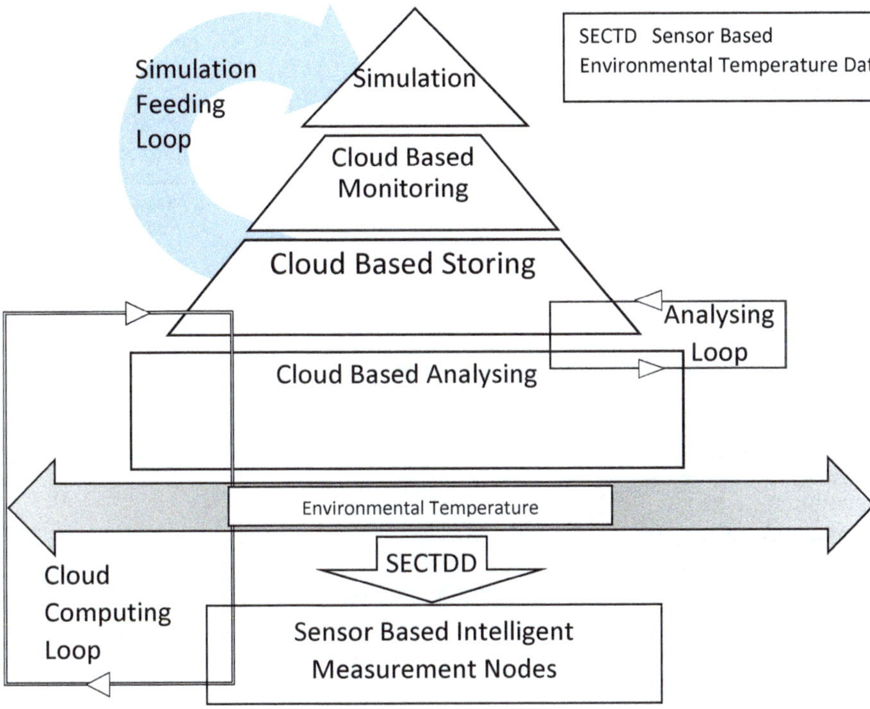

Fig. 1. Block diagram of the designed system

The intelligent sensor-based measurement system used for collecting the environmental temperature to define exact parameter, time, and warrant pre-adjustment process stability under measurable and correlated with temperature parameters. The proposed metrology methodology noted as a part of intelligent measurement and waste free manufacturing. This metrology methodology is taking a position on the quality-based cycle as given in Fig. 2.

For the study 1048575 temperature data was collected. The features of the temperature sensor [Ardumotive: 'Humidity and Temperature', http://www.ardumotive.com/how-to-use-dht-21-sensor-en.html, accessed August 2018] that used when data are collected:

(a) Max sensible temperature: 80 °C
(b) Min sensible temperature: −40 °C
(c) Temperature Sensitivity: ±1 °C
(d) Sample period: 2 s

All that collected temperature data's distribution is given in Figs. 3 and 4.

Depending on the collected data, the empirical distribution function was used for the explaining the distribution of the collected data. An empirical distribution function can be defined as a distribution function linked to the empirical measure of a sample data set. The cumulative distribution function (CDF) is a step function that jumps up to

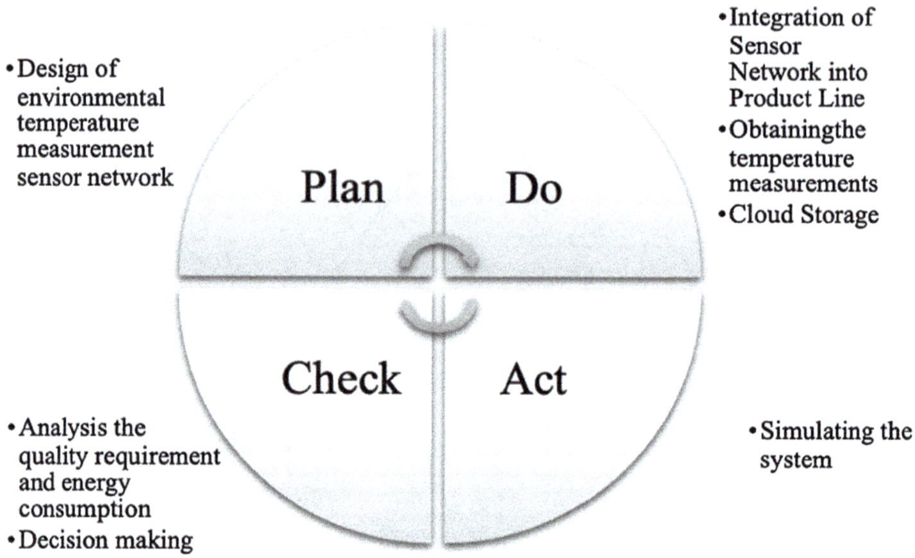

Fig. 2. Total quality management

Column Labels	Count	Cumulative	% Cumulative
	302	302	0,00028801
1	12	314	0,000299454
2	143	457	0,00043583
3	12	469	0,000447274
4	21	490	0,000467301
10	1	491	0,000468255
13	1	492	0,000469208
19	5	497	0,000473977
20	33713	34210	0,032625229
21	241098	275308	0,262554419
22	319251	594559	0,567016189
23	227920	822479	0,784377846
24	175835	998314	0,952067329
25	38058	1036372	0,988362301
26	11788	1048160	0,999604225
27	105	1048265	0,999704361
28	170	1048435	0,999866485
29	7	1048442	0,999873161
30	14	1048456	0,999886513
40	4	1048460	0,999890327
47	2	1048462	0,999892235
50	4	1048466	0,999896049
2528	1	1048575	1
General Sum	1048575		

Fig. 3. Distribution of the temperature data

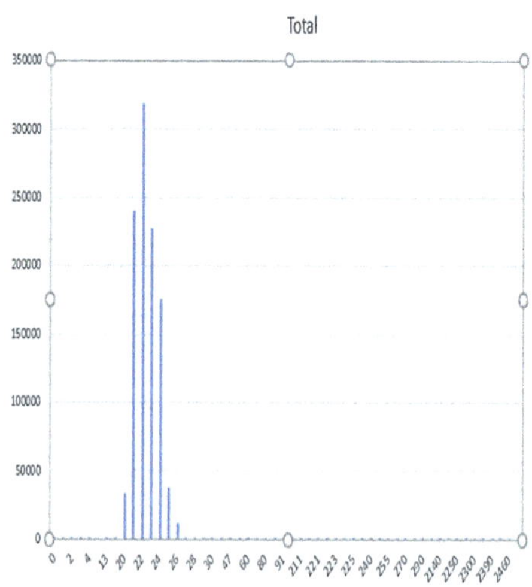

Fig. 4. Histogram form of the collected data

1/n at each of the n data points. The value of the determined data at any given time is the portion of the determined variable whose observation value is less than or equal to the determined value.

The empirical distribution function is an estimate of the cumulative distribution function, which is formed from the determined data. There are several random outcomes to determine the rate of convergence to the cumulative distribution function, which forms the basis of the empirical distribution function.

X is the random variable of the CDF F(x) = P(X $\leq$ x), and the series x_1, …, x_n be n random variables sampled from X [6]. The empirical distribution function is F(x):

$$F(x) = \frac{sampleset \leq x}{n} = \frac{1}{n}\sum_{i=1}^{n} I(x_i \leq x)$$

where I (x) is the indicator function [7].

With this formula description, a fixed value of x, I(xi $\leq$ x) is equal to a random value. Considering the fact that, the random value is equal to variance of F(x). This process provides the realistic temperature data for the simulation. For the increasing the success of the simulation the sample set is extended with the latest collected factory environment temperature data. With this updating the sample set design not only the improving the simulation success but also improve the TQM structure.

3 Process Management Modeling

Modeling the modules as nodes of implementation to collect and evaluate the data have enabled to design our project flexible in a manufacturing environment. The designed nodes are integrated within the operational process management model. Using the process management variables such as work force, cost, time and resources, we can assess the future scenarios. The possibilities and verification of the operational data can be checked the conformity to the required standards. These advantages will be used for any other Small and medium-sized enterprises (SMEs), which are in transition to the Industry 4.0 applications.

Managing the processes using the model designed, a knowledge database will be kept for further improvement and decision-making. Understanding the bottlenecks of the processes and making optimal changes will be based on quantified analysis. Realistic and precise results will provide the advantage to the organizations that keeps quality improvement as their main strategy.

4 Conclusion

The work conducted with the scope of this study, the determined quality standards based determined risks in production lines have either been fully eliminated or reduced to acceptable ‘low risk’ levels with the simulation system which is improved via continuously updating sample set. It has been evaluated that the proposed simulation model can benefit all Small and medium-sized enterprises (SMEs) by reducing production waste and increasing energy efficiency, provided that the data continuously obtained from the production line. This proposed cloud-based continuously updating simulation system will provide companies the advantage of cost-effectiveness, which will surely increase their competing power.

Acknowledgements. We thank TÜBİTAK (The Scientific and Technological Research Council of TURKEY) for encouraging us in our study.

References

1. Taherizadeh, S., Jones, A.C., Taylor, I., Zhao, Z., Stankovski, V.: Monitoring self-adaptive applications within edge computing frameworks: a state-of-the-art review. J. Syst. Softw. **136**, 19–38 (2018). ISSN: 0164-1212
2. Ai, Y., Peng, M., Zhang, K.: Edge computing technologies for Internet of Things: a primer. Digit. Commun. Netw. **4**(2), 77–86 (2018). ISSN: 2352-8648
3. Oliff, H., Liu, Y.: Towards Industry utilizing data-mining techniques: a case study on quality improvement. Procedia CIRP **63**, 167–172 (2017). ISSN: 2212-8271
4. Durakbasa, M.N., Bauer, J.M., Bas, G., Kräuter, L.: Novel trends in the development of autonomation and robotics in metrology for production in the future. IFAC-PapersOnLine **49**(29), 6–11 (2016)
5. Sen, Ö., Durakbasa, N., Erdol, H., Berber, T., Bas, G., Sevik, U.: Implementation of digitalization in food industry. In: DAAAM International Scientific Book 2017, Vienna, Austria, pp. 091–104 (2017). ISBN 978-3-902734-12-9. ISSN 1726-9687
6. http://bjlkeng.github.io/posts/the-empirical-distribution-function, 11 June 2019, 22:39
7. https://www.statisticshowto.datasciencecentral.com/empirical-distribution-function/, 11 June 2019, 23:47
8. Yerofeyev, S., Ipatov, O., Markov, S., Potekhin, V., Sulerova, A., Shkodyrev, V.: Adaptive Intelligent Manufacturing Control Systems, pp. 1016–1024 (2016)

Effect of Shape of Cutting Edge on Face Milled Surface Topography

Csaba Felhő(✉), Antal Nagy, and Janos Kundrák

Institute of Manufacturing Science, University of Miskolc, Miskolc, Hungary
{csaba.felho,antal.nagy,janos.kundrak}@uni-miskolc.hu

Abstract. For improved machining efficiency, in addition to the increase in material removal performance, the operating conditions and the assurance of the roughness values specified by the manufacturers for the surface of the component must be considered. In this article the topography of surfaces created with the same cutting data is analyzed with different geometry inserts. It is shown how the 2D and 3D roughness parameters can be estimated in different cases. Then the theoretical (estimated) and actual values from cutting operations are inspected, and the differences between them are discussed. Finally, the different versions are ranked based on the smallest achievable roughness values, and it was found, that the round insert gives the lowest roughness values.

Keywords: Face milling · Surface roughness · Theoretical roughness · Insert geometry · 3D roughness

1 Introduction

Today, many manufacturing processes are characterized by growing demands for cutting technologies. Since the quality of the machined surface, especially its topography, is of decisive importance for the wear and life of the parts encountering friction, the desired roughness characteristics have to be ensured during machining. This is also valid for the face milled surfaces we examine; these processes are widely used in the manufacture of various machine parts since we can produce relatively large flat surfaces economically and in good quality. The estimation of the surface roughness expected with the application of the exact cutting data has long been of interest to researchers, as this makes the cutting processes more predictable on the basis of surface roughness. Previously, this was done with a large number of cutting experiments, later using different methods and proper validation. Researchers have presented several different methods for determining surface roughness [1]. Basically, these methods can be divided into three major groups. The first of them is the use of analytical models or numerical calculation methods, where the surface roughness is determined by analytical or numerical calculations based on the tool geometry and the kinematics of the particular process. The general feature of this method is that it is often unable to take into account the dynamic effects of cutting. The other large group is the use of experimental methods, whose essence is to measure the value of the roughness characteristics under given conditions during cutting experiments, then to store the data in a knowledge base. The next time the same cutting parameters are set, guide values can be

N. M. Durakbasa and M. G. Gençyılmaz (Eds.): ISPR 2019, LNME, pp. 525–534, 2020.
https://doi.org/10.1007/978-3-030-31343-2_46

chosen to achieve the appropriate surface roughness characteristics. The third major group is the use of soft computing techniques (artificial intelligence, artificial neural networks, genetic algorithms), where a computer program gives an estimate of the expected roughness after being taught with appropriate data (real roughness results). In the following, some of the research on the surface topography of the face milling process is presented.

Munoz-Escalona and Maropoulos [2] described a geometric modeling method based on the generation of residual traces on machined surfaces in the case of milling with square inserts. The model was validated with real experiments, where the parameters changed were the cutting speed, the feed rate per tooth, the axial depth of cut and the tool edge radius. It was found that the method can be used to estimate roughness with an accuracy of about 98%. Wu and Yin [3] created a model which uses regression analysis to execute roughness estimation by varying speed, per-feed rate, and axial and radial depths. In addition, they developed a genetic algorithm based on the results of the measured data. Karkalos et al. [4] investigated the effect of three technological parameters – the depth of cut, cutting speed and feed rate – on the roughness of a machined surface made from Ti-6Al-4V aluminum alloy. They used several advanced procedures such as Response Surface Methodology, Analysis of variance (ANOVA), Box Behnken Design experiments and Artificial Neural Networks. Moaz et al. [5] used the Finite Element Method to estimate the roughness of milled titanium alloy surfaces. They found that there is a direct link between the force component in the feed direction (Ff) and the surface roughness, and developed a method for estimating roughness with finite element method based on the force component. Miko [6] using statistical methods on the basis of cutting experiment data, established a correlation established to estimate the expected roughness, in the case of milling steep free-form surfaces. Savkovic et al. [7] using two methods – the factorial experiment design and the fuzzy logic method – modeled the roughness of the surface with different input parameters (cutting speed, feed rate per tooth and depth of cut) for milling semi-solid aluminum alloy. Demircioglu and Durakbasa [8] compared optical and stylus-type roughness measurement methods in turning, milling and grinding, using statistical methods.

In the present article, partly as a continuation of our previous research results [9, 10], we examined the effect of tool edge geometry on the surface topography in case of face milling. First we performed cutting experiments with the selected cutting data and with three inserts of different geometries, and measured the 2D and 3D roughness parameters. After that, the theoretical roughness values were determined under the same conditions. In the research described in this article, we used CAD modeling of the machined surfaces to determine the theoretical roughness characteristics. The essence of this is to create a model of the machined surface in a CAD system, and then to translate its points with the help of an interface software into an evaluation software of the roughness measuring device (Fig. 1). Thus, real and theoretical roughness are evaluated on the same basis using the same software, which facilitates comparison of results [11]. For face milling, the geometry of the tool (cutting insert) is displayed on the surface in the form of a periodic profile where the distance of each trace corresponds to the feed rate per tooth fz. It was observed in [11] that the height of the imprints in the center line of the tool is the largest, and it decreases as we move away from the centerline.

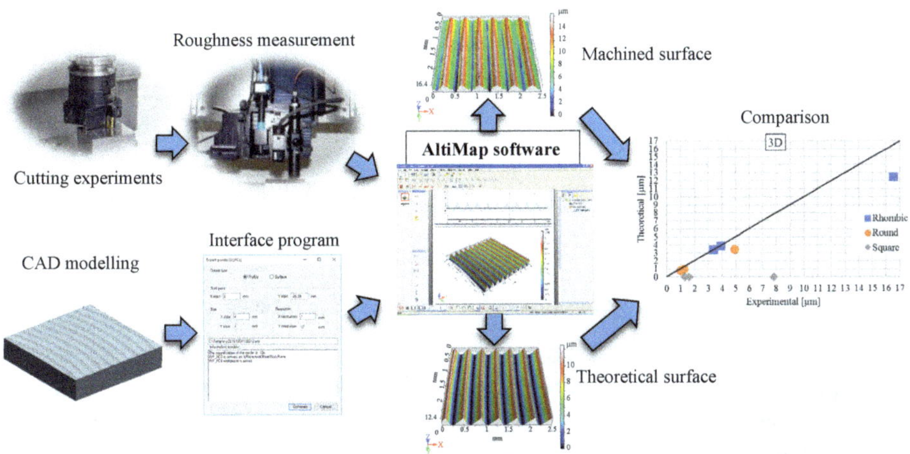

Fig. 1. Flow chart of examining surface topography (roughness characteristics)

2 Experiments

The aim of the experiments is to analyze and compare the values of the roughness parameters of face milled surfaces made with three types of cutting tool geometry.

2.1 Experimental Conditions

Experiments were performed with a Perfect Jet MCV-M8 vertical machining center. We created the workpiece surfaces with three different types of cutting edge geometry (Table 1). The specimens were made of unalloyed steel of normalized C45 (1.0503) grade with a tensile strength of 580 MPa and a hardness of 207 HBW [12]. The machined flat surfaces were 58 mm long and 50 mm wide, as illustrated in Fig. 2. The workpieces were milled with symmetrical setting.

Table 1. List of cutting inserts

Symbol	Insert code	D_t [mm]	Insert type	Details
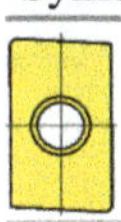	R215.44-15T308 M-WL	80	Rhombic 90°	$\kappa_r = 90°$; $\gamma_o = 0°$; $\alpha_o = 11°$; $r_\varepsilon = 0.8$ mm
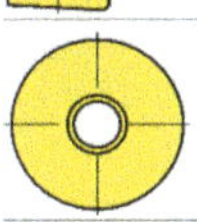	RCKT1204M0-PM	68	Round	$\gamma_o = 0°$; $\alpha_o = 7°$
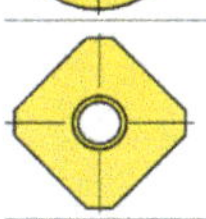	SEKN 1203 AFTN JC5030	63	Square 45°	$\kappa_r = 45°$; $\gamma_o = 0°$; $\alpha_o = 20°$; 0.85 × 45°

During the experiments the depth of cut was set to ap = 0.8 mm, the feed rate to fz = 0.4 mm/rev. and the cutting speed to vc = 200 m/min.

2.2 Method of the Examination

The topography was examined by analyzing the theoretical and real values of the 2D and 3D roughness values and by studying the relationship between the two types of values.

2.3 Analysis of the Roughness of the Face Milled Surfaces

The milled surfaces are analyzed based on 2D and 3D recordings of the surface topographies and roughness parameter values. The surface roughness was measured on an AltiSurf 520 surface roughness measuring device using a CL2 confocal chromatic sensor. We used AltiMap software to evaluate the data. The same system of analysis of the topography is used for the evaluation of the surface made with face milling experiments; therefore it is possible to validate the data obtained during theoretical modeling with the same roughness profile and values.

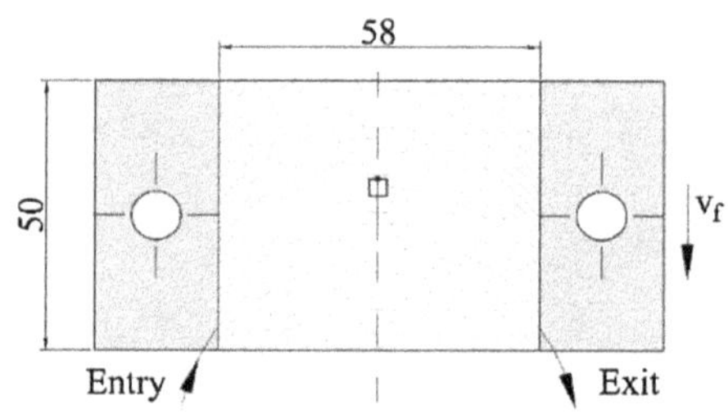

Fig. 2. Place of the measurements (mm)

The 2D profiles, recorded in the center line of the milling head in the feed direction, were measured according to ISO 4288:1998 and ISO 3274:1998. The 3D topographies were measured and evaluated according to ISO 25178-2:2012. The roughness was examined in the symmetry plane, because the maximal values can be found here, and the roughness measurements were carried out parallel to the feed direction according to the standard surface roughness procedure. The length of the 2D measurements was determined according to ISO 4288, the area of the 3D measurements was 2.5 × 2.5 mm each. In this work three kinds of 2D (Ra, Rz and Rq) and their appropriate 3D (Sa, Sz, Sq) roughness parameters were investigated that are commonly used in practice.

2.4 Determination of the Roughness Values Based on the Theoretical Profiles

The theoretical profile (topography) of the surface was determined with the cutting data used for the cutting experiments, which was created with a previously developed CAD model [13]. Then the points of the (theoretical) surface (x, y, z coordinates) are transferred to professional surface topography analysis software using the interface program, which will perform the evaluation according to the standard two- or three-dimensional roughness parameters to the theoretical values. The resulting surface points can be scanned into the AltiMap surface roughness evaluation software for evaluation of the desired two- or three-dimensional surface roughness parameters and visualization of roughness profiles.

One of the great advantages of this method is that we can generate 2D roughness profiles and 3D surfaces in any measurement position and direction. The novelty of the method using the applied principle is the possibility to analyze the complex surface topography created by any tool and any combination of tool and workpiece movement. Another great advantage is that other factors can be added to the model relatively easily. In this study we also examined the possibility of modeling different edge geometry of the inserts.

The theoretical surfaces created by modeling are shown in Fig. 3, and it can be seen, that the square insert with a bs wiper edge section (which is longer than the feed per tooth value) produces a flat surface.

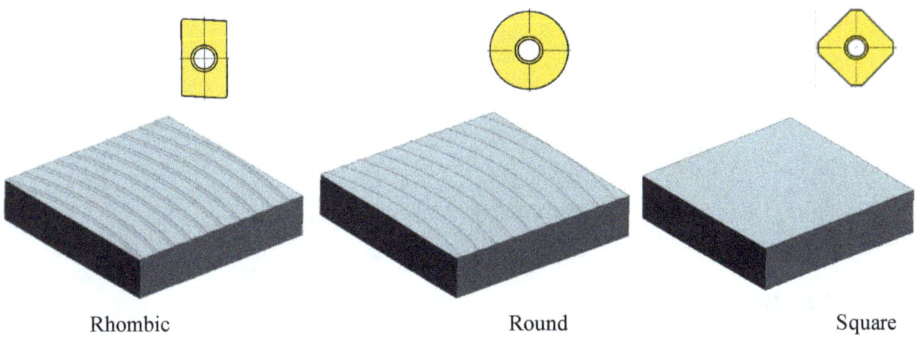

Fig. 3. 3D theoretical profiles created by modeling

2.5 Results of Experiments

The examinations were carried out with the three different cutting inserts and 2D and 3D roughness parameters were analyzed. The roughness values measured on the milled surface in the symmetry plane are shown in Table 2. Of the measurement results, Ra, Rz, Rq, Sa, Sz, Sq are shown in this paper. The roughness values determined from the profiles are summarized in Table 3.

Table 2. Roughness values measured in the symmetry plane

Experimental	Ra	Rz	Rq	Sa	Sz	Sq
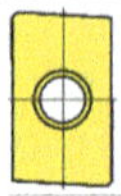	3.591	15.263	4.088	3.383	16.535	3.915
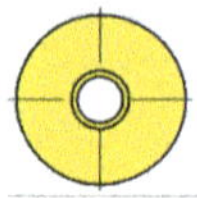	0.949	4.731	1.141	0.999	4.907	1.208
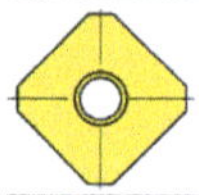	1.256	8.566	1.594	1.292	7.810	1.606

Table 3. Theoretical roughness values

Theoretical	Ra	Rz	Rq	Sa	Sz	Sq
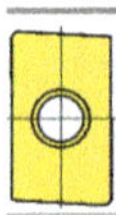	3.172	11.846	3.641	3.447	12.504	3.918
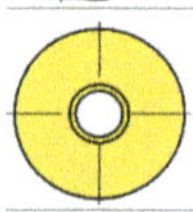	0.802	3.150	0.936	0.853	3.489	0.990
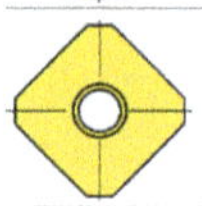	(0)	(0)	(0)	(0)	(0)	(0)

3 Discussion

Comparing the topography of the surfaces with different edge geometry inserts, it can be concluded that the roughness values differ significantly with the applied cutting data. The smallest values for each 2D parameter are for the surface milled with the round insert. Those are followed by the square insert, whose values at Ra are 1.32 times, at Rz 1.81 times, and at Rq 1.40 times higher. The highest values were found on the surface milled with the rhombic insert, where the difference compared to the circular insert was 3.78 times at Ra, 3.22 times at Rz, and 3.59 times at Rq.

In the examination of the theoretical profile, it is immediately obvious that no useful data was gained for the square shape. This turned out to be due to the following reason: when an insert with a chamfer parallel to the surface is used (here, the square insert), when the width of the chamfer bs is greater than the feed per tooth fz, the values of roughness equal 0, as indicated in Table 3. In the theoretical roughness values, the circular insert yielded the second smallest roughness values, while compared to the circular insert the Ra of the rhombic shape was 3.96 times higher, the Rz was 3.76 times higher, and the Rq was 3.89 times higher. When looking at these ratios we can conclude that the results of modeling compare well with the measured values.

The profile diagrams of the cut surface (Figs. 4 and 5) show the differences in the measured values for the round and the rhombic-shaped inserts. It can be seen that, despite the parallel section of the square insert, there is no better roughness than for the round shaped insert. Consideration should be given to the future investigations to measure roughness values for different κr' values of the tool edge. In the other two cases examined, the theoretical 2D profile follows the shape of the measured profile very well, usefully displaying the effects of microscopic material removal.

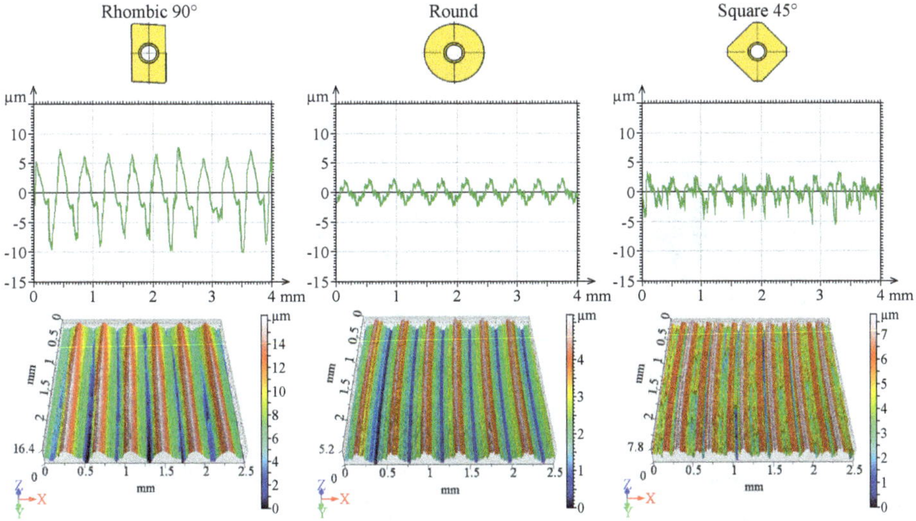

Fig. 4. Profile diagrams of the milled surfaces

Regarding the analysis of 3D surfaces, it can be stated that the values of roughness characteristics are more or less in line with the values measured in 2D profiles, the tendency of change is similar. However, 3D measurements give a much more spectacular sight of the nature of the cut surface, as shown in Figs. 4 and 5. We found that if the circular shape is used, the tool imprints can be easily seen on the surface with a relatively regular pattern, while the roughness parameters are the smallest of the three applied inserts. Therefore, bearing in mind the minimization of roughness, it seems more appropriate to use a radial edge instead of a large parallel edge section bs, which is often difficult to accurately set.

Our findings are well illustrated in Figs. 6 and 7, where a comparison of the theoretical and real roughness characteristics can be seen. Both the 2D and 3D roughness parameters for the circular insert have the lowest values and the theoretical values agree with this.

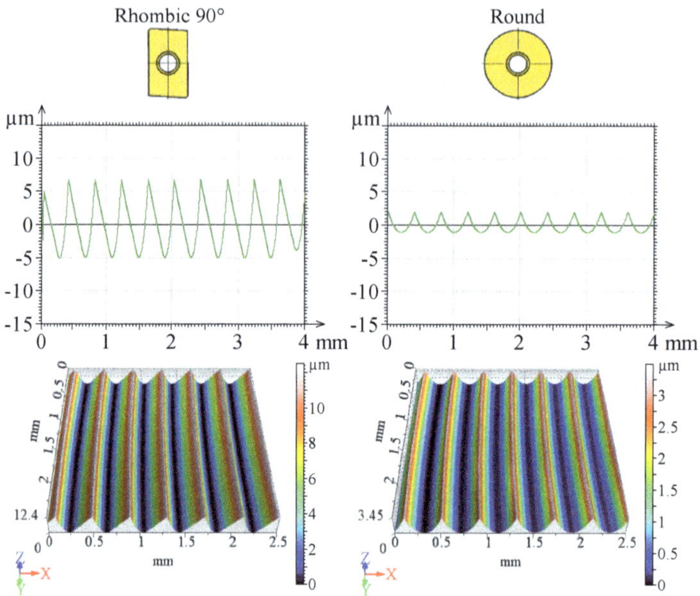

Fig. 5. Theoretical profile diagrams

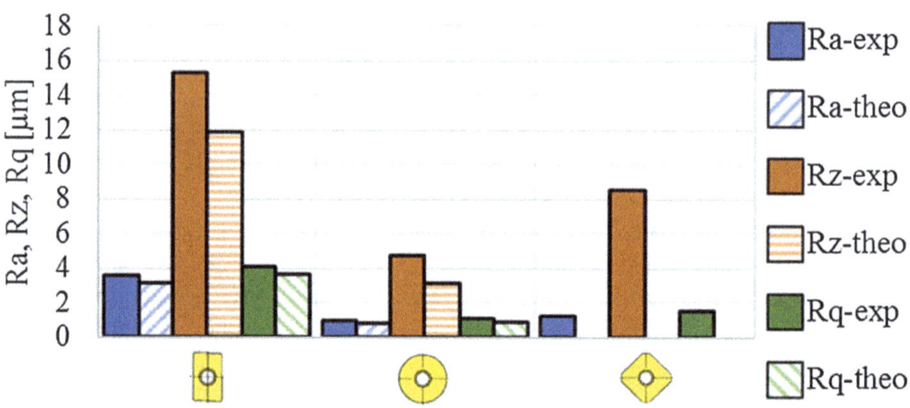

Fig. 6. Experimental and theoretical 2D roughness parameters when applying different inserts

Figure 8 shows the ratios of experimental and theoretical pairs of values. Most of the points are very close to the values that are considered ideal; only the points with the highest roughness values, Rz and Sz, show a greater deviation.

Based on the results it can be stated that the developed theoretical roughness modeling method is suitable for characterization of both 2D profiles and 3D surfaces. The obtained theoretical roughness data can be used to estimate the value of the surface roughness characteristics expected with the application of the given parameters. It is also apparent from the results that not all edge geometries have the same precision in the approximation, and it is often necessary to consider other factors and errors as well.

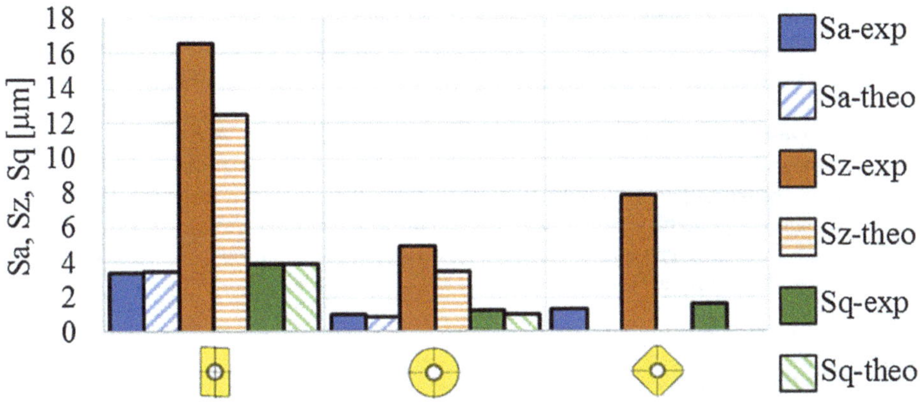

Fig. 7. Experimental and theoretical 3D roughness parameters when applying different inserts

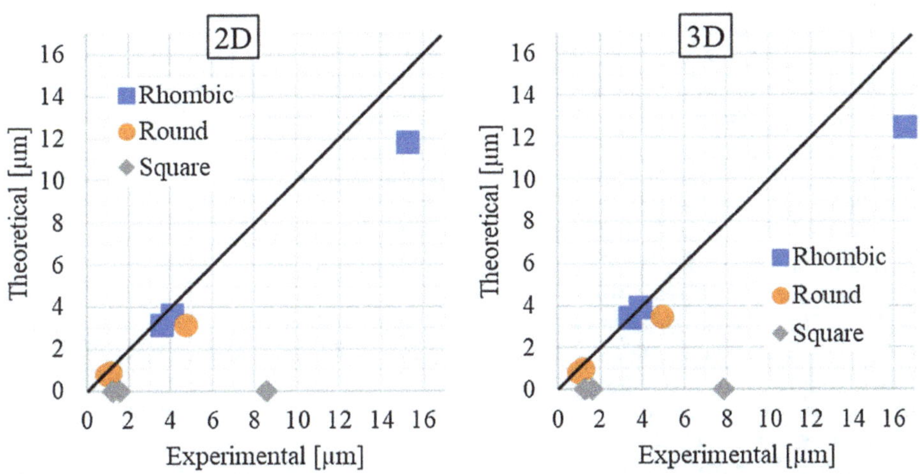

Fig. 8. Comparison of theoretical and real roughness values

4 Conclusions

In our work we have analyzed the topography of surfaces face milled with different cutting edge geometries. After carrying out cutting experiments, we compared the 2D and 3D roughness parameters of the surface and found that there was a significant difference in their values. The most favorable (i.e. the lowest roughness) is milling with round inserts, followed by the square insert, with the rhombic insert yielding the worst (highest) roughness.

Then we examined the extent to which the roughness parameters calculated using the previously developed CAD model follow the effect of edge geometry. We found that it is not advisable to apply the model to inserts working with the edge parallel to the cut surface, since when the length of the tool chamfer is greater than the feed per tooth, the theoretical value of the roughness parameters will always be equal to zero. The other edges follow the changes well and confirm that the CAD model is able to determine the expected roughness of the surface.

Acknowledgement. The authors greatly appreciate the support of the National Research, Development and Innovation Office – NKFIH (No. of Agreement: K 116876). The described study was carried out as part of the EFOP-3.6.1-16-00011 "Younger and Renewing University – Innovative Knowledge City – institutional development of the University of Miskolc aiming at intelligent specialization" project implemented in the framework of the program Szechenyi 2020. Both grants are gratefully acknowledged.

References

1. van Luttervelt, C.A., Childs, T.H.C., Jawahir, I.S., Klocke, F., Venuvinod, P.K., Altintas, Y., Armarego, E., Dornfeld, D., Grabec, I., Leopold, J., Lindstrom, B., Lucca, D., Obikawa, T., Shirakashi, Sato, H.: Present situation and future trends in modelling of machining operations progress report of the CIRP working group 'modelling of machining operations'. CIRP Ann. Manuf. Technol. **47**(2), 587–626 (1998)
2. Munoz-Escalona, P., Maropoulos, P.G.: A geometrical model for surface roughness prediction when face milling Al 7075-T7351 with square insert tools. J. Manuf. Syst. **36**, 216–223 (2015)
3. Wu, X., Yin, X.: Surface roughness analysis and parameter optimization of mold steel milling. Procedia CIRP **71**, 317–321 (2018)
4. Karkalos, N.E., Galanis, N.I., Markopoulos, A.P.: Surface roughness prediction for the milling of Ti–6Al–4V ELI alloy with the use of statistical and soft computing techniques. Measurement **90**, 25–35 (2016)
5. Moaz, H.A., Basim, A.K., Ansari, M.N.M., Bashir, M.: FEM to predict the effect of feed rate on surface roughness with cutting force during face milling of titanium alloy. HBRC J. **9**, 263–269 (2013)
6. Miko, B.: Surface quality prediction in case of steep free form surface milling. Key Eng. Mater. **686**, 119–124 (2016)
7. Savkovic, B., Kovac, P., Mankova, I., Gostimirovic, M., Rokosz, K., Rodic, D.: Surface roughness modeling of semi solid aluminum milling by fuzzy logic. J. Adv. Technol. Eng. Stud. **3**(2), 44–56 (2017)
8. Demircioglu, P., Durakbasa, M.N.: Investigations on machined metal surfaces through the stylus type and optical 3D instruments and their mathematical modeling with the help of statistical techniques. Measurement **44**, 611–619 (2011)
9. Felho, C., Kundrák, J.: Comparison of theoretical and real surface roughness in face milling with octagonal and circular inserts. Key Eng. Mater. **581**, 360–365 (2014)
10. Kundrák, J., Felho, C.: 3D roughness parameters of surfaces face milled by special tools. Manuf. Technol. **16**(3), 532–538 (2016)
11. Felho, C.: Investigation of surface roughness in machining by single and multi-point tools. Ph.D. thesis, Otto von Guericke University Magdeburg, Germany (2014)
12. C45/1.0503 Non-alloy quality steel - equivalent, chemical composition, properties. http://www.steelnumber.com/en/steel_composition_eu.php?name_id=152
13. Felho, C., Karpuschewski, B., Kundrák, J.: Surface roughness modelling in face milling. Procedia CIRP **31**, 136–141 (2015)

Examination of Surface State-Change on Diamond Burnished Aluminium Components

Viktoria Ferencsik and Gyula Varga(✉)

Institute of Manufacturing Science, University of Miskolc, Miskolc, Hungary
{ferencsik.viktoria, gyula.varga}@uni-miskolc.hu

Abstract. In the quality assurance of machine components finishing processes have important roles and the post-machining cold forming methods such as burnishing, rolling, shot peening and others have occupied a main place in industry. This work refers to the study of the ball-burnishing process which is a particular finishing treatment that provides good surface finish, high compressive residual stress and increased hardness in the surface layer. These characteristics are key for the fatigue life improvement of the component. In the present study, burnishing technique was used to investigate the effect of input parameters of the process (number of passes, feed rate, burnishing force) on selected surface layer features like micro-hardness and residual stress of the low alloyed aluminium components. For plan and execute the experiments we used the Taguchi type full factorial experimental design method by which empirical formulas can be created easily. The measurement of the surface micro-hardness was executed with Wilson Instruments Tukon 2100B measuring equipment, while the analyzation of the changing of tension conditions was performed by X-ray diffraction method. The measured results were evaluated by the comparison of a special correlation formula to determine the optimal combination level of the different parameters in the given interval.

Keywords: Burnishing · Plastic deformation · Micro-hardness · Residual stress · Full factorial experimental design

1 Introduction

At present, the manufacturing industry is focusing much attention on dimensional accuracy and surface finish. A surface, machined by conventional machining processes such as turning and milling, consists of inherent irregularities produced by the cutting tool, or a finer structure due to tearing of the material during machining. So final surface quality depends on largely the applied finishing treatments but reaching the required surface quality (required by the component drawing) requires great care when choosing the conditions for these processes. The reason of it is that different processes and/or their kinematics, the pairing of the tool and workpiece material, etc. or differences in technological data will result different surface quality. The different peculiarities of material removal should also be taken into account because other factors playing role, for example, when machining brittle materials (ceramics) [1] or using multipurpose [2] or abrasive tools [3, 4].

N. M. Durakbasa and M. G. Gençyılmaz (Eds.): ISPR 2019, LNME, pp. 535–544, 2020.
https://doi.org/10.1007/978-3-030-31343-2_47

Depending on the applied technique, there may already be significant differences in surface quality in the machining process, the nature and magnitude of the residual stresses [4], but this difference is also evident, for example, in the change of the micro-hardness of the surface layer [3].

There are many finishing treatments used in mechanical industry like polishing, grinding, lapping or honing and one of the most competent chipless finishing process is the burnishing [5–7].

Ball-burnishing is defined as a technological operation that consists of plastically deforming surface irregularities by the action of the force exerted by a ball [8].

In the past, burnishing was utilized only for smoothing of shafts and bores. After 1950, this method was applied in Germany and the former Soviet Union for work hardening of railway wagon axles and automotive crankshafts. Then the usage range becomes more extensive: inner and outer surfaces of hydraulic components, bearings, sealing surfaces, filets, etc. New materials, new tools, new attachments made possible the usage of burnishing on CNC machines [9].

Several studies deal with the theoretic and experimental examination of burnishing process, for example Luo et al. dealt with diamond burnishing of components made from aluminium alloy type LY12 and bronze type H62 [10, 11]. They created empirical formulas and experienced that from the point of view micro-hardness of the burnished surface the depth of burnishing is the most effective parameter which is followed by the burnishing force, feed rate and the speed of burnishing. Among the burnishing force, feed rate and burnishing speed Yu and Wang took into account the effect of nose radius of the tool, furthermore the rigidity of the tool spring to the surface roughness when executed experiments on aluminium alloy [12].

In this experiment, we analysed the influence of the number of passes (i), the feed rate (f) and the burnishing force (F).

2 Burnishing of Outer Cylindrical Surfaces

The burnishing process is realized by crushing the surface of materials without machining and it consists of pressing ball into the surface of the workpiece and imparting a feed motion at the same [13, 14].

The ball of which the material can be hardened steel, carbide, mineral ceramics and natural or artificial diamond, acts as tool surface layer, for the given normal force which gives high specific pressure, when this burnishing pressure exceeds the yield strength of the component, localized plastic deformation at the surface will take place in 0.01–0.2 mm thick layer in the subsurface area. This action leads to the spreading out of the valleys; causing an improvement of surface roughness. The grain structure is condensed, producing a hard surface with superior load-carrying and wear-resistance characteristics [15]. Figure 1 [16, 17] shows the mechanism of this phenomenon with the movement of burnishing tool over the workpiece.

The range of this deformation and as the hardening layer of the material is quite high, and changes in the structure of the surface layer occur at a significant depth.

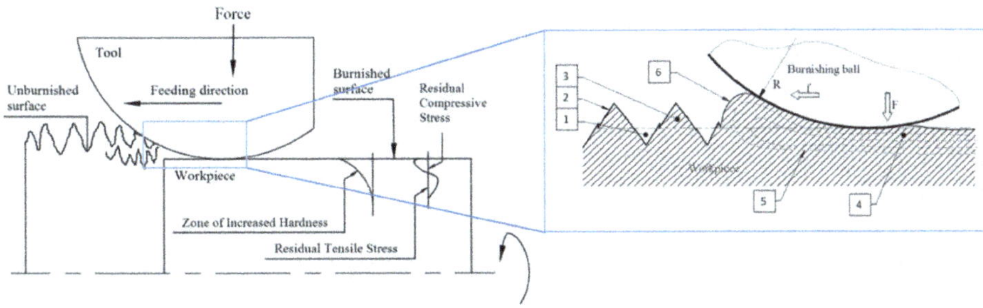

Fig. 1. Schematic drawing of a burnishing process [16, 17]. 1 - valley of roughness, 2 - surface profile, 3 - peak of roughness, 4 - zone of material deformation, 5 - zone of plastic and elastic deformation, 6 - wave of material in front of the burnishing tool

3 Experimental Conditions

3.1 Material

The quality and hardness of the workpiece to be burnished can vary widely. We have chosen low-alloyed aluminium for this experiment, due to its widespread use in industrial practice.

Figure 2 shows the geometry of the specimen influenced by the chosen experimental design method and the risk of bending during machining. On the basis of this, the test specimen is clamped with three jaw chuck and held up by standing peak.

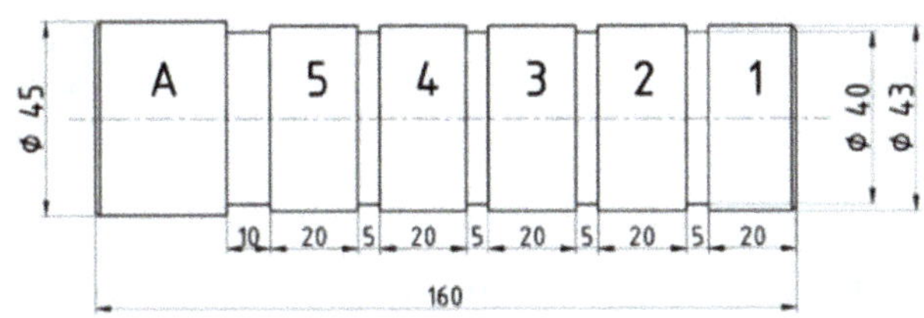

Fig. 2. Draft of the specimen

Before diamond burnishing, the specimens were smoothing turned with 0.2 mm/rev than 0.15 mm/rev feed rates.

3.2 Experimental Methodology

Many experiments involve the study of the effects of two or more factors. In general, factorial designs are most efficient for this type of experiment. By a factorial design, so in each complete trial or replication of the experiment all possible combinations are investigated [18]. For the design and implementation of the experiment, we have chosen the Taguchi type full factorial experiment design method [19, 20] with which empirical formulas can be efficiently created.

The aim of this method is to determine the function relationship between the dependent variable and the independent variables. Each independent variable, called factor can take more values, which are called levels.

The experimental work is conducted on a CNC lathe machine type OPTIMUM OPTIturn L-Series 440 at the Institute of Manufacturing Science at the University of Miskolc. A PCD-spherical (r = 3.5 mm) burnishing tool was used with a manual feed small amount of lubricating oil and 30 m/min speed was set. The main advantage of using numerically controlled machine is its flexibility. It enables the machining and burnishing operations to be accomplished easily in a sequential order. Any change in the burnishing conditions (such as speed, force, ball diameters, number of tool passes and feed) can easily be adjusted [15].

The matrix of Taguchi type full factorial experimental design is shown in Table 1, which contains the pre-experimental testing parameters for both the natural and the transformed dimensions. The numerical determination of the parameters was done based on preliminary experiments and extensive literature review.

Table 1. Burnishing parameters

No.	Parameters of burnishing			Transformed parameters		
	i [ø]	f [mm/rev]	F [N]	x_1	x_2	x_3
1	1	0.001	10	−1	−1	−1
2	3	0.001	10	+1	−1	−1
3	1	0.005	10	−1	+1	−1
4	3	0.005	10	+1	+1	−1
5	1	0.001	20	−1	−1	+1
6	3	0.001	20	+1	−1	+1
7	1	0.005	20	−1	+1	+1
8	3	0.005	20	+1	+1	+1

3.3 Measuring of Surface Micro-Hardness

Measuring of the surface micro-hardness of the specimens in three points in 120°, were executed with Wilson Instruments Tukon 2100B measuring equipment in Vickers hardness. The principle of this measurement technique is that how a material is subject to plastic deformation by using a standard force. During the measurement, a 136° diamond pulley is pressed with a specific force, 10 N in our experiments for 10 s on the surface to be measured. In the course of the evaluation, using a CCD camera, weigh the traces of the imprint and the average of the two is calculated by the device own software calculating the impression surface [21].

3.4 Measuring of Residual Stress

In a structural material or component, residual stresses exist in the object without the application of services or other external loads. They affect a part similarly to externally applied stresses. In materials because of the macroscopic residual stress the atomic cores in the lattice points are put of their balance. In the point of view of crystallography, it means that the parameters of the lattice are changing. Distance between lattice planes are changed because of the residual stress, these distances can be measured, so the values of the stress can be calculated from them [22].

So during measuring of residual stress, the wave-length of the X-ray is known and the dislocation of the Bragg-angle is measured that is caused by the changing of lattice planar distances. The formula (1) was applied what is called Bragg-equation [23]:

$$n\lambda = 2d_{hkl}\sin\Theta, \tag{1}$$

where:

n	integer determined by order given
λ	wave-length of X-ray
d_{hkl}	spacing between the planes in the atomic lattice
Θ	angle between the incident ray and the scattering planes

Residual stress distributions in cold formed sections are dependent primarily on the plastic deformation in the manufacturing process. Thus, it is necessary to know the manufacturing process before measuring the residual stress. In this investigation the measuring process was executed with an X-ray diffraction measuring machine type Stresstech Xstress 3000 G3R, which realizes non-destruction test and complex shapes can be measured providing rotation of the measuring head is not restricted. Figure 3 illustrates the measuring process which was done on 3 points at 120° in tangential and axial direction too because it was supposed that the direction of the processing has influence on the changing of the stress condition.

Fig. 3. The residual stress measuring process

4 Results

For evaluation of measured data improvement ratios were introduced, which are shown in formula (2)–(3):

$$\rho_{\sigma/HV} = \frac{|x_b|}{|x_t|} \tag{2}$$

$$\rho\% = (\rho - 1) \cdot 100\%, \tag{3}$$

where:

ρ_σ Improvement ratios of residual stresses (σ) and micro-hardness (HV). This is a dimensionless ratio, which textures the changes occurring because of manufacturing,

x_t Residual stress and micro-hardness remain after turning,

x_b Residual stress and micro-hardness remain after burnishing,

$\rho\%$ The percentage value of the improvement ratio.

The higher the value of $\rho\%$, the greater the improvement due to burnishing.

The measured data and the calculated improvement ratios are summarized in Table 2.

Table 2. Measured values and calculated improvement ratios

No.	σ_t [MPa]		$\rho_{\sigma t}$ [%]	σ_a [MPa]		$\rho_{\sigma a}$ [%]	Vickers hardness [HV]		ρ_{HV} [%]
	Turned	Burnished		Turned	Burnished		Turned	Burnished	
1	−9.5	−187.0	1868.4	74.5	−217.0	391.3	165.7	172.3	4.02
2	−9.5	−142.3	1397.9	74.5	−202.7	372.1	155.7	168.7	8.34
3	−9.5	−194.1	1943.2	74.5	−300.3	503.1	168.3	170.3	1.19
4	−9.5	−183.5	1831.6	74.5	−307.9	513.3	157.0	172.0	9.55
5	−9.5	−152.8	1508.4	74.5	−212.9	385.8	156.7	161.0	2.76
6	−18.5	−115.1	522.2	67.0	−167.1	349.4	149.0	167.0	12.08
7	−18.5	−198.7	974.1	67.0	−308.9	561.0	160.7	173.0	7.67
8	−18.5	−132.5	616.2	67.0	−198.9	396.9	178.3	179.0	0.38

With the application of full factorial experiment design method empirical formulas (4–6) were created and these calculations, such as the axonometric figures (Figs. 4, 5 and 6) were prepared using "MathCAD 15.0" software.

$$\rho_{\sigma_t} = 2.0463 \cdot 10^3 + 10.637 \cdot i + 1.597 \cdot 10^5 \cdot f + 8.324 \cdot F + 1.126 \cdot 10^4 \cdot i \cdot f - 29.101 \cdot i \cdot F - 1.858 \cdot 10^4 \cdot f \cdot F + 3.361 \cdot i \cdot f \cdot F \tag{4}$$

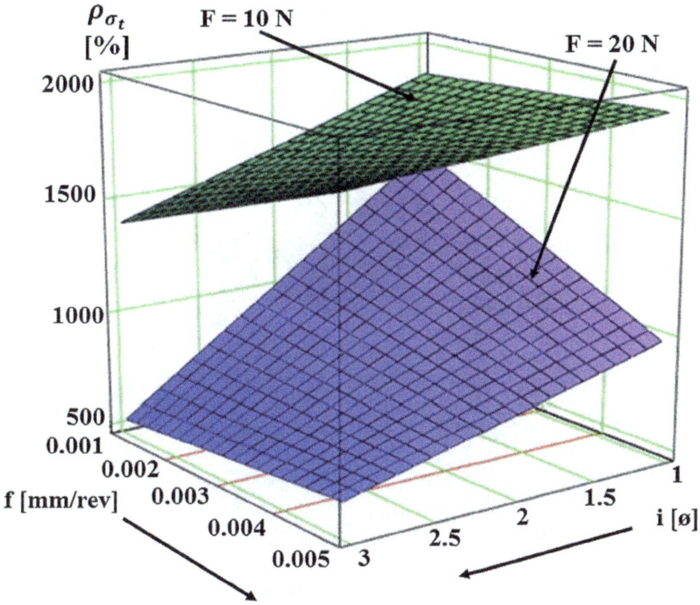

Fig. 4. Changing of improvement ratio of tangential residual stress

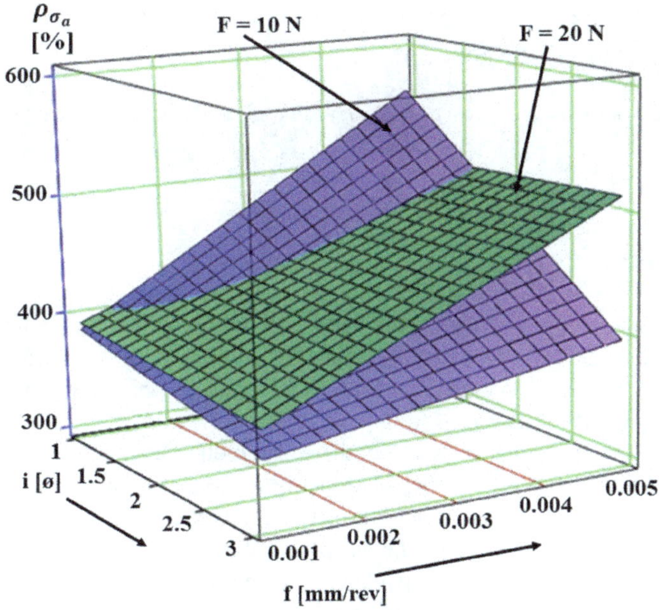

Fig. 5. Changing of improvement ratio of axial residual stress

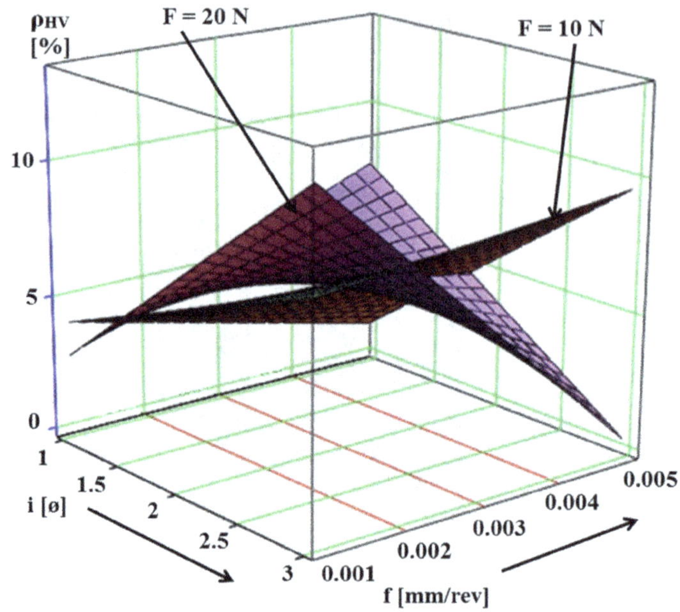

Fig. 6. Changing of improvement ratio of surface micro-hardness

$$\rho_{\sigma_a} = 409.0125 - 24.312 \cdot i - 1.121 \cdot 10^4 \cdot f - 3.239 \cdot F + 2.331 \cdot 10^4 \cdot i \cdot f + 1.104 \cdot i \cdot F + 3.549 \cdot 10^3 \cdot f \cdot F - 1.964 \cdot 10^3 \cdot i \cdot f \cdot F \quad (5)$$

$$\rho_{HV} = 11.3488 - 3.426 \cdot i - 5.729 \cdot 10^3 \cdot f - 0.828 \cdot F + 3.086 \cdot 10^3 \cdot i \cdot f + +0.508 \cdot i \cdot F + 451.625 \cdot f \cdot F - 258.125 \cdot i \cdot f \cdot F \quad (6)$$

5 Conclusions

The paper deals with the experimental analysis of diamond burnishing process on external cylindrical surfaces when the material of the workpiece was low-alloyed aluminum. Experimental parameters that were taken in consideration: number of passes, feed rate and burnishing force.

The aim of the experiments was how these parameters have effect on the changing of stress condition and surface micro-hardness. The experiments were executed and evaluated on the base of Taguchi type full factorial design experimental design method. The evaluation was more visible by the use of the dimensionless improvement ratios of tangential/axial residual stresses (ρ_t/ρ_a) and micro-hardness (ρ_{HV}), which were formed from the ratio of pre- and post-burnished results. After determining the empirical formula the 3D demonstration of it made the evaluation more picturesque.

On the base of the present research work it can be stated:

- Perpendicular to the direction of manufacturing, so in the case of tangential residual stress, the improvement ratio was the highest: $\rho_{\sigma t} = 1943\%$, while in axial direction the maximum value of this ratio was $\rho_{\sigma a} = 503.1\%$ with the application of same burnishing parameters (f = 0.005 mm, F = 10 N and single number of passes),
- The number of passes have more effect on $\rho_{\sigma t}$ and the explanation of it can be that the restructuring of dislocation can be realized in deeper in the subsurface layer,
- The above mentioned phenomenon is the interpretation of that the highest ρ_{HV} was caused by the application of triple number of passes because cold plastic forming and thus the hardening can be realized more efficiently as the tool and the surface of the workpiece are contacted with more power and greater machining time,
- Overall, it can be determined as many and more times as possible within reasonable limits, the tool is in contact with surface of workpiece the greater the degree of improvement will be due to burnishing technique.

In the future we would like to investigate the effect of number of passes on surface roughness at diamond burnished surfaces with the same parameter settings.

Acknowledgment. "The described study was carried out as part of the EFOP-3.6.1-16-00011 "Younger and Renewing University – Innovative Knowledge City – institutional development of the University of Miskolc aiming at intelligent specialisation" project implemented in the framework of the Szechenyi 2020 program. The realization of this project is supported by the European Union, co-financed by the European Social Fund."

"Project no. NKFI-125117 has been implemented with the support provided from the National Research, Development and Innovation Fund of Hungary, financed under the K_17 funding scheme."

References

1. Horvath, M., Kundrak, J., Mamalis, A.G., Gyani, K.: On the precision grinding of advanced ceramics. Int. J. Adv. Manuf. Technol. **20**(4), 255–258 (2002)
2. Felho, C., Kundrak, J.: Characterization of topography of cut surface based on theoretical roughness indexes. Key Eng. Mater. **496**, 194–199 (2012)
3. Kundrak, J., Mamalis, A.G., Markopoulos, A.P.: Finishing of hardened boreholes: grinding or hard cutting? Mater. Manuf. Process. **19**(6), 979–993 (2004)
4. Mamalis, A.G., Kundrak, J., Gyani, K.: On the dry machining of steel surfaces using superhard tools. Int. J. Adv. Manuf. Technol. **19**(3), 157–162 (2002)
5. Dzierwa, A., Markopoulos, A.P.: Influence of ball-burnishing process on surface topography parameters and tribological properties of hardened steel. Machines **7**(11), 1–3 (2019)
6. Shetty, D., Kumar, M., Prashed, K., Kumar, B.: Studies on effect of ball burnishing parameters on ferrous and non-ferrous materials with and without presence of abrasive particles. Int. J. Eng. Trends Technol. **48**(7), 378–381 (2017)
7. Ravenkar, G.D., Shetty, R., Rao, S.S., Gaitonde, V.N.: Analysis of surface roughness and hardness in ball burnishing of titanium alloy. Measurement **58**, 256–268 (2014)
8. Travieso-Rodríguez, J.A., Dessein, G., Gonzáles-Rojes, H.A.: Improving the surface finish of concave and convex surface using a ball-burnishing process. Mater. Manuf. Process. **26**(12), 1494–1502 (2011)

9. Neagu-Ventzel, S., Luca, L., Cioc, S., Schuster, M., Pilar, J., Marinescu, I.: FE 3D Burnishing model, p. 1–8. https://pdfs.semanticscholar.org/3339/77830bfed355badff0e2e0179027a6f953b4.pdf?_ga=2.153165176.119141369.1554545287-1125775818.1554545287. Year of download 20198
10. Luo, H., Liu, J., Wang, L., Zhong, Q.: The effect of burnishing parameters on burnishing force and surface micro-hardness. Int. J. Adv. Manuf. Technol. **28**, 707–713 (2006)
11. Luo, H., Liu, J., Wang, L., Zhong, Q.: Investigation of the burnishing process with PCD tool on non-ferrous metals. Int. J. Adv. Manuf. Technol. **25**, 454–459 (2005)
12. Yu, L., Wang, L.: Effect of various parameters on the surface roughness of an aluminium alloy burnished with a spherical surfaced polycrystalline diamond tool. Int. J. Mach. Tools Manuf **39**, 459–469 (1999)
13. Basak, H., Sonmez, F.: Examination of the effect of burnishing apparatus on surface roughness and hardness in burnishing process. Indian J. Eng. Mater. Sci. **24**, 115–122 (2017)
14. Mahajan, D., Tajane, R.: A review on ball burnishing process. Int. J. Sci. Res. Publ. **3**, 1–8 (2013)
15. Dabeer, P.S., Purohit, G.K.: Determination of surface roughness by ball burnishing process using factorial techniques. Int. J. Mech. Aerosp. Ind. Mechatron. Manuf. Eng. **7**(7), 1528–1531 (2013)
16. Vereschaka, A.A.: Improvement of working efficiency of cutting tools by modifying its surface properties by application of wear-resistant complexes. In: Advanced Materials Research, vol. 712–715, pp. 347-351. Trans Tech Publications (2013)
17. Dzionk, S., Scibiorski, B., Przybylski, W.: Surface texture analysis of hardened shafts after ceramic ball burnishing. Materials **12**(204), 1–15 (2019)
18. Montgomery, C.: Design and Analysis of Experiments, 7th edn. International Student Version, Arizona State University (2009)
19. Taguchi, G.: System of Experiment Design, p. 143. UNIPUB, Kraus International Publications, White Plains (1984)
20. Fridrik, L.: Chosen chapters from the topics of experimental design of production engineering. Műszaki Könyvkiadó, Budapest (1987). (In Hungarian)
21. Wilson® Instruments Tukon™ 2100B: Vickers/Knoop Hardness Tester, 11 July 2019. https://www.bergeng.com/mm5/downloads/hardness/WilsonInstruments_Tukon2100B.pdf
22. Mertinger, V., Sólyom, J.: Examination of residual stress by X-Ray diffraction. Study aid, University of Miskolc, pp 3–39 (2017)
23. Fitzpatrick, M.E., Fry, A.T., Holdway, P., Kandil, F.A., Shackleton, J., Suominen, L.: Determination of residual stresses by X-ray diffraction. Measurement Good Practice Guide, No. 52, p. 68 (2005)

Supply Chain Management

A Literature Review on Closed Loop Supply Chains

Aziz Kemal Konyalıoğlu(✉) and İlke Bereketli Zafeirakopoulos

Department of Industrial Engineering, Galatasaray University, Istanbul, Turkey
konyalioglua@itu.edu.tr, ibereketli@gsu.edu.tr

Abstract. Management of a supply chain is a complex process to model and optimize. Nowadays it is getting even more complex, due to environmental and economic problems we face. Supply chains require higher efficiency, energy and material recovery, and working with less cost. To fulfil these requirements, the concept of closed loop supply chain arose. There exist studies about closed loop supply chains in the literature and each of them proposes different methods to implement. In this study, it is aimed to put forward a literature review about closed loop supply chains, see which methods are used, and make a classification among them. Furthermore, the study will give a perspective about the most preferred methods and specified topic in closed loop supply chains in details.

Keywords: Supply chain management · Closed loop supply chain · Literature review · Production systems

1 Introduction

Since the economic and environmental parameters in supply chains change by the time, new approaches arise to implement in its management. Closed loop supply chain (CLSC) concept is one of them, and is getting more and more importance every year. CLSC can be defined as a complex task to recover of returned products or extended as recovering or collecting Inderfurth and Teunter (2001). It consists of both forward and backward flow of a supply chain. The chain starts with raw material suppliers and continues with production, finished good delivery to the customers and then collection of end-of-life products via third party companies or directly from consumers to manufacturers, suppliers or waste treatment partners, and finally the loop of material flow is closed. This leads to a sustainable system that may survive for many years, deals with environmental issues, and brings high efficiency to the supply chain. For designing a closed loop supply chain there are different methods and optimization models proposed in the literature.

In this study, it is aimed to investigate the studies from the literature about closed loop supply chain between the years 1999 and 2017, to see how the approaches to closed loop supply chain evolved in time. The study includes four main parts: an introduction part to give a perspective, a literature review part for examining the topics and methods in closed loop supply chain, an analysis part in order to classify the studies by their topics, years and methods and finally a conclusion and discussion part to debate the results of the study.

N. M. Durakbasa and M. G. Gençyılmaz (Eds.): ISPR 2019, LNME, pp. 547–556, 2020.
https://doi.org/10.1007/978-3-030-31343-2_48

2 Literature Review

Jayaraman et al. (1999) used a 0–1 integer programming for not only a location and distribution problem of remanufactured products but also their optimal amounts and transshipment costs in a closed loop supply chain. Krikke et al. (2003) studied closed loop supply chain of a Japanese refrigerators company by using their research and development data. Environmental worries lead new designs of products and process. Usually design of a product and logistics are dealing separately but at this study a model of combination of these topics is suggested. Different scenarios are used for process centralizing, product design alternatives, quality and quantities of returned products and environmental policies.

Min et al. (2005) used a heuristic approximation in a Lagrangian relaxation manner for obtaining a solution of multi echelon and commodity closed loop supply chain design problems. The important side of the solving the problem proposed by Min et al. is to practically solve the center allocation problems located in a closed loop supply chain by explaining the reverse and forward logistics activities and by not neglecting the backward flows.

Schultmann et al. (2006) studied reverse logistics in the view of automotive industry. They considered end of life vehicle in Germany to conduct a closed loop supply chain model using Tabu Search Algorithm. A system dynamics approach is presented by Kumar and Yamaoka (2007) by considering Japanese automotive sector. They used system dynamics for the reverse flow of the cars to find out the optimal scenario in terms of different recovery options.

Georgiadis and Besiou (2008) used a system dynamics approach for electrical and electronic equipment closed loop supply chains by taking into consideration just one single producer and single product closed loop supply chains in order to see independencies between key factors of the manufacturing system. They have also concluded that even if an environment of closed loop supply chain is generally suitable for forward logistics, recoverable and recyclable products environment are suitable for reverse logistics so that the design of closed loop logistic system developed by Georgiadis and Besious is supply driven and recycle-driven system.

Amaro and Barbosa-Póvoa (2009) have evaluated the effects of uncertainty while optimizing closed loop supply chain planning through different structures. To have an efficient supply chain, the product portfolio demands and prices have been evaluated by Amaro and Barbosa-Póvoa, and a mixed integer linear programming in a multi-period form has been constructed. The solution procedure of this mixed integer linear programming is a Branch and Bound algorithm to reach to the optimal solution. A real case study is conducted in a pharmaceutical company even if a centralized supply chain problem is not very convenient for the combination of stochastic demand and economical flows.

Raw material suppliers, producers, retailers, customers and recovery facilities are included in the closed loop supply chain model of Yang et al. They used variational inequalities theory for optimizing the equilibrium state of the network (2009). Pan et al. (2009) studied capacitated dynamic lot sizing problems in closed loop supply chain considering returned products. The market demand consists of both new produced goods and also remanufactured ones which are recovered from returned products. Manufacturing, remanufacturing and disposing capacities are limited and backlogging is forbidden. Under these conditions a model is suggested by solving it via dynamic programming algorithms. Different scenarios were conducted under capacitated and incapacitated conditions.

Pishvaee and Torabi (2010) studied a bi-objective probabilistic model by using mixed integer programming. The model includes both forward and backward flow of a supply chain and it copes with uncertainties. The solution is reached by using fuzzy approach.

Also, Kannan et al. (2010) studied genetic algorithm to solve mixed integer linear programming model. With legal issues of being environmental not only producing green but also collecting end of life products is a must today. So reverse logistics are very important to manage for returning products. These returned products will be faced with different recovery options such as recycling, remanufacturing or reusing. This process flow has to be linked to a logistics network. The proposed model in this study is designing a model for lead-acid batteries to produce new ones because these battery wastes are hazardous, it is completely restricted for disposal. A multi echelon, multi period, multi product closed loop supply chain network model is suggested considering material, production, distribution and recovery flows. Genetic algorithm results are compared to GAMS software solutions that results were sufficient and high quality.

Pishvaee et al. (2011) studied closed loop supply chain in terms of robustness. Robust supply chain design is needed because of important changes in business such as demand or costs. The study suggests a robust optimization model for uncertainty of the inputs for a closed loop supply chain. A mixed integer linear programming model is used for the model design and then the model is presented under robust optimization theory and robustness verification is done by other robust models. Wei and Zhao (2011) studied pricing decisions using fuzziness in a closed loop supply chain. They used a combination of game theory and fuzzy theory to find optimal wholesale price in a retail competition and different scenarios were evaluated. Winkler (2011) made a linkage between a sustainable supply chain and a closed loop supply chain which is directly related to each other.

On the other hand, Zarandi et al. (2011) developed and designed a closed supply chain for distribution network design problems which are reverse flows are going into forward flows by using a fuzzy goal programming approach in an interactive manner. By the aid of fuzzy goal programming, Zarandi et al. concluded that the logistics activities, especially designed as a closed loop supply chains, have a time-manageable property and network design can make backward flows possible while dividing into sub-backward problems.

Another study of a closed loop supply chain is done by Oulgu and Wong (2012) in automotive industry. They presented fuzzy rule-based system by using a software called Visual Basic.Net. The automotive company is used as a case study and results may be a road map for the future studies. Furthermore, there exists another study used a hybrid system for production planning by Kenné et al. (2012). They designed a model under machine constraints such as failures and repairs. It is a single production that includes both manufacturing and remanufacturing operations. Remanufacturing is a better option than disposing end of life products. The proposed model aims to minimize holding and backlog costs. For solution they used stochastic dynamic programming.

Mitra (2012) investigated two-echelon closed loop supply chains considering correlated demands and returns and also inventories. Stochastic and deterministic models for two-echelon systems, including reverse and forward logistics have been generalized by cost structures. Numerical example is done in the study and it is assumed that demands are suitable for normal and gamma distributions. In the study, at the end, an inventory management and revenue management have been combined by the aid of vehicle routing modelling.

Wei and Zhao (2013) evaluated fuzzy closed loop supply chain for backward flow decisions. Produced new goods or remanufactured ones are collected by producers, retailers or third-party companies. By using game theory and fuzzy theory together, companies' strategies are evaluated.

Wei and Zhao (2013) have tried to decide the reverse channel choices by using fuzzy Closed Loop supply chains, including important elements of a manufacturing system such as suppliers, retailers, used products. Also, they have investigated the third-party logistics activities in order to see how these activities affect closed loop supply chain. The first goal of the study is to understand three different used product collection behaviors by taking into consideration fuzzy theory and game theory. Fuzziness is used to comprehend the customer demand fluctuations and manufacturing costs by a numerical analysis. The main assumptions in the study are symmetry assumption which implies the manufacturer, retailer, and third party have vice-versa information. Furthermore, Subramanian et al. (2013) used a priority based simulated annealing for closed loop supply chain network design problems. A one-period and one-product closed loop supply chain under a deterministic demand and stochastic product return has been constructed and heuristic based Vogel's approximation has been used to reach to the initial solution by taking into consideration a recursive cost function. Implementation of sustainability in this closed loop supply chain problems has been discussed.

Another mixed integer programming study is conducted by Amin and Zhang (2013a, b). They introduced a two phased model of supplier selection using fuzzy approach and a multi objective mixed integer linear programming for selecting refurbishing site selection. They studied under uncertain demand and return conditions in 2013. A closed loop supply chain model was designed and a multi objective facility location model is presented (Amin and Zhang 2013a, b). Özceylan and Paksoy (2013a, b) used mixed integer linear programming for closed loop supply chain. This model includes both two ways of the supply chain with multi products and periods. The model helps to determine locations and retailers and also gives the optimal amount of transported goods and disassembled products.

Amin and Zhang (2013a, b) configured a closed loop supply chain including multiple customers, products, remanufacturing subcontractors and suppliers in which a three-stage model consisting of evaluation and network configuration with allocation has been tried to comprehend. Furthermore, the first stage of the study by Amin et al. is to construct a qualify function deployment model to understand the relationship between customer requirements and process requirements by the aid of fuzzy set theory in order to deal with fuzziness and uncertainty. The second stage of the study is to model a stochastic mixed integer nonlinear programming model in which demand is stochastic to allocate suppliers and other parameters of this closed loop supply chain.

Devika et al. (2014) studied a sustainable closed loop supply chain model again by using mixed integer programming. This study not only concerns economic issues but also considers environmental effects of the whole supply chain system. Another study of Özceylan et al. (2014) used nonlinear mixed integer programming for a closed loop supply chain network. This model aims to optimize tactical and strategic decisions of the chain. Strategic decisions are gathered from the quantities of the goods in the forward and backward flows. Tactical ones are concerning reverse flow for disassembly line balancing. The objective function is minimizing the cost for every step such as procurement, remanufacturing, disassembly or transportation.

Ramezani et al. (2014) designed a closed loop supply chain network under fuzziness by considering multiproduct, closed loop supply chain interaction and multiperiod properties in which maximizing the profit and quality, minimizing the delivery time are the main goals of the model. Mathematical modelling, which is used in the study, is a fuzzy mixed integer linear programming as constraints are not certain and coefficients are not deterministic. Crisp optimization values and fuzzy optimization values are compared at the end of the study and the numerical example has been used to show if the model is appropriate for this closed loop supply chain. The proposed model by Ramezani et al. provides a reverse logistics and a forward logistics strategy for not only suppliers but also for retailers.

Designing a closed loop supply chain is an NP-hard problem that a new hybrid model was suggested by Soleimani and Kannan (2015). They used particle swarm optimization and generic algorithm together for designing a closed loop supply chain model for large scale networks. Planning and designing variables were included in the new model. These two popular meta-heuristic algorithms helped to improve results when they are taken individually by genetic algorithm and swarm optimization. This model is used for an Iranian hospital furniture manufacturer as a case study.

Another hybrid model is conducted by Keyvanshokooh et al. (2016), that robust and stochastic optimization were combined. Their objective is to maximize profit as a mixed integer linear program. The hybrid model of robust stochastic modelling is covering up uncertainties of transportation costs and demand and returned products sets. Latin Hypercube Sampling method is used for transportation cost scenarios and stochastic Benders decomposition is presented to solve this model.

Talaei et al. (2016) modelled a carbon-efficient closed loop supply chain network design problem by the robust fuzzy way. They have used a mixed integer programming model for solving facility location and allocation problems of a closed loop supply chain network under fuzziness. The main reason why they have used fuzzy programming is that carbon efficient closed loop supply chain problems can be considered under uncertain conditions by aiming to minimize the total cost while minimizing carbon emission.

Tseng et al. (2015) tried to assess the features of a closed loop supply chain in the view of sustainability concept. They have used a balanced scorecard measurement system financially and no financially in a closed loop supply chain hierarchy. To model this system, fuzzy Delphi method and analytical network process have been used to understand the interdependency through measurements. The result of their model provides a support system for the stakeholders and management strategy in sustainable closed loop supply chains.

Zhalechian et al. (2016) designed a sustainable closed loop location routing inventory supply chain network under a stochastic environment. A new mathematical modelling, which is a multi-objective mathematical modelling for designing a sustainable closed loop supply chain, introduced by Zhalechian et al. has advantages as it includes allocation and routing decisions and green factors. Combining the stochastic and deterministic variables in the model make the study approaches to a real-life case as it also includes a sensitivity analysis. They mainly concluded that the size of closed loop supply chain has a big effect on the costs and inventory management.

Soleimani et al. (2017) have used a fuzzy multi-objective modelling for a CLSC network design and developed a genetic algorithm for solving their model. As demand is not crisp, they used a fuzzy approach for demand modelling.

3 Analysis of the Literature Review

40 articles were evaluated in terms of closed loop supply chain. They all used different methods and gave different suggestions by considering both forward and backward material, goods, service flows. Figure 1 shows years of the inspected articles.

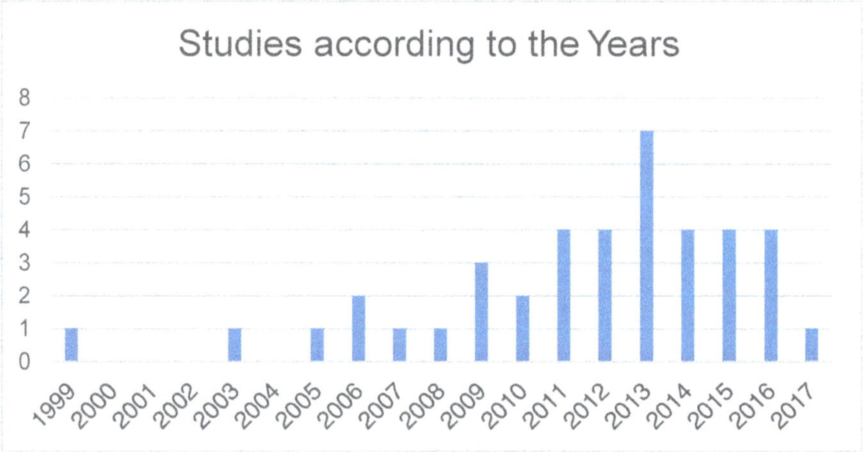

Fig. 1. Literature studies by years

In the Fig. 1, it has been seen that the studies through the years change and fluctuate but generally the number of studies about closed loop supply chains attempts to increase. 2013 is the year that the highest number of studies has been published.

Another graph in Fig. 2 shows the methodology differences in between 40 articles which are conducted from the literature review.

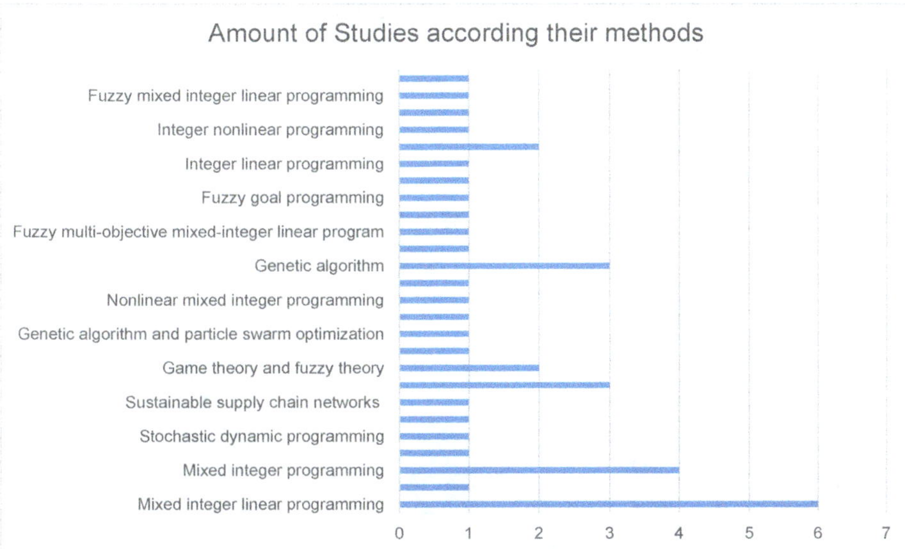

Fig. 2. Amounts of different methods that are used in the articles

In the Fig. 2, it is seen that the most used method is mixed integer linear programming in closed loop supply chain themed studies. Furthermore, mixed integer programming, including nonlinear and linear programming, is also widely used and genetics algorithms are designed for closed loop supply chain problems.

It is also understood that fuzzy logic is also widely used in closed loop supply chain problems compared to other deterministic problems. In Fig. 3, it can be seen that generally non-fuzzy methods are used in the articles.

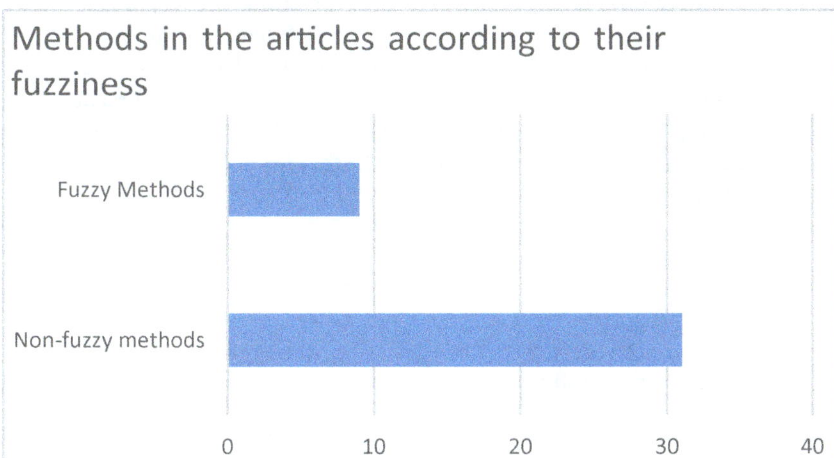

Fig. 3. Amount of studies according to their fuzziness

Figure 4 shows the articles according to their sectors. As seen in the Fig. 4, logistic activities are the area that CLSC design and applications are the most-used one. Automotive and Electric and Electronic Equipment (EEE) sectors are respectively the second and the third one.

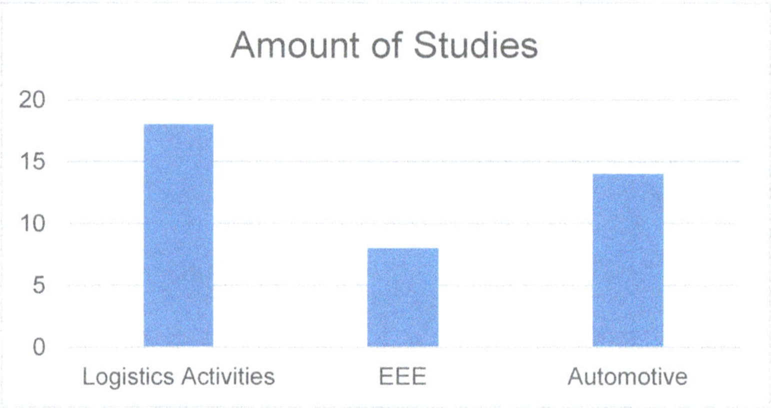

Fig. 4. Amount of studies according to sectors

4 Conclusion

In this study, it is aimed to review studies about closed loop supply chain in the past 20 years. The years, sectors and methodologies based on the studies have been grouped to make an analysis. As the environmental concerns and green activities increase, the importance of closed loop supply chain, especially green CLSC is growing. In the studies which are reviewed, it is seen that there are many different methods to optimize or design a closed loop supply chain. Not only deterministic approaches but also stochastic approaches have been used for modelling closed loop supply chains. Furthermore, generally mathematical approaches are used to understand it. The studies are increased as of the year 2013. The cases are mostly chosen from the logistics, electric and electronic equipment and automotive sectors. It can be seen that logistic activities sector is the area that CLSC is mostly used. For further research, the amount of studies can be increased to see other sectors.

References

Amaro, A.C.S., Barbosa-Póvoa, A.P.F.: The effect of uncertainty on the optimal closed-loop supply chain planning under different partnerships structure. Comput. Chem. Eng. **33**(12), 2144–2158 (2009)

Amin, S.H., Zhang, G.: A multi-objective facility location model for closed-loop supply chain network under uncertain demand and return. Appl. Math. Model. **37**(6), 4165–4176 (2013a)

Amin, S.H., Zhang, G.: A three-stage model for closed-loop supply chain configuration under uncertainty. Int. J. Prod. Res. **51**(5), 1405–1425 (2013b)

Devika, K., Jafarian, A., Nourbakhsh, V.: Designing a sustainable closed-loop supply chain network based on triple bottom line approach: a comparison of metaheuristics hybridization techniques. Eur. J. Oper. Res. **235**(3), 594–615 (2014)

Georgiadis, P., Besiou, M.: Sustainability in electrical and electronic equipment closed-loop supply chains: a system dynamics approach. J. Clean. Prod. **16**(15), 1665–1678 (2008)

Jayaraman, V., Guide Jr., V.D.R., Srivastava, R.: A closed-loop logistics model for remanufacturing. J. Oper. Res. Soc. **50**(5), 497–508 (1999)

Kannan, G., Sasikumar, P., Devika, K.: A genetic algorithm approach for solving a closed loop supply chain model: A case of battery recycling. Appl. Math. Model. **34**(3), 655–670 (2010)

Kenné, J.P., Dejax, P., Gharbi, A.: Production planning of a hybrid manufacturing–remanufacturing system under uncertainty within a closed-loop supply chain. Int. J. Prod. Econ. **135**(1), 81–93 (2012)

Keyvanshokooh, E., Ryan, S.M., Kabir, E.: Hybrid robust and stochastic optimization for closed-loop supply chain network design using accelerated Benders decomposition. Eur. J. Oper. Res. **249**(1), 76–92 (2016)

Krikke, H., Bloemhof-Ruwaard, J., Van Wassenhove, L.N.: Concurrent product and closed-loop supply chain design with an application to refrigerators. Int. J. Prod. Res. **41**(16), 3689–3719 (2003)

Kumar, S., Yamaoka, T.: System dynamics study of the Japanese automotive industry closed loop supply chain. J. Manuf. Technol. Manag. **18**(2), 115–138 (2007)

Min, H., Ko, H.J., Park, B.I.: A Lagrangian relaxation heuristic for solving the multi-echelon, multi-commodity, closed-loop supply chain network design problem. Int. J. Logist. Syst. Manag. **1**(4), 382–404 (2005)

Mitra, S.: Inventory management in a two-echelon closed-loop supply chain with correlated demands and returns. Comput. Ind. Eng. **62**(4), 870–879 (2012)

Olugu, E.U., Wong, K.Y.: An expert fuzzy rule-based system for closed-loop supply chain performance assessment in the automotive industry. Expert Syst. Appl. **39**(1), 375–384 (2012)

Özceylan, E., Paksoy, T.: A mixed integer programming model for a closed-loop supply-chain network. Int. J. Prod. Res. **51**(3), 718–734 (2013a)

Özceylan, E., Paksoy, T.: Fuzzy multi-objective linear programming approach for optimising a closed-loop supply chain network. Int. J. Prod. Res. **51**(8), 2443–2461 (2013b)

Özceylan, E., Paksoy, T., Bektaş, T.: Modeling and optimizing the integrated problem of closed-loop supply chain network design and disassembly line balancing. Transp. Res. Part E Logist. Transp. Rev. **61**, 142–164 (2014)

Pan, Z., Tang, J., Liu, O.: Capacitated dynamic lot sizing problems in closed-loop supply chain. Eur. J. Oper. Res. **198**(3), 810–821 (2009)

Pishvaee, M.S., Torabi, S.A.: A possibilistic programming approach for closed-loop supply chain network design under uncertainty. Fuzzy Sets Syst. **161**(20), 2668–2683 (2010)

Pishvaee, M.S., Rabbani, M., Torabi, S.A.: A robust optimization approach to closed-loop supply chain network design under uncertainty. Appl. Math. Model. **35**(2), 637–649 (2011)

Ramezani, M., Kimiagari, A.M., Karimi, B., Hejazi, T.H.: Closed-loop supply chain network design under a fuzzy environment. Knowl. Based Syst. **59**, 108–120 (2014)

Schultmann, F., Zumkeller, M., Rentz, O.: Modeling reverse logistic tasks within closed-loop supply chains: an example from the automotive industry. Eur. J. Oper. Res. **171**(3), 1033–1050 (2006)

Soleimani, H., Kannan, G.: A hybrid particle swarm optimization and genetic algorithm for closed-loop supply chain network design in large-scale networks. Appl. Math. Model. **39**(14), 3990–4012 (2015)

Soleimani, H., Govindan, K., Saghafi, H., Jafari, H.: Fuzzy multi-objective sustainable and green closed-loop supply chain network design. Comput. Ind. Eng. **109**, 191–203 (2017)

Subramanian, P., Ramkumar, N., Narendran, T.T., Ganesh, K.: PRISM: PRIority based SiMulated annealing for a closed loop supply chain network design problem. Appl. Soft Comput. **13**(2), 1121–1135 (2013)

Talaei, M., Moghaddam, B.F., Pishvaee, M.S., Bozorgi-Amiri, A., Gholamnejad, S.: A robust fuzzy optimization model for carbon-efficient closed-loop supply chain network design problem: a numerical illustration in electronics industry. J. Clean. Prod. **113**, 662–673 (2016)

Tseng, M., Lim, M., Wong, W.P.: Sustainable supply chain management: a closed-loop network hierarchical approach. Ind. Manag. Data Syst. **115**(3), 436–461 (2015)

Wei, J., Zhao, J.: Pricing decisions with retail competition in a fuzzy closed-loop supply chain. Expert Syst. Appl. **38**(9), 11209–11216 (2011)

Wei, J., Zhao, J.: Reverse channel decisions for a fuzzy closed-loop supply chain. Appl. Math. Model. **37**(3), 1502–1513 (2013)

Winkler, H.: Closed-loop production systems—a sustainable supply chain approach. CIRP J. Manufact. Sci. Technol. **4**(3), 243–246 (2011)

Zarandi, M.H.F., Sisakht, A.H., Davari, S.: Design of a closed-loop supply chain (CLSC) model using an interactive fuzzy goal programming. Int. J. Adv. Manuf. Technol. **56**(5–8), 809–821 (2011)

Zhalechian, M., Tavakkoli-Moghaddam, R., Zahiri, B., Mohammadi, M.: Sustainable design of a closed-loop location-routing-inventory supply chain network under mixed uncertainty. Transp. Res. Part E Logist. Transp. Rev. **89**, 182–214 (2016)

Inderfurth, K., Teunter, R.: Production planning and control of closed-loop supply chains (No. EI 2001-39) (2001)

Evaluation of Wireless In-flight Entertainment System in Aviation Sector

Elif Altın(✉) and İlke Bereketli Zafeirakopoulos

Industrial Engineering, Galatasaray University, Istanbul, Turkey
eliffaltin@gmail.com, ibereketli@gsu.edu.tr

Abstract. Supply chain management, which require doing better with less cost and with maximum quality, is one of the most important subjects in the developing aviation world. The aviation industry has many district regulations, complex products and systems. In order to deal with that, aviation companies endeavour to follow recent technologies, to develop new innovation projects and to associate them in supply chain operations effectively. Moreover, airline companies purpose to focus on sustainable development of their systems. Sustainability on the system performance can be supported by adding values to the supply chain operations. Innovation projects contribute to value-adding activities that can support sustainable supply chain management. One of the breakthrough projects in the airline industry is Wireless In-Flight Entertainment (WIFE) system. The project is represented as the most innovative version of in-flight entertainment systems. It enables connectivity of passengers on the internet by their own portable electronic devices (PED). Additionally, the system provides reduction of size, weight and energy consumption. The system also offers passenger satisfaction, cost saving, additional revenue and functional efficiency for airlines. The aim of the study is to evaluate the WIFE system in respect of economic aspect during the management of the supply chain operations. Cost-benefit analysis is envisaged to be employed to compare two different systems in the same research efficiently. The research is planned to be implemented in Turkish Airlines Technic that is a maintenance, repair and overhaul (MRO) company.

Keywords: Supply chain management · Cost-benefit analysis · Aviation industry · Wireless inflight entertainment (WIFE)

1 Introduction

There is a world-wide request for more efficient systems that demands low cost and less resource consumption; also presents high value and quality during supply chain processes. Those systems require innovation by promoting better sustainable performance of products.

Supply chain in aviation industry requires management of different product components that are produced by various suppliers in different countries. Therefore, airline companies reveal their significant efforts to achieve sustainable development.

With growing attention to sustainable development, airline companies focus on product and process innovation activities to achieve green and sustainable aviation. As a result of adopting innovative practices systematically by means of project

N. M. Durakbasa and M. G. Gençyılmaz (Eds.): ISPR 2019, LNME, pp. 557–569, 2020.
https://doi.org/10.1007/978-3-030-31343-2_49

management standards, better environmental performance of products and processes can be gained under industrial competitiveness.

When targeting to a more sustainable universe, aviation industries highly prefer light weight components and solutions. One of the important projects for reducing the environmental impacts of airline companies is Wireless In-flight Entertainment (WIFE) systems. WIFE system provide savings in many areas through reduction of size, weight and energy consumption and it is designed to influence the passenger's comfort level in a positive way.

Wireless in-flight entertainment is growing into one of the most advanced areas of the in-flight experience. Besides, WIFE system has been providing in-flight internet connectivity to passengers with the rule of following necessary instructions. Additionally, airlines can also provide a chance of using portable electronic devices (PEDs) during the flight.

Sustainable supply chain management requires right analysis of economic situation of the systems. The aim of the study is to evaluate WIFE project in the scope of Supply Chain Management by depending on economic dimension of the system in detail at Turkish Airlines Technic.

2 Literature Review

The research was advanced with all studies with keywords "supply chain management, sustainability, aviation industry, wireless inflight entertainment (WIFE) and cost benefit analysis. The study was improved with the support of previous researches. In the literature review, new product designing, sustainability, sustainable supply chain and in-flight entertainment systems were evaluated under economic, environmental and social aspects. According to the reviews, it has seen that there is no study conducting cost benefit analysis of WIFE systems in aviation industry. The main aim of the study is to fill this gap and contribute to the literature (Table 1).

Table 1. The literature survey of supply chain management in aviation industry and in flight entertainment systems

Author	Title	Objective of the study	Methods used	Topic of the study	Impact category
(Görener et al. 2017) [1]	A hybrid type-2 fuzzy based supplier performance evaluation methodology: The Turkish Airlines technic case	Aiming to improve effectiveness of sustainable supplier selection in a MRO company	Fuzzy Logic AHP TOPSIS	Supplier Performance Evaluation (SPE)	Sustainable supply chain management

(*continued*)

Table 1. (*continued*)

Author	Title	Objective of the study	Methods used	Topic of the study	Impact category
(Keivanpour et al. 2015) [2]	End-of-life aircraft treatment in the context of sustainable development, lean management, and global business	Analysing retired aircrafts in respect of different aspects	Optimization, Lean management	Directing retired aircraft models	Environmental Social Economic
(Kim et al. 2016) [3]	First-Class in-Flight Services and Advertising Effectiveness: Antecedents of Customer-Centric Innovativeness and Brand Loyalty In the United States (US) Airline Industry	Examining effectivity of in-flight services from the aspect of customer experience	A qualitative analysis	In-flight entertainment	Customer experience
(Liu 2007) [4]	In-flight entertainment system: State of the art and research directions	Aiming to introduce future in-flight systems guiding	Literature review	In-flight entertainment systems	Customer focused
(Moreira 2014) [5]	Integrating Eco-design and PLM in the Aviation Completion Industry: A Case Study	Examining benefits of product life cycle and eco-design activities	A qualitative analysis	Product life and eco-design management	Environmental Economic
(Morgado 2016) [6]	Accessible aircraft seat and console controls for passengers with disabilities through mobile devices	Introducing a in-flight entertainment model for passengers with disabilities	A qualitative analysis	In-flight entertainment	Passenger focused
(Phu et al. 2009) [7]	Towards An In-flight Infrared Peer-to-Peer Media Streaming Network	Analysing more technological versions of in-flight entertainment systems	A qualitative analysis	In-flight entertainment	Quality Reliability Technology

(*continued*)

Table 1. (*continued*)

Author	Title	Objective of the study	Methods used	Topic of the study	Impact category
(Raj and Srivastava 2018) [8]	Sustainability performance assessment of an aircraft manufacturing firm	Proposing a model to develop sustainability performance of an aircraft company	Fuzzy Best Worst Multi Criteria (FBWM) decision-making	Sustainability of an aircraft manufacturing company	Environmental Social Economic
(Tsai and Lai 2018) [9]	A green approach to the weight reduction of aircraft cabins	Evaluation of impacts on implementation of green activities in aircraft cabins	Activity-Based Costing (ABC), the Theory of Constraints (TOC), a mathematical programming model	Green Aviation	Environmental

3 In-flight Entertainment System

All airlines are concerned about customer satisfaction and keeping up with competitors. Therefore paying a particular attention to in-flight systems is considered as one of the main aspires. Airlines are perpetually seeking new and innovative ways to boost the in-flight expertise because advanced, user friendly in-flight amusement technologies can invariably be in demand.

The first purpose of the in-flight entertainment system (IFE) is to provide passengers comfort with food and beverages, especially when they see nothing else than the sky. The enclosed environment of the aircraft can cause discomfort or even problems to passengers [4]. The second purpose of IFE is to reduce stress level of passengers intelligently and effectively, and to make them enjoy their flight to the last.

With the increasing demands of the passengers, the competition of airlines and the advancement of technology, the airline companies started to offer more sophisticated services in where electronic devices played a major role. This has led to the in-flight entertainment system being a multi-purpose system, with the exception of catering and physical comfort. Moreover, in-flight entertainment systems provide passengers with the opportunity to follow their work throughout the flight, watching movies, listening to music, and playing games.

Passenger travel demand is expected to increase 5 percent/year over the next 20 years, and a total of 56,000 new aircrafts will be needed by 2040 to accommodate the projected demand [10]. Therefore the use of the in-flight entertainment will be one of the most requested factors of attracting customers for aviation companies in the coming years.

There are two types of in-flight entertainment systems that are classified as wired and wireless.

3.1 Wired In-flight Entertainment System

Today, modern in-flight entertainment systems present astonishing number of films, television series, songs, games and moving-maps. They are considered with embedded hardware (touch screens, seat remote controls, earphones) Fig. 1.

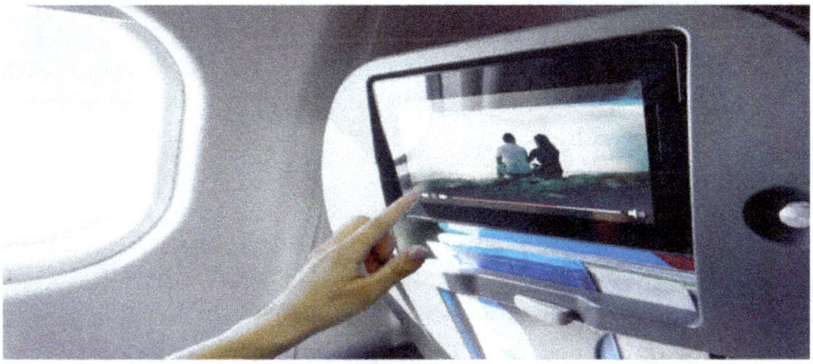

Fig. 1. Wired IFE system [11]

With the helping of analysis based on related information on Turkish Airlines Technic Inc., it is seen that there are some limitations of wired in-flight systems as below:

(1) *Setting up of the System:* Installation of the system is a laborious and long-term process. Also, the existing systems are difficult and costly to install, and to maintain is a disadvantage in terms of capturing and developing technology.
(2) *Expense:* The installation of an IFE is expensive. In addition, the weight of aircraft components also has a large impact on cost. The difference between the wired in-flight entertainment system and the wireless in-flight entertainment system is evident on the seat side. In the wireless in-flight entertainment system, all the components of the seat side are reduced to the passenger's own device.
(3) *Serviceability:* The use of wired systems for passengers is not very useful due to sensitivity of the screens. Passengers can be more comfortable while using their own personal devices.

3.2 Wireless In-flight Entertainment System

Technology is ever-changing so with the rise in laptops, smartphones, tablets e-readers and MP3 players, airline companies have initiated to provide in-flight entertainment that can stream content to passengers' own devices. In this system, customers can access to the published movies, music, reading and various entertainment content by a web page for computers and by a developed application for smartphones and tablets. After the application is installed, the in-flight entertainment system will be used. Moreover, airlines are willing to present on board Wi-Fi connectivity that gives chances for passengers to encounter an even more personalised system. This type of new innovative project is called wireless in-flight system.

According to understanding obtained from Turkish Airlines Technic Inc., the two systems are compared in terms of customer pleasure, cost, maintenance and weight. It can be said that the wireless in-flight entertainment system has many advantages over wired in-flight entertainment system. These are as follows:

(1) *Passenger satisfaction:* Passengers can have access to many kinds of films, music and games streamed to their portable electronic devices (PEDs).
(2) *Lighter aircraft with reduced fuel burn:* The biggest difference between the wired and the wireless in-air entertainment system is on the seat side. Many seat side components are not used therefore it presents a lighter aircraft and thanks to the decrease in aircraft, fuel burn is reduced.
(3) *A very short installation time:* No installation is required on the seat side. This can absolutely shorten the delivery time of the system.
(4) *A much lower initial investment costs:* No exterior hardware and special demand are needed. It enables the system to be initiated easily with a low cost.
(5) *Less maintenance costs:* Since the wireless system will not be a seat side or will consist of only a power unit, it will generate less maintenance costs than the wired system.
(6) *Being passenger oriented:* In wired systems, the devices are manufactured according to the firms' rules so it can be hard to understand all the critical details of the systems. In addition to this, the response time of the systems is very slow. Otherwise, in the wireless in-flight entertainment system, the passengers can use their own devices to enjoy the entertainment system by easily connecting to the system.
(7) *Ancillary revenues for airlines:* Airlines can obtain ancillary revenues by enabling additional services such as advertising and online shopping.
(8) *Operational efficiencies for airlines:* Contents can be updated more frequently and content integration lead times and costs are eliminated.

4 Methodology: Cost Benefit Analysis

Cost benefit analysis (CBA) is a systematic method to evaluate investment options in terms of cost and benefits such as ongoing cost-reduction and labour saving [13]. CBA is the most substantially used project evaluation methodology to compare projects by observing their benefits and costs in order to make decision on an investment plan. If the comparison is made among several projects, the project which is the most beneficiary is chosen. Otherwise, if the decision is made for a project, the project that has net benefits more than the net costs is resolved as profitable.

The purpose of using CBA is choosing the right project that brings maximum profit. While managing the CBA for any project, all costs and benefits should be converted to a common monetary units to advance the basis for comparison.

There are general steps which must be followed in CBA in the following order:

identification of costs and benefits,
measurement of costs and benefits,
comparison of costs and benefits, and
project selection.

$$\text{Cost Benefit Ratio } C/B = \frac{\sum_{t=0}^{n} \frac{C_t}{(1+i)^n}}{\sum_{t=0}^{n} \frac{B_t}{(1+i)^n}} \tag{1}$$

where C is cost, B is benefits, t is year, and i is the discount rate.

If C/B < 1, Benefits exceed the costs. The project should be permitted to proceed.
If C/B > 1, Costs exceed the benefits. The project should not proceed.
If C/B = 1, Costs equal the benefits which means the project should be permitted to proceed.

5 Modelling

For constructing the model, firstly the criteria affecting the selection of the more beneficial system are determined. The criteria are identified as economical inputs and economical outputs for both systems. Secondly, under each inputs and outputs, costs and benefits are determined. Thirdly, measurement of costs and benefits are generated. Later, comparison of present value (PV) of costs and benefits between the two systems is constructed. Finally, the most profitable system is found.

6 Application

The application of cost benefit analysis between two different in-flight entertainment system is implemented in Turkish Airlines Technic Inc. The most profitable system is determined by using qualitative and quantitative data generated by real values and assumptions.

Turkish Airlines Technic Inc. is part of the brand community of Turkish Airlines A.O. The brand, which is one of the leading aviation service suppliers in the world, provides its customers with its expert team; offers high quality and comprehensive maintenance, repair, modification and reconfiguration services. The company operates in two locations in Istanbul, at Istanbul Airport and Sabiha Gökçen Airport. In addition to engineering and maintenance activities, the company provides services in many areas of component maintenance and replacement support, design, certification and production.

6.1 Cost Benefits Analysis of Wired IFE System

As a common calculation for wireless and wired IFE systems, material weight, that are used in a narrow body aircraft, are evaluated by collecting real data separately. Secondly, fuel costs that are occurred for the two systems are calculated by assuming the fuel cost $108 for a kg of weight. The details are shown in Table 2.

Table 2. Material weight-fuel cost calculation for the two IFE systems

	Wireless IFE	Wired IFE
Material weight	45 kg	750 kg
Fuel cost	$4.860	$81.000

It is assumed that the two different systems have the same amount of sales in five years as in Table 3.

Table 3. Sales amounts for the two systems

Years	2018	2019	2020	2021	2022
Sales amounts	1	3	5	7	9

For five years, costs and benefits that can occur for wired system are presented in Table 4. Below cost estimations are generated by hypothesis based on some values taken from company employees.

Table 4. Cost distribution for wired IFE for five years

Wired IFE system	Years				
	2018	2019	2020	2021	2022
Labor cost	$1.000.000	$1.500.000	$2.000.000	$2.500.000	$3.000.000
Material cost	$1.000.000	$3.000.000	$5.000.000	$7.000.000	$9.000.000
Indirect cost	$3.000.000	$3.500.000	$3.500.000	$3.000.000	$3.000.000
Fuel cost	$81.000	$243.000	$405.000	$567.000	$729.000
Benefit	$2.000.000	$6.000.000	$10.000.000	$14.000.000	$18.000.000

Table 5 shows that in 2021, the net benefit will begin to progress positively. Otherwise, in respect of the net cumulative benefits, it can be said that the system will not totally bring net profits until the end of the year 2022.

Table 5. Calculation of cumulative benefits and costs for wired Ife for five years

Wired IFE system	Years				
	2018	2019	2020	2021	2022
Present value of costs	$5.233.430	$8.243.000	$10.587.379	$12.316.901	$14.394.263
Cumulative costs	$5.233.430	$13.476.430	$24.063.809	$36.380.709	$50.774.973
Present value of benefits	$2.060.000	$6.000.000	$9.708.738	$13.196.343	$16.472.550
Cumulative benefits	$2.060.000	$8.060.000	$17.768.738	$30.965.081	$47.437.630
Net benefit	−$3.173.430	−$2.243.000	−$878.641	$879.442	$2.078.287
Net cumulative benefits	−$3.173.430	−$5.416.430	−$6.295.071	−$5.415.629	−$3.337.342

The Fig. 2 below shows that in year 4, PV value of benefits and costs are estimated approximately (Fig. 4).

Fig. 2. Wireless IFE system [12]

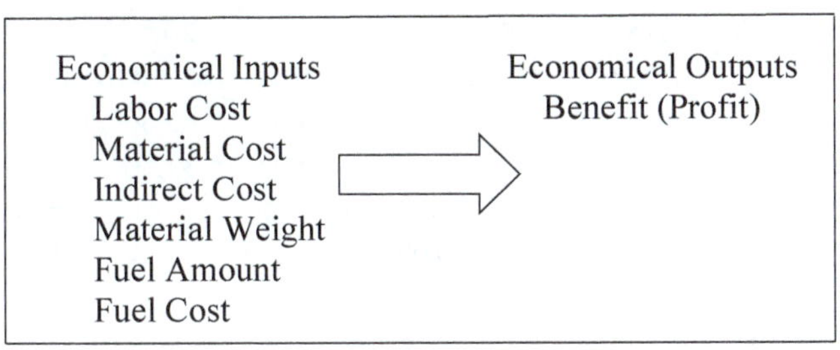

Fig. 3. Economical inputs-outputs

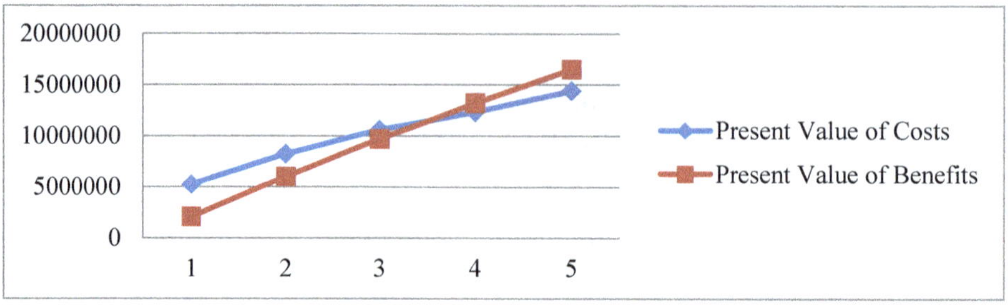

Fig. 4. PV of costs and benefits for wired system for five years

In Table 6, C/B for wired system is evaluated and it demonstrates that the ratio is more than one. It means that the system is not beneficial and is not providing profits in five years.

Table 6. Ratio of cost benefit for wired system

Total present value of costs	$50.774.973
Total present value of benefits	$47.437.630
Ratio of cost and benefit (C/B)	1,0703522

6.2 Cost Benefit Analysis of Wireless IFE System

For five years, costs and benefits that can occur are demonstrated in Table 7. Based on some values taken from company employees, cost estimations are made by hypothesis for five years.

Table 7. Cost distribution for wireless IFE for five years

Wireless IFE system	Years				
	2018	2019	2020	2021	2022
Labor cost	$400.000	$800.000	$1.000.000	$1.500.000	$2.000.000
Material cost	$400.000	$1.200.000	$2.000.000	$2.500.000	$3.000.000
Indirect cost	$1.000.000	$1.500.000	$1.500.000	$1.500.000	$1.500.000
Fuel cost	$4.860	$14.580	$24.300	$34.020	$43.740
Benefit	$1.000.000	$3.000.000	$5.000.000	$7.000.000	$9.000.000

Table 8 shows that in 2020 the system will start to make net profits progressively and in 2022 according to the net cumulative benefit, the system will continue to bring increasing profits.

Table 8. Calculation of cumulative benefits and costs for wireless IFE for five years

Wireless IFE system	Years				
	2018	2019	2020	2021	2022
PV of costs	$1.752.291	$3.414.580	$4.877.961	$5.687.642	$6.446.019
Cumulative PV of costs	$1.752.291	$5.166.871	$10.044.832	$15.732.474	$22.178.493
PV of benefits	$1.030.000	$3.000.000	$4.900.000	$6.598.171	$8.236.274
Cumulative PV of benefits	$1.030.000	$4.030.000	$8.930.000	$15.528.171	$23.764.445
Net benefit	−$722.291	−$414.580	$22.039	$910.529	$1.790.255
Net cumulative benefit	−$722.291	−$1.136.871	−$1.114.832	−$204.303	$1.585.952

The Fig. 3 below shows that in year 3, PV value of benefits and costs are balanced almost equally (Fig. 5).

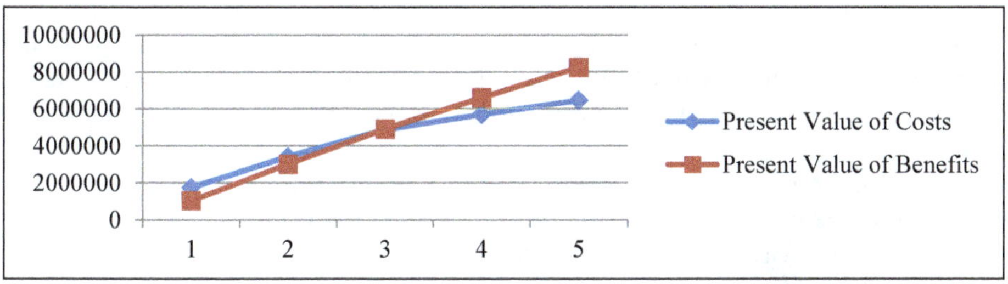

Fig. 5. PV of costs and benefits for wired system for five years

In Table 9, C/B for wireless system is calculated and found that the ratio is less than one. It means that the system is profitable and is bringing benefits in years.

Table 9. Ratio of cost benefit for wireless system

Total Present Value of Costs	$22.178.493
Total Present Value of Benefits	$23.764.445
Ratio of Cost and Benefit (C/B)	0,933263663

7 Conclusion

Nowadays, there is an ongoing competition between airline companies to be able to survive in the industry. Those companies can progress their development by giving a special attention to their customers. They must follow general aviation innovations, regulations and requirements of the industry to reach their main aims.

Following and creating innovations requires a systematic improvement in supply chain processes because supply chain management supports to provide efficient products with low cost and high quality that those especially contribute much to aviation industry.

In addition to this, sustainability of the industry can be supported by presenting simplified systems with small size, low weight and energy consumption. One of the modern examples is Wireless in-flight system (WIFE). In this study, this system is evaluated by applying cost benefit analysis. The main goal of the study is to compare Wired and Wireless IFE systems to demonstrate the economic advantages of the WIFE system over wired system.

Firstly, economical inputs and outputs are determined and taken into consideration as costs and benefits. Secondly, costs and benefits are measured for the two systems during five years. Later, cost benefit analysis is implemented for the two systems. According to the application, wireless system starts to bring net profit earlier than wired system does. In addition to this, cost benefit ratio of wireless system is found smaller than wired system and it means that wireless system will contribute to the company much positively in the long term.

As a summary, the analysis mainly shows that economically, wireless IFE system is more advantageous than wired IFE system. The wireless system requires less weight, labour, material, indirect and fuel cost therefore those factors support a more beneficial system.

The main limitation of this study is that some of the data are generated by assumptions. The main contribution of this research is that it fills a gap in the literature by evaluating the wireless in-flight entertainment systems by using cost benefit analysis. Finally, it is found economically more favourable than traditional wired systems.

References

1. Görener, A., Ayvaz, B., Kuşakcı, A.O., Altınok, E.: A hybrid type-2 fuzzy based supplier performance evaluation methodology: the Turkish Airlines technic case. Appl. Soft Comput. J. **56**, 436–445 (2017)
2. Keivanpour, S., Ait Kadi, D., Mascle, C.: End of life aircrafts recovery and green supply chain (a conceptual framework for addressing opportunities and challenges). Manag. Res. Rev. **38**(10), 1098–1124 (2015)
3. Kim, S., Kim, I., Hyun, S.S.: First-class in-flight services and advertising effectiveness: antecedents of customer-centric innovativeness and brand loyalty in the United States (US) airline industry. J. Travel Tour. Mark. **33**(1), 118–140 (2016)
4. Liu, H.: In-flight entertainment system: state of the art and research directions. In: Second International Workshop on Semantic Media Adaptation and Personalization (SMAP 2007), pp. 241–244 (2007)
5. Moreira, N., Aït-Kadi, D., Vieira, D.R., Romero, A., de Santa-Eulalia, L.A., Wang, Y.: Integrating eco-design and PLM in the aviation completion industry: a case study, pp. 169–180 (2014)

6. Morgado, M.R., José, M.A., Rocha, A.M., Fleury, A.L., Lopes, R.D., Zancul, E.S.: Accessible aircraft seat and console controls for passengers with disabilities through mobile devices. In: Proceedings of the International Symposium on Consumer Electronics, ISCE, pp. 63–64 (2016)
7. Phu, T.V., Huynh, V.A., Bretscheider, T.R.: Towards an in-flight infrared peer-to-peer media streaming network. In: ITNG 2009 - 6th International Conference on Information Technology: New Generations, pp. 652–656 (2009)
8. Raj, A., Srivastava, S.K.: Sustainability performance assessment of an aircraft manufacturing firm. Benchmarking **25**(5), 1500–1527 (2018)
9. Tsai, W.H., Lai, S.Y.: Green production planning and control model with ABC under industry 4.0 for the paper industry. Sustainability **10**(8), 2932 (2018)
10. The ICAO 2013, «International Civil Aviation Organization» website (2013). https://www.icao.int/
11. Patel, N.: The Einfochips homepage (2018). https://www.einfochips.com/blog/top-emerging-trends-of-the-global-in-flight-entertainment-market/
12. The Gogoair website (2019). https://www.gogoair.com/
13. David, R., Ngulube, P., Dube, A.: A cost-benefit analysis of document management strategies used at a financial institution in Zimbabwe: a case study. SA J. Inf. Manag. **15**(2), Art. #540, 10 p. (2013). https://doi.org/10.4102/sajim.v15i2.540

Supply Chain Management in Healthcare: A Literature Review

Tuğçe Beldek, Aziz Kemal Konyalıoğlu, and Hatice Camgöz Akdağ(✉)

Management Engineering Department, Istanbul Technical University, Istanbul, Turkey
{beldek, konyalioglua, camgozakdag}@itu.edu.tr

Abstract. Supply chain management becomes more important in the developing world in order to provide a material flow with an optimum cost and time generally. Many sectors focus on supply chain management optimization and qualifications to construct a better network of suppliers for their end consumers. Especially in healthcare sector, not only for pharmaceutical products but also for hospital materials, supply chain management gets importance for service quality and patient satisfaction. Thus, there are many studies about supply chain management in healthcare to emphasize its importance. In this study, it is aimed to put forward a literature review of supply chain management in healthcare in order to give a perspective by expressing the important cases and researches which are parallel to the latest studies. Furthermore, this literature review will assess a perspective in order to understand how to manage a complex supply chain in healthcare sector by investigating studies in the literature.

Keywords: Supply chain management · Pharmaceutical companies · Distribution · Healthcare sector

1 Introduction

Supply chain management (SCM) can be defined as a set of companies which transfers the product to the end user via suppliers, product assemblers, merchandisers and transportation companies which are the parts of a supply chain [1]. Another definition of a supply chain can be extended as the flow of goods, services and information starting from raw materials and going finally to the end user [2]. The importance of Supply Chain Management continues to grow and to be introduced in the literature [3]. Supply chain, especially in multi-vendor supply chains, which can be globally or locally, may be difficult in terms of evaluation of performance as it is hard to define performance measurement criteria in a supply chain [4]. Furthermore, supply chain management provides an energy between local and global companies to integrate the process excellence of management and to manage the interaction between all members of a supply chain [5].

In healthcare sector, supply chain management also gains importance because of logistics, pharmaceutical products and patient satisfaction to improve service quality compared to other sectors. Furthermore, the change in healthcare sector has been very

N. M. Durakbasa and M. G. Gençyılmaz (Eds.): ISPR 2019, LNME, pp. 570–579, 2020.
https://doi.org/10.1007/978-3-030-31343-2_50

rapid during the last years. Many healthcare organizations are aware of the importance of applying a better technique and method in order to adopt their supply chain management strategies in spite of the fact that the methods and techniques pose problems in terms of industrial settings [6]. Improving healthcare quality in a public or private sector depends directly on its supply chain management efficiency.

In this study, it is aimed to investigate the studies in the literature about supply chain management in healthcare between the years 2000 and 2018. The study includes four main parts: an introduction part to give a perspective and brief introduction about supply chain management in healthcare, a literature review part for examining the topics and methods in supply chain management in healthcare, an analysis part in order to classify the studies by their topics, years and methods and finally a conclusion and discussion part to debate the results of the study.

2 Literature Review

Kumar, Ozdamar and Zhang (2008) have investigated a cost reduction technique about the medical suppliers from a case study in Singapore. Furthermore, they have concluded that some applications of just-in-time (JIT), reengineering, outsourcing reductions may be cost-efficient even if information technologies (IT) implementations start from a high initial cost because of the deficiency of experts. Information technologies at the initial stage of identifying suppliers can be beneficial in terms of cost reduction [7]. Another study explained by Meijboom, Schmidt-Bakx and Westert (2011) emphasized the organizational lacks affecting the cure of patients and taking the inputs from multiple healthcare suppliers in order to solve by the practices of Supply Chain Management. Also, they provided a literature survey including industrial processes in healthcare. They have mainly focused on the lead time of suppliers, integration and suitable IT practices [8].

Another study by Kumar, Swanson and Tran (2009) focused on the radio frequency identification devices (RFID) which enable healthcare supply chains more effective and reductive in terms of costs and the current RFID systems in healthcare supply chain are too costly to apply even if there are some applications of RFID systems to be installed a cost-efficient way [9]. On the other hand, Attaran (2012) has expressed the important achievement factors and the latest challenges in implementing RFID systems within healthcare supply chain management and the positive and negative sides to apply RFID systems in supply chains have been identified in terms of cost and processes [10]. Uthayakumar and Priyan (2013) have proposed an inventory model for the distribution which combines the pharmaceutical and hospital supply chains. The model they proposed has taken into consideration the lead time, time and space availability, customer service levels and by the aid of this model, they determined an optimal proposition for lead time and available lot size by illustrating a numerical example [11]. Settanni, Harrington and Srai (2017) have evaluated Pharmaceutical Supply Chains in terms of production and delivery models. Also, they have categorized the modellings and concluded that the definitions of Pharmaceutical Supply Chains are extremely based on production centralized definitions and have deficiencies to reflect patient consumptions [12].

Aronsson, Abrahamsson and Spens (2011) have analyzed healthcare supply chains in terms of agility and lean manufacturing. They tried to understand how the process of healthcare supply chains can be agile and lean by establishing a supply chain orientation and what is needed for by applying an empirical analysis by the aid of agile and lean philosophies in supply chain management in healthcare in order to develop the effectiveness of healthcare supply chains [13]. Shah, Goldstein, Unger and Henry (2008) aimed to investigate the work design by its improvement for a supply chain management in healthcare in order to better transfer a better patient treatment service quality given that the patient care services cannot be easily predictable and there are many private and public companies which are independent. These companies are generally collaborated and in a rhythm in terms of supply chain process. They have taken four organizations, that are independents, for defining their different supply chain processes in healthcare industry and for understanding their supply chain mechanisms [14].

Sinha and Kohnke (2009) identified the lack between increasing demand and current high-quality supply in healthcare industry, effectiveness of cost and time which is a problem in developing and developed countries. They have proposed a framework consisting of 3A, which are affordability, awareness and access to inform healthcare supply chain managers and in order to integrate the continuous improvement as a definition of quality and technological improvements [15].

Cook et al. (2001) took into consideration of service industry if the supply chain management is as affective as in the manufacturing industry or not. They showed improvement of the quality, reduction of costs and lead times which are results of a high performed supply chain at the healthcare area [16]. Another study explains the increase at a pharmaceutical supply chain efficiency. In this paper, Australian e-commerce project provides an information system integration to health care sector and gets improved levels of data as a result [17]. Chandra and Kachhal (2004) served a supply chain model that contains e-commerce integration. They also showed different methods such as optimization and simulation for inventory and purchasing policies [18]. Another e-commerce study is done by Kim (2004), which includes statistical results to find out effects of B2B e-commerce in health care sector. He conducted a survey and evaluated results to show internet-based integration improves supply chain management in health care sector [19].

McKone-Sweet et al. (2005) studied supply chain management implementation to healthcare sector. They found out some constraints such as education, team operations for procurement, all actors in the chain, lack of leadership, uncertain incentives, data collection and evaluation. The main problem is both leaders' and managers' supply chain knowledge in practice which directly effects supply chain performance [20]. Kim made a practical study at a hospital, which shows nearly 30% of decrease in inventory levels by improving its supply chain management system. This improvement includes purchasing operations and inventory policies of pharmaceuticals. A transparent system was designed to optimize inventory levels at every step of the supply chain. By this system, demand forecasting is more accurate and ordering process is linked with the online system. Decrease in inventory resulted with decrease in costs [21]. Langabeer (2005) mentioned about the difficulties of supply chain management in health care. He analyzed the current situation of management systems and seek for the reasons of technology usage scarcity. He gives advices for future studies to adopt new technologies to improve health care supply chain [22].

Zheng et al. (2006) studied at a different area: e-adoption of healthcare supply chain. They drew a framework in the light of English National Health Service that considers health, business and supply in terms of every organization to use e-commerce [23]. A common supply chain model is given by Baltacioglu et al. (2007), for service sector. Capacity, demand, customer and supplier relationship, service and order operations management are main elements of the chain and included in the proposed model. This model is applicable for also healthcare sector [2]. Kitsiou et al. (2007) studied healthcare supply chain to analyze its information system to define every element such as entities, data and information flows, processes etc. They also suggest new technological alternatives to improve health care supply chain management system while adding value [24].

Sousa et al. (2011) built a dynamic programming for optimizing the global supply chain problem of a pharmaceutical company as a case study. The aim of the study was to maximize net present value of the company by considering costs of distribution and production which are located in different places and tax rates. Also, they have built this model by dividing into two problem which are primary and secondary subproblems [25]. Rahimnia and Moghadasian (2010) examined the leagility of supply chain managements in healthcare sector, especially in hospitals. They have conducted a case study to evaluate the perspective of leagility in hospitals on behalf of a professional supplier since the situation of current system requires an agile and lean system for hospitals in terms of delivery systems [26]. Walker et al. (2008) studied the important factors of green supply chain management and initiatives in the private and public sector such as external and internal barriers to implement the environmental concepts. They have also studied a private healthcare company, as a case study, in order to understand how the legislations, affect the suppliers of a hospital and concluded that hospitals generally choose local providers to contribute to the local economy [27].

Bhakoo and Choi (2013) investigated the implementation and reaction of IOS in hospitals by considering different parts of a supply chain including distributors, manufacturers and healthcare sector. They concluded that the implementation of IOS is very complex and the reaction of internal pressures can be different in healthcare supply chains [28]. Kwon et al. (2016) discussed the importance of supply chain management in healthcare sector in terms of costs of patients compared to the rate of readmission. They also investigated the three main strategic areas in order to increase the profit of suppliers and to improve supply chain process and concluded that the quality of healthcare service quality depends on the supply chain surplus [29].

Kogan et al. (2014) studied the collaboration between healthcare supply chain members and analyzed how the effects can change the relationships of providers in healthcare sector. Also, they concluded that the uncertain demand and operational costs may block the consolidation of providers and even if they consolidate each other, the net profit of medical providers may not change [30]. Ishii et al. (2017) explained that the integration of supply chain in healthcare sector can provide a cost reduction and improve healthcare quality by examining three healthcare departments with the aid of experts in supply chain. They concluded that a more effective supply chain management in hospitals can mostly give a healthier academic life for those who work in hospitals [31].

Imran et al. (2018) tried to build a multiperiod model of pharmaceutical supply chain by assuming that the amount of medicines coming from providers is not certain. On the other hand, time, quality and cost, called as business triad, are the three objectives of the model and they supported their model by a numerical example by concluding that the required satisfaction has a different combination of these three objectives [32]. Syahrir et al. (2015) have prepared a report to make a survey and analyze the topics of supply chain management in healthcare and disaster by taking into considerations inventory-control management, service quality, operations research topics in healthcare and information technologies [33]. Zepeda et al. (2016) studied on the logistics services and uncertain demand for hospitals, which are two main factors considering the effects of internal institutional regulations on inventory costs by the data of the State of California. They concluded that logistics services have strong effects in local systems' infrastructures if they are weak [34]. Also, Rossetti (2011) examined the factors affecting pharmaceutical supply chain management and logistic evolution in healthcare sector have been investigated in detail based on industry dynamics [35].

Moons et al. (2019) conducted a literature review to evaluate the hospitals' supply chain logistics performances by the latest studies done in this area [36]. Dobrzykowski et al. (2014) also studied a literature review to list operations and supply chain management in healthcare sector studies between the years 1982 and 2011, which are used a structured analysis [37]. A different literature review study conducted by Narayana et al. (2014) gives important aspects on managerial researches and insights about pharmaceutical supply chain concept by a content analysis which focuses on performance improvement in supply chain of hospitals [38]. Chen et al. (2013) used an empirical test to evaluate hospitals' supply chain performance. They used a research model by taking into consideration the performance factors of hospitals in terms of IT integration of hospitals and suppliers, the base of trust and the flow of knowledge and also, the effects of these factors are very important for the supply chain performance in a hospital as they are interrelated factors [39]. Another literature study conducted by Bishara (2006) mentioned the crucial factors of pharmaceutical supply chain management by a literature survey in the light of the definition "cold supply chain" [40].

Bagchi et al. (2014) studied the supply competency effects on foreign direct investments by using Boolean-based logical analysis by considering the environment of supply, infrastructure, and capacity. They advised strategic planners to what to focus on in order to attract the foreign direct investors and eased the decision-making process in this global competitive environment [41]. Kelle et al. (2012) proposed a multi-item inventory model for a hospital in terms of pharmaceutical supply chain by considering three main strategies which are operational, tactical and strategical. They conducted their case study in an individual care unit for the inventory management at a local warehouse by evaluating service levels, ordering systems and optimal allocations [42]. As a different study of inventory control systems in hospital conducted by Stecca et al. (2016), they tried to understand the optimization of inventory by focusing on the central of pharmacy and the optimization of inventory costs of logistics of pharmaceutical products with the aid of mixed integer linear programming level by level [43].

3 Analysis

In the analysis section, the 43 articles which have been reviewed are classified according to their years, their usage of operations research techniques or modellings

Figure 1 shows the article number of supply chain management in healthcare according to their years. In the Fig. 1, it has been seen that the article number has an intention to increase between years 2000 and 2006. Furthermore, the number of articles fluctuates in other years.

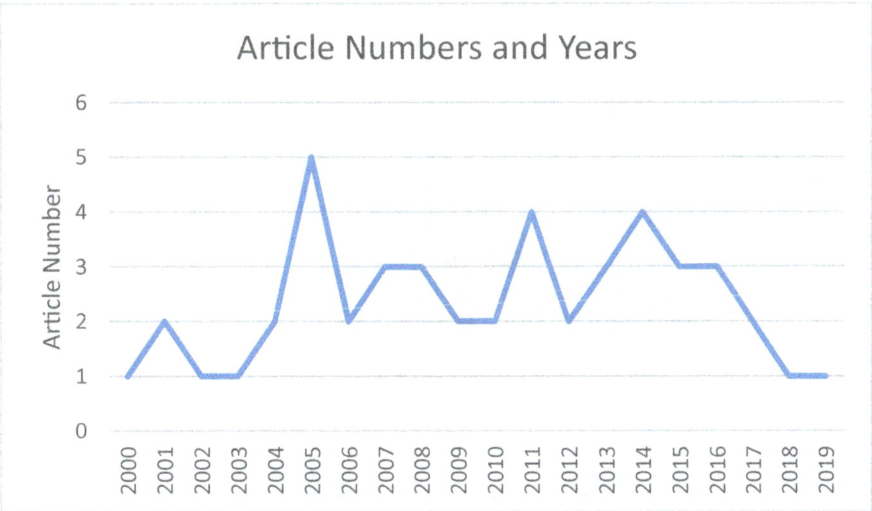

Fig. 1. The number of articles according to the years

Figure 2 shows that the number of articles according to the existence of mathematical modelling in the articles reviewed.

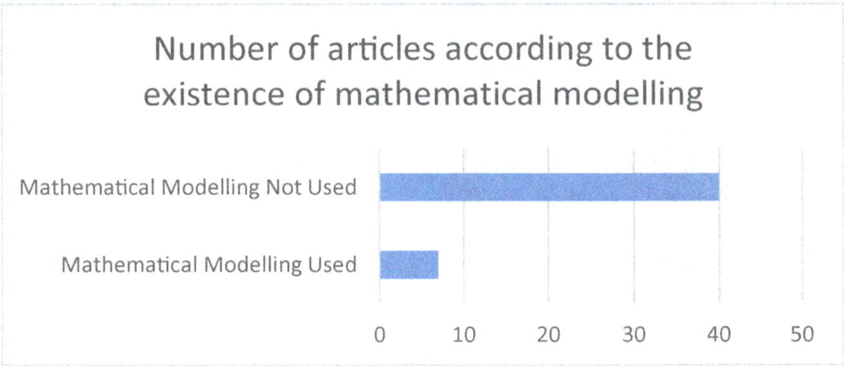

Fig. 2. The number of articles according to existence of mathematical modelling

The articles in which any mathematical modelling is not used have a high percentage compared to the articles that mathematical modelling is used.

On the other hand, there are two types in healthcare supply chain management articles, which are pharmaceutical SCM and hospital SCM specifically. Figure 3 shows the rates of reviewed articles in terms of pharmaceutical SCM and Hospital SCM. It has been seen that Hospital Supply Chain management articles have a bigger percentage than Pharmaceutical SCM.

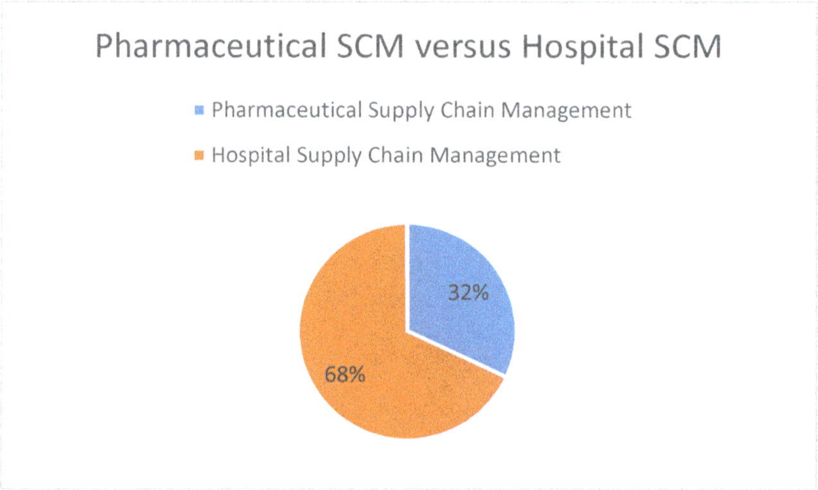

Fig. 3. The Pharmaceutical SCM and Hospital SCM article percentages in the reviewed articles

4 Conclusion

Understanding the structure of supply chain management for healthcare industry is not an easy topic and its value increases nowadays. There are many studies in the literature that supply chain management in healthcare industry in which this study examined 43 articles about pharmaceutical and hospital supply chain management. Firstly, the big gap about articles in the literature is that mathematical modelling in the literature. Nearly 15% of the papers examined have not any mathematical modelling for supply chain management. Models might help inventory, cost and other operation factors' optimization in terms of operations research.

Another classification can be considered as the supply chain focus area: evaluation of pharmaceutical or hospital SCM. Pharmaceutical SCM articles are less than hospitals' studies. Medicine sector is as important as the hospital operations so that the ratio of the studies about pharmaceutical SCM and hospital SCM might increase to improve the efficiency and reduce the costs of health care supply chain. Last but not least there are not many studies about industry 4.0 applications in health care systems attached to supply chain management evaluation in order to adapt the latest challenges of industry 4.0.

References

1. Londe, La, Bernard, J., Masters, James M.: Emerging logistics strategies: blueprints for the next century. Int. J. Phys. Distrib. Logistics Manag. **24**(7), 35–47 (1994)
2. Baltacioglu, T., Ada, E., Kaplan, M.D., Yurt And, O., Cem Kaplan, Y.: A new framework for service supply chains. Serv. Ind. J. **27**(2), 105–124 (2007)
3. Giunipero, L.C., Brand, R.R.: Purchasing's role in supply chain management. Int. J. Logistics Manag. **7**(1), 29–38 (1996)
4. Hervani, A.A., Helms, M.M., Sarkis, J.: Performance measurement for green supply chain management. Benchmarking Int. J. **12**(4), 330–353 (2005)
5. Lambert, D.M., Cooper, M.C.: Issues in supply chain management. Ind. Mark. Manage. **29**(1), 65–83 (2000)
6. De Vries, J., Huijsman, R.: Supply chain management in health services: an overview. Supply Chain Manag. Int. J. **16**(3), 159–165 (2011)
7. Kumar, A., Ozdamar, L., Ning Zhang, C.: Supply chain redesign in the healthcare industry of Singapore. Supply Chain Manage. Int. J. **13**(2), 95–103 (2008)
8. Meijboom, B., Schmidt-Bakx, S., Westert, G.: Supply chain management practices for improving patient-oriented care. Supply Chain Manage. Int. J. **16**(3), 166–175 (2011)
9. Kumar, S., Swanson, E., Tran, T.: RFID in the healthcare supply chain: usage and application. Int. J. Health Care Qual. Assur. **22**(1), 67–81 (2009)
10. Attaran, M.: Critical success factors and challenges of implementing RFID in supply chain management. J. Supply Chain Oper. Manage. **10**(1), 144–167 (2012)
11. Uthayakumar, R., Priyan, S.: Pharmaceutical supply chain and inventory management strategies: Optimization for a pharmaceutical company and a hospital. Oper. Res. Health Care **2**(3), 52–64 (2013)
12. Settanni, E., Harrington, T.S., Srai, J.S.: Pharmaceutical supply chain models: a synthesis from a system view of operations research. Oper. Res. Perspect. **4**, 74–95 (2017)
13. Aronsson, H., Abrahamsson, M., Spens, K.: Developing lean and agile health care supply chains. Supply Chain Manage. Int. J. **16**(3), 176–183 (2011)
14. Shah, R., Goldstein, S.M., Unger, B.T., Henry, T.D.: Explaining anomalous high performance in a health care supply chain. Decis. Sci. **39**(4), 759–789 (2008)
15. Sinha, K.K., Kohnke, E.J.: Health care supply chain design: toward linking the development and delivery of care globally. Decis. Sci. **40**(2), 197–212 (2009)
16. Cook, J., DeBree, K., Feroleto, A.: From raw materials to customers: supply chain management in the service industry. SAM Adv. Manage. J. (2001)
17. Bhakoo, V., Chan, C.: Collaborative implementation of e-business processes within the health-care supply chain: the Monash Pharmacy Project. Supply Chain Manage. Int. J. **16**(3), 184–193 (2011)
18. Chandra, C., Kachhal, S.K.: Managing health care supply chain: trends, issues, and solutions from a logistics perspective. In: Proceedings of the Sixteenth Annual Society of Health Systems Management Engineering Forum, pp. 20–21, February 2004
19. Kim, S.M.: An empirical investigation of the impact of electronic commerce on supply chain management: a study in the healthcare industry (2004)
20. McKone-Sweet, K.E., Hamilton, P., Willis, S.B.: The ailing healthcare supply chain: a prescription for change. J. Supply Chain Manage. **41**(1), 4–17 (2005)
21. Kim, D.: An integrated supply chain management system: a case study in healthcare sector. In: International Conference on Electronic Commerce and Web Technologies, pp. 218–227. Springer, Heidelberg, August 2005

22. Langabeer, J.: The evolving role of supply chain management technology in healthcare. J. Healthcare Inf. Manage JHIM **19**(2), 27–33 (2005)
23. Zheng, J., Bakker, E., Knight, L., Gilhespy, H., Harland, C., Walker, H.: A strategic case for e-adoption in healthcare supply chains. Int. J. Inf. Manage. **26**(4), 290–301 (2006)
24. Kitsiou, S., Matopoulos, A., Manthou, V., Vlachopoulou, M.: Evaluation of integration technology approaches in the healthcare supply chain. Int. J. Value Chain Manage. **1**(4), 325–343 (2007)
25. Sousa, R.T., Liu, S., Papageorgiou, L.G., Shah, N.: Global supply chain planning for pharmaceuticals. Chem. Eng. Res. Des. **89**(11), 2396–2409 (2011)
26. Rahimnia, F., Moghadasian, M.: Supply chain leagility in professional services: how to apply decoupling point concept in healthcare delivery system. Supply Chain Manage. Int. J. **15**(1), 80–91 (2010)
27. Walker, H., Di Sisto, L., McBain, D.: Drivers and barriers to environmental supply chain management practices: Lessons from the public and private sectors. J. Purchasing Supply Manage. **14**(1), 69–85 (2008)
28. Bhakoo, V., Choi, T.: The iron cage exposed: Institutional pressures and heterogeneity across the healthcare supply chain. J. Oper. Manage. **31**(6), 432–449 (2013)
29. Kwon, I.W.G., Kim, S.H., Martin, D.G.: Healthcare supply chain management; strategic areas for quality and financial improvement. Technol. Forecast. Soc. Chang. **113**, 422–428 (2016)
30. Kogan, K., Leu, J., Chernonog, T.: Healthcare supply chain operations: Why are doctors reluctant to consolidate? Oper. Res. Health Care **3**(3), 101–115 (2014)
31. Ishii, L., Demski, R., Lee, K.K., Mustafa, Z., Frank, S., Wolisnky, J.P., Cohenh, D., Khanna, J., Ammerman, J., Khanuja, H.S., Unger, A.S., Gould, L., Wachter, P.A., Stearns, L., Werthman, R., Pronovost, P.: Improving healthcare value through clinical community and supply chain collaboration. Healthcare **5**(1–2), 1–5 (2017)
32. Imran, M., Kang, C., Ramzan, M.B.: Medicine supply chain model for an integrated healthcare system with uncertain product complaints. J. Manuf. Syst. **46**, 13–28 (2018)
33. Syahrir, I., Vanany, I.: Healthcare and disaster supply chain: literature review and future research. Procedia Manuf. **4**, 2–9 (2015)
34. Zepeda, E.D., Nyaga, G.N., Young, G.J.: Supply chain risk management and hospital inventory: effects of system affiliation. J. Oper. Manage. **44**, 30–47 (2016)
35. Rossetti, C.L., Handfield, R., Dooley, K.J.: Forces, trends, and decisions in pharmaceutical supply chain management. Int. J. Phys. Distrib. Logistics Manage. **41**(6), 601–622 (2011)
36. Moons, K., Waeyenbergh, G., Pintelon, L.: Measuring the logistics performance of internal hospital supply chains–a literature study. Omega **82**, 205–217 (2019)
37. Dobrzykowski, D., Deilami, V.S., Hong, P., Kim, S.C.: A structured analysis of operations and supply chain management research in healthcare (1982–2011). Int. J. Prod. Econ. **147**, 514–530 (2014)
38. Narayana, S.A., Pati, R.K., Vrat, P.: Managerial research on the pharmaceutical supply chain–a critical review and some insights for future directions. J. Purchasing Supply Manage. **20**(1), 18–40 (2014)
39. Chen, D.Q., Preston, D.S., Xia, W.: Enhancing hospital supply chain performance: a relational view and empirical test. J. Oper. Manage. **31**(6), 391–408 (2013)
40. Bishara, R.H.: Cold chain management–an essential component of the global pharmaceutical supply chain. Am. Pharm. Rev. **9**(1), 105–109 (2006)

41. Bagchi, P., Lejeune, M.A., Alam, A.: How supply competency affects FDI decisions: some insights. Int. J. Prod. Econ. **147**, 239–251 (2014)
42. Kelle, P., Woosley, J., Schneider, H.: Pharmaceutical supply chain specifics and inventory solutions for a hospital case. Oper. Res. Health Care **1**(2–3), 54–63 (2012)
43. Stecca, G., Baffo, I., Kaihara, T.: Design and operation of strategic inventory control system for drug delivery in healthcare industry. IFAC-PapersOnLine **49**(12), 904–909 (2016)

Supply Chain Orientation, ERP Usage and Knowledge Management in Supply Chain

Arafat Salih Aydiner[1(✉)], Mehmet Fatih Acar[2], Selim Zaim[3], and Dursun Delen[4(✉)]

[1] Istanbul Medeniyet University, Istanbul, Turkey
arafat.aydiner@medeniyet.edu.tr
[2] Havatek Makina A.Ş., İzmir, Turkey
facar@havatek.com.tr
[3] Istanbul Şehir University, Istanbul, Turkey
selimzaim@sehir.edu.tr
[4] Oklahoma State University, Stillwater, USA
dursun.delen@okstate.edu

Abstract. The principal aim of this study is to investigate the direct and indirect impacts of Knowledge Management (KM) and Enterprise Resource Planning (ERP) usage with the mediating effect of Supply Chain Orientation (SCO) on operational performance drawing on a sample of 200 Turkish manufacturing companies. Universal Structure Modelling (USM) were applied to examine the direct and indirect effects of variables on performance. In the path analysis seven hypothesis were evaluated with USM. Data analysis reveals that SCO affects operational performance positively with a degressive nonlinear relationship. Also it is found that there is a positive degressive nonlinear relationship between operational performance (OPER) and financial performance (FIN). The study shows that all relationships other than above are linear. In addition, no support has been found between ERP usage relations with OPER.

Keywords: Supply Chain Orientation · Knowledge Management · ERP · Operational performance

1 Introduction

Many companies buy and sell different items and equipment at the same time as they manufacture goods. Raw materials or equipment is needed for production in corporations, and produced items are sold at a profit. Supply chain management can organize and arrange these processes. Therefore, supply chain is one of the most popular topics for managers and academicians in today's highly competitive market environment. Many firms try to gain competitive advantage by improving their supply chain capabilities [1]. Even though, there are a lot of research related to supply chain management in the literature, supply chain orientation and its combination with knowledge and enterprise resource planning is not common. Supply Chain Orientation (SCO) can be defined as the motivation of organizations to manage supply chain relations with their contractors [2].

N. M. Durakbasa and M. G. Gençyılmaz (Eds.): ISPR 2019, LNME, pp. 580–590, 2020.
https://doi.org/10.1007/978-3-030-31343-2_51

Knowledge Management (KM) is also a new concept for organizations, and it has been accepted by both academics and managers. Knowledge can be seen as a soft power that provides competitive advantage for its users. Knowledge plays an important role in competition, and KM provides a means to get it.

ERP systems are integrated software programs, which allow the execution of many internal and external functions in enterprises through a combined system. ERP systems provide a better communication of functions within the enterprise through the efficient use of corporate resources and supply chain to facilitate better decision-making when implemented successfully.

The purpose of this thesis is to investigate the effects of Knowledge Management and ERP Usage on business performance, taking into account Supply Chain Orientation (SCO). In the literature, a positive relationship between SCO and business performance is discussed. In this study, it is shown that KM and ERP have an even stronger positive impact on performance with the contribution of SCO. Additional factors of SCO, Knowledge Management and ERP Usage are also discussed in this study. Thus, in this research, the effects of Knowledge Management (KM) and ERP on business performance, especially in terms of supply chain capabilities, are investigated in relation to the SCO.

The study is organized as follows: first the necessary literature are reviewed, later research model and hypothesis are proposed, then research method is explained, lastly research analysis and the conclusion are discussed.

2 Literature Review

Supply chain is a very popular concept in today's market environment. Sometimes, supply chains can be seen as only logistics activities, but this idea understates the importance of the concept. All processes, from buying of raw materials to after-sale services, are elements of the supply chain. According to many authors, it includes all firm functions such as: logistics, manufacturing, purchasing, marketing, promotion, sales, R&D and product design [3, 4].

In contrast to the past, the classical concept of supply chain management (SCM) (i.e. functional integration) has expanded in recent years, and the importance of cooperation among supply chain members is increasing to improve competitiveness of the supply chain [5]. To satisfy the SCM activities (e.g. efficient consumer response, quick response, forecasting and replenishment) efficiently, collaborative activities such as joint forecasting and planning, information sharing, joint inventory management are needed [5] (Min and Mentzer 2004). Additionally, many authors [5, 6] emphasized several key elements of successful SCM, including information sharing, cooperation, long term relationships, mutual risk and reward sharing. Christopher (1992) suggested that real competition is supply chain against supply chain, rather than company against company [2]. To manage the supply chain operations effectively, managers should concentrate on supply chain activities inside the firm. Mentzer et al. (2001) underlined the importance of this condition and called it a supply chain orientation (SCO) [7]. According to Min and Mentzer (2004), it is not possible to conduct efficient supply chain management among not supply chain oriented firms [5]. SCO can be defined as 'the implementation by an organization of the systemic, strategic implications of the tactical activities involved in managing the various flows in a supply chain [5].

Knowledge has been one of the most popular debate topics in philosophy since old eras [8]. Knowledge can be defined as, 'information plus the causal links that help to make sense of this information' and Knowledge Management is'a process that establishes and clearly articulates such links [9, 10]. Also Alavi and Leidner [8], defined knowledge as "information possessed in the mind of individuals: it is personalized information (which may or may not be new, unique, useful or accurate) related to facts, procedures, concepts, interpretations, ideas, observations and judgments". Knowledge Management (KM) is a key point for organizations to use their intangible resources efficiently. It is determining and administration of knowledge to help organization's processes and improves the innovation activities inside a firm. In the last years, integration of information systems with Knowledge Management to generate, transfer and apply it in organizations has been also discussed [8]. A survey of KPMG showed that nearly half of the organizations faced with problems when their key employees left. Additionally, the different survey indicated that most of firms thought some of necessary knowledge is embedded inside them but the essential problems are to find, store and use it [8, 11–13]. Knowledge Management have three goals; (1) making knowledge clear and stating the importance of it in a firm, (2) establishing a knowledge-intensive culture with stimulating behaviors like knowledge sharing, (3) establishing a knowledge infrastructure, both of technical (i.e. e-mail system, servers etc.) and social (i.e. collaborating, meetings) ones [8].

Different Enterprise Resource Planning (ERP) software packages were widely used by companies in different sectors. Material Requirement Planning (MRP) from the 1970s and Material Requirement Planning II (MRP II) from the 1980s are prior versions of ERP [1]. ERP is used because of pressure from competitors or requests from suppliers or customers in the supply chain to provide linkage [1]. Su and Yang [1], defined ERP as, 'is an integrated enterprise computing system that is designed to automate the flow of material, information, and financial resources among all functions within an enterprise on a common database.' There are many benefits of ERP if it is fully applied in business organization, including having faster transactions, reduced cycle time, better financial management, and making tacit knowledge explicit [1]. Moreover, Davenport [14], also pointed out the benefits of ERP, such as reduced cycle time, improving information flow, and rapid formation of financial information.

3 Research Model and Hypotheses

Previous studies show that there is a positive relationship between SCO and supply chain management (SCM) [15, 16]. Min et al. [16], emphasizes that SCO has positive effect on firm performance and a strong antecedent of SCM operations. Thus, it is asserted that Supply Chain Orientation has a positive impact on firm performance, meaning that firms who focus on supply chain operations have a competitive advantage over that do not use SCO.

H1: Supply Chain Orientation (SCO) is positively related with operational performance.

Sivakumar and Roy [26], studied about knowledge redundancy and supply chain performance. According to their definition, the direct effect of Knowledge Management on operational performance is positive and with the mediating effect of SCO, KM has a stronger positive effect on performance.

H2a: Knowledge Management (KM) is positively related with operational performance.
H2b: The impact of KM mediated by SCO is stronger than KM's direct effect on operational performance.

ERP facilitates greater managerial control, rapid decision-making, and reduction of operational costs. Furthermore, in the research of Su and Yang [1], a positive relationship between ERP and SCM competences was shown. Therefore, the direct effect of ERP on operational performance is positive, and secondly, with a mediating effect of SCO, ERP has positive effect on performance.

H3a: ERP Usage is positively related to operational performance.
H3b: The impact of ERP Usage mediated by SCO is stronger than ERP's direct effect on operational performance.

Sedera and Gable et al. [17], showed the relationship between KM competence and ERP. They suggested that there is a positive and significant relationship between KM competence and ERP success. Xu et al. [18], also argued the correlation between ERP and KM.

H4: The impact of ERP Usage on operational performance mediated by KM and SCO simultaneously is stronger than all aforementioned ERP Usage's affects.

There is a direct relationship between operational performance and financial performance because operational performance refers to efficiency in the activities of a firm [19].

H5: Operational performance is positively related with financial performance.

To analyze these hypotheses with the nonlinearity among variables, universal structural equation modelling (USM) was applied to the proposed model. The research model is illustrated below in Fig. 1.

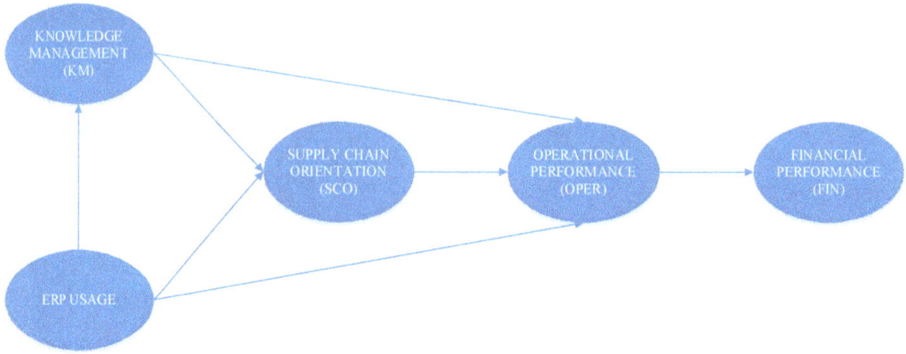

Fig. 1. Proposed model

4 Research Methodology

4.1 Survey Instrument

Questionnaire is created to evaluate SCO, KM, ERP usage, OPER and FIN. The original version of the questionnaire was in English. This questionnaire was translated into Turkish. The Turkish version was back translated by taking approval of a group of experts that the two versions were comparable. The questionnaire was pre-tested several times to ensure that there is no problem related to the wording, format, and sequencing of questions.

The questionnaire consists of 69 latent and demographic variables. Respondents were asked to evaluate each of the constructs based on 5-point Likert scale. The statements are rated according to following scale except performance questions (1 = strongly disagree, 2 = disagree, 3 = neither agree/nor disagree, 4 = agree, 5 = strongly agree). The performance questions are rated with the scale as follows: (1 = Very Bad; 2 = Bad; 3 = Undecided; 4 = Good; 5 = Very Good).

4.2 Sample

The sample was collected from manufacturing companies. These companies were selected from the members of Industrial Zones and Chambers of Commerce from the most industrialized cities of Turkey. In addition, the companies were chosen according to their ERP usage which was published on their web site. The respondents for the survey required to be managers whom are related to supply chain operations, including production planning engineers, warehouse managers, and procurement managers

Approximately 800 surveys were sent to different companies, and nearly 250 firms responded to our questionnaire. After eliminating several surveys because of missing data, the sample has been reduced to 200 surveys. Thus, overall response rate was 25%, which is considered satisfactory for subsequent analysis. The respondent companies are from different sectors such as the automotive, chemical, construction, and food industries.

5 Analysis and Results

Universal Structure Modeling (USM) is a structural equation modeling that captures nonlinear relationships between variables in the model [20]. USM can be seen as complementary method for covariance based structural equation modeling (CVSEM). USM does not require normality assumptions, and it provides the use of formative scales. Additionally, USM can capture nonlinearity and interactions with sample sizes of less than 250 cases [20]. Thus, the data analysis was completed by applying unidimensionality tests for each variables.

5.1 Unidimentionality Test of Variables

The validity and reliability of three variables were also evaluated by testing unidimensionality three ways: principal component analysis, Cronbach's α and Dillon-Goldstein's ρ. All the Cronbach's α values were closer or greater than the threshold value of 0.70. When considering principal component analysis, the first eigenvalue score of the manifest variable of each variable was greater than one, and the second score was lower than one. Therefore, each variable could be thought as unidimensional [21, 22]. Lastly, Dillon-Goldstein's ρ value for each variable is larger than 0.70, so this also gave extra support for unidimensionality for all variables as seen in Table 1.

Table 1. Unidimensionality tests

Variable	Number of ind.	Cronbach's α	Dillon-Goldstein's ρ	First eigen.	Second eigen.
SCO	4	0.7419	0.8393	1.4158	0.4241
KM	3	0.8616	0.9185	1.6011	0.2888
ERP usage	2	0.6216	0.8459	1.6384	0.5811

5.2 Neusrel Results

To analyze the nonlinearity among variables, universal structural equation modelling (USM) was applied to the proposed model. The goodness of fit index was calculated as 0.53 after three iterations. Moreover, when R2 values are compared, Neusrel results are bigger than linear PLS results. These are shown below as Table 2.

Table 2. Comparison Of R2 values and goodness of fit index

R^2	KM	SCO	OPER	FIN
Neusrel	0.2105	0.6917	0.5286	0.2772
Linear	0.2054	0.5981	0.3889	0.2716

GOF -Goodness of Fit: 0.53
Moreover, factor score represents the correlation coefficients between factors and latent variables (LV). Factor weights are 'weights which are used to calculate an LV using this set of manifest variables (MVs) [23]. Factor scores refer to correlation coefficients between manifest and latent variables. In addition to these, Cronbach α values, Average Explained Variance (AEV) scores and composite reliability scores were closer or larger than the critical threshold values 0.7, 0.5 and 0.6, respectively [24, 25]. These calculations are shown in Tables 3 and 4 respectively.

Table 3. Measurement values for USM

	Cronbach's Alpha	AEV average explained variance	Composite reliability
KM	0.88	0.80	0.95
SCO	0.76	0.58	0.76
ERP	0.62	0.72	0.90
OPER	0.73	0.48	0.63
FIN	0.81	0.72	0.91

Table 4. USM estimation results

Latent variable	Manifest variable	Factor scores
ERP	Modules	0.79
	Utility	0.90
KM	Generation	0.92
	Storage	0.84
	Usage	0.92
SCO	Credibility	0.71
	Benevolence	0.76
	Commitment	0.78
	TMS	0.80
OPER	Delivery on time	0.71
	Forecasting accuracy	0.70
	Lead time	0.49
	Service after sale	0.70
	Average inventory level	0.81
FIN	Profit	0.85
	Revenue	0.85
	Return on investment	0.85

Furthermore, Neusrel gives nonlinear relationships between variables. According to the results, there are three possible nonlinearities. However, when significance levels are considered, the number of nonlinear relationships decreased by two, since the significance level for ERP and SCO relationship is greater than 0.05. In conclusion, there are nonlinear relationships between SCO and OPER and OPER and FIN. The USM results with related coefficients and significance levels are shown in Fig. 2.

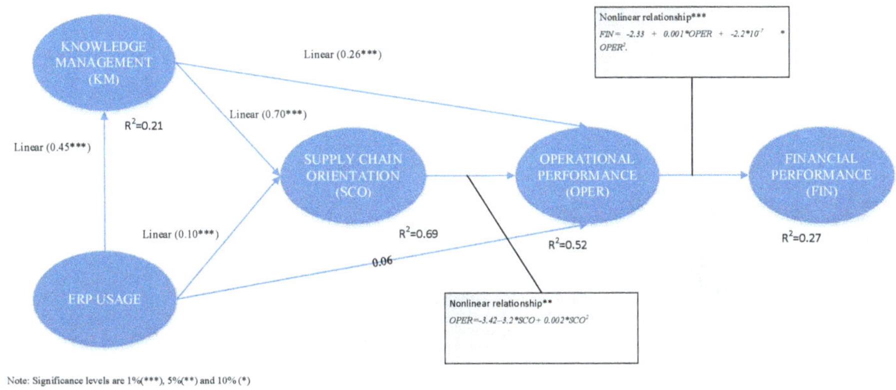

Fig. 2. Results of USM

The formula between OPER and SCO is:
OPER = −3.42–3.2 * SCO + 0.002 * SCO2
and the formula between OPER and FIN is:
FIN = −2.33 + 0.001 * OPER + −2.2*10 − 7 * OPER2.

There are quadratic relationship between OPER and SCO and OPER and FIN. The leverage factors of two models are 0.17 and 0.75 respectively. It shows whether the model is degressive or progressive. 'Degressive means that a change in the effected variable is higher for low values of the causing variable than for high values. The opposite is true for progressive functions - the effect of high values is higher as with low values [23]. A leverage factor lower than 1 refers to a degressive function and higher than 1 refers to progressive function. Therefore, both relationships have degressive functions.

The first formula refers that ratio of the effect of SCO on OPER is decreasing in high values. It is an expected result, because if there is no supply chain orientation in the organization, the settlement of it can bring many benefits in the beginning period, however in the following times increasing rate of operational performance will decrease. Since, there are many determinative factors except SCO that affect the operational performance, it means that to improve efficiency, capability of SCO is limited. Moreover, the second formula shows the same relationship between OPER and FIN. A change in the financial performance is higher for low values of operational performance. This situation is expected normally, when companies transform from inefficient situation to efficient one in their operations, their financial performances increase significantly. However, in the next times the increasing ratio of financial indicators will decrease, because financial performance is not affected just operational performance. Nonlinear relationships are also shown in Figs. 3 and 4.

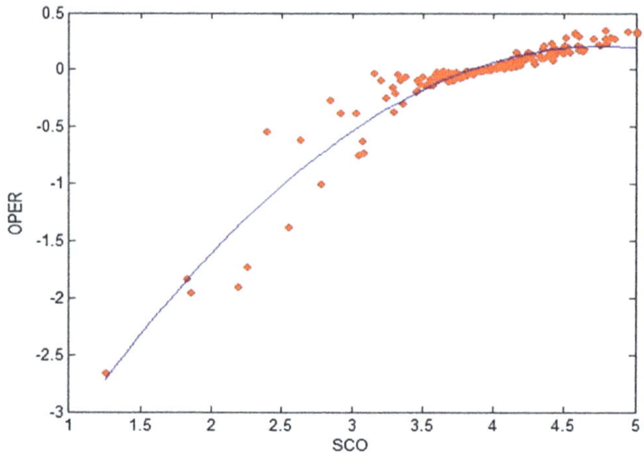

Fig. 3. Relationship between SCO and OPER

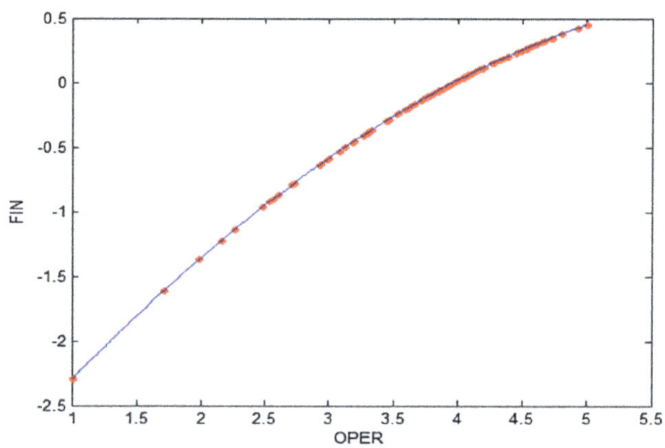

Fig. 4. Relationship between OPER and FIN

6 Conclusion

In this research, the effects of KM, ERP Usage and SCO on operational and financial performance were investigated. The analysis USM showed that SCO affects operational performance positively. Furthermore, the direct effect of KM on operational performance and indirect effects on financial performance (over operational performance) is positively significant. In addition to this, ERP Usage had no significant impact on operational performance.

The USM shows that the effect of SCO on OPER is decreasing in high values, means degressive relationship. This is an expected situation, since the establishment of SCO can bring many benefits in the beginning period, but later rising rate of operational performance will decrease. There are many elements except SCO that affect the operational performance, it means that capabilitiy of SCO is limited to raise efficiency.

Additionally, there is a same relationship between OPER and FIN. A change in the financial performance is higher for low values of operational performance. This situation is expected normally, when companies transform from inefficient situation to efficient one in their operations, their financial performances increase significantly. However, in the next times the increasing ratio of financial indicators will decrease, because financial performance is not affected just operational performance.

Although it is an intensive study, this research has some limitations. Since only manufacturing companies were considered, it is very hard to collect data from large number of companies, so the sample size of this research is not sufficient to apply covariance-based structural equation modelling, therefore, the proposed model in this research can be analyzed using covariance-based structural equation modelling (AMOS or LISREL) with restrictions. In addition to this, the service sector was not considered in this paper. In future research, it may be also added and results can be compared with those of the manufacturing sector.

References

1. Su, Y.F., Yang, C.: A structural equation model for analyzing the impact of ERP on SCM. Expert Syst. Appl. **37**(1), 456–469 (2010)
2. Christopher, M.L.: Logistics and Supply Chain Management. Pitman Publishing, London (1992)
3. Bechtel, C., Jayaram, J.: Supply chain management: a strategic perspective. Int. J. Logistics Manage. **8**(1), 15–34 (1997)
4. Min, S.: A market orientation in supply chain management. PhD Dissertation, The University of Tennessee (2001)
5. Min, S., Mentzer, J.T.: Developing and measuring supply chain management concepts. J. Bus. Logistics **25**(1), 63–99 (2004)
6. Cooper, M.C., Ellram, L.M.: Characteristics of supply chain management and the implications for purchasing and logistics strategy. Int. J. Logistics Manage. **4**(2), 13–24 (1993)
7. Mentzer, J.T., DeWitt, W., Keebler, J.S., Min, S., Nix, N.W., Smith, C.D., Zacharia, Z.G.: Defining supply chain management. J. Bus. Logistics **22**(2), 1–25 (2001)
8. Alavi, M., Leidner, D.E.: Review: knowledge management and knowledge management systems: conceptual foundations and research issues. MIS Q. 107–136 (2001)
9. Sarvary, M.: Knowledge management and competition in the consulting industry. Calif. Manage. Rev. **41**(2), 95–106 (1999)
10. McGinnis, T.C., Huang, Z.: Rethinking ERP success: a new perspective from knowledge management and continuous improvement. Inf. Manage. **44**(7), 626–634 (2007)
11. Hackbarth, G.: The impact of organizational memory on IT systems. In: Hoadley, E., Benbasat, I. (eds.) Proceedings of the Fourth Americas Conference on Information Systems, pp. 588–590 (1998)
12. Von Krogh, G.: Care in knowledge creation. Calif. Manage. Rev. **40**(3), 133–153 (1998)
13. Cranfield University: The Cranfield/Information Strategy Knowledge Survey: Europe's State of the Art in Knowledge Management. The Economist Group (1998)

14. Davenport, T.: Putting the enterprise into the enterprise system. Harvard Bus. Rev. **76**(4), 113–121 (1998)
15. Hult, G.T., Ketchen, D.J., Adams, G.L., Mena, J.A.: Supply chain orientation and balance scorecard performance. J. Manag. Issues **20**(4), 526–544 (2008)
16. Min, S., Mentzer, J.T., Ladd, R.T.: A market orientation in supply chain management. J. Acad. Mark. Sci. **35**(4), 507–522 (2007)
17. Sedera, D., Gable, G.G.: Knowledge management competence for enterprise system success. J. Strateg. Inf. Syst. **19**(4), 296–306 (2010)
18. Xu, L., Wang, C., Luo, X., Shi, Z.: Integrating knowledge management and ERP in enterprise information systems. Syst. Res. Behav. Sci. **23**(2), 147–156 (2006)
19. Coelli, T.J., Prasada Rao, D.S., O'Donnell, C.J., Battese, G.E.: An Introduction to Efficiency and Productivity Analysis. Springer, Boston (2005)
20. Buckler, F., Hennig-Thurau, T.: Identifying hidden structures in marketing's structural models through universal structure modeling: an explorative Bayesian neural network complement to LISREL and PLS. Mark. J. Res. Manage. **4**(2), 49–68 (2008)
21. Chin, W.W.: The partial least squares approach for structural equation modeling. In: Marcoulides, G.A. (ed.) Modern methods for business research Mahwah, pp. 295–336. Lawrence Erlbaum Associates, NJ (1998)
22. Zaim, H., Tatoglu, E., Zaim, S.: Performance of knowledge management practices: a causal analysis. J. Knowl. Manage. **11**(6), 54–67 (2007)
23. Buckler, F.: Manual. Neusrel 5.0. Cologne, Germany (2013)
24. Nunnally, J.Y., Bernstein, I.H.: Psychometric Theory. McGraw-Hill (1994)
25. Bagozzi, R.P., Yi, Y.: On the evaluation of structural equation models. J. Acad. Mark. Sci. **16**(1), 74–94 (1988)
26. Sivakumar, K., Roy, S.: Knowledge redundancy in supply chains: a framework. Supply Chain Manage. Int. J. **9**(3), 241–249 (2004)

University-Industry Collaboration

Entrepreneurial Competences of University Students in the Digital Age: A Scale Development Study

Mahmut Tekin[1], Deniz Baş[2], Tahsin Geçkil[3], and Özdal Koyuncuoğlu[3](✉)

[1] Faculty of Economic and Administrative Sciences, Selçuk University, Konya, Turkey
mahtekin@selcuk.edu.tr
[2] Vocational School of Health Services, Kırklareli University, Kırklareli, Turkey
deniz.bas@klu.edu.tr
[3] Faculty of Applied Sciences, Necmettin Erbakan University, Konya, Turkey
{tgeckil,okoyuncuoglu}@erbakan.edu.tr

Abstract. Advances in technology continue to affect our lives day by day. The new technological developments gaining momentum with the 4th Industrial Revolution have an impact on the working life, employment and unemployment in the 4th industrial revolution as in every industrial revolution. Every new technology that has emerged may adversely affect some sectors and business areas and cause unemployment. The ongoing industrial revolution is expected to contribute to the formation of new businesses and new jobs. This change necessitates the preparation of the individuals for the needs of the new age as well as the equipping of individuals for entrepreneurship activities. There is no doubt that the dynamos of the preparatory process will be universities. This situation increases the importance of monitoring the developments in entrepreneurship and innovation in the field of technology and in gaining competence in this field for university students.

As a requirement of competition, state authorities care about the role of universities in entrepreneurship and have serious expectations. Public authorities support and promote the universities in this area significantly. The questions such as "How are the students, who are the main target groups of universities, taken into account in the university entrepreneurship ecosystem?" and "What is the impact of universities on students through training, incentives and support activities in the field of entrepreneurship?" have become a topic to be studied in the field of higher education. While looking for answers to these questions, determining entrepreneurial competences of university students has emerged as an important requirement. The aim of this methodological study is to develop a tool to measure the entrepreneurial competencies of university students based on Kailer's Entrepreneurship Competence Model. For the study, a sample was used consisting of 415 students from Selcuk University Faculty of Economics and Administrative Sciences and Konya Technical University Faculty of Engineering and Natural Sciences in Konya, Turkey. Content validity, construct validity and reliability studies were conducted at the development stage of Entrepreneurial Competences Scale of Students. The Data were analyzed on the computer using SPSS and LISREL package programs. According to the results of

N. M. Durakbasa and M. G. Gençyılmaz (Eds.): ISPR 2019, LNME, pp. 593–604, 2020.
https://doi.org/10.1007/978-3-030-31343-2_52

EFA, 25-item and two-dimensional Entrepreneurial Competency Scale was developed. Cronbach's alpha coefficient of the scale was calculated as .90 and the percentage of variance explanatory was measured as 42.28%. Subsequent CFA results showed that the scale had good fit values (X2/df: 2.39, RMSEA: .073, CFI: .94, NNFI: .94). It is concluded that the scale developed in this study is a valid and reliable measurement tool that can be used to measure the entrepreneurship competencies of university students. It can be stated that the Entrepreneurship Competence Scale of University Students will be important for the universities that want to improve their student-based entrepreneurial-innovative features, public athorities and for the whole economy at macro level.

Keywords: Industry 4.0 · Student entrepreneurship · Entrepreneurial knowledge · Entrepreneurial motivation · Entrepreneurial competencies scale

1 Introductıon and Theoretical Framework

As a result of globalization, technological, economic and environmental developments, higher education institutions need to keep up with new developments to survive in global competition for a long time. The most important of these new developments is Industry 4.0 which is called the Fourth Industrial Revolution. The main philosophy of Industry 4.0 includes the creation of intelligent systems based on the use of cyber-physical systems. The interconnection of objects and services through the internet and the creation of intelligent systems with artificial intelligence capabilities and algorithms will result in direct and indirect effects multi-dimensionally [1]. Therefore, the Industry 4.0 process represents a process that creates its own economy and fundamentally changes the established value chains.

In this new process, increase in productivity, higher added value, cost reduction, zero inventory production, better quality, faster and environmentally friendly production, individual production and customer satisfaction and the need for qualified manpower will increase [2]. The transformation of production and related functions into intelligent systems and much more intensive use of automation will lead to less employment and cause workforce differentiation in demographics. Industry 4.0, which will differentiate the demographic of the workforce, will lead to the elimination of labor-intensive business models and at the same time identify a series of new labor profiles with advanced competencies. These new job descriptions will not only be limited to engineering areas, but will also apply to all areas of health, law, business and social sciences [1].

Digitalization and the systems it supports affect all business functions from production to marketing. Moreover, this process seems to be affective not only in enterprise level but also in a much wider area. The issues at macro level such as growth, employment, education, investment environment, entrepreneurship can be given as examples. One of the issues that the concept of Industry 4.0 is expected to affect is the emergence of new business models and the establishment of new appropriate initiatives.

In addition to the existing forms of work, the emergence of different applications and new business areas and occupations in the business life will be discussed. However, it is seen that there is not enough study in the literature about this subject.

The concept of entrepreneurship, whose foundations are based on Schumpeter's theory of entrepreneurship, is one of the main factors creating economic development in the region, thus contributing to the development of the region. According to Schumpeter (1934), entrepreneurship and innovation are both important elements of national economies, and entrepreneurs are the main reason for economic development. Schumpeter sees entrepreneurs as a process of creative destruction. According to him, there is no entrepreneurial activity without innovation and there is no driving force of capitalism and capitalist gains without entrepreneurial activity [3]. According to Tekin (2006), entrepreneurship is one of the most important elements of free competition or free market economy. Free competition system will not occur properly without entrepreneurial activities in a country In this context, entrepreneurship is important in many respects [4].

When the development of universities is examined, it is seen that the last generation of universities are entrepreneurial-innovative universities. This led to the use of the knowledge and encouraging the transfer of knowledge and technology as well as the entrepreneurial thought of the students and academists [5].

It is seen that competencies-based evaluations including performance evaluation, recruitment and career development have been used in human resources management since 1980s. It is important to define the competencies related to entrepreneurship in recruitment, performance evaluation and training processes to monitor and develop intercorporate entrepreneurship. Entrepreneurship intentions require entrepreneurship competencies. This requires entrepreneurial knowledge (transferable knowledge), entrepreneurship skills (experience-non-transferable knowledge) and motivation (attitude and intention). Kailer (2005) states that entrepreneurship competence is necessary for individuals to start a business and recommends the Entrepreneurship Competency Model developed by himself [6].

In this study, a scale that measures the entrepreneurship competencies of university students was developed based on Kailer's Entrepreneurship Competency Model. According to the Entrepreneurship Competency Model of Kailer [6], based on the Bochum Competence Model [7], the implementation of an individual's intention to establish a business or take over a business requires an individual entrepreneurial competence. Individual entrepreneurship competence includes fund of knowledge and motivation about entrepreneurship. The knowledge about entrepreneurship is divided into two as explicit knowledge and implicit knowledge. In other words, fund of knowledge expresses knowledge and skills. The motivation of the enterprise refers to the basic attitude of the individual towards self-employment and the intention for enterprise. Therefore, entrepreneurship will lead to the planning of the initiative activities or the implementation of the planned intervention activities together with fund of knowledge (knowledge and skill) and enterprise motivation (attitude and intention). The developed scale is based on these sub-dimensions.

The measurement process in social sciences is an important concept covering a wide area. The measurement process and the period are one of the concepts that form the basis of science. People, events and processes can be examined through observation which is a scientific method. The data obtained from these observations must be defined in a way that everyone can understand to use them scientifically. For this purpose, scales are developed to be used scientifically according to the results of the measurement process [8].

2 Problem and Purpose of the Study

To develop entrepreneurship competencies of university students and to gain widespread currency, it is important to measure them, that is to say to evaluate them quantitatively. The aim of this research which starts with the question "Can a valid and reliable measurement tool be developed to assess student entrepreneurship competencies?" is to develop Entrepreneurship Competencies Scale for measuring students' entrepreneurship competencies.

3 Method of the Study

Type of the Study: This study, which was planned as a methodological type, was based on the content validity, construct validity and reliability analyzes.

Population and Sample: Population consists of the students of Selçuk University Faculty of Economics and Administrative Sciences and Konya Technical University Faculty of Engineering and Natural Sciences. The sample consists of 415 participants including 113 from Selçuk University and 302 from Konya Technical University. The number of items in the measurement tool was selected as the baseline to determine the sample size. For factor analysis, it is stated that 4:1 to 10:1 ratio of sample size is sufficient per item [9–11]. In this study, this ratio is about 9:1 for EFA and 7:1 for CFA. Convenience sampling method was used for sampling method.

Data Collecting Tools: In the Explanatory Factor Analysis stage, the data were collected by Entrepreneurship Competencies Scale (candidate scale) which was developed by the researchers and reduced to 27 items as a result of content validity analysis. To test the external criterion validity of the developed scale, the Individual Innovativeness Scale which was adapted to Turkish by Sarıoğlu-Kemer and Altuntaş (2017) was used as a parallel form [12].

The data were collected by two-factor Entrepreneurship Competencies Scale with 25 items developed as a result of Explanatory Factor Analysis (EFA) during Confirmatory Factor Analysis (CFA) (find enclosed). The scale was 5-point Likert type and the responses were scored as: 1 = I never agree, 2 = I don't agree, 3 = I neither agree nor disagree 4 = I agree and 5 = I strongly agree. There are no items in the scale to reverse the score. The lowest possible score in the scale is 25 and the highest score is 125. As the scores obtained from the scale increase, the level of entrepreneurship competencies increases as well.

Collecting the Data: The data for EFA were collected by the researcher (second author) between the dates of 21 February and 07 March 2019 from university students.

Data for test-retest were collected between 28–29 March 2019. At the EFA stage, the participants were asked to write a nickname in the survey saying that the same study would be repeated soon in two pilot classes. The survey form was given to the participants again in re-test period and they were asked to write the same nickname. First and second surveys with the same nicknames were matched. 21 days were left between the two applications for the test-retest. Half of the data collected during the EFA stage was used in the CFA phase.

Data Analysis: The data were analyzed by calculating Content Validity Index in the content validity stage. The data collected at the EFA stage were analyzed in IBM SPSS Statistics 25 program by means of mean, standard deviation, pearson product-moment correlation, Chronbach's Alpha reliability coefficient and Explanatory Factor Analysis. The data obtained during the CFA phase were evaluated in LISREL 9.3 programs by Confirmatory Factor Analysis validator factor analysis [11].

Ethical Authorisation of the Study: At the beginning of the study, ethics committee approval was received from Social Science and Humanities Research Association and Publication Ethics Committee. In addition, written permission was obtained from the rectorships of the universities to collect data, and the related faculty deans and department chairs were informed. The necessary explanations about the research were given to the students to be included in the study in the classroom environment and it was informed that the participation in the research would be on a voluntary basis. Data were collected from the volunteers by means of questionnaires. The questionnaires were distributed to the participants and the forms were filled in under the supervision of two researchers of the study in the classroom and later the forms were collected.

4 Results and Discussion

4.1 Validity and Reliability Studies

The validity study of Entrepreneurship Competencies Scale was carried out by (1) content validity, (2) construct validity and (3) criterion validity.

(1) Content validity was performed using Lawshe technique [13]. At this stage, a draft scale consisting of 32 items was presented to 10 experts and a 27-item candidate scale was obtained by calculating the content validity index [14, 15].

(2) Construct validity was tested with Explanatory Factor Analysis. Explanatory Factor Analysis was performed from the data obtained from 302 participants from Faculty of Engineering and Natural Sciences of Konya Technical University and 113 participants from Faculty of Economics and Administrative Sciences of Selçuk University. 40.9% of the participants were female and 59.1% were male. The fact that the Kaiser-Meyer-Olkin (KMO) value before the factor analysis was .88 and the Bartlett's Test of Sphericity value was 3553.87 indicates that the sample size was very good [11]. As a result of the factor analysis, a 25-item two factor structure with an eigen-value of more than 1 was obtained expressing a 42.28% of the total variance [9].

The first factor of Student Entrepreneurship Competencies Scale was called Entrepreneurial Knowledge (Knowledge + Skill) and the second factor was called Entrepreneurial Motivation (attitude + intention). The first factor consists of 14 items and the factor loadings vary between .64 and .49. The items collected in this factor represent the knowledge (knowledge + skills) dimension in the field of entrepreneurship. So, this factor is called Entrepreneurial Knowledge. There are 11 items in the second factor whose factor loadings vary between .83 and .49. When these 11 items were analyzed, it was found to be related to the dimension of Entrepreneurial Motivation of Entrepreneurship Competencies Model and this factor was called Entrepreneurial Motivation (see Table 1).

Table 1. Factor structure of entrepreneurship competencies scale and factor loads of materials (n = 240)

Scale items	Items	Factors/Dimensions	
		Factor I Entrepreneurial knowledge	Factor II Entrepreneurial motivation
1	When I graduate from college, I think I will have enough knowledge to start my own business	.53	
2	When I have an idea, I know where to start	.64	
3	In terms of legal status, I know the advantages and disadvantages of business types (individual companies, capital companies)	.54	
4	I think I have enough knowledge to discuss with my peers about entrepreneurship	.63	
5	I have knowledge about eh esctor I am interested in	.56	
6	I have experience about eh esctor I am interested in	.49	
7	I can build a team to realize my business idea	.64	
8	I can handle everything with my team mates	.65	
9	I can say I'm good at team work	.57	
10	I can say that I am skillful in bargaining, sales and presentation	.60	
11	I can guide them about what to do when I am with my friends	.55	
12	I am skilled at managing a team	.53	
13	I'm not having trouble making my ideas into reality	.59	
14	When I encounter a problem, I can easily solve it	.55	
15	Owning my own business is my greatest ideal		.72
16	Working in the public sector is not for me		.49
17	Risks must be taken to achieve something in life		.57
18	Doing the right job at the right time is necessary for success		.55
19	I prefer to do my own work by implementing my own idea instead of working for someone else		.77

(*continued*)

Table 1. (*continued*)

Scale items	Items	Factors/Dimensions	
		Factor I Entrepreneurial knowledge	Factor II Entrepreneurial motivation
20	It's not for me to work as a paid staff in a job that belongs to someone else		.70
21	My career goal is to become an entrepreneur		.75
22	I make every effort to establish and maintain my own business		.78
23	I seriously think about setting up my own business		.83
24	One day I will definitely start my own business		.80
25	I'm ready to do anything to be an entrepreneur		.75
Eigenvalue		6.29	5.13
Variance Explained by Factors (%)		23.28	19.00
Total Variance Explained (%)		42.28	
Kaiser Meyer Olkin Value		.88	
Bartlett test (sd: 351, p: .000)		3553.87	

(3) In the criterion validity test, scale scores are correlated with some external criteria (similar scales) [13, 16]. The correlation coefficient between the two scale scores is calculated after both scales are applied to the same sample group. In this study, it was determined that there is a positive, strong and highly statistically significant relationship between the Individual Innovation Scale used as a similar scale and the Entrepreneurship Competencies Scale that is being developed ($p < .001$, Table 2).

Table 2. The relationship between entrepreneurial competencies scale and individual innovation scale

Scales	Number (n)	r	p
Entrepreneurship Competencies Scale	52	.70	.000
Individual Innovation Scale	52		

Reliability: In reliability test of this study, (1) internal consistency and (2) Invariance tests were used. (1) internal consistency: Item analysis and Chronbach's alpha coefficient were calculated in Internal Consistency. In item analysis, it is recommended that total scale and the correlation of the items in the scale should be above .30 [13]. This value is considered as .40 to keep the reliability high in this study. 1.st and 2nd items were excluded from the scale whose total item Pearson correlation coefficient was below .40. Factor analysis was performed with 25 items left in the scale. As a result of factor analysis, a 25-item and 2-factor scale was formed. The total item correlations of Entrepreneurship Competencies Scale are between .49 and .83. Cronbach's α coefficient is a measure of the homogeneity of the internal consistency of the items in the scale [16]. The total and factor Cronbach's alpha coefficients of the scale are given in Table 3.

Table 3. Total entrepreneurial competences scale and Cronbach's alpha coefficients for factors (N = 240)

Scale and its sub-dimensions	Number of items	Items (New item no)	∝
Entrepreneurial Competences Scale (Total)	25	All items	.90
Factor 1: Entrepreneurial knowledge (Knowledge + Skill)	14	2, 3, 5, 6, 7, 8, 9, 10, 11, 12, 13, 14, 15, 16 (1–14)	.86
Factor 2: Entrepreneurial motivation (Attitude + Intention)	11	17, 18, 19, 20, 21, 22, 23, 24, 25, 26, 27 (15–25)	.90

As shown in Table 3, in the analysis of internal consistency which is one of the reliability indexes of Entrepreneurial Competences Scale and its sub-dimensions, Cronbach's alpha reliability coefficient was .90 for the whole scale, .86 in the sub-dimension of Entrepreneurial knowledge and .90 in the sub-dimension of Entrepreneurial Motivation. Cronbach's α coefficient, which may be sufficient on a Likert-type scale, is considered to be sufficient for test reliability [17], and this value should be as close to 1 as possible [16]. Usually, the reliability coefficient (Cronbach α) is considered to be "excellent" when it is around .90, very good when it is .80 and sufficient when it is around .70 [9]. If the alpha coefficient is less than .40, the measuring tool is not reliable [18]. Accordingly, it is seen that the whole internal consistency of the Entrepreneurship Competencies Scale is excellent, the internal consistency of the first factor is very good and the second factor is excellent.

(2) Invariance: Test-retest reliability was used for invariance test. The scale was applied to a group of 52 people twice for the test-retest three week intervals. The difference between the mean scores obtained from the first and second measurement of Entrepreneurship Competencies Scale applied to the students 3-week intervals was compared with the t-test in the dependent groups. In addition, the compatibility between the scale scores in two measurements was assessed by intraclass correlation coefficient (SKK = ICC) (Table 4).

Table 4. Test-retest mean scores and the relationship of entrepreneurship competencies scale and sub-dimensions (n: 52)

Scale and sub-dimensions	First test $\bar{x} \pm$ SS	Last test $\bar{x} \pm$ SS	t*	p	ICC	p
Total scale	89.52 ± 10.77	88.83 ± 11.05	1.162	.251	.96	**.000**
Factor 1: Entrepreneurial knowledge	48.79 ± 6.10	48.81 ± 6.98	.041	.968	.93	**.000**
Factor 2: Entrepreneurial motivation	40.73 ± 7.67	40.02 ± 7.70	1.924	.060	.97	**.000**

* T test in dependent groups: degree of freedom: 51

As shown in Table 4, when comparing the mean scores obtained from test-retest as reliability analysis to evaluate the time-invariance of Entrepreneurship Competencies Scale and two Sub-Dimensions, it was found that there was no statistically significant difference between the mean scores of the two measurements made 3-week intervals ($p > .05$, Table 4). When the intra-class correlation coefficient (two-way random effect, consistency model) was examined to test the consistency between the first and recurrent test scores of student Entrepreneurship Competencies Scale and two sub-dimensions, it was seen that the reliability coefficient was 96% in the total scale score, 93% in the sub-dimension of Entrepreneurial Knowledge, and 97% in the sub-dimension of Entrepreneurial Motivation and the rates were highly significant ($p < .001$, Table 4). It was found that time-invariance reliability of the scale was high and there was excellent and high consistency between the recurrent measurements at the level of 93% and 97%.

4.2 Confirmatory Factor Analysis (CFA)

CFA is the stage of verification of the model resulting from EFA in the simplest term. The coefficients (factor loadings) of all items in the entrepreneurship Competencies Scale with their size were found between .35 and .91 To test compliance competency as a result of CFA, X^2/df was examined and it was found as 2.39. The ratio between 0–2 is a perfect compliance and it is good compliance when it is between 2–3 [11] (Fig. 1).

Other common good compliance tests are Root Mean Square Error of Approximation (RMSEA), Standardized Root-mean-Square Residual (SRMR), Comparative Fit Index (CFI), Non-Normed Fit Index (NNFI), Goodness of Fit Index, (GFI), Adjusted Goodness of Fit Index (AGFI) [19, 20]. The fact that the RMSEA is equal to or less than 0.080 and the p value is less than .05 (statistically significant) shows a good compliance [19, 20]. If RMSEA value of the entrepreneurship Competencies Scale developed in this study is (<.80), (.073) model fit is considered good.

The fact that SRMR compliance index is less than .10 [20], CFI, NNFI values are equal to or greater than 0.90 (.95 and higher perfect compliance) [19, 20] and AGFI is equal or bigger than 0.80 indicates a good compliance. When GFI value equals to or is greater than 0.85 [11], it indicates compliance. It was found that there was a perfect fit according to SRMR (.068), CFI (.94), NNFI (.94) values obtained from the Entrepreneurship Competencies Scale.

Economic and social development at national and international levels is largely on the shoulders of entrepreneurs. A significant number of successful entrepreneurs of this age start their activities after getting a good university education. Some of them set up their business by leaving the workplace after working as an employee or going on to work - Hybrid entrepreneur - until starting their business. It is also seen that some of them set up their business after becoming the partner of the workplace or taking over that place. In today's conditions, transforming the business idea into an initiative and sustaining that initiative requires significant competencies. While parts of these requirements such as personality traits, socio-economic conditions including family and friends are formed by factors outside the university, a significant portion of them can be gained during university education. This study was based on Kailer's (2005) Entrepreneurship Competence Model. According to Kailer, entrepreneurship competence consists of entrepreneurship knowledge (explicit knowledge, and experience as

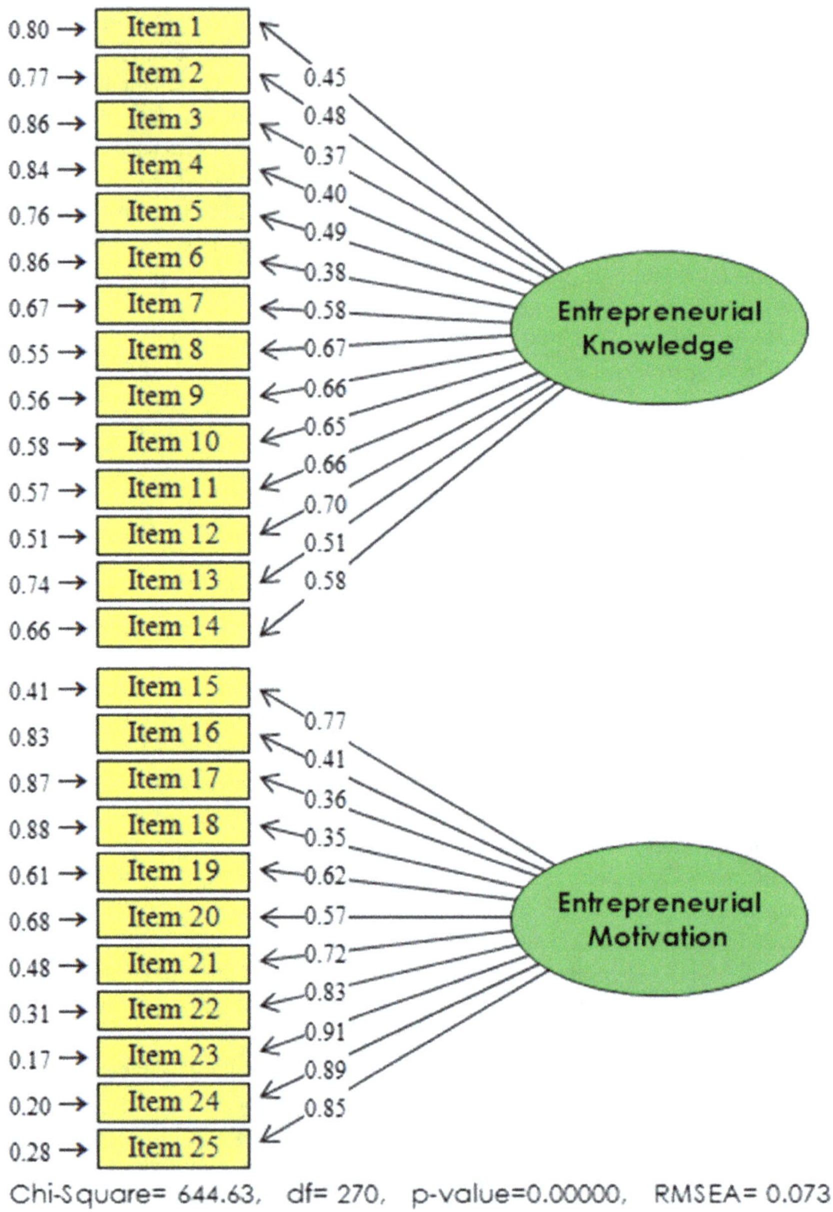

Fig. 1. Results of confirmatory factor analysis of entrepreneurship competencies scale: path coefficients and error variances

known implicit knowledge) and entrepreneurship motivation (entrepreneurship attitude and entrepreneurship intention). It is very important for universities to evaluate the results of their education and training activities well in terms of measuring the effectiveness of their work and the requirements. Although there are other scales developed in the international literature on the same subject, this study was carried out because it was believed that developing a new scale based on Kailer's Entrepreneurship Competencies Model would be beneficial.

This study was conducted to obtain a reliable and valid measurement tool to measure entrepreneurship competencies of university students and to bring them into international entrepreneurship literature. Entrepreneurship competency scale is a self-report scale with two sub-dimensions of 25 items. The scores to be taken from the scale vary between 25 and 125 and the higher the scores, the more entrepreneurship competencies increase. The total Chronbach-alpha reliability coefficient of the scale was .90 and the reliability coefficient of all dimensions was above .86. The scale can be used as a whole and since the dimensions have a high Chronbach 'alpha value, each dimension can be used alone. Similar scale correlation coefficient was .70 and test retest coefficient was .96. The total variance explanation percentage of the scale was measured as 42.28. Item's factor loading of the scale ranged between .49 and .83. As result of CFA analysis, it was seen that RMSEA was 0.073, SRMR < 0.068, Chi-Square/df value was 2.39, Comparative Fit Index (CFI) was 0.94, Non-Normed Fit Index (NNFI) was 0.94, Goodness of Fit Index (GFI) was 0.84 and Adjusted Goodness of Fit Index (AGFI) was 0.80.

In conclusion, the results showed that the Entrepreneurial Competencies Scale is a valid and reliable measurement tool that can be used to measure the students' entrepreneurial competencies. The fact that the data were collected from the universities in Konya only is the limitation of this study. The scale can be used by universities and researchers to assess students' competence gains. In addition, it can be used as pre-test and post-test measurement to evaluate the programs for the development of entrepreneurship competencies.

References

1. Gürsoy, G.: 40endustri40.com (2016). http://40endustri40.com/endustri-4-0-veyuksekogretim. Accessed 30 Apr 2019
2. Tüsiad Sanayi 4.0 Raporu. Türkiye'nin Küresel Rekabetçiliği İçin Bir Gereklilik Olarak Sanayi 4.0 – Gelişmekte Olan Ekonomi Perspektifi, The Boston Consulting Group. http://www.tusiad.org/indir/2016/sanayi-40.pdf. Accessed 30 Apr 2019
3. Schumpeter, J.A.: Theory of Economic Development. Harvard University Press, Cambridge (1934)
4. Tekin, M.: Girişimcilik ve Küçük İşletme Yöneticiliği. Güney Ofset, Konya (2006)
5. Koyuncuoğlu, Ö., Tekin, M.: Türkiye'de Girişimci Ve Yenilikçi Üniversitelerin Gömülü Teoriye Göre Değerlendirmesi ve Bir Model Önerisi", Selçuk Üniversitesi Sosyal Bilimler Enstitü Dergisi, vol. 41, pp. 16–31, Nisan 2019
6. Kailer, N.: Evaluation of entrepreneurship education at universities. Ibw-Mitteilungen, 3, pp. 1–11 (2005)
7. Staudt, E., Kriegesmann, B.: Kompetenzentwicklung: Neue Aufgaben für die Umsetzung von Innovationen. In: Wagner, K.-R. (eds.) Mitarbeiterbeteiligung, Gabler Verlag, Wiesbaden, pp. 138–151 (2002)
8. Şeşen, H.: Ölçek Geliştirme, Lefke Üniversitesi, İ.İ.B.F. Kıbrıs. Available: https://slideplayer.biz.tr/slide/2607750/. Accessed 15 May 2019

9. Özcan, K., Balyer, A.: Liderlik Oryantasyon Ölçeğinin Türkçeye uyarlanması. Mersin Üniversitesi Eğitim Fakültesi Dergisi **9**(1), 136–150 (2013)
10. Geçkil, T., Tikici, M.: Örgütsel Demokrasi Ölçeği Geliştirme Çalışması. Amme İdaresi Dergisi **48**(4), 41–78 (2015)
11. Çokluk, Ö., Şekercioğlu, G., Büyüköztürk,Ş.: Sosyal Bilimler İçin Çok Değişkenli İstatistik SPSS ve Lisrel Uygulamaları. 3. Baskı, Pegem Akademi, Ankara (2014)
12. Sarıoğlu-Kemer, A., Altuntaş, S.: Adaptation of The Individual Innovativeness Scale in Nursing Profesion: Turkish Validity - Reliability Study. Hemşirelikte Eğitim ve Araştırma Dergisi **14**(1), 52–61 (2017)
13. Lawshe, C.H.: A quantitative approach content validity. Pers. Psychol. **28**, 563–575 (1975)
14. Gözüm, S., Aksayan, S.: Kültürlerarası ölçek uyarlaması için Rehber II: Psikometrik özellikler ve kültürlerarası Karşılaştırma. Hemşirelik Araştırma Dergisi **1**(5), 3–14 (2003)
15. Polit, D.F., Beck, C.T.: The content validity index: are you sure you know what's being reported? Critiqueand recommendations. Res. Nurs. Health **29**, 489–497 (2006)
16. Tezbaşaran, A.: Likert Tipi Ölçek Geliştirme Kılavuzu", Psikologlar Derneği Yayınları, pp. 27–36, Ankara 1996
17. Kurnaz, M.A., Yiğit, N.: Fizik Tutum Ölçeği: Geliştirilmesi, Geçerliği ve güvenirliği. Necatibey Eğitim Fakültesi Elektronik Fen ve Matematik Eğitimi Dergisi **4**(1), 29–49 (2010)
18. Akgül, A.: Tıbbi Araştırmalarda İstatistiksel Analiz Teknikleri – SPSS Uygulamaları. 3. Baskı, Ankara 2005
19. Şimşek, Ö.F.: Yapısal Eşitlik Modellemesine Giriş Temel İlkeler ve LISREL Uygulamaları. Cem web Ofset, Ankara (2007)
20. Harrington, D.: Confirmatory Factor Analysis. Oxford University Press, New York (2009)

Evaluation of Entrepreneurial-Innovative University Conditions and Activities from Students' Point of View in the Context of Industry 4.0

Mahmut Tekin[1], Özdal Koyuncuoğlu[2(✉)], Tahsin Geçkil[2], and Deniz Baş[3]

[1] Faculty of Economic and Administrative Sciences, Selçuk University, Konya, Turkey
mahtekin@selcuk.edu.tr
[2] Faculty of Applied Sciences, Necmettin Erbakan University, Konya, Turkey
{okoyuncuoglu,tgeckil}@erbakan.edu.tr
[3] Vocational School of Health Services, Kırklareli University, Kırklareli, Turkey
deniz.bas@klu.edu.tr

Abstract. One of the issues frequently encountered in the digitalization of higher education is the transformation of universities to the University 4.0 with the Industry 4.0. University 4.0 refers to the characteristics that universities must have in the Industry 4.0 era. The adaptation of the workforce that will be most affected by the Industry 4.0 process and training it for the new competency needs will be carried out by universities and research centers to a great extent. The evaluation of the studies to be carried out within this framework in a wide range, not only in the fields of engineering but also in the fields of social sciences, medicine and even law, will accelerate the integration process of the Industry 4.0 at universities. In this way, with the change of the objectives and learning outcomes of the academic programs developed based on competency, it is naturally expected that the industry will have an impact on the educational programs and techniques. As a requirement of competition, state authorities care about the role of universities in entrepreneurship and have serious expectations. Public authorities support and promote universities in this area significantly. It is observed that universities have recently put forward their entrepreneurial and innovative university feature in marketing activities. However, how much are the students who are the real target group of the universities taken into consideration in the university entrepreneurship ecosystem? The questions such as "How can a university be measured and assessed as an entrepreneurial-innovative university?" have become a subject to be studied in the field of higher education. While looking for answers to these questions, evaluation of entrepreneurial-innovative university conditions and activities from the university students' point of view in the context of Industry 4.0 and University 4.0 emerged as an important requirement. The aim of this methodological study is to develop a tool to measure the internal environmental conditions and activities of the university students and entrepreneurial-innovative university based on Koyuncuoglu and Tekin (2019)'s Entrepreneurial University Model. For the study, a sample was used consisting of 528 students from Selçuk University

N. M. Durakbasa and M. G. Gençyılmaz (Eds.): ISPR 2019, LNME, pp. 605–618, 2020.
https://doi.org/10.1007/978-3-030-31343-2_53

Faculty of Economics and Administrative Sciences and Konya Technical University Faculty of Engineering and Natural Sciences, in Konya Province in Turkey. Content validity, construct validity and reliability studies were conducted at the stage of development of the Entrepreneurial-Innovative University Conditions Scale. Data were analyzed on the computer using SPSS and LISREL package programs. According to the results of the explanatory factor analysis (EFA), 31-item and three-dimensional Entrepreneurial-Innovative University Conditions Scale was developed. The Cronbach's alpha coefficient of the scale was calculated as .95 and the explanatory percentage was calculated as 51.29%. As a result of confirmatory factor analysis (CFA), it was seen that the scale had good fit values (X2/df: 2.47, RMSEA: .075, CFI: .96, NNFI: .96).

It was concluded that the scale developed in this research is a valid and reliable measurement tool that can be used to measure the internal environmental conditions and activities of the Entrepreneurial-Innovative universities. It can be said that the entrepreneurial-innovative university conditions scale will be important for the universities which want to improve their student-oriented entrepreneurial-innovative features, public authorities and for the entire economy in macro level.

Keywords: Industry 4.0 · Higher eduction · Entrepreneurial-innovative universities · Scale development study · Entrepreneurial-innovative university conditions scale

1 Introduction and Theoretical Framework

In the age of digitalization, it could be stated that providing graduates with qualifications of the age and qualified labor to sectors is in the authority of universities. The construction of universities, which can manage the digital era by developing global and local strategies with international organizations, governments and industry cooperation, is now being discussed in the context of higher education [1].

As a matter of fact, at the meeting of the Council of Higher Education at the date of 01.03.2018, "From Higher Education to Industry: Qualified Human Power Workshop" was held to debate and reconstruct the current situation of higher education in Turkey. The chairman of YÖK, Prof. Dr. M.A. Yekta Saraç who made an opening a speech at the workshop underlined the importance of the graduates from 4-year undergraduate programs and 2-year vocational schools to enhance Turkey's economic development and the international competitiveness; and for qualified manpower needed by the private sector and the industry. This situation, together with the digital economy created by "Industry 4.0" revolution, shows the importance of assessing the existing university structure in Turkey according to the wave of "technological unemployment" all over the world. In addition, it is also important to equip students with skills related to the jobs of future professions in higher education [2]. Ankara Chamber of Industry (ACI) President Nurettin Özdebir stated that there is a supply-demand dispute in the labor market and stated that university or vocational college graduates can not find jobs. He also said that industrialists complained about not being able to find qualified employees in accordance with their needs. The main reason for this is that the needs of the industry are quickly changing in today's global competitive environment and the educational programs of schools can't follow the developments as it should be. It was emphasized in the workshop that the sectors with high value added which produce intensive products by means of knowledge and technology need creative individuals who are equipped with basic sciences and mathematics, developed hand skills, and are able to adapt to changing conditions [3].

It is expressed by OECD reports that Education 4.0, which is considered to be necessary for a country to integrate with Industry 4.0, is implemented by countries with high automation levels [4]. It is obvious that Education 4.0 is a must to compete with developed countries economically and socially for the development of the country. At the beginning, there was a transition from the process focused on narration and memorizing (Education 1.0) and the computer and internet-oriented process (Education 2.0) to knowledge production process (Education 3.0); and now the world transits to the stage of innovation and production (Education 4.0). In the Education 4.0 approach, it is stated that constructivist education systems will be applied in general and a learning process based on 3 areas will be applied moving beyond Bloom taxonom [5]. With this process, (1) 3Rs regulating comprehension (Recalling, Relating, Refining), (2) 3Is encouraging to search (Inquiring, Interacting, Interpreting) and (3) and Result-oriented 3Ps (Participating, Processing-Presenting) are aimed to be achieved. It is pointed out that it is necessary to use the approaches such as visual learning, personalized education systems, game and scenario based learning, project based problem solving and augmented reality approaches to equip students with these skills [6].

When the development of universities is examined, it is seen that the last generation of universities are entrepreneurial-innovative universities. This has led to the promotion of the use of knowledge, that is to say the transfer of knowledge and technology as well as the entrepreneurial thought of students and academics [7].

In this study, it was tried to develop a scale which enabled the evaluation of university conditions and activities from the students' point of view based on the Entrepreneurial University Model of Koyuncuoğlu and Tekin (2019). Some authors call entrepreneurial and innovative universities as Entrepreneur Friendly Universities and point out the paradigm shift here [8, 9]. In their study, Tekin et al. show that the answer should be sought to the question "which university provides the best conditions for entrepreneur candidates?" rather than the question "which university has the most output?".

The evaluation about the entrepreneurial and innovative features of universities has been made by TUBITAK every year since 2012 in Turkey. As the Entrepreneurial and Innovative University Index implemented by TUBITAK is more output-oriented, a student-based and community interactive university evaluation model with balanced input and output activities was developed by Koyuncuoğlu and Tekin (2019). Koyuncuoğlu and Tekin (2019) argued that the inputs and outputs of the entrepreneurial and innovative universities aiming to be student-oriented and community interactive at the same time should be in balance. In this context, Koyuncuoğlu and Tekin recommend entrepreneurial university model which is sustainable for universities. In their study, Koyuncuoğlu and Tekin (2019), without any reference to any model, reached the theoretical structure consisting of 3 themes and 14 sub-categories of these themes including *strategic entrepreneurship orientation* (internal environmental conditions), *potential activation and development* (input) and *projects and commercialization* (output).

This study was built on the first two themes of the model developed by Koyuncuoğlu and Tekin (2019) and it was designed to measure the entrepreneurial and innovative characteristics of universities from the students' point of view. The theme of *strategic entrepreneurship orientation* in the model developed by Koyuncuoğlu and Tekin refers to the internal environmental conditions of the university and *potential activation and development* theme refers to the input activities of the model.

The measuring process in social sciences is an important concept covering a wide area. The measuring process is one of the concepts that form the basis of science. People, events and processes can be examined through observation which is a scientific method. To use the data obtained from these observations scientifically, they must be defined in a way that everyone can understand. For this purpose, scales are developed to be used scientifically according to the results of the measurement process [10].

2 Problem and Purpose of the Study

It is important that universities in Turkey can be measured and evaluated quantitatively to develop and extend their entrepreneurial and innovative activities. The aim of this research which starts with the question "Can a valid and reliable measurement tool be developed to evaluate the entrepreneurial and innovative features of universities from the students' point of view?" is to develop "Student-based Entrepreneurial and Innovative University Conditions Scale" so as to measure the entrepreneurial and innovative conditions and activities of the universities in Turkey in the context of Industry 4.0.

3 Method of the Study

3.1 Type of the Study

This study, which is planned as a methodological type, is based on the content validity, construct validity and reliability analyzes.

3.2 Population and Sample

Population consists of the students of Selçuk University Faculty of Economics and Administrative Sciences and Konya Technical University, Faculty of Engineering and Natural Sciences. The sample consists of 528 participants including 131 from Selçuk University and 397 from Konya Technical University. The number of items in the measurement tool was selected as the baseline to determine the sample size. For factor analysis, it is stated that 4:1 to 10:1 ratio of sample size is sufficient per item [11–13]. In this study, this ratio is about 10:1 for EFA and CFA. Convenience sampling method was used for sampling method.

3.3 Data Collecting Tool

The data were collected by Student-based Entrepreneurial and Innovative University Conditions Scale which was developed by the researchers. Student-Oriented

Entrepreneurial and Innovative University Conditions Scale consists of three sub-dimensions and 31 items. The scale was 5-point Likert type and the responses were scored as: 1 = I never agree, 2 = I don't agree, 3 = I neither agree nor disagree 4 = I agree and 5 = I strongly agree. The lowest possible score in the scale is 31 and the highest score is 155. The high scores indicate that the entrepreneurial-innovative conditions of the university are increasing/better.

3.4 Collecting the Data

The data for EFA were collected by the researcher (fourth author) from university students between the dates of 07 March and 21 March 2019. The data for test-retest were collected between 04–05 April 2019. At the EFA stage, the participants were asked to write a nickname in the survey saying that the same study would be repeated soon in two pilot classes. The survey form was given to the participants again in re-test period and they were asked to write the same nickname. First and second surveys with the same nicknames were matched. 14 days were left between the two applications for the test-retest. Half of the data collected during the EFA stage was used in the CFA phase.

3.5 Data Analysis

The data were analyzed by calculating Content Validity Index in the content validity stage. The data collected at the EFA stage were analyzed in IBM SPSS Statistics 25 program by means of mean, standard deviation, pearson product-moment correlation, Chronbach's Alpha reliability coefficient and Explanatory Factor Analysis. The data obtained during the CFA phase were evaluated in LISREL 9.3 programs by Confirmatory Factor Analysis validator factor analysis [13].

3.6 Ethical Authorisation of the Study

At the beginning of the study, ethics committee approval was received from Social Science and Humanities Research Association and Publication Ethics Committee. In addition, written permission was obtained from the rectorships of the universities to collect data, and the related faculty deans and department chairs were informed. The necessary explanations about the research were given to the students to be included in the study in the classroom environment and it was informed that the participation in the research would be on a voluntary basis. Data were collected from the volunteers by means of questionnaires. The questionnaires were distributed to the participants and the forms were filled in under the supervision of two researchers of the study in the classroom and later the forms were collected.

4 Results

4.1 Validity and Reliability Studies

The validity study of Student-based Entrepreneurial and Innovative University Conditions Scale was carried out by (1) content validity, (2) construct validity and (3) criterion-dependent validity.

(1) *Content Validity:* Content validity was performed using Lawshe technique [14]. At this stage, a draft scale consisting of 38 items was presented to 10 experts and a 37-item candidate scale was obtained by calculating the content validity index [15, 16].

(2) *Construct Validity:* Construct validity was tested with Explanatory Factor Analysis. Explanatory Factor Analysis was made with the data obtained from 131 participants from Selçuk University Faculty of Economics and Administrative Sciences (business administration 67, department of economics 51, international trade 13), and 397 participants from Konya Technical University Faculty of Engineering and Natural Sciences (industrial engineering 87, computer engineering 87, civil engineering 141, electrical and electronic engineering 62, chemical engineering 3, mechanical engineering 8, environmental engineering 5, geomatic engineering 3 and food engineering 1). 40.9% of the participants were female and 59.1% were male. The fact that the Kaiser-Meyer-Olkin (KMO) value before the factor analysis was .94 and the Bartlett Sps Test Sphericity value was 5950.79 indicated that the sample size was very good [13]. As a result of the factor analysis, a 31-item three-factor structure with an eigen-value of more than 1 was obtained expressing a 51.29% of the total variance. Variance rates are considered to be sufficient in the range of 40% to 60% in social sciences [11].

The first factor of the entrepreneurial-innovative university conditions scale is "University Internal Environmental Conditions", the second factor is "Awareness and Training Activities" and the third factor is "Incentive and Support Activities". The first factor consists of 7 items and the factor loadings vary between .77 and .42. The items collected in this factor represent the internal environmental conditions of the university in an entrepreneurial university model. For this reason, this factor is called "University Internal Environmental Conditions". There are 13 items in the second factor whose factor loadings vary between .69 and .43. When these 13 items were analyzed, it was found to be related to the dimension of Awareness and Training Activities of entrepreneurial university model and this factor was called "Awareness and Training Activities". There are 11 items in the third factor whose factor loadings vary between .78 and .43. When looking into these 11 items, it was found to be related to the dimension of "Incentive and Support Activities" of entrepreneurial university model and this factor was called "Incentive and Support Activities" as seen in Table 1.

The test of criterion-dependent validity couldn't be carried out in this study because the subject of measurement of entrepreneurial and innovative universities is extremely dynamic and literature is still developing and there is no parallel scale which can be used for criterion-referenced validity together with the candidate scale (such a scale could not be reached during the planning and conducting the research).

Table 1. Explanatory factor analysis results of the entrepreneurial-innovative university conditions scale (n: 264)

Item no	Items	Factor		
		1	2	3
1	There is an entrepreneurial culture at the university	.76		
2	University management gives importance to entrepreneurship	.76		
3	The university management approaches the entrepreneurship issue strategically	.75		
4	University management provides the necessary conditions (education, consultancy, mentoring, background etc.) to the students who want to be entrepreneurs	.77		
5	My personal communication network has expanded through activities at the university	.62		
6	The university has support units such as entrepreneurship center, technology transfer office	.42		
7	There are physical infrastructures at the university such as technopark, technocity, cyberpark and incubation center	.52		
8	The university regularly organizes activities such as competitions, project markets, career days, seminars, conferences, panels, publicity etc	.44	.57	
9	Entrepreneurship speakers are often invited to the university		.59	
10	Student clubs often organize activities related to entrepreneurship		.52	
11	Entrepreneurship is given particular importance in the official website of the university		.64	
12	In the social media accounts of the university (facebook, instagram etc.), there is active information/guidance on entrepreneurship		.59	
13	Adequate contact meetings are held at the university about entrepreneurship activities		.69	
14	The university makes efforts to increase the motivation of entrepreneurship		.69	
15	The university offers courses related to entrepreneurship (entrepreneurship culture, innovation, brand management, business plan and commercialization, technology management, innovation management etc.)		.41	
16	The university offers a rich variety of courses in the field of entrepreneurship		.57	
17	There are practices related to entrepreneurship (case study, workshop, field work etc.)		.49	
18	KOSGEB certificate is given as a result of the entrepreneurship courses I take at my university		.54	
19	The university offers entrepreneurship education incrementally (1. General entrepreneurship education, 2. Entrepreneurship training at the beginning of the work, 3. Presentation of entrepreneurship training modules after starting work)		.60	.43
20	When I wanted to turn my ideas into real, I learned at the university where to start		.52	

(*continued*)

Table 1. *(continued)*

Item no	Items	Factor		
		1	2	3
21	The university encourages and supports students about entrepreneurship			.52
22	The university supports students' social activities			.47
23	The university encourages students to participate in events			.48
24	The university allows students to make use of the network services (investment angels, student clubs, etc.)			.54
25	Academicians who give entrepreneurship courses encourage students to become entrepreneurs			.64
26	I can easily reach the university's entrepreneurship support units (technocity, TTO, entrepreneurship center etc.)			.68
27	The university's entrepreneurship support units produce fast and effective solutions			.70
28	Students who have an idea or project can exchange ideas with faculty/entrepreneurship support staff			.78
29	Students who have an idea or project help to canalize the instructors/entrepreneurship support staff			.71
30	The consultancy centers of the university (technology transfer office, technology development center, etc.) offer useful services (consultancy, mentoring, network, etc.) to students with an idea or project			.68
31	When information or consultancy is requested, the university's entrepreneurship support units are making a return		.43	.65
	Eigen value	6.75	6.75	5.48
	Variance explained by factors (%)	18.25	18.23	14.80
	Total variance explained (%)	51.29		
	Kaiser Meyer Olkin	.94		
	Barlett test (sd: 666, p: .000)	5950.79		

(3) *Reliability:* In reliability test of this study, (1) internal consistency and (2) Invariance tests were used. *Internal Consistency:* Item analysis and Chronbach's alpha coefficient were calculated in internal Consistency. In item analysis, it is recommended that total scale and the correlation of the items in the scale should be above .30 [15]. This value is considered as .40 to keep the reliability high in this study. A total of 6 items including 6, 9, 10, 22, 23 and 30 were excluded from the scale whose total item Pearson correlation coefficient was below .40. Factor analysis was performed with 31 items left in the scale. As a result of factor analysis, a 31-item and 3-factor scale was formed. The total item correlations of the Entrepreneurial-Innovative University Conditions Scale are between .41 and .78. Cronbach's α coefficient is a measure of the homogeneity of the internal consistency of the items in the scale [17]. The total and factor Cronbach's alpha coefficients of the scale are given in Table 2.

Table 2. Student-based entrepreneurial-innovative university conditions scale and Cronbach alpha reliability coefficients of the sub-dimensions (n: 528)

Scale and its sub-dimensions	Number of items	Items (Ne item no)	∝
Entrepreneurial-innovative university conditions scale (total)	31	All items	.95
Factor 1: University Internal Environmental Conditions	7	1, 2, 3, 4, 5, 6, 7	.87
Factor 2: Awareness and Training Activities	13	8, 9, 10, 11, 12, 13, 14, 15, 16, 17, 18, 19, 20	.90
Factor 3: Incentive and Support Activities	11	21, 22, 23, 24, 25, 26, 27, 28, 29, 30, 31	.91

As shown in Table 2, in the analysis of internal consistency which is one of the reliability indexes of Entrepreneurial-Innovative University Conditions Scale and its sub-dimensions, Cronbach's alpha reliability coefficient was .95 for the whole scale, .87 in the sub-dimension of University Internal Environmental Conditions, .90 in the sub-dimension of Awareness and Training Activities and .91 in the sub-dimension of Incentive and Support Activities. Cronbach's α coefficient, which may be sufficient on a Likert-type scale, is considered to be sufficient for test reliability [18], and this value should be as close to 1 as possible [17]. Usually, the reliability coefficient (Cronbach α) is considered to be "excellent" when it is around .90, very good when it is .80 and sufficient when it is around .70 [11]. If the alpha coefficient is less than .40, the measuring tool is not reliable [19]. Accordingly, it is seen that the whole internal consistency of the Entrepreneur-Innovative University Conditions Scale is excellent, the internal consistency of the first factor is very good, and the second and the third factor's internal consistency is excellent.

Invariance: Test-retest reliability was used for invariance test. The scale was applied to a group of 54 patients twice for the test-retest three-week intervals. The difference between the mean scores obtained from the first and second measurement of Entrepreneurial-Innovative University Conditions Scale applied to the students 2-week intervals was compared with the t-test in the dependent groups. In addition, the conformity between the scale scores in two measurements was assessed by intraclass correlation coefficient (ICC) (Table 3).

Table 3. Test-retest mean scores and the relationship of Entrepreneurial-Innovative University Conditions scale and their sub-dimensions (n: 54)

Scale and sub-dimensions	First test $\bar{X} \pm SS$	Last Test $\bar{X} \pm SS$	t	p	ICC	p
Total scale	90.56 ± 13.33	88.58 ± 17.55	1.302	.199	.86	.000
Factor 1: University Internal Environmental Conditions	20.56 ± 4.03	20.21 ± 4.92	.667	.508	.79	.000
Factor 2: Awareness and Training Activities	36.00 ± 7.14	35.27 ± 7.82	1.082	.284	.88	.000
Factor 3: Incentive and Support Activities	34.00 ± 5.16	33.10 ± 6.78	1.271	.209	.78	.000

t: t test in dependent groups: degree of freedom = 51
ICC: Intraclass Correlation Coefficient

As shown in Table 3, when comparing the mean scores obtained from test-retest as reliability analysis to evaluate the time-invariance of Entrepreneurial-Innovative University Conditions Scale and sub-dimensions, it was found that there was no statistically significant difference between the mean scores of the two measurements made 2-week intervals ($p > .05$, Table 3). When the intra-class correlation coefficient (two-way random effect, consistency model) was examined to test the consistency between the first and recurrent test scores of student-based Entrepreneurial-Innovative University Conditions Scale and sub-dimensions, it was seen that the reliability coefficient was 86% in the total scale score, 79% in the sub-dimension of University Internal Environment Conditions, 88% in the sub-dimension of Awareness and Training Activities and in advanced level as .78% in the sub-dimension of Incentive and Support Activities and the rates were highly significant ($p < .001$, Table 3). It was found that time-invariance reliability of the scale was high and there was excellent and high conformity between the recurrent measurements at the level of 78% and 88%.

4.2 Confirmatory Factor Analysis (CFA)

CFA is the stage of verification of the model resulting from EFA in the simplest term. The coefficients (factor loadings) of all items in the Entrepreneurial-Innovative University Conditions Scale with their size were found between .36 and .87. To test compliance competency as a result of CFA, X^2/df was examined and it was found as 2.47. The ratio between 0–2 is a perfect compliance and it is good compliance when it is between 2–3 [13] (Fig. 1).

Other common good compliance tests are Root Mean Square Error of Approximation (RMSEA), Standardized Root-mean-Square Residual (SRMR), Comparative Fit Index (CFI), Non-Normed Fit Index (NNFI), Goodness of Fit Index, (GFI), Adjusted Goodness of Fit Index (AGFI) [20, 21]. The fact that the RMSEA is equal to or less than 0.080 and the p value is less than .05 (statistically significant) shows a good compliance [20, 21]. The fact that RMSEA value of the Entrepreneur-Innovative University Conditions Scale developed in this study was found to be (< .80) showed that (.075) model fit was good. The fact that SRMR compliance index is less than .10 [21], CFI, NNFI values are equal to or greater than 0.90 (.95 and higher perfect compliance) [20, 21] and AGFI is equal to or bigger than 0.80 indicates a good compliance. It was observed that there was a perfect compliance compared to SRMR (.064), CFI (.96), NNFI (.96) values obtained in Entrepreneurial-Innovative University Conditions Scale.

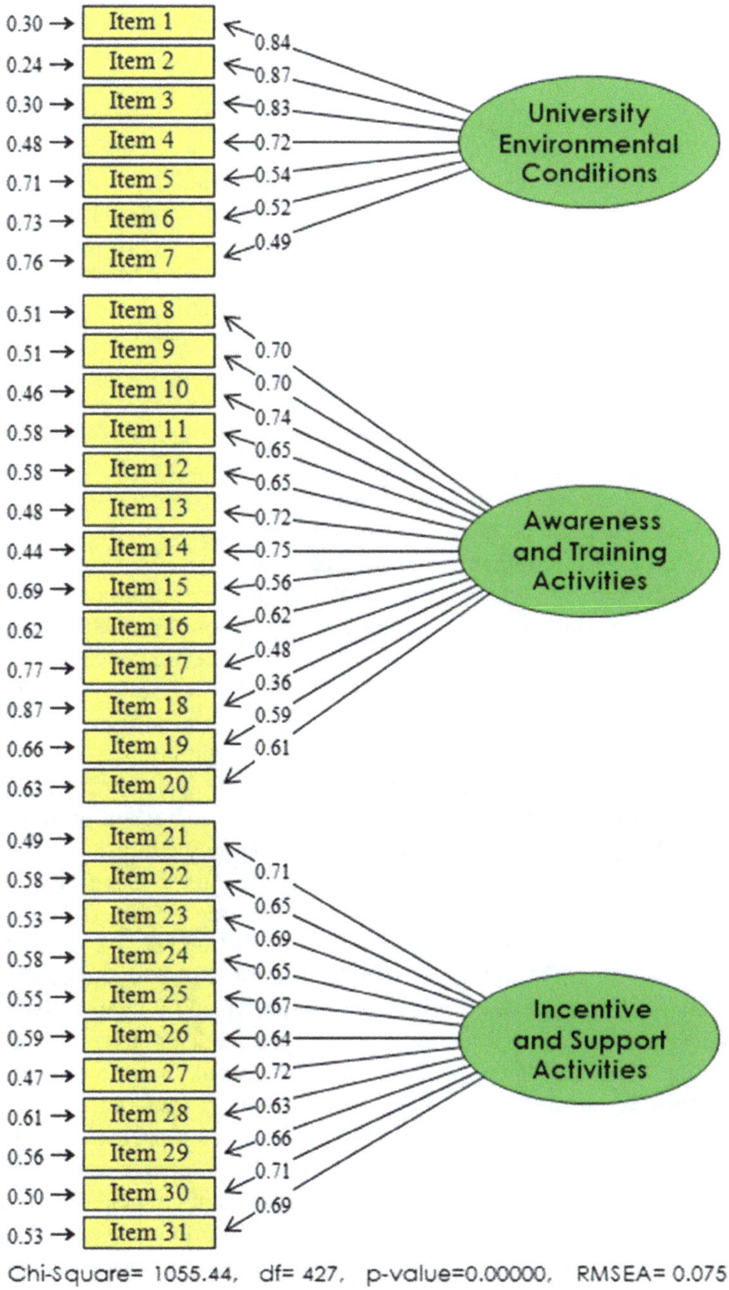

Fig. 1. Confirmatory factor analysis results of the entrepreneurial-innovative university conditions scale: path coefficients and error variances

5 Conclusion

Economic and social development at national and international levels is largely on the shoulders of entrepreneurs. A significant number of successful entrepreneurs of this age start their activities after getting a good university education. In today's conditions, it will not be wrong to say that turning the idea of business into an enterprise and maintaining that enterprise depends on the entrepreneurship conditions and activities of the university where entrepreneur candidates are educated. This study is based on Koyuncuoğlu and Tekin's (2019) Entrepreneurial University Model. According to Koyuncuoğlu and Tekin, Entrepreneurial Universities consist of the internal environmental conditions of the university (the attitude of the university), the input dimension (awareness, training, incentive and support activities) and the output dimension (projects and commercialization). It is very important for universities to evaluate the results of their education and training activities in terms of measuring the effectiveness of their work and the requirements. This study was conducted because it was believed that developing a scale based on the Entrepreneurial University Model of Koyuncuoğlu and Tekin would be beneficial as there is no other scale developed in the international literature that evaluates entrepreneur-innovative universities from the perspective of students. The aim of this study is to obtain a reliable and valid measurement tool for the evaluation of entrepreneur-innovative university conditions from the perspective of students and to bring them into the international entrepreneurship literature. Entrepreneur-innovative university conditions scale is a scale with three sub-dimensions of 31 items. The scores to be taken from the scale vary between 31 and 155 and indicate that the entrepreneur-innovative conditions and activities of the university are good from the students' point of view. The total Chronbach-alpha reliability coefficient of the scale was .95 and the reliability coefficient of all dimensions was above .87. The scale can be used as a whole and since the dimensions have a high Chronbach' alpha value, each dimension can also be used alone. The test retest coefficient was .86. Total variance explanation of the scale was measured as 51.29%. The item factor loadings of the scale ranged between .41 and .78. As a result of CFA analysis, it was seen that RMSEA was 0.075, SRMR < 0.064, Chi-Square/df value was 2.47, Comparative Fit Index (CFI) was .96 and Non-Normed Fit Index (NNFI) was .96.

Universities also have important duties to increase students' interest in technology and to encourage their studies in this field. Student technology communities will play an important role in raising the awareness of university students in the field of technology. The fact that universities organize special education programs in the field of technology and their physical infrastructures such as techno parks open their doors to the students will encourage the students to take initiatives in the field of technology. In addition, universities should make the best of the opportunities offered by digital transformation and consider digital technologies as a key element. Universities are advised to integrate digital technologies into their infrastructure and increase their digital capabilities to support innovation and entrepreneurship.

In conclusion, the results showed that the Entrepreneur-Innovative University Conditions Scale is a valid and reliable measurement tool that can be used to measure the entrepreneurial and innovative conditions and activities of the universities from students' point of view. The fact that criterion-referenced validity was not performed and the data were collected from the universities in Konya only is the limitation of this study. Entrepreneurial-innovative university conditions scale is a scale filled with three sub-dimensions of 31 items based on self-report. The scale can be used by universities and researchers to evaluate the entrepreneurial and innovative conditions and activities of universities. In addition, it can be used as a pre-test and post-test measurement tool to evaluate the effectiveness of the programs for the development of the entrepreneurial-innovative conditions and activities of universities.

References

1. Aybek, Y.: Üniversite 4.0'a geçiş süreci: kavramsal bir yaklaşım. Açıköğretim Uygulamaları ve Araştırmaları Dergisi **3**(2), 164–176 (2017)
2. YÖK: The YÖK (2018). http://www.yok.gov.tr/yuksekogretimden-endustriye-nitelikli-insan-gucu-calistayi
3. ACI: The Ankara Chamber of Industry (2016). http://www.aso.org.tr/wp-content/uploads/2016/09/1-20160927MILLIYET_ANKARA_SF_1.pdf
4. OECD Programme for International Student Assessment. http://www.oecd.org/pisa/publications. Accessed 30 Apr 2019
5. Gomaratat, S.: Subject: learning productivity. In: Sinlarat, P. (ed.) 10 Ways of Progressive Learning Encouraging/Facilitating the Ability of the Learner of 21st Century. Education Science, Dhurakit Bandit University, Bangkok (2015)
6. Nedeva, V., Dineva, S.: New learning innovations with web 4.0. In: Proceedings 7th International Conference on Virtual Learning ICVL, pp. 316–321
7. Koyuncuoğlu, Ö., Tekin, M.: Türkiye'de Girişimci Ve Yenilikçi Üniversitelerin Gömülü Teoriye Göre Değerlendirmesi ve Bir Model Önerisi. Selçuk Üniversitesi Sosyal Bilimler Enstitü Dergisi **41**, 16–31 (2019)
8. Tekin, M., Geçkil, T., Koyuncuoğlu, Ö.: Entrepreneurial universities index: a scale development study. Macrothink Inst. Bus. Econ. Res. **7**(2), 70–81 (2017)
9. Tekin, M., Geçkil, T., Koyuncuoğlu, Ö., Tekin, E.: Girişimci Dostu Üniversiteler İndeksi ve Bir Model Geliştirilmesi. Selçuk Üniversitesi Sosyal Bilimler Enstitüsü Dergisi **39**, 138–150 (2018)
10. Şeşen, H.: Ölçek Geliştirme, Lefke Üniversitesi, İ.İ.B.F. Kıbrıs. https://slideplayer.biz.tr/slide/2607750/. Accessed 03 May 2019
11. Özcan, K., Balyer, A.: Liderlik Oryantasyon Ölçeğinin Türkçeye uyarlanması. Mersin Üniversitesi Eğitim Fakültesi Dergisi **9**(1), 136–150 (2013)
12. Geçkil, T., Tikici, M.: Örgütsel Demokrasi Ölçeği Geliştirme Çalışması. Amme İdaresi Dergisi **48**(4), 41–78 (2015)
13. Çokluk, Ö., Şekercioğlu, G., Büyüköztürk, Ş.: Sosyal Bilimler İçin Çok Değişkenli İstatistik SPSS ve Lisrel Uygulamaları, 3. Baskı, Pegem Akademi, Ankara (2014)
14. Lawshe, C.H.: A quantitative approach content validity. Pers. Psychol. **28**, 563–575 (1975)
15. Gözüm, S., Aksayan, S.: Kültürlerarası ölçek uyarlaması için Rehber II: Psikometrik özellikler ve kültürlerarası Karşılaştırma. Hemşirelik Araştırma Dergisi **1**(5), 3–14 (2003)
16. Polit, D.F., Beck, C.T.: The content validity index: are you sure you know what's being reported? critique and recommendations. Res. Nurs. Health **29**, 489–497 (2006)
17. Tezbaşaran, A.: Likert Tipi Ölçek Geliştirme Kılavuzu1, pp. 27–36. Psikologlar Derneği Yayınları, Ankara (1996)

18. Kurnaz, M.A., Yiğit, N.: Fizik Tutum Ölçeği: Geliştirilmesi, Geçerliği ve güvenirliği. Necatibey Eğitim Fakültesi Elektronik Fen ve Matematik Eğitimi Dergisi **4**(1), 29–49 (2010)
19. Akgül, A.: Tıbbi Araştırmalarda İstatistiksel Analiz Teknikleri – SPSS Uygulamaları, 3. Baskı, Ankara (2005)
20. Şimşek, Ö.F.: Yapısal Eşitlik Modellemesine Giriş Temel İlkeler ve LISREL Uygulamaları. Cem web Ofset, Ankara (2007)
21. Harrington, D.: Confirmatory Factor Analysis. Oxford University Press, New York (2009)

Capstone Projects

A Framework of Route Selection for Hazardous Goods Transportation in Tunnels

Ayça Türkmen, Berna Özçınar, Gökay Aydoğdu, Tilbe Adsız(✉), and Melisa Özbiltekin

International Logistics Management, Yaşar University, Bornova, Turkey
aycaturkmen1997@gmail.com, bernaozcnar@gmail.com, gokayaydgd@gmail.com, adsiztilbe@gmail.com, melisa.ozbiltekin@yasar.edu.tr

Abstract. Due to the increase in globalization, higher customer expectations and environment volatility, supply chains are exposed to risks more easily. Therefore, supply chain risk management become a crucial issue for companies and countries in last decades. Supply chain risk management (SCRM) involves all type of risks such as supplier risks, locational risks, natural risks. Hazardous goods transport is a very destructive type of transport which causes very risky accidents. Hazardous goods transport risks can have serious consequences for the environment, property or people. The importance of these substances in daily life and industrialization and the increasing demand for these products are important for minimizing and assessing the risks associated with Dangerous Goods Transportation (DGT). Assets that may be affected by these risks are classified under 5 main criteria in this research. As a result of the Analytic Hierarchy Process (AHP) method, it is aimed to help the selection by establishing a risk-based superiority between alternative ways and tunnels while selecting the route in the pilot area.

Keywords: DGT · AHP · SCRM · Road transportation · Tunnel

1 Introduction

Recently, increasing competition encourages firms to develop their supply chains. The supply chain represents the complete path from the first supplier to the final supplier. A smooth supply chain flow can create competitive advantage for companies. Moreover, logistics activities in supply chain management are the crucial parts that need to be considered and developed. Risk is the unexpected occurrence of unexpected events. The companies or countries can face with risks at any stages of supply chain. It may cause structural problems in the part of supply chain operations. Furthermore, risk management is more important than past years because of the risks and potential results they face with are getting bigger. The risk impacts must be managed if companies or countries want to survive crisis that occurs after the risk occurs [1].

Traffic accidents are the most important risk in road transport. It can be caused by driver's misconduct, the condition of the vehicle, the adversely weather conditions, the

N. M. Durakbasa and M. G. Gençyılmaz (Eds.): ISPR 2019, LNME, pp. 621–635, 2020.
https://doi.org/10.1007/978-3-030-31343-2_54

traffic density, etc. However, the most dangerous traffic accidents are caused by hazardous goods transportation. The best way to protect against these destructive effects is to establish and maintain a continuous monitoring system.

Moreover, hazardous substances are dangerous materials for human health and safety, animals and property depending on their content. Therefore, regulations such as ADR are designed to prevent accidents involving people, property or the environment as much as possible. Although regulations are arranged, it is not enough to avoid hazardous goods accidents caused by human errors and the results may be catastrophic.

Road traffic intensity has increased over the years and also the amount of transportation of hazardous goods has increased. When transporting hazardous materials, depending on their impact level, they can have devastating consequences on the population and the environment as in the example when an accident occurs in densely populated areas.

In order to give comprehensive detailed information about the transportation of hazardous substances and the level of risk, decision-makers consider the following steps in detail to highlight the importance of controlling such transports and analyzing the relevant risks. The ways in which they define risk are performed by analyzing past events, collecting ideas, or directly analyzing them.

In the following sections, problem statement is explained. Then, literature review about risk management and some regulations about hazardous materials transportation, methodology of the study, data collection are described in detail.

2 Problem Statement

This study is focused on risk of hazardous goods transportation. Hazardous goods transportation is risky not only open road but also in tunnels. Moreover, hazardous goods transportation is risky for human and environment. Recently, with improvement in industrialization need for hazardous materials is growing, so traffic volume of hazmat by road transportation is increasing. Hazardous materials can cause extensive damage due to their characteristic features. The number of people who are exposed to risk and the number of inhabitants in the vicinity may be subject to considerable loss according to the results of the risk.

These risks come out during the transportation of hazardous material by roads, such as loss of life and property and environmental disasters. International agreements try to provide reducing risk to lowest level that occurs during transportation in tunnels or roads. By analyzing these risks, study has tried to develop a framework aimed at minimizing damage and losses. In order to accomplish this aim, firstly the masses and structures that could be affected determined and the hierarchy between them has tried to establish. For this purpose, a questionnaire which includes the effects of hazardous materials during the accident is prepared in this study. Moreover, after collected data from the questionnaire, expert opinions were analyzed. The results are obtained with the AHP method to find the prioritize of the criteria.

By using AHP and Scale methods, the route which minimizes the risk is tried to determine with comparing alternative ways of Bayraklı Tunnel that has chosen as pilot region to make hazardous goods transportation safer.

3 Literature Review

In this study, literature review consists of six parts as the supply chain risk management, NATO management of transportation of hazardous goods, ADR which is related regulations on transportation of hazardous goods, information about nine class of hazardous materials, regulations about safety transport on tunnels and the way of minimizing risks with the help of Supply Chain Risk Management (2007), NATO Transport of Dangerous Goods (2010) and Transport Safety in Tunnels books (2001).

Firstly, supply chain risk management is analyzed with by considering literature.

3.1 Supply Chain Risk Management

The risks to the supply chain are interrupted or deteriorated due to unpredictable events [1]. The risks in the supply chain include cost fluctuations, material shortages, supplier financial problems, failures, natural and man- made disasters. Logistics activities in supply chain management are important activities that should be taken into consideration and developed because of the risks they face. Precautions should be taken for future uncertainties and approaches should be determined by identifying the risks in the best way [1]. Risk management processes in the supply chain include risk planning, risk identification, risk analysis, managing risk, and monitoring risk [1]. Using these steps, creating a risk management matrix is very important for the future of the supply chain.

In the following part, there are some information about transportation of hazardous goods by considering NATO transport of hazardous goods books.

3.2 NATO Transport of Dangerous Goods

The United States Department of Transportation defines dangerous good (DG) or hazardous material as "any substance or material capable of causing harm to people, property, and the environment" The United Nations divides them into nine classes according to their physical, chemical and nuclear characteristics [2].

According to manufacturers, consumers and developers, accidents and failures have some reasons. These reasons can be listed as;

- Poor quality projects
- Poor quality operational tests
- Poor quality monitoring of operations
- Human error
- Terrorist actions
- Failure to obey the traffic rules to train on time [2].

Therefore, besides the production, it is important to give importance transportation of the goods. Generally, hazardous substances are used in places other than where they are produced, so each type of substance should be considered separately in defining the risks that may occur during transport. These risks are caused by the failure of logistics activities or human errors in planning process. Hazardous substances are mostly transported by road. Hazardous substance transportation process can organize in 3 areas such as at the place of production, at the destination or on the way.

DGT accidents are generally classified as low probability–high result events [2]. From the point of government, the government is responsible for risk-reducing routes. From the carrier perspectives, carrier companies prefer routes which one is less costly. Cost–benefit analysis can be used when deciding on the route selection problem. Measures productivity or the effects of resource management, considering all potential impacts on society and the economy.

In the following part, ADR which is related to transportation of hazardous goods is explained in detail.

3.3 ADR (European Agreement Concerning the International Carriage of Dangerous Goods by Road)

It is a European Convention on the Transportation of Dangerous Goods on International Roads. It is an international treaty prepared in Geneva on September 30, 1957 and entered into force in 1968. Turkey has signed agreements on 22 March 2010.

ADR is a directive that ensures safe and orderly transport of hazardous substances without causing harm to human health and the environment. ADR provide additional requirements within the carrier except the transported load and vehicle. These requirements are the conditions for the construction of vehicles, special training for the driver and the enterprises involved in transportation [2].

In the next part, nine class of hazardous materials are explained.

3.4 Nine Class of Hazardous Materials

Hazardous materials are divided into nine classes which each class has its own characteristics. In determining the rules for the transport of hazardous substances, provisions are set separately for each product. Each class is represented by individual labels, while separate rules and provisions are set for each class in packaging, labeling, handling and marking. Hazardous materials can be classified according to their risk levels and structural characteristics [2].

In the following part, transport safety in tunnels are explained.

3.5 Transport Safety in Tunnels

Traffic density, population and environmental factors cause major problems in tunnels. In developing technical processes related to tunnels, the problems were encountered during the passage of hazardous goods are not taken into consideration. A risk involving hazardous goods in the tunnel can lead to costly consequences such as human life, environmental and tunnel damage. On the other hand, prohibiting tunnels for vehicles which transports hazardous goods, creates economic losses for carrier companies. As a result, the density of the population passes through the regions, which increases the risk.

It should be noted that there are three major hazards during categorization that can cause injury or cause serious damage to the tunnel structure; explosions, exposure of toxic gases or volatile toxic fluids, fires [3]. Furthermore, there are five tunnel categories; A (no restriction on the carriage of hazardous goods), B (restrictions for the carriage of hazardous goods which may lead to a very large explosion), C (restrictions for the carriage of hazardous goods which may lead to a very large explosion, a large explosion or a large toxic release), D (restrictions for the carriage of hazardous goods which may lead to a very large explosion, a large explosion, a large toxic release or to a large fire), E (restriction for the carriage of all hazardous goods other than UN 2919, 3291, 3331, 3359 and 3373 and all other hazardous goods in accordance if the quantities carried exceed 8 tones total gross mass per transport unit) [3].

For reducing risks of transportation of hazardous materials, finding an alternative route can be a solution.

Alternative route can be found depending on the desired conditions;

- Shortest path.
- Minimum impact for population.
- Minimum event accident.
- Minimum environmental risk.

In the next part, some information about minimizing risks with the help of risk management processes for transportation of hazardous goods are given.

3.6 Minimizing Risk

Recently, risk management and risk minimization become a hot topic in real world. Many researchers seek to find true and safety route in their article. By considering hazmat risks, risk will be at more high level and impact area will have serious consequences. Hazmat solutions is complicated because regulation and natural impact effect researchers. These studies were focused especially on choosing safe transportation route by using many method [4]. The recommended route depends critically on the model adopted for the quantification of risk, as shown by [5]. For minimizing risk, it depends on a few circumcenters such as time, route, citizen population and tunnel allowance. Tunnel authorities need to make informed decisions as to whether such transport should be permitted in certain tunnels and, if so, how this can be achieved in a safe and practicable manner [5].

Moreover, by considering literature review, reduction of hazardous material transportation risk can be achieved in many different ways [6]. Some of these risk reduction measures, such as driver training and regular vehicle maintenance, have little connection to Operations Research (OR) [6]. According to Ümit et al. (2003) Road transportation of hazmat is special since it requires consideration of various actors who pursue different goals. An environmental risk analysis model is developed to help assess environmental risks for fuel products shipments [7].

In the following part, methodology of the study is explained.

4 Methodology

Methodology of the study consists of two main calculation; determining risk probability and risk impact in tunnels and roads. Therefore, firstly, risk probability of roads and tunnels are determined with the help of accident rate data. Then, weights of AHP is determined and a scale is prepared. Firstly, calculation of risk probability on roads and tunnels are explained.

4.1 Risk Probability

To calculate risk probability on tunnels and roads caused by hazardous goods transportation, firstly, accident rate of pilot roads and traffic volumes are tried to obtain from the government. After obtaining data, the accident rate of pilot destinations is calculated and risk probabilities of them are found. Therefore, following equation is used to calculate accident rate of roads.

$$Risk = Risk\ Possibility \times Risk\ Impact \tag{1}$$

$$Accident\ Rate = \frac{X(Accident\ Rate)}{Traffic\ Volume} \tag{2}$$

4.2 AHP

In this study, AHP method is used to measure the impact of the risk. AHP consists of 2 different calculations including weights and scale.

(1) *AHP and Criteria Selection:* The analytic hierarchy process (AHP) was developed by Thomas H. Saaty in 1977 [5]. The main purpose of the AHP method is to define and compare the superiority of the criteria and sub-criteria in a systematic way. The best result that we want to achieve in this information is obtained by comparing.

As a first step in the AHP, the problem is defined. In the direction of the intended use, the transmitter determines the criteria and sub-criteria and forms a hierarchical structure. Based on these criteria, alternative options are determined. Moreover, the criteria are obtained from the reviewed articles and studies and after that experts are approved these criteria to apply the study. These people are expert of dangerous good transportation systems in private firms. The criteria reference table is given in Table 1.

Table 1. Criteria reference table

Criteria	Author(s)
People in the impact area	Bubbico et al. [8]; Chen et al. [9]; Fabiano et al. [10]; Frank et al. [11]; OECD [12]
Environmental impact – water environment, soil environment, featured areas	Bubbico et al. [8]; Erkut et al. [6]; Fabiano et al. [10]; Huang et al. [13]; Topuz, et al. [14];
Emergency action speed	AFAD et al. [14]; Bubbico et al. [8]; Chen et al. [9]; Ümit et al. [7]; OECD [12]
Proximity to settlement	Özyağcı et al. [15]

By considering literature review, main and sub criteria are defined to find weights with AHP to find prioritization of the hazardous materials on water, soil, featured areas, human and emergency cases and to find best road for hazardous goods transportation. The criteria table is given Table 2.

Table 2. Criteria table

Criteria	Sub-criteria
Water environment (A)	Stream (A1)
	Dam (A2)
	Sea (A3)
	Lake (A4)
	Irrigation areas (A5)
Soil environment (B)	Forest areas (B1)
	Natural protected areas (B2)
	Farming areas (B3)
Featured areas (C)	
Human (D)	Number of people to be affected (D1)
	Proximity to settlement (D2)
Emergency case (E)	Proximity to hospital (E1)
	Proximity to fire stations (E2)
	Proximity to law enforcement officers (E3)

In order to compare each criterion among each other, binary comparison decision matrices are created. When creating these matrices, 1–9 significance scale developed by Saaty is used [4].

The scales are used for questionnaire of the study. The scale table is given in Table 3.

Table 3. Scale table of AHP [20]

Intensity of importance	Definition	Explanation
1	Equal importance	Two factors contribute equally to the objective
3	Somewhat more important	Experience and judgement slightly favor one over the other
5	Much more important	Experience and judgement strongly favor one over the other
7	Very much more important	Experience and judgement very strongly favor one over the other. Its importance is demonstrated in practice
9	Absolutely more important	The evidence favoring one over the highest possible validity
2, 4, 6, 8	Intermediate values	When compromise is needed

Firstly, decision maker evaluates criteria according to scale to prioritize criteria. Secondly, normalized matrices are formed. The value of each column is obtained by dividing the value of the respective column separately. Then the average of the value of each column is calculated. As a result of the matrix calculation, importance criteria are found for each criterion [4].

After the binary comparison decision matrices are created, the next step is the calculation of priority or weight vectors. A common method for calculating priority vectors is: The normalized matrix is obtained by dividing each column value by the respective column sum and moving from the normalized matrix; averages of each row values are taken; these values are the weight of importance for each criterion. The priority vector is created with these weights. To measure whether the decision maker is consistent when comparing between factors, for each generated comparison matrix, it is necessary to calculate the Consistency Ratio (C.R.). In this calculation, rationality index (R.I.) numbers are used depending on the alternative number of n. If the value found in the calculations is less than 0.10, it is concluded that the comparison matrix generated is consistent. Otherwise, the comparison matrix is inconsistent and must be rearranged [2]. Rationality index is shown in Table 4.

Table 4. Rationality index

n	1	2	3	4	5	6	7	8	9	10	11	12	13	14	15
RI	0	0	0,58	0,9	1,12	1,24	1,32	1,41	1,45	1,49	1,51	1,48	1,56	1,57	1,59

The following equation is used for consistency analysis.
CI = Consistency Indicator RI = Rationality Indicator CR = Consistency Ratio

$$CI = \frac{\lambda - n}{n - 1} \tag{3}$$

$$CR = \frac{CI}{RI} \tag{4}$$

Furthermore, the consistency ratio is calculated in order to measure whether the decision maker is consistent during the comparison. If the result is below 0.10, the matrix is consistent, and if a big result is obtained, a re- evaluation may be required. Lastly, the importance weights and alternatives are multiplied by their weight and the priority value of each alternative is found. The sum of these values must be equal to 1. The best value is the best alternative [4].

(2) *Creation of Scale to Alternative Roads:* Priority scoring table is created to find assigned reviewers determine preliminary scores for each of the scored review criteria. After that a preliminary score for the overall impact and impact score should reflect the reviewer's overall evaluation. Scoring table should consider the full range of the rating scale and the scoring descriptors in assigning preliminary and final scores. The impact score for an application is based on assessments of each individual reviewer of the scored criteria multiple additional criteria regarding the protection and inclusion of human, animal care and welfare; bio-hazards, and criteria specific to the funding opportunity.

While creating scoring table for all classes, it is not appropriate to work with all classes at the same time because of variability of substance change while substance in same class. It is a challenge to decide one specific substance and one class. Therefore, class 2 which is gases [16] is selected. For instance, LPG has an explosive gas spectrum between 1.8% and 9.5% in the air which is more suitable than other common gaseous fuels [17]. Therefore, when the LPG explosive, it will affect soil environment more than water environment, for that soil environment coefficient range is narrower marked. Also, emergency response time is most critic value about in the scoring table because of distance from the hospital, Fire Department, Police Center and Military Centers' proximity is key point. Moreover, the basic distance calculation is used to get average time of arrival in emergency situations [16]. Furthermore, impact rate on human scored according to the number of people SQ/Km to be affected and the proximity to the settlement [18]. Although the final process of prioritization is performed by doing the AHP pair-wise comparison, the calculation of particular hazard risk priority must be conveniently made by an individual set of equations.

The scores are obtained from the literature and expert opinions [19]. The scoring table is given in Table 5.

Table 5. Priority score of each risk level for sub-criteria

Criteria	Risk level	Description	Description of priority (KM-time-ETC)	Normalized priority score
Water environment (A)	Extreme	Scored according to the distance from the stream, dam, sea, lake, irrigation areas	1.00–0.90 = 0/900 M	1
	High		0.75–0.89 = 1.0/1.9 km	0.75
	Medium		0.50–0.74 = 2.0/2.9 km	0.5
	Low		0.25–0.49 = 3.0/4.0 km	0.25
	No effect		0.00 = >4.0 km	0
Soil environment (B)	Extreme	Scored according to the distance from the forest areas, natural protected areas, farming areas	1.00–0.90 = 0/540 M	1
	High		0.75–0.89 = 539/950 M	0.75
	Medium		0.50–0.74 = 940 M/1,45 km	0.5
	Low		0.25–0.49 = 1,44/2,0 km	0.25
	No effect		0.00 = >2,0 km	0
Featured areas ©	Extreme	The building cost is scored according to the high valuable places (tunnel's) and the distance to the private areas and the number of places	Cost >500.000$ per 1,0 km	1
	High		Cost 300.000$–499.999$ per 1,0 km	0.75
	Medium		Cost 100.0000–299.999$ per 1,0 km	0.5
	Low		Cost <99.000$ places per 1,0/2,0 km	0.25
	No effect		0 places per > 2,0 km	0
Human (D)	Extreme	Scored according to the number of people to be affected and the proximity to the settlement	>10000 humans per 1 Sq/Km	1
	High		>7000 and <9999 human per 1 Sq/Km	0.75
	Medium		>1000 and < 6999 human per 1 Sq/Km	0.5
	Low		<999 human per 1 Sq/Km	0.25
	No effect		None human per 1 Sq/Km	0
Emergency case (E)	Extreme	According to the hospital, Fire department, police center and military centers' proximity, it is calculated according to the average time of arrival in emergency situations (without considering the traffic) (100 km/hour)	50 min or above	1
	High		30–49 min	0.75
	Medium		11–29 min	0.5
	Low		Under 10 min	0.25
	No effect		Respond promptly in 5 min	0

Normalized priority scores are found that risk levels have been established depending on the specified criteria. Each criterion is divided into effect radii that vary depending on the characteristics of the hazardous goods carried. 1 has been determined as the extreme, and 0 has no effect. The results were in this range, as the score ranges discussed were between 0 and 1.

In the following part, implementation of the study is explained in detail. The implementation of the study consists of finding weights of criteria with the help of AHP, finding scores of the alternative roads with the help of scale.

5 Implementation

The aim of this study is to determine the pollution they create during transportation of materials belonging to nine hazardous materials classes and to determine alternative routes in order to prevent the destruction of water resources, vegetation, agricultural lands and forests as a result of a possible accident [9]. By considering literature review, the environmental elements face with risks consist of water environment, soil environment, featured areas, human and emergency cases.

In the study, firstly, a questionnaire is prepared to weight of criteria with the help of AHP.

5.1 Data Collection – Questionnaire

The AHP method is used to find priority weights of environmental criteria when carrying out environmental risk analysis for hazardous goods transportation. Therefore, the questionnaire is developed by considering literature. The questionnaire consists of 27 questions that compare main and sub criteria for all classes of hazardous materials. The study involves 9 experts who are in hazardous materials and working as experts in various institutions to prioritization of the criteria. Only data of second class taken into consideration. In addition, the questionnaire is made with face to face interviews and via e-mail in order to find the priority weights of the factors, a hierarchical model has been created and a double comparison matrix of the criteria and sub-criteria has been formed as a result of the evaluations obtained from the experts. A scale is formed, and points are given according to the importance of roads in the pilot area to be used in the project.

5.2 Weights of Criteria with AHP

The main aim is to use AHP in the study is to evaluate the risks by class-based evaluation and to determine an alternative route usage according to the most risk-bearing criteria. Moreover, hazardous substances that emerge accidentally to the water cause contamination of water resources above and below ground. Furthermore, the soil may contain hazardous substances which are harmful to living things. Erosion also facilitates the spread of hazardous substances. In addition, AHP is used to determine priority weight of environmental criteria when carrying out environmental risk analysis [13].

As a result of the questionnaire conducted with 9 experts, the data are entered the Excel program. As required by the AHP calculation, firstly individual-based AHP is applied to the questionnaires and the consistency rates are calculated. For this purpose, the consistency of participants of the survey is calculated to get consistent weights at the end of the study. This ratio is expected to be less than 0.1 for the questionnaire to be considered consistent according to the AHP rule [1]. As a result of these calculations, Expert 1 and Expert 6 are observed to be inconsistent and the study is evaluated with the remaining 7 questionnaires. The weights of each criteria are given in Table 6.

Table 6. Weights and consistency analysis of sub-criteria

	Weights		
Consistency rate of main criteria	Main criteria	Sub-criteria	Consistency rate of sub-criteria
0,068	Water environment (0,442)	Stream (0,253)	0,0106
		Dam (0,222)	
		Sea (0,178)	
		Lake (0,191)	
		Irrigation areas (0,156)	
	Soil environment (0,268)	Forest areas (0,532)	0,002
		Natural protected areas (0,287)	
		Farming areas (0,181)	
	Featured areas (0,147)		
	Human (0,086)	Number of people to be affected (0,667)	0
		Proximity to settlement (0,333)	
	Emergency case (0,057)	Proximity to hospital (0,666)	0,025
		Proximity to fire stations (0,225)	
		Proximity to law enforcement officers (0,109)	

5.3 Choosing Pilot Region and Calculation of Scale

In this study, Bayraklı Tunnel is selected as pilot location. Bayraklı Tunnels, which are formerly Karşıyaka Tunnels, are two highway tunnels and it is between Karşıyaka and Bornova. Twin-tube tunnels of 320 and 317 m (1,050 and 1,040 ft) (Bayraklı-1) and 1,865 and 1,876 m (6,119 and 6,155 feet) in length (Bayraklı-2) carry three traffic lanes in both directions.

Hazardous substance carriers are not allowed to use the tunnel. Instead, it is expected to use one of the given paths from these vehicles:

- Bornova exit-D550-Altınyol-Girne Boulevard-Karşıyaka Entrance
- Bayraklı Street No: 1620/39 Street 1620/41 Street 1609-Kubilay Street- Onur Neighborhood Entry
- Bayraklı Tunnel's E87 Highway- Çiğli Intersection

The following locations have been selected as the pilot region due to the following features; closest path from Çiğli Intersection from all the highway in Izmir. Moreover, it is decided to analyses these three roads (D550- Altınyol, Onur Neighborhood and Bayraklı Tunnel's- E87). Bayraklı Tunnel's is on the E87 highway and total distance of this road is 12,7 km and total daily traffic volume is nearly 38,250. Furthermore, Altınyol is on the D550 highway and total distance of this road is 13,6 km and total daily traffic volume is nearly 51,789 and Onur Neighborhood is 18.8 km and total daily traffic volume is 39,041.

Scale score of each risk level for sub-criteria is based on rational and heuristic approaches. When the water environment is calculated, the score of the closest sea distance in multiple location in those roads are tried to find. Firstly, Altınyol is near the sea. Therefore, average distance from sea is calculated 403 M in four different point cross head in Altınyol which they are in Park Içi Road 450 M, Gediz Electricity 97 M, Izban Turan 875 M and Alaybey Shipyard 190 M. However, Bayraklı Tunnel's (3,4 km) and Onur Neighborhood (3,35 km) are all the time far away from seaside. Therefore, Altınyol gets 1 point and Bayraklı Tunnel's and Onur Neighborhood gets 0,25 point. As it is known that soil environment coefficient range is narrower than water environment. Altınyol road pass by 6 biggest forest in Izmir. However close range is lower than other two roads. When the average of those six forests are calculated, the result is 1.39 km far away from Altınyol and it gets 0,5 point. However, Bayraklı tunnels (0,626 km) and Onur Neighborhood (0,35 km) pass by in three forest but they are at close quarters. When average of those two roads to distance from forest, Bayraklı Tunnels gets 0,75 point and 1,00 point for Onur Neighborhood. Onur Neighborhood get 1,00 point because of past by between two largest forest which is Nezih Dağdelen Park Area (50 M) and Basın Yayın Forest (400 M). Moreover, similar calculation is made for other criteria. After scoring all roads according to scale, the accident rate of the roads is founded, and the risky road is determined.

5.4 Accident Rate of Roads

In this study, firstly it is tried to find road by road accident rate of three destination (Bayraklı Tunnel's, Altınyol and Onur Neighborhood). However, General Directorate of Highway published only all roads average accident rate and it is calculated as 52.628 in 2018. To get data about different roads needs special permissions from the government. Therefore, it is not possible to collect data because of bureaucratic reasons. For this reason, the study can be applied different roads when there is accident rate of alternative roads. For the study, the possible calculations are given.

If the general accident rate was X = 4385,66 (52.628/12), which route is most having accident rate.

If it is accepted that Bayraklı Tunnel is X weight, Altınyol is accepted as 2X weights and Onur Neighborhood is 1.5X weight because of the road condition, traffic volume and usage of those roads. It is known the roads traffic volume as mentioned in Scale section. Calculation will be made with using Eq. 2 which is mentioned in methodology section.

Therefore, Bayraklı Tunnel's accident rate will be 0,11, Altınyol 0,22 and Onur Neighborhood 0,17. After obtaining accident rate, risk score is calculated using. Equation 5 calculated using (AHP Weight * Scale Score n)* (Accident Rate);

Bayraklı Tunnels;

$$(0,25 * 5,534 * 0,11) + (0,75 * 5,504 * 0,11) + (1 * 5,22 * 0,11) \\ + (0,5 * 5,11 * 0,11) + (0,25 * 5,14 * 0,11) = 1,6029$$

Altınyol;

$$(1 * 5,534 * 0,22) + (0,5 * 5,504 * 0,22) + (0,75 * 5,22 * 0,22) \\ + (1 * 5,11 * 0,22) + (0,5 * 5,14 * 0,22) = 4,3738$$

Onur Neighbourhood;

$$(0,25 * 5,534 * 0,17) + (1 * 5,504 * 0,17) + (0,25 * 5,22 * 0,17) \\ + (1 * 5,11 * 0,17) + (0,75 * 5,14 * 0,17) = 2,9168$$

The results of risk scores are shown in Table 7.

Table 7. Sum of risk scores

	Bayraklı Tunnel's (0,11)	Altınyol (0,22)	Onur neighborhood (0,17)
(AHP weight n* scale score n) * (accident rate)	0,152185	1,21748	0,235195
	0,45408	0,60544	0,93568
	0,5742	0,8613	0,22185
	0,28105	1,1242	0,8687
	0,14135	0,5654	0,65535
Sum	1,602865	4,37382	2,916775

At the end of the calculations, the results are found as follows; 0.065 for Altınyol, 0.023 for Bayraklı Tunnel's and 0.048 for Onur Neighborhood. These values show the prioritization of selected roads and tunnels. Therefore, these scores are low. According to results, Bayraklı Tunnels have been chosen as the least risky route in hazardous goods transportation.

6 Conclusion

SCRM is a crucial issue for supply chain management due to increasing type of risks in supply chains. In addition, one of the biggest risks occur in transportation process. Transportation of hazardous materials are becoming important issue due to lack of management in road regulations. Especially, transportation of hazardous materials has risks in tunnels and open roads. For this purpose, in this study, it is aimed to find the damage of the accident of hazardous goods transportation on tunnels and open routes and also to find best alternative route for hazardous good transportation. Therefore, opinion of 9 experts are obtained for questionnaire and the data are used in AHP method to find prioritization of the criteria. As a result of all calculations, it was aimed to select the least risky route among the results obtained by multiplication with accident rate statistic. Bayraklı Tunnels' score is 0,024, Onur Neighborhood's score is 0,048 and Altınyol's score is 0,066.

According to result it can be said Bayraklı Tunnels are the less risky route option in the pilot research. However, in Turkey according to regulation transporting hazardous goods in tunnels is forbidden. In this research, it is aimed to show that opening tunnels to traffic in accordance with the regulations in European Agreement Concerning the International Carriage of Dangerous Goods by Road which Turkey already signed in 2010 but did not apply, minimize the transportation risks and catastrophic results.

Acknowledgment. The study is prepared by Ayça Türkmen, Berna Özçınar, Meriç Ulucaağaç, Tilbe Adsız, Gökay Aydoğdu, Ezgi Güneş in the context of senior design project in 2019.

References

1. Waters, D.: Supply Chain Risk Management. British Library Cataloguing-in-Publication Data, London (2007)
2. Mauro Benza, C.B.: Technical and Functional Standards to Provide a High-Quality Service for Dangerous Good Transport on Road. NATO, Paris, France
3. ADRBOOK: Adrbook: Adrbook.Com (2017)
4. Pala, O.: Bulanik Analitik Hiyerarşi Prosesi Ve Meslek Seçiminde Uygulanması. Dokuz Eylül Üniversitesi, Sosyal Bilimler Enstitüsü Dergisi **18**(3), 427–444 (2016)
5. Erkut, E., Verter, V.: Modeling of transport risk for hazardous materials. Oper. Res. **46**(5), 625–642 (1998). https://doi.org/10.1287/opre.46.5.625
6. Erkut, E., Tjandra, S., Verter, V.: Hazardous materials transportation. In: Barnhart, C., Laporte, G. (eds.) Handbook in OR & MS, vol. 14 (2007)
7. Ümit, H., Kara, B.Y.: Büyük Ölçekli Bir Tehlikeli Madde Taşımacılığının Coğrafi Bilgi Sistemleri Tabanlı Analizi: Shell Türkiye Örneği. Yöneylem Araştırma Dergisi, no. 15 (2003)
8. Bubbico, R., Di Cave, S., Mazzarotta, B., Silvetti, B.: Preliminary study on the transport of hazardous materials through tunnels. Accid. Anal. Prev. **41**, 1199–1205 (2019)
9. Chen, H., Chena, J., Tanb, J., Zhangb, J.: Intelligent ventilation and emergency evacuation of underwater tunnel under fire conditions. Procedia Eng. **166**, 379–388 (2016)

10. Fabiano, B., Curro, F., Palazzi, E., Pastorino, R.: A framework for risk assessment and decision-making strategies in dangerous good transportation. J. Hazard. Mater. **93**, 1–15 (2002)
11. Frank, W.C., Thill, J.C., Batta, R.: Spatial decision support system for hazardous material truck routing. Transp. Res. Part C: Emerg. Technol. **8**(1–6), 337–359 (2002)
12. OECD: Safety in Tunnels. PIARC (2001)
13. Huang, B.: A GIS-AHP method for HAZMAT route planning with consideration of security. Environ. Inform. Arch. **2**, 818–830 (2004)
14. Tehlikeli Madde Taşımacılığı Kazaları, AFAD. https://www.afad.gov.tr/upload/Node/3911/xfiles/tasimacilik_kazalari-son.pdf. Accessed 13 Sept 2014
15. Özyagcı, S.: Tehlikeli Maddelerin Karayolu İle Taşınması. İstanbul Teknik Üniversitesi Fen Bilimleri Enstitüsü, Yayımlanmamış Doktora Tezi, İstanbul (2008)
16. Lin, M.C., Canny, J.F.: A fast algorithm for incremental distance calculation. In: Proceedings 1991 IEEE International Conference on Robotics and Automation, pp. 1008–1014. IEEE (1991)
17. Waghulade, R.B., Patil, P.P., Pasricha, R.: Synthesis and LPG sensing properties of nano-sized cadmium oxide. Talanta **72**(2), 594–599 (2007)
18. Sattayaprasert, W., Hanaoka, S., Taneerananon, P., Pradhananga, R.: Creating a risk-based network for hazmat logistics by route prioritization with AHP: case study: gasoline logistics in Rayong, Thailand. IATSS Res. **32**(1), 74–87 (2007)
19. Karabulut, S., Öcalir-Akünal, E.V.: Karayolu Ile Tehlikeli Madde Taşimaciliği Için Coğrafi Bilgi Sistemi Destekli Çevresel Risk Analizi. Gazi Üniversitesi Mühendislik-Mimarlı Fakültesi Dergisi **30**(3), 351–359 (2015)
20. Erkut, E., Ingolfsson, A.: Catastrophe avoidance models for hazardous materials route planning. Transp. Sci. **34**, 165–179 (2000)

A Hybrid Flow Shop Scheduling Problem

Ayşegül Eda Özen, Gülce Çini, Merve Çamlıca, Nilay Çınar, Hasan Bahtiyar Soydan, Levent Kandiller, and Hande Öztop(✉)

Department of Industrial Engineering, Yaşar University, İzmir, Turkey
ayseguleda.ozen@gmail.com, gulce.cini@gmail.com, mervecmlca@gmail.com, nilayycinar@gmail.com, hasanbahtiyar96@gmail.com, {levent.kandiller, hande.oztop}@yasar.edu.tr

Abstract. Hybrid flow shop environment generally refers to the flow shop with multiple parallel machines per stage. Hybrid flow shop scheduling problem (HFSP) is a complex combinatorial optimization problem that came across in many real-life problems. In this study, a real-life HFSP of a lubricant company is considered, where the aim is to minimize total weighted completion time of the jobs. Apart from classical HFSPs, the studied problem has additional constraints such as machine eligibility, sequence-dependent setup times and machine capacities. Due to the additional constraints in the system, a novel mixed integer linear programming model is proposed for the studied HFSP with three stages. As the problem is NP-hard, two constructive heuristic algorithms and an improvement heuristic algorithm are also developed. The performance of the proposed heuristic algorithms is evaluated by comparisons with the optimal results obtained from the mathematical model. The extensive computational results show that proposed heuristic algorithms find near optimal results in reasonable computational times. Sensitivity analysis is also performed for the weight parameter of the problem, which indicates that the proposed heuristic algorithms also perform very well for different weight parameter values. Finally, the proposed heuristic algorithms are integrated into a user-friendly decision support system using Microsoft Excel VBA interface to provide an efficient scheduling tool for the company.

Keywords: Hybrid flow shop scheduling · Total weighted completion time · Sequence-dependent setup times · Machine eligibility · Heuristic algorithm

1 Introduction

An efficient and applicable scheduling system is very important for the companies in order to decrease their costs by improving their production processes; provide accurate and realistic due dates for their customers and use existing resources more efficiently. This paper considers a real-life hybrid flow shop scheduling problem (HFSP) of a lubricant company in Izmir, Turkey. The company produces three different types of main products as automotive lubricants, industrial lubricants and auto-care products. These products have wide variety and each product has a different bill of materials, which makes company's production planning more challenging. Currently, in the

N. M. Durakbasa and M. G. Gençyılmaz (Eds.): ISPR 2019, LNME, pp. 636–649, 2020.
https://doi.org/10.1007/978-3-030-31343-2_55

company, production planning is based on estimations and experience of the planners. As the current production planning does not follow a standard and efficient methodology, capacity gaps and financial losses may occur in the system. Therefore, the focus of this study is to minimize total weighted completion time of the jobs in a hybrid flow shop with machine eligibility, sequence-dependent setup times and machine capacity restrictions. Creating an efficient scheduling tool for the company is also one of the main purposes of this study.

Ruiz and Vázquez-Rodríguez [1] presented a detailed literature review on exact, heuristic and metaheuristic methods that have been proposed for the HFSP. They discuss solution approaches for various variants of the HFSP, considering different assumptions, constraints and objective functions. Another extensive literature review is also provided by Ribas et al. [2]. In the literature, sequence-dependent setup times and machine eligibility constraints are studied in several studies [3–9]. The makespan minimization for two-stage hybrid flow shop with dedicated machines and additional constraints is studied by Chikhi and Abbas [3]. Lin and Liao [4] proposed a heuristic algorithm for a real label sticker manufacturing problem, which corresponds to a two-stage HFSP with sequence-dependent setup times and machine eligibility. Ruiz et al. [5] have studied a complex HFSP in the ceramic tile industry, which includes many additional restrictions such as unrelated parallel machines, sequence-dependent setup times, machine eligibility and machine release dates. A genetic algorithm was proposed to minimize makespan for the HFSP with unrelated parallel machines, sequence-dependent setup times and machine eligibility by Ruiz and Maroto [6]. Yaurima et al. [7] also presented a genetic algorithm for the HFSP with unrelated machines, sequence-dependent setup times, availability constraints and limited buffers. Later, Pan et al. [8] proposed nine heuristic algorithms to minimize the makespan for the HFSP with sequence-dependent setup times. Yu et al. [9] also developed a genetic algorithm for the HFSP with unrelated machines and machine eligibility restrictions to minimize the total tardiness. Recently, a mathematical model and several metaheuristic algorithms were proposed for the HFSP considering makespan and total flow time criteria by Öztop et al. [10].

In this study, a novel mixed integer linear programming (MILP) model is proposed for the real-life 3-stage HFSP due to the additional constraints in the system. As the HFSP is known to be NP-hard [11], two constructive heuristic algorithms and an improvement heuristic algorithm are also proposed for the problem, based on well-known Shortest Processing Time (SPT) rule. The performance of the proposed heuristic algorithms is evaluated by comparisons with the optimal results obtained from the mathematical model on 50 real instances. A user-friendly decision support system has also been created by integrating the proposed heuristic algorithms.

The rest of the paper is organized as follows. The formal problem definition is provided in Sect. 2. In Sect. 3, the proposed MILP model and heuristic algorithms are presented, respectively. Computational results of the proposed solution approaches are presented in Sect. 4. Section 5 explains the developed decision support system. Finally, Sect. 6 provides the conclusions and future suggestions.

2 Problem Definition

The hybrid flow shop, which is also called as flexible flow shop, is a special variant of the generalized flow shop. In the standard HFSP, a set of n jobs must be processed on m ($m > 1$) serial stages following the same route, where each stage includes at least one parallel identical machines and each job must be assigned to the one of these machines in each stage. Note that, at least one of the stages include more than one machine. As mentioned in the previous section, we consider the real-life HFSP of a lubricant company, which has many additional constraints.

In the company, the production process consists of three main stages as blending, dropping and filling, respectively. In the first stage, base oil and additives are blended in the eligible blending machines based on blending codes of the products. As ingredients of the products are different for each blending code, there are several machine groups based on blending codes. Note that, capacities of the machines are different within each machine group. After the blending stage, samples of blended products are sent to the quality control process and the approved products proceed to the next stage, i.e., dropping stage. Similar to the blending stage, machine eligibility and capacity limitations also exist in the dropping stage. After the dropping stage, the ready products for the filling process are transferred to the filling stage via pipelines. In the filling stage, there are 3 different filling machines according to the packaging material type. Note that, during the transition from the dropping machines to the filling machines, pipe cleaning process is needed between different types of the products. Furthermore, mould change process is needed between different sizes of packaging materials in each filling machine.

As mentioned above, in blending and dropping stages, products are assigned to the eligible machines according to their blending codes. There are also machine capacity constraints in these two stages. In the filling stage, each product is assigned to an eligible filling machine according to its packaging material type. Note that, in the filling stage, there are also sequence-dependent setup times between different products due to the exchange of mould and pipeline cleaning processes. These additional constraints differentiate the studied problem from the standard HFSP, further complicating the structure. In the studied HFSP, it is assumed that all machines are available initially and there is no initial setup time for the jobs. It is also assumed that, the quality control durations between stages are included in the processing times of the jobs in the corresponding stages. Consequently, the aim of the problem is to minimize total weighted completion time of the jobs, where each job has a priority weight based on its profit. In order to determine the weights of the jobs, detailed ABC analyses are performed based on profits of the jobs, where the products are classified in three main groups, as A, B and C according to their profits.

3 Solution Methodology

3.1 Mathematical Model

The proposed MILP model is explained below for the studied real-life HFSP with machine eligibility, sequence-dependent setup times and machine capacity constraints, where the objective is to minimize total weighted completion time of the jobs.

Sets and Indices

i, q: index of job
j: index of blending code
k: index of machine
I: Set of jobs including the dummy job, $I = \{0, \ldots, n\}$
J: Set of blending codes
F: Set of machines in filling stage
B_j: Set of machines for blending code j in blending stage
D_j: Set of machines for blending code j in dropping stage
F_i: Set of machines for job i in filling stage

Parameters

w_i = Weight of job i
px_i = Processing time of job i in blending stage
py_i = Processing time of job i in dropping stage
pz_i = Processing time of job i in filling stage
s_{iq} = Setup time when job q is processed immediately after job i
N_k = Capacity of machine k in blending stage
A_k = Capacity of machine k in dropping stage
v_i = Volume of job i
Q = Sufficiently large integer

Decision Variables

$$X_{ijk} = \begin{cases} 1, \text{ if job } i, \text{ which has blending code } j, \text{ is blended in machine } k \text{ in blending stage} \\ 0, \text{otherwise.} \quad i \in I, j \in J, k \in B_j \end{cases}$$

$$Y_{ijk} = \begin{cases} 1, \text{ if job } i, \text{ which has blending code } j, \text{ is processed in machine } k \text{ in dropping stage} \\ 0, \text{otherwise.} \quad i \in I, j \in J, k \in D_j \end{cases}$$

$$Z_{ijk} = \begin{cases} 1, \text{ if job } i, \text{ which has blending code } j, \text{ is filled in filling machine } k \text{ in filling stage} \\ 0, \text{otherwise.} \quad i \in I, j \in J, k \in F_i \end{cases}$$

$$ox_{iqk} = \begin{cases} 1, \text{ if job } i \text{ precedes job } q \text{ in blending machine } k \\ 0, \text{otherwise.} \end{cases}$$

$$oy_{iqk} = \begin{cases} 1, \text{ if job } i \text{ precedes job } q \text{ in dropping machine } k \\ 0, \text{otherwise.} \end{cases}$$

$$oz_{iqk} = \begin{cases} 1, \text{ if job } i \text{ immediately precedes job } q \text{ in filling machine } k \\ 0, \text{otherwise.} \end{cases}$$

cx_i = Completion time of job i in blending stage, $cx_i \geq 0$
cy_i = Completion time of job i in dropping stage, $cy_i \geq 0$
cz_i = Completion time of job i in filling stage, $cz_i \geq 0$

Based on the above definitions, the MILP formulation of the problem is given as follows.

$$\text{Minimize } \sum\nolimits_{i \in I/\{0\}} w_i \, cz_i \tag{1}$$

$$\sum\nolimits_{k \in B_j} X_{ijk} = 1 \qquad \forall\, i \in I/\{0\}, \forall\, j \in J \tag{2}$$

$$\sum\nolimits_{k \in D_j} Y_{ijk} = 1 \qquad \forall\, i \in I/\{0\}, \forall\, j \in J \tag{3}$$

$$\sum\nolimits_{k \in F_i} Z_{ijk} = 1 \qquad \forall\, i \in I/\{0\}, \forall\, j \in J \tag{4}$$

$$Q(2 - X_{ijk} - X_{qjk} + ox_{iqk}) + cx_i - cx_q \geq px_i \qquad \forall i, q \in I/\{0\}, \forall j \in J, \forall k \in B_j | i < q \tag{5}$$

$$Q(3 - X_{ijk} - X_{qjk} - ox_{iqk}) + cx_q - cx_i \geq px_q \qquad \forall i, q \in I/\{0\}, \forall j \in J, \forall k \in B_j | i < q \tag{6}$$

$$Q(2 - Y_{ijk} - Y_{qjk} + oy_{iqk}) + cy_i - cy_q \geq py_i \qquad \forall i, q \in I/\{0\}, \forall j \in J, \forall k \in D_j | i < q \tag{7}$$

$$Q(3 - Y_{ijk} - Y_{qjk} - oy_{iqk}) + cy_q - cy_i \geq py_q \qquad \forall i, q \in I/\{0\}, \forall j \in J, \forall k \in D_j | i < q \tag{8}$$

$$Q\left(1 - oz_{iqk}\right) + cz_q - cz_i \geq pz_q + s_{iq} \qquad \forall i, q \in I/\{0\}, \forall k \in F_i | i \neq q \tag{9}$$

$$\sum\nolimits_{q \in I} oz_{0qk} = 1 \qquad \forall k \in F \tag{10}$$

$$\sum\nolimits_{q \in I} oz_{q0k} = 1 \qquad \forall k \in F \tag{11}$$

$$\sum\nolimits_{i \in I | i \neq q} oz_{iqk} = \sum\nolimits_{j \in J} Z_{qjk} \qquad \forall q \in I, \forall k \in F \tag{12}$$

$$\sum\nolimits_{q \in I | i \neq q} oz_{iqk} = \sum\nolimits_{j \in J} Z_{ijk} \qquad \forall i \in I, \forall k \in F \tag{13}$$

$$cx_i \geq px_i \qquad \forall i \in I/\{0\} \tag{14}$$

$$cy_i - cx_i \geq \sum\nolimits_{k \in D_j} Y_{ijk} py_i \qquad \forall i \in I/\{0\}, \forall j \in J \tag{15}$$

$$cz_i - cy_i \geq \sum\nolimits_{k \in F_i} Z_{ijk} pz_i \qquad \forall i \in I/\{0\}, \forall j \in J \tag{16}$$

$$v_i X_{ijk} \leq N_k \qquad \forall i \in I/\{0\}, \forall j \in J, \forall k \in B_j \tag{17}$$

$$v_i Y_{ijk} \leq A_k \qquad \forall i \in I/\{0\}, \forall j \in J, \forall k \in D_j \tag{18}$$

The objective function (1) minimizes the total weighted completion time of the jobs. Constraint sets (2), (3) and (4) ensure that each job is assigned to an eligible machine in each stage. Constraint sets (5) and (6) prevent any two job operations from being overlapped on the same blending machine. Similarly, constraint sets (7) and (8) prevent any two job operations from overlapping on the same dropping machine. The constraint set (9) guarantees that the starting time of a job on a filling machine must be greater than or equal to the finishing time of the preceding job on that machine, regarding the sequence dependent setup time in between. Constraint sets (10) and (11) ensure that the dummy job 0 is the first and the last job in the job sequence on each filling machine. The constraint set (12) states that each job must be immediately preceded by another job on a filling machine. Similarly, the constraint set (13) ensures that each job must be immediately succeeded by another job on a filling machine. Constraint set (14) guarantees that the completion time of a job in the blending stage is greater than or equal to its processing time in that stage. Constraint sets (15) and (16) ensure that the starting time of a job in a stage is greater than or equal to its release time from the previous stage. Constraint sets (17) and (18) state that each job is assigned to a machine without exceeding the machine's capacity.

The proposed MILP model is solved for 5 instances with varying job sizes from 5 to 100, on IBM ILOG CPLEX Optimization Studio with CPLEX 12.8. The computational time of the mathematical model for different job sizes are shown in Table 1. As shown in Table 1, the solution time of the mathematical model increases exponentially as the job size increases. Note that, the entry '-' in the table means that the mathematical model cannot find the optimal solution within the computational time. This result is expected, as the studied problem has already been proven as NP-hard [11]. Consequently, heuristic algorithms have been proposed for the problem in the following subsection.

Table 1. Results of the mathematical model for different job sizes

Instance	Number of jobs	Number of blending machines	Number of dropping machines	Number of filling machines	Optimal solution	Computational time (sec.)
1	5	5	4	3	1221.1	0.27
2	10	5	4	3	2660.7	1.16
3	15	5	4	3	4180.7	66.13
4	20	5	4	3	6139.4	10846.5
5	100	5	4	3	–	43925.8

3.2 Heuristic Algorithms

Due to the NP-hardness of the studied problem, two constructive heuristic algorithms are developed based on well-known Shortest Processing Time (SPT) rule. According to the SPT rule, jobs are sorted in increasing order of their total processing times and they are assigned to the available machines in this order. In the first constructive heuristic algorithm, SPT rule is applied for each stage by considering only the processing times at that stage. While, in the second algorithm, SPT rule is applied for each stage by considering the total processing times at that stage and the following stages. Namely, SPT rule is applied: for the first stage by considering the total processing times of all (blending, dropping and filling) stages; for the second stage by considering the total processing times of dropping and filling stages; and for the third stage by considering the processing times of that stage (filling). In both heuristic algorithms, after the jobs are sorted according to the SPT rule, they are scheduled to the eligible machines regarding their blending codes, ready times from previous stage, weights and volumes. The general steps of the proposed heuristic algorithms are listed below. Note that, only difference between the two algorithms is the aforementioned processing time calculations for SPT rule.

1. *Blending Stage*
   - 1.1. Group the jobs according to their blending codes.
   - 1.2. For each group, schedule the jobs to the eligible machines according to their blending codes as below:
     - 1.2.1. Sort the jobs in decreasing order of their weights.
     - 1.2.2. Sort the jobs with same weight according to the SPT rule among themselves.
     - 1.2.3. In line with the order in step 1.2.2, schedule the jobs to the available machines regarding their blending codes and machine capacity limitations.
2. *Dropping Stage*
   - 2.1. Group the jobs according to their blending codes.
   - 2.2. For each group, schedule the jobs to the eligible machines according to their blending codes as below:
     - 2.2.1. Sort the jobs in increasing order of their ready times from previous stage.
     - 2.2.2. According to the order in step 2.2.1, schedule the first job for each machine regarding the blending codes of jobs and the machine capacities.
     - 2.2.3. Calculate the ready time of each machine according to the completion time of the last job on the machine.
     - 2.2.4. Determine the machine m_{min}, which has the minimum ready time.
       - 2.2.4.1. According to the order in step 2.2.1, if the earliest starting time of the next job is greater than or equal to the ready time of m_{min}, schedule the next job to the machine m_{min} regarding its ready time from previous stage.
       - 2.2.4.2. According to the order in step 2.2.1, if the earliest starting time of the next job is less than the ready time of m_{min}:
         - 2.2.4.2.1. Sort the jobs, which have smaller ready time than the ready time of m_{min}, in decreasing order of their weights.
         - 2.2.4.2.2. Sort the jobs with same weight according to the SPT rule among themselves.
         - 2.2.4.2.3. Schedule the first job in this order to the machine m_{min} regarding the ready time of m_{min}.
     - 2.2.5. If all jobs are not scheduled, go back to step 2.2.3.
3. *Filling Stage*
   - 3.1. Group the jobs according their packaging material types.
   - 3.2. For each group, schedule the jobs to the eligible machine according to their packaging material types as below:
     - 3.2.1. Sort the jobs in increasing order of their ready times from previous stage.
     - 3.2.2. Schedule the first job to the machine according to the order in step 3.2.1.

3.3. Calculate the earliest starting time of each job that may follow the last job on the machine, regarding the sequence-dependent setup time in between.
   3.3.1. For the jobs which has an earliest starting time that is greater than its ready time:
      3.3.1.1. Sort these jobs in increasing order of their earliest starting times, and then sort them in decreasing order of their weights.
      3.3.1.2. Sort the jobs with highest weight according to the SPT rule among themselves.
      3.3.1.3. Schedule the first job in this order to the machine.
   3.3.2. For the jobs which has an earliest starting time that is less than or equal to its ready time:
      3.3.2.1. According to the order in step 3.2.1, schedule the next job to the machine regarding its ready time and the sequence-dependent setup time in between.

In order to improve the performance of the proposed heuristic algorithms, additional improvement heuristic algorithm is also employed after the aforementioned constructive heuristics. In this study, as an improvement heuristic, insertion local search is applied to the solutions obtained by the above heuristic algorithms. In the insertion local search, a job is removed from its current position and inserted into all possible positions except its current position. Then, the removed job is inserted into the best improving position. This procedure is repeated for all jobs.

4 Computational Results

In this section, computational results are presented for the proposed heuristic algorithms. In order to evaluate the performance of the heuristic algorithms, 50 real instances with 15 jobs are obtained from the company, which have 4 blending machines, 4 dropping machines and 3 filling machines. The proposed MILP model in Sect. 3 is coded on IBM ILOG CPLEX Optimization Studio and all instances are solved optimally with CPLEX 12.8. All proposed heuristic algorithms are coded in Microsoft Visual Basic for Applications (VBA) Software and all instances are solved by heuristic algorithms within few minutes. The relative percent deviations (RPD) from the optimal results are reported for each heuristic algorithm, which is calculated as below:

$$RPD = \frac{Heuristic\,Solution - Optimal\,Solution}{Optimal\,Solution} \times 100 \tag{19}$$

4.1 Comparison of Constructive Heuristic Algorithms

Table 2 reports the RPD values for each constructive heuristic algorithm. As shown in the table, the average RPD value of the first constructive heuristic algorithm is 9.97%, while it is 11.32% for the second constructive heuristic algorithm over 50 instances. "Best Heuristic Algorithm" column of the table reports the best RPD values among these two constructive heuristic algorithms. As shown in the table, the average RPD value of the best constructive algorithm is 9.26%. Note that, only difference between the two constructive heuristics is the aforementioned processing time calculations for SPT rule. As the first constructive heuristic slightly performs better than the second one, these processing time calculations slightly affects the performance of the algorithm.

Table 2. Comparison of constructive heuristics

Constructive heuristic algorithms							
Instance	RPD (%)			Instance	RPD (%)		
	Heuristic algorithm 1	Heuristic algorithm 2	The best heuristic algorithm		Heuristic algorithm 1	Heuristic algorithm 2	The best heuristic algorithm
1	8.42	6.63	6.63	26	2.15	0.07	0.07
2	5.47	5.86	5.47	27	8.78	12.73	8.78
3	12.28	12.28	12.28	28	7.19	9.26	7.19
4	13.03	12.47	12.47	29	26.17	22.90	22.90
5	1.94	2.51	1.94	30	23.78	35.60	23.78
6	2.69	4.44	2.69	31	9.99	9.99	9.99
7	4.43	4.43	4.43	32	9.03	9.03	9.03
8	8.04	8.34	8.04	33	11.05	11.05	11.05
9	8.38	4.32	4.32	34	15.92	15.92	15.92
10	9.21	9.60	9.21	35	8.86	2.88	2.88
11	12.62	12.62	12.62	36	7.58	7.33	7.33
12	4.23	5.68	4.23	37	6.69	6.69	6.69
13	18.24	26.14	18.24	38	8.78	8.96	8.78
14	4.18	4.18	4.18	39	1.03	14.47	1.03
15	24.90	24.42	24.42	40	28.54	31.53	28.54
16	9.74	14.44	9.74	41	17.26	24.05	17.26
17	6.26	4.39	4.39	42	7.77	13.20	7.77
18	10.74	12.77	10.74	43	6.97	3.10	3.10
19	7.92	20.30	7.92	44	20.56	24.49	20.56
20	2.23	6.14	2.23	45	5.80	16.86	5.80
21	3.88	2.98	2.98	46	23.62	22.93	22.93
22	6.50	6.46	6.46	47	8.73	11.77	8.73
23	3.29	3.84	3.29	48	2.94	4.25	2.94
24	2.43	2.17	2.17	49	13.40	7.12	7.12
25	8.64	9.39	8.64	50	15.95	12.95	12.95
				Average	**9.97**	**11.32**	**9.26**

4.2 Comparison of Constructive Heuristic Algorithms with Insertion Local Search

As mentioned before, an insertion local search is applied to the aforementioned constructive heuristic algorithms in order to improve the solution quality. Table 3 reports the RPD values for each constructive heuristic algorithm with insertion local search, for the same 50 real instances in the previous subsection. As shown in the table, after applying the insertion local search procedure, the average RPD value of the first constructive heuristic algorithm is decreased to 5.45%, while it is decreased to 5.12% for the second constructive heuristic algorithm. When the best heuristic algorithm is chosen for each instance, the average RPD value is reduced to 4.78%, as shown in the "Best Heuristic Algorithm" column of the table. Hence, it can be said that insertion local search enhances the solution quality significantly.

Table 3. Comparison of constructive heuristics with insertion local search

Constructive heuristic algorithms with insertion local search							
Instance	RPD (%)			Instance	RPD (%)		
	Heuristic algorithm 1	Heuristic algorithm 2	The best heuristic algorithm		Heuristic algorithm 1	Heuristic algorithm 2	The best heuristic algorithm
1	7.70	5.90	5.90	26	2.10	0.10	0.10
2	5.50	5.90	5.50	27	1.40	4.50	1.40
3	12.30	12.30	12.30	28	7.20	8.90	7.20
4	4.90	8.10	4.90	29	9.80	6.70	6.70
5	1.50	1.50	1.50	30	15.20	15.20	15.20
6	2.70	3.60	2.70	31	10.00	10.00	10.00
7	2.50	2.00	2.00	32	3.00	3.00	3.00
8	8.00	8.30	8.00	33	7.00	7.00	7.00
9	6.70	4.30	4.30	34	4.20	4.20	4.20
10	3.10	2.70	2.70	35	7.40	2.90	2.90
11	11.00	11.00	11.00	36	3.70	3.70	3.70
12	4.20	5.70	4.20	37	4.20	4.20	4.20
13	7.00	5.70	5.70	38	5.00	6.20	5.00
14	2.80	2.80	2.80	39	1.00	2.60	1.00
15	10.60	10.00	10.00	40	5.70	4.80	4.80
16	6.20	6.20	6.20	41	1.90	1.90	1.90
17	2.10	1.50	1.50	42	4.30	3.70	3.70
18	5.00	3.60	3.60	43	4.50	3.10	3.10
19	2.60	2.70	2.60	44	14.20	14.20	14.20
20	0.10	0.10	0.10	45	4.90	4.90	4.90
21	3.30	3.00	3.00	46	7.80	7.70	7.70
22	6.50	4.90	4.90	47	1.30	1.30	1.30
23	3.00	3.00	3.00	48	1.70	1.00	1.00
24	2.20	2.20	2.20	49	13.00	3.70	3.70
25	4.00	5.70	4.00	50	6.90	8.00	6.90
				Average	**5.45**	**5.12**	**4.78**

4.3 Comparison of Proposed Solution Approaches with the Current System of the Company

Another 15 real instances are also obtained from the company in order to compare the current system with the proposed solution approaches. Note that, these instances have varying job sizes from 15 to 24, where there are 21 blending machines, 10 dropping machines and 3 filling machines in each instance. All these 15 real instances are optimally solved by MILP model and the feasible results are obtained by using heuristic algorithms. Table 4 reports the improvement percentage for each solution approach against the current system. Note that, the "Heuristic Algorithm" column reports the best results among the results obtained by the two heuristic algorithms with insertion local search. As shown in the table, 27.80% improvement is observed in the total weighted completion time of the jobs in average, by using the developed heuristic algorithms. On the other hand, the optimal results are 31.67% better than the results of the current system, in average.

Table 4. Comparison of proposed solution approaches with the current system of the company

Instance	Improvement (%)	
	Heuristic algorithm	Optimal solution (MILP)
1	45.88	49.82
2	49.17	56.82
3	10.41	12.32
4	49.63	77.64
5	37.30	37.30
6	55.02	55.79
7	27.29	32.88
8	4.92	4.92
9	44.24	44.24
10	23.02	27.95
11	0.89	0.89
12	2.08	5.22
13	4.08	4.08
14	60.00	60.00
15	3.14	5.11
Average	**27.80**	**31.67**

4.4 Sensitivity Analysis

Sensitivity analysis is also performed for the weight parameter of the problem. The aim of this analysis is to observe that how the performance of the solution approaches will be affected by the change of the weight parameter. For this analysis, 15 instances with 15 jobs are obtained from the company, which have 4 blending machines, 4 dropping machines and 3 filling machines. Table 5 reports the RPD values and computational times of the solution approaches for both weighted and unweighted form of the problem. As shown in Table 5, the proposed solution approaches perform robustly for both unweighted and weighted form of the problem, in terms of both solution quality and time. Note that, the average RPD value is 5.71% for the weighted form of the problem, while it is 3.97% for the unweighted form of the problem over 15 instances.

Table 5. Sensitivity analysis

Instance	MILP Computational time (sec.)		Heuristic algorithm Computational time (sec.)		RPD (%)	
	Weighted	Unweighted	Weighted	Unweighted	Weighted	Unweighted
1	31.03	20.79	98.16	53.16	2.30	3.31
2	30.73	20.01	86.17	40.27	10.75	6.24
3	27.32	20.42	76.41	41.39	7.31	2.36
4	11.19	27.27	94.04	30.60	3.29	2.14
5	3.19	12.21	157.52	36.53	3.68	1.26
6	6.76	12.01	84.62	31.45	3.10	2.53
7	25.15	3.92	40.72	33.19	2.68	1.82
8	26.28	5.84	40.17	32.47	4.85	8.19
9	28.77	16.25	35.35	38.12	1.51	4.29
10	8.00	12.52	32.59	33.27	3.67	2.27
11	7.30	8.95	34.55	33.45	4.38	3.01
12	8.04	5.16	29.95	29.29	26.42	13.76
13	31.70	7.84	39.59	34.27	3.98	0.62
14	9.70	19.73	37.27	30.23	6.53	1.66
15	7.64	20.56	40.30	32.69	1.17	6.15
Average	**17.52**	**14.23**	**61.83**	**35.36**	**5.71**	**3.97**

5 Decision Support System

A user-friendly decision support system (DSS) is created by embedding proposed heuristic algorithms in Microsoft Visual Basic for Applications (VBA) Software. In the developed DSS, the user is able to work in a practical way by entering only product codes and units. When the product codes and units are entered by the user, other information about the product automatically appears on the page. In order to provide flexibility for the user, the user can also edit, delete and add the input data easily. The main page of the DSS is shown in Fig. 1.

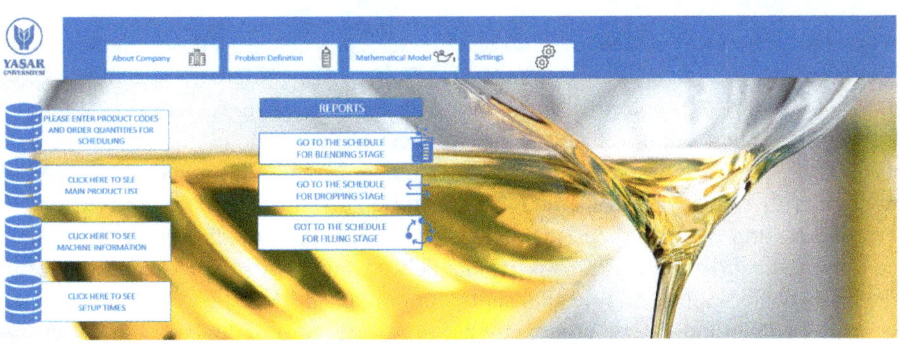

Fig. 1. Main page of the decision support system

The developed DSS employs both of the two heuristic algorithms and it offers several options to see different solution alternatives. The options include the first heuristic algorithm, the second heuristic algorithm and the best option among these two heuristic algorithms. Note that, insertion local search is included in all heuristic algorithms. After choosing the solution approach, the DSS reports the results to the user with detailed lists and Gantt Charts. Solution report screen of the DSS is shown in Fig. 2. As shown in Fig. 2, the starting time, the processing time and the completion time of each job at each stage can be seen in these reports as well as detailed Gantt Charts.

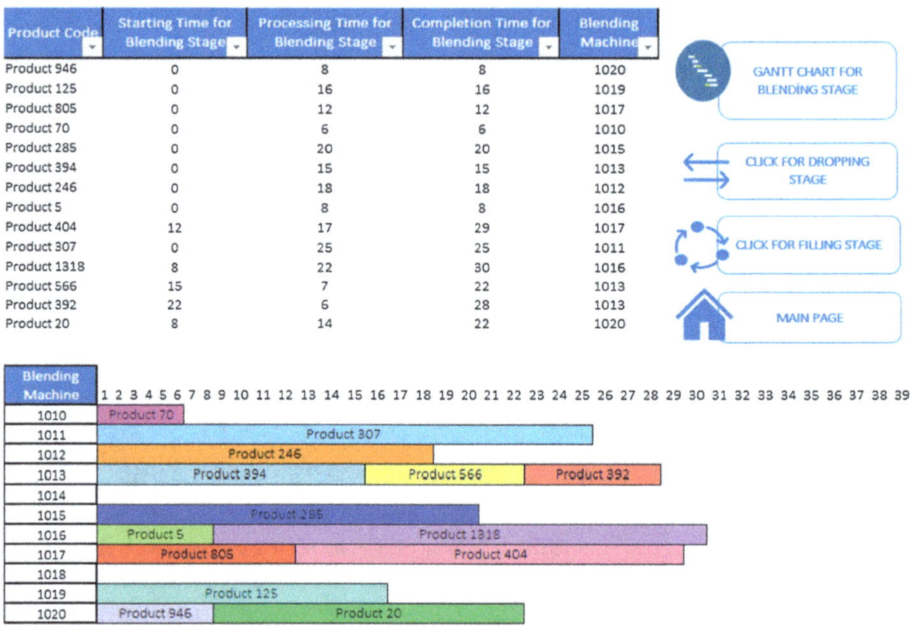

Product Code	Starting Time for Blending Stage	Processing Time for Blending Stage	Completion Time for Blending Stage	Blending Machine
Product 946	0	8	8	1020
Product 125	0	16	16	1019
Product 805	0	12	12	1017
Product 70	0	6	6	1010
Product 285	0	20	20	1015
Product 394	0	15	15	1013
Product 246	0	18	18	1012
Product 5	0	8	8	1016
Product 404	12	17	29	1017
Product 307	0	25	25	1011
Product 1318	8	22	30	1016
Product 566	15	7	22	1013
Product 392	22	6	28	1013
Product 20	8	14	22	1020

Fig. 2. Solution report of the decision support system

6 Conclusions

In this study, a real-life HFSP of a lubricant company is considered, where the aim is to minimize total weighted completion time of the jobs. Different from the standard HFSP, the studied problem has additional limitations such as machine eligibility, sequence-dependent setup times and machine capacities. A novel mixed integer linear programming model is proposed for the studied HFSP with three stages. As the problem is NP-hard, two constructive heuristic algorithms, which are based on SPT rule, and an improvement heuristic algorithm are also developed. The performance of the proposed heuristic algorithms is evaluated by comparisons with the optimal results. The extensive computational results show that proposed heuristic algorithms find near optimal results in reasonable computational times. Based on 50 real instances, average RPD value is 9.26% for the best constructive heuristic. After applying insertion local search to the constructive heuristics, the average RPD value is further decreased to

4.78%. When the current system is compared with the heuristic algorithms, it can be seen that there is 27.80% improvement on the total weighted completion time of the jobs. Finally, the proposed heuristic algorithms are integrated into a user-friendly decision support system using Microsoft Excel VBA interface, to provide an efficient scheduling tool for the company. Note that, the developed decision support system reports the results to the user with detailed lists and Gantt Charts.

As a future work, the decision support system can be made more professional by integrating to the current resource management system of the company. Furthermore, metaheuristic algorithms can be developed for the studied problem employing the proposed constructive heuristics. The developed metaheuristic algorithms can also be embedded into the decision support system. Different objectives or constraints can also be considered for the studied problem in future studies.

Acknowledgments. This study is supported by TUBITAK (The Scientific and Technological Research Council of Turkey) under the program of "2209-B Undergraduate Research Projects for Industrial Applications Fellowship Program".

We would like to thank Deniz Türsel Eliiyi for her guidance and enlightenment. Additionally, we would like to thank İ. Çağcan Çatıkkaş and Alp Arslan for their support in this study. Finally, we are grateful to the company and Dilek Turan for their co-operation and support.

References

1. Ruiz, R., Vázquez-Rodríguez, J.A.: The hybrid flow shop scheduling problem. Eur. J. Oper. Res. **205**(1), 1–18 (2009)
2. Ribas, I., Leisten, R., Framinan, J.M.: Review and classification of hybrid flowshop scheduling problems from a production system and a solutions procedure perspective. Comput. Oper. Res. **37**, 1439–1454 (2010)
3. Chikhi, N., Abbas, M.: Makespan minimization for two-stage hybrid flow shop with dedicated machines and additional constraints. In: 9th International Conference on Modeling, Optimization & Simulation – MOSIM 2012, France, pp. 123–133 (2012)
4. Lin, H.-T., Liao, C.-J.: A case study in a two-stage hybrid flow shop with setup time and dedicated machines. Int. J. Prod. Econ. **86**(2), 133–143 (2003)
5. Ruiz, R., Serifoglu, F.S., Urlings, T.: Modeling realistic hybrid flexible flowshop scheduling problems. Comput. Oper. Res. **35**(4), 1151–1175 (2008)
6. Ruiz, R., Maroto, C.: A genetic algorithm for hybrid flowshops with sequence dependent setup times and machine eligibility. Eur. J. Oper. Res. **169**(3), 781–800 (2006)
7. Yaurima, V., Burtseva, L., Tchernykh, A.: Hybrid flowshop with unrelated machines, sequence-dependent setup time, availability constraints and limited buffers. Comput. Ind. Eng. **56**(4), 1452–1463 (2009)
8. Pan, Q.K., Gao, L., Li, X.Y., Gao, K.Z.: Effective metaheuristics for scheduling a hybrid flowshop with sequence-dependent setup times. Appl. Math. Comput. **303**, 89–112 (2017)
9. Yu, C., Semeraro, Q., Matta, A.: A genetic algorithm for the hybrid flow shop scheduling with unrelated machines and machine eligibility. Comput. Oper. Res. **100**, 211–229 (2018)
10. Öztop, H., Tasgetiren, M.F., Eliiyi, D.T., Pan, Q.K.: Metaheuristic algorithms for the hybrid flowshop scheduling problem. Comput. Oper. Res. **111**, 177–196 (2019)
11. Gupta, J.N.D.: Two-stage hybrid flow shop scheduling problem. J. Oper. Res. Soc. **39**(4), 359–364 (1988)

An Open Vehicle Routing Problem for Daily Shipment Plan of a Local Bedding Company in Turkey

Bartu Arslan, Özlem Erdin, Deniz Gülce Dağıstanlıoğlu, Ozan Bolel, Çağlar Su, Banu Y. Ekren, and Cansu Yurtseven(✉)

Industrial Engineering, Yasar University, Izmir, Turkey
bartuarslan@yahoo.com, ozlemerdin96@gmail.com, gulcedd09@gmail.com, ozanbolel@outlook.com, caglarsu21@gmail.com, {banu.ekren, cansu.yurtseven}@yasar.edu.tr

Abstract. This study deals with a daily shipment plan for a bedding company in Turkey. We develop a decision support system optimizing total travel distance of vehicles completing daily shipment plans. In detail, the developed models both assign the shipment goods to capacitated vehicles as well as minimize the total travel distance of these vehicles by considering an open vehicle routing algorithm. The decision support system is coded by using the MS excel-macro and a user friendly tool is developed for users' convenience. In the developed tool, there are two different solution approaches: a mixed integer mathematical programming and a heuristic solution approaches. For the small size plans (i.e. less than 25 demand points) mathematical modelling guaranteeing the optimal solution can be utilized. For the large size cases the heuristic algorithm providing a near-optimal result promptly can be utilized. The macro-based tool is designed such a flexible way that the user can make any revisions on the provided plan whenever necessary.

Keywords: Open vehicle routing · Daily shipment plan · Capacitated vehicle routing · Optimization · Decision support system

1 Introduction

During the last decade, in the logistics area, the competition has increased among the distributors for fast delivery of goods. New technologies such as delivery optimization and fleet management software emerged as the key advantage in this competition. The overall supply chain cost can be reduced significantly by focusing on the transportation cost. Transportation cost can be reduced significantly if manual delivery planning process is entirely replaced or augmented by the kinds of optimization software. In logistics, transition from manual dispatching planning to computer aided dispatching plan incorporating integrated route optimization and planning can create a domino effect on positive changes in total transportation cost. The first prerequisite for this efficient transition is a true information system. Then, the development of proper optimization algorithms is critical to manage the shipment plan, efficiently.

N. M. Durakbasa and M. G. Gençyılmaz (Eds.): ISPR 2019, LNME, pp. 650–660, 2020.
https://doi.org/10.1007/978-3-030-31343-2_56

Shipment plan is a set of decision processes mostly considering the creation of shipment orders, assignment of these orders to vehicles, routing of these vehicles, and delivery of the orders to customers on time. Developing integrated models considering these multiple processes is significant in the optimization process of this operation. Otherwise, an independently solved model may violate the other's objective and may not provide an optimal plan.

To improve the efficiency of a shipment plan, the solution should address and fulfil the following requirements: work-load balancing, which vehicle should follow which route to fulfil the customer demands, effectively segmenting of customers and products in a given region, utilization of vehicles effectively to maximize the area coverage, tracking of vehicle's route plan and dynamically change it on need-basis, utilization of vehicle space, tracking the performance, etc. In this paper, we solely focus on assignment of orders to vehicles and routing of these vehicles based on daily demand profile of the studied company. We develop both mathematical and a heuristic algorithm for this purpose and integrate these algorithms in a macro-based decision support system tool. We also develop a user-friendly interface in this tool for the users' convenience.

The studied problem is presented by following these sections: In Sect. 2, we define the studied problem in detail. In Sect. 3, a literature survey covering the regarding problems is presented. In Sect. 4, we explain our solution approaches and methodologies. In Sect. 5, the developed algorithms are tested by considering different size scenarios. In Sect. 6, the developed decision support system tool is presented. In Sect. 7, a conclusion summary is given for this study.

2 Problem Definition

In the studied company, shipment which is the last step in the post-production process is carried out with certain stages. These stages are: creation of a daily shipment orders, confirmation process of products ready for the shipment, vehicle and routing plans, vehicle assignment and delivery of the products to the customer addresses. First, a daily shipment order list is obtained from the customers. Then, the due dates and the prices of these products are checked out and sent to the regarding departments for approve. Before the approval process, orders are controlled for existence of any missing or incorrect information. After the confirmation process is completed, order lists are taken from the planning department. The received product list is sorted manually by the due date priority. After the product is checked in the warehouse, the distribution plan starts to be created by the experience of the planning order. Since, no optimization procedure is followed in the plan process, so many mistakes may happen resulting with high cost.

In any case, important resources such as time and money, should be used optimally. Since lack of the systematic and scientific methods in the current shipment policy of the company, the current experience-based policy does not guarantee an optimal solution. By utilizing optimization algorithms, time, money and labour costs can be minimized. Therefore, customer satisfaction can be increased, and important matters like vehicle selection, and route assigning can be balanced.

The bedding company has a single distribution depot in the studied region. The customer profile varies depending on the sale. In this study, we aim to assign customer demands to the limited number of vehicles which is five. The vehicles have capacity constraints. Then, we optimize the routing plan of these five vehicles. However, if the number of vehicles vary in time, since it is defined as an input variable in the model, the company can still use the developed tool. The vehicles are assumed to be same with loading capacities of 17 m^3 products in total. In this problem, the main inputs are as follows; a daily shipment list, number of products based on types, delivery addresses of the customers, requested shipment date, volume of product types, vehicle capacities based on types and number of vehicles.

3 Literature Review

In this section, we provide a review for the related papers of the studied problem. The studied problem includes two steps one of which is the assignment of products to the vehicles, and second is routing of these vehicles. Therefore, our attention was also given to the vehicle routing problem (VRP) methodologies. However, we also provide some related studies on the assignment procedure. The first article that is encountered was a study from Jouber [1]. In that study, the basic VRP requires several assumptions, and those assumptions can be eliminated by adding new constraints. As those constraints are added, new VRP variants come into existence.

For this study, three VRP variants are reviewed. The first one is vehicle routing problem with pickup and delivery (VRPPD). Whenever the same vehicle satisfies both pickup and delivery demand, vehicle routing problem with pickup and delivery emerges [2]. Since this is not in accordance with our problem, literature research is continued with other VRP variants. The next variant was capacitated vehicle routing problem (CVRP). When many articles are observed, a study of Fisher and Jaikumar [3] has so many similarities with our problem in terms of the problem that they deal with. In that study, the vehicle routes must satisfy several constraints due to the factors like fixed vehicle capacities. In our study, the vehicle capacities are fixed as well. However, we realized that our case does not need vehicles to return to the depot unlike this study. This difference forced us to research more VRP types by taking that need into account. That was the point where this study's scope narrowed with open vehicle routing problem (OVRP). Open vehicle routing problem is a type of vehicle routing problem where trucks start with the depot and end the route with one of the customers [4]. According to the article "A General Heuristic for Vehicle Routing Problems", CVRP is a very similar methodology to OVRP, and an OVRP can be solved as an asymmetric CVRP [5].

Galic et al. [6] presented a case study of a distributor of fast moving consumer goods modelled like a mix fleet multi depot. They developed computer-aided system for optimization for delivery planning every working day without exception for more than four years.

As we benefit from several articles about main methodology and some constraints, our research continued with assignment problem in mind. This study requires an assignment problem and a vehicle routing problem to be solved as integrated. The study by Fisher and Jaikumar was very beneficial in this regard, since this study also presents an assignment problem. Therefore, new constraints related to the assignment problem are added to our study.

Based on the article mentioned by Seker et al. [7], customer demands had been assigned to the trucks at the initial point, and then, total travelling distances between depot and customers had been calculating based on the OVRP rules [7].

At the final point of this study, since it is observed that it takes too long for mathematical model to generate solutions with larger data sets, we searched for new solution methodologies. As a result of these searches, Clarke and Wright savings algorithm [8], a sequential insertion algorithm, a parallel insertion algorithm, cluster first route second and route first cluster second heuristics, and the nearest neighbour algorithms were studied.

4 Solution Methodology

For the shipment plan, first, we develop a mixed integer programming model based on an OVRP. Second, we develop a heuristic algorithm for large size cases. The assumptions of the studied system are summarized below:

4.1 Assumptions

- Volume each each demand is pre-known.
- Address information of customer is pre-known.
- Vehicle capacities are same and equal to 17 m^3.
- Total number of vehicles are sufficient to satisfy the total demand.
- Setup time for the shipment process is ignored.
- Each vehicle starts its trip at the depot and ends at a customer address node (i.e., open vehicle routing).
- Product shapes of the demands are ignored, only their volumes are considered.

4.2 Mathematical Formulation

(1) *Parameters:*

K	Total number of vehicles
N	Set of travelling nodes
d_{ij}	Distance between customer i and customer j, $j = 2, \ldots, N$; $i = 1, \ldots, N$
Q_i	Order volume of customer i, $i = 2, \ldots, N$
C_k	Capacity of vehicle k, $k = 1, \ldots, K$

(2) *Decision Variables:*

$$U_i = \text{Variable used for subtour elimination}$$

$$x_{ijk} = \begin{cases} 1, & \textit{if vehicle k travels from customer i to customer j} \\ 0, & \textit{otherwise} \end{cases}$$

$$y_{ik} = \begin{cases} 1, \textit{ if order of customer i is delivered by vehicle k} \\ 0, \textit{ otherwise} \end{cases}$$

(3) *Mathematical Model:*

$$\text{Minimize } \sum_k^K \sum_i^N \sum_j^N d_{ij} x_{ijk}$$
Subject to

$$\sum_k^K y_{ik} = K \qquad i = 1 \tag{1}$$

$$\sum_k^K y_{ik} = 1 \qquad i = 2, \ldots, N \tag{2}$$

$$\sum_i^N x_{ijk} = y_{jk}, \qquad j = 2, \ldots, N \quad k = 1, \ldots, K \tag{3}$$

$$\sum_j^N x_{ijk} \le y_{ik}, \qquad i = 1, \ldots, N \quad k = 1, \ldots, K \tag{4}$$

$$\sum_i^N Q_i y_{ik} \le C_k, \qquad k = 1, \ldots, K \tag{5}$$

$$U_i + 1 = U_j + M\left(1 - x_{ijk}\right), \qquad i = 2, \ldots, N \quad j = 2, \ldots, N \quad k = 1, \ldots, K \tag{6}$$

$$x_{ijk} = 0 \; or \; 1, \qquad i = 1, \ldots, N \quad j = 2, \ldots, N \quad k = 1, \ldots, K \tag{7}$$

$$y_{ik} = 0 \; or \; 1, \qquad i = 1, \ldots, N \quad k = 1, \ldots, K \tag{8}$$

$$U_i \ge 0, \qquad i = 2, \ldots, N \tag{9}$$

The objective function of the mathematical model minimizes total travelling distance of the vehicles. Constraint (1) defines that all vehicles must begin from the depot while the first node is considered as depot node. Constraint (2) ensures of assigning a vehicle to each customer; constraints (3) and (4) assure that if the vehicle k is assigned to the customer i, the same vehicle must again transport between i and j customers. Constraint (5) is the capacity constraint of each vehicle. Constraint (6) is subtour elimination constraint. Constraints (7) and (8) show that every x_{ijk} and y_{ik} are binary decision variables. Constraint (9) is non-negativity constraint of subtour decision variable.

4.3 Heuristic Method

In the cases that mathematical model takes too much time to generate a solution for large size instances, a heuristic method is utilized whose details are summarized in this section. According to the literature review, the nearest neighbour algorithm (NNA) is used widely for heuristic approach that is fast and easy to implement for VRPs [9]. For capacitated OVRP this algorithm can also used as shown by Fig. 1.

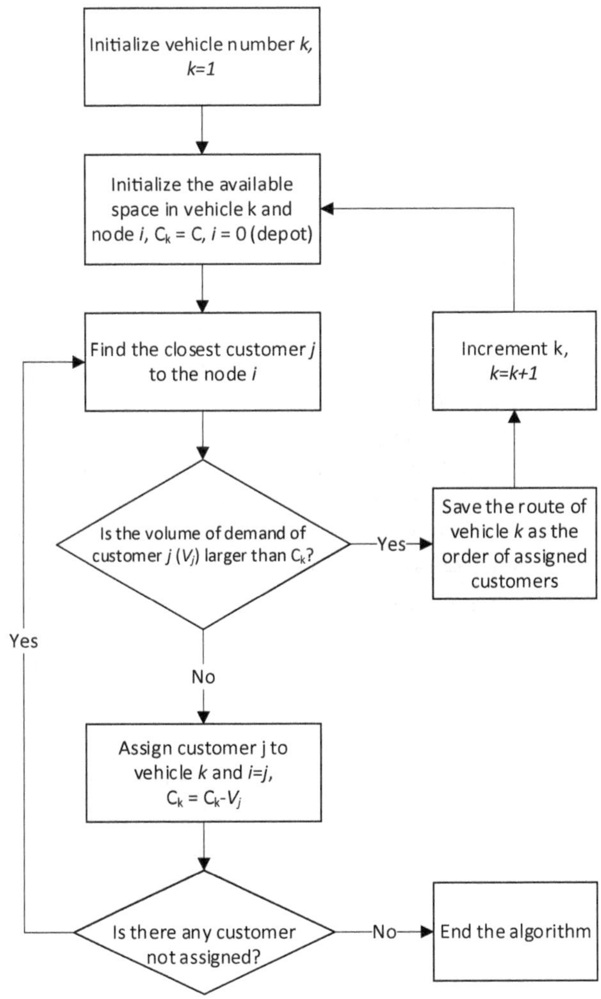

Fig. 1. Nearest neighbour algorithm flow chart

Nearest neighbour method is a tour based constructive algorithm which only considers to minimize traveled distance or total cost. Since it relies on fewer constraints, it is easy to implement for OVRP. In addition, its effortless nature makes it reliable and speed wise efficient especially while dealing with larger data sets. The steps of this algorithm is outlined below and flow chart of the heuristic method is given by Fig. 1.

1. Start the route with the depot node.
2. Add the closest remaining node to the last added node in the route while not exceeding the truck capacity.
3. Delete the last added node from the distance matrix.
4. If the truck capacity reaches the limit and there is at least one node in the matrix then, stop the algorithm and start with a new vehicle and go to step 1; otherwise go to step 5.
5. End the algorithm.

Note that in Fig. 1, k, C_k, i, and V_j represent the vehicle number, vehicle capacity, node number (demand point), volume of demand of customer j, respectively.

5 Computational Results

5.1 Runtimes

First, the developed mathematical model is tested for different size of the problem. These designs are shown in Table 1. Each is solved by CPLEX 12.6.3 on i7 6700 processor with 32 GB of RAM. The run time of the results for the tried scenarios are also summarized in Table 1. The first column, shows the number of customers to be visited. The second column is the number of vehicle scenarios. In the third column, the solution time averages are summarized. Seven replications are completed for all customer (node) whose results are given in Table 1. Namely, the distance and the demand volume input variables are changed by creating random values from uniform distribution for these scenarios and repeated for seven different replications. Once again, the average solution times of those repeats are illustrated in the third column of this table.

Table 1. Solution time for different number of nodes and vehicles

Number of nodes	Number of vehicles	Solution time (sec.)
8	3	2.5
10	4	4.6
15	4	5.9
20	5	18
25	5	34.6
30	5	834.9
35	5	4125
40	5	18000+

Figure 2 shows how the run time changes based on the size of the problem. When the number of nodes increases, the decision variables also increase. The number of decision variables is shown as number of outputs in Fig. 2. From Fig. 2, it can be observed that the run time increases exponentially with the size of the problem. The mathematical model may be seen as reasonable in terms of the run time, for up to 35 nodes (i.e., 4,125 s.). However, the daily shipment request of the company usually exceeds the 40 nodes. Therefore, a heuristic algorithm is also studied for such, more customer node cases.

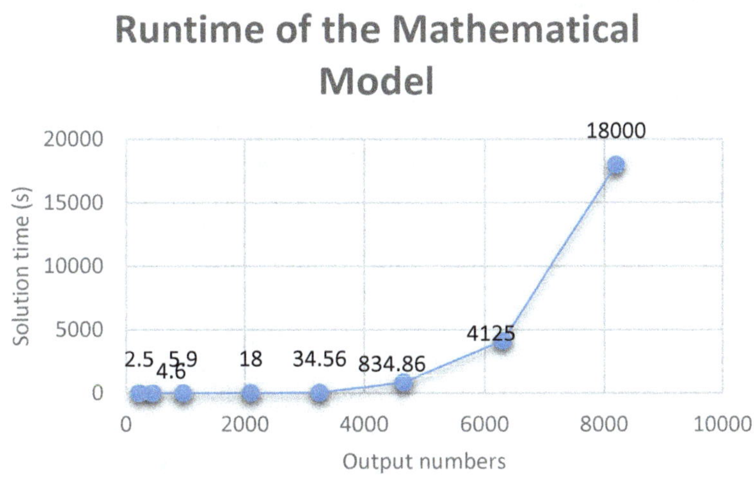

Fig. 2. Runtime of the mathematical model

5.2 Implementation of the Optimization Models for the Real Case

The considered optimization models are tried for a real set of customer data requested from the company. The both models' results are compared with the current vehicle routing plan and summarized in Table 2. The sample data shows total of 331 km is travelled with 22 customers including the depot. Then, using our decision support system, two solution methods are solved with the same data. Numerical solutions are given in Table 2.

Table 2. Comparison of methods for real case

	Total distance (km)	Improvement (%)
Current system	331	–
Mathematical model	226	31.8
Heuristic method	245	27.2

According to Table 2, the current travel distance is 331 km. This distance is 226 km. with the mathematical model solution and 245 km. with the heuristic algorithm solution. The improvement obtained by the heuristic algorithm is 27.2% while it is

31.8% by the mathematical model. Thus, we believe that both our solution approaches achieved a significant improvement over the current routing approach in the company.

6 Developed Decision Support System

We designed and developed a user-friendly decision support system (DSS) by embedding both mathematical and heuristic-based algorithms for the solution of a daily shipment problem of a bedding company in Turkey. The main screen of the decision support system has four main buttons as shown in Fig. 3. The first button is used to receive the daily orders from the company database. This data base includes the customer delivery addresses as well. The user can also edit the demand data provided by the system upon pressing the second button. By the third button, the user can solve the problem either by using mathematical model or by using the heuristic algorithm. On submit, it generates a solution with routes and vehicle plans. Finally, the last button opens the shipment plan report for every vehicle.

Fig. 3. Decision support system main screen

Bing Maps API is integrated into the DSS to find out the real coordinates of the addresses and the coordinates are used to create the distance matrix and visualize the routes.

The developed DSS also has some other screens to provide different options to utilize for the user. In Fig. 4, some of them can be observed. For instance, there is a window for editing the current customer data and removing the any product from the shipment plan. In addition, total shipment volume, customer shipment volume, total vehicle number can also be seen in the same screen.

After solving the problem, the result is visualized in a screen as shown in Fig. 5. On the left side, there are two maps. The upper map shows all the customer locations. The lower map shows the route of the selected vehicle besides the customer locations. The current selected vehicle can be changed by the user via the route dropdown. In the middle part, all vehicles can be seen with assigned customer addresses. The user can

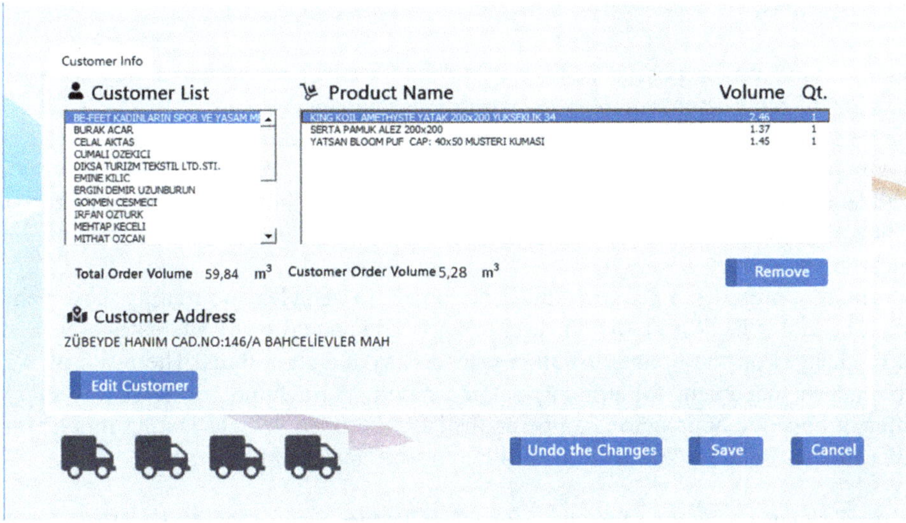

Fig. 4. Edit order list user form

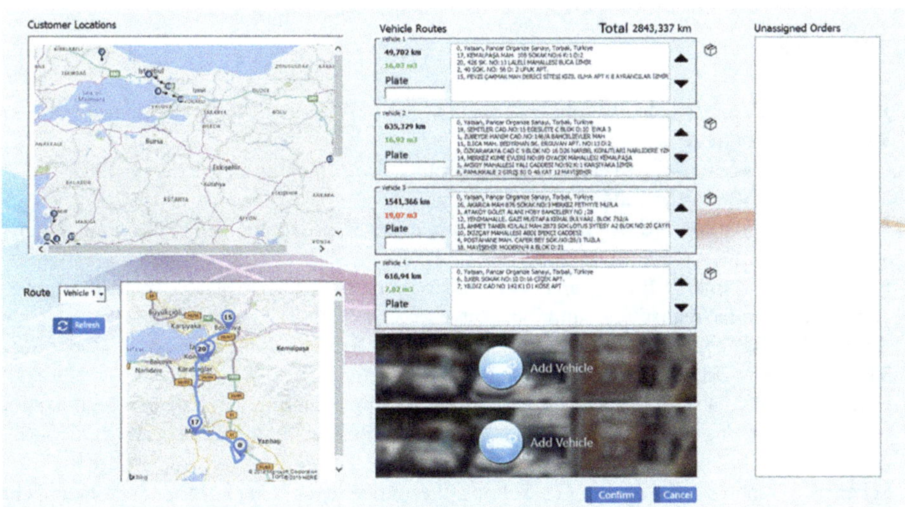

Fig. 5. Shipment plan user form

also observe the number of routes, customer addresses, total distance to be travelled, product volume, total volume of assigned products, and plate number of the vehicles. Customers can be dragged and dropped between vehicles, and if it is desired, route of the vehicle can be changed with the up and down buttons in order to further improve the routing plan. If necessary, new vehicles can also be added via the add vehicle button. In addition, at the right side, the user can also drag and drop the customers from vehicles to unassigned orders.

7 Conclusion

In this study, we propose a flexible user-friendly excel macro-based tool for optimization of daily shipment plan of a bedding company in Turkey. In the tool, we integrate two optimization procedures that can be used alternatively based on the size of the problem. The mathematical model guarantees the optimal solution for the shipment planning. However, when the number of nodes increases, this solution procedure may become inefficient in terms of the run time. In this case, the user can use the heuristic algorithm.

The tool provides a plan, assigning the orders to vehicles and routing those vehicles. The tool is flexible so that the user can also add and drop any addresses as well as make changes on the optimal solution provided by the algorithms. The result plan is written in a document for printing out the report. As a future work of this study, different heuristic approaches can be applied for the open vehicle routing problem in order to contribute the literature with different solution methodologies.

Acknowledgment. We would like to thank our colleagues Batuhan Taş and Atakan Türkyılmaz, and our company supervisor Süleyman Çolak for their kind support and contribution to this study.

References

1. Joubert, J.W.: The Vehicle Routing Problem: Origins and Variants. PhD Thesis, University of Pretoria etd, South Africa (2007)
2. Gajpal, Y.: Algorithms for vehicle routing problem with pickup and delivery, PhD Thesis, McMaster University, Hamilton, Ontario, May 2008
3. Fisher, M.L., Jaikumar, R.: A Generalized Assıgnment Heuristic for Vehicle Routing, August 1979
4. Picka, K.H., Ashjari, B., Ziaeifar, A., Nickbeen, P.: Open vehicle routing problem optimization under realistic assumptions. Int. J. Res. Ind. Eng. **3**, 46–55 (2014)
5. Pisinger, D., Røpke, S.: A general heuristic for vehicle routing problems. Comput. Oper. Res. **34**, 2403–2435 (2007)
6. Galić, A., Carić, T., Fosin, J.: The case study of implementing the delivery optimization system at a fast-moving consumer goods distributor. Promet Traffic Transp. **25**(6), 595–603 (2013)
7. Şeker, Ş., Güneri, A.F., Ermiş, G.: Solving the multi depot vehicle routing problem with assignment algorithm. In: 13th International Research/Expert Conference "Trends in the Development of Machinery and Associated Technology", Hammamet, pp. 225–228, October 2009
8. Toth, P., Vigo, D.: The Vehicle Routing Problem. Society for Industrial and Applied Mathematics, Philadelphia (2002)
9. Daanish, A.R., Naick, B.K.: Implementation of charging station based electric vehicle routing problem using nearest neighbour search algorithm. In: 2nd IEEE International Conference on Intelligent Transportation Engineering (ICITE), September 2017

Building a Decision Support System for Vehicle Routing Problem: A Real-Life Case Study from Turkey

Ayşenur Doğan, İrem Bilici, Osman Kaan Demiral, Mehmet Serdar Erdoğan, and Özgür Kabadurmuş(✉)

International Logistics Management, Yasar University, Izmir, Turkey
adogan96@icloud.com, irembilici@outlook.com, kaan.dmrl.kd@gmail.com, {mehmet.erdogan, ozgur.kabadurmus}@yasar.edu.tr

Abstract. One of the most costly operations in logistics is the distribution of goods. Inefficient vehicle routes increase distribution costs, especially for companies performing distribution operations daily. Vehicle Routing Problem (VRP) addresses this inefficiency and optimizes the distribution routes of vehicles. In this study, we developed a decision support system to solve the Vehicle Routing Problem with Time Windows and Split Delivery and applied it to a real-life case company. The data of the problem were obtained by a real logistic company, which is one of the leading Turkish logistics companies located in Izmir, Turkey. The company distributes goods to the customers located in various cities in Turkey and currently does not use any decision-making tool to optimize the routes of its trucks. We formulated the mathematical model as Mixed Integer Linear Programming (MILP) and solved it by using IBM OPL CPLEX. Our proposed decision support system clusters the customers into geographical groups and then optimizes the routes within the clusters. The results of the decision support system can be manually adjusted by the decision maker to fine-tune the routes. We demonstrated the efficiency of our proposed methodology on the regional distribution of the company. The results of the study showed that our proposed model decreases the total distribution distance by 16% and total distribution time by approximately 13%.

Keywords: Vehicle Routing Problem · Mixed Integer Linear Programming Model · Optimization · Time windows · Heterogeneous Fleet

1 Introduction

In today's global world, rapidly developing new technologies, increasing customer expectations and intercompany competitions necessitate the improvements in the supply chain operations, such as transportation, distribution and logistics. Optimization studies based on operations research and mathematical programming techniques have been conducted for effective management in these fields. As explained by Hillier and Lieberman [1], one of the most well-known NP-hard problems among these fields is the

N. M. Durakbasa and M. G. Gençyılmaz (Eds.): ISPR 2019, LNME, pp. 661–675, 2020.
https://doi.org/10.1007/978-3-030-31343-2_57

Vehicle Routing Problem (VRP), which has a direct impact on the effectiveness and cost of the operations of the company.

The VRP was first proposed by Dantzig and Ramser [2] and can be explained as a combinatorial optimization problem aiming to find the optimal routes of the vehicles that distribute customer demand between geographical locations from one or more depots as shown in Fig. 1. Therefore, researchers approach the problem from two different perspectives: mathematical programming formulations to optimally solve the problem and heuristic methods to obtain the near-optimum results.

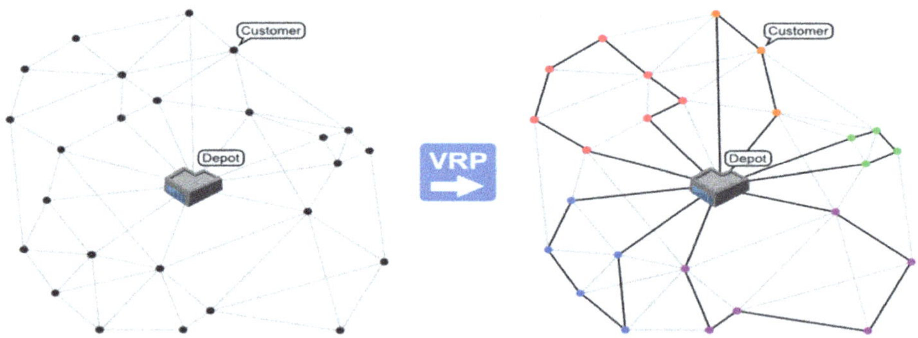

Fig. 1. Illustration of the Vehicle Routing Problem

The main purpose of the VRP is to reduce the total transportation cost by minimizing distances and making effective routes. Transportation cost can be divided into two categories: fixed costs (vehicle investment, depreciation, and driver related costs) and variable costs (fuel cost). With using VRP algorithms, companies can minimize not only variable cost by reducing distances but also fixed costs by reducing the number of vehicles. However, real-life practices of VRP may require many constraints that are needed to be considered. Vehicle-related constraints include the number of vehicles, the capacity of vehicles (generally in weight or volume), and working periods of drivers. Besides, the constraints related to customers are time windows required by customers service or satisfying the priority relations among customers. Lastly, precedence relations between pairs of locations and traffic & road conditions may be included in real-life applications.

These complex constraints of the real-life distribution applications define the variants of the VRP as presented in Fig. 2. The most common VRP variant is the Capacitated Vehicle Routing Problem (CVRP) [3], which includes distribution vehicles with certain capacities and a different capacity for each vehicle. Each customer must be served once, and every vehicle tour must start and end at the same depot. The CVRP aims to minimize the distribution cost by taking these constraints into account. The distribution company may have more than one depot in Multiple Depot Vehicle Routing Problems (MDVRP), and each depot should be assigned to a vehicle within the fleet. Also, the customers should be assigned to a depot according to their proximity. Another other variant is the Split Delivery Vehicle Routing Problem (SDVRP). The SDVRP allows dividing large quantities of orders into smaller quantities and delivering at different times with different vehicles. The objective is to minimize the total cost (fixed cost and variable cost). In the Vehicle Routing Problem with Time Windows (VRPTW), the customer's request must be fulfilled within a certain period of time and the objective is to minimize the total cost by distributing all customers within their required hours. Also, the Heterogeneous Fleet VRP (HVRP) delivers customer demand from the central depot to the customer set by taking into account the vehicle capacity constraints of the fleet of various vehicles with different properties (e.g., cost or capacity). In VRP with Pick-up and Delivery (VRPPD), the possibility of customers returning goods is also considered. Therefore, vehicles must have an available space for the returned cargo. In this problem, the vehicle capacity and route are intended to be optimized to prevent increased travel distances or a need for more vehicles. The Dynamic VRP (DVRP) allows visits to the customers must be sequenced along the routes. In real-life problems, customer data and demand are generally uncertain, and this uncertainty can cause additional costs, such as the unplanned return of a vehicle to a replenishment depot. The Stochastic Customer (VRPSC) and Stochastic Demand (VRPSD) VRP variants are to solve problems in real-life by considering the uncertainty of the data. The VRP with Crew Scheduling deals with the assignment of tasks to crew duties so that each duty is performed within a set of working rules (schedules and capabilities) and the total labor costs are also minimized. In VRP with two-three-dimensional loading (2L-VRP, 3L-VRP), loading constraints are also considered. The Periodic VRP is another variant of the VRP, which consists of several periods and assumes that each customer must be visited at a specified frequency and dedicated combinations of visiting periods.

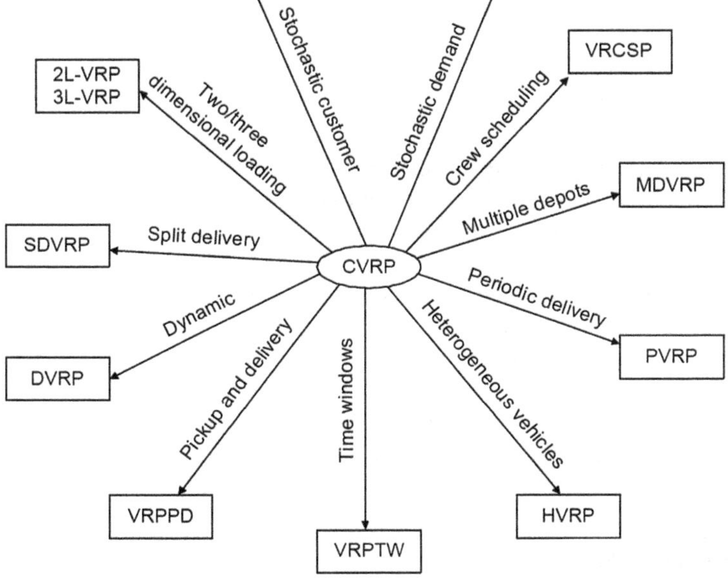

Fig. 2. Extensions of the VRP [4]

VRP algorithms are used by various companies in real life. UPS developed On-Road Integrated Optimization and Navigation (ORION) program to provide the optimized routes for drivers. According to UPS Pressroom [5], the program helps UPS drivers to find the optimal way for delivery operations. ORION's decisions can change instantaneously with changing conditions, such as traffic or accidents. As of 2016, UPS estimated that it has optimized 55,000 routes, saved 10 million gallons of fuel, reduced 100,000 tons CO^2 and saved the company between $300–400 million in costs. FedEx also has a VRP software, which is called GroundCloud[1]. It optimizes the routes and significantly reduces distribution costs. In addition, there is an organization called Routific[2], which provides VRP solutions to the companies. Routific optimizes the last mile delivery operations by considering time windows, vehicle capacities, delivery types, priority stops, traffic, driver speeds, driver shift times, and breaks.

In this study, we developed a decision support system for a real-life VRP application. Our proposed decision support system is designed for a cargo company located in Izmir, Turkey. The proposed methodology first clusters the customers into groups according to their geographical regions. Then, the VRP problem is solved for each cluster. The mathematical model of the VRP problem presented herein uses many realistic aspects of the distribution requirements of the company: time windows, heterogeneous fleet, and split delivery. The results of the initial model can be manually adjusted by the decision maker to fine-tune the results. To demonstrate the efficiency of our proposed methodology, we applied the decision support system on the regional distribution of the company. The results showed that our proposed methodology reduces the total distribution distance and time.

This paper is organized as follows. The relevant literature on the Vehicle Routing Problem variants is summarized in Sect. 2. The problem definition is explained in Sect. 3. The proposed decision support system methodology is presented in Sect. 4. The case study and results are presented in Sect. 5. Conclusion and future work are summarized in Sect. 6.

2 Literature Review

In this section, we review the extensions of the Vehicle Routing Problem in the literature including heterogeneous fleet, time windows, split delivery, backhauls and mixed backhauls, multi-depot and stochastic demand variants.

Karagül and Güngör [6], Ceschia et al. [7], Kritikos and Ioannou [8], Belmecheri et al. [9] examined the Heterogeneous Fleet VRP to solve the case of a fleet of vehicles characterized by different capacities. As Karagül and Güngör [6] mentioned, the heterogeneous fleet of vehicles can be explained as a fleet that has various capacities and costs and can be used in various delivery activities. Ceschia et al. [7] developed a Tabu Search heuristic with a multi-day planning horizon, a complex carrier-dependent cost for vehicles, and the possibility of leaving orders unscheduled with using a tabu list and

[1] https://www.gogroundcloud.com/.

[2] https://routific.com/.

punishment method. The results of these techniques are close to real-life representation and competitive with the best methods of the literature. Kritikos and Ioannou [8] worked on Heterogeneous Fleet Vehicle Routing with Overloads and Time Windows (HFVRPTW), which has not been previously studied in the literature. They modeled the problem by integrating the total trip load into the objective function and solved it by a Sequential Insertion heuristic that employs a penalty function allowing capacity violations but limiting them to a predefined upper bound. As another extension, HVRP with mixed backhauls and time windows (HVRPMBTW) is investigated by Belmecheri et al. [9], which was rarely studied in the literature. They considered linehaul customers, where some customers need deliveries from a central depot, while others have some goods to be picked up and centralized at the same depot (backhaul customers) considering delivery times and different vehicle types. A Particle Swarm Optimization approach with a Local Search was successfully applied to minimize the total distance traveled and weighted by the variable costs of the vehicles used. Salhi et al. [10] introduced the Fleet Size and Mix Vehicle Routing Problem with Backhauls (FSMVRPB) and proposed a Set Partitioning Problem (SPP) based heuristic. In their paper, the Fleet Size and Mix VRP and the VRP with Backhauls have been investigated separately. Unlike VRP, FSMVRP considers a heterogeneous vehicle fleet. In Vehicle Routing Problem with Backhauls (VRPB), the fleet can consider pick-ups after the deliveries are made and no route is allowed with only pick-up (backhaul) customers. The Set Partitioning based heuristics on FSMVRPB offers the most appropriate solution in terms of solution quality and calculation time by efficiently reducing the number of routes in the pool. The Multi-Period Fleet Size and Mix Vehicle Routing Problem with Stochastic Demands was defined as an extension of the FSMVRP by Pasha et al. [11]. In this problem, the customer demands vary from period to period, and different routes and different vehicles are identified for serving customers in each period. Pasha et al. [11] showed that the uncertain customer demands can cause an unplanned return to the depot for refilling, which increases the cost and that happens usually before serving the last customer on the route. Another variant of the VRP with mix split delivery and fleet size using Mixed Integer Programming (MIP) and Constraint Programming (CP) was considered by Maheo et al. [12]. In split delivery VRP, vehicles can visit customers more than once. Applying split delivery can be useful when customer demands exceed the vehicle capacity. Belfiore and Yoshizaki [13], Bertoli et al. [14], and Chu et al. [15] examined VRP variants using split deliveries and these articles showed that splitting deliveries can decrease variable transportation cost and the fixed costs. Bae and Moon [16] studied the multi-depot vehicle routing problem with time windows (MDVRPTW), which is a practical and challenging problem in logistics and supply chain management. They optimized the routes of service vehicles that are used for delivery and installation of electronics to minimize fixed costs of the depots and the delivery and installation vehicles as well as the expenses related to travel distances and labor. Along with a mixed integer programming model, they also developed a heuristic and a genetic algorithm to obtain near-optimal solutions for large problems.

As Laporte [17] summarized, the VRP solution methods can be divided into two: (1) exact and (2) heuristic methods. According to the literature review, we observed that exact algorithms can only solve relatively small problems, but approximate (heuristic) algorithms like Tabu Search, Genetic Algorithms or Particle Swarm can be very effective. For example, Particle Swarm Optimization (PSO) with a Local Search as proposed by Belmecheri et al. [9] yielded high-quality solutions on small problems and improved the results of the large benchmark problems.

Similar to our study, there are also many real-life case studies in the literature. Amini et al. [18] studied the VRPTW for a real-life company which is responsible for providing enough safe potable water in Tehran. They used the Particle Swarm Optimization algorithm to minimize the total distance. In Onut et al. [19], HVRP was used to optimize the routes of a company that operates in the LPG sector in Turkey. Similar to our real-life case study herein, the company of Onut et al. [19] determines the routes manually, and therefore, Onut et al. [19] provided a significantly better solution for the company. Another study with fleet size and mixed vehicle routing problem was performed by Karagül and Güngör [6] to make the routing decision of tourists coming to the hotels in Alanya region from Antalya airport. The purpose of their study is to minimize the traveling distances between the airport and the hotels at the minimum cost. Various solution methods were used by Karagül and Güngör [6] such as savings algorithm, sweep algorithm, random permutation alignment, standard genetic algorithms and random search algorithms. As a result, they reduced the cost by 35% by using a random search algorithm. Another case study belongs to Panapinun and Charnsethikul [20], in which they developed a daily truck routing and scheduling software to minimize total distance and improve the efficiency of the perishable distribution for the food company located in Bangkok. As a result of their study, the total distance was reduced between 11% and 27% thanks to their Vehicle Routing and Scheduling Software (VRASS). In another case study, Worwa [21] presented a mathematical formulation to optimize the school transportation system in one of the Polish junior high schools. The results of the study showed that small-scale problems can be solved by location-allocation-routing method.

3 Problem Definition

Our study focuses on the cargo distribution in the Aegean Region, which also includes some southern Marmara cities and Mugla province. The goal of the company is to find the best distribution routes to minimize the total distribution cost, which includes fixed and variable costs. The vehicle capacities are fixed since the company has its own fleet. Also, the company aims to have high truck utilization rates.

The company has a single depot to serve the Aegean Region in Turkey, which is located in Izmir. This is the only depot in the Aegean region. In most situations, a vehicle must return to the depot on the same day. In rare occasions, the vehicle is allowed to return in two days due to a large number of deliveries in a large geographical region. The customer demands become available at the end of the day, and the vehicle routes are determined daily.

In the current practice, an operation specialist is responsible for the determination of all routes in Turkey. The company has more than 1,000 trucks and at least 10% of them are used in cargo distribution, whereas the rest are used in FTL (full truckload) operations. Even though there are a large number of trucks to route and many customers located in diversified geographical regions to distribute, the vehicle routing problem is manually solved by an appointed employee. With using his past experience, the routing expert makes the routing decisions every evening so that the vehicles are loaded and departed in the morning. After obtaining the routes, additional iterations are performed to manually adjust the routes. Obviously, this manual process of route assignments by a single employee is far from optimal. It requires an intensive amount of manual labor and has a large margin of error. Figure 3 illustrates the current practice of the company.

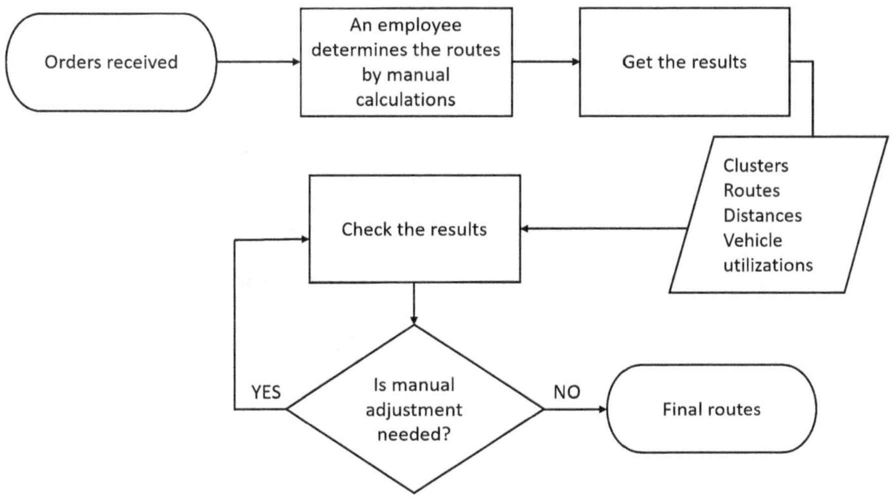

Fig. 3. The company's current practice to solve VRP

To solve this problem more effectively we proposed a decision support system as presented in Sect. 4 in detail. The mathematical model of the problem is presented in the next section.

3.1 Mathematical Model

We adopted the mathematical model of Belfiore and Yoshizaki [13] to solve a real-life heterogeneous fleet vehicle routing problem with time windows and split deliveries (HFVRPTWSD).

The parameters of the model are given below:

N	Set of customers, where $N = \{1, 2, \ldots, \mathrm{n}\}$
n	The number of customers
Node 0	Central depot
T	Set of different vehicles, where $T = \{1, 2, \ldots, \mathrm{K}\}$
t_{ij}	Travel time between nodes i and j
d_{ij}	Distance between nodes i and j
q_i	Demand of customer i

a_k Capacity of vehicle k
f_k Fixed cost of vehicle k
g_k Variable cost of vehicle k
e_i Earliest delivery time required by customer i
l_i Latest delivery time required by customer i
s_i Service time of customer i

The decision variables of the model are:

$$x_{ij}^{k} = \begin{cases} 1 & \textit{if a customer j is supplied after customer i by a vehicle of type k}, \\ 0 & \textit{otherwise} \end{cases}$$

b_i^k = The time at which service begins at customer i by a vehicle of type k,

y_i^k = *Fraction* of customer i's demand delivered by a vehicle of type k.

The objective function and the constraints are given below.

$$\min z = \sum_{k=1}^{K} f_k \sum_{j=1}^{n} x_{0j}^{k} + \sum_{i=0}^{n} \sum_{j=0}^{n} \sum_{k=1}^{K} c_{ij}^{k} x_{ij}^{k}. \tag{1}$$

s.t.

$$\sum_{j=1}^{n} x_{0j}^{k} = 1; k = 1, \ldots, K, \tag{2}$$

$$\sum_{i=0}^{n} x_{ip}^{k} - \sum_{j=0}^{n} x_{pj}^{k} = 0; p = 0, \ldots, n; k = 1, \ldots, K, \tag{3}$$

$$\sum_{k=1}^{K} y_i^k = 1, i = 1, \ldots, n, \tag{4}$$

$$\sum_{i=1}^{n} q_i y_i^k \leq a_k; k = 1, \ldots, K, \tag{5}$$

$$y_i^k \leq \sum_{j=0}^{n} x_{ji}^{k}; i = 1, \ldots, n; k = 1, \ldots, K, \tag{6}$$

$$b_i^k + s_i + t_{ij} - M_{ij}\left(1 - x_{ij}^{k}\right) \leq b_j^k; i = 1, \ldots, n; j = 1, \ldots, n; k = 1, \ldots, K, \tag{7}$$

$$e_i \leq b_i^k \leq l_i; i = 1, \ldots, n; k = 1, \ldots, K, \tag{8}$$

$$y_i^k \geq 0; i = 1, \ldots, n; k = 1, \ldots, K, \tag{9}$$

$$b_i^k \geq 0; i = 1, \ldots, n; k = 1, \ldots, K, \tag{10}$$

$$x_{ij}^k \in \{0, 1\}; i = 0, \ldots, n; j = 0, \ldots, n; k = 1, \ldots, K. \tag{11}$$

The objective function (1) is to minimize the fixed vehicle and travel costs. Constraints (2) and (3) define that each vehicle leaves from the depot, arrive at a customer, leaves again and finally arrives at a depot. Constraint (4) ensures that the total demand of each customer is satisfied. Constraint (5) makes certain that the total vehicle capacity is not exceeded. Constraint (6) ensures that the demand of each customer can only be compensated if a vehicle goes to that customer. Constraint (7) adjusts the minimum time for beginning service and ensures that there are no subtours. Constraint (8) defines the time windows. Constraints (9) and (10) assure that decision variables y_i^k and b_i^k are positive, and Constraint (11) defines decision variable x_{ij}^k as binary.

4 Methodology

In this section, the proposed decision support methodology is presented. The model has been formulated as a Mixed-Integer Linear Programming (MILP) model and solved in IBM OPL CPLEX. However, the model is not able to find any feasible solution within 24 hours when the number of customers is above 50. Therefore, we have clustered the customers by employing *k*-means clustering method and optimized the vehicle routing problem within each cluster.

Clustering is the process of partitioning a set of data elements into subsets [22]. Each subset is a cluster and elements in a cluster are similar to one another while they are dissimilar to the elements in the other clusters. *K*-means clustering method divides a set of data into *k* clusters so that each cluster contains at least one element. The number of clusters (denoted as "*k*") must be given in advance in *k*-means clustering method. Each cluster has a center that is the mean value of all elements in that cluster. The aim of the *k*-means algorithm is to minimize total within cluster error which is the total distance of all elements to their cluster centers. The procedure of *k*-means clustering method is given below:

1. Determine *k* cluster centers.
2. Assign each element to the cluster with the nearest center.
3. Re-compute the cluster centers.
4. Repeat steps 2 and 3 until cluster memberships, cluster centers or the error sum of squares do not change.

To determine the number of clusters in our problem, we have created a formula according to a total load of customer orders and minimum vehicle load. Equation (12) shows the calculation of a number of clusters. We divide the total customer load to 10 because the company aims to achieve a minimum vehicle load of 10 tons for their 15-ton vehicles. If only one vehicle is used for each cluster, an average of 10 tons can be loaded to the vehicles with this formula.

$$k = \left\lfloor \frac{Total\ Customer\ Load}{10} \right\rfloor \tag{12}$$

Upon finishing the clustering process, vehicle routing problem is optimized within each cluster by using CPLEX and the routes are obtained. If the decision maker is not satisfied with the resulting routes, she can adjust the number of clusters and reiterate the process until she is satisfied. Then, the decision maker can make manual adjustments in the obtained routes. In this process, the decision maker can discard a route, insert or swap customers between and within routes if necessary. Note that, most of this process is automated and the decision maker can make adjustments without making any calculations by hand. Figure 4 illustrates our proposed decision support system.

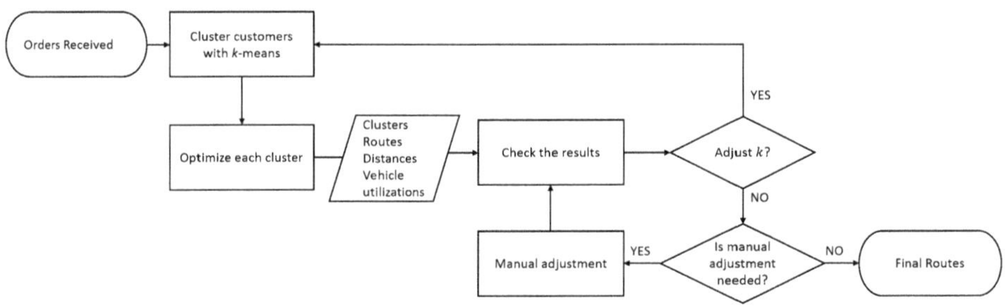

Fig. 4. The proposed decision support system

5 Results

The proposed model has been tested on a real-life case study. The real-life distribution data set was obtained from a logistics company located in Izmir, Turkey. The identity of the company is not revealed due to the company's request. However, this Izmir based company entered into logistics sector almost 50 years ago. Since the 1960s, the company has become one of the leading companies in the logistics and personnel transportation market in Turkey.

In the data set, the locations, demands, and time windows of 66 customers in Aegean region of Turkey are defined. The company also provided information on the vehicle types and their capacities. The routing decisions made by the company's expert were also obtained to compare our results. The model has been formulated as a Mixed-Integer Linear Programming model and solved in IBM OPL CPLEX with an Intel(R) Core(TM) i7-4510U CPU @ 2.00 GHz 2.60 GHz 8 GB RAM computer. The maximum solution time limit of the model is chosen as two hours. Since the distribution decisions are made daily, applying a time limit more than 2 hours would not be practical.

In our problem, the total customer load is 52.2, thus, the number of clusters ("k") is selected as 5 ($\lfloor 52.2/10 \rfloor = 5$). Figure 5 shows the comparison of the clustering of the company expert and k-means clustering. According to Fig. 5, the clustering of the company and our clustering are different. The clusters of customers 1, 2, 3, 4, 5, 6, 8, 42, 56, and 57 have changed.

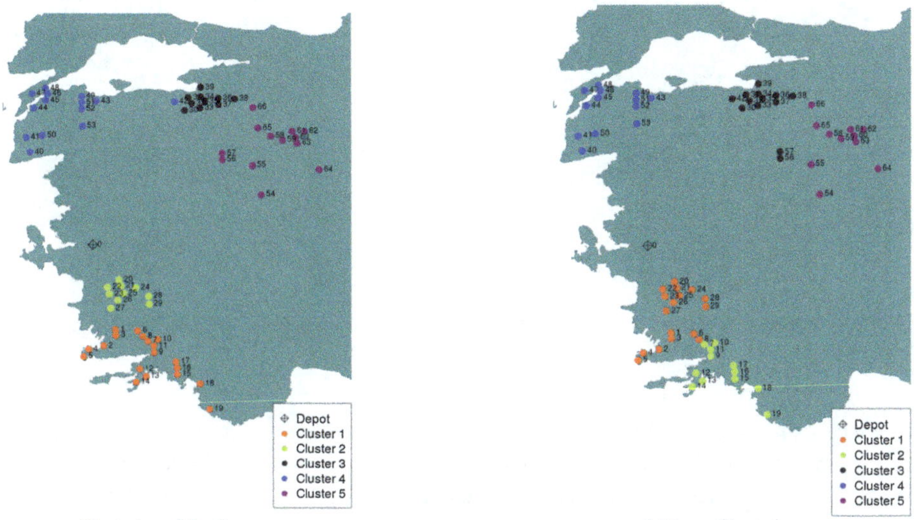

Clustering of the Company

k-Means Clustering

Fig. 5. Comparison of company's clustering and *k*-means clustering

After solving the VRP for each cluster as a separate problem, we have obtained six different routes for five clusters. Figures 6 and 7 show the comparison of the routes generated by the company and our routes. According to Figs. 6 and 7, both the company and our decision support system found three routes in the North Aegean region. On the other hand, the company generated two routes, while our method generated three routes in the South Aegean region.

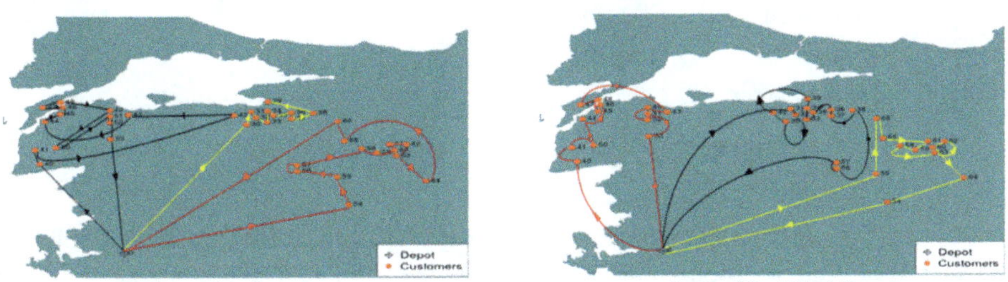

Routes of the Company

Routes of Our Methodology

Fig. 6. Comparison of the routes of the company and routes obtained by CPLEX in north Aegean region map

Our proposed method generated two routes in one of the clusters of the south Aegean region since the capacity of one vehicle is not enough to fulfill the demands of all customers in that cluster. While one of the vehicles visit 16 customers, the other one visits a single customer (customer 26).

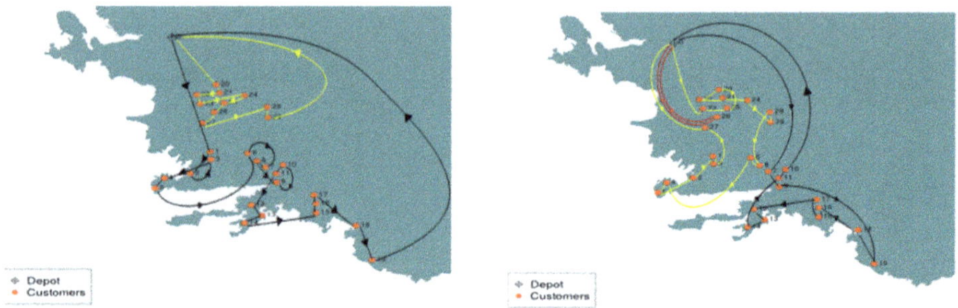

Routes of the Company Routes of Our Methodology

Fig. 7. Comparison of the routes of the company and routes obtained by CPLEX in south Aegean region map

Since using one additional vehicle for one customer is not a good decision, manual adjustments should be made as suggested in our proposed methodology. The first adjustment is to cancel the delivery of "Customer 26" today and deliver it on the next day. The second adjustment is to deliver the cargo of "Customer 26" with an available vehicle serving to the nearest cluster. Figure 8 illustrates the new routes after the two manual adjustments that we have suggested.

Manual Adjustment 1 Manual Adjustment 2

Fig. 8. The Routes after the Manual Adjustments

Table 1 presents the comparison of the total distances of the company's solution and our suggested solutions. According to Table 1, the first solution that we have obtained provides 12.27% reduction in the total delivery distance, however, an additional vehicle is used in that solution. Manual Adjustment 1 provides 16.35% decrease in total distance. However, it may cause customer dissatisfaction since the delivery is scheduled to be carried out the next day. Manual Adjustment 2 provides 16.3% reduction in distance, which is comparable to the reduction of Manual Adjustment 2.

Table 1. Results related to total distance

Solution method	Number of vehicles	Total distance (km)	Distance improvement (%)
Company solution	5	5137.36	
Our initial solution	6	4507.22	12.27
Manual adjustment 1	5	4297.15	16.35
Manual adjustment 2	5	4297.15	16.30

Table 2 presents the comparison of the distribution times of the company's solution and our suggested solutions. According to Table 2, all solutions that we have suggested have stable route durations that are around 12 hours while the route durations that the company expert found are highly unstable that are ranging from 6.42 to 21.03 hours. We provided 9.30% reduction in the total delivery time in our initial solution. Manual Adjustments 1 and 2 reduce the total delivery times by 13.04% and 12.87%, respectively.

Table 2. Results related to distribution times (in hours)

Cluster	Company's solution	Our initial solution	Manual adjustment 1	Manual adjustment 2
1	6.42	12.70	10.07	10.07
2	16.35	11.96	11.96	12.07
3	10.78	12.60	12.60	12.60
4	21.03	12.65	12.65	12.65
5	15.65	13.79	13.79	13.79
Total	**70.23**	**63.69**	**61.06**	**61.18**
Improvement (%)	–	**9.30%**	**13.04%**	**12.87%**

Therefore, as the Manual Adjustment 2 satisfy the customer demand and significantly reduces both delivery distance and time, it is chosen as the best option and suggested to the company. We have also suggested one additional adjustment to the company that could reduce the delivery cost. A smaller vehicle with a capacity of 10.2 tons, called as MAN TGL 12.220, can be used in the delivery of the two routes in the north Aegean region since the vehicle loads in these routes are less than 10 tons.

As for the solution times, CPLEX was not able optimally solve the vehicle routing problems within the clusters within two hours of solution time limit. The optimality gaps reported by CPLEX were 50% on average.

6 Conclusion and Future Work

In this study, we have developed a decision support system to optimize the distribution operations of one of the biggest Turkish logistics companies located in İzmir, Turkey. In the current practice of the company, an employee makes all the routing decisions manually. Our proposed methodology first clusters the customers into k clusters using k-means clustering method. Then, the route of each cluster is optimized separately using our proposed mathematical formulation. In our study, we formulated a Heterogeneous Fleet Vehicle Routing Problem with Time Windows and Split Delivery problem as a Mixed Integer Linear Programming model and coded in IBM OPL CPLEX. After obtaining the routes for each cluster, the decision maker can adjust the number of clusters (k) and reiterate the clustering algorithm and VRP optimization process. In addition, the decision maker can manually adjust the routes by discarding a route, inserting or moving customers between and within routes, if necessary. Our contribution is providing a decision support system to optimize routes of a real logistics company. Note that, our proposed decision support system is an automated process and it is significantly more efficient to operate than the company's current practice.

The proposed decision support system eliminates human error and provides an improved VRP solution to the decision maker of the company. It also significantly reduces the total distribution distances and delivery times. To show its effectiveness, we applied our methodology to a real-life distribution case study of the company. We have obtained 12.27% of reduction in total distance and 9.3% of reduction in total distribution time in our initial solution. However, we presented two manual adjustments to further improve the initial solution. The first manual adjustment discarded one of the underutilized routes and reduced the total distribution distance by 16.35% and total distribution time by 13.04%. The second manual adjustment moved the customer of the underutilized route to the nearest cluster and reduced the total distribution distance by 16.3% and the total distribution time by 12.87%.

The current study has only examined one case study. For future work, this study can be extended by applying our methodology to different case studies (İzmir city distribution or entire Turkey distribution) within the company. Also, metaheuristic algorithms can be applied to solve the problem faster. Finally, the solution methodology can be implemented in the Enterprise Resource Planning (ERP) system of the company to streamline and fully automate the decision-making process.

Acknowledgment. Burak Can Özaslan, Nazmihan Öterbülbül and Emre Can Kayadelen also helped this study during their senior design project at Yasar University in 2019.

References

1. Hillier, S.F., Lieberman, J.G.: Introduction to Operations Research, 7th edn. McGraw-Hill International Editions, New York (1995)
2. Dantzig, G.B., Ramser, J.H.: The truck dispatching problem. Manage. Sci. **6**, 80–91 (1959)
3. Belachgar, K.: Vehicle Routing Problem with Distance Constraints and Clustering. Al Akhawayn University, Morocco (2017)

4. Wen, M.: Rich vehicle routing problems and applications. Ph.D. Thesis, Technical University of Denmark, Denmark (2010)
5. ORION Backgrounder (2018). https://www.pressroom.ups.com/pressroom/ContentDetailsViewer.page?ConceptType=Factsheets&id=1426321616277-282
6. Karagul, K., Gungor, I.: A case study of heterogeneous fleet vehicle routing problem: touristic distribution application in Alanya. Int. J. Optim. Control: Theor. Appl. (IJOCTA) **4**(2), 67–76 (2014)
7. Ceschia, S., DiGaspero, L., Schaerf, A.: Tabu search techniques for the heterogeneous vehicle routing problem with time windows and carrier-dependent costs. J. Sched. **14**, 601–615 (2011)
8. Kritikos, M.N., Ioannou, G.: The heterogeneous fleet vehicle routing problem with overloads and time windows. Int. J. Prod. Econ. **144**(1), 68–75 (2013)
9. Belmecheri, F., Prins, C., Yalaoui, F., Amodeo, L.: Particle swarm optimization algorithm for a vehicle routing problem with heterogeneous fleet, mixed backhauls, and time windows. J. Intell. Manuf. **24**, 775–789 (2013)
10. Salhi, S., Wassan, N., Hajarat, M.: The fleet size and mix vehicle routing problem with backhauls: Formulation and set partitioning-based heuristics. Transp. Res. Part E: Logist. Transp. Rev. **56**, 22–35 (2013)
11. Pasha, U., Hoff, A., Hvattum, L.M.: The multi-period fleet size and mix vehicle routing problem with stochastic demands. In: European Congress on Computational Methods in Applied Sciences and Engineering, pp. 121–146. Springer, Cham, May 2015
12. Maheo, A., Urli, T., Kilby, P.: Fleet size and mix split-delivery vehicle routing. arXiv preprint arXiv:1612.01691 (2016)
13. Belfiore, P., Yoshizaki, H.T.: Heuristic methods for the fleet size and mix vehicle routing problem with time windows and split deliveries. Comput. Ind. Eng. **64**(2), 589–601 (1974)
14. Bertoli, F., Kilby, P., Urli, T.: Vehicle routing problems with deliveries split over days. J. Veh. Routing Algorithms **1**, 1–17 (2018)
15. Chu, J.C., Yan, S., Huang, H.J.: A multi-trip split-delivery vehicle routing problem with time windows for inventory replenishment under stochastic travel times. Netw. Spatial Econ. **17**, 41–68 (2017)
16. Bae, H., Moon, I.: Multi-depot vehicle routing problem with time windows considering delivery and installation vehicles. Appl. Math. Model. **40**(13–14), 6536–6549 (2016)
17. Laporte, G.: The vehicle routing problem: An overview of exact and approximate algorithms. Eur. J. Oper. Res. **59**(3), 345–358 (1992)
18. Amini, S., Javanshir, H., Tavakkoli-Moghaddam, R.: A PSO approach for solving VRPTW with real case study. Int. J. Recent Res. Appl. Stud. **4**, 118–126 (2010)
19. Onut, S., Kamber, M.R., Altay, G.: A heterogeneous fleet vehicle routing model for solving the LPG distribution problem: a case study. J. Phys.: Conf. Ser. **490**(1). IOP Publishing (2014)
20. Panapinun, K., Charnsethikul, P.: Vehicle Routing and Scheduling Problems: A Case Study of Food Distribution in Greater Bangkok. Kasetsart University, Bangkok (2005)
21. Worwa, K.: A case study in school transportation logistics. Res. Logistics Prod. **4**(1), 45–54 (2014)
22. Huang, Z.: Extensions to the *k*-means algorithm for clustering large data sets with categorical values. Data Min. Knowl. Discov. **2**(3), 283–304 (1998)

Developing a Spare Parts Demand Forecasting System

Elif Özbay, Banu Hacialioğlu, Büşra İlayda Dokuyucu, Hakan Şahin, Mehmet Mukan Saçlı, Merve Nur Genç, Efthymia Staiou(✉), and Mert Paldrak

Industrial Engineering, Yaşar University, Izmir, Turkey
eozbayelif@gmail.com, banu91001l@gmail.com, ilaydaadokuyucu@gmail.com, hakansahin1212@hotmail.com, mukansacli@gmail.com, mervengenc@gmail.com, {effi.staiou,mert.paldrak}@yasar.edu.tr

Abstract. The focus of this study is on developing a decision support system (DSS) in order to forecast spare parts demand for a company producing high technology products in Turkey. The company is one of the world's leading original design manufacturers in the field of consumer electronics and white goods. Accurate forecasts of customer demand for preliminary products and spare parts play an important role in order to reduce costs and increase customer satisfaction. Currently, the company's forecasting system is based on personnel experience and a statistical approach, which lacks the ability of capturing demand data behaviour. The approach followed results in an increased forecasting error, thus increases production costs, results in lack of spare parts, and decreases customer satisfaction. The aim of this project is to develop a DSS to minimize the forecasting error; therefore help the company develop a policy for optimizing the stock levels kept, reducing costs and increasing customer satisfaction. In order to understand the behaviour of customer demand of spare parts, the company's television products are chosen for the pilot study, since these products are highly influenced by rapid technological changes and changes in the product models. The spare parts are classified into different groups, using ABC analysis, in order to develop a forecasting model for each group. In the solution methodology part, three different statistical methodologies for the forecasting process were respectively studied; Winter's, Double Exponential Smoothing and Moving Average Methods. Winter's Method is used for the data which exhibit trend and seasonality, Double Exponential Smoothing is used for the data which exhibit trend and Moving Average Method is used for the data which exhibit stationary behaviour. In the DSS developed, the above-mentioned methodologies are coded, using Excel VBA programming language, historical data's behaviour is analysed and forecasts for future spare parts demand are made. The forecasting results are compared based on the minimum error (PAE), to decide upon which is the most appropriate forecasting methodology to use according to the specific spare parts past data behaviour.

Keywords: Forecasting · Decision Support System · Winter's Method · Exponential Smoothing Method · Spare parts

N. M. Durakbasa and M. G. Gençyılmaz (Eds.): ISPR 2019, LNME, pp. 676–691, 2020.
https://doi.org/10.1007/978-3-030-31343-2_58

1 Introduction

Spare parts forecasting is a vital issue for high technology production companies aiming to achieve high customer satisfaction. In this paper, a spare parts demand-forecasting problem is addressed, dealing with the Spare Parts Planning and Procurement (SPPP) Department of a company producing high technology products in Turkey. The main focus of the project is on the development of a Decision Support System that forecasts the amount of spare parts to be ordered, considering statistical methods according to historical data behaviour. In order to better analyse the behaviour of customer demand on spare parts, the company's television products are chosen for the pilot study, since these products are highly influenced by rapid technological changes and changes in the product models. Thus, the behaviour of customer demand on television spare parts is analysed according to the spare parts codes.

The company emphasises on customer satisfaction. Therefore, meeting the customer's requirement on spare parts on time is a primary goal. Although the legal period to meet customer demand is claimed to be 20 working-days, the company aims at meeting the customer demands within 3 days. Moreover, if the demand of a customer cannot be met within the legal period, the company is committed to provide a new product to satisfy the customer. Therefore, inventory management is a vital issue for the company. In addition to this, optimizing the safety stock levels will be easier with reliable demand estimations. Extra costs, such as inventory costs, may arise when the demand for spare parts is not forecasted accurately and customer satisfaction is adversely affected because the demand is not met on time. As a result, the spare parts' forecasting error needs to be minimized for any specific time.

In the following sections, the problem definition, literature review, modelling and solution methodology, verification and validation, a description of the Decision Support System developed, the results and the conclusions of the paper are presented.

2 Problem Definition

2.1 Current System

In the company, the Spare Parts Planning and Procurement (SPPP) Department is responsible for forecasting spare parts demands. This department works under Customer Services Directory (CSD). SPPP Department supports the Service Operations for all the products of the company. The workflow of the customer requesting a service process from the SPPP department is demonstrated in Fig. 1. In order to forecast the demand level of the following periods in the current system, the required data are obtained from the Enterprise Requirements Planning (ERP) system of the company and other Information Technology applications, such as SAP (Systems Applications and Products), CRM (Customer Relationship Management), and BI (Business Intelligence), which keeps the needed quantitative information about sales and purchases. The data, extracted in an Excel sheet, include the amount of orders delivered to the local Service points of the company. Using these data, SAP forecasts the demand of spare parts and offers an estimation for the following month of spare parts demand. Since this approach of SAP does not consider the statistical behaviour of spare parts demand, the responsible manager, for each product group, evaluates the forecasted values estimated using SAP and might change them according to his/her experience. In addition to the monthly forecasts, a safety stock exists, for covering a period of 20 days demand for each spare part. The stock level is checked continuously and when it goes under the safety stock level, SAP immediately notifies the department to order new spare parts.

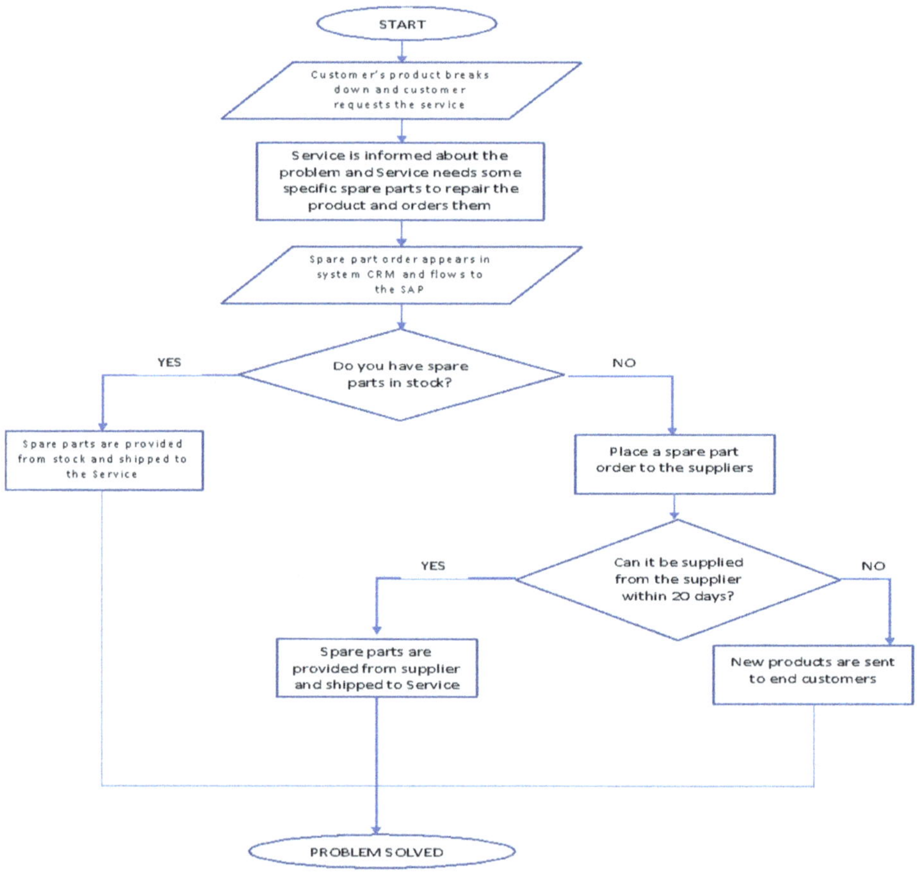

Fig. 1. SPPP department workflow

2.2 Observation and Symptoms about Current Forecasting System

The current forecasting process of SPPP Department combines experience with a statistical approach which does not take into consideration the data behaviour of the demand such as seasonality and trend. Therefore, the demand estimation may not be accurate as desired. It leads to forecasts that need a review by the personnel resulting in time delays. Furthermore, operational blindness may occur since they use experience to focus on the problem from one perspective. Because of the lack of the appropriate statistical methodology, forecasting error is undesirably high. Thus, surplus and shortage occurs which leads to increases in cost of keeping stock or decreases in customer satisfaction. After conducting an analysis on the current system, it is observed that the difference between spare parts requirements, from code to code, are considerably large. Under the analysis made, symptoms of the problem are summarized below.

Inaccurate forecasting systems which are mostly based on subjective factors may create bullwhip effect. Based on forecasted values, SPPP stocks spare parts by taking into consideration production plans. The company's current system results in a forecasting error higher than the desired level. When demand of spare parts is overestimated, the capacity of the warehouses will be set based on overestimated demand. This situation will cause unutilized capacity, inefficient production/purchasing of these spare parts, unnecessary manpower and excess stock which generates extra costs. As a result, the company will be able to satisfy the spare parts requirements of the customers, but the excess inventory of spare parts will be obsolete in the case of production of main products' is terminated when they go out of fashion. When they underestimate the demand for spare parts, the plant needs to change its capacity requirement production plans rapidly in order to satisfy the excess demand of the customer. Under this condition, bottlenecks may occur, and workers may have to work overtime. If the spare parts are replenished from other companies, the replenishment may take longer than expected and it may cause an additional transportation cost in order to supply necessary spare parts to prevent spare parts shortage. These symptoms require the usage of the appropriate statistical and scientific forecasting approaches.

2.3 Problem Definition

After the observation of the current system and symptoms, it is realised that the historical behaviour of the data is not taken into consideration in the selection of the statistical methodology used for forecasting spare parts demand. In addition, a decision support system, which can use different statistical approaches and minimize the error of forecasting does not exist. As a result, the accuracy of the existing forecasting approach is weak, resulting in an increased forecasting error. Furthermore, the absence of an accurate forecasting DSS results in surpluses and shortages in spares parts, which leads to increased stock keeping costs, decreased customer satisfaction and increased levels of renewing products.

The main target is to develop a decision support system with the most appropriate statistical methods, based on the demand data's structure, which will result in minimizing the forecasting error for each spare part.

Furthermore, more accurate forecasting will help the company to develop a policy for optimizing the stock levels kept, reducing costs and increasing customer satisfaction. The output of the decision support system will provide the future demand forecasted values, per spare part code, for a number of desired periods ahead.

The main success criterion for the DSS is the decrease in the forecasting error. As a result, the following success criteria can be reached as well:

- Achieving the target of 95% customer satisfaction,
- Increasing fill rate (rate of satisfied demand),
- Decreasing renewing rate (number of products renewed due to lack of spare parts).

3 Literature Review

For selecting the most appropriate forecasting methodology, considering the behaviour of the data (stationary, trend, seasonality) and aiming to minimize the error of estimation, a literature review was performed from many local and foreign sources. Similar studies on this subject were investigated, statistical time series forecasting methods and artificial neural network methods were examined.

Artificial Neural Networks is a process mechanism that works by learning and processing information like the human brain. In the literature review, there are many prediction studies using artificial neural networks. However, the duration of the learning of artificial neural networks is long and requires a lot of experimentation to train the network [1].

Forecasting is a tool used for predicting future demand based on past demand information. Forecasting methods are divided in two parts: subjective and objective forecasting methods. On one hand, subjective methods depend on either individual or group opinions. There are several techniques used for these methods such as sales force composites, customer surveys, jury of executive opinion and the Delphi method [2]. On the other hand, objective methods make forecasts based on historical behaviour of the data and a model. Objective methods are divided into two categories, namely Casual methods and Time Series methods. Casual methods use data from sources other than the series being predicted, while time series methods depend on past values of the variable being predicted. Time series methods are used to capture three types of data behaviours such as; stationarity, trend and seasonality [2].

For stationary time series there are two common methods for forecasting which are Moving Average and Single Exponential Smoothing. Moving Average is used to analyse data points by creating a series of averages of different subsets of the full data set. It simply takes the average of last n periods as the forecast of the next period. Averaging methods are suitable for stationary time series data where the series is in equilibrium around a constant value with a constant variance over time and they can be easily computed. Moving Average's disadvantage occurs when n is large enough to make forecasts less sensitive to changes. It does not forecast trendy data series well, requires much historical data and for quarterly data, a four-quarter moving average, MA (4), eliminates or averages out seasonal effects. Single Exponential Smoothing method only requires three pieces of data: the forecast for the most recent period, the actual value for that period and the value of the smoothing constant which is a weighting factor that reflects the weight given to the most recent data values. Moreover, observed data is the sum of two or more components and the random error which is the difference between the observed value and the exact value. However, Single Exponential Smoothing method neglects the random variation. In addition, forecasts are not accurate when seasonal or cyclical variations are present in the data [2].

According to Nahmias and Olsen [2], Linear Regression and Double Exponential Smoothing via Holt's method can be applied in the trend-based series. Linear regression is used to fit a straight line to a set of data. The aim of this method is to construct a model that predicts the dependent variable (Y) based on change of one or more independent variables (X). When applying this method to the forecasting problem, the independent variable corresponds to time and the dependent variable to the data to be forecasted. The model is shown as $Y = a + bX$. The values of a and b provide best fit of the data. Therefore, Least Square Method is used to compute the values of a and b. The disadvantages of this method are its complexity and its requirement of great historical data. Xie, Hong and Wohlin [3] points that, Double Exponential Smoothing technique is useful for predicting the number of failures in the next time period. However, sometimes it is more useful to provide an estimate of failure intensity. It can be calculated as the reciprocal of the number of the failures. According to Nahmias and Olsen [2], Double Exponential Smoothing designed to track time series with linear trend. The method requires specification of two smoothing constants, α for stationary, and β for the trend. There is no parameter to handle the seasonality.

According to Nahmias and Olsen [4], Decomposition and Winter's Methods are the methods used in estimation of seasonal and trendy series. There are 3 equations for Winter's Method and these equations have different smoothing constants which are α, β and γ. The constant of α is used for base level of time series, β for trend and γ for the seasonality. There are two steps in this method; the initialization and the updating step, respectively. After these steps are completed, final forecast is made in period t for any future period $t + T$. As the Winter's Method covers the seasonality and trend in the data, the number of formulas is high when comparing this method with other time series methods. This method has the advantage of being easily updated when new data are obtained. Classical Decomposition method is a process, which disassembles the components of time series, and it is being used as a starting procedure of most forecasting methods. Decomposition method formulates the time series as a function of 4 components [5]. The main advantage of classical decomposition is that it is simple and clear to apply, and it provides early detection of the components. The serious difficulty of the method occurs when seasonal indexes are low which causes error to increase because of the appliance of the procedure.

Croston [6] and Box, Jenkins and Reinsel [7] (ARIMA) methods are proper to forecast the intermittent data. The two most common linear time series models are Autoregressive (AR) and Moving Average (MA) Models. The statistical combination of these two models generates Autoregressive Moving Average (ARMA) and Autoregressive Integrated Moving Average (ARIMA). ARMA (p, q), where p is AR and q are MA, is mostly appropriate for univariate time series. ARIMA (p, d, q) where p is AR, d is differencing, and q is MA, is the extended version of ARMA which also includes non-stationarity [8].

4 Modelling and Solution Methodology

4.1 Data Analysis

The company provides its customers a wide range of products and supplies spare parts for these products. The company provided the amount of spare parts data related to the TV products. Data cover a period of 42 months, 166 weeks between years 2016–2019. In the data there are spare parts orders received from the services during a period of 3.5 years, 12.276 different spare part types, 52 different product groups (video card, led bar, accessories etc.). The reason why there are 52 different product groups for TV spare parts is that television is a product with advanced technology. Due to developing technology and rapidly changing trends, product variety and thus the number of spare parts is constantly increasing. It was also observed that this situation increased the complexity of the data.

In order to make the data useful and understandable, the data were classified, extracted and analysed using ABC analysis according to the demand level. The spare parts that belong to the first 20% of the total number of orders are called A class, following 30% and 50% are called B and C classes respectively. As a second categorization, for 42 months, spare parts orders from services are classified according to their characteristics. Spare parts are divided into 5 groups as; New-born, dead, inactive, rare and regular demanded spare parts.

The names of the groups and the conditions of classifications are shown in Table 1. In addition, ABC analysis results are shown in Table 2.

Table 1. Explanation of classification groups

Class	Condition	Quantities
Dead	Less than 50 demand in 3 years	9822
New-born	No demand before the January 2017	456
Inactive	No demand after the December 2017	116
Rare	No demand Minimum 15 weeks in total 156 week	1553
Regular	There is demand in every week	99

Table 2. ABC analysis

Class	Number of spare parts
A	8
B	75
C	12193

The company has recorded 4 different categories of demand for spare parts. These categories are called “Sales Document Type”. The first type of document is listed as “orders opened by the service to address customer demand directly”, and secondly “orders opened for spare parts in services’ own stock”. The third and fourth document types have been created to record the spare parts orders that were made when there are discounts in the periodic spare part prices. The company offers discounts its spare parts prices twice a year in May-June and September-October. The spare part orders in the first and second order document types show mostly a continuous and regular distribution of data. However, in the third and fourth types, there is an intermittent distribution of data.

4.2 Modelling and Formulation

In this project, the behaviour of historical customer demand data is examined, and trend is observed in most of document type 1 sales. Document type 2 sales are nearly all stationary. Moreover, it is deduced that some of them also behave seasonally. With this information, it is decided to model Double Exponential Smoothing for demand with trend but no seasonality, Winter's Method for demand with seasonality and Moving Average for data that is stationary. These three methods are clarified in detail in the following section.

(1) Winter's Method: This method is used when the data have seasonality and trend. There are 3 equations for this method and these equations may have different smoothing constants which are α, β and γ; α for base level of time series, β for trend and γ for the seasonality. This method is divided into two parts. These parts are respectively; initialization of starting values and updating these values period by period.
For the initialization step, summation of actual values (V_i) for each season i, is calculated to identify the relationship between seasons. According to V_i, initial values G_O, S_O and c_t are calculated using the below formulations. G_O represent initial trend, S_O represents the level at a current time and c_i represent seasonal factors for each season i. Formulas are as follows;

$$G_0 = \frac{(V_m - V_1)}{(m * m)}$$

$$S_0 = \frac{V_m}{m}$$

$$c_i = \left(\frac{D_i}{S_0}\right), \; \forall i < m \; and \; i < 0$$

For updating step, St, Gt and ct, trend, level and seasonal index values at period t respectively, should be calculated.

$$S_t = a\left(\frac{D_t}{c_{t-n}}\right) + (1-a)(S_{t-1} + G_{t-1})$$

$$G_t = \beta(S_t - S_{t-1}) + (1-\beta)(G_{t-1})$$

$$c_t = \gamma\left(\frac{D_t}{S_t}\right) + (1-\gamma)c_{t-n}$$

After calculating St, Gt and ct, the forecasted value, Ft can be calculated in period t for any future period t + T.

$$F_t = S_t + kG_t + c_{t+T}$$

(2) Double Exponential Smoothing Method: Double Exponential Smoothing Method is used when data are trendy. The difference of this method from Winter's Method

is that it does not consider the seasonality. The formulations are shown below. The parameters; L_t is for level, a is the smoothing constant for level, T_t is for trend, γ is the smoothing constant for trend, Y_t is the actual value, F_t is the forecasted value and k is the number of steps to forecast ahead, at time t.

$$L_t = \alpha Y_t + (1 - \alpha)[L_{t-1} + T_{t-1}]$$

$$T_t = \gamma[L_t - L_{t-1}] + (1 - \gamma)T_{t-1}$$

$$F_t = L_{t-1} + k * T_{t-1}$$

(3) Moving Average: The forecast for the next period is the average of the last n observations of the time series. Moving Average method is used if there is no or negligible amount of trend. The formulation is given below, where n is the number of periods to average and D_i is the demand level in period i.

$$F_i = \frac{\sum_{j=1}^{n} D_{i-n}}{n}$$

(4) Error Calculation: The difference between the forecast and actual values is referred to as the forecast error. Thus, one requirement for a good forecast is that its forecast error should be close to zero. For this reason, we measure the accuracy of forecasts over time. There are many ways of measuring the accuracy of forecasts such as; Mean Absolute Deviation (MAD), The Mean Squared Error (MSE), The Root Mean Square Error (RMSE), The Mean Absolute Percentage Error (MAPE) and The Percentage Absolute Error (PAE). Mean absolute deviation (MAD) and Mean squared error (MSE) are two common measures of forecast accuracy during n periods. Mean absolute percentage error (MAPE) gives the average percentage error.

In this project, PAE (Percentage Absolute Error) is used as an error measurement. When MAPE was chosen as a criterion, it gave undefined results if zero demand occurs in a period because its formulation contains demand level as a denominator. In the PAE formulation, the sum of the differences between demand and forecast is divided by the sum of demands. For this reason, the only failure that may occur is the situation where all the actual data consist of zero demands, then PAE results in a division by zero failure [9]. The formulation of PAE is shown below where Demand_i stands for demand at period i, Forecast_i for forecasted value at period i and N for total number of periods;

$$PAE = 100 * \frac{\sum_{i=1}^{N} |Demand_i - Forecast_i|}{\sum_{i=1}^{N} Demand_i}$$

In addition to these, α, β and γ parameters have an influence on the results. Because of this, when the DSS system is running, these parameters are optimized to increase the accuracy of forecasts. Usually, in literature, this optimization is being made only once for all the data, but in this case, it ignores the possible changes in the parameters' values over time. In order to avoid this, in our study the data are divided into parts with cycle/2 data points in them, where cycle is defined as the complete set of data with all seasons included, for example a year is a cycle with 4 seasons included. After that, each part of data can have its localized parameters α, β and γ and Excel Solver works to minimize PAE by changing the values of those local α, β and γ parameters for each part of data. Rather than optimizing parameters only once, the parameters are optimized locally in every half cycle in order to react to behaviour changes. This responsiveness increases the accuracy of forecasts.

5 Verification and Validation

After choosing the appropriate statistical methods to be applied in the DSS, Winter's, Double Exponential Smoothing and Moving Average Methods were coded in Excel VBA. The results of the DSS when applying all three methods for forecasting, using sample sales data, were compared with the outputs of Minitab. It is observed that the results were very close to each other and therefore the model of the DSS is validated.

As seen in Figs. 2, 3 and 4, Winter's Method with seasonal data, Double Exponential Smoothing with trendy data and Moving Average with stationary data, respectively, caught the behaviour of the data and gave good results as planned.

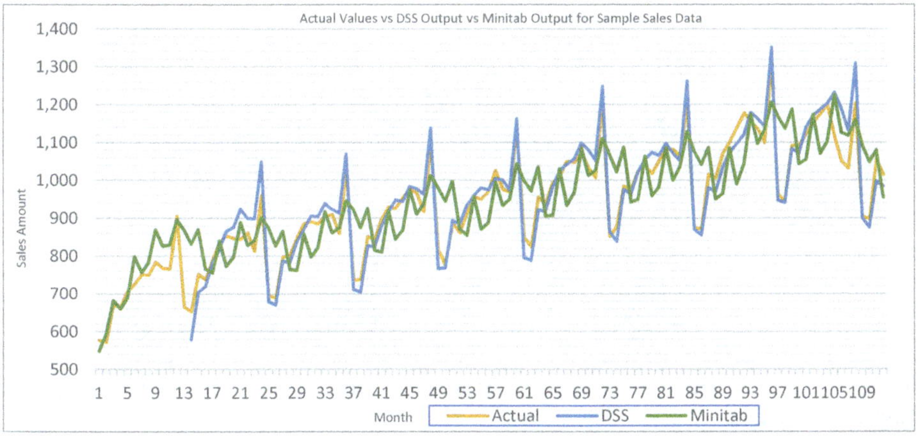

Fig. 2. Winter's method comparison with actual data, DSS output and minitab output for sample sales data

For Winter's Method, DSS forecasted with 3.00% PAE and Minitab forecasted with 8.07% PAE. For Double Exponential Smoothing Method, DSS forecasted with 8.06% PAE and Minitab forecasted with 7.98% PAE. Finally, for Moving Average Method, DSS forecasted with 8.48% PAE and Minitab forecasted with 8.48% PAE. Each parameter, α, β and γ, in Winter's and in Double Exponential Smoothing Method, are set equal to 0.2 respectively and n is set equal to 2 in the Moving Average Method.

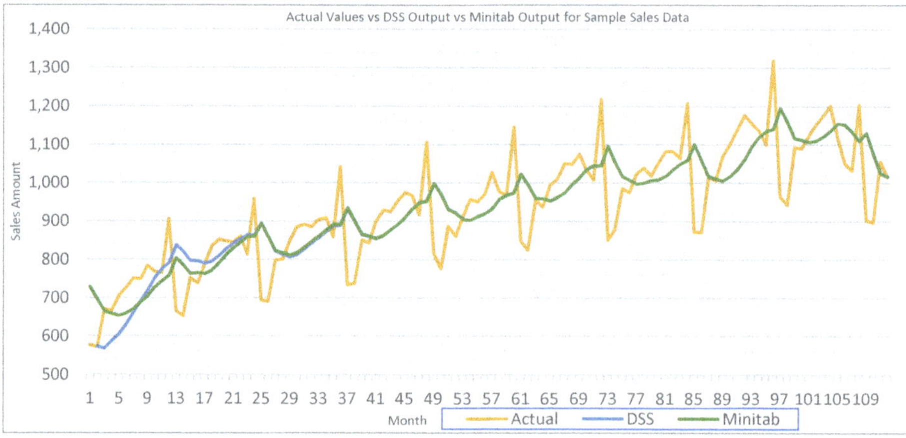

Fig. 3. Double Exponential Smoothing Method comparison with actual data, DSS output and minitab output for sample sales data

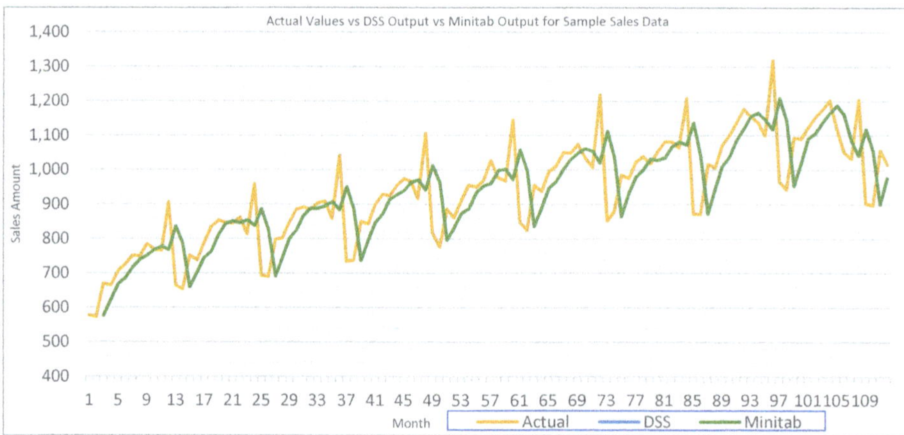

Fig. 4. Moving average method comparison with actual data, DSS output and minitab output for sample sales data

The company provided spare parts demand historical data for 166 weeks' (between years 2016–2019). Since the data for the rare, new-born, inactive and dead classes were not enough for the model to recognize and analyse their behaviour, forecasting for these classes could not be implemented. The DSS is used for forecasting spare parts demand only for the regular class. Finally, the PAE of the current forecasting system used by the company and our project's PAE were compared.

The DSS used for its analysis data of a period of 3.5 years. The first year was used for the initialization part and the remaining years were used for the updating part.

In Table 3, the PAE of the forecasts for a sample of nine spare parts codes are illustrated, for sales document type 1, presenting the PAE results when Winter's,

Double Exponential Smoothing and Moving Average methods are used respectively, highlighting the method giving the smallest error. In Table 4, the same results are presented for sales document type 2.

Table 3. Sales document type 1 results

Sales Document Type 1			
Material Code	Winter's Method PAE	Double Exponential Smoothing PAE	Moving Average (Company Forecast) PAE
1	10.98	11.69	73.75
2	9.67	10.6	64.37
3	14.08	15.54	70.39
4	21.32	16.19	77.65
5	25.18	22.61	70.75
6	4.48	3.79	77.19
7	46.04	28.27	17.87
8	31.24	26.53	16.79
9	49.89	44.26	26.14

Table 4. Sales document type 2 results

Sales Document Type 2			
Material Code	Winter's Method PAE	Double Exponential Smoothing PAE	Moving Average (Company Forecast) PAE
1	43.98	47.46	46,83
2	25.47	29.2	76,84
3	19.32	12.1	100
4	46.81	30.77	88,62
5	38.53	40.16	24,94
6	51.24	49.82	27,76

As seen in the results presented above, depending on the spare parts demand behaviour, different methods give the smallest error values. When examined in more detail, it was observed that for some material codes the seasonality of the year that was used in the initialization part of the Winter's Method is not similar with the one appearing in the following years. In these cases, Winter's Method resulted in an increased PAE compared to the methods that are less complex.

6 Decision Support System

This project was carried out to create a DSS for forecasting spare parts demand, by using the appropriate statistics methodology and by minimizing the forecasting error. In the DSS, three forecasting methodologies; Winters's Method, Double Exponential Smoothing and Moving Average Methods, are used for making the spare part demand forecasts and are modelled in Microsoft Excel VBA language. The forecasts can be made weekly and monthly according to the choice of the user. In addition, a simple and user-friendly interface is prepared in Excel. The homepage of the system is as shown in Fig. 5.

Fig. 5. Homepage of the DSS

When the program starts, the homepage appears and the user can choose one of the operations, by clicking on the buttons located at the right top corner: "CATEGORIZATION", "ABC ANALYSIS" and "FORECAST". If the user chooses one of the first two operations, a separate spreadsheet is opened containing the results of the report.

When the user selects the "FORECAST" operation, the user form shown in Fig. 6 appears. The user can enter the material code and the time period for which she/he wants to forecast; the program retrieves the historical data of the selected spare part from Microsoft Access Database and makes an accurate forecast with all three methods.

The results of the 3 methods are presented in the user form shown in Fig. 7. After that, the user can make use of the forecasted value with the least PAE.

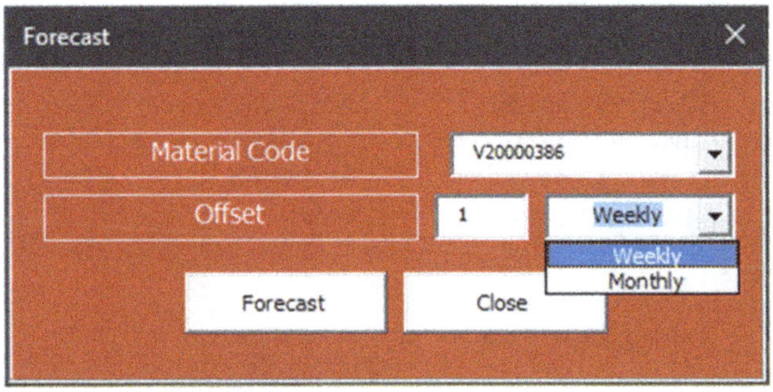

Fig. 6. Forecasting page

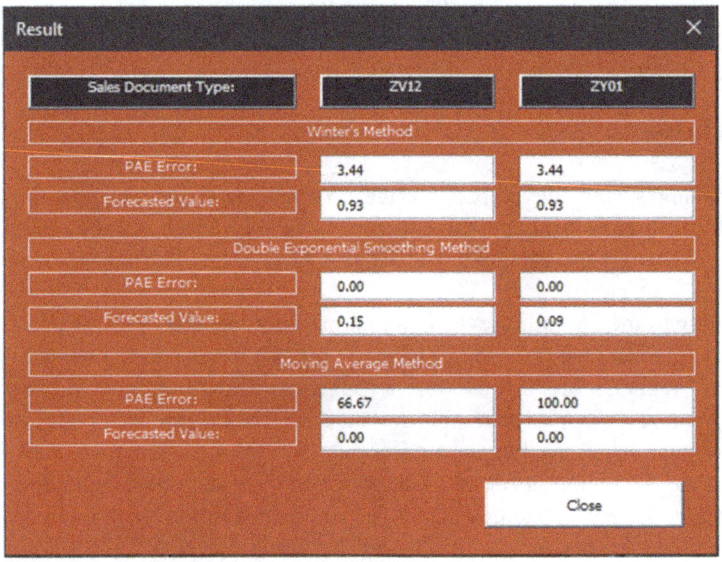

Fig. 7. Result page

7 Results

After modelling and developing the DSS which uses the three methods described above (Winter's Method, Double Exponential Smoothing and Moving Average), 99 regular parts are forecasted with weekly data, and the results are recorded. It took around 40 min for the DSS to run and calculate all forecasted values for all spare part codes.

For sales document type 1 as shown in Table 5, Winter's and Double Exponential Smoothing Methods gave better results for most of the regular spare parts and, led to an improvement rate of 47.97% on the average. For sales document type 2 as shown in Table 6, because of the stationary behaviour of the data, Moving Average method gave better results for most of the spare parts; thereby it led to an improvement rate of 3.9% on the average.

Table 5. Error calculations for sales document type 1

Sales document type 1						
Class	Number of spare parts	Winter's method PAE	Double exponential smoothing PAE	Moving average (company forecast) PAE	DSS output PAE	Improvement percentage
Regular	99	44.14	45.9	64.42	33.52	47.97

Table 6. Error calculations for sales document type 2

Sales document type 2						
Class	Number of spare parts	Winter's method PAE	Double exponential smoothing PAE	Moving average (company forecast) PAE	DSS output PAE	Improvement percentage
Regular	99	43.33	42.44	25.84	24.83	3.9

8 Conclusion

The forecasting system developed for the company is based on detecting the most appropriate methodology to be used according to the data behaviour of the specific spare part. To solve the problem, historical data is received from the company and analysed.

According to the literature review and taking into consideration the analysis of data behaviour, it is decided to use 3 methods for the forecasting process. These are respectively, Winter's Method, Double Exponential Smoothing and Moving Average. Winter's Method is appropriate for the data which exhibit trend and seasonality, Double Exponential Smoothing is appropriate for the data which exhibit trend and Moving Average Method is for the data which exhibit stationary behaviour.

A Decision Support System is developed to forecast future spare parts demand. The above-mentioned methodologies are coded by using Excel VBA programming language and compared based on the error to decide upon which is the most appropriate one to use.

To measure the performance of the models, PAE error calculation is used. For optimizing parameters, PAE output is used as an objective function and the parameters are used as decision variables. Excel Solver tool gives different values to the parameters in forecasting process to minimize PAE.

Three years data are provided by the company for this project. The data received are classified according to their demand structure. After that, it is seen that the Decision Support System cannot work properly for all classes. Because of that, in this project, only regular class is examined. The project results in forecasting accuracy improvements up to an average rate of 47.97%, for example in regular class spare parts of sales document type 1. As time progresses, the amount of data used by the DSS will increase. In this way, the forecast results are expected to improve.

Additionally, the data which are used for the forecasting process are kept on and retrieved from Microsoft Access Database, resulting in dynamic forecasting. Thus, data recording is made in a more regular and sustainable way and VBA's efficiency and working speed has increased.

Furthermore, although only television data have been examined in this study, since the pilot product of this project is television, the Decision Support System is ready and can be used for all the other products of the company.

References

1. Karahan, M.: İstatistiksel Tahmin Yöntemleri: Yapay Sinir Ağları Metodu ile Ürün Talep Tahmini Uygulaması. Doktora tezi, Selçuk Üniversitesi (2011)
2. Nahmias, S., Olsen, T.L.: Production and Operations Analysis, 7th edn. WaveLand Press, Long Grove (2015)
3. Xie, M., Hong, G.Y., Wohlin, C.: A study of exponential smoothing technique in software reliability growth prediction. Qual. Reliab. Eng. Int. **13**, 247–353 (1997)
4. Nahmias, S., Olsen, T.L.: Winters's method for seasonal problems. In: Nahmias, S., Olsen, T. L. (eds.) Production and Operation Analysis, pp. 85–86. Waveland Press, Long Grove (2015)
5. Hyndman, R.J., Athanasopoulos, G.: Forecasting: Principles And Practice, 2nd edn. Monash University, Melbourne (2018)
6. Croston, J.: Forecasting and stock control for intermittent demands. Oper. Res. Q. **23**(3), 289–303 (1972)
7. Box, G.E.P., Jenkins, G.M., Reinsel, G.C.: Time Series Analysis; Forecasting and Control, 3rd edn. Prentice Hall, Englewood Cliffs (1994)
8. Montgomery, D., Jennings, C.L., Kulahci, M.: Introduction to Time Series Analysis and Forecasting. Wiley, New York (2008)
9. Adalı, E., Ataş, Y., Baykan, O.M., Güldoğan, İ., Koral, E.: Son Ürün Talep Takip Ve Tahminleme Projesi. Lıcense thesıs, Project Yasar Unıversıty (2015)

Linking Container Shipping Company Service Attributes to Shipper Satisfaction and Loyalty

Gülümser Gür[1], Ecem Ervansel[1], Ege Taylan Er[1], Aylin Caliskan[2], and Yucel Ozturkoglu[1(✉)]

[1] International Logistics Management, Yaşar University, Izmir, Turkey
gulumser.gur@hotmail.com, ecemrvnsl@gmail.com, egetaylaner@gmail.com, yucel.ozturkoglu@yasar.edu.tr
[2] Transportation Services, Yaşar University, Izmir, Turkey
aylin.caliskan@yasar.edu.tr

Abstract. In container shipping industry, due to the strategic alliances and vessel sharing, the services given to the customers have become the same each other. High competition remains but differentiation among carriers now totally depends on understanding customers and getting their loyalties. Therefore, the aim of this study is to find how and which service attributes of a container shipping company affect the shipper satisfaction and loyalty. Additionally, because there are thousands of shippers with different characteristics, such as product, organizational structure, trade region, and capacity, in this study it is also aimed to find the basic differences among shippers toward container shipping service attributes. Structural equation modelling (SEM) applied to assess data collected by a questionnaire survey. Data assessment achieved through hypotheses testing and with the help of some other statistical analyses such as ANOVA and T-Test. In this study, the factors affecting the choice of carriers were determined by using the literature. Factors are divided into sub-headings such as rates, operation, and customer service. Data analyses were conducted by using LISREL and SPSS.

Keywords: Customer satisfaction · Customer loyalty · Structural equation modelling · B2B marketing · Container shipping

1 Introduction

Globalization today affects all business and supply chains and gives value and importance to the logistics sector and services. The two main components of globalization are international ocean transportation and information and communication technologies [1]. Shipping, a global service industry, is the lifeblood of international trade. Thanks to shipping, intercontinental trade, bulk transport of raw materials and import/export of affordable food and finished goods can be made. According to some research, 90% of international trade is provided by seaway [2]. Container lines have an important place in international trade and global supply chains. Due to the density in the sector, container lines face high competition and overcapacity problems. These problems complicate the work of the companies carrying the container and ensure that

N. M. Durakbasa and M. G. Gençyılmaz (Eds.): ISPR 2019, LNME, pp. 692–704, 2020.
https://doi.org/10.1007/978-3-030-31343-2_59

the companies are more customers focused. In order to survive in this competitive environment, the quality of transportation must be ensured [3]. According to Ting and Tzeng [4], optimal capacity and route planning are also used in various capacity utilization methods.

With the standard of services provided by the companies, it became more important to understand the characteristics and needs of the customers. Besides, carriers have begun to adopt new service offerings, such as point-to point transportation [5]. In addition to increasing competition, high cost of the sector and the demands of the carriers are the other reasons that make difficult the work of the container transportation companies. The slowdown in global economic growth since 2009 has affected the business environment of container carriers [6]. Since 2016, the container shipping market has been stagnant due to low freight rates [6]. As a solution to these problems, container lines began to use bigger vessels, and initiated cooperation to reduce costs and use business resources more efficiently [3]. The capital intensive structure of the sector, which is one of the problems faced by the container lines, has led to restrictive business practices such as alliances [2]. In some parts of the world, strategic alliances have replaced liner conferences, where it is not allowed to allocate carriers or determine freight rates [7].

Logistics companies should provide a service to meet their own profitability and the needs of their customers at the same time and in addition to long-term customer relations and efficient management of profitable customers is also necessary. Since the container lines are based on standard service, they have limited opportunities to allocate their main services from port to port. While the carriers compete against their strategic partners, Balcı [3] believes that the basic product characteristics such as transit time, sailing date and damage performance have become the same. It is important to understand the characteristics and needs of the customers, to develop new services, and to create differentiation in order to maintain customer relations. There exists a variety of shippers with different characteristics, such as cargo, demographics, capacity, or different targets or service requirements. Furthermore, the demand for shippers is different from each other, so if it is not carefully planned, marketing efforts created to meet the needs of the customer cannot exceed any profit from marketing.

The purpose of this study is to find out how container shipping companies service characteristics affect shipper satisfaction and loyalty. Also in this study, it is aimed to find the main differences between the shippers regarding the container transport service qualities because there are thousands of shippers with different characteristics such as product, organizational structure, trade zone, and capacity.

2 Literature Review

2.1 Customer Loyalty in Container Shipping Industry

Indicates that more customers will be retained in the future when the loyalty of existing customers increases. If a company's customer loyalty is strong, it reflects on the firm's economic returns because it provides a stable cash flow in the future [8].

There are important points for a customer to purchase a product or service again. One can be defined as the customer's satisfaction with a certain discount or offer, and another as the emotional commitment of a customer to the product or service. Businesses are more interested in customer loyalty efforts. Connecting customers to a business and continuing to prefer that business creates a competitive advantage for the business. Because the cost of retaining the old customer is higher than the cost of acquiring new customers [9].

According to researches, the cost of acquiring a new customer is 6 times more than the cost of protection of the old customer [10]. The increase in the percentage of loyal customers increases the profitability of the firm. Hançer [11], an increase of 5% in customer loyalty indicates that the profit increased from 25% to 80%.

2.2 The Relationship Between Customer Satisfaction and Customer Loyalty

In the sector where there is intense competition, companies need to keep their existing customers and create customer loyalty. The relationship between customer satisfaction and loyalty has been an important topic in many studies. Thanks to the literature review, it was revealed that customer satisfaction was the main factor in ensuring customer loyalty. There are other factors affecting customer loyalty. According to the European Customer Satisfaction Index Model, these factors are determined as the business image, customer expectations, perceived service quality, perceived product quality, perceived value and customer satisfaction [10].

Customer loyalty to the company is the most important contribution of customers who are satisfied. Other contributions;

- Goes to purchase more products.
- It also buys from other products produced by the company.
- Positive thoughts about the products produced by the enterprise and the business contribute to the positive image of the business.
- It is less sensitive to the brands and products of competitors [12].

Measuring the level of customer satisfaction is one of the issues that businesses emphasize. There are many different methods to measure this. In this study, the survey method will be used to measure the level of satisfaction.

2.3 Customer Value and Customer Retention

Customer satisfaction is the result of the evaluation between the expectations before the purchase and the outputs obtained after the purchase [13]. Customer satisfaction or dissatisfaction occurs according to the positive, negative or equivalent of this difference. Customer satisfaction and customer value are different from each other and important concepts for customer relationship management. Customer value is the result of pre-purchase and post-purchase evaluations and preferences of target customers. On the other hand, customer satisfaction occurs after purchase. Customer value requires a strategic approach to determine the difference between service and expectation, while customer satisfaction requires a tactical focus on good service delivery [14]. Customer value, which is of great importance in marketing, increases with quality and service but falls with the price [15].

Customer retention is the customer's desire to maintain business relations with existing companies [15]. Customer retention is more important than acquiring new customers because research shows that the cost of acquiring a new customer is about 5 times more than the cost of holding a current customer [15]. Customer retention has a direct impact on customer loyalty. Customer retention has a direct impact on customer loyalty. The greater the loyalty to the customer, the easier it is for the customer to retain the customer.

3 Problem Statement

Customer satisfaction in the sector has been an important issue with the intensity of competition in the service sector. As the customer becomes the most important link in the value chain, the roles within the distribution channel have changed and become the most important element of the customer distribution channel. Firms should make better understand their customers, differentiate their needs and differentiate their services, or develop new services because customer satisfaction is the most important issue that will enable the company to differentiate and take place in the logistics sector.

Many logistics firms face problems in order to take place in the sector. The biggest problem to satisfy the company's customer is the inability to find empty containers. Normally, there are several options to find empty containers. One of these options is the export-based return of empty containers to the export area. For example, an export container going to England is full of return to Izmir. This helps to eliminate trade imbalances. The company does not prefer these methods because the capacity of a firm on a ship is limited, so different shipping companies share ships. Since this limited space is usually loaded with full containers, there is no room for empty containers. The other option is to provide empty containers from different destinations. In particular, empty containers can be brought from Egypt, but collecting empty containers from different destinations is costly. Another option, leasing a container, is not an easy solution because of the constraints such as the target and time period brought by leasing companies. Lack of equipment can be removed by buying a new container but buying a new container cannot solve the problem permanently because the new container will go to the destination and stay there again. Moreover, stocking containers in warehouses are another method of finding equipment. This method is also not preferred because a container must be exported before the expiration date and the container has a maximum storage period of 6 months. Therefore, the possibility of storage is limited.

The reason why companies attach importance to the empty container issue is that they cannot meet customer needs due to this issue. New customers and existing customers will help the company to find a burden, maintain its burden, generate revenue and fill its place on board. When the customer is in such an important position, the firm has to reject the demands of some customers due to problems. This creates customer dissatisfaction. In recent years, with the similarity of the company services in the sector, customer satisfaction and loyalty has become an important element to highlight the firms in the competitive environment.

4 Methodology

Structural equation modelling is a multivariate statistical analysis technique that is used to analyze structural relationships. This strategy is the mix of factor investigation and numerous relapse examinations, and it is utilized to dissect the auxiliary connection between estimated factors and dormant builds. This method preferred because it estimates the multiple and interrelated dependence in a single analysis.

As mentioned earlier, the factors that are effective in the carrier selection of the carrier were first determined by literature review. Factor list is as in Table 1 and listed under 7 main headings. Then, the hypotheses needed to develop the model were created by the determined variables. With the help of hypotheses, the model was determined by determining the relationships with customer satisfaction and loyalty. At the same time, factors have helped to create surveys. Thanks to the survey, the relationships of the different customer groups with the company, their loyalty to the company, their demands and demands from the company have been revealed. By analysing the results of the survey, customer satisfaction and customer loyalty will be measured.

When using this model, the target is to protect existing customers, ensure customer satisfaction and increase customer loyalty. As a result, the purpose of the research is to link the service characteristics of a container transport company with customer satisfaction and loyalty.

[16] and [17] found in their study that there is a positive relationship between rate and customer satisfaction. Similarly, [18] and [19] indicated the relationships between rate and customer satisfaction. Thus, Fig. 1 shows the following hypothesis in this study.

H1 :Rate is positively associated with the customer satisfaction.

H2 :Rate is positively associated with the customer loyalty.

H3 :Operation is positively associated with the customer satisfaction.

H4 :Operation is positively associated with the customer loyalty.

H5 :Value added is positively associated with the customer satisfaction.

H6 :Value added is positively associated with the customer loyalty.

H7 :IT is positively associated with the customer satisfaction.

H8 :IT is positively associated with the customer loyalty.

H9 :Scheduling is positively associated with the customer satisfaction.

H10 :Scheduling is positively associated with the customer loyalty.

H11:Customer service is exactly connected with the customer satisfaction.

H12 :Customer service is exactly connected with the customer loyalty.

H13 :Other services is positively associated with the customer satisfaction.

H14 :Other services is positively associated with the customer loyalty.

H15: Customer satisfaction is positively associated with the customer loyalty.

Fig. 1. Hypotheses

4.1 Population and Sample

The survey was applied to companies using container lines as envisaged. The field study was planned according to the return received and the survey was conducted by the company employees in different districts of İzmir. Areas of application: transport, international transport, hydraulic-electricity, construction, electricity, electricity, marble, food, textile, brush-sanding, chemistry, metal, mining. As was targeted, answers were received not only from Izmir but also from various provinces. The target is to get close to 200 questionnaire responses. 144 responses were obtained and the data required for the analysis were collected.

4.2 Survey Development and Data Collection Process

Customer satisfaction in the sector has become an important issue with the intensity of competition in the service sector. The customer has become the most important link in the value chain and the roles in the distribution channel have changed, becoming the most important element of the customer distribution channel. In the literature surveys, problems in the sector were revealed and the variables affecting customer satisfaction were determined and a questionnaire was formed within the framework of these variables.

The research area of this survey, which is based on the service quality of the container lines, is the companies that use container lines. In order to determine the service quality of container lines, it is aimed to reach many companies using this line. These companies are planned to be from different sectors and from different business dimensions. In addition, each company has different expectations from the cooperation line. This variability has expanded the field of research and included not only Izmir but also different cities such as Istanbul and Mersin. Every place or every customer has different needs/expectations. While visiting companies, it was targeted to conduct interviews with the export managers/sales managers of the firms to collect the data and these data formed the framework of the project. Before applying the project, companies in the sector were searched and a list was created. The sectors planned to be implemented are transport companies, food and textile, automotive, construction, chemistry. The survey was applied to companies using container lines as envisaged. After the planning of the companies to be discussed, mail was sent first. The field study was planned according to the return received and the survey was conducted by the company employees in different districts of İzmir. The questionnaire was sent by e-mail and was made by one-to-one interviews with company visits. 144 responses were obtained and the data required for the analysis were collected.

After the data were collected, the analysis phase was started. Statistical analysis for social sciences (SPSS) was used to make analysis. Initially, the data were organized and prepared for analysis. Data were analyzed by factor analysis and the results were interpreted. The predicted research model was created as a result of the analysis. Data averages and standard deviations were calculated. The frequency distributions of the service characteristics in the survey were analyzed. As a result, the purpose of the research is to link the service characteristics of a container transport company with customer satisfaction and loyalty.

Pre-Testing of the survey: The preliminary test was conducted in order to be able to give analysisable data. Pre-test is composed of two stages as qualitative and quantitative. A focus group was formed from the companies in the qualitative test. There are 20 companies in the determined group. As a result of these interviews, some questions were eliminated from the survey. After the questions were eliminated, the questionnaire was continued and factor analysis was performed after 50 responses were obtained. As a result of factor analysis, some questions were eliminated.

5 Findings

5.1 Descriptive Statistics

According to the results, 60% of the responses belong to the freight forwarders and the rest is manufacturing companies. This result is fare because most of the customers of a liner shipping company is freight forwarders. 42% of the respondents have managerial level positions. 35% of the companies' ages are between 10–24 years and 31% of them 1–9 years. Respondents indicated that they mostly (42%) use MSC as a container shipping company. Middle East constitutes the 34% of cargo destination while Northern Europe constitutes 28%.

5.2 Factor Analysis

After applying factor analysis, we obtained 3 important values; Kaiser Meyer Olkin (KMO) and Bartlett's test, Total Variance Explained and Rotated Component Matrix. KMO and Bartlett's test are prerequisites for making a factor analysis. KMO explains whether the data set is sufficient for analysis. The value of KMO should be greater than 0.50. As this ratio increases, the data set is getting better for making factor analysis. At the following table KMO score of our factor analysis is 0,792 and the sample size was found sufficient for making factor analysis. Before the factor analysis there were 30 variables. To obtain the appropriate factor analysis, varimax was selected as the rotation method and 40 iterations was selected for convergence. After the factor analysis was used 23 variables remained. Accordingly, 7 factors were obtained and these factors explained 69,898 of the total variance and cumulative percentage were found as 69,898. While the factor analysis was improved, the interest variable, the eco friendly variable, the damage rate variable, the direct service variable, the container condition variable, the customs work variable and the repayment variable were extracted. Because these variables lied heavy on either on more than one factor or low. As a result of the factor analysis, our foreseen research model has become the following as seen in the Fig. 2.

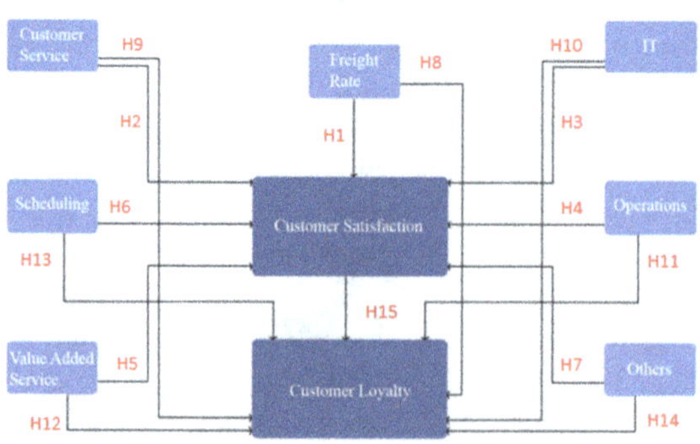

Fig. 2. Research model

5.3 ANOVA and T-Test

ANOVA test which is used for three or more variables and t test table which is used for two independent variables were used to find out if survey results are significant. In other words, they helped us to figure out if we need to reject the null hypothesis or accept the alternate hypothesis. Significant differences between the averages of groups were found out in the six questions of first part of survey.

The factors, which the firms in different cities were considering while evaluating the ship line, were evaluated. The evaluation that is for schedule performance of ship line differs according to the city where is located of the firm participating in the survey ($F = 2.729$, $p = 0.016$). In other words, the Ho hypothesis that is for one-way analysis of variance was rejected because the significance value of schedule is less than 0.05. For this reason, there is a statistically significant difference between the averages of schedule groups with 95% confidence. The evaluation that is for loyalty of firm against ship line differs according to the year that the firm which participating in the survey has been operating ($F = 3.023$, $p = 0.020$). For customer service performance of ship line differs according to the annual TEU amount of firms participating in the survey ($F = 3.071$, $p = 0.012$). The Ho hypothesis that is for one-way analysis of variance was rejected because the significance value of customer service is less than 0.05.

5.4 Confirmatory Factor Analysis (CFA)

Confirmatory factor analysis (CFA) is a multivariate factual method that is utilized to test how well the deliberate factors speak to the quantity of builds. In Confirmatory factor analysis (CFA), analysts can indicate the quantity of elements required in the information and which estimated variable is identified with which inactive variable.

Confirmatory factor analysis (CFA) is an apparatus that is utilized to affirm or dismiss the estimation hypothesis. CFA ought to be seen as a for the most part dis-confirmatory method. That is, CFA encourages the factual dismissal—or, best case scenario, a provisional maintenance—of a particular hypothesis in regards to the factor (s) in charge of the watched relations in the information. In the event that, then again, the examiner's expectations are a generally ungrounded investigation of relations recommended by the information, traditional exploratory factor examination is the more fitting methodology.

In most researches, factors has been reviewed and processed one by one. But in studies like we are working on, we have to review factors jointly. In this case, we are using CFA as method. CFA is optimal for comparison the real data with the hypotheses and factors that we obtained before collecting the real data.

With the rule for factors to be less than 1 and to be closest to 0, they can be accepted as consistent. Primarily, for freight rate and operation we can say that all three sub factors are consistent. Secondly, in value added services, packing has been resulted 1.15 so our first inconsistent sub factor is packing. For information technology results are consistent again. Schedule contains two sub factors which are close to being inconsistent however still schedule can be considered has consistent. On the rest of factors are consistent with real data which has been collected. Measurement model results Table 1 was created based on Standardized loadings and T values table.

Table 1. Measurement model results

Construct	Indicator	Standardized loading	Significance (t-value)
Freight rate	Flexibility	0.60	6.70
	Payment	0.61	6.83
	Freight rate	0.55	6.04
Operation	Container availability	0.69	7.63
	Shipment time	0.64	7.12
	Inland haulage	0.39	4.13
Value added services	Packing	1.15	7.15
	Warehouse	0.70	5.92
Information tech.	EDI	0.41	4.46
	Online booking	0.79	8.51
	Tracking	0.58	6.37
Schedule	On time	0.88	12.55
	Transit time	0.92	13.39
	Departure calendar	0.56	7.05
Customer service	Knowledgeable employee	0.61	7.88
	Salesperson	0.81	11.55
	Polite	0.81	11.52
	Professional	0.85	12.38
	Problem solving	0.58	7.41
	Reply	0.80	11.33
	Rapid documentation	0.82	11.70
Other services	Container types	0.59	6.10
	Port connectivity	0.84	7.67
Satisfaction	Relationship	0.90	13.78
	Comparison	0.92	14.37
	Performance of line	0.88	13.29
	Suitability of line	0.74	10.13
Loyalty	First choice	0.82	11.62
	Willingness	0.80	11.21
	Positive comment	0.83	11.90
	Change	0.72	9.71
	Partner	0.83	11.89

Standardized loads show the correlations between latent variables which it relates each observed variables. The freight rate factor is explained mostly by the payment variable. The variability of operation factor is mostly container availability; the variability of value added services factor is the maximum of packing, the variability of customer service factor is most explained by the observed professional variable. Loyalty rate factor is explained mostly by positive comment and partner.

5.5 Structural Equation Modeling (SEM)

Structural equation modelling is, seemingly, a standout amongst the most prevalent factual procedures available to quantitative social researchers. The fame of structural equation model has prompted the formation of an academic diary committed explicitly to structural equation model also. What's more, there are an expansive number of programming programs that consider advanced and profoundly adaptable displaying. Research has also shown how growth curve models could also be incorporated into the structural equation modeling framework utilizing existing structural equation modeling software. The advantages of the structural equation modeling framework to growth curve modeling is its tremendous flexibility in specifying models of substantive interest.

In studies, that as many sub factors to work on, SEM has big opportunity against other methods. SEM is more economic in perspective of time and exertion. Using the SEM you can analyze a single model unlike the regression analyzes. This prevent the need of applying the analyzes several times. In summary, structural equation modeling shows the factor loadings and error variances to the researcher. In addition, the main consideration is to evaluate the appropriateness of the model. As shown in Fig. 3 the results of research that we processed with SEM.

The RMSEA value indicates whether the model conforms to the default model. Another value used for model fit is RMSEA. This value is limited to zero as the lower limit. The value is between 0 and 0.05, indicating that the model has a good fit; the value between 0.05 and 0.08 indicates that the model has acceptable compatibility. The RMSEA value is between 0.08 and 0.10, while the model fit is low, it can be said that the model is not acceptable for values higher than 0.10. The RMSEA value in this study is 0.067, which indicates that the model is acceptable.

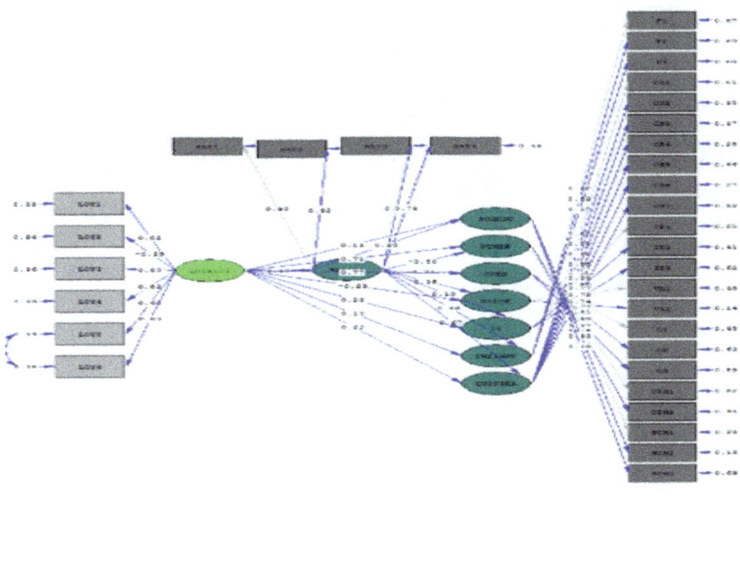

Fig. 3. Structural model

The modification indices suggested to add an error covariance between Loy5 and Loy6 for this reason error covariance was added in there, and Loy2 is removed. Thus, Chi-Square is decreased and CFI value got closer to 1.

We performed Structural equation modeling (SEM) using LISREL to statistically analyze the data and test the proposed conceptual model and hypotheses. To evaluate the goodness of model fit, we used chi square and degree of freedom ($\chi 2/df$), comparative fit index (CFI), root mean square error of approximation (RMSEA). The results showed an acceptable fit for our hypothesized model because based on the recommended cut-off values [20], all of the fit indices were in acceptable level ($\chi 2/df$ = 1,6514; CFI = 0.97; RMSEA = 0.067) as seen in Table 2.

Table 2. Structural model results

Path	Standardized loading	Significance (t-value)	Interpretation
Freight rate → Satisfaction	0.58	7.87	Significant
Freight rate → Loyalty	0.10	0.31	Non-significant
Operation → Satisfaction	0.22	5.04	Significant
Operation → Loyalty	0.77	6.53	Significant
Value added services → Satisfaction	0.38	5.31	Significant
Value added services → Loyalty	0.23	0.72	Non-significant
Information technology → Satisfaction	0.10	0.38	Non-significant
Information technology → Loyalty	0.29	3.42	Significant
Schedule → Satisfaction	0.26	2.19	Significant
Schedule → Loyalty	0.13	0.95	Non-significant
Customer service → Satisfaction	0.67	7.59	Significant
Customer service → Loyalty	0.71	7.12	Significant
Other services → Satisfaction	0.50	5.42	Significant
Other services → Loyalty	0.53	5.76	Significant
Satisfaction → Loyalty	0.87	6.79	Significant

χ^2 = 791.04; p = 0.00000; df = 479; RMSEA = 0.067; CFI = 0.97

The results of the data analysis show that eleven hypotheses of the model are supported and four are not supported as seen in the Fig. 4.

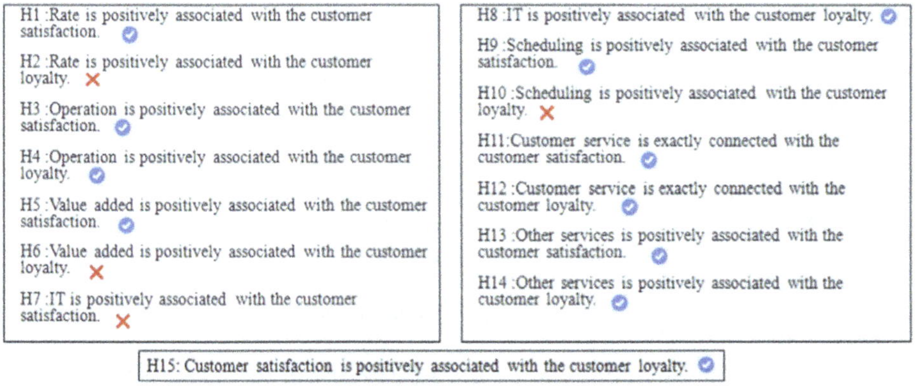

Fig. 4. Hypothesis Results

6 Conclusion

Today, transportation is an important part of global trade. With the growth of the logistics sector, container transportation companies entered into a big competition [21]. This competition has brought a lot of problems. In particular, commercial imbalances revealed an important problem such as not finding empty containers. Firms need to differentiate themselves because they are faced with similar services. In order to achieve this differentiation, it is necessary to be customer-focused because a significant link of the chain is the customer. Companies should understand the needs of the customer and develop their services accordingly. The aim of this study is to understand the wishes of the customers and to measure the company's customer satisfaction and loyalty. While carrying out this study, the criteria in the carrier selection were determined by using the literature. A questionnaire was developed and applied to many firms in the sector. Survey results were analyzed to measure customer satisfaction and loyalty levels. SPSS and LISREL were used to analyze data which have been collected. With factor analysis, the responses are grouped together to form a smaller number of factors and model was developed. ANOVA and t Test were used to figure out if we need to reject the null hypothesis or accept the alternate hypothesis. Significant differences between the averages of groups were found out in the six questions of first part of survey. CFA was used to analyze for comparison the real data with the hypotheses and factors that we obtained before collecting the real data. We performed Structural equation modeling (SEM) using LISREL to statistically analyze the data and test the proposed conceptual model and hypotheses. According to the structural equation modelling results, customers are mostly satisfied with the customer service attributes of a shipping line (%67). Customers are more loyal according to the operations of the shipping line (%77). Information Technologies used by a shipping line doesn't impact the satisfaction level of customer. Value added services, freight rate and schedule of a shipping line do not impact the loyalty of the customer.

Acknowledgment. The study is prepared by Gulumser Gur, Ecem Ervansel, Ege Taylan Er, Efe Yagci, Selin Aydin, Elif Kiran in the context of senior design project at Yasar University in 2019.

References

1. Stiglitz, J.E.: Making Globalization Work. W.W. Norton and Company, Inc., New York (2006)
2. Haralambides, H.E.: Structure and operations in the liner shipping industry. In: Handbook of Transport Modelling, pp. 761–775. Pergamon-Elsevier Science (2007)
3. Balci, G., Cetin, I.B.: Market segmentation in container shipping services: a qualitative study. Manag. Res. Rev. **40**(10), 1100–1116 (2017)
4. Ting, S.-C., Tzeng, G.-H.: Ship scheduling and cost analysis for route planning in liner shipping. Marit. Econ. Logist. **5**(4), 378–392 (2003)
5. Heaver, T.D.: The evolving roles of shipping lines in international logistics. Int. J. Marit. Econ. **4**(3), 210–230 (2002)
6. Chen, K.K., Chiu, R.H., Chang, C.T.: Using beta regression to explore the relationship between service attributes and likelihood of customer retention for the container shipping industry. Transp. Res. Part E Logist. Transp. Rev. **104**, 1–16 (2017)
7. Sheppard, E.J., Seidman, D.: Ocean shipping alliances: the wave of the future? Int. J. Marit. Econ. **3**(4), 351–367 (2001)
8. Reichheld, F.F., Sasser, J.W.: Zero defections: quality comes to services. Harv. Bus. Rev. **68**(5), 105–111 (1990)
9. Lin, H.H., Wang, Y.: An examination of the determinants of customer loyalty in mobile commerce contexts. Inf. Manag. **43**, 271–282 (2006)
10. Öztürkoğlu, Y., Çalışkan, A.: Deniz Taşımacılığında Broker Seçimi Kararını Etkileyen Kriter Skorlarının Belirlenmesi ve Alternatiflerin Değerlendirilmesi. Dokuz Eylül Üniversitesi Denizcilik Fakültesi Dergisi **8**(1), 31–61 (2016)
11. Hançer, M.: Konaklama Endüstrisinde Müşteri Sadakati: Anlam, Önem, Etki Ve Sonuçlar. Seyahat ve Turizm Araştırmaları Dergisi, Cilt 3, no. 1–2 (2003)
12. Baytekin, P.: Toplam Kalite Hedefinde Müşteri Memnuniyetinden Müşteri Sadakatine. Yeni Düşünceler Dergisi **1**(1), 41–52 (2005)
13. Eggert, A., Ulaga, W.: Customer perceived value; a substitute for satisfaction in business markets. J. Bus. Ind. Mark. **17**(2/3), 107–118 (2002)
14. Onaran, B., Bulut, Z.A., Ozmen, A.: A study to investigate the effect of customer value on customer satisfaction, brand loyalty and customer relationship management performance. Bus. Econ. Res. J. **4**(2), 1–37 (2013)
15. Kotler, P.T., Keller, K.L.: Marketing Management. Prentice Hall, Upper Saddle River (2012)
16. Kannan, V., Bose, S.K., Kannan, N.G.: An evaluation of ocean container carrier selection criteria: An Indian shipper's perspective. Manag. Res. Rev. **34**(7), 754–772 (2011)
17. Lu, C.S.: The impact of carrier service attributes on shipper–carrier partnering relationships: a shipper's perspective. Transp. Res. Part E Logist. Transp. Rev. **39**(5), 399–415 (2003)
18. Brooks, M.R.: Understanding the ocean container market—a seven country study. J. Hist. Econ. Thought **22**(1), 39–49 (1995)
19. Wen, C.H., Huang, J.Y.: A discrete choice model of ocean carrier choice. J. East. Asia Soc. Transp. Stud. **7**, 795–807 (2007)
20. Hu, L.T., Bentler, P.M.: Cutoff criteria for fit indexes in covariance structure analysis: conventional criteria versus new alternatives. Struct. Equ. Model. Multidiscip. J. **6**, 1–55 (1999)
21. Çalışkan, A., Kalkan, M., Ozturkoglu, Y.: City logistics: problems and recovery proposals. Int. J. Logist. Syst. Manag. **26**(2), 145–162 (2017)

Lot Size Optimization of Purchased Products

Merve Başay, Yağız Can Çelgin, Naz Çelik, Merve Eyüpoğlu, Furkan Koç, Pırıl Köksal, Ege Özgültekin(✉), Damla Kizilay, and M. Arslan Örnek

Department of Industrial Engineering, Yasar University, Bornova, Turkey
basaymerve@gmail.com, yagizcelgin96@gmail.com, nzclk96@gmail.com, eyupoglumerve@hotmail.com, furkan_koc_92@hotmail.com, piril.koksal@hotmail.com, ege.ozgul35@gmail.com, {damla.kizilay,arslan.ornek}@yasar.edu.tr

Abstract. In Turkey, Pınar Et is one of the leading companies in meat sector since 1985. Within the scope of senior year project, this paper considers the optimization of lot sizes of auxiliary and packaging materials, purchased by the company. Aim of the study is the minimization of total inventory costs, which include unit purchasing, setup, and holding costs subject to a service level. For this purpose, multi-item dynamic lot sizing models are formulated with time-varying demand, deterministic lead time, with backorders allowed. The first model is formulated by considering Joint Replenishment of items supplied from the same suppliers. The second model is formulated to reflect Vendor-Based minimum lot sizes and rounding quantities to the replenishment quantity. Although these models produce an optimal solution in a reasonable time, a Wagner-Whitin based heuristic algorithm is developed as an additional solution to the company. All of the models run according to Rolling Horizon (RH) approach, which updates the existing plan at the beginning of each planning period in order to cope with changes in demand and other parameters. As a result of the study, potential savings in holding cost and reduction in planning duration are obtained and are reported.

Keywords: Multi-item dynamic lot sizing · Wagner-Whitin algorithm · Minimum order size · Joint replenishment · Rolling horizon

1 Introductıon

This study is conducted to improve the current material planning system of Pınar Et. In this paper, the problem related to two activities, namely planning and purchasing, is considered. In a proper manner, detailed studies is done by identifying current system and symptoms of the problem. After the literature review is completed, it is determined that multi-item dynamic lot sizing optimization should be used to minimize total inventory cost and improve the planning process. As a result, two dynamic mathematical models are formulated with the time-varying deterministic demand, dynamic cost components, deterministic lead time properties when the shortage is allowed under a certain service level. Also, Wagner-Whitin based heuristic is developed in Visual

N. M. Durakbasa and M. G. Gençyılmaz (Eds.): ISPR 2019, LNME, pp. 705–719, 2020.
https://doi.org/10.1007/978-3-030-31343-2_60

Basic for Applications (VBA) to present an additional solution method and prevent dependency on the software of IBM ILOG CPLEX. This study aims to minimize the total inventory costs and to determine the optimal lot sizes, ordering frequencies, and timings. Two proposed mathematical models and Wagner-Whitin based heuristic algorithm are solved with actual data by IBM ILOG CPLEX and VBA, and solutions are obtained, and are presented. Furthermore, conclusions and future research directions are provided in this paper.

Pınar Et works with various suppliers and customers. The Company mainly procures lamb, calf, and turkey as a raw material. Also, spices, additives, packages, foils, etc. are provided as auxiliary and packaging materials from different suppliers for each raw material. According to the bill of materials, inventory levels, and lead times, a weekly production schedule is generated through the SAP by the planning department. The Production process starts with raw material entry, and after many automated steps, they turn into processed meat. Finally, all finished products are packed and distributed.

2 Problem Definition

To determine the customer demand, next four-months sales are forecasted. According to the forecasts, the planning department determines the material requirements for all semi-products and finished-products. After receiving the material requirements, the purchasing process is done for all raw materials among the suppliers who have contracted by the purchasing department beforehand.

Because of the real-life constraints, the planning department is not usually able to reach exact required quantities. The reason for that is the purchasing department desires to order in larger quantities than planned, taking into account minimum lot sizes of the suppliers and long lead times. Also, shortages are generally undesirable because the Company aims to have service level as high as possible for customer satisfaction. On the other hand, ordering more than the planning department requires results in keeping the inventory levels high. As a consequence, inventory holding costs increase. Shortly, the Company is in pursuit of a scientific methodology to improve the planning and purchasing procedure. Therefore, in this study, it is aimed to formulate a material planning model for determining of minimum lot sizes of each auxiliary and packages material.

After the comprehensive analysis of the current system and observed symptoms, the principal goal of this study is set as to use a systems approach to develop an optimization model and systematic solution approaches that deliver smart plans considering all factors to improve the planning process. In this way, the balance between the planning and purchasing departments related to ordering policy is ensured. Furthermore, minimum lot sizes are optimized, and the total cost is expected to decrease.

Performance measures of this system are the total cost and service fill rate. In this study, the optimization of these performance measures is the primary objectives. It is crucial to meet the demand on time and thus to keep the service level as high as possible while considering other parameters.

Outputs of this study are optimal lot sizes, ordering frequencies, and timings. Additionally, a user-friendly decision support system is developed for this study, and is made available to be used of in the Company. The decision support system gives out a report, including the optimal values and the total expected inventory cost of the previous period while considering the lead time and some other constraints. They are detailed in section "Limitations and Assumptions."

3 Literature Review

Inventory management is crucial for many businesses at all times. Companies can minimize the possible risks that will arise during the production process and optimize productivity by effectively tracking and managing inventory. Also, it is noticed that problem is deterministic and time-varying demand and deterministic delivery time when the data are examined. As a result, the dynamic lot sizing model is the most suitable one for the problem.

When the demand rate varies with time, it is wrong to assume that the correct strategy is always to use the same replenishment quantity. According to Axsäter [1], variations in demand rate depend on different reasons. As an example, the item might be used as a component or subassembly. Also, seasonal variations in demand can be caused by variations in demand rate. However, it still considered deterministic demand, and all variations are known in advance. Additionally, if the demand rate varies with time, namely dynamic, modeling time categorization has to be chosen as a limited period. This limited period is known as the planning horizon. When dealing with the dynamic lot sizing, some assumptions have to be made. These are explained as follows: Firstly, demand is known in each period, and it is assumed that periodical demands actualize at the beginning of the period. Secondly, delivery is made as a batch, and the whole batch is delivered at the same time. Thirdly, there is no initial stock, and cost components are fixed ordering cost, the variable cost, and holding a cost. Also, the variable cost does not depend on the replenishment quantity; in particular, there are no discounts in either the unit purchase cost or there are no discounts in the unit transportation cost. Finally, backorders are not allowed, and there is no lead time. Deciding on replenishment quantity of a single item in each period with considering cost minimization objective is the only issue.

Solution algorithm of dynamic lot sizing model, which is called Wagner-Whitin is introduced by Harvey M. Wagner and Thomson M. Whitin in 1958 [2]. It is an algorithm for finding the optimal solution by dynamic programming. In addition to that, some of the assumptions of this algorithm which are related to backorders, cost components, initial stocks are relaxed for this study.

Formulated dynamic lot sizing model will run according to a rolling horizon approach. According to Şahin et al. [3], rolling horizon systems iteratively solve a series of linked short-term stationary lot-sizing problems for which demand is known in advance. Rolling through time, the firm executes a subset of the earliest replenishment decisions, and after waiting for the duration of the re-planning periodicity (RP), the system updates demand information and then initiates the next planning cycle. In each subsequent planning cycle, orders within the frozen interval are not subject to change, but orders in the free interval may be rescheduled. Policy decision variables governing the implementation of rolling horizon planning systems include the choice of lot-sizing rule, RP and the lengths of the planning horizon and frozen interval. The length of the frozen interval and RP may be either time or order based.

In 2009, a new metric for measuring instability in multi-item multi-level planning systems is written by Kabak and Ornek [4]. New metric composed of four sub-instability measures is proposed to measure multi-item and multi-level schedule instability under rolling schedules. It defines the instability between two successive planning cycles and divides it into a number of sub-instability measures. According to writers, the main advantage of the proposed metric is that it makes an independent comparison of alternative production systems possible and that it facilitates the application of nervousness dampening procedures.

There are many types of inventory models dealing with multi-item environments [5]. The objective of these models is usually to minimize total inventory cost while satisfying demand. In such systems, cost minimization can be achieved by coordinating the replenishment of several purchased items. The joint replenishment problem (JRP) deals with the problem of coordinating the replenishment of a group of items that may be jointly ordered from a single supplier [6]. Since the replenishment of multi-items is a general context within the operation environment, companies try to find models to make decisions on inventory management that result in efficient JRP solutions. Furthermore, according to Salameh et al. [7], the JRP guarantee total cost reduction (ordering cost, transportation cost) compared to single-item replenishment. In addition to that, according to Goyal [6], joint replenishment models are more likely to be applied in real life than single item replenishment models.

There are many cost parameters to be calculated in this study. In real production environments and also in studied company Pınar Et, estimation of cost parameters, especially backorder cost, is difficult.

Backorder is associated with not being able to fill an order [8]. If an item is demanded and cannot be met due to a shortage, various costs such as administration, price discounts for late deliveries, material handling, transportation and loss of goodwill that may affect the sales, in the long run, will occur. Since shortage costs are so difficult to estimate, it is widespread to replace them with a suitable service constraint [1]. It is also challenging to determine an adequate service level, but the company provides this. As a result, the backorder cost is not included in the objective function of the mathematical model. However, the provided service level is added to the model as a constraint in order to allow backorder to a certain level.

4 Problem Modellıng and Formulations

4.1 Solution Approach

After analysis of the current system in Pınar Et and literature review is completed, it is decided to use multi-item dynamic lot sizing optimization considering lead times and backorders.

The first step is to determine the critical items to focus on this study. As the product range of the Company is wide, the number of components, which are auxiliary and packaging materials in the study, used in these products is many. At first, various 948 of materials are considered by the Pınar Et. After performing the ABC analysis, the most valuable, significantly consumed. A class 46 materials are selected. First, a code is written in Python, and the results are obtained as the system sorted the items. Finally, those materials are discussed with Pınar Et for the final decision. Some materials which are called as joker materials are added by the Company. Although these materials do not have high consumption values, Pınar Et has problems with these items and offered us to study on them as well.

4.2 Limitations and Assumptions

Some limitations and assumptions are made to formulate the mathematical model while determining the solution methodology. All the assumptions are listed as follows:

- Demand rate is deterministic and time-varying for all items.
- Periodical demands actualize at the beginning of each period.
- Items are ordered and delivered in batch.
- Lead time is deterministic and fixed.
- Time-based model is chosen as the planning horizon (periodic review).
- Planning horizon assumed as 20 weeks.
- Backlogging is allowed.
- Cost parameters are item dependent.
- Ordering cost does not depend on the replenishment quantity, and in particular, there are no discounts in unit purchasing cost.
- Setup cost is affected by only logistics cost.

4.3 Parameters, Sets and Decision Variables

Input parameters, decision variables and relevant sets of the problem are determined and listed below.

<u>Parameters:</u>

t, u: time index
i, j: item index
d_t^i: demand rate of item i at time t
c_t^i: unit cost of item i at time t
k_t^i: fixed ordering (setup) cost of item i at time t
h_t^i: unit inventory holding cost of item i at time t
l: lead time of item i
T: length of the planning horizon
J: the number of items
M^i: a big number for item i, calculated as sum of the demands of each item for each period
I_0^i: initial inventory level of item i
r_t^i: inventory quantity of item i, which is ordered in previous planning horizon and received at time t of current planning horizon, scheduled receipts

Sets:

$u, t \in$ U = {1, …, T}: the set of time

$i, j \in$ N {1, …, J}: the set of items

Decision Variables:

- $X^i_{u,t} = \begin{cases} 1, & \text{if item } i \text{ is ordered at time } u, \text{arrived at time } t \\ 0, & \text{otherwise} \end{cases}$
- $P^i_t = \begin{cases} 1, & \text{if backorder occurs for item } i \text{ at time } t \\ 0, & \text{otherwise} \end{cases}$
- $Q^i_{u,t}$: replenishment quantity of item i, ordered at time u, arrived at time t.
- I^i_t : on hand inventory of item i at the end of time t
- B^i_t : backorder quantitiy of item i at time t

4.4 Conceptual Model

In this study, models, that are compatible with the rolling horizon approach, are formulated to cope with an uncertain environment, that is common for today's industries [13]. In this model, 20-week planning is considered, and at the beginning of each week, the plan is updated. Concordantly, the mathematical model aims to minimize total inventory system cost, which includes the sum of the setup, variable, and holding costs. In addition, inventory balance equations are written for first period, middle periods, and last period. When writing these equations, ordering quantity, inventory level, demand, scheduled receipts, and backorder are taken into account. As the backorder is allowed under certain service level and if there is an inventory, then backorder does not occur.

4.5 Model Formulation

(1) *Base Model:*

$$min \sum_{t=1}^{T}\sum_{i=1}^{J}\left(c^i_{(t-l^i)}.Q^i_{(t-l^i),t} + k^i_{(t-l^i)}.X^i_{(t-l^i),t}\right) + \sum_{t=1}^{T-1}\sum_{i=1}^{J}(h^i_t.I^i_t) \tag{1}$$

subject to

$$Q^i_{(t-l^i),t} + I^i_0 - d^i_t + B^i_t + r^i_t = I^i_t, \quad t = 1 \parallel t - l_i \geq 1, \quad \forall i \in N, \tag{2}$$

$$Q^i_{(t-l^i),t} - d^i_t + I^i_{(t-1)} + B^i_t - B^i_{t-1} + r^i_t = I^i_t \quad t = 2,\ldots,T-1 \parallel t - l_i \geq 1, \quad \forall i \in N, \tag{3}$$

$$Q^i_{(t-l^i),t} - d^i_t + I^i_{(t-1)} - B^i_{(t-1)} + r^i_t = 0, \qquad t = T \parallel t - l_i \geq 1, \qquad \forall i \in N, \tag{4}$$

$$1 - \left(\sum_{t=1}^{T}\sum_{i=1}^{J} B^i_t / \sum_{t=1}^{T}\sum_{i=1}^{J} d^i_t\right) \geq 0,98, \qquad \forall t \in U, \qquad \forall i \in N, \tag{5}$$

$$X^{i}_{(t-l^i),t}.M^i \geq Q^{i}_{(t-l^i),t}, \qquad \forall t \in U, \qquad \forall i \in N, \tag{6}$$

$$P^i_t.M^i \geq B^i_t, \qquad \forall t \in U, \qquad \forall i \in N, \tag{7}$$

$$\left(1 - P^i_t\right).M^i \geq I^i_t, \qquad \forall t \in U, \qquad \forall i \in N, \tag{8}$$

$$X^i_{u,t}, P^i_t \in \{0, 1\}, \qquad \forall t \in U, \qquad \forall i \in N, \tag{9}$$

$$I^i_t, Q^i_{u,t}, B^i_t \geq 0, \qquad \forall t \in U, \qquad \forall i \in N, \tag{10}$$

The objective of the multi-item dynamic lot-sizing problem, shown in Eq. (1), is to minimize total cost. Total cost includes the sum of the variable, setup, and inventory carrying costs. Equations (2), (3) and (4) show the inventory balance equations for period t = 1, t = 2 to T-1 and t = T, respectively. Equation (2) indicates that inventory level of item *i* at the end of period 1, which means the beginning of any planning horizon, is depend on replenishment quantity, beginning inventory, backorder quantity, scheduled receipts and demand of the period. In addition to Eq. (2), backorder quantity of previous period is considered in Eq. (3) while determining the inventory level of any period *t*. Equation (4) denotes that, an inventory level of item *i* at the end of last period, which means the end of any planning horizon, depends on replenishment quantity, scheduled receipts, beginning inventory, backorder quantity of previous period and demand of last period. No holding inventory in the last period is defined by the system in Eq. (4). The model measures the service level in terms of the average proportion of demand that is satisfied over time. Equation (5) indicates that at least 98% of the time no backorder occurs; therefore, the service level is greater than or equal to 98%. Equation (6), forces $x^i_{t-l^i}$ to be 1 if there is ordering for item *i* at time $t - l^i$ and $x^i_{t-l^i} = 0$ implies that there is no ordering at time $t - l^i$, which means $Q^i_{(t-l^i),t} = 0$. Equations (7–8), force P^i_t to be 1 if there is backorder for item i in period t. Finally, Eq. (9–10) define the domains of the variables and their sign restrictions.

The model as mentioned earlier is formulated in the most classical form. In addition to that, two different versions are formulated. First one is "Joint Replenishment Policy," and the second one is "Vendor-Based Minimum Lot Sizing Policy." All models cover a 20-week planning horizon.

Many companies order a group of items simultaneously, rather than individually. This is known as the joint replenishment. We have extended the classical rolling horizon model to achieve better inventory management by coordinating the replenishment of several purchased items, and to reduce the time spent on the order tracking in the planning department. The proposed joint replenishment model optimizes ordering quantities for each product to reduce fixed costs, e.g., by shipping several products on one truck while partially meeting demand.

(2) *Joint Replenishment Policy:* The changes that is made in the model to make joint replenishment and additionally defined parameters, set, decision variable are presented below.

Parameters and Set:

A:	group index, based on lead time interval and city information
$k_t^{i,A}$:	fixed ordering (setup) cost, special for item i which is in group A
l^A:	maximum lead time in group A, valid for all items in group.
$A \in G = \{1, \ldots, \lvert G \rvert\}$:	the set of groups

Decision Variables:

$Q_{u,t}^{i,A}$: replenishment quantity of item $i$which is in group A, ordered in time u, arrived at time t

$$X_{u,t}^{i,A} = \begin{cases} 1, & \text{if item i which is in group A is ordered at time u, arrived at time t} \\ 0, & \text{otherwise} \end{cases}$$

$$min \sum_{t=1}^{T} \sum_{i=1}^{J} \left(c^i_{(t-l^A)} . Q^{i,A}_{(t-l^A),t} + k^{i,A}_{(t-l^A)} . X^{i,A}_{(t-l^A),t} \right) + \sum_{t=1}^{T-1} \sum_{i=1}^{J} (h^i_t . I^i_t) \quad (1)$$

subject to

$$Q^{i,A}_{(t-l^A),t} + I^i_0 - d^i_t + B^i_t + r^i_t = I^i_t, \quad t = 1 \parallel t - l_i \geq 1, \quad \forall i \in A, \quad \forall A \in G \quad (2)$$

$$Q^{i,A}_{(t-l^A),t} - d^i_t + I^i_{(t-1)} + B^i_t - B^i_{t-1} + r^i_t = I^i_t, \quad t = 2, \ldots, T-1 \parallel t - l_i \geq 1, \quad \forall i \in A, \forall A \in G \quad (3)$$

$$Q^{i,A}_{(t-l^A),t} - d^i_t + I^i_{(t-1)} - B^i_{(t-1)} + r^i_t = 0, \quad t = T \parallel t - l_i \geq 1, \quad \forall i \in A, \quad \forall A \in G \quad (4)$$

$$X^{i,A}_{(t-l^A)} \geq X^{j,A}_{(t-l^A)}, \quad \forall t \in U, \quad \forall i,j \in A, i \neq j, \quad \forall A \in G \quad (5)$$

$$X^{i,A}_{(t-l^A)} \leq X^{j,A}_{(t-l^A)} \quad \forall t \in U, \quad \forall i,j \in A, i \neq j, \quad \forall A \in G \quad (6)$$

$$1 - \left(\sum_{t=1}^{T} \sum_{i=1}^{J} B^i_t / \sum_{t=1}^{T} \sum_{i=1}^{J} d^i_t\right) \geq 0,98, \quad \forall t \in U, \quad \forall i \in N, \quad (7)$$

$$X^{i,A}_{t-l^i} . M^i \geq Q^{i,A}_{(t-l^i),t}, \quad \forall t \in U, \quad \forall i \in N, \quad (8)$$

$$P^i_t . M^i \geq B^i_t, \quad \forall t \in U, \quad \forall i \in N, \quad (9)$$

$$\left(1 - P^i_t\right) . M^i \geq I^i_t, \quad \forall t \in U, \quad \forall i \in N, \quad (10)$$

$$X^{i,A}_{u,t}, P^i_t \in \{0, 1\}, \quad \forall t \in U, \quad \forall i \in N, \quad (11)$$

$$I_t^i, Q_{u,t}^{i,A}, B_t^i \geq 0, \qquad \forall t \in U, \qquad \forall i \in N, \tag{12}$$

Different from the first model, in the above-mentioned model, instead of the cost of placing a separate order for each product, the cost of making joint replenishment $k_t^{i,A}$ is used. The change is shown in Eq. (1). In addition to that, group index A and group lead time parameter lA, which are used in the objective function and Eqs. (2, 3, 4), affect replenishment quantity. Equations (5) and (6) ensure making joint replenishment for items which are in the same group. Equations (7–12) are same with "Base Model".

(3) *Vendor-Based Minimum Lot Sizing Policy Model*: The base model determines the lot sizes without the restrictions imposed by the suppliers. We also extend the model to reflect real-life conditions by adding vendor-based minimum lot size quantities and rounding quantities to the model. In a real production environment, the manufacturer of each purchased product has its minimum order quantity.

The changes that have been made in the model to make vendor-based minimum lot sizing and additional parameters and decision variables defined are provided below.

Parameters:

p^i: minimum order size of item i
s^i: fixed replenishment quantity when item i is ordered more than the minimum order size

Decision Variables:
R_u^i: A multiple of fixed replenishment quantity

$$min \sum_{t=1}^{T} \sum_{i=1}^{J} \left(c^i_{(t-l^i)} \cdot (p^i + s^i.R^i_{t-l^i}) + k^i_{(t-l^i)}.X^i_{(t-l^i),t} \right) + \sum_{t=1}^{T-1} \sum_{i=1}^{J} (h_t^i.I_t^i) \tag{1}$$

subject to

$$(p^i + s^i.R^i_{t-l^i}) + I_0^i - d_t^i + B_t^i + r_t^i = I_t^i, \quad t = 1 \; \| \; t - l_i \geq 1, \quad \forall i \in N, \tag{2}$$

$$(p^i + s^i.R^i_{t-l^i}) - d_t^i + I^i_{(t-1)} + B_t^i - B^i_{t-1} + r_t^i = I_t^i, \qquad t = 2, \ldots, T-1 \; \| \; t - l_i \geq 1, \quad \forall i \in N, \tag{3}$$

$$(p^i + s^i.R^i_{t-l^i}) - d_t^i + I^i_{(t-1)} - B^i_{(t-1)} + r_t^i = 0, \qquad t = T \; \| \; t - l_i \geq 1, \quad \forall i \in N, \tag{4}$$

$$R_u^i \geq 0, \text{integer}, \quad \forall t \in U, \quad \forall i \in N, \tag{5}$$

$$1 - \left(\sum_{t=1}^{T} \sum_{i=1}^{J} B_t^i / \sum_{t=1}^{T} \sum_{i=1}^{J} d_t^i\right) \geq 0,98, \qquad \forall t \in U \qquad \forall i \in N, \tag{6}$$

$$P_t^i.M^i \geq B_t^i, \qquad \forall t \in U, \quad \forall i \in N, \tag{7}$$

$$\left(1 - P_t^i\right).M^i \geq I_t^i, \quad \forall t \in U, \quad \forall i \in N, \tag{8}$$

$$X_{u,t}^i, P_t^i \in \{0, 1\}, \quad \forall t \in U, \qquad \forall i \in N, \tag{9}$$

$$I_t^i, B_t^i \geq 0, \quad \forall t \in U, \quad \forall i \in N, \tag{10}$$

In this model, the replenishment quantity of item i is affected by the minimum lot size and rounding quantity of that product. The change in total unit cost is shown in the objective function. In inventory balance Eqs. (2, 3, 4) replenishment quantity is written as $\left(p^i + s^i \cdot R_{t-l_i}^i\right)$. Also, Eq. (5) defines the domains of the variables and sign restrictions. Equations (6–10) are the same with "Base Model".

(4) *Wagner - Whitin Based Heuristic:* Since the purchasing of the IBM ILOG CPLEX would create an extra cost for the Company, Excel Solver is used to solving these models. However, the computation time of solving the real-life problem has taken too long the problem could not be solved with Excel Solver, which has a limit of 200 decision variable. Thus, a Wagner-Whitin based heuristic is developed in Visual Basic for Applications (VBA).

The objective function of the heuristic is the minimization of total inventory cost, which includes variable, joint setup, and a holding cost. While deciding to the replenishment of items, the lead time is still considered. However, backorder is not allowed. Also, it works joint replenishment approach in order to provide a reduction in setup cost., pseudocode of the developed heuristic method is represented in Fig. 1, below.

The proposed heuristic algorithm is a Wagner-Whitin based algorithm. However, instead of using single item to calculate progress, it treats each group as a single item. First, a group's cost matrix is calculated from any period to each period. The cost matrix includes costs of variable and holding when ordered at one period to another. Then starting from the last period, the algorithm compares whether ordering at that period or proceeding without any calculation (same with Wagner–Within algorithm). This process continuous until the lead time of the group concerned. For example, if the last period is 20 and the lead time of the group is 6, the algorithm works from period 20 to period 6. The weeks from the start to the week during the supply period are considered to be covered by the on hand inventory. The cost incurred during these periods is only the cost of holding.

```
Step 1 Initialization
Step 2  Recursion
        FOR each group
                Calculate costMatrix values
                FOR t1=periods from last one to groups lead time
                        FOR t2=t1+1 to last period
                                IF best_t1 is greater than costMatrix_t1,t2 + best_t2
                                        SET best_t1 as costMatrix_t1,t2 + best_t2
                                        SET bestTo_t1 to t2
                                ENDIF
                        ENDFOR
                ENDFOR
        ENDFOR
                FOR t1=periods from groups lead time to first period
                        FOR every item in group
                                        SET best_t1 as best_t2+ Total demand from first period to
current calculated period
                                        SET bestTo_t1 to t2
                                ENDIF
                        ENDFOR
                ENDFOR
ENDFOR

FOR each period t
        FOR every item i
                Fill order_i,t structure according to bestTo_t
                Shift items of order_i,t as lead time to back
        ENDFOR
ENDFOR
```

Fig. 1. Pseudocode of Wagner-Whitin Based Heuristic

5 Quantitative Results

In this section, numerical results of the proposed models are examined, and fundamental analyses are given.

5.1 Verification and Validation

In order to test the accuracy of the proposed mathematical models, demand and cost information of five products for ten weeks are obtained from the company and solved. It is observed that the problem outputs results as desired by the company, and the models responded quickly and effectively while calculating the objective function and considering constraints of the models. Solution time of the models and their closeness to optimal solution are shown in Table 1 below. Additionally, the closeness to optimal solution of Wagner-Whitin Based Heuristics is compared with Joint Replenishment Model. Optimal solutions are obtained with mathametical models. After the sample data set had given consistent results, it is decided to develop these models since they promised the optimal solution. Finally, the problem of “Lot Size Optimization of Purchased Products” is solved with different approaches. The minimization of the total inventory cost as much as possible in accordance with the specified objective function is obtained when the proposed models solved by "IBM ILOG CPLEX" and "Visual Basic for Applications (VBA)" with actual data provided by the company.

According to the objective function, replenishment quantities of each item for each period are obtained. Optimal ordering timings and order frequencies are also determined considering lead times of items. On hand inventories and backorder, quantities are found by inventory balance equations, which are shown in Eqs. (2, 3, 4), of a mathematical model. Finally, solutions of related decision variables for the joint replenishment model are represented in Table 2, for vendor-based minimum lot size model in Table 3 and Wagner-Whitin based heuristic in Table 4. Furthermore, costs obtained by three optimization models and also the cost of the heuristic algorithm are represented in Fig. 2 below.

In order to take feedbacks from Pınar Et, the results are shared with related departments while validating the models. Finally, solutions and improvements of the study are validated by the company after the outputs are found consistent.

Table 1. Solution time and closeness to optimal solution

Solution methods	Solution time (sec)	Gap to optimality(%)
Base model	8,07	0
Joint replenishment model	9,12	0
Vendor based min lot size model	28,75	0
Wagner-Whitin based heuristics	15,00	2

Table 2. Solutions of joint replenishment model

	Replenish. Qty.	Backorder Qty.	Replenish. Qty.	Backorder Qty.	Replenish. Qty.	Backorder Qty.	Replenish. Qty.	Backorder Qty.
Week/Item	10		11		12		13	
Item1	0	0	7740	0	5521	0	5521	0
Item2	0	0	0	0	0	0	0	0
Item3	2192	0	2149	0	2149	0	2149	0
Item4	0	0	65	0	65	0	65	0
Item5	0	0	0	0	0	0	0	0
Item6	0	0	4440	0	3314	0	3314	0
Item7	0	0	0	0	0	0	0	0
Item8	0	0	0	0	0	0	0	0
Item9	3874	0	5486	0	5486	0	5486	1783
Item10	0	0	0	0	0	0	0	0

Table 3. Solutions of vendor based minimum lot sizing model

	Replenish. Qty.	Backorder Qty.	Replenish. Qty.	Backorder Qty.	Replenish. Qty.	Backorder Qty.	Replenish. Qty.	Backorder Qty.
Week/Item	10		11		12		13	
Item1	0	0	22000	0	0	0	0	0
Item2	0	0	0	0	0	0	0	0
Item3	2250	0	2250	0	4050	0	0	0
Item4	0	0	0	0	200	0	0	0
Item5	0	0	0	0	0	0	0	0
Item6	0	0	13250	0	0	0	0	0
Item7	0	0	0	0	0	0	0	0
Item8	0	0	0	0	0	0	0	0
Item9	0	0	11250	0	0	0	11250	0
Item10	0	0	0	0	0	0	0	0

Table 4. Solutions of Wagner-Whitin based heuristic

	Replenishment quantities			
Week/Item	10	11	12	13
Item1	5190	5521	5521	11043
Item2	0	0	0	0
Item3	2192	2149	2149	2149
Item4	102	65	65	130
Item5	0	0	0	0
Item6	2543	3314	3314	6629
Item7	0	0	0	0
Item8	0	0	0	0
Item9	3874	5486	5486	10971

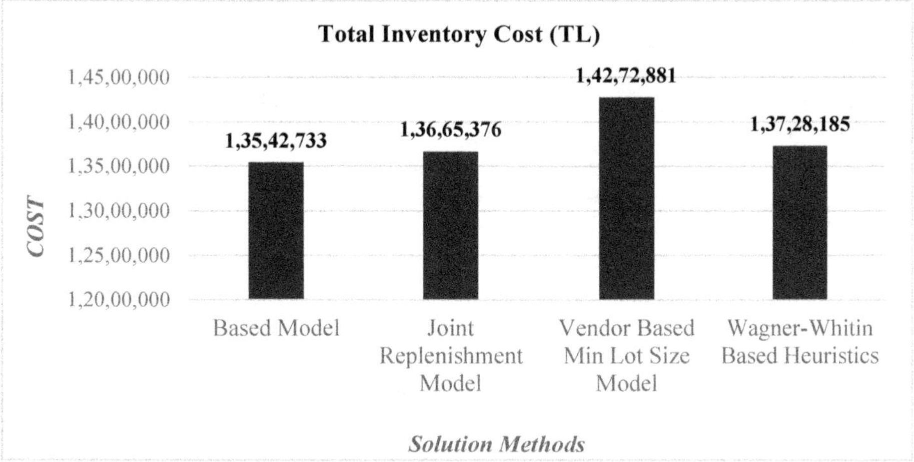

Fig. 2. Total inventory cost of solution methods

5.2 Comparison and Quantification of Improvements

The most important contribution of this study to Pınar Et is to provide structured 5-months purchase plans for 46 -selected- items. Decision Support System (DSS) is developed, which prepares purchase plans considering demand, lead time, and service level, also provide on-hand inventory quantities to users for each period. In addition to that, users are able to view alternative plans and their outputs with graphical illustrations and matrices. Different solution methodologies -plans- offer flexibility to users in decision making according to purchase strategy. Ordering quantities of all items which should be ordered in the current week are reported in PDF form and it can be sent by e-mail. Also, dynamic DSS allows changes in cost parameters, demands, lead times, item names and suppliers, supplier locations, vendor-based minimum lot sizes and rounding quantities and initial inventory levels of each item.

Moreover, while the planning of these 46 items takes at least 1 h in the company, this system offers an optimal solution and alternative solutions, at 4 min at most. Hence, time allocated for planning is reduced by 93.3% with this system. Not only time is shortened, but also workload is decreased substantially.

One more advantage of this user-friendly decision support system is to enable procurement planners to keep track of inventory systems efficiently with less effort. Additionally, designed DSS may provide purchase plans for any company.

As the setup and holding cost parameters are difficult to calculate in real life systems, the company is not used to take into account of these costs while deciding lot sizes. Now, in addition to the purchasing cost, these two costs are taken into consideration, and a balance between them is provided while making purchasing decisions. Also, the cost of logistics, accepted as a setup cost, has been actively used in the planning process for the first time in the company by this project. Since there is no buffer stock anymore, and reduction in on-hand inventory is provided, a 28.9% decrease is achieved in the inventory held. This decrease corresponds to 36.223 TL per month, resulting 434.678 TL gain annualy.

6 Conclusion and Future Research Directions

The determination of optimal lot sizes of purchased products is a common and inevitable problem that trouble the production planners and purchasing employees most. In consideration of reducing total inventory cost, a multi-item dynamic lot-sizing problem is studied in this paper. Three different approaches are proposed. Firstly, a joint replenishment model is an offer for the Company to reduce the total setup cost. The second model, which is vendor-based minimum lot sizes model, is formulated to optimize the current system in the Company. We suggest that these two models should be used according to a rolling horizon approach in order to keep the system updated in terms of demand, costs, etc. Lastly, Wagner-Whitin based heuristic is developed in order to prevent the extra cost of installation of IBM ILOG CPLEX. After the formulation of mathematical models and heuristic, validation is done, and solutions are found consistent by the Company. As a result, the problem of planning and purchasing departments is solved with scientific methodology. Also, user-friendly DSS is presented to improve the planning and purchasing procedure. In addition to that, approximately 29% cost reducing provided in total holding cost potentially. For further studies, it is foreseen that; improvements will be measured better when this plan contains all purchased items of the Company. In the future, this study can be extended, considering another two different problems of the Company. First one is supplier selection according to the supplier's score, and the other one is vehicle routing in joint replenishment.

Acknowledgment. This study is supported by The Scientific and Technological Research Council of Turkey (TUBITAK) in the program of "2209-B Undergraduate Research Projects for Industrial Applications Fellowship Program". We would like to express our special appreciation and thanks to our advisors at Yaşar University; especially Dr. Onder Bulut for assisting this difficult study, with constructive comments that greatly improved the paper. Also we would like to acknowledge the studied company Pınar Et for their cooperation and providing necessary data during this educational process.

References

1. Sven, A.: Inventory Control. Springer, New York (2006)
2. Wagner, H.M., Whitin, T.M.: Dynamic version of the economic lot size model. Manage. Sci. **5**, 89–96 (1958)
3. Sahin, F., Narayanan, A., Robinson, P.: Rolling horizon planning in supply chains: review, implications and directions for future research. Int. J. Prod. Res. **51**, 5413–5436 (2012)
4. Kabak, K.E., Ornek, A.M.: An improved metric for measuring multi-item multi-level schedule instability under rolling schedules. Comput. Ind. Eng. **56**(2), 691–707 (2009)
5. Aksoy, Y., Selcuk Erenguc, S.: Multi-item inventory models with co-ordinated replenishments: a survey. Int. J. Oper. Prod. Manag. **8**(1), 63–73 (2013)
6. Khouja, M., Goyal, S.: A review of the joint replenishment problem literature: 1989–2005. Eur. J. Oper. Res. **186**(1), 1–16 (2008)
7. Salameh, M.K., Yassine, A.A., Maddah, B., Ghaddar, L.: Joint replenishment model with substitution. Appl. Math. Model. **38**(14), 3662–3671 (2014)
8. Backorder Costs: Investing Answers Building and Protecting Your Wealth through Education Publisher of The Next Banks That Could Fail. https://investinganswers.com/financial-dictionary/businesses-corporations/backorder-costs-4856

Measuring Service Quality in the Logistic Sector by Using Servqual and Best Worst Method

Delfin Kolat(✉), H. Ajlan Kökçü, Mete Kiranli, Melisa Özbiltekin, and Yücel Öztürkoğlu

International Logistics Management, Yaşar University, Izmir, Turkey
delfinkolat@gmail.com, kokcuajlan@gmail.com,
mete.kiranli@outlook.com,
{melisa.ozbiltekin,yucel.ozturkoglu}@yasar.edu.tr

Abstract. In this study, it is aimed to analyze the logistics service quality provided by freight forwarder logistics company in Izmir, Turkey. As in every sector, competition between companies is increasing day by day in the logistics sector. Therefore, all companies in the service sector must provide good and quality service in the face of increasing competition. In addition to the cost factor to attract limited customers in the market, logistics companies can also come to the fore with the service quality they offer. The service quality of a business depends on what the customer expects from that service and the extent to which their expectations are met. In the study, firstly, the expectations of the customers are analyzed from the quality service perspective and secondly, the quality of the company, which serves for the same quality indicators, has been tried to be measured. To achieve these aims, service quality (SERVQUAL) method will be used. Also, the Best Worst Method (BWM) will be used to find the weight of each logistics service quality dimensions with a mathematical model.

Keywords: Freight forwarder · Service quality · SERVQUAL · Best Worst Method

1 Introduction

Service is not a tangible item and it is an experience; service quality is connected to the customer satisfaction. Satisfaction level with the service is prior for customer's perception of quality ([1, 2]). There are three characteristics of service, heterogeneity, intangibility and inseparability [3]. Most services are cannot be counted measured stocked tasted and confirmed ahead of time of sale to guarantee the quality [4]. It is not easy to measure the service quality. Moreover, the benefits and satisfactions offered directly with sale or supplied with the sale of these products [5]. Moreover, another definition is service sector is the sector in which all kinds of beneficial work and activity are produced besides tangible material products [6].

Recently, importance of companies in the service sector and the share of logistics sector are increasing day by day. In related to the increase, the number of logistics

N. M. Durakbasa and M. G. Gençyılmaz (Eds.): ISPR 2019, LNME, pp. 720–731, 2020.
https://doi.org/10.1007/978-3-030-31343-2_61

companies increases. Consequently, in the logistics sector, competition between companies is increasing. With the rapid increase, forwarders in the sector face with some problems and have taken a step forward to communicate directly with the customer according to the logistic companies and to respond more quickly to customer requests and expectations. Therefore, to increase competitiveness the role of forwarders is increased, and service quality concepts have gained importance. All companies in the service sector like hotels needs to provide better and quality service against increasing competition [7].

Service quality is a measure of how well the level of service delivered meets the customer requirements [8]. Researches shown that measuring the service quality provides benefits and increasing the market share. Moreover, it also allows the company to take part in a competitive environment [9]. It is observed that the companies that give the importance to the measurement of service quality are more successful when the customer-based strategy is applied [10].

Especially, in Turkish economy, share of industry in Turkey is 38%, while the share of trade is almost 45%. This ratio is almost 17% for the service sector and service sector has the third highest share in Turkey's economy [11]. Furthermore, in the economy of Turkey, there are sub-sectors of service sector such as transportation and storage, administrative and support service activities, accommodation and food service activities, information and communication, professional, scientific, technical activities, and others. The share of transportation and storage activities is almost 37% as having the largest share in Turkey's economy [11].

Therefore, companies in the service sector measure their service quality to respond to the needs and demands of their customers and thus to highlight themselves in the sector. In this study, the main aim is to conduct research on the quality of service expected and perceived by customers who get service from the Logistics Company using SERVQUAL and BWM. With the study, it is aimed to examine whether the customers who get service from the logistics company have a difference in the expected service quality and perceived service quality.

2 Problem Statement

Each service has a unique concept without a certain standard that each customer can perceive differently [7]. There are many definitions of service in the literature. According to American Marketing Institute, services are activities that provide satisfaction and benefits associated with the sale of a good or offered for sale [8]. Since the second half of the twentieth century, the two main sectors which are agricultural and industrial sector have left behind the service sector, in both developed and developing countries. For this reason, the services sector has gained increasing importance in the development process of the economy. In addition, the fact that the services sector is an indicator of the level of prosperity and development in the countries, the increase in human needs parallel to the developing technology and the different types of services and the large increase in the human-weighted employment volume clearly reveal the importance of this sector [12].

In a highly competitive environment, companies develop strategies to protect their existing customers and to gain new customers. While the competition among the companies are increasing, it is not enough to reduce costs. Therefore, it has become important to measure the quality of services in order not to lose their customers. Service quality affects customer satisfaction and determines whether the service meets certain criteria [13]. Service quality consists of perceived quality and expected quality. Perceived quality can be defined as the customer's decision about the general position and excellence of the service they receive, while the expected quality explains expectations about the service they receive [14].

In addition, the measurement of service quality in the sector is more difficult than measuring the quality of production. There are certain concrete data in the production sector but the expectations of the customers in the service sector may vary. That is why to fill the gaps between customer perceptions and expectations about the service received is essential for customer satisfaction. In this study, we aimed to identify the service quality of the third-party logistics company's by using the SERVQUAL method. Quality of services is measured by many different methods. These methods are SERVQUAL, Servperf, Grönroos, the critical incent technique and Benchmarking. SERVQUAL Method is one of the most common method in these techniques to measure service quality ([15, 16, 17]).

The purpose of the study is to measure the service quality perceived by customers of the logistics company during the service they receive. In the study, firstly it is aimed that determining the dimensions of logistics service quality and their importance level. Secondly, determining the SERVQUAL points of service provided by a logistics company providing transportation services and comparing these points according to perceived and expected service quality. Lastly, analyzing the results based on demographic characteristics of the target audience by using SERVQUAL scores and BWM.

In the following section, methodology of the study is explained in detail.

3 Methodology

The methodology of the study includes two main parts as SERVQUAL Method and Best Worst Method (BWM). In the first part, a questionnaire is prepared by considering literature review to measure service quality of the logistics company. The survey consists of three parts; 5 demographic questions, 22 questions based on expectations, 22 questions based on perceptions. The survey study is conducted on 110 people who are the customers of the logistics company and workers of the logistics company via by e-mail or face-to-face between 22 February and 5 April 2019. After collected data, the results are analyzed with SERVQUAL Method and BWM.

In the following part, the SERVQUAL Method is explained in detail.

3.1 SERVQUAL Method

Quality means to something whether it is great or not [18]. Service quality is a consumer's overall view of the related inferiority/superiority of the company and its services [19]. Service quality conditions affect the competitive advantage of a logistics company, which brings market share and profitability [20].

By considering literature review, it can be seen that there are some studies maintains that the quality of service can be defined from different expectations of users towards the services provided with the perception of the service received [21]. Furthermore, the service quality is the benefit of the service received after the purchase of the service and it expresses whether they are satisfied with the service [22]. The concept of service quality has a great importance to meet the customer needs and expectations [22].

Factors determining service quality; the environment and appearance where the service is presented, the timing of the service, the expertise of the service providers, the continuity of the service, the reliability, reliability, and flexibility [23]. Moreover, factors that separates service from goods are; having characteristic features such as abstraction (lack of physical existence), inseparability (production and consumption are inseparable), variability, heterogeneity and intolerance [23]. SERVQUAL method as service quality through the gap among the customers' perception and expectation of company's service quality performance. Service quality is consisting of expected and desired quality [17]. By considering literature review, lots of researches have defined that SERVQUAL method is the most proper and true model to be used in determining the quality of the services offered ([15–17]).

An instrument to measure service quality based on their exploratory research regarding consumer assessment of service quality. The instrument, SERVQUAL, consists of 22 statements for assessing consumer perceptions and expectations of service quality in service and retailing organizations. Respondents are asked to indicate the degree to which they agree or disagree with a statement on a 7-point scale. The 22 items are the result of a 2-stage process, beginning with 97 items representing 10 determinants of service quality. The first stage is focused on condensing the number of items based on discriminating ability and examining dimensionality and reliability. The purpose of the second stage is to examine the dimensionality of the scale and to establish the reliabilities of its components. This is accomplished by analyzing data from four independent samples (bank, credit card company, repair and maintenance company, and long-distance telephone company) [24].

SERVQUAL method is a comprehensive method of measuring the perceived and expected service quality of its customers [3]. In the following years, the scale was updated because of the gradual implementation of the service sectors in different business sectors. The first scale prepared was 10-dimensional and then reduced to 5. Then, this model is intended to be applied in different service sectors [25]. Furthermore, SERVQUAL scale as 5 sensory activity dimensions such as tangible, reliability, responsiveness, assurance and empathy [9].

In the following part, Best Worst Method is explained in detail.

3.2 Best Worst Method (BWM)

Best Worst Method (BWM) is the one of the multi criteria decision-making methods. This method developed by Rezai [27]. The method is used to evaluate a set of alternatives with respect to a set of decision criteria. The BWM has a unique structure. The major aspect of BWM, in contrast to most existing MCDM methods, are: (i) requires less comparison data; (ii) it leads to more consistent comparisons, which means that it produces more reliable outcomes [28]. It develops the comparison system in a very structured way reducing the existence of inconsistency. Its structured way of comparison is more consistent with rational human thinking, main attribute of the method that has been confirmed by the respondents within the case company. In the following section, the steps of conducting the Best-Worst Method (BWM) are explained, and the requirements of conducting BWM are also presented [29].

The steps of BWM are explained in below.

S1: A set of decision criteria are identified. In this step, a set of criteria $\{c_1, c_2, c_3, \ldots, c_n\}$ is chosen to make a decision.
S2: The decision maker selects the best criterion and worst criterion among the available set of criteria, which is determined in step 1.
S3: The decision maker carries out pairwise comparisons between the best criterion and other criteria.
S4: The decision maker carries out pairwise comparisons between the worst criterion and other criteria.
S5: The maximum absolute differences are minimized to obtain unique solution weights.

Let us define model decision variables and parameters and variables as follows:

w_B : weights of the best criteria
w_W : weights of the worst criteria
w_j : weights of criteria
a_{Bj} : vector ofbest criterion with other criteria j
a_{jW} : vector ofworst criterion with other criteria j
ε_j : the maximum absoulute differences

This problem is transferred to the following mathematical linear model;

$$
\begin{aligned}
&\min \varepsilon_j \\
&\text{s.t.} \\
&\left|w_B - a_{Bj} w_j\right| \leq \varepsilon \\
&\left|w_j - a_{jw} w_w\right| \leq \varepsilon \\
&\sum_j jW_J = 1 \\
&w_j \geq 0, \quad \text{for all j}
\end{aligned}
$$

The proposed mathematical model is coded using AMPL and solved by CPLEX 9.1 on a computer with a Pentium IV 2.8-GHz processor and 1 GB of RAM.

In the following part, findings and results of the study is explained.

4 Findings and Results

After the results of the survey are obtained, demographic results are analyzed and SPSS analysis is interpreted based on SERVQUAL Method, Reliability analysis, Data analysis, T-tests and ANOVA analysis respectively.

4.1 Demographic Results

The survey which consists three parts and 110 respondents is prepared to measure service quality of a logistics company in Turkey with SERVQUAL and BWM.

Demographic results are given in Fig. 1. According to demographic results, it can be seen that the majority of participant in the company receive sea transportation services with the percent of 32%. Furthermore, 26% of the participants of the survey received air transportation services. Moreover, 25% of the participants of the survey receive freight transportation services. In addition, 17% of the survey participant of the customer buying other type of services i.e. forwarding services, intermodal and multimodal transport services.

Types of service	%	Sectors	%	Working Years	Number of company
Sea Transportation	32	Logistics	27	1	10
Air Transportation	26	Others	18	2	27
Freight Transportation	25	Food	16	3	26
Others (Forwarding services,intermodal, multimodal etc.)	17	Textile	16	4	21
		Automotive	10	5 and more	26
		Electronics	5		
		Agriculture	5		
		White Appliances	3		

Core Activity	%
Production	39
Service	36
Industry	12
Others(natural stone,husbandry etc.)	9
Agriculture	3

Fig. 1. Demographic results

Moreover, 27% of customers which has the highest portion of the participant of the survey have been working with the company for 2 years. Furthermore, 26% of survey participants have been working together with the company for 3 years, and the other 26% is working for 5 and more years. In addition, it is obtained that the highest portion of the participant of the survey work in core activity is production with percent of 39%. Moreover, 36% of the participants of the survey work in service core activity.

Furthermore, the majority of participant in the sector survey serve in logistics with the percent of 27%. Moreover, 18% of the participants of the survey serve in other industries i.e. plastics, natural stone. Furthermore, 16% of the participant survey serves in textile and 16% in food industries with same percent.

The majority of participant in the transport sector receive sea transportation with the percent of 11%. Moreover, 8% of the participants of the survey received freight transportation. Furthermore, the highest portion of participant of the survey work in sales and marketing department with the percent of 26%. Moreover, 24% of the survey participant work in logistics department in their company.

4.2 Results and Interpretation

In the following section, SPSS analysis interpretation is explained.

(1) *SERVQUAL Method*

The SERVQUAL scale is based on measuring the perception of the performance of the customer from the service and the actual performance of the service as a result of the consumption experience [13].

Reliability refers to the similarity of independent but comparable measurements of a concept, property or object using the same method [29]. For this purpose, reliability analysis is performed to test the reliability of the SERVQUAL model. The Cronbach Alpha coefficient for the scale is calculated as 0.901. Moreover, the fact that the alpha coefficient between 0 and 1 is close to 1 is an indicator that the scale is such a reliable scale [30].

The item-total statistics are given in Table 1.

Table 1. Item-total statistics

	Scale mean if item deleted	Scale variance if item deleted	Corrected item-total correlation	Squared multiple correlation	Cronbach's alpha if item deleted
Tangible	16,550	5,132	,573	,391	,914
Reliability	16,417	4,525	,812	,667	,866
Responsiveness	16,615	4,365	,813	,710	,865
Assurance	16,508	4,572	,819	,751	,865
Empathy	16,487	4,438	,758	,629	,878

As a result of the reliability analysis, factor coefficient is found to be 0.865, Alpha coefficient is 0.878, reliability coefficient Alpha coefficient is 0.866, Alpha factor of confidence factor is 0.865 and alpha coefficient of concrete properties factor is 0.914.

In the following part, data analysis is explained.

4.3 Data Analysis

For this study, SPSS software is used to analyze the data. Quality factor is firstly determined by factor analysis. The result of the Kaiser-Meyer-Olkin test, which is an index comparing the magnitude of the correlation coefficients between the variables and the size of the partial correlation coefficients, is 0.794. This result is quite acceptable since being close to 1.

In the following part, determining the perceptions of the Logistics company customers on service quality is analyzed.

4.4 Determining the Perceptions of the Logistics Company Customers on Service Quality

The perception and expectation levels of customers regarding the different dimensions of service quality are presented in Table 2. It is observed that Perception scores are below the expectation scores in all dimensions without exception from the values in the average column of the table.

The results of paired samples statistics are given in Table 2.

Table 2. Paired samples statistics

		Mean	N	Std. deviation	Std. error mean
Pair 1	Perceived tangible	4,0945	110	0,59274	0,05652
	Expected tangible	4,5564	110	0,42178	0,04021
Pair 2	Perceived reliability	4,2273	110	0,61382	0,05853
	Expected reliability	4,6205	110	0,42396	0,04042
Pair 3	Perceived responsiveness	4,0295	110	0,65554	0,0625
	Expected responsiveness	4,6114	110	0,46381	0,04422
Pair 4	Perceived assurance	4,1364	110	0,59781	0,057
	Expected assurance	4,5327	110	0,51136	0,04876
Pair 5	Perceived empathy	4,1568	110	0,6695	0,06383
	Expected empathy	4,4932	110	0,60811	0,05798

However, the following hypotheses are tested by using the paired sample t-test in order to determine whether the difference is coincidental (not based on any reason) or an indication of a problem in service quality.

H0 = There is no significant difference between the expectations and perception levels of the service quality of the customers of the logistics company.

H1 = There is a difference between the expectations and perception levels of the service quality of the logistics company customers.

The quality of service points that the Logistics company customers have received from the enthusiasm, empathy, reliability, trust and concrete characteristics are given in the Average column of Table 3. The values in the significance column of the table show that these differences are statistically significant at the significance level of $p < 0.05$.

Table 3. Paired samples statistics

		t	df	Sig. (2-tailed)
Pair 1	Perceived tangible - expected tangible	−6,982	109	,000
Pair 2	Perceived reliability - expected tangible	−5,596	109	,000
Pair 3	Perceived responsiveness - expected responsiveness	−8,002	109	,000
Pair 4	Perceived assurance - expected assurance	−5,279	109	,000
Pair 5	Perceived empathy - expected empathy	−4,179	109	,000

Based on these findings, H0 hypothesis is rejected and it is concluded that the Logistics customers' perceptions about service quality are lower than their expectations and consequently service quality of banks could be considered as insufficient.

In the following part ANOVA analysis is applied.

4.5 ANOVA Analysis

For this study, ANOVA analysis is applied. One-way ANOVA determines whether the perceptions of service quality dimensions and satisfaction were significantly different depending on the differences in the sample's demographic characteristics. In Table 4, the results of the ANOVA analysis is given. The analysis includes the relationship between perceptions and expectations of service quality's dimensions.

H0: = There is no significant difference between the expectations and perception levels of the service quality based on types of services for the logistics company.

H1 = There is a difference between the expectations and perception levels of the service quality based on types of services for the logistics company.

Table 4. ANOVA analysis based on service type

Based on service type	Tangible	Reliability	Responsiveness	Assurance	Empathy
Significant	0,533	0,771	0,686	0,346	0,504

ANOVA analysis is applied based on types of service. The types of services can be listed as road, air, sea and other. When it is examined the Table 4, in terms of service type of the logistics company, H0 hypothesis is rejected. There is no difference in H0a, H0b, H0c, H0e (tangible, reliability, responsiveness and empathy values), but there is a difference in H0d (assurance values). H0d is accepted. Significant values in the table were evaluated according to $p > 0.5$.

Furthermore, another ANOVA analysis is applied based on main activities. These main activities are production, industry, agriculture, service and others.

H0 = There is no significant difference between the expectations and perception levels of the service quality based on main activities for the logistics company.

H1 h = There is a difference between the expectations and perception levels of the service quality based on main activities for the logistics company.

Table 5. ANOVA analysis based on main activity

Based on main activity	Tangible	Reliability	Responsiveness	Assurance	Empahty
Significant	0,004	0,565	0,688	0,871	0,911

When it is examined the Table 5, in terms of service type of the logistics company, in terms of main activities of the logistics company, H0 hypothesis is rejected there is no difference in H0b, H0c, H0d, H0e (assurance, reliability, responsiveness and empathy values).There is a difference in H0a (tangible values). Significant values in the table are evaluated according to $p > 0.5$.

Moreover, weighted SERVQUAL method for the logistics company is calculated as follows;

Weighted SERVQUAL score = (reliability * 0.2098992 + empathy * 0.202 + assurance * 0.201 + tangibles * 0.196714 + responsiveness * 0.19073). Therefore, overall Weighted SERVQUAL score is calculated as 4.130.

4.6 Best Worst Method

As mentioned before, the proposed mathematical model is coded using AMPL and solved by CPLEX 9.1 on a computer with a Pentium IV 2.8-GHz processor and 1 GB of RAM. By solving the Best Worst Method model, the weights of different criteria and sub-criteria are determined. Table 6 shows the weights of the main criteria because of responses received from the respondents in the surveys. By solving the BWM model, the weights of different criteria and was obtained. Table 6 shows the weights of the main criteria based on responses received from the respondents in the survey.

Table 6. The weights of the main criteria

	Tangible	Reliablity	Responsiveness	Assurance	Empathy	ε_j
Mean	0,09	0,255	0,358	0,152	0,145	0,056
Median	0,055	0,288	0,310	0,165	0,112	0,182
Std	0,048	0,155	0,187	0,092	0,065	0,085

According to the mean values, responsiveness has been selected the most important criterion. The tangible is the least important criteria based on the BWM results. Based on the median scores, again responsiveness is the most important criteria. ε_j represents the consistency of the comparisons. The ε_j values are very close to zero. This means that there is a high consistency of the comparisons and a high reliability of the results.

5 Conclusion

The study subject is determined as the measurement of service quality. Many techniques are searched for measuring service quality and as a result of the researches, SERVQUAL technique is decided. Starting from the preparation process of the report, literature review is done from different sources and the study is supported by reference from many articles. As a result of the interviews conducted with the company, the compliance of the survey with the customers is determined and the questionnaire is presented to the customers. At the end of the survey, firstly, demographic questions are analyzed which are explained in Findings Section.

After collected survey data, the SPSS analysis is applied by calculating the difference between perception and expectation of the respondents. Afterwards, the analyzes in the research model are applied to the data of the scale. To determine the reliability of the study, the reliability analysis (calculation of Alpha coefficient) is made to measure the internal consistency of the data. In this process, reliability of the scale is found to be 0.925 and the analysis is continued as this value is within acceptable limits. Then, analyzes are performed on SERVQUAL scale. The service quality level offered by the firm on the one hand can be measured by using the SERVQUAL scale and the total quality level of the service offered by the company can be evaluated thanks to the equally weighted and weighted SERVQUAL scores calculated on the other hand. Moreover, to find optimal dimensions, BWM is applied in the study. Simple formulas are used to find the overall scores of the importance of the dimensions. BWM has several important attributes that make it a strong and user-friendly method compared to most multi-criteria decision-making methods. It requires less data; results in more reliable results and, because it does not use fractional numbers, it is easier to understand by the decision-makers. In this study, it is found that the equally weighted and weighted SERVQUAL scores are negative based on the quality dimensions of the firm. Accordingly, it can be interpreted that the company cannot meet customer expectations. Means that, the perception of customer quality related to the service provided in the company is low.

Acknowledgment. The study is prepared by Delfin Kolat, H.Ajlan Kökçü, Mete Kıranlı, Nazlı Ocak, Sinan Dağdelen, Fulya Ünlü in the context of senior design project in 2019.

References

1. Brady, M.K., Cronin, J.J.J.: Some new thoughts on conceptualizing perceived service quality: a hierarchical approach. J. Mark. **65**, 34–49 (2001)
2. Falk, T., Hammerschmidt, M., Schepers, J.J.L.: The service quality-satisfaction link revisited: exploring asymmetries and dynamics. J. Acad. Mark. Sci. **38**, 288–302 (2010). http://dx.doi.org/10.1007/s11747-009-0152-2288
3. Parasuraman, A., Zeithaml, V.A., Berry, L.L.: A conceptual model of service quality and its implications for future research. J. Market. **49**(4), 41–50 (1985)
4. Zeithaml, V.A.: How consumer evaluation processes differ between goods and services. In: Donnelly, J., George, W. (eds.) Marketing of Services, pp. 186–190. American Marketing, Chicago (1981)
5. Tenekecioğlu, B.: Makro Pazarlama, p. 159. MET Basim Yayim ve Organizasyon, Eskişehir (1992)
6. Belgin, Ö.: Hizmet Sektöründe Verimlilik. Bilişim Dergisi, Yil **38**, 113–114 (2010)
7. Filiz, Z., Kolukisaoğlu, S.: SERVQUAL Yöntemi Ve Bir Hizmet İşletmesinde Uygulamasi. Dumlupinar Üniversitesi Sosyal Bilimler Dergisi, 253–266 (2014)
8. Ozturkoglu, Y., Esendemir, E.: ERP software selection using IFS and GRA methods. J. Emer. Trends Comput. Inf. Sci. **5**(5), 363–370 (2014)
9. Parasuraman, A., Zeithaml, V.A., Berry, L.L.: Refinement a Reassessment of SERVQUAL scale. J. Retail. **67**, 420–450 (1991)
10. Whiteley, R.: The Customer-Driven Company, pp. 44–47. Addison-Wesley, Boston (1991)
11. TÜİK, Yillik Sanayi ve Hizmet İstatistikleri (2015). http://www.tuik.gov.tr/PdfGetir.do?id=21528
12. ÖZSAĞIR, A., Aliye, A.K.I.N.: Hizmetler Sektörü İçinde Hizmet Ticaretinin Yeri Ve Karşilaştirmali Bir Analizi. ElektronikSosyalBilimlerDergisi, **41**(41) (2012)

13. Kidak, L.B., Nişancı, Z.N., Burmaoğlu, S.: Sağlik Hizmetlerinde Kalite Ölçümü: Kamu Hastanesi Örneği (Quality Measurement in Health Services: a Public Hospital Case). Yönetim ve Ekonomi: Celal Bayar Üniversitesi İktisadive İdari Bilimler Fakültesi Dergisi **22**(2), 483–500 (2015)
14. Şekerkaya, A.K.: Bankacilik Hizmetlerinde Algilanan Toplam Kalite Ölcümü, p. 87. Sermaye Piyasası Kurulu, Ankara (1997)
15. Rauyruen, P., Miller, K.E., Barrett, N.J.: Relationship Quality as a Predictor of B2B Customer Loyalty **60**(1), 21–31 (2007)
16. Banomyong, R., Supatn, N.: Selecting logistics providers in thailand: a shippers'. Perspective **45**(3), 419–437 (2011)
17. Baki, B., Sahin Basfirinci, C., Murat A.R.I., Cilingir, Z.: An application of integrating SERVQUAL and Kano's model into QFD for logistics services: a case study from Turkey. Asia Pac. J. Market. Logistics, **21**(1), 106–126 (2009)
18. Zuraidah: New Oxford English-English-Malay Dictionary. Oxford FajarSdn. Bhd, Malaysia, p. 458 (2007)
19. Bitner, M.J., Hubbert, A.R.: Encounter satisfaction versus overall satisfaction versus service quality: the consumer's voice. In: Rust, R.T., Oliver, R.L. (Eds) Service Quality: New Directions in Theory and Practice, Sage Publications, Thousand Oaks, vol. 34(2), pp. 72–94 (1994)
20. Morash, E.A., Ozment, J.: Toward management of transportation service quality. Logistics Transp. Rev. **30**(2), 115140 (1994)
21. Munusamy, J., Cselliah, S., Hor, W.M.: Service quality delivery and its impact on customer satisfaction in the banking sector in malaysia. Int. J. Innov. Technol. Manag. **1**(4), 399–400 (2010)
22. Devebakan, N., Aksarayli, M.: Sağlık İşletmelerinde Algılanan Hizmet Kalitesinin Ölçümünde SERVQUAL Skorlarinin Kullanimi ve Özel Altinordu Hastanesi Uygulaması. Dokuz Eylül, Sosyal Bilimler Enstitüsü Dergisi **5**(1), 38–54 (2003)
23. Özveren, Y.S.: Müşteri Memnuniyetive Hizmet Kalitesi İlişkisi: Mersin İlindeki 4 ve 5 Yıldızlı Oteller Örneği (YayınlanmamışYüksek Lisans Tezi), pp. 531–533. Mersin Üniversitesi, Mersin (2010)
24. Bulgan, U.: Kütüphanecilik sektöründe hizmet kalitesinin ölçümü ve bir üniversite Kütüphanesi uygulaması. Yükseklisans tezi, Beykent Üniversitesi Sosyal Bilimler Enstitüsü, Türkiye, 21–25 (2002)
25. Karaman, D., Yumurtaci Aydoğmuş, H., Karakoç, A.: Sigorta Hizmeti Alan Müşterilerin BakışAçisiyla İlişki Pazarlaması. Uluslararasi Multi disipliner Çalışmaları Kongresi (International Multidisciplinary Studies Congress), 46 (2017)
26. Ersöz, S., Pınarbaşı, M., Türker, K.A., Yüzükırmızı, M.: Hizmet Kalitesinin Servqual Metodu ile Ölçümüve Sonuçlarin Yapısal Eşitlik Modelleri ile Analizi: Öğretmen Evi Uygulaması. Int. J. Eng. Res. Dev. **1**(1), 19–27 (2009)
27. Rezaei, J.: Best-worst multi-criteria decision-making method. Omega **53**, 49–57 (2015)
28. Rezaei, J.: Best-worst multi-criteria decision-making method: some properties and a linear model. Omega **64**, 126–130 (2016)
29. Rezaei, J., Nispeling, T., Sarkis, J., Tavasszy, L.: A supplier selection life cycle approach integrating traditional and environmental criteria using the best worst method. J. Cleaner Prod. **135**, 577–588 (2016)
30. Dalgakiran, A.B., Öztürkoğlu, Y.: Scale and relationship analysis for turkish furniture sector/türkiye mobilya sektörüne yönelik ölçek ve ilişki analizi. Bus. Manag. Stud.: An Int. J. **5**(1), 147 (2017)

Optimization of Costs in Empty Container Repositioning

Mehmet Yasin Göçen[1(✉)], Öykü Çağlar[1], Elif Ercan[2], and Damla Kizilay[1]

[1] Department of Industrial Engineering, Yasar University, Bornova, Turkey
gocenmehmet@outlook.com, oykucaglar12@gmail.com, damla.kizilay@yasar.edu.tr

[2] Department of Industrial Engineering, Dokuz Eylul University, İzmir, Turkey
elif.ercan@deu.edu.tr

Abstract. In this study, the empty container repositioning problem in a liner shipping company is considered. The research includes necessary information about the company and maritime transportation, highlighting the importance of empty container storage. As the worldwide containerized trade increases, the container traffic increases. Consequently, the surplus containers are repositioned to locations where they are required, which causes high costs. Therefore, in the existing system, the company deals with higher prices. In this study, the storage location of the empty containers, which are coming from the vessels to the ports, are determined according to the different cost policies of the warehouses. Thus, our problem is defined as an empty container repositioning problem. The primary objective of this study is to build an operations-research based decision-making system in order to minimize the total cost, regarding changing prices of warehouses and transportation cost of containers. In the decision-making process, this study will help the company to overcome the complexity about repositioning empty containers regarding the container size, storage day of the containers, changing inventory holding costs of the warehouses, and the transportation costs of the containers from port to the corresponding warehouses. Taking into consideration all these parts of the problem, this study will eventually help in reducing inventory holding and transportation costs of the empty container storage. As a solution methodology, the mixed integer linear programming model is proposed, and optimal solutions are obtained in a concise time.

Keywords: Empty container repositioning · Mixed integer linear programming model · Decision support system · Cost reduction · Optimization

1 Introduction

Maritime transport is defined as the transportation of people and products from one place to another. As the new port terminal investments are made, the operational efficiency of the existing terminals, and the importance of maritime transport is increasing. Today, approximately 90% of world trade is carried out by the sea. One of the important advantages of maritime transport is that it is cheaper than other

N. M. Durakbasa and M. G. Gençyılmaz (Eds.): ISPR 2019, LNME, pp. 732–747, 2020.
https://doi.org/10.1007/978-3-030-31343-2_62

transportation methods such as 3.5 times more economical when compared to the railroad, 7 times according to the road and 22 times less expensive than airline transportation. For this reason, maritime transport is carried out in different ways. One of them is Tramp Transportation. The critical thing in this type of transportation is the existence of the cargo. Therefore, shipowners would like to direct their ships to areas where there is more likely to be a load. The other is Liner Transportation. In liner transportation, the arrival, departure time, and transfer ports of the vessels are predetermined. Vessels come to these ports in the days and hours determined and perform the discharge operations. Due to the predetermined timelines, the port of the ship is specified.

In the 1950s, container transportation has been developed as a means of transportation. It became an important sector for the realization of lowering cost. The main components of container transportation are cargo, container, container ships, and ports. Container transportation in Turkey began to appear in the 1980s. Container handling rate in Turkey is the average of all the time in the world due to Turkey's strategical location. In general, making the transport as soon as possible is an essential detail while minimizing costs is controversial. Since it is more accessible, the nearest port is selected.

Container transportation prevents the physical effects that can come from outside during the transportation of the goods and allows to carry more load in one go. It is also a kind of transportation that has fewer harms for nature. Since there is intense competition within the container market, how to transfer empty containers effectively and efficiently at the right time is a critical decision at the operational stage. The commercial imbalance is the most significant cause of the problem of where and how to transfer. This imbalance creates an empty container surplus in some areas, and some regions create an empty container scarcity. Empty container assignment is usually performed randomly by the experienced operators. However, since each port and container tariffs vary, it is crucial to ensure this transportation with minimum cost.

This study is performed in a liner shipping company, which is located in Izmir, Turkey. The company has contracted with a software development company which runs software operations only for the liner shipping company. This liner shipping company is one of 67 companies and has 7300 employees; therefore, it is Turkey's largest shipping company and has several port operations. The company has 14 offices currently in Turkey. Apart from that, it works in 25 countries, 62 of them are workplaces. These countries are Greece, Italy, Monaco, France, Spain, Portugal, Germany, Morocco, Algeria, Tunisia, Egypt, Lebanon, Libya, Bulgaria, Romania, Ukraine, Russia, Georgia, Azerbaijan, Kazakhstan, Turkish Republic of Northern Cyprus, China, Ghana, and Nigeria. The liner shipping company is partnering with 10 successful companies (Hapag Lloyd, Yang Ming, and Tarros SPA). There are also representative relations with these ten companies (Grimaldi Lines, ACL, Marguisa Line, DAL, UAFL Ethiopian Shipping). The company has foreign subsidiaries outside Turkey. That is why it is an international company. The company in Turkey contributes to foreign trade. Turkey's highest-volume container port is Marport. It has an annual capacity of 2.3 million twenty-foot equivalent unit (TEU). This liner shipping company has five empty container depots in Izmir. There are many types of container which comes to Izmir Port. The container types are given in Table 1.

Table 1. Types of containers

Size	Door Openings mm Feet/Inches		Internal dimensions mm Feet/Inches		
Feet/Inches	Width	Height	Length	Width	Height
20 dry container	2330–2340	2275–2280	5880–5890	2330–2350	2380–2390
40 dry container	2330	2275–2280	12024–12032	2330–2350	2380–2390
40 high cube dry container	2330–2340	2585–2583	12032	2330–2350	2680–2695
20 flat rack container	–	–	5700–5900	2335–2350	2250–2276
40 flat rack container	–	–	11700–11980	2378	1960
40 platform container	–	–	11700–11980	2380	1955
40 high cube reefer container	2290–2294	2569	11590	2290	2544–2547
20 Iso tank	–	–	–	–	–
40 pallet wide	2395–2400	2274–2279	12095	2422	2387
45 high cube pallet Wide	2363–2410	2500–2584	13532	2426	2694
Feet/Inches	Width - Length (Roof Opening)		Length	Width	Height
20 open top container	2200–2230	5396–5757	5880–5897	2330–2350	2329–2350
40 open top container	2190	11500–11871	12007–12030	2330–2350	2329–2367

The liner shipping company uses the shipping plan of its sub-firm. The company has 71 years of experience in international transportation. Its sub-firm conducts regular transportation between the Black Sea, Mediterranean and West African ports with a total of 41 flagships. The aim of the sub-firm is managing decisions about the transformation between port, which are unloading and demand points or warehouses. Its companies are divided into shipowner and ship management, port operations, automotive services, maritime transport and agency, bunkering services, logistic services, and other services.

A software development company which has contracted with the liner shipping company started to work as the data innovation division in 1984, but in 1997, it was restructured. It was established within the liner shipping company to provide IT management to other companies.

A software development company attaches importance to its customers and employees in management procedures. In addition, the company considers creativity and working culture. It helps its customers to achieve success and cares about their satisfaction, customer-focused work. According to the software development company, business partners are its customers. The company meets the needs of its customers and examine their business processes. It attaches importance to the specialization of their employees in every subject and contributes to the training and development of each employee. Also, it is open to innovation and creativity. By making inferences from their mistakes and working more effectively, they increase the quality. The company considers the decisions of each employee and ensures their participation.

This study tries to optimize the cost of the empty container repositioning, including both storage and transportation costs. The rest of the paper is organized as follows: In Sect. 2, the symptoms in the system are explained, and concerning those symptoms, the problem is defined. Section 3 presents the related literature with the leading articles on empty container management problems. Formulation of the mathematical model with the all parameter and decision variables explanations are given in Sect. 4, which presents the solution methodology of this study. The formulated model is verified and validated for the real data obtained from the company. These verification and validation procedure are explained in Sect. 5. For the collaborated company, a decision support system (DSS) is developed and presented in Sect. 6 with computational results of the formulated model. Finally, the conclusion and future work are stated in Sect. 7.

2 Problem Definition

Empty containers management in the maritime logistics sector has always been an issue that needs to be improved. Once the vessel arrives at the container port terminal, the import containers are unloaded from the vessel, and the export containers are loaded onto the vessel. For an export order, an empty container is transported to the customer where it is filled, and the full container is then delivered at the departure terminal. For an import order, the flow is usually reverse, i.e., a full container must be delivered from the arrival terminal to the customer. In the existing system, there are two different transportation methods for empty containers. In the first method, for import orders, once the customer receives his/her import container(s), these full containers need to be emptied at the container port terminal and the goods need to be transported to the customers' depot usually by the customers' trucks. Once the load is moved out from the import container, the container becomes empty and needs to be cleaned (if necessary) and then, be transported to one of the available warehouses by the liner shipping company's trucks and stored there for a while. In the second method, import orders are retrieved by the customers from the terminal and emptied at the customer's depot. Then those empty containers are transported from customer to one of the available warehouses by the customer's trucks. Those two methods are shown in Fig. 1.

As it is shown in Fig. 1, the company has five warehouses. In the first method, emptied containers are checked in the unloaded area of the terminal, whether they are damaged or not. Damaged containers are directly transferred to Sütçüler Warehouse, which is shown by red arrow because there are no other warehouse options for the damaged containers. Other empty containers which are not damaged, are moved to appropriate warehouses as shown by the black arrows. For those containers, the company is responsible for their transportation costs. In the second method, after customers take their containers from the port with their trucks, they empty the containers in their container unloading area, and each container is controlled by customers or agencies whether it is damaged or not. Damaged containers are directly moved to Sütçüler Warehouse, which is shown by the red dashed arrow. Remaining empty containers which are not damaged will be transferred to the suitable warehouses by the customers as shown by the dashed black arrows. Therefore, the transportation cost of those containers belongs to the customer. However, in both methods, it is the company's responsibility to check the availability of the warehouses and to pay the storage cost of the empty containers according to the pricing policies of the assigned warehouses.

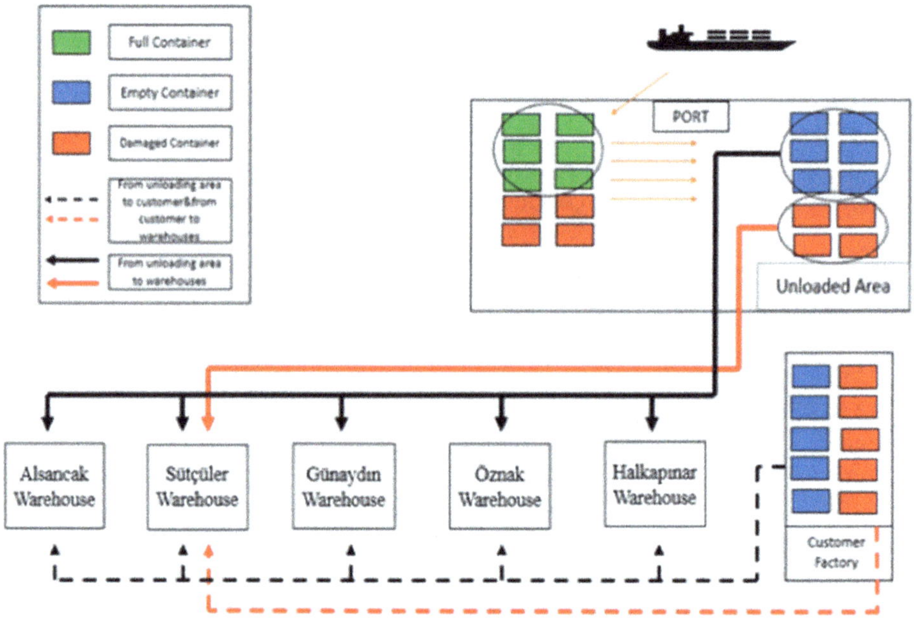

Fig. 1. Empty container flow in the existing system

Depending on the container sizes, the number of containers stored, and the total storage days of the containers, the company has several contracting schemes, incurring different pricing policies for each candidate warehouse as given in Table 2. The company has to pay two types of costs; gate cost and daily storage cost. As shown in the table, every warehouse has a gate cost, which is paid for each container. For the first, second, third, and fourth warehouses, the price for gate cost is \$15 per container. For the fifth warehouse, gate cost is \$30 per container. Daily storage cost is the price paid for each container daily according to their storage days. For the first, second, and third warehouse, the price is \$2 per day per container. For the fourth warehouse, the daily storage cost is increasing as the storage day of the container increases. If a container stays at the fourth warehouse between 0–7 days, then the company pays \$2 per day. If containers are stored in that warehouse for more than 7 days, then the company starts to pay \$2.5 for each additional day until day 14. If the container stays in the fourth warehouse between 14–21 days, then the company pays \$3 per extra day. If the storage day of the container is more than 21 days, then \$4 is paid for those additional days.

The company has many kinds of container types for maritime transport. They are standard, high cube, open top, flat rack, platform, reefer, and ISO tank. They have different size of each type of container. The dimensions of each container are 20 or 40 TEU. The term "free pool" is defined for some of the warehouses. "free pool" indicates that the specified number of containers depending on their TEUs can stay on that warehouse without paying for the daily storage costs. If the number of assigned containers to those warehouses exceeds the specified free pool amount, then the company starts to pay for the daily storage cost for the excess containers. The free pool is available for warehouse 1, warehouse 3 and warehouse 4 and it is restricted for 100 containers of 20 TEU's or 50 containers of 40 TEU's. The term "free day" indicates that if the storage day of the containers is less than the specified free day amount, then the company does not pay for the daily storage cost for those containers. On the other hand, if the storage day of the container is greater than the free day amount, then the company starts to pay the daily storage cost for the excess days of each container. Free days policy is valid for second and third warehouses and limited to 7 days.

Table 2. Pricing policies of warehouses

	Gate cost	Daily storage cost	Free pool	Free days
Warehouse 1 (Halkapınar)	$15	$2	100 units of 20 TEU or 50 units of 40 TEU	N/A
Warehouse 2 (Günaydın)	$15	$2	N/A	7
Warehouse 3 (Sütçüler)	$15	$2	100 units of 20 TEU or 50 units of 40 TEU	7
Warehouse 4 (Alsancak)	$15	$2.0 (0–7) $2.5 (7–14) $3.0 (14–21) $4.0 (21 +)	100 units of 20 TEU or 50 units of 40 TEU	N/A
Warehouse 5 (Öznak)	$30	$0	N/A	N/A

The third warehouse has both free pool and free days policies; therefore, both conditions should be checked if a container is assigned to that warehouse. If the assigned container is in the limits of free pool, then its storage day should be checked. If the storage day does not exceed the free day, then the company does not pay for the daily storage cost. Otherwise, the company should pay $2 for each exceeding day of each container. If the assigned container exceeds the limits of the free pool, then the company pays the daily storage cost for that container regardless of the free day policy. The fifth warehouse does not have any daily storage cost. However, its gate cost is more than the other warehouses. In Table 2, the warehouses that do not have any free pool or free day policies are shown as N/A, which means not applicable.

Moreover, these warehouses can be either within the container port terminal or can be located at the outside. Hence, the transportation cost and time changes depending on where the container is transported. The names of the warehouses, distances of warehouses from the port of Izmir, in which the company operates, are given in Table 3 in terms of kilometers (km).

Table 3. Distances between the warehouses of the company and the Port of Izmir

Name of the warehouses	The distance from port to warehouses	Empty/Full container storage
1. Alsancak warehouse	2 km	✓
2. Halkapınar warehouse	2 km	✓
3. Sütçüler warehouse	30 km	✓(Extra: damaged containers)
4. Günaydın warehouse	5.3 km	✓
5. Öznak warehouse	2 km	✓

Currently, the company assigns the empty containers to the warehouses according to the experiences of the operators manual. These manual assignments take a very long time due to the manual calculation of the costs. Also, according to the interviews with the managers, in these assignments, the storage day of the containers are ignored, which has a significant effect on the storage costs. Therefore, the current planning process is not systematic and causes unnecessary expenses to the company. In the following fishbone diagram, the symptoms of the problem are defined concerning the current container-warehouse assignments.

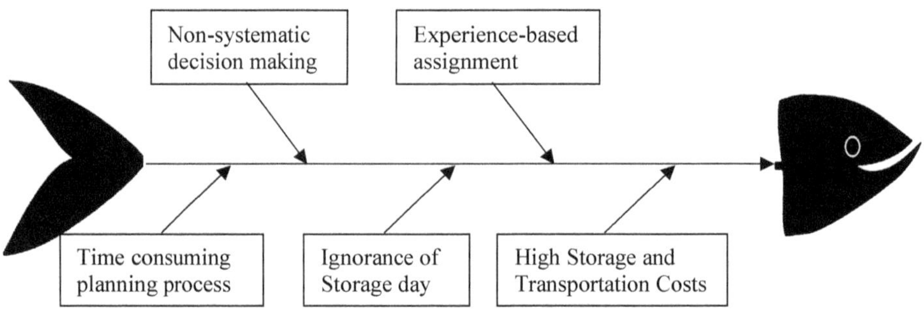

Fig. 2. Fishbone diagram

Figure 2 shows the issues of the current system through the fishbone diagram. The specified problems result in the necessity of repositioning empty containers. Therefore, to satisfy the inventory constraints for empty containers, the company has requirements for where to store the empty containers.

Our project's aim is minimizing the total cost of empty containers while satisfying the assignment of the containers to the warehouses. First performance measure and the main objective is to minimize the total cost, including the transportation cost of the empty containers and inventory holding cost at the warehouses. Empty container repositioning plays a vital role in the liner shipping industry. It does have an economic impact on the stakeholders in the container logistics chain; meanwhile, it has an environmental and sustainability effect on the society because of the decrease of empty container movements will also decrease fuel consumption and reduce crowding. In this case, we will deal with empty containers' management.

3 Literature Review

In this section, a literature review is done related to empty container repositioning problem.

Song and Carter [1] focus on empty container repositioning in liner shipping. The objective of this article is to recognize essential components that affect purge container development. They mention the reduction of empty container movements, which has impact reducing fuel consumption, congestion, and emissions. They propose a deterministic linear programming model as a solution method. Epstein et al. [2] built a decision support system to make thousands of container repositioning decisions automatically. Since the decision-making process is complex, and there are uncertainties, their decision support system (ECO) helped both researchers and the company to overcome the complexity of the problem. Song and Dong [3] consider cargo routing and empty container repositioning problem. The objective of this study is to minimize the total cost in the planning horizon, and they propose a stochastic linear programming model. Bandeira et al. [4] propose a mathematical model for transshipment operations of both empty and full containers. Moon et al. [5] mention the empty container repositioning problem for companies. The study aims to minimize total cost, which includes folding/unfolding cost, inventory cost, container purchasing cost, and repositioning cost. They formulate a mathematical model and propose heuristic algorithms. Shintani et al. [6] develop the design of container liner shipping service networks by explicitly considering empty container repositioning. Both ship and container deployment problems are discussed in this study. Thus, the problem is formulated as a two-stage problem. They aim to find the sequence of calling ports, the required number of ships for each size category and the cruising speed with maximum profit objective for the liner shipping company. The proposed problem is solved by a heuristic based on a genetic algorithm. Hjortnaes et al. [7] propose a multi-commodity model which considers the flows of both damaged and non-damaged containers separately. The proposed model gives the repositioning of containers in the network, and the objective of the study is to minimize total cost.

4 Mathematical Model and Solution Methodology

In this section, the formulated mixed integer linear programming model is introduced. Our solution method consists of two steps. In the first step, the data is pre-processed, and the damaged containers are sent to the Sütçüler Warehouse directly. Then the proposed model is solved for undamaged containers, and the model determines the assignment of the empty containers to the warehouses optimally.

The mathematical model is written regarding five different warehouse policies. Assumptions of the mathematical model are explained below:

- We assume that inspection on emptied containers is carried out, and the model considers only undamaged empty containers as an input.

- Every warehouse has a maximum capacity level for each type of container. However, we assume that the capacity of each warehouse is sufficient enough to store all the containers. Our planning period is restricted with each upcoming vessel to the port. Therefore, there is always enough space at the warehouses to store all the empty containers discharged from the vessel.

Sets, parameters, decision variables, and the mathematical model formulation are given below. The objective function and constraints are explained in detail.

(1) *Sets and the Parameters*

N : Set of containers
M : Set of warehouses
N^v : Set of containers transferred from port to warehouses by the company's trucks
g_j : Gate cost of warehouse $j, j \in M$
c_j : Unit transportation cost of each container from port to warehouse j, $j \in M$
d_i : Total number of storage days for container $i, i \in N$
ds_j : Daily storage cost of warehouse $j, j \in M | j \neq 4$
$ds_{j,k}$: Increasing daily storage cost of warehouse j based on the number of storage days in interval k $k|k = 1, 0 \le d_i < 8;$ $k = 2, 8 \le d_i < 15;$ $k = 3, 15 \le d_i < 22;$ $k = 4, 22 \le d_i,$ $j = 4$
$freedays_j$: Storage day limit of warehouse j for free storage
$freepool_j$: Container limit of warehouse j for free storage

$$TEU_i = \begin{cases} 1, \text{ if container } i\text{'s TEU is 20} \\ 2, \text{ if container } i\text{'s TEU is 40} \end{cases}$$

(2) *Decision Variables*

$$x_{i,j} = \begin{cases} 1, \text{ if container } i \text{ is assigned to warehouse } j \\ 0, \text{ otherwise} \end{cases}$$

$$w_{i,j} = \begin{cases} 1, \text{ if container } i \text{ excesses free pool limit for warehouse } j \\ 0, \text{ otherwise} \end{cases}$$

$$fp_{i,j} : \begin{cases} 1, \text{ if container } i \text{ is 20 TEU and in the free pool of warehouse } j. \\ 2, \text{ if container } i \text{ is 40 TEU and in the free pool of warehouse } j. \\ 0, \text{ if container } i \text{ exceeds the free pool limit of warehouse } j. \end{cases}$$

$y_{i,j}$: The amount of days exceeding the number of free days of the container i assigned to warehouse j

(3) *Mathematical Model*

Minimize Z =

$$\sum_{i \in N} \sum_{j \in M} g_j * x_{i,j} +$$
$$\sum_{i \in N^v} \sum_{j \in M} c_j * x_{i,j} +$$
$$\sum_{i \in N} \sum_{j \in M | j \neq 2,5} d_i * w_{i,j} * ds_j +$$
$$\sum_{i \in N} \sum_{j \in M | j \neq 1,4,5} y_{i,j} * ds_j +$$
$$\sum_{i \in N | d_i < 8} d_i * w_{i,4} * ds_{4,1} + \sum_{i \in N | 8 \leq d_i < 15} (7 * w_{i,4} * ds_{4,1}) + ((d_i - 7) * w_{i,4} * ds_{4,2}) +$$
$$\sum_{i \in N | 15 \leq d_i < 21} (7 * w_{i,4} * ds_{4,1}) + (7 * w_{i,4} * ds_{4,2}) + ((d_i - 14) * w_{i,4} * ds_{4,3}) +$$
$$\sum_{i \in N | 22 \leq d_i} (7 * w_{i,4} * ds_{4,1}) + (7 * w_{i,4} * ds_{4,2}) + (7 * w_{i,4} * ds_{4,3}) + ((d_i - 21) * w_{i,4} * ds_{4,4})$$

Subject to

$$\sum_{j \in M} x_{i,j} = 1, \qquad \forall i \in N \tag{1}$$

$$x_{i,j} * TEU_i \leq fp_{i,j} + (w_{i,j} * TEU_i), \qquad \forall i \in N, \forall j \in M | j \neq 2,5 \tag{2}$$

$$\sum_{i \in N} fp_{i,j} \leq freepool_j, \qquad \forall j \in M | j \neq 2,5 \tag{3}$$

$$x_{i,j} * d_i \leq freedays_j + y_{i,j}, \qquad \forall i \in N, j = 2 \tag{4}$$

$$x_{i,j} * d_i \leq freedays_j + y_{i,j} + (M * w_{i,j}), \qquad \forall i \in N, j = 3 \tag{5}$$

The objective function minimizes the gate and daily storage cost of containers, as well as the transportation cost of containers from port to each warehouse. The first equation of objective function describes gate cost of the warehouses for each container, while the second equation of the objective function indicates the unit transportation cost of containers that sent to warehouses from the port by the company's trucks. The company pays only unit transportation cost for those containers that transported from the port. The transportation costs of the empty containers that are carried by the customers belong to the customer. The third part of the objective function defines the daily storage cost for containers which exceed the free pool limits of the first, third, and fourth warehouses. The free pool is not valid for warehouse two and warehouse five. The fourth equation calculates the daily storage cost for the containers which exceed the free day limit of the warehouses, and this policy is valid for second and third warehouses. The fifth equation of the objective function is defined for warehouse four and calculates, increasing daily storage cost, regarding the storage days of the containers.

According to the constraint (1), each container can only be assigned to a single warehouse. It is assumed that all warehouses have enough capacity to store all incoming containers within the planning period. Constraints (2) and (3) are written for warehouses with free pool policy. The constraint (2) indicates that if the container is inside $\left(fp_{i,j} = TEU_i\right)$ or outside $\left(w_{i,j} = 1\right)$ the free pool limit of its assigned warehouse, and according to the TEU volume of each container and computes the number of containers which exceed the free pool limit $\left(w_{i,j} = 1\right)$. If the containers do not exceed the free pool limit, then the value of $w_{i,j}$ equals to zero. The restriction (3) ensures that the containers counted in the free pool limit of their assigned warehouses do not exceed the given free pool limit of those warehouses. For the containers which are outside the free pool limit, the company has to pay the daily storage cost, which is calculated in the objective function. Free pool limit should be less than the determined number. Constraints (4) and (5) are written for warehouses that have free days policy. Constraint (4) states that if the storage days of a container is greater than the specified free days, then the value of $y_{i,j}$ will be equal to the exceeding storage days of that container for the second warehouse. If the storage day of a container is within the free day limit, then the value of $y_{i,j}$ will be zero. Daily storage cost for those containers that exceed free day limit is calculated in the objective function. Constraint (5) is written for the fourth warehouse, which has both free pool and free days policies. If the container assigned to this warehouse is within the free pool $\left(w_{i,j} = 0\right)$, then its storage days are checked if it is exceeding the free day limit $\left(y_{i,j} > 0\right)$ or not $\left(y_{i,j} = 0\right)$, and the company pays for its daily storage cost accordingly. If the container assigned to this warehouse exceeds the free pool limit $\left(w_{i,j} = 1\right)$, the company pays for its daily storage cost, and there is no need to check if it is in the free day limit or not. Constraints in accordance with the policy of each warehouse have been determined, and the storage and transportation cost of the containers have been minimized in the objective function.

5 Verification, Validation and Computational Results

At first, the formulated mathematical model is verified with a toy data set. Toy data is generated randomly, and it is shown in Table 4. This table also shows the example of the inputs in our problem. In our toy data set, there are 10 empty containers which are indicated as "ID" in the first column. Then, their TEU volume is generated uniformly between U(1,2) and shown as "TEU" in the second column. Storage days of the containers are created uniformly between U(2,50) to see the effect of the free day policy of the warehouses. Finally, the last column indicates the origin of the empty containers. If the empty container comes from the port by the company's trucks, then the origin value equals to one. If the empty container comes from the customer by the customer's vehicles, then the origin value equals to two. We should keep in mind that the company pays the transportation cost of the containers if they are transferred by the company's trucks. Therefore, the "origin" value of the containers are generated uniformly between U(1,2).

Table 4. Input for toy data

ID	TEU	Storage Day	Origin
1	1	48	1
2	2	40	2
3	1	6	2
4	2	27	1
5	2	12	2
6	2	5	2
7	2	23	2
8	1	32	1
9	1	19	1
10	2	8	1

Due to the small number of containers are included in the toy data, free pool, and free day limits of the warehouses are given accordingly. Free pool limit of warehouses 1, 3, and 4 are given as 4 containers for 20 TEU sizes, or 2 containers for 40 TEU sizes. Free pool limit of the warehouses 2 and 3 are given as 7 days as stated in the real data. Transportation costs of the empty containers from port to warehouses are determined regarding the distances, as shown in Table 3. Trucks can carry one container at a time independent from TEU size of the containers; therefore, unit transportation cost of all containers are the same but differs for each warehouse. Table 5 presents the unit transportation costs of each warehouse, calculated with respect to the distances.

Table 5. Unit transportation costs of the warehouses

Name of the warehouse	Transportation cost
1. Alsancak warehouse	$ 0.5
2. Halkapınar warehouse	$ 0.5
3. Sütçüler warehouse	$ 7.5
4. Günaydın warehouse	$ 1.3
5. Öznak warehouse	$ 0.5

The presented toy data is solved by IBM ILOG CPLEX Optimization Studio v12.8 optimally within seconds. Also, all the alternative assignments are performed manually, and the cost of each assignment is calculated. Result of the optimal solution and the costs of all alternatives are compared to verify the developed mathematical model. The optimal solution of the toy example is shown in Table 6, indicating the assignment of each container to one of the five warehouses concerning their TEU sizes, storage days, and the distances of the warehouses from the port. In the table, "1" indicates that the container in that row is assigned to the warehouse in the corresponding column. Otherwise, the cell has "0" value. At the bottom row, the assignment costs of each warehouse are calculated.

Table 6. Optimal results of toy data

ID	Warehouse-1	Warehouse-2	Warehouse-3	Warehouse-4	Warehouse-5
1	1	0	0	0	0
2	0	0	0	1	0
3	0	1	0	0	0
4	1	0	0	0	0
5	0	1	0	0	0
6	0	1	0	0	0
7	0	0	0	1	0
8	1	0	0	0	0
9	1	0	0	0	0
10	0	1	0	0	0
Cost	$ 60	$ 72	$ 0	$ 30	$ 0

According to the optimal solutions of the toy problem, none of the containers is assigned to warehouse 3 and warehouse 5 due to the high transportation costs and high gate cost of those warehouses. Containers, whose storage days are less than 7 and less than the other container's storage days are assigned to the warehouse 2 because of its free day policy. Containers whose storage days are significant, are preferred to be assigned to the warehouses 1 and 4 respecting their free pool policy. Those assigned containers do not exceed the free pool limit, so the company does not pay for the daily storage cost.

At the validation process, the empty container repositioning problem is solved with different sizes of real instances by IBM ILOG CPLEX Optimization Studio v12.8. In the instance sets, the size of the containers is taken as 50, 100, 150, 200, respectively. The optimal solution is found for each instance set in a concise time. The runtimes of these instances are listed as follows:

- Runtime for 50 containers: 8 s 97 ms.
- Runtime for 100 containers: 7 s 71 ms.
- Runtime for 150 containers: 8 s 45 ms.
- Runtime for 200 containers: 8 s 75 ms.

These CPU results indicate that the formulated model to solve the problem is adequate to address large-sized instances, and optimal results can be obtained in a very short time. According to the interviews with the port managers, up to 200 empty containers are derived from a single incoming vessel. This means that our largest case has 200 containers. The inputs of the real data are obtained from the liner shipping company, and the actual policies of the warehouses are used. Free pool limit of the warehouses 1, 3 and 4 are 100 containers of 20 TEU size or 50 containers of 40 TEU size. Free day limit of the warehouses 2 and 3 are 7 days. Unit transportation costs are taken as stated in Table 5. The following Fig. 3 summarizes the results of the optimal assignments of real data for 200 containers getting an average of 10 different instances.

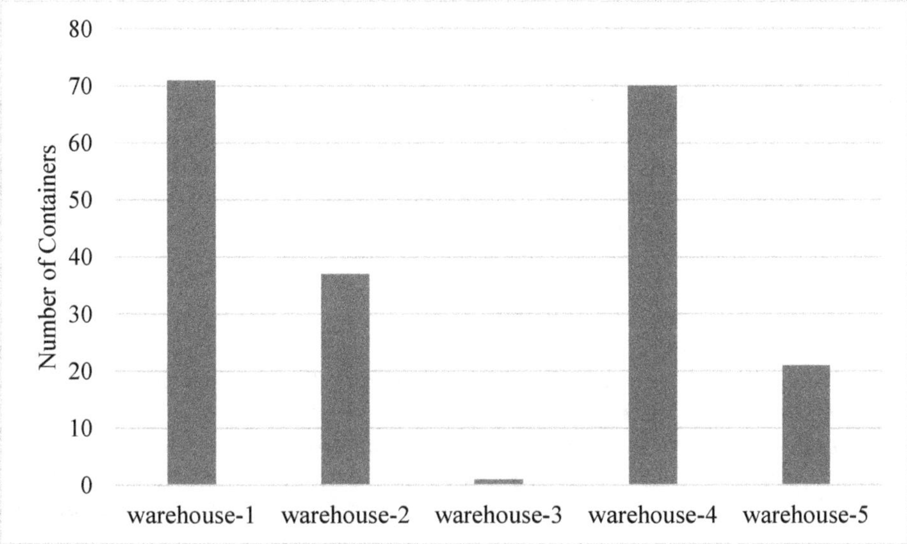

Fig. 3. Number of containers assigned to each warehouse

The graph shows that warehouses 1 and 4 are the most preferable due to their high free pool limits, and low transportation costs. From all the instances, 35% and 36% of the incoming empty containers are assigned to warehouses 1 and 4 respectively. Warehouse 2 has free day policy, and its transportation cost is low, so the containers that have smaller storage days are assigned to this warehouse. Nearly 19% of the empty containers are assigned to warehouse 2. Warehouse 5 do not have any daily storage cost, but its gate cost is twice that of other warehouses. Therefore, the small number of containers are assigned to that warehouse, indicating 10% of all containers. Warehouse 3 is the farthest warehouse to the port, and it has both free day and free pool policies, which causes more daily storage costs for the containers. Thus, this warehouse rarely used by the company.

6 DSS and Implementation

A decision support system (DSS) is developed for the company to provide easy implementation of the proposed solution methodology. DSS is developed using Microsoft Excel VBA application. User-friendly and straightforward interfaces are developed regarding the requirements of the users in the liner shipping company.

Our DSS has two primary interfaces, such as input and output. In the input part, a list of upcoming containers is required from the user. This list includes the identification, TEU size, storage day, and the origin of the container. Using combo boxes, the user can select the corresponding size and origin. The user should enter the storage day of the container as an integer value. Also, unit transportation costs of the warehouses can be changed by the users as the fuel oil prices changes. Furthermore, users can change the free pool and free day limits of the corresponding warehouses according to the yearly contracts with the liner shipping company.

After entering all the input values, the optimization program can be called by the DSS by clicking on the "solve" button. Then, the problem is solved optimally within seconds, and the users can monitor the solution. In the output part of the DSS, users can control which containers are assigned to which warehouses. Also, warehouse usage rates are shown by the graphs that are created by the DSS. Thus, the users can decide whether to have a contract with that warehouse or not.

7 Conclusions and Future Work

In this study, the empty container repositioning problem is considered, and the assignment of empty containers to the available warehouses are provided with the minimum costs. The problem environment is researched in detailed and related literature is reviewed carefully. This study is conducted in a liner shipping company. The problem is defined according to the observations during the assignment processes and the interviews with the managers. First of all, the symptoms are defined, and then the problem statement is determined. A mathematical model is formulated to solve the specified model regarding the number of containers, container specifications, and the different storage policies of the warehouses. Containers have three different specifications, such as the size of the containers, total storage days of the containers, and the origin of the containers. Warehouses have two types of costs, such as gate cost and daily storage cost. Regarding all the specifications of the containers and the warehouse policies, the optimal assignments are done, and the optimal solutions are obtained.

In the current planning system of the company, the empty container and warehouse assignments are implemented by the workers according to their experiences. However, these experience-based assignments take a very long time due to the manual calculations of the costs. Moreover, the current assignments cause high storage costs and unnecessary expenses to the company. The optimal assignments can be obtained by employing the proposed model through the developed user-friendly DSS. The proposed model provides 34.7% of cost improvement compared to the experience-based assignments. Besides the cost reduction, the developed model provides time reduction in the planning process. While experience-based assignments take one hour in a day, the developed model can be solved optimally within seconds. Therefore, the proposed approach will help the company to save their time and cost. Moreover, through the DSS, all the assignments and the utilization of the warehouses are shown by graphs and tables. User can monitor them and can change the policies to see the different cases. This will help the company on the contract that they sign with the warehouses.

As a future study, capacity constraints of the warehouses can be added to the model. Also, the model can be turned into a dynamic model considering the current stock levels of the warehouses and leaving containers from the warehouses. Also, more containers can be regarded as coming by more than one vessel dynamically.

Acknowledgment. This study is supported by The Scientific and Technological Research Council of Turkey (TUBITAK) in the program of "2209-B Undergraduate Research Projects for

Industrial Applications Fellowship Program." We would like to thank our colleagues İdil Efe, Doruk Kızıltunç, Volkan Helvacı, Koray Çetin, and Barış Atay for their help and contribution to the study.

References

1. Song, D.P., Carter, J.: Empty container repositioning in liner shipping. Marit. Policy Manag. **36**(4), 291–307 (2009)
2. Epstein, R., Neely, A., Weintraub, A., Valenzuela, F., Hurtado, S., Gonzalez, G., Angulo, G.: A strategic empty container logistics optimization in a major shipping company. Interfaces **42**(1), 5–16 (2012)
3. Song, D.P., Dong, J.X.: Cargo routing and empty container repositioning in multiple shipping service routes. Transp. Res. Part B: Methodol. **46**(10), 1556–1575 (2012)
4. Bandeira, D.L., Becker, J.L., Borenstein, D.: A DSS for integrated distribution of empty and full containers. Decis. Support Syst. **47**(4), 383–397 (2009)
5. Moon, I., Do Ngoc, A.D., Konings, R.: Foldable and standard containers in empty container repositioning. Transp. Res. Part E: Logistics Transp. Rev. **49**(1), 107–124 (2013)
6. Shintani, K., Imai, A., Nishimura, E., Papadimitriou, S.: The container shipping network design problem with empty container repositioning. Transp. Res. Part E: Logistics Transp. Rev. **43**(1), 39–59 (2007)
7. Hjortnaes, T., Wiegmans, B., Negenborn, R.R., Zuidwijk, R.A., Klijnhout, R.: Minimizing cost of empty container repositioning in port hinterlands, while taking repair operations into account. J. Transp. Geogr. **58**, 209–219 (2017)

Periodic Route Optimization for FMCG Distributors

Onur Çopur, Mert Yıldız, Simru Göven, Ali Övünç Güneri, Alper Berke Yavuz, Mahmut Ali Gökçe, and Cansu Yurtseven(✉)

Industrial Engineering, Yaşar University, Bornova, Turkey
onurcopur12@gmail.com, mrtyldz.1909@gmail.com, simrugoven@gmail.com, alperberkeyavuz@gmail.com, a.ovuncguneri@yahoo.com, {ali.gokce, cansu.yurtseven}@yasar.edu.tr

Abstract. Izmir based Information Technology (IT) solution provider develops supply chain management software tools for many customers from fast moving consumer goods sector (FMCG). Customers of the FMCG sector needs to be visited a certain number of times in a given period by sales representatives. The company must decide which customers must be visited in which sequence by each sales representative while obeying visit frequency and time windows requirements. These decisions have significant impact on total cost. For this reason, finding an optimal route for every sales representative for each day of the planning horizon is important. This practically challenging and technically important problem can be described as periodic multiple depot traveling salesman problem with time windows (PMDTSPTW). We propose a novel mathematical model for the optimal solution of this problem. The proposed model minimizes the total distance traveled by sales representatives by deciding which sales representative will visit which customer on which day while following the time windows to collect demand data. The proposed model is applicable to any company from FMCG sector. The solution approach in this study is implemented and tested with real life including coordinates and location data of customers and sales representatives from Turkey's largest beer distributor. The proposed model is solved using IBM ILOG CPLEX Optimization Studio version 12.8. The results show significant improvement over the current situation. To ensure efficient usage of the proposed approach, a user-friendly decision support system (DSS) is constructed and implemented.

Keywords: Periodicity · Travelling salesman problem · Time windows · Mathematical programming · Decision support system

1 Introduction

The purpose of this study is to propose a novel mathematical model for the solution of periodic multi-depot traveling salesman problem with time windows (PMDTSPTW) and to present the implementation of the solution in a real-life case study with the data provided by one of the biggest beer distributor in Turkey. This work focuses on route optimization problem of sales representatives in fast moving consumer goods (FMCG)

N. M. Durakbasa and M. G. Gençyılmaz (Eds.): ISPR 2019, LNME, pp. 748–759, 2020.
https://doi.org/10.1007/978-3-030-31343-2_63

sector. In the FMCG sector, big retailers and manufacturers need sales representatives (SR) and distribution vehicles to sell and distribute their goods to large number of customers. Each customer should be visited a certain number of times in during a planning horizon, within the given time windows by sales representatives. Therefore, finding routes with minimum cost for the sales representatives is a relevant and frequently studied problem in the FMCG sector.

This problem is practically important, technically challenging and forms a new case for literature. As the data of the problem gets larger, difficulty of optimization increases exponentially. Proposing a good solution approach for such a complex case is challenging. The problem studied belongs to one of the largest Information Technology (IT) solution provider which is a software company headquartered in Izmir and has several customers in FMCG sector. The customers' request is a routing module to optimize the visits of sales representatives from seven starting points (houses of SR's) to customers with minimum travelling distance. Sales representatives visit local markets and supermarkets for collecting the orders. The case study presented with the proposed solution contains 869 customers and 7 sales representatives, which leads a complicated and dense problem. This study specifically focuses on routing optimization of SR's and PMDTSPTW is used to apply real world conditions as much as possible.

2 Problem Definition

The problem can be classified as periodic multi-depot travelling salesman problem with time windows (PMDTSPTW) and it is a variant of TSP. To clarify, the visit frequency requirements of customers makes the problem periodic and different starting points of each route makes the problem multi-depot. In addition to that, each customer has (a) specific time window(s) for accepting the sales representatives. The company has to assign each sales representative customers to visit with routing information and collect demand. After collection of demand, the orders are assigned to the delivery vehicles which distribute the goods. Currently, the last part of problem, which covers assignment of vehicles and distribution of the orders, can be solved by the company to their satisfaction. However, route determination for sales representatives is done by intuitive methods based on experience, which can be quite far from the optimal in complex cases. There are customers to be visited at certain times for a given number of times in a week, while each sale representative starts and ends the day at his/her own house. This makes the problem periodic and multi-depot. The solution proposed in this study may enable customer visits to be performed with less sales representatives than appointed. To the best of our knowledge, there are a limited number of studies in the literature and sources related with this type of problem. Therefore, the two motivations of this study are satisfying requests of the company and contributing the literature.

In order to solve the problem, following input data are required: number of visits per period for each customer, time spent at each customer, work shifts of sales representatives, earliest and latest possible arrival time for each customer, and a distance matrix that shows the distances between each pair of node. So far, the company provided the real coordinates of the customers and sales representatives for testing and implementation of the proposed solution method. The data used in the case study belong to one of the biggest beer distributors in Turkey. This distributor is located in the center of Izmir and has 869 customers in total. Among the 869 customers, 621 of them need to be visited twice a week and the rest is needed to be visited only once a week. There are seven (7) sales representatives employed by this distributor. According to this data set, a distance matrix will be fetched by using Bing Maps Restful API on Python 3. The Bing Maps Distance Matrix API provides travel time and distances for a set of origins and destinations. The distances and times returned are based on the routes, calculated by the Bing Maps Route API. Distance matrices can be calculated for driving, walking and public transit routes. For this case, it is assumed that all distances between customers are driving distances. A distance matrix can be requested that has up to 2500 origins-destinations pairs which is calculated by multiplying the number of origins, by the number of destinations. However, in this problem, the distance matrix size should be 869 * 869. A python code is developed to create the whole distance matrix by creating a sliding 50 * 50 matrix inside the 869 * 869 size matrix. If a new customer is added to the system, or if it is desired to remove, updating the data will be easier with the API code. To fetch the data from Bing Maps API, Json and urllib libraries are used.

In addition, on average, a sales representative works 10 h a day, spends 10–15 min at each customer and visits 35–40 customers in a day. Currently, routing operations are made based on personal experience of the employees. Therefore, total traveled distance by sales representatives and the workload balance of sales representatives can be compared with the previous status as a performance measure. In this problem, main objective is minimizing the total traveled distance by sales representatives within working hours. They start their routes from different locations (considered as a multi-depot problem) for collecting orders from different customers who are needed to be visited according to their visit frequency. This problem is easy to solve for small data sets. However, with 869 customers mathematical model takes too much time to generate a solution for large size instances.

3 Literature Review

The problem studied is the combination of three different extensions of vehicle routing problem (VRP). These are Periodic VRP (PVRP), Multi-depot VRP (MDVRP) and VRP with time windows (VRPTW). Since, TSP is quite similar to VRP and the difference between them is that in VRP, vehicles have a limited capacity which pick-up and deliver the goods to the customers. The extensions of these vehicle routing problems are frequently studied in FMCG sector and known as challenging problems. Because of the similarities to our problem, we have reviewed many articles. The articles are examined in detail below and sorted by the relevance.

Chu et al. [2] proposed a periodic capacitated arc routing problem. This study contains linear programming model, metaheuristic and lower bounds, also includes task assignments and determination of vehicle trips while focusing on minimizing the total cost.

Thangiah [10] represented a study about vehicle routing problems with time windows. This study contains Hybrid Genetic Algorithms, Simulated Annealing and Tabu Search Heuristic. With similar to our problem, this study focuses on minimizing the distance traveled to service the customers. In addition this study includes decisions about the number of vehicles.

Pacheco et al. [7], studied about optimizing vehicle routes in a bakery company. In this case, delivery dates are flexible. This study is related to our problem with same objective function; minimizing the total distance travelled for the daily routes over the week. Unlike our problem, this study has flexibility at delivery dates.

Alinaghian [1] proposed a navel heuristic algorithm for a periodic vehicle routing problem. This paper also focuses on client visits over a given time horizon. The represented model also optimizing the routes in each time period. In addition, this paper represents a new heuristic algorithm based on data mining.

Tan et al. [9] proposed a heuristic method for vehicle routing problem with time windows. Objective of this article is to achieve the minimum transportation cost including travelled distance while considering the penalties. Also a novel genetic algorithm clustering method is proposed in this paper.

Clarke and Wright [3] represented a study about vehicle scheduling. In this case there is a central depot and delivery points. This study focuses on finding the optimum routes under the conditions of varying truck capacities. To mention this study finds optimum or near-optimum routes for the problem.

Paletta [8] developed a new heuristic algorithm for the periodic traveling salesman problem. This new algorithm has smaller average percentage error than existing algorithms. This paper is also considered as a traveling salesman problem and objective is to minimize the total distance traveled.

Coene et al. [4] studied on a periodic vehicle routing problem. This paper is about a variant of a periodic vehicle routing problem. Objective function of this study is to minimize the sum of the costs of all routes over the planning horizon.

Mancini [6] proposed a solution approach for a multi depot multi period vehicle routing problem. In this study there is a heterogeneous fleet. The aim of this study is to minimize the total delivery cost. This fleet contains vehicles with different capacity. Similar to our problem, this paper also has mixed integer programming (MIP) solution approach.

The most relevant article with our study is proposed by Cordeau et al. [5]. They developed a base mathematical model and suggested minor changes to solve the periodic vehicle routing problem, the periodic traveling salesman problem, and the multi-depot vehicle routing problem. Besides the mathematical model, they also proposed a tabu search heuristic. The proposed solution for this study based on computational experiments carried out on instances taken from the literature indicates that the proposed method outperforms existing heuristics for all three problems. All relevant articles are mentioned in Table 1.

There are a number of contributions to literature from this study. To the best knowledge of the authors, the novel mathematical model is the first of its kind for a periodic routing problem of multiple sales representatives as described, taking into account not only time windows but also considers traffic conditions while minimizing the total distance traveled and ensuring the workload balance. Also, the model is verified and validated and tested with real life data, as presented in Sect. 5. In the literature, there are problems with similar aspects to our study, but there is no problem that meets our case.

Table 1. Literature table

	Periodic	Multi depot	Time windows	Vehicle routing problem	Traveling salesman problem
Alinaghian, M., 2014	✓			✓	
Chu et al., 2004	✓			✓	
Clarke, G., Wright, J.W., 1964				✓	
Coene et al., 2010	✓			✓	
Cordeau et al., 2010	✓	✓		✓	
Mancini, S., 2015	✓	✓		✓	
Pacheco et al., 2012				✓	
Paletta, G., 2002	✓				✓
Tan, K.C. et al., 2001			✓	✓	
Thangiah, S.R., 1999			✓	✓	
Our Study	✓	✓	✓		✓

4 Modeling and Solution Methodology

In this section, a novel mixed integer mathematical programming model is proposed for PMDTSPTW. Indices, sets, parameters and decision variables of the problem are shown in Table 2.

Table 2. Indices, sets, parameters and decision variables

Indices, Sets, Parameters	
i	Starting node index
j	Destination node index
k	Sales representative index
l	Day index
r	Possible visits index
I, J	Set of customers $I, J = \{M+1, \ldots, N\}$
H	Set of dummy indices to ensure that a sales representative arrives at a customer on a given day it also leaves that customer on the same day. $H = \{M+1, \ldots, N\}$
K	Set of sales representatives $K = \{1, \ldots, M\}$
L	Set of days of week (period), 1 represents Monday, 6 represents Saturday. $L = \{1, \ldots, 6\}$
R	Set of all possible visit combination
C_i	Set of allowable visit combination for customer $i \in I, C_i \in R$

(*continued*)

Table 2. (*continued*)

M	Number of sales representatives
N	Number of customers
Er_i	Earliest service time for each customer i
La_i	Latest service time for each customer i
V	Average Speed of vehicle
d_{ij}	The distance from node $i \in I$ to $j \in J$
a_{rl}	$\begin{cases} 1, \text{iff day } l \in L \text{ belongs to visit combination } r \in C_i \\ 0, \text{otherwise} \end{cases}$
WE	Work ending hour in terms of minutes
MTD	Morning traffic duration
ETD	Evening traffic duration
S_i	Service duration of customer $i \in I$
Decision Variables	
x_{ijkl}	$\begin{cases} 1, \text{iff sales representative } k \in K \text{ visits customer } j \in J \text{ immediately} \\ \text{after visiting customer } i \in I \text{ during day } l \in L \\ 0, \text{otherwise} \end{cases}$
y_{ir}	$\begin{cases} 1, \text{iff visit combination } r \in C_i \text{ is assigned to customer } i \in I \\ 0, \text{otherwise} \end{cases}$
w_{ikl}	Service start time of customer $\forall i \in I$ by sales representative $\forall k \in K$ on day $\forall l \in L$
u_{ikl}	Sequence index variable defined for subtour elimination $\forall i \in I, \forall k \in K, \forall l \in L$
x_{ikl}	Sequence index variable defined for subtour elimination $\forall i \in I, \forall k \in K, \forall l \in L$
y_{ikl}	Sequence index variable defined for subtour elimination $\forall i \in I, \forall k \in K, \forall l \in L$
z_{ikl}	Sequence index variable defined for subtour elimination $\forall i \in I, \forall k \in K, \forall l \in L$
$\boldsymbol{Cf_{il}}$	$\begin{cases} \mathbf{0.85, if\ w_{ikl} \leq 90\ or\ w_{ikl} \geq 390} \\ \mathbf{1, otherwise} \end{cases}$

4.1 Mathematical Model

$$\min z = \sum_{i \in I \cup K} \sum_{j \in J \cup K} \sum_{k \in K} \sum_{l \in L} d_{ij} x_{ijkl} \tag{0}$$

Subject to

$$\sum_{r \in C_i} y_{ir} = 1 \qquad \forall i \in I \tag{1}$$

$$\sum_{j \in J} \sum_{k \in K} x_{ijkl} = \sum_{r \in C_i} a_{rl} y_{ir} \qquad \forall i \in I, \forall l \in L \tag{2}$$

$$\sum_{i \in I} x_{ihkl} = \sum_{j \in J} x_{ihkl} \qquad \forall h \in H, \forall k \in K, \forall l \in L \tag{3}$$

$$\sum_{i\in I}\sum_{k\in K} x_{ijkl} \leq 1 \qquad \forall j \in J, \forall l \in L \tag{4}$$

$$\sum_{j\in K} x_{ijkl} = 1 \qquad \forall j \in J, \forall k \in K, \forall l \in L \tag{5}$$

$$\sum_{j\in K} x_{ijkl} = 1 \qquad \forall i \in I, \forall k \in K, \forall l \in L \tag{6}$$

$$\begin{gathered} w_{ikl} + S_i + \left(\frac{d_{ij}}{V * Cf_{il}}\right) * 60 - M\left(1 - x_{ijkl}\right) \leq w_{jkl} \\ \forall i \in I \cup K,\ \forall j \in J \cup K,\ \forall k \in K,\ \forall l \in L \end{gathered} \tag{7}$$

$$Er_i \leq w_{ikl} \leq La_i \qquad \forall i \in I, \forall k \in K, \forall l \in L \tag{8}$$

$$w_{ikl} \leq MTD + (WE - ETD) * x_{ikl} \tag{9}$$

$$w_{ikl} \geq (WE - ETD) * (1 - y_{ikl}) \tag{10}$$

$$w_{ikl} \geq MTD * (1 - z_{ikl}) \tag{11}$$

$$w_{ikl} \leq (WE - ETD) + (MTD * z_{ikl}) \tag{12}$$

$$x_{ikl} + y_{ikl} + z_{ikl} = 2 \tag{13}$$

$$cf_{il} \leq 0.85 * z_{ikl} + (1 - z_{ikl}) \tag{14}$$

$$u_{ikl} + 1 \leq u_{jkl} + M\left(1 - x_{ijkl}\right) \qquad \forall i \in I, \forall j \in J, \forall k \in K, \forall l \in L \tag{15}$$

$$x_{ijkl} \in \{0, 1\} \qquad \forall i \in I, \forall j \in J, \forall k \in K, \forall l \in L \tag{16}$$

$$y_{ir} \in \{0, 1\} \qquad \forall i \in I, \forall j \in J, \forall k \in K, \forall l \in L \tag{17}$$

$$u_{ikl} \geq 0 \qquad \forall i \in I, \forall k \in K, \forall l \in L \tag{18}$$

$$w_{ikl} \geq 0 \qquad \forall i \in I, \forall k \in K, \forall l \in L \tag{19}$$

The objective function minimizes the total distance travelled by sales representatives (0). Constraint (1) ensures that exactly one feasible visit combination must be assigned to each customer. Constraint (2) indicates that each customer is visited only on the days corresponding to the assigned combination. Constraint (3) ensures that when sales representatives arrive at a customer on a given day it also leaves that customer on the same day. Constraint (4) restricts that each customer must be visited by at most one sales representative on each day. Constraint (5) indicates that each sales representative starts the tour from their homes on each day. Constraint (6) ensures that each sales representative goes back to their homes at the end of every day. Constraints (7–8) ensures that the arrival time of two consecutive customers does not collide with each other considering the duration of service and travel between customers and limiting the sales representatives to arrive at a specific time. Constraints (9–14) ensure that velocity constant should be 0.85 if w_{ikl} is in morning traffic duration or evening traffic duration. Constraint (15) is for subtour elimination. Constraints (16–19) are sign restrictions.

5 Computational Results

Due to the size and complexity of the real data received from the company, a general coordinate scan was performed to test the accuracy of the data. The mathematical model is solved by IBM ILOG CPLEX version 12.8 optimization software with different sizes of data for the purpose of monitoring performance and accuracy control. The results of MIP model are shown in Table 3.

Table 3. Results of MIP by IBM ILOG CPLEX 12.8

Number of SR[b]	Number of customers	Number of customers with single visit frequency	Number of customers with double visit frequency	Objective function (Km)	Solution time (Min)
2	30	9	21	17.74	10.37
2	40	12	28	19.35	25.48
2	50	15	35	20.32	52.29
2	60	18	42	21.80	92.54
2	70	21	49	34.97	123.45
2	80	24	56	37.65	240.00[a]
2	90	27	63	43.85	240.00[a]
2	120	36	84	54.12	240.00[a]

[a]Cut off point (4 h) results.
[b]SR = Sales Representatives

Table 3 presents the results of the optimization model for varying number of customers. These cases use the real life data from the case study. The results reported in Table 3 show that the problem a steep increase in solution time as the problem size gets larger. While for the cases with number of customers 30, 40, 50, 60, 70 have optimal results, the best results for cases with number of customers 80, 90 and 120 are best solutions reported by CPLEX after the cutoff point of 4 h applied. Even though problems larger than 70 customers are run without any time limit, the best computer in hand, 2.8 GHz Core i7 CPU and 16 GB RAM, turned memory error and stopped before reaching to an optimum solution. For this reason, we have added cutoff points for CPLEX and the results returned for this restricted times. It seems that for real large size problems can only be solved by dividing the problem into the sub-problems and/or by employing heuristic methods. The mathematical model results prove that the model runs smoothly. The solution founded by MIP on CPLEX checked manually and the routes satisfy all constraints which mean the method is verified and considered as valid.

6 Decision Support System

Decision Support System (DSS) is an information system that allows people to get support from computer technology during decision making. Dense and complex data used while problem solution phase can be modeled and reduced to understandable levels. To mention briefly, decision support system is a combination of data, information, software, analysis and mathematical model. DSS helps people to understand the complex systems and solution methodology of these systems. Due to such benefits, in this study a user-friendly decision support system is constructed.

There are three main parts of DSS; "Main Page", "Solve" and "PDFPrint". "Main Page" screen displays general information and explanations about DSS, what operations DSS performs and how it performs, and information such as printouts. DSS main screen is enriched with vehicles and route visuals to better represent the purpose of the project visually. There are two buttons on the "Main Page" screen; "Start" and "Help" button, which can be seen in Fig. 1.

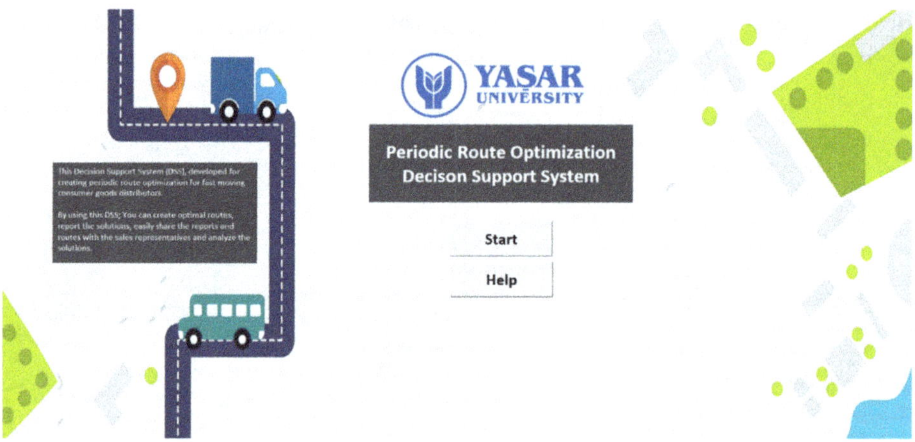

Fig. 1. DSS Main Page

Clicking the "Start" button displays the "Solve" page of DSS, see Fig. 2. "Solve" page is divided into three separate parts. Simply, there are inputs on the left side and the outputs in the middle of the "Solve" page. There is also "Google Maps" area on the right side of the page, which visualizes optimal routes found at the end of the solution. "Add" button adds sales representative or customer to the system. "Remove" button to removes sales representative or customer from the system. There is one extra button, "Clear", which helps user to clear all sales representatives or customers from the system with just one click.

To see the solution, sales representative name and day are selected from the "Output" screen. The solution is displayed on the screen when "Bring the Result" button is

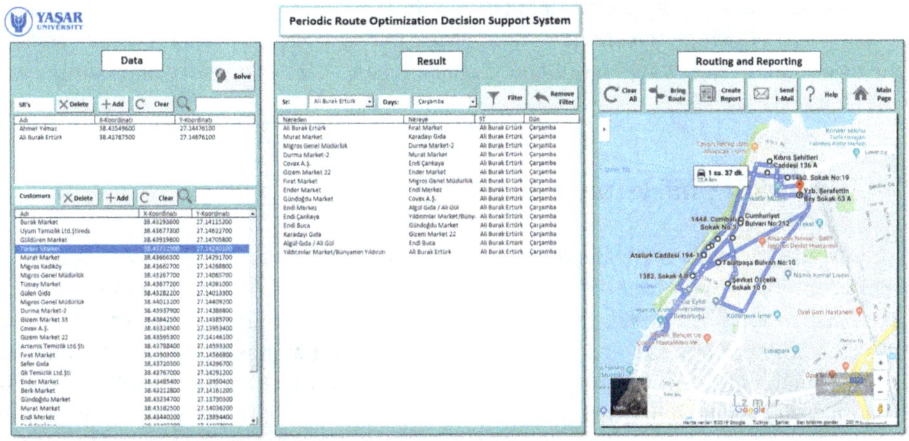

Fig. 2. DSS system menu

pressed. The solution screen displays where the sales representative comes from and where the sales representative goes, the name of the sales representative, the specific day and arriving and departure time of the customer. When the "Detailed PDF Report" button is clicked, system generates a detailed PDF report which is visualized in Fig. 3.

Weekly Periodic Route Optimization Day Report			
Date	21.05.2019		Sales Representatives sholud visit customers according to below order.
Sales Representative			
Destination	**Day**	**Sales Representative**	**Notes**
Ali Burak Ertürk	Wednesday		
Fırat Market	Wednesday		
Migros Genel Müdürlük	Wednesday		
Durma Market-2	Wednesday		
Murat Market	Wednesday		
Karadayı Gıda	Wednesday		
Gizem Market 22	Wednesday		
Ender Market	Wednesday		
Endi Merkez	Wednesday		
Algül Gıda / Ali Gül	Wednesday		
Endi Buca	Wednesday		
Gündoğdu Market	Wednesday		
Covax A.Ş.	Wednesday		
Endi Çankaya	Wednesday		
Yıldırımlar Market/Bünyamin Yıldırım	Wednesday		
Total Visits	15	**Approved By**	Övünç

Fig. 3. DSS report output

When all transactions are completed, the detailed report is sent to sales representatives by just clicking on a single button, "Mail to Sales Representatives", and sending them as e-mail. In this way, loss of communication and time loss between processes are prevented.

7 Conclusion and Future Work

In this study, a mathematical model is proposed for periodic multi-depot traveling salesman problem with time windows (PMDTSPTW). Firstly, a general system analysis is performed. To explain briefly, the technically challenging and practically very important problem studied and to mention, given data belongs to one of the biggest beer distributor in Izmir. Companies associated with FMCG sector needs route optimization to distribute their goods. The coordinate's data given by company used to find the distance matrix. Bing Maps Restful API is the tool that found and creates this matrix on Python 3.6. As it is mentioned before the goal of this study is to minimize the total traveled distance by sales representatives. According to customers' visit frequency multiple routes, for multiple sales representatives who start to their routes from different depots (houses) founded for a planning horizon. While these routes founded a certain time window considered. The mixed integer programming (MIP) method is used.

There are important contributions both to company and literature from this study. To the best knowledge of the authors, the novel mathematical model is the first of its kind for a periodic routing problem of multiple sales representatives as described, taking into account time windows and also traffic conditions while the aim is minimizing the total distance traveled and ensuring the workload balance. In the literature, since there are problems with similar aspects to our study but there is no problem that meets our case, a contribution to the literature is made.

The main contribution to the company by this study can be identified as a solution method that has been created which increases the efficiency, minimizes the total traveled distance and ensures the workload balance. In the old system, the routes created based on the experience of the sales representatives are formed with the help of scientific methods in the new system. The decision support system is developed for a routing system based on scientific methods. With this mathematical model and user friendly decision support system, new system can be adapted to any company from real life. The decision support system created with the Excel VBA interface enables the finding, visualization and reporting of optimal routes for sales representatives. The reporting process is an additional feature of the decision support system that enables the sales representatives to receive their daily routes by e-mail. The summary report consists of destination, day, sales representative name, total number of visits and notes. In accordance with the observed examples, the total distance traveled was reduced to a minimum as intended.

Finally, as future work; there may be different types of sales representatives such as pedestrians, motor driver sales representatives. Another possibility is that if a sales representative finishes his/her route and working hours remain, another route can be assigned to him/her. In addition, a lunch break constraint can be added to implement the problem. For the solution method for cases with really large number of customers, a detailed heuristic approach must be developed, tested and implemented.

Acknowledgement. This study is supported by TUBITAK (The Scientific and Technological Research Council of Turkey) in the program of "2209-B Undergraduate Research Projects for Industrial Applications Fellowship Program". We would like to thank Ege Can Erdoğan, Doğacan TANIŞ and Zafer Yapıcıel for their support and contributions to this study.

References

1. Alinaghian, M.: A Navel Heuristic Algorithm for the Periodic Vehicle Routing Problem (2014)
2. Chu, F., Labadie, N., Prins, C.: The Periodic Capacitated Arc Routing Problem linear programming model, metaheuristic and lower bounds. J. Syst. Sci. Syst. Eng. **13**, 423–435 (2004). https://doi.org/10.1007/s11518-006-0174-y
3. Clarke, G., Wright, J.W.: Scheduling of vehicles from a central depot to a number of delivery points. Oper. Res. **12**, 568–581 (1964)
4. Coene, S., Arnout, A., Spieksma, F.: On a periodic vehicle routing problem. J. Oper. Res. Soc. **61**, 1719–1728 (2010). https://doi.org/10.1057/jors.2009.154
5. Cordeau, J.-F., Gendreau, M., Laporte, G.: A Tabu Search heuristic for periodic and multi-depot vehicle routing problems. Networks **30**, 105–119 (1997). https://doi.org/10.1002/(sici)1097-0037(199709)30:23.3.co;2-n
6. Mancini, S.: A real-life multi depot multi period vehicle routing problem with a heterogeneous fleet: formulation and adaptive large neighborhood search based matheuristic. Transport. Res. Part C (2015). https://doi.org/10.1016/j.trc.2015.06.016
7. Pacheco, J., Alvarez, A., García, I., Ángel-Bello, F.: Optimizing vehicle routes in a bakery company allowing flexibility in delivery dates. J. Oper. Res. Soc. **63**, 569–581 (2012). https://doi.org/10.1057/jors.2011.51
8. Paletta, G.: The period traveling salesman problem: A new heuristic algorithm. Comput. Oper. Res. **29**, 1343–1352 (2002). https://doi.org/10.1016/s0305-0548(01)00035-1
9. Tan, K.C., Lee, L.H., Zhu, K.Q., Qu, K.: Heuristic methods for vehicle routing problem with time windows. Artif. Intell. Eng. **15**, 281–295 (2001)
10. Thangiah, S.R.: A hybrid genetic algorithms, simulated annealing and tabu search heuristic for vehicle routing problems with time windows. In: Chambers, L. (ed.) Practical Handbook of Genetic Algorithms Complex Structures, vol. 3, pp. 347–381. CRC Press (1999)

Scheduling of Test Operations at Bosch Thermotechnology R&D Laboratory

Pelin Akçay, Mustafa Doğa Çetin, Batu Ekmekçi(✉), Beste Yıldız, Levent Kandiller, and Damla Kızılay

Department of Industrial Engineering, Yasar University, İzmir, Turkey
oyaakcay@hotmail.com, dogactnn@gmail.com, batubaker@gmail.com, beste427@hotmail.com, {levent.kandiller,damla.kizilay}@yasar.edu.tr

Abstract. In Bosch's Research and Development (R&D) laboratories, several performance tests are executed to check whether the targeted quality and reliability values of the products are attained. Each product has to go through several tests on several test machines with different processing times. The specified problem in Bosch Thermotechnology entirely fits the Flexible Job Shop Scheduling (FJSS) Problem with machine and operation dependent setup times. The most crucial factor while performing these schedules of the products is to complete them as soon as possible to provide customer satisfaction as well as test machine utilization. Hence, our objective function is to minimize the makespan. Additionally, a decision support system (DSS) for the company on MS Excel is developed.

Keywords: Flexible Job Shop Scheduling · Sequence dependent setup times · Greedy heuristics

1 Introduction

In 1886, Robert Bosch founded the "Workshop for Precision Mechanics and Electrical Engineering" in Stuttgart. This was the birth of today's globally operating company. Right from the start, it was characterized by innovative strength and social commitment [1]. Bosch Thermotechnology is the world's leading supplier of energy efficient heating systems, hot water solutions, and cooling systems, founded in 1910. The Bosch Thermotechnology division is responsible for all activities involving heating technology and hot water solutions. The division has several major international and regional thermotechnology brands and supplies people with state-of-the-art technologies worldwide. In times of scarce energy resources, the topic of efficient heat generation is of existential importance. Innovations from Bosch Thermotechnology have been setting trends for a decade [2]. The vision of Bosch Thermotechnology Turkey, focusing on eco-friendly and efficient energy use, is to create and share value and to become a market and customer-focused leader in heating, ventilating and air conditioning (HVAC) both in Turkey and the other countries it operates in all around the world. In technology and services, Bosch is one of the leading suppliers in Turkey, 'Mobility Solutions', 'Energy and Building Technology', 'Industrial Technology' and 'Durable

N. M. Durakbasa and M. G. Gençyılmaz (Eds.): ISPR 2019, LNME, pp. 760–774, 2020.
https://doi.org/10.1007/978-3-030-31343-2_64

Consumer Goods' five separate centers. Bosch Group products and services are designed to improve the quality of life through innovative, beneficial solutions. In Turkey, they produce in Manisa, Bursa, Tekirdağ, Kocaeli, and Istanbul. There are three Research & Development (R&D) centers in Bursa, Manisa, and Tekirdağ. Bosch Thermotechnology Heating and Air Conditioning Industry Trade Inc. is located in Manisa Organized Industrial Zone No: 14 Part 2. The facility has a closed area of 25,900 m^2 on a land of 109,000 m^2. Bosch Thermotechnology Manisa Plant's leading services are wall-hung and floor-standing boilers, heat pumps and ventilation, air-conditioning and commercial products. Bosch Thermotechnology produces 551 types of boilers at its Manisa factory for exporting to 41 countries in Europe, the Caucasian Region, the Middle East, the Gulf Region, Latin America, and the Far East. The main components used in production are manufactured in the factory. Approximately 70% of the raw materials and components used in the manufacturing process are supplied by domestically based subsidiary industries. Considering its international sales, the Bosch Manisa plant is Turkey's 39th largest exporter and holds the titles for best Research & Development (R&D) Center and export leadership for its sector [2].

Manisa Bosch Thermotechnology Research & Development (R&D) center was established in 2010 and had 85 employees. In the last three years, 55 million TL has been spent on R&D work, and in the 8th year of its establishment, it was honored with five awards. R&D finds solutions that will make a difference in terms of productivity, quality, and reliability for companies and customers. It also compares competitive solutions by considering customers' feedback, defining the most appropriate engineering solutions and monitoring the level of development. The main objectives of the R&D center are ensuring the continuity of competence, developing university-industry cooperation, producing energy-efficient and environment-friendly innovative products [3].

The R&D center houses specialized laboratories for various purposes. These are an electrical laboratory, functional laboratory, sound room, wind room, ERP, system laboratory, and approval laboratory. Many different tests are performed in these laboratories. These test machines are divided into five groups: A, B, C, D, E. Functional tests are performed in functional, approval, and ERP laboratories. In the wind laboratory, emission tests are carried out under certain conditions, taking the changing weather conditions into account. In the climatization laboratory from −30° to 80°, under different conditions of the device to the environmental conditions are tested. In the electrical laboratory, software and related electrical tests are performed. The audio level of the boiler and component are measured in the audio laboratory. In the system laboratory, tests are performed in different systems to measure the efficiency of the boilers. The main objective of these laboratories is to extend the life of the device to adapt it to Bosch's quality standards called TT-VAN and to find immediate solutions to the problems that may be encountered in the future.

In the job shop scheduling production environment, each job has its route. In parallel machine scheduling problems, demanded jobs can be processed in any of the test machines. Flexible job shop scheduling (FJSS) problems arise with the addition of machines parallel to the job shop production environment. Each of these environments has workstations with similar machines. In the case of flexible job shop scheduling, each operation of each job can be processed on at least one machine [4, 5], meaning there are alternative machines for each operation. It reflects the production environment where it is possible to perform tests on more than one test machine. In FJSS problems, it is very important to schedule the operation to be performed on the machines and to process them in the most appropriate way to the main objective. In Bosch Thermotechnology Research and Development (R&D) Laboratory, boilers (jobs) are subjected to many different tests (operations). These tests are carried out on various test stands (machines). The aim is to minimize the maximum completion time (makespan) of the tests of the given set boilers. The problem mentioned in Bosch Thermotechnology is an FJSS problem.

In the following section, the problem is defined based on the system analysis carried out in the R&D Laboratory. Consequently, the related literature is explained in Sect. 3. Section 4 covers model formulation and solution techniques. The next section explains our Decision Support System (DSS) developed for the company. Finally, we conclude and address future research directions.

2 Problem Definition

Based on the observations, the symptoms directly affecting the efficiency of the test machines were collected. The first symptom is shown numerically in Fig. 1, which depicts the observed test machine-capacity utilization over a four-week period. Firstly, there was a noticeable difference in usage between the test machines. It is observed that the disproportionate machine-test assignment due to manual test machine assignment is performed. The most demanded test machines are A and B while other test machines are also capable of performing certain common operations. For instance, test machine C can undertake the same operations that test machines A and B can perform. However, its utilization is relatively low, creating a variance.

The second prominent symptom of the current system arises from the usage of discrete periods. In R&D laboratories, test requests are planned for certain periods. Each day consists of two periods, and each period consists of four hours. Tests are assigned to the test machines by an authorized engineer, taking into account these four-hour periods determined by the engineers. However, this system is not a preferred system for effective time usage. For example, if the processing time of an operation of a job exceeds the specified period by a little amount, then the consecutive period will also be occupied, blocking the assignment of another test. This results in a significant loss of time in the test machine, and it will greatly affect the results of the test machine utilization. Figure 2 illustrates this effect with numerical values. The processing time of Operation 1 is defined as six hours during which the first four hours completely cover a single period, and the second period is used for the remaining two hours. This means that the test machine does not operate for two hours of the second period, meaning half of the period remains idle. As seen in the numerical example, the duration of the periods determined in the planning phase, machine utilization rates, operators' operating time significantly affects planning time. These listed reasons indicate a lack of decision support systems in which the duration of the periods can be changed in each case requested by the user.

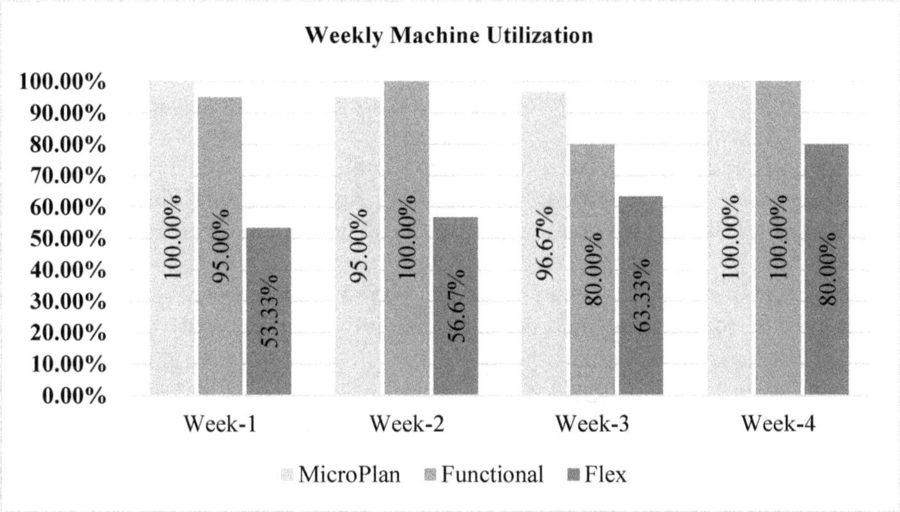

Fig. 1. Weekly machine utilization

Other observed symptoms can be classified into four main headings given in Fig. 3 as material, method, people, and time. Method-based symptoms are due to the use of manual assignment and manual scheduling methods. The absence of a standard process in the system and the high number of test requests make the planning difficult with the manual methods available in the current system. In addition, all planning and assignment procedures must be carried out manually in case of an urgent test request. This increases the workload of the authorized engineer and causes a delay in response to the demands and makes it difficult to reach the most appropriate planning solution.

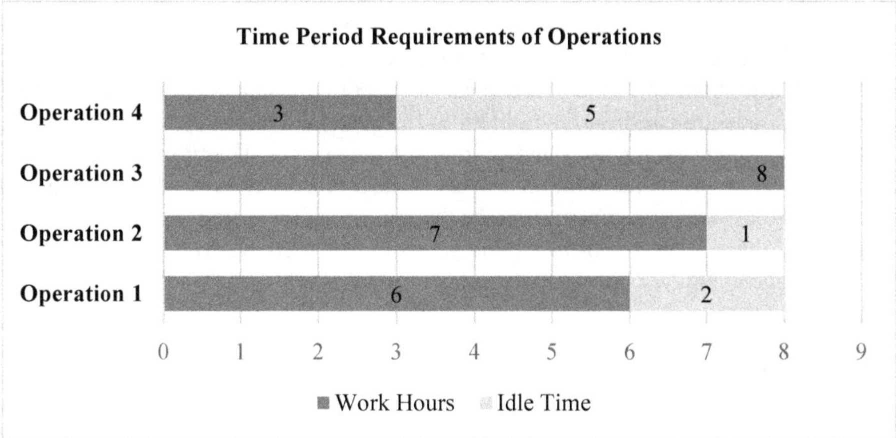

Fig. 2. Time period requirements of operation

Some problems arising from model differences of test machines are also available in the system. For example, each machine's model, the time of purchase due to the depreciation, and the installation time for the tests are different from each other. Therefore, there is no standard procedure for all existing test machines. The other problem is the lack of communication. Due to the complexity of the current system, there is a communication problem between the semi-authorized engineers and operators. This disconnection may extend the completion time of the tests. For example, if an operator finishes the test assigned to him earlier than the given period, and does not inform the authorized engineer, no further tests can be assigned to the test machine. This causes the operator to remain idle for that period, leading test machines cannot be used at the desired efficiency, and the upcoming jobs are accumulated.

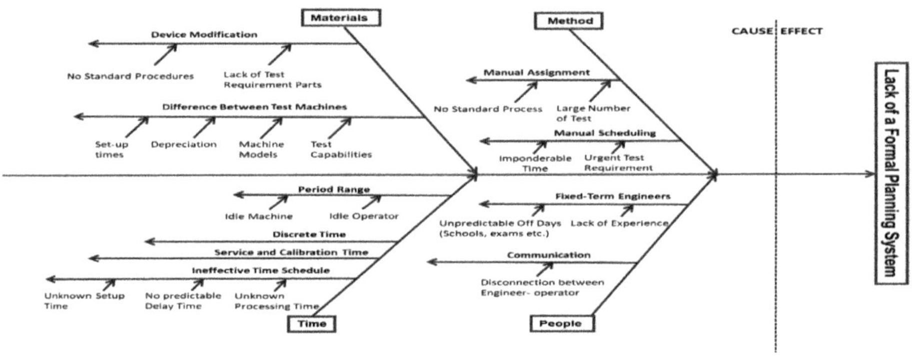

Fig. 3. Fishbone diagram

In Bosch Research & Development (R&D) laboratories, the scheduling and assignment of the jobs are carried out manually. Because of this manual scheduling, some engineers or operators cannot find available test machines to perform their tasks. This situation leads to delayed jobs to the customers. Furthermore, test machines are not used efficiently. In light of the observations, the symptoms are indeed the main source of the problem. Human-based manual scheduling and lack of scientific optimization methods are causing the problem in the system. This study aims to solve the problem using scientific methods and to improve the system. The primary objective of this study is to minimize the total completion time and then compare the utilization of the test machines. Other crucial criteria are the minimization of flow times, setup times, and delays. All these key performance indicators (KPIs) were determined together with Bosch's management.

3 Literature Review

A literature review was made from various sources to determine the information and research results related to the subjects and sub-topics discussed. In this process, it has been observed that key performance criteria are defined to measure the success of the solution methods used in scheduling problems. Key performance criteria were used as the objective function in scheduling problems. Scheduling problems according to performance criteria can be studied under three main headings, such as completion time, delivery time, and test machine usage cost. In light of the current Bosch system, only time-based studies have been examined in this study.

We consider first the review papers and surveys listed. There were two different scheduling problems at this stage, namely the flow shop scheduling and the job shop scheduling problems. However, the differences between the two main parts were examined with the symptoms, and then it is concluded that our problem was the Job Shop Scheduling (JSS) Problem. The first article that we examined for JSS was by Fattahi, Bidgoli and Samouei [6] to compare our developed heuristic by applying Tabu Search algorithm. The second article that we studied is to develop solution with the Tabu Search algorithm using a proposed algorithm technique for the JSS problem by Nakandhrakumar, Seralathan, Azarudeen, and Narendran [7]. However, in later stages, the classical JSS problem did not meet our assumptions and symptoms. So, it is decided that the Flexible Job Shop Scheduling (FJSS) problem, which is a more specialized type of JSS problem, would be the most appropriate guidelines for the project. A similar problem, resource-constrained project scheduling, has the following review studies: Kandiller [8], Herroelen [9], Özdamar and Ulusoy [10], Kolischa and Padmanb [11], Hartmann and Briskorn [12], Fattahi, Bidgoli and Samouei [6], Nakandhrakumar, Seralathan, Azarudeen and Narendran [7].

Finally, after deciding that the problem was FJSS and the main goal was to minimize the makespan, we focused on several FJSS articles having mathematical models. The similarities and differences between the models used in these articles were examined whether or not they could be integrated into the current Bosch's system. We considered the works of Kaya and Fığlalı [5], Wang, Cai, Li and Liu [13], Mehrabad and Fattahi [14], Demir and İşleyen [15].

The most suitable models for the current system was selected [15]. The five different formulations of flexible job shop scheduling problem were found to be the most suitable formulations to the current system. The number of constraints, binary variables, and continuous variables in these model formulations was calculated for size comparisons given in Table 1.

Table 1. Comparison of 5 candidate models from literature

	Model 1	Model 2	Model 3	Model 4	Model 5
Type of model	Sequence-position variable based	Precedence variable based	Precedence variable based	Sequence-position variable based	Time indexed
Time representation	Continuous	Continuous	Continuous	Continuous	Discrete
Complexity parameters	A, B, C, D	A, B, C	A, B, C	A, B, C, D	A, B, C, U
# Of constraints	4(A * B) + 3(A * B * C * D) + 2(A * B * C) + (C * D)	A + 3(A * B) + 2 (A * B * C) + 2(A^2 * B^2 * C)	2(A * B) + 2(A * B * C) + 3(A^2 * B^2 * C)	3(A * B) + (A * B * C) + 3(A^2 * B^2 * C)	2(A * B * U) + (A * B * C) + (C * U) + (A * B * C * U) + (A * B) + 2(A^2 * B^2 * C * U)
# Of binary variables	(A * B * C) + (A * B * C * D)	(A * B * C) + (A^2 * B^2 * C)	(A * B * C)	(A^2 * B^2 * C) + (A ^ 2 * B^2)	(A * B * C) + (A^2 * B^2 * C) + (A * B * C * U)
# Of continuous variables	2(A * B) + (C * D)	A + 2(A * B * C)	(A * B)	(A * B) + (A * B * C)	–

Model 1 [16] and Model 4 [17] are sequence-position variable based model, Model 2 [18] and 3 [19] are precedence variable-based, and Model 5 [20, 21] is time-indexed. In all our models, A defines the number of jobs, B defines the number of operations, and C defines the number of test machines. In Model 1 and 4, there is D parameter that test machines for a sequence of assigned operation to that model. In Model 5, U parameter is used for the number of periods that only Model 5 has. All decision variables, parameters, and constraints of the candidate models were examined, and necessary changes were made to adapt them to the current system. These changes were made for two different timing methods. These time methods are classified as discrete time and continuous time. Based on the Bosch's consideration, we have formulated a discrete time model getting benefit from Model 5 [20, 21].

4 Model Formation

The following formulation is based on [15]. The notation used is as follows:

Indices

i, h: index of jobs (1, …, n)
j, g: index of operations (1, …, J_i)
k: index of test machines (1, …, m)
u: index of time periods (1, …, T)

Parameters

p_{kij}: processing time of operation j of job i (O_{ij}) on test machine k
M_{ij}: eligible machine set for O_{ij}
M: a large number

Decision Variables

Cmax: makespan

$$v_{ijk} = \begin{cases} 1, & \text{if } O_{ij} \text{ is performed on machine } k \\ 0, & \text{otherwise} \end{cases}$$

$$z_{ijhgk} = \begin{cases} 1, & \text{if operation } O_{ij} \text{ precedence operation } O_{hg} \text{ on machine } k \\ 0, & \text{otherwise} \end{cases}$$

$$W_{ijku} = \begin{cases} 1, & \text{if } O_{ij} \text{ is processed by machine during period } u \\ 0, & \text{otherwise} \end{cases}$$

Constraints

$$\left(\sum\nolimits_{k \in Mij} W_{ijku}\right) * u \leq C_{max} \qquad \forall i, u, j = J_i \qquad (1)$$

$$\sum_u W_{ijku} = p_{kij} * v_{ijk} \qquad \forall i,j \quad \forall k \in M_{ij}. \tag{2}$$

$$\sum_i \sum_j W_{ijku} \leq 1 \qquad \forall k, u \tag{3}$$

$$p_{kij} * \left(W_{ijku} - W_{ijku+1}\right) + \sum_{r=u+2}^{T} W_{ijkr} \leq p_{kij} \qquad \forall i,j,k,u = 1,\ldots,T-2 \tag{4}$$

$$\sum_{k \in Mij} v_{ijk} = 1 \qquad \forall i,j \tag{5}$$

$$\sum_j \sum_k W_{ijku} \leq 1 \qquad \forall i, u \tag{6}$$

$$p_{kij} * W_{hgku} \leq \sum_{r=1}^{u-1} W_{ijkr} + M * \left(1 - Z_{ijhgk}\right) \qquad \forall i \leq h, \quad \forall j,g,u = 2,..,T \quad \forall k \in M_{ij} \cap M_{hg} \tag{7}$$

$$\sum_{k \in Mij} p_{kij} * v_{ijk} * \sum_{k \in Mij} W_{ij+1ku} \leq \sum_{r=1}^{u-1} \sum_{k \in Mij} W_{ijkr} * v_{ijk} \quad \forall i,j = 1,\ldots,J_{i-1}, \quad \forall u = 2,\ldots,T \tag{8}$$

$$W_{ijku} \in \{0,1\} \qquad \forall i,j,k,u \tag{9}$$

$$Z_{ijhgk} \in \{0,1\} \qquad \forall i \leq h, \ \forall j,g, \ \forall k \in M_{ij} \cap M_{hg} \tag{10}$$

$$v_{ijk} \in \{0,1\} \qquad \forall i,j,k \tag{11}$$

$$C_{max} \geq 0 \tag{12}$$

Constraint (1) determines the makespan. Constraint (2) defines the number of periods (processing time) of if assigned to machine k. Constraint (3) ensures that in each period of each machine, only one operation can be performed. Constraint (4) guarantees that each operation will not be interrupted before it is finished. Constraint (5) ensures that an operation is performed on one and only one machine. Constraint (6 and 7) guarantees that operation O_i and operation O_{hg} cannot be done at the same time on any machine in the set $M_{ij} \cap M_{hg}$. Constraint (8) presents the precedence relationship of two consecutive operations. Constraints (9), (10), and (11) are binary variables, whereas (12) is nonnegativity. The most important KPI in this problem is the maximum time of completion of the tests (*Cmax*). Thus a classical objective function to the above formulation is

$$Min\ Cmax \tag{13}$$

Test machine and operation based sequence dependent setup times have been added to the process timelines of each machine in order to solve complexity of the system. Thus, for each operation, it is not necessary to define parameters in order to determine the setup times on the corresponding test machine and include these parameters into the constraints.

In order to calculate the size and complexity of our model, which is Model 5 [20, 21] discrete time model in Table 1, the data of the toy problem obtained from the company has been put into the formulations and necessary calculations have been made. In our Model 5 [20, 21], A has 6 jobs, B has 19 operations, C has 7 test machines, and U has 10 periods for each day (each period has 45 min). When we calculated the number of constraints and binary variables, we found out 1.830.682 constraints and 99.750 binary variables. Flexible job shop scheduling problems are classified as NP-Hard problems. As seen from the results, many constraint and binary variables are required even for the small-scale problem. After deciding our variables for Model 5, we used IBM ILOG CPLEX 12.8 in order to solve this toy example. As seen, even for such a small-scale problem it did not solve the problem in a reasonable time by using CPLEX.

An example problem is shown in Table 2, has six jobs, nineteen operation, and seven machines exist. Each job has a different number of operations, such as Job 1 has three operations. Process time for each operation for specific machines is shown in the Table 2.

Table 2. Problem data

Title	Job	Operation	Machine 1	Machine 2	Machine 3	Machine 4	Machine 5	Machine 6	Machine 7	Due date
Op11	1	1	1	1	1	2	3	2	5	2
Op12	1	2	2	2	2	1	3	1	1	
Op13	1	3	2	2	2	1	1	5	1	
Op21	2	1	3	3	3	2	2	1	4	5
Op22	2	2	2	2	2	3	4	5	1	
Op23	2	3	2	2	2	4	4	1	2	
Op24	2	4	1	2	3	1	2	3	1	
Op25	2	5	2	2	5	6	4	1	1	
Op31	3	1	2	2	2	1	3	2	1	4
Op32	3	2	1	1	1	2	3	2	1	
Op33	3	3	3	3	3	1	2	2	1	
Op41	4	1	2	2	2	5	4	1	2	9
Op42	4	2	3	3	3	2	3	2	3	
Op51	5	1	3	3	3	2	3	2	3	7
Op52	5	2	1	1	1	2	2	1	1	
Op61	6	1	4	4	4	2	3	2	3	20
Op62	6	2	3	3	3	2	2	5	2	
Op63	6	3	1	1	1	2	2	2	2	
Op64	6	4	1	1	1	2	1	2	1	

Having a high number of constraints and variables in the problem, it was not possible to find the optimal solution in a reasonable time. In fact, this is an NP-Hard problem [22]. For this reason, a heuristic algorithm has been developed which gives near optimal solutions in reasonable computational times. The algorithm is shortest processing time (SPT) based greedy construction algorithm. Our heuristic consists of three main parts in a loop. Firstly, heuristic search eligible operations and machines (in the beginning, all machines are eligible). Eligible operations are added to the list with their minimum machine processing time. Then we sort operations in increasing order. Secondly, the algorithm searches condition if the best candidate operation is available for minimum processing time machine for candidate operation. If machine is eligible (ready time = 0) then we assign selected operation to the best machine, otherwise we assign candidate operation to any eligible machine at time t. We update machine's ready time as assigned operation's processing time. Finally, we update time as minimum ready time of machines' and mark assigned operation as planned operation. At the end of final step, we return first step until there are no operations exist. This algorithm is encoded in Microsoft Excel using the Visual Basic Application (VBA) (Fig. 4).

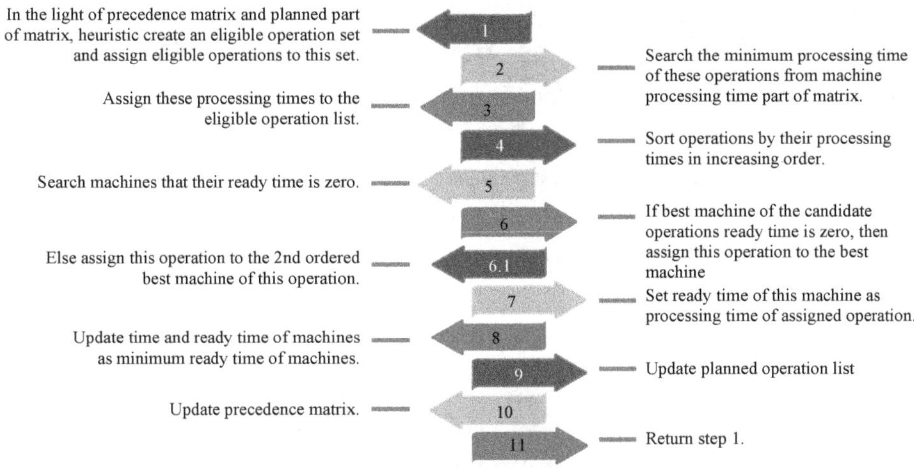

Fig. 4. Heuristic algorithm steps diagram

By using our example data gathered from Bosch's current system, manual planning outputs are compared and validated. Figures 5 and 6 visualize the solutions of the same instance for the exact model and the heuristic approach respectively, while the Fig. 7 represents the current plan of the firm. For each test machine, machine utilization is calculated. For each job, lateness, tardiness, and flow time are calculated and shows in Table 3. When this table is examined, it is seen that the best results are given by the developed CPLEX model as expected.

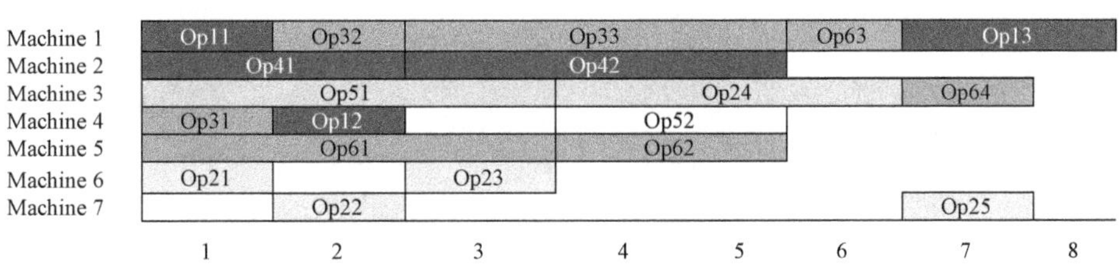

Fig. 5. Heuristic solution Gantt chart

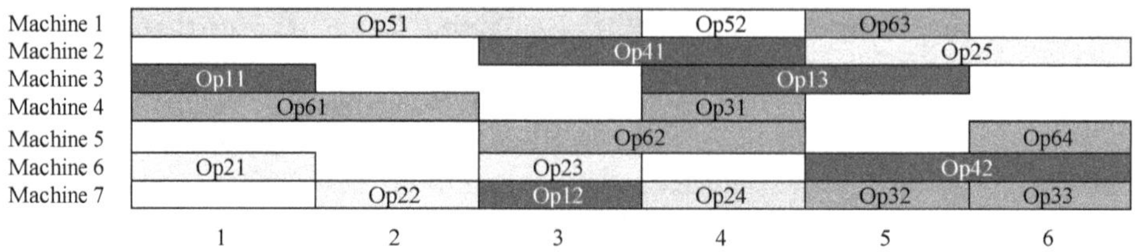

Fig. 6. Optimal (CPLEX) solution Gantt chart

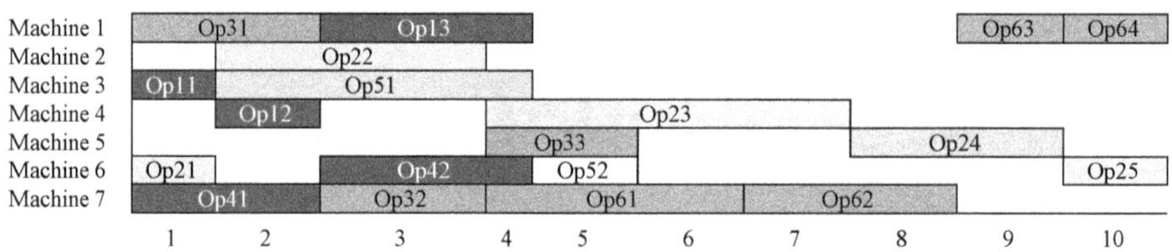

Fig. 7. Current system BOSCH's planning Gantt chart

Table 3. Comparison of the solutions 1

		Bosch's outputs	Cplex outputs	Heuristic outputs
Machine utilization for each machine	Machine 1	60,0%	83,3%	100,0%
	Machine 2	20,0%	66,7%	62,5%
	Machine 3	40,0%	50,0%	87,5%
	Machine 4	50,0%	50,0%	50,0%
	Machine 5	40,0%	50,0%	62,5%
	Machine 6	50,0%	66,7%	25,0%
	Machine 7	80,0%	83,3%	25,0%
Tardiness for each job	Job 1	0	0	0
	Job 2	3	0	0
	Job 3	1	0	1
	Job 4	0	0	0
	Job 5	0	0	0
	Job 6	0	0	0
Lateness for each job	Job 1	0	0	−1
	Job 2	3	−1	0
	Job 3	1	−1	1
	Job 4	−2	−2	−1
	Job 5	−4	−4	−3
	Job 6	−3	−4	−3
Flow time for each job	Job 1	4	4	3
	Job 2	10	6	7
	Job 3	5	3	5
	Job 4	4	4	5
	Job 5	4	4	5
	Job 6	7	6	7

KPI based comparisons are given in Table 4 numerically. The optimal result for these identified KPIs was obtained from the developed CPLEX model. However, even a small-scale data problem such as a toy example takes a long time, such as four hours to solve the problem. This reduces the applicability of CPLEX modelling. As shown in the table, the CPLEX result for small-scale data is solved in four hours while developed heuristic model solves it in eleven seconds. This solution method reduces the time loss and gives a close result to the optimal result at a reasonable time. This advantage increases the applicability of heuristic model in real-life problems.

Table 4. Comparison of the Solutions 2

Time	2 h	4 h	11 s
	Bosch's outputs	Cplex outputs	Heuristic outputs
Makespan (CMAX)	10	6	8
Machine utilization	48.57%	64.28%	58.92%
Average tardiness	0.66	0	0.16
Average lateness	8.33	2	1.16
Average flow time	4.66	4.5	5.33

5 Decision Support System

From the DSS's welcome page "Job-Operation Input Page" can be selected to modify the data of the tests in the system, to add new tests to the system and display the current tests. Job Operation Input Page leads to a database of tests which all possible tests are stored.

After pressing the situated "Job-Operation Input Page" button in Fig. 8, the following window will appear. (The data below is a representation. The jobs and operations will be opened as empty.) The data to be run for heuristic will be entered in the forms (USERFORM) on this page.

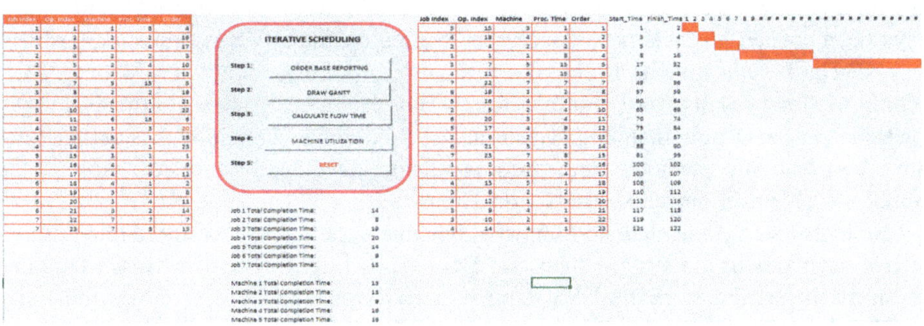

Fig. 8. Output page

After the problem is solved, the above page will be opened for outputs. The test machine, where each operation's operation is handled, the operation times of these operations, the total completion times of the jobs, the total completion times of the test machines and the test machine controls are on the output page. The Gantt chart located to the right of the page is for operations in Fig. 8.

6 Conclusion and Future Work

The problem observed in the Bosch Research and Development (R&D) laboratories is the manual assignment of test machine and tests. The primary objective of this study is to minimize the completion time of the planned tests without exceeding the determined delivery time and using the test machines efficiently. With the help of a developed decision support system (DSS), it is aimed to compare the efficiency analysis of test machines and to improve the system. In this study, improved techniques and codes can be applied to the similar systems, making the study suitable to use not only for the target companies but also for the others, which makes it innovative and global. During the study, delivery of the tests requested by the customer or engineer within the given time, keeping the machine utilization high and minimizing the delays have been taken into consideration. Duration of the periods used in the company have been specified as four hours, and this decision has caused the periods not to be used efficiently. With the improved system, periods are set as forty-five minutes, and the total daily test planning period is divided into ten periods.

The mathematical models of the flexible job shop scheduling problem have been researched, and five models that are suitable for the nature of the problem have been identified as candidate models. The constraints, parameters, and decision variables of the selected models were examined. Various constraints have been added to ensure that the models are matched to the system, and necessary changes have been made on the decision variables. The two models, which are closest to the current system, use two different types of the periods, are updated to the improved system and coded in the CPLEX. A toy problem has been created by using the small-scale data to compare results between the current planning method and developed one at the end of the project. In order to be able to interpret and compare these results, dynamic Gantt Charts have been created by making the necessary coding on Excel VBA. Since the problem was NP-Hard it was difficult to find the optimal result of these problems. With CPLEX coding, optimal results were obtained for the toy problem. However, in real-life systems such as the current planning system in Bosch, finding a solution takes quite a long time, and heuristic methods were used. With our developed heuristic it solved the Bosch's problem in eleven seconds with Cmax being eight time periods.

As future work, it is planned to have a dynamic system (DSS) reactive to a number of test machines in the system such that DSS. The current system in Bosch contains seven machines, therefore the DSS is made accordingly. To increase the applicability of DSS, it can be updated for other companies. In order to compare the DSS results' effectiveness, there will be future studies on different methods. In the future, it is planned to solve the improved system using Ant Colony heuristic method. There are a limited number of flexible job shop scheduling problem examples of real-life problems in the literature. This study is expected to contribute to in the FJSS problem literature in that sense.

Acknowledgements. We are indebted to our team members Hakan Elmas, Sehercan Yılmaz and Tuğçe Akboyun for their contribution. We appreciate our Bosch advisors Güler Öztürk, Hüseyin Özçimen and Alper Atias for their guidance and support throughout the project.

References

1. Bosch global (2018). https://www.bosch.com/our-company/our-history/
2. Bosch Thermotechnology (2018). https://www.bosch-thermotechnology.com/corporate/en/company/history.html
3. Manisa, T.T. (2017). https://www.bosch-thermotechnology.com/tr/en/residential/knowledge/bosch-termoteknoloji/bosch-tt-turkey/
4. Ekos Group website. https://www.ekosinerji.com/mp-include/uploads/2016/07/retim-sistemlerinde-cizelgeleme.pdf
5. Kaya, S., Fığlalı, N.: A review on meta-heuristic approaches to solving flexible job shop scheduling problem. Sak. Univ. J. Sci. **20**(2), 223–244 (2016). https://doi.org/10.16984/saufenbilder.12029
6. Bidgoli, M.M., Fattahi, P., Samouei, P.: An improved tabu search algorithm for job shop scheduling problem trough hybrid solution representations. J. Qual. Eng. Prod. Optim. **3**(1), 13–26 (2018). https://doi.org/10.22070/jqepo.2018.1360.1035
7. Nakandhrakumar, R.S., Seralathan, S., Azarudeen, A., Narendran, V.: Optimization of job shop scheduling problem using tabu search optimization technique. Int. J. Innov. Res. Sci. Eng. Technol. **3**(3), 1241–1244 (2014)
8. Kandiller, L.: Project management: management engineering programme, pp. 2–55 (2014)
9. Herroelen, W.S.: Resource-constrained project scheduling: the state of the art. J. Oper. Res. Soc. **23**(3), 261–275 (1972). https://doi.org/10.1057/jors.1972.48
10. Özdamar, L., Ulusoy, G.: a survey on the resource-constrained project scheduling problem. IIE Trans. **27**(5), 574–586 (1995). https://doi.org/10.1080/07408179508936773
11. Kolischa, R., Padmanb, R.: An integrated survey of deterministic project scheduling: model-data-algorithm integration. Omega **29**(3), 249–272 (2001). https://doi.org/10.1016/S0305-0483(00)00046-3
12. Hartmann, S., Briskorn, D.: A survey of variants and extensions of the resource-constrained project scheduling problem. Eur. J. Oper. Res. **207**(1), 1–14 (2010). https://doi.org/10.1016/j.ejor.2009.11.005
13. Wang, L., Cai, J., Li, M., Liu, Z.: Flexible job shop scheduling problem using an improved ant colony optimization. Sci. Program. **2017**, 1–11 (2017). https://doi.org/10.1155/2017/9016303
14. Mehrabad, S.M., Fattahi, P.: Flexible job shop scheduling with tabu search algorithms. Int. J. Adv. Manuf. Technol. **32**(5–6), 563–570 (2007)
15. Demir, Y., İşleyen, K.S.: Evaluation of mathematical models for flexible job-shop scheduling problems. Appl. Math. Model. **37**(3), 977–988 (2013). https://doi.org/10.1016/j.apm.2012.03.020
16. Wagner, H.M.: An integer linear programming model for machine scheduling. Nav. Res. Logist. Q. **6**, 131–140 (1959)

17. Kim, K.-H., Egbelu, P.J.: Scheduling in a production environment with multiple process plans per job. Int. J. Prod. Res. **37**, 2725–2753 (1999)
18. Manne, A.S.: On the job-shop scheduling problem. Oper. Res. **8**, 219–223 (1960)
19. Gao, J., Gen, M., Sun, L.: Scheduling jobs and maintenances in flexible job shop with a hybrid genetic algorithm. J. Intell. Manuf. **17**, 493–507 (2006)
20. Thomalla, C.S.: Job shop scheduling with alternative process plans. Int. J. Prod. Econ. **71**, 125–134 (2001)
21. Gomes, M.C., Barbosa-Povoa, A.P., Novais, A.Q.: Optimal scheduling for flexible job shop operation. Int. J. Prod. Res. **43**, 2323–2353 (2005)
22. Sirkeci, E.: Esnek Atölye Tipi Çizelgeleme Problemi İçin Çözüm Yaklaşimlari: Savunma Sanayinde Bir Uygulama. MsC Thesis, Gazi Üniversitesi Fen Bilimleri Enstitüsü, Ankara, Turkey (2015)

Standardization of Packaging Materials for Various Products and Cost Optimization in Packaging

Özge Ceyhan, Merve Türk, Hafize Helvacı, Damla Yüksel(✉), and Nejat Kutup

Department of Industrial Engineering, Yasar University, İzmir, Turkey
ozgeceyhan6@gmail.com, merveturkk59@gmail.com, hafizehelvaci1243@gmail.com, {damla.yuksel, nejat.kutup}@yasar.edu.tr

Abstract. This project is related to the standardization of the packaging of variable-size products, which belongs to base, headboard, marquise and cupboard (nightstand) product categories, for a bed products company. The company cares importance to special size demands of their customers and produces products in a wide range of sizes. This causes complexity in the stock-tracking of cardboard plates, which are the most important packaging materials in the packaging process, and high waste of cardboards. Therefore, this study aims to improve the packaging system of the company and to reduce the packaging costs by decreasing the amount of waste related to the materials used for packaging. The main objective of the study is to create a standardized packaging plan by considering all product sizes the company has produced in the last three years in the product categories of base, headboard, marquise and cupboard. To achieve this goal, after the completion of literature search, forecasts for the next year's sales were done, and a mathematical model that has two conflicting objectives was formulated. Then, the mathematical model was solved by using IBM ILOG CPLEX program for each product category and optimal package sizes were determined. A useful Decision Support System (DSS) which enables to prepare weekly packaging plans that reduce the extra packaging costs and the amount of waste has been created at the end of the project.

Keywords: Optimization · Packaging problem · Standardization · Mathematical model · Sales forecasting · Waste minimization

1 Introduction

This study is conducted in a local bedding company which was founded in 1974 as a manufacturer of bed products and has 60.000 m^2 production plant in Torbalı, Turkey. The production capacity of the company is 700 mattresses per day. The company, which has high market share among its competitors, produces personalized bed products and has a wide range of products. The company has 377 headboards, 45 coffer base, 31 electro base, 30 drawers base and 112 cupboard types. Although they produce different sizes of all products, the customers may order in any size of the products.

N. M. Durakbasa and M. G. Gençyılmaz (Eds.): ISPR 2019, LNME, pp. 775–790, 2020.
https://doi.org/10.1007/978-3-030-31343-2_65

Therefore, this significantly increases the number of product size. Thus, the waste occurs in the packaging process rises significantly. For example, the fact that the suitable cardboard size for the product is not in the inventory increases the use of cardboard. So, this study aims to determine the dimensions of the packaging cardboards which will minimize the waste occurring in the process.

The company uses different raw materials due to the product variety. Raw materials are consumed in different stages of production. Some of those creates a semi-finished product such as sewn fabric, furnished product and packaging semi-finished products. The main raw materials used for furnished products such as headboards, bases and marquises are skeleton, sponge, cardboard, fabric and yarn on the production process. Semi-finished sewn fabric is composed of different types of fabric and yarn. Packaging semi-finished product consists of packaging materials such as cardboard, bubble pack, stretch and polyethylene profile material. Moreover, the company supplies semi-finished base and headboard skeletons from suppliers. At the same time, different labels can be used in products to represent the brand of the good, since the company makes order-oriented production and uses custom labels for the products it produces.

The production flow in the company starts with the order arrival as shown in Fig. 1. Then, skeleton and fabric are supplied from suppliers after the order is received. Each raw material supplied is subjected to quality control tests. After the preparation of fabric, semi-finished products of each order are collected in one area, which is called kit area, and produced simultaneously during the production process. Every finished product also passes a quality control test. After all these operations, the only remaining process is packaging and labelling. Packaged and labelled products are sent to the shipping area to be delivered to customers.

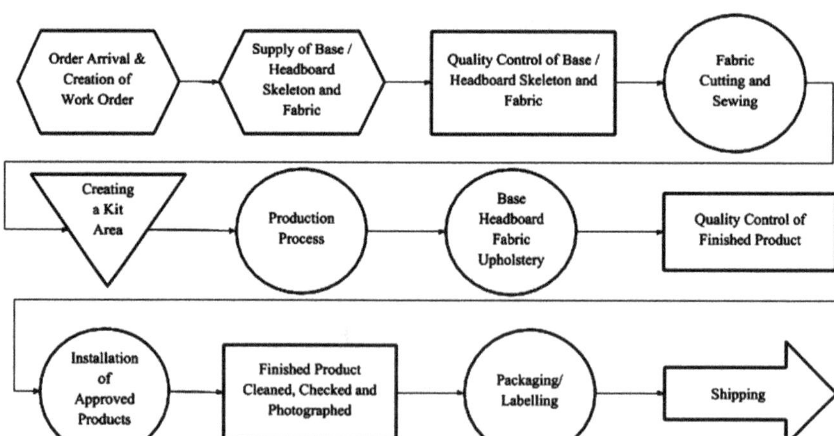

Fig. 1. Production flow

As seen in Fig. 1, packaging is the last step of the products before leaving the company. Therefore, it has a great importance on production flow. Also, because of the different product sizes due to the specials orders' of the customers, the waste caused by packaging materials, especially by cardboards, increases significantly in the packaging process. Consequently, the cost of packaging rises directly. So, the main goal of this study is to standardize the dimensions of the packaging cardboards which will minimize the waste occurring in the packaging process.

The detailed problem definition is given in Sect. 2 and the literature review is represented in Sect. 3. Section 4 explains the mathematical model and Sect. 5 discusses the computational results. Finally, conclusion and future work are presented in Sect. 6.

2 Problem Definition

The company uses two main packaging methods for export and domestic market. In the domestic market, a semi-packaging is used. On the other hand, in the products which create 90% of exported goods, full packaging is applied to prevent damage, since protection of export products is more difficult. It causes the usage of extra packaging materials. Another point is that the company arranges these extra packaging materials according to wishes of the customers. Hence, there is no product-based standardization in packaging because of this flexibility. However, this study only considers the export products which are fully packed.

Moreover, since the company cares importance to the special demands of their customers, variable sizes of products are produced. The product diversity has an important role on packaging process in the company. The packaging structure is affected by the customer-based variation of product sizes. For instance, there are at least 13 sizes of width, which are 70, 75, 80, 90, 100, 120, 140, 150, 160, 180, 200, 210 and 220 cm, for electro base and drawer base product categories and at least 4 sizes of height, which are 190, 200, 210 and 220. Although the company has a much more variety of sizes in its products than the other bed product companies in Turkey, variety of sizes are still quite high due to the acceptance of the customer-based orders. It creates complexity for stock tracking of packaging materials, especially cardboard. Cardboards are purchased from suppliers as cut and ready to be used in packaging process of the company. However, workers in the packaging area bend and glue some of the cardboards to shape them three dimensional as can be seen in Fig. 2. The packaging operation performs manually in the company, almost no machine is being used during the process. Another significant thing is fully packed products are packaged with more than two pieces of cardboards, not a single whole piece of cardboard. In Fig. 3, fully packed bases are shown.

Workers arrange and select the size of the cardboard according to size of the product before the packaging process. Firstly, the product is wrapped with protective materials such as bubble pack. The bubble pack is taped to the product manually. As a product protection, a polyethylene profile material can be wound around the edges of the product or a corner protector material also can be used. Following that, appropriate cardboard sizes are chosen and cardboards to be used brought from the stock area and arranged on the packaging table. The product is placed in cardboards on the packaging table. Styrofoam blocks are used for the spaces between the cardboard and the product due to the shape of the product. In Fig. 4, how a Styrofoam block is used in the packaging of headboards is seen.

Fig. 2. Shaped cardboards

Fig. 3. Fully packed bases

After the materials used in the package are fixed, the top cardboards are closed, and the product is packed. The product packaged with cardboard is sometimes wrapped with plastic foil on a separate table if a customer desires and made prepared for shipment after the hooping operation. In the half-packed products, different cardboard sizes are used, and each side of the packaged product is not covered with cardboard. An example of half-packed product can be seen in Fig. 5. Also, generally, two pieces of cardboard are used for products to be half-packed. In the packaging process, even though an appropriate size of cardboards is being chosen, the cardboards are folded manually according to the product sizes and the remaining parts are cut.

Fig. 4. Styrofoam blocks in the packaging

Fig. 5. A semi-packed headboard

As the number of different sizes of package increases, box inventory increases. Increase in variety of package size leads to increase in handling and purchase costs. The company may have to purchase small amounts of boxes with higher price. The absence of a standard packaging system in the company causes an excess or lack of stock in packaging materials such as cardboards. Besides, it makes inventory control harder. In the company, there is difficulty in stock tracking of cardboards. Therefore, the cardboard size in stock can be cut and arranged according to the desired size from time to time, when there is no stock in the proper size of cardboard. The packaging system in the company wastes packaging materials in significant amounts. Furthermore, the company's use of expensive C-BC dopel cardboards increases the importance of this situation.

For these reasons, it is observed that the packaging cost constitutes a certain part of the total production cost of the company. Also, another symptom determined through observation is that although the packaging is supported by C-BC dopel cardboards and protective materials, the products are damaged somehow, and this is the most important part of customer complaints. The company wants to make sure that the complaints are not related with packaging quality.

This study focuses on the standardization of the packaging system of the company by creating efficient packaging plans for headboards, cupboards and bases, which include coffer bases, drawers' bases and electro bases, that decrease waste in materials and eliminate the packaging differences between domestic market and export products if it will be less costly, and to set out a similar plan for a determined interval of packaging measures. At the end, the cost in packaging materials will be reduced by decreasing waste in the system. It is aimed to standardize the packaging by creating common packaging sizes in each product group separately. In addition, the inventory of cardboards is aimed to be minimized by decreasing the number of different package sizes. Thus, monitoring of inventory will be more controllable. Therefore, instead of using different package sizes for different sizes of a product, one packaging size will be used for a group of sizes. Moreover, a flow chart for the packaging process is shown in Fig. 6.

In other respects, some numerical data were needed and acquired from the company. These are past sales data of related products, bill of material for packaging of these products and unit material costs of raw materials which are used in packaging of products. Firstly, the given data are divided into product-based categories such as bases, headboards, marquises and cupboards. Bill of materials for packaging of each product category is given by the company for unique sizes of products. To calculate the packaging costs of each product of each category, unit costs of raw materials have been obtained. Moreover, in the mathematical model, forecasted demand values for each product is used as parameters and these data are acquired by means of past sales data of the company.

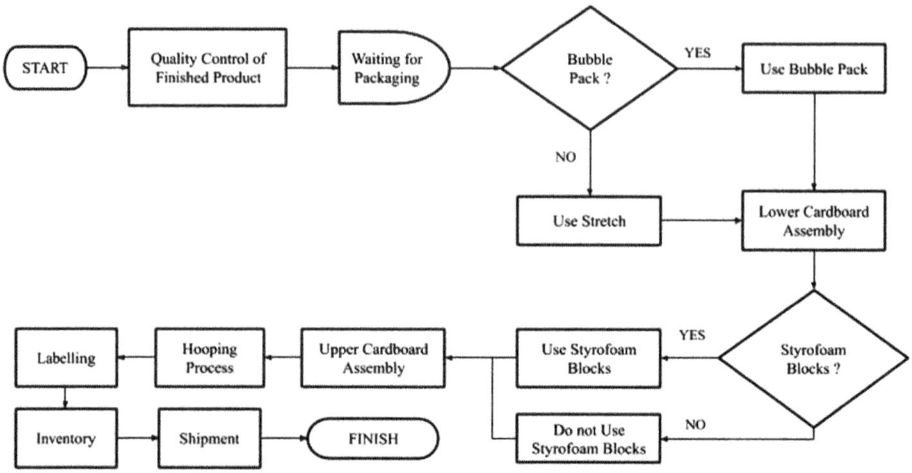

Fig. 6. Flow chart of packaging process

3 Literature Review

In this literature research, packaging problem that can be observed on each production process, attracts the attention of mathematical community for many years. These problems occur directly or as a sub-problem in other problems. Mathematicians attempt to find an appropriate solution which provides waste reduction on packaging materials and saving cost. Cost reduction and amount of package material usage will be reduced in production facilities when the appropriate solution method for this packaging problems are found. Egeblad's (2008) study is distinguishing different between 1DCSP and 1DBPP. While cutting stock problems concern with the few item sizes (homogeneous), bin packing is used number of more items (heterogenous). In terms of 2D, based on the thesis of Egeblad (2008), develop the solution benefit from 2D knapsack problem. According to this method, determine the between variance of product size of height and weight and usage material size and height. Until the whole area is filled, to select product suitable for package. Thirdly, this study is examined in terms of 3D, it is aimed to reduce number of bins. Egeblad (2008), the size of the produced product has changed to fit the package. We used the explanation for 1D to determine the type of approach in the solution method of the problem. According to description of 2D used identified the variation, add the constraints to reduce the variations. However, the studies for multidimensional are not used because this does not apply to our problem definition.

Martello et al. (2002) study the orthogonal packaging of a given series. In such problems, rectangular boxes consist of loading a minimum number of rectangular containers. For the resolution of the problem, algorithms such as exact three-dimensional problems are mentioned. The theoretical and practical performances of the lower boundaries were compared during the solution of the problem.

Variable sized bin packing problem (VSBP) is one of the other problems related to packaging in the literature. In this problem, which is different from the classical bin packing problem, has a finite number of different box sizes and there is no constraint about supplying bins of different sizes. The aim is to pack different pieces into bins while minimizing the total space used in the packaging process. The objective also includes the minimization of the total cost of packing since it is assumed that the cost of the bin is directly proportional to the its size. Friesen and Langston (1986), study three heuristics for the variable sized bin packing problem. These are called Next-Fit using largest bins only (NFL), First-Fit decreasing using largest bins, at end repack to smallest possible bins (FFDLR) and First-Fit decreasing using largest bins but shifting as necessary (FFDLS).

Haouari and Serair (2009) study the problem of one-dimensional variable bin packaging. Bin packaging problems constitute a class of optimization problem in a wide variety of applied areas including packaging, cutting and timing. In this paper, algorithms for variable size packaging problem which are generalization of one-dimensional packaging problem are investigated. This problem, minimum cost of many items has unequal size and cost involves placing a set. Different bin types are used in modeling and it has been tried to ensure that the total weight of all items loaded in each item has a weight, does not exceed a thousand capacity. In this study, it was tried to obtain a minimum set of packaging cost. One-dimensional variable packaging has proposed and analyzed six intuitive experimental performances for the problem. Although this research is related to the purpose of our problem, it has been studied one dimensional variable bin and the solution methods not related.

In different industrial sectors, since packing is one of the most important resource allocation problems, computer technology is important role in optimization problems for developing the utilization of resources. It is undeniable that optimal loading space utilization saves transportation cost and resources, not only space. Wu et al. (2017) study a new useful packing problem which can be categorized three-dimensional irregular packing with variable-size cardboards problem, three-dimensional variable-size bin packing problem and the single container loading problem. Searching a good solution for the problem is difficult because the sub problems are NP-hard. Their paper considers the three sub problems as how to pack irregular fasteners into variable-size cardboards such that the utilization of cardboards is highest firstly. Second problem is related to three-dimensional variable-bin packing problem and finally different products are loaded into container or shipping. For this reason, it includes a single container loading problem. For the solution the main aim is to maximize the filling rate of the container and they use a novel algorithm for 3D irregular fasteners packing problem and 3D bin packing problem.

Ilıcak (2003) studies on bi-objective bin problems which the objective functions are contradictory each other. This problem involves assigning items with different weights to boxes that have identical capacity. The main objectives of the problem are to minimize the total deviation from the box capacity or minimizing the maximum over-deviation while minimizing the total number of boxes used. The aim of this study is to create a procedure for an effective solution of all two-objective problems, by using the results of the scheduling theory, constructive and curative upper limiting heuristic methods. In this study, more than one product can be assigned to a box, such that it does not exceed the capacity. On the contrary, in our problem, each product will be assigned to a single box. Despite this difference, while examining the thesis, similarities were found between the thesis and our problem. Since the cardboard stock area in the factory has a capacity, we also aim to minimize the number of different sizes of cardboard used in this study as well. As targeted in this study, we also aim to minimize the amount of deviation between the size of the cardboard and the product to be used for packaging in our problem. As a result, this literature has been used to determine the objective function and constraints of mathematical model in line with all these.

There exist several packaging problems that have been studied before. Some of these examine the problem from the point of view of efficient space utilization. One of these is the study made by Richard C. Wilson that investigates packaging of similar types of products which differ in size in 1965. The study includes minimization of the total cost of having a specific number of different box sizes in the system by considering the purchasing cost of each box size and the cost of warehouse space that arises due to box sizes. Wilson (1965) states that if each product is packaged in its own size, warehouse space costs, shipping costs and cardboard box expenses can be limited. On the other hand, inventory holding, handling and purchase costs can be decreased in case the usage of one box size for all sizes. Wilson (1965) reports that the savings achieved by the usage of different box sizes for each size of products are balanced out by the higher price of boxes purchased in small quantities, the additional order of many different box sizes and inventory and handling costs. In fact, since the number of variables and constraints are quite a lot in the study, after deciding on solving the mathematical model of the problem as stated by integer programming is unrealistic at that time, a heuristic procedure is developed. At first, a set of candidate box sizes is generated. Following that, an incremental cost method is used to decrease the first set of box sizes. Forecasted sales values that are used as the annual activity of the products plays a significant role in the determination of final box sizes and the reduction in the number of box sizes is acquired while it does not cause an increase in the total cost of space at the end of the study.

4 Modelling and Solution Methodology

Solution methodology of the project is carried out for standardizing packaging structures of bases, headboards, marquises and cupboards product categories of the company. The methodology aims to determine the optimal package sizes for each product size in each product category in the system. After the completion of the literature review, the problems and solutions that may be relevant to the project have been decided so that the similarities and differences with the problem have been analyzed extensively.

Ilıcak's (2003) master thesis presents bi-objective bin packing problems. One of the problems, which aims to minimize total deviation over bin capacity and minimize number of bins, has two conflicting objectives. Also, during the definition of the decision variables and some constraints, which are constraint (2) and constraint (3), of the problem, it has been taken advantage of the master thesis of Ilıcak's. In addition, based on the idea of over deviation used in the master thesis, total deviation of the package sizes from product sizes has been formulated in the objective function of the mathematical model.

Wilson's (1965) study is about a packaging problem that selects the optimum number and the sizes of boxes which minimize the total system cost. In the problem statement, the forecasted demand activity of each product is used as a parameter in the objective function. Based on Wilson's study, in the objective function of the mathematical model, forecasted demand values of each product were used as a weighting function on the total deviation.

To be able to obtain parameters of the mathematical model, it has been decided to analyze last three years of sales data of each product category and to make a forecast for the next year's sales data. Simple Linear Regression method has been used as a forecasting method. The method is applied on the sales data of last three years in order to acquire the forecasted demands for each product category after the data organization step. Forecasted demand values are quite important for the mathematical model which is formulated for determining optimal package sizes for each product size in each product category and the total number of different package sizes to be used at the end.

The objective function of the mathematical model aims to minimize the total number of different package sizes to be used and the total waste occurred in the system at the same time. Consequently, it has been noticed that the problem contains more than one objective. Since it is difficult to achieve all the conflicting objectives of this problem simultaneously, it has been decided to use the two objectives with their own weights as a single objective function. Thus, the weighted sum of these two objectives is obtained in the modelling. The aim is to minimize the total number of different package sizes to be used in each category at the end and the difference between the package size and the product size by considering 3 dimensions. The strategy followed is that the same package size is used for a group of products which have approximate sizes. The model determines package sizes proposed to be used by weighting forecasted demand values of each product. Products that have high forecasted demand values are assigned to packages whose dimensions close to their own size due to the definition of the model. The sets, indices and parameters are presented below.

Sets and Indices

i Products in different size $i \in N$

j Packages in different size $j \in M$

d Dimension index, $d = 1, 2, 3;$ 1 refers to width, 2 refers to depth, 3 refers to height

Parameters

z_{id} The size of dth dimension of product i

a_{jd} The size of dth dimension of package j

f_i The forecasted demand of product i

w_1 The weight of total number of different packages to be used

w_2 The weight of total waste occurred

A set of package sizes is given to the model as if each product had its own package.

Decision Variables

y_j 1 if package j is selected, 0 otherwise

x_{ij} 1 if product i is assigned to package j, 0 otherwise

The mathematical model created according to the definitions given above is as follows:

$$\text{Min } \mathbf{w_1}.\sum_{j=1}^{M} y_j + \mathbf{w_2}.\sum_{i=1}^{N}\sum_{j=1}^{M} f_i.x_{ij}.\left(\left(a_{j1} - z_{i1}\right) + \left(a_{j2} - z_{i2}\right) + \left(a_{j3} - z_{i3}\right)\right) \quad (1)$$

Subject to;

$$\sum_{j=1}^{M} x_{ij} = 1 \qquad \forall i \in N \quad (2)$$

$$x_{ij} \le y_j \qquad \forall i \in N, j \in M \quad (3)$$

$$a_{j1} - z_{i1} \ge -M\left(1 - x_{ij}\right) \qquad \forall i \in N, j \in M \quad (4)$$

$$a_{j2} - z_{i2} \ge -M\left(1 - x_{ij}\right) \qquad \forall i \in N, j \in M \quad (5)$$

$$a_{j3} - z_{i3} \ge -M\left(1 - x_{ij}\right) \qquad \forall i \in N, j \in M \quad (6)$$

$$\sum_{j=1}^{N} x_{ij} \ge y_j \qquad \forall j \in M \quad (7)$$

$$x_{ij}, y_j \in \{0, 1\} \qquad \forall i \in N, j \in M \quad (8)$$

While writing the objective function, we have inspired from Ilıcak's (2003) master thesis. The thesis concerns with two bi-objective bin packing problems. One of them, which is Classical Bin Packing problem, aims to minimize the total deviation over capacity. In our project, the aim of the mathematical model is that while minimizing the total number of different package sizes to be used, on the other hand, acquiring proper product size and package size assignments which will minimize the total difference between the product sizes and the package sizes assigned to them by considering three dimensions (1). Moreover, in Wilson's (1965) study, forecasted demand activity of products is used as a weighting function on the annual costs of cardboards. In our model, forecasted demand of a product size is used to determine the importance of the size of the product. If the forecasted demand is high, the size gain importance and the model choose the package size which can create a minimum space between the product and the package. Weights are assigned to the goals clarified in the objective function of the model in line with their importance levels and two objectives of the model are expressed in a single objective function. Also, values of w_1 and w_2 have been determined as subjective to indicate the importance levels of the goals. While the total waste will be a very large number compared to the total number of packages, w_2 is considered to be 1, and w_1 has been studied by trying high numbers to give importance to the total number of packages. Constraint (2) of the model indicates that every product should be assigned to exactly one package. Constraint (3) ensure that if a product is assigned to a package, then the package is used. Constraints (4), (5) and (6) state that if a product is assigned to a package, then the difference between the package and the product size should be greater than or equal to zero. These constraints prevent assignment of the product to a smaller package size. Constraint (7) of the model indicates that if there is no product size assigned to a package size from the set of packages, then the package size is not selected by the model in the solution. Finally, Constraint (8) indicates the domain of decision variables.

5 Computational Results

The developed a mathematical model has been worked on IBM ILOG CPLEX program by using all different product sizes, which are obtained from the last 3 years of sales data, in each category and forecasted demand values of each product size. When the last 3 years of sales data of the company are examined, the number of different product sizes observed in each product category is shown in Table 1. 504 different product sizes in the bases category, 1385 different product sizes in the headboards category, 72 different product sizes in the marquises category and 4 different product sizes in the cupboards category have been used as parameters for the product sizes in the mathematical model.

Table 1. Total number of different product sizes

Product category			
Bases	Headboards	Marquises	Cupboards
504	1385	72	4

The model has been run separately for each product category of the company and results that show assignment of package sizes to each product size and give the value of total number of different package sizes to be used are obtained. Due to the idea of weighted sum of two objectives, the mathematical model gives different solutions, as the value of w_1 changes. Therefore, in each product category, different w_1 values have been used to get different results of the model and the total difference between the product and package sizes, which refers to total deviation, and the total number of different package sizes used in each solution have been reported. Line graphs were plotted according to results of the mathematical model. These graphs that can be seen in Figs. 7, 8 and 9, show the relationship of number of cardboard variety and the total deviation for different w_1 values for the product categories of bases, headboards and marquises.

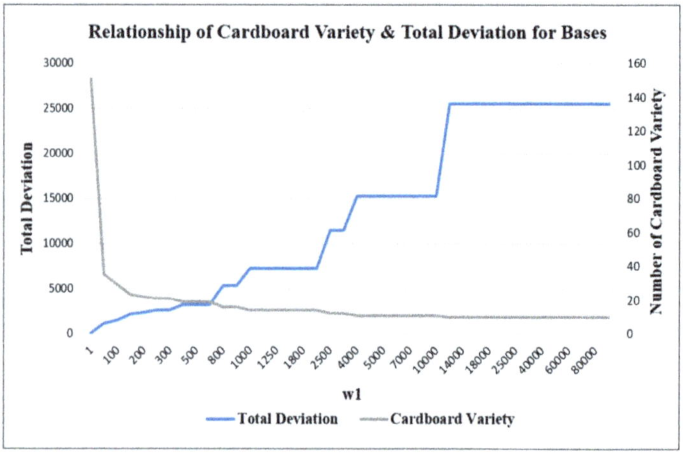

Fig. 7. Relationship of the total cardboard variety and the total deviation for bases

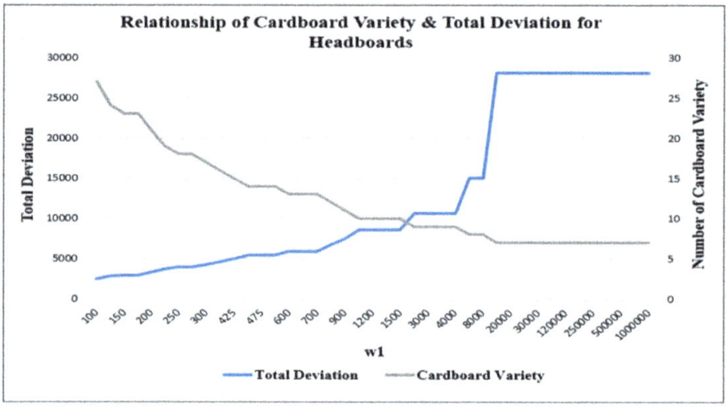

Fig. 8. Relationship of the total cardboard variety and the total deviation for headboards

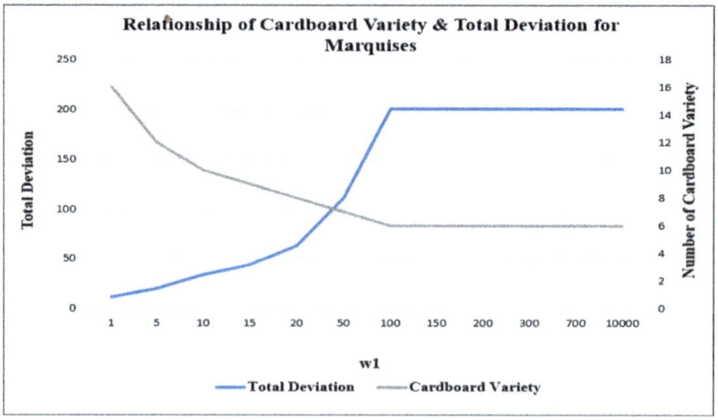

Fig. 9. Relationship of the total cardboard variety and the total deviation for marquises

By looking at Figs. 7, 8 and 9, it can be concluded that when the number of cardboard variety increase, the total deviation that refers to the total difference between products and the assigned package sizes decreases automatically. This inverse relationship between the number of cardboard variety and the total deviation also supports that the problem has two conflicting objectives.

Moreover, in the product category of cupboard, there are just 4 different product sizes the company produces. When the mathematical model was run with different w_1 values, the model gives the same solution that shows the same product size and the package size assignment. As a result, 4 different sizes of products are packaged by 3 different packages as minimum.

To make a comparison with the current system, the total waste occurred in each solution with different w_1 values has been calculated by subtracting the total product area from the total area of optimal package size assigned to the product. Thus, we have attained the waste levels for all solutions with different w_1 values in headboards, bases and marquises product categories. Furthermore, since we have the old package size information for all product sizes of bases, headboards and marquises, the waste level of the old packaging system the company used have also been calculated. The different results that obtained by using the mathematical model have been compared with the old packaging system of the company in terms of the total amount of waste and the total number of different packages used. If the total amount of waste calculated and the total number of different package sizes used in the result of the model are lower than the values of the old packaging system, then the result is acceptable for the company. As a result, in each product category, w_1 values which give better results in the total amount of waste and the total number of packages used have been assumed as optimal solutions of the model. After the analysis of the results and the comparison step, values of w_1 and the number of cardboard varieties are in Table 2 have been accepted as values that give optimal solutions of the mathematical model. Percentage improvements in total waste and total cardboard packaging cost have also been calculated based on assignments of optimal number of cardboard varieties in each product category and percentage improvements can be seen in Tables 2 and 3.

Table 2. Results of optimal solutions and percentage improvements in waste

Category	w1	Number of cardboard variety	Total waste level of old packaging system (cm^2)	Total waste level of new packaging system (cm^2)	Percentage improvement (%)
Base	12000	10	13078413	9657168	26
Headboard	1100	8	5627898	4113180	27
Marquise	20	8	120188	67826	44
Cupboards	Noneffective	3	–	–	–

Total cost of old packaging system includes the total cost of cardboard plates used by the company for all products in each category. In addition, to be able to calculate the total cost of new packaging system, the unit cost of cardboard has been calculated by

Table 3. Results of optimal solutions and percentage improvements in total cardboard packaging cost

Category	w1	Number of cardboard variety	Total cost of old packaging system (₺)	Total cost of new packaging system (₺)	Percentage improvement (%)
Bases	12000	10	19296	18436	4.46
Headboards	1100	8	6876	6144	10.64
Marquises	20	8	297	146	50.93
Cupboards	Noneffective	3	–	–	–

using the current cardboard cost information of the company. Following that, cost of standard cardboard sizes to be used in the new system have been obtained by assuming the same quality of cardboard will be used in the new packaging system with the same unit cost. Thus, total cost of new packaging system has been calculated and reported as can be seen in Table 3.

On the other hand, although standard cardboard sizes were obtained in the cupboard product category, the waste and cost comparisons with the old packaging system could not be made due to lack of data. Another important thing in cupboard category is that as the value of w_1 change, neither the product-package size assignment results nor the total number of cardboard varieties used varies. Therefore, w_1 is stated as noneffective in the tables.

When the mathematical model is run with the w_1 values shown in Tables 2 and 3 in each product category, the package sizes to be used in the packaging of all product sizes in related product categories are selected as in Table 4. Thus, in the bases category 10 different cardboard sizes, in the headboards category 8 different cardboard sizes, in the marquises category 8 different cardboard sizes and in the cupboards (nightstands) category, 3 different cardboard sizes are used by the model in total.

Table 4. Standard package sizes determined for each product category

	Product category			
	Base	Headboard	Marquise	Cupboard
Standard package sizes	102 × 24 × 222	122 × 14 × 122	102 × 52 × 23	52 × 48 × 33
	102 × 25 × 202	162 × 14 × 122	152 × 42 × 31	52 × 52 × 31
	112 × 24 × 202	182 × 19 × 122	162 × 57 × 27	53 × 42 × 35
	122 × 24 × 192	202 × 14 × 142	172 × 57 × 25	
	142 × 23 × 222	202 × 14 × 222	182 × 42 × 31	
	202 × 20 × 222	202 × 19 × 202	182 × 47 × 28	
	202 × 23 × 212	222 × 19 × 182	182 × 48 × 28	
	222 × 15 × 222	222 × 22 × 172	202 × 42 × 29	
	222 × 20 × 202			
	92 × 25 × 212			

6 Conclusion and Future Work

This project has been conducted to minimize the amount of cardboard waste in the packaging process by determining the standard package sizes required in accordance with the demands of the existing products. The current packaging system of the company and system requirements have been analyzed and the mathematical model most suitable for the observed problem has been developed. In order to make the necessary comparisons in the packaging plans obtained from the model for each product category, firstly the sales data of the previous 3 years were divided into product categories and the waste and packaging cost analysis of these products were made for the packaging system used by the company. These data were compared with the packaging cost and total waste in the newly developed standard packaging system. Following that, it has been concluded that it is appropriate to use a total of 8 different cardboard sizes for headboards, 10 different cardboard sizes for bases, 8 different cardboard sizes for marquises and 3 different cardboard sizes for cupboards. According to these results, declines obtained in waste level are 26% in the base category, 27% in the headboard category and 44% in the marquise category. As a direct result of the improvements in the amount of waste, the observed improvements in the packaging cost are 4.46% in the base category, 10.64% in the headboard category and 50.93% in the marquise product category. Thus, the old packaging system which is based on employee experience has been improved by applying mathematical methods and transformed into a system where resources are used more efficiently. In addition, as a result, a useful Decision Support System (DSS) has been created which eliminates dependency on labor force and reduces extra packaging costs and the amount of waste in the packaging system of the company. The user-friendly DSS created with Excel VBA interfaces shows the optimal package sizes for products in the weekly production plan and the total cardboard requirements and it also makes tracking the inventory of the cardboards easier for the user. If the result of the project is desired to be developed and improved, studies on the quality of the cardboard used by the company can be done in order to reduce the total cost of cardboard usage.

As future work, to evaluate the quality of alternative cardboards and to decide on if the currently used cardboard type should be changed, cardboard quality control scheme can be analyzed. Therefore, the cost comparison for different type of cardboards can be studied in the future. Also, improvement of consumption of materials and cost optimization can be done for all packaging materials used in the process. Thus, our work covers standardization of all packaging materials. The percentage of cost improvement and the efficiency of amount of materials used will increase.

Acknowledgment. This study is supported by TUBITAK (The Scientific and Technological Research Council of Turkey) in the program of "2209-B Undergraduate Research Projects for Industrial Applications Fellowship Program". We would like to thank our colleagues Ayça Serra Bayhan, Tuğçe Alkan, Fatoş Nur Direkli and Ekin Aydın for their help and contribution to the study.

References

Egeblad, J.: Heuristics for Multidimensional Packing Problems (2008). http://hjemmesider.diku.dk/~jegeblad/phd.pdf

Friesen, D.K., Langston, M.A.: Variable sized bin packing. SIAM J. Comput. **15**(1), 222–230 (1986). https://doi.org/10.1137/0215016

Haouari, M., Serairi, M.: Heuristics for the variable sized bin-packing problem. Comput. Oper. Res. **36**(10), 2877–2884 (2009). https://doi.org/10.1016/j.cor.2008.12.016

Ilıcak, I.: Bi-objective bin packing problems. Master's thesis (2003). https://etd.lib.metu.edu.tr/upload/2/1079987/index.pdf

Lodi, A., Martello, S., Monaci, M.: Two-dimensional packing problems: a survey. Eur. J. Oper. Res. **141**(2), 241–252 (2002). https://doi.org/10.1016/S0377-2217(02)00123-6

Wilson, R.C.: A packaging problem. Manag. Sci. **12**(4), 135–145 (1965). https://doi.org/10.2307/2627807

Wu, H., Leung, S.C.H., Si, Y.-w., Zhang, D., Lin, A.: Three-stage heuristic algorithm for three-dimensional irregular packing problem. Appl. Math. Model. **41**, 431–444 (2017). https://doi.org/10.1016/j.apm.2016.09.018

Sustainability Evaluation of Textile Warehouses from Social and Environmental Perspective

Burcu Dilaver, Hasan Tahsin Çokoğlu, Meriç Yamanlar, Yeşim Deniz Özkan-Özen, and Yiğit Kazançoğlu(✉)

Department of International Logistics Management, Yaşar University, Izmir, Turkey
dilaverburcu@outlook.com, hasan.cokoglu@gmail.com, yamanlarmeric@gmail.com, {yesim.ozen, yigit.kazancoglu}@yasar.edu.tr

Abstract. Growing complexity in supply chains has increased the importance of warehousing activities more and more with each passing year. This study focuses on textile warehouses, where fluctuations in demand is high, and sustainability concerns are growing due to high production and consumption rates. With this view, aim is to present social and environmental sustainability criteria for textile warehouses and present the relationship between them in order to guide sustainability transformation of textile warehouses. Social and environmental aspects of sustainability are considered as a core subject based on company needs that are determined in the light of meetings performed with company experts, and the literature review made on environmental and social sustainability perspective. After that, interviews were held with the company to specialize the criteria for textile warehouses. Finally, fuzzy DEMATEL method was used to present the importance order and causal relationship between criteria for implications. At the end of the implementations, metrics were determined for each criterion, and recommendations were made to create a roadmap for the company.

Keywords: Environmental sustainability · Social sustainability · Warehouse · Textile · Fuzzy DEMATEL

1 Introduction

In contemporary supply chains, holding inventory and warehousing activities have a crucial position [1]. As a result of increasing complexity in logistic network, warehouse performance analysis has also become important. Warehousing have a critical importance on the success or failure of the supply chains [2]. Textile warehouses have also a great impact to satisfy the demand properly.

Also, with the awareness of environmental problems in the supply chain, companies turned to performance evaluations and GSCM (Green Supply Chain Management) to survive in an increasingly competitive environment [3]. Because of that sustainable

N. M. Durakbasa and M. G. Gençyılmaz (Eds.): ISPR 2019, LNME, pp. 791–810, 2020.
https://doi.org/10.1007/978-3-030-31343-2_66

warehouses and production areas growing day by day to provide sustainability with its three dimensions: economic, environmental and social.

In this context, this study especially focuses on social and environmental sustainability of textile warehouses and presenting some wide world criteria used for measuring performance of the warehouse. From this point of view, this study aims to give a road map to textile warehouses to increase the sustainability of the operations and to build a sustainable brand image. At the same time to make a sustainable performance evaluation, metrics are given to measure sustainable warehouse performance of the company. To do so, one of multi criteria decision making method (MCDM) fuzzy DEMATEL is used in this study to give the importance order of the social and environmental criteria separately, and to present causal relationship between them.

In the following sections firstly, literature review related to warehouse sustainability criteria, and environmental sustainability of the warehouse specially for textile industry are examined. Additionally, proposed methodology is given with steps, and after that implementation of the study and implications are specified respectively.

2 Literature Review

Literature review of this study includes sustainable warehouse performance management criteria. The subject of sustainability is examined under three main headings in the literature. These headings are; social, environmental and economic criteria. However, economic criteria were not evaluated in this literature review. Though financial duties are crucial to business, society regularly has various desires for organizations [4]. Through expanded requests on solid financial performance of the supply chains, associations are presently considered in charge of the environmental and social performance of their suppliers and accomplices [5]. Additionally, social and environmental performance of companies have a significant effect on sustainable brand image. Therefore, social and environmental criteria considered in this study. Literature review were concentrated upon sustainability criteria special to warehouses. All these studies are summarized with the definitions of each in Table 1.

Literature review of this study focused on social sustainability criteria and environmental sustainability criteria. Sustainability is one of the key elements of sustainable success of the companies. To start with social sustainability, there are some major performance criteria that could be given as an example such as Strong Leadership, Teamwork and Employee Satisfaction which are examined in many researches. For instance, Strong Leadership defined as the leadership skills to impact workers to follow one's lead to decisions ([6–9]). Another important social sustainability criterion is Teamwork. It is one of the crucial parts of social sustainability according to studies ([6, 10–12]).

Moreover, due to increase in the environmental consciousness, environmental sustainability became crucial for supply chain. Eco-friendly and cost-effective practices are the key stones for the industry. Legal regulations are also a driving force through providing environmental sustainability. Because of these, importance of environmental sustainability performance of the supply chain is increased. Sustainability criteria such as Recycling [13–16], Reuse [14–17] and Disposal [18–21] are one of the most

Table 1. Literature review

Main criteria	No	Sub criteria	Definition	Author's
End of life activities	1	Disposal	It is the name given to the process of getting rid of the materials that have reached the end of their useful life or the residual material	[18–21]
	2	Reuse	To use the object for the same purpose or for any other purpose, without changing the structure of the object	[14–17]
	3	Re manufacturing	The products used are products that are as reliable as new products according to the original equipment manufacturers (OEM) specification and customer perspective	[13, 16, 21]
	4	Recovery	It is the process to regain a product or material that cannot be used	[14, 15, 17]
	5	Reconditioning	Generally, it requires the main components to be constructed according to the operating conditions expected to be lower than the original model	[14, 15, 17]
	6	Refurbishment	Service or renewal of old or damaged equipment to work or make it look better	[13, 14, 22, 23]
	7	Recycling	It is the name given to all activities where waste materials are collected, accumulated, processed and put into use at the stage of production of a new product	[13–16]
Waste	8	Water waste	Any kind of water used in any part of the production or affected by people's use is waste water	[20, 24, 25]
	9	Solid waste	Commercial and industrial waste (Plastic waste, Paper/card waste, Tins and metals, Ceramics and glass)	[20, 21, 26, 27]
	10	Liquid waste	These wastes, both as domestic waste and industrial waste, contain contaminated water, organic liquids, washing water, waste detergents and rain water	[20, 25, 28, 29]
	11	Gaseous waste	Evolution of unwanted gases as a result of some reactions in production, most reactions can occur as HCl, CO2, H2, NH3, HCl	[20, 25, 28, 30]
	12	Hazardous	It contains all types of garbage which are flammable, toxic, corrosive and reactive	[20, 21, 31]

(*continued*)

Table 1. (*continued*)

Main criteria	No	Sub criteria	Definition	Author's
Emission	13	Air	Air pollution is a mix of particles and gases that can reach harmful concentrations both outside and indoors	[32–35]
	14	Green house gas emission	Methodology for calculation and declaration of energy consumption and GHG emissions of transport services	[36–39]
	15	Noise emission	One form of pollution on which the efforts of the individual can have a direct effect	[40–42]
	16	CO2 emission	CO2 gas into the atmosphere leads to global warming and deteriorating effects	[32–34, 43]
	17	CH4	Methane is the second largest greenhouse gas affecting global warming and accounts for about 16% of anthropogenic greenhouse gas emissions	[39]
	18	Particulate	Particulate matter is made of and particles and arises from vehicle exhaust, industrial emissions, road dust, and windblown soil	[44, 45]
Energy & resource consumption	19	Water	A percent of total consumption of recycled water over the current water consumption	[46–48]
	20	Electricity	Consumption of electricity presented in kilowatt used per hour	[49–51]
	21	Oil	Number of oil barrels consumed	[52, 53]
	22	Coal	Tons of coal consumed	[54, 55]
	23	Renewable energy	A percent of using renewable energy with respect to total energy used	[46, 50, 56]
	24	Material	The most appropriate usage of material, energy, and other tangible natural assets	[49, 51, 57]
	25	Fuel	Amount of fossil fuel consumption per tonne consumed	[51, 52, 56, 58]

(*continued*)

Table 1. (*continued*)

Main criteria	No	Sub criteria	Definition	Author's
Pollution	26	Air	Release of different air pollutants in the atmosphere and environment causes many of other environmental issues like greenhouse effect, ozone layer destruction, smog formation, etc.	[46, 51, 59]
	27	Water	One or more substances have built up in water to such an extent that they cause problems for animals or people	[46, 59]
	28	Dust	Pollution arise with dust particles in atmosphere	[55]
Corporate culture	29	Workforce diversity	The effectiveness of a companies on the workforce in relation to gender, ethnicity, and disability	[46, 60–62]
	30	Workforce engagement	A possession of the connection among a corporation and its workers	[46, 63]
	31	Social responsibility	The plan that company be supposed to stability profit-making movements with activities that advantage civilization	[6, 64–66]
	32	Employee turnover	Proportion of workers who leave the company and replace those who leave	[6, 67–69]
	33	Discipline management	Definite policy of behaviour to be followed by the administrator & moreover a variety of techniques for managing resource professionally	[70–72]
Health and safety	34	Number of accidents per year	Amount of accident within one year in the workplace	[6, 46, 73]
	35	Physical hazard	Condition of injury to the worker, even if there is contact or not	[6, 73]
	36	Safeguarding against work intensification to promote job control	Applying to prevent the raised focus and commitment requested of workers at work, generally experienced as time pressure, expanded pace and work overload	[74–76]

(*continued*)

Table 1. (*continued*)

Main criteria	No	Sub criteria	Definition	Author's
Training	37	Health and safety orientation training	It is an organization's business to give health and safety training to its personnel's, so they can be comfortable with hazards in their workplace and how to prevent them. The level of employees prepared in health and safety must be estimated and revealed	[46, 77, 79]
	38	Employee training	To develop the abilities of an individual employee, effectiveness of staff training programs	[79–81]
	39	Cultural awareness training for employees	Entity of cultural awareness training for employees, education and awareness enterprises	[80, 81]
Well - being workers	40	Employee satisfaction	Satisfaction with the decision-making process, percent of employees who report complete job satisfaction, effectiveness of keeping personnel feedback. When employees would be more developed, they would be more satisfied with the job	[80, 81, 83]
	41	Salary rates	Internal and external salary ratios, average staff turnover by classification	[80, 81]
	42	Employee development plans	Percent of staff development business plans. This would help in increasing the skills thus employee would be able to develop career with their own realistic career plan	80, 81, 83]
	43	Career management	Practices and strategies providing workers to sustain positive inclusion in career, family and individual roles over the life standards	[84–86]
	44	Work–life balance	Satisfaction and impression of achievement in meeting work and non-work status life	[87, 88]

(*continued*)

Table 1. (*continued*)

Main criteria	No	Sub criteria	Definition	Author's
Working conditions	45	Employment stability	A business ventures effect on work chances within the company, the stability there of as well as estimating the fairness of indemnity	[67, 68]
	46	Teamwork	The combined act of a group, particularly when active and productive	[6, 10–12]
	47	Labour productivity	It is the ratio of the outputs that each worker creates at a specific time (usually one hour) in a workplace	[7, 48, 89]
	48	Strong leadership	Leadership is the skills to impact workers to follow one's lead to decisions	[6],[7–9]
	49	Ergonomics	The examine of people's yield in their working medium	[90–93]
	50	Participation	Different things to different people depending upon their objectives and expectations	[6, 46, 63]

investigated environmental sustainability criteria in the literature. To give an example recycling is an important criterion for reducing waste by collecting, accumulating, processing waste materials and putting them into use at the stage of production of a new product.

In Table 1, summary of the literature review is presented by including social and environmental warehouse sustainability criteria with their brief definitions.

In the following section, sustainability criteria for the BR retailing are determined according to the literature review presented above.

3 Sustainability Criteria for Br Retailing

In the literature review, criteria for social and environmental sustainability for warehouses were investigated. According to the results of research, 28 criteria have identified for environmental sustainability also for social sustainability 22 criteria found and already explained in Table 2. The criteria found in the literature review were reviewed with the experts in the company and then specified for the textile warehouses. At the end of the interviews, 12 social sustainability criteria, and 9 environmental sustainability criteria were selected for textile warehouses.

3.1 Social Criteria

Workforce Stability & Turnover (S1): To keep the chances of success, stability and compensation justice within the company and to keep the rate of entry-exit to the minimum.

Teamwork (S2): It is a community of individuals who have a common approach to what they are responsible for and who are co-ordinated and complementary to each other.

Employee Efficiency (S3): It is the ratio of the outputs that each worker creates at a specific time (usually one hour) in a workplace.

Strong Leadership (S4): Leadership is the skills to impact workers to follow one's lead to decisions and behaviours in the field.

Ergonomics (S5): It is to adjust the working conditions to ensure the safety and comfort of the workers inside the warehouse.

Workforce Participation (S6): To meet the expectations and objectives of the employees with such things as teamwork and premiums. At the same time to ensure the link between the union between people and the company-employee.

OHSAS Training (S7): It is an organization's business to give health and safety training to its personnel. Employees are not allowed to work without compulsory occupational health and safety training.

Employee Training and Development (S8): Seminars and trainings to improve the skills and cultural awareness of all staff. As a result, a realistic and individual career plan can be created.

Employee Satisfaction (S9): Ensuring success and satisfaction monitoring through meetings to ensure workers' satisfaction at the same level in both their work and social lives.

Social Responsibility (S10): The plans that company be supposed to stability profit-making movements with activities that advantage civilization.

Continuous Monitoring & Discipline Management (S11): The company policy of employee's behaviour to be controlled by the manager. Moreover, a variety of techniques for monitoring resource professionally.

Accidents and Physical Dangers (S12): All personal injuries and accident risks in the workplace. Operators of vehicles and equipment can be cause injury to pedestrians so there are certain walkways in the warehouse.

3.2 Environmental Criteria

Disposal (E1): It is the name given to the process of getting rid of the materials that have reached the end of their useful life or the residual material.

Reuse (E2): To use the object again for the same purpose or for any other purpose, without changing the structure of the object.

Recovery (E3): It is the process to regain or repair a product or material that cannot be used.

Recycling (E4): It is the name given to all activities where waste materials are collected, accumulated, processed and put into use at the stage of production of a new product.

Solid Wastes (E5): Commercial and industrial wastes (Plastic and paper/ card waste used to protect and pack textile products)

Textile Dust (E6): Dust particles from textile products into the atmosphere.

Material (E7): The most appropriate usage of material, energy, and other tangible natural assets.

Electricity (E8): Electricity consumption in kilowatts used in all activities of textile warehouse in one hour.

Fuel (E9): Amount of fossil fuel consumption per tonne consumed.
The methodology that is used in this project is described in the next section.

4 Proposed Methodology: Fuzzy DEMATEL

4.1 DEMATEL

In order to explain the structure of complex causal relations with the help of digraphs and matrices, the Research Institute of Geneva developed the method of 'Decision making and evaluation laboratory (DEMATEL)'. According to [94], if DEMATEL is considered as a structural modelling method, the greatest benefit is to analyse the cause and effect associations between the components of a system.

The Fuzzy set theory was found in 1965 by Zadeh to understand and manage uncertainty in decision making. By following below steps, fuzzy DEMATEL can be implemented.

Step 1. In the first step, a fuzzy linguistics scale is used to evaluate the causal relationships between factors. In order to measure the uncertainty and error in human evaluations, linguistic terms such as "No, Very Low, Low, High, Very High" can be used.

Table 2. The fuzzy linguistic scale

Linguistic terms	Triangular fuzzy numbers
Very high influence (VH)	(0.75, 1.0, 1.0)
High influence (H)	(0.5, 0.75, 1.0)
Low influence (L)	(0.25, 0.5, 0.75)
Very low influence (VL)	(0, 0.25, 0.5)
No influence (No)	(0, 0, 0.25)

Step2. The first direct relationship matrix is obtained by binary comparison. The first fuzzy direct relationship matrix Z is made by influencing the relationships between the components in an n x n matrix. Therefore, the direct-relation matrix is established as i.e. $Z = [Z_{ij}]_{n \times n}$. where Z is a n × n non-negative matrix; Z_{ij}. represents the direct impact of factor i on factor j.
Step3. The normalized direct relationship matrix X, i.e. $X = [X_{ij}]$ and $0 \leq X_{ij} \leq 1$ can be obtained from the Eqs. (1) and (2) in which all the basic diagonal is present with 0.

$$X = s.Z, \tag{1}$$

$$s = \frac{1}{max \leq i \leq n \sum_{j=1}^{n} Z_{ij}} \quad i, j = 1, 2. \ldots, n \tag{2}$$

Step 4. By using the Eq. (3) where I is shown as the identity matrix, total relationship matrix T is obtained.

$$T = x(I - x)^{-1} \tag{3}$$

Step 5. Row and column totals are shown as D and R in the T matrix with the Eqs. (4) and (5).

$$T = t_{ij,} i, j = 1, 2. \ldots, n$$

$$D = \sum_{j=1}^{n} t_{ij} \tag{4}$$

$$R = \sum_{i=1}^{n} t_{ij} \tag{5}$$

Where, D and R denote the sum of rows and the sum of columns, respectively.

In this study, one of the defuzzification methods, Converting Fuzzy data into Crisp Scores (CFCS) is used. The CFCS method is grounded on the practice for defining the left and right points by the fuzzy min and the fuzzy max, is determined by the weighted average by the total score membership functions.

In the following section, fuzzy DEMATEL implementation is presented.

5 Implementation of the Study

This study conducted in BR retailing, a subsidiary of Ran Apparel and Boyner Stores. Fuzzy DEMATEL implementation is conducted separately for social and environmental criteria. Causal relation and importance order of social criteria are implemented between 12 criteria. Causal relation and significance order of environmental criteria are implemented between 9 criteria. 4 experts are participated in the survey for social criteria. 2 experts are participated in the survey for environmental criteria. Result of social and environmental implementation is presented in Table 5.

5.1 Implementation of Social Criteria

As mentioned in the previous sections, 12 social sustainability criteria were defined for textile warehouse, which are; Workforce Stability & Turnover (S1), Teamwork (S2), Employee Efficiency (S3), Strong Leadership (S4), Ergonomics (S5), Workforce Participation (S6), OHSAS Training (S7), Employee Training and Development (S8), Employee Satisfaction (S9), Social Responsibility (S10), Continuous Monitoring & Discipline Management (S11), Accidents and Physical Dangers (S12).

In order to solve the equations of fuzzy DEMATEL method, all the formulas and linguistic assessments of participants entered into Microsoft Excel. After the deviation using the CFCS method and the collection of these evaluations, the initial direct relationship matrix and the normalized direct relationship matrix are generated. In the next step of the method, the total relationship matrix is generated and is shown in Table 3.

Table 3. Total relation matrix

T	S1	S2	S3	S4	S5	S6	S7	S8	S9	S10	S11	S12	D
S1	0,4125	0,4962	0,5066	0,4744	0,3496	0,4713	0,4296	0,4434	0,5794	0,2258	0,4751	0,4979	**5,3618**
S2	0,5285	0,4654	0,5698	0,5434	0,3449	0,5322	0,4148	0,4983	0,6329	0,2785	0,4867	0,5076	**5,8032**
S3	0,4859	0,5334	0,4555	0,5433	0,3642	0,5026	0,3745	0,4797	0,6106	0,2362	0,4748	0,4891	**5,5499**
S4	0,5660	0,6036	0,6219	0,5074	0,4066	0,5813	0,4517	0,5590	0,6803	0,2798	0,5665	0,5515	**6,3758**
S5	0,4082	0,4084	0,4493	0,3990	0,2527	0,4122	0,3664	0,3721	0,5065	0,1965	0,3770	0,4247	**4,5731**
S6	0,5084	0,5393	0,5625	0,5481	0,3521	0,4370	0,4081	0,5045	0,6241	0,2758	0,4665	0,4808	**5,7072**
S7	0,4241	0,4433	0,4714	0,4351	0,3398	0,4139	0,3154	0,4186	0,5346	0,2087	0,4573	0,4788	**4,9409**
S8	0,5156	0,5068	0,5162	0,5159	0,3244	0,4854	0,4329	0,4118	0,6155	0,2617	0,4959	0,4724	**5,5545**
S9	0,6066	0,6183	0,6003	0,6080	0,4190	0,5899	0,4548	0,5723	0,6023	0,2976	0,5559	0,5661	**6,4913**
S10	0,3337	0,3856	0,4051	0,3579	0,2218	0,3448	0,2764	0,3686	0,4288	0,1537	0,3219	0,3329	**3,9313**
S11	0,4807	0,5012	0,5032	0,5241	0,3107	0,4634	0,3990	0,4934	0,5898	0,2222	0,4049	0,4979	**5,3908**
S12	0,5462	0,5500	0,5560	0,5408	0,4083	0,5424	0,4731	0,4688	0,6471	0,2352	0,5374	0,4560	**5,9611**
R	**5,8165**	**6,0517**	**6,2179**	**5,9973**	**4,0943**	**5,7766**	**4,7967**	**5,5905**	**7,0520**	**2,8715**	**5,6200**	**5,7557**	

Considering the data set in Table 3, Social Implementation Results of Casual Diagram is shown in Fig. 1.

Fig. 1. Casual diagram for social criteria

Finally, in order to establish and analyse the structural model, D+R and D−R values are calculated The causal diagram is constructed by mapping the data set D+R and D−R and presented in Figure 1. According to the results, cause group includes: S10, S5, S4, S12, and S7, while the effect group was composed of S8, S6, S11, S2, S1, S9, and S3. Moreover, increasing order of importance level of the criteria is found as S9, S4, S2, S3, S12, S6, S1, S8 S11, S7, S5, and S10. Results of the fuzzy DEMATEL application are summarized in Table 4 and discussions related to these results are made in the following section.

5.2 Implementation of Environmental Criteria

Environmental sustainability criteria for textile warehouses were presented as; Disposal (E1), Reuse (E2), Recovery (E3), Recycling (E4), Solid Wastes (E5), Textile Dust (E6), Material (E7), Electricity (E8), Fuel (E9).

After the deviation of the CFCS method and the collection of these evaluations, the first direct relationship matrix and the normalized direct relationship matrix were generated respectively. Then, the total relationship. Then, the total relationship matrix is generated and is shown in Table 4.

Table 4. Total relation matrix for environmental

T	E1	E2	E3	E4	E5	E6	E7	E8	E9	D
E1	0,1593	0,3836	0,4119	0,3491	0,2972	0,2283	0,2554	0,1971	0,1881	2,4700
E2	0,2663	0,3906	0,4099	0,5759	0,5081	0,2822	0,3531	0,3468	0,3030	3,4359
E3	0,2282	0,3387	0,2363	0,3042	0,2811	0,2631	0,2765	0,2719	0,2886	2,4886
E4	0,2935	0,5562	0,4978	0,3711	0,4377	0,3097	0,3574	0,3498	0,3096	3,4828
E5	0,2649	0,5264	0,4668	0,5379	0,3217	0,3061	0,3241	0,3179	0,3011	3,3669
E6	0,1194	0,1167	0,1097	0,1100	0,1258	0,0658	0,1423	0,1395	0,1362	1,0653
E7	0,2517	0,5865	0,5205	0,5309	0,5088	0,2444	0,2890	0,3668	0,3501	3,6487
E8	0,0471	0,0903	0,0819	0,0840	0,0792	0,0475	0,1470	0,0657	0,1193	0,7620
E9	0,0330	0,0573	0,0526	0,0542	0,0506	0,0338	0,0746	0,1013	0,0434	0,5006
R	1,6634	3,0463	2,7874	2,9173	2,6101	1,7809	2,2193	2,1568	2,0394	

The data which is given in Table 4 shows that the Environmental Implementation Results of the Causal Diagram in Fig. 2.

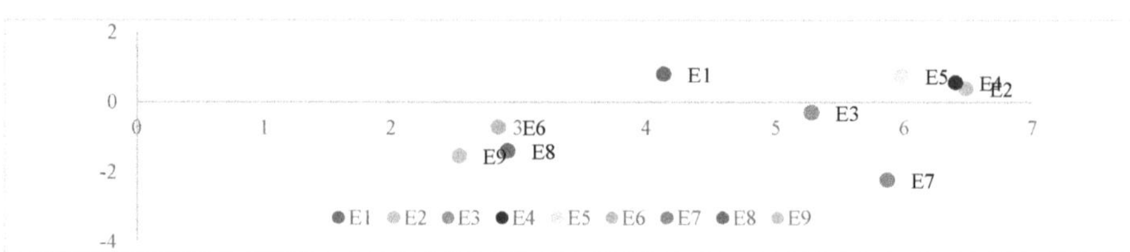

Fig. 2. Casual diagram for environmental

Finally, D+R and D−R values are calculated and the causal diagram is created by mapping the D + R and D data set. Figure 2 According to the results, the cause group comprises: E1, E5, E4 and E2, the effect groups E3, E6, E8, E9 and E7. In addition, the level of importance of the criteria is E2, E4, E5, E7, E3, E1, E8, E6 and E9. The results of the application of Fuzzy DEMATEL are summarized in Table VII and the discussions about these results are given in the next section.

Table 5 is presented summary of social and environmental implementation results.

Table 5. Summary of the implementation results

Social results			Environmental results		
Importance order	Cause	Effect	Importance order	Cause	Effect
Employee Satisfaction (S9)		x	Reuse (E2)	x	
Strong Leadership (S4)	x		Recycling (E4)	x	
Teamwork (S2)		x	Solid Wastes (E5)	x	
Employee Efficiency (S3)		x	Material (E7)		x
Accidents and Physical Dangers (S12)	x		Recovery (E3)		x
Workforce Participation (S6)		x	Disposal (E1)	x	
Workforce Stability & Turnover (S1)		x	Electricity (E8)		x
Employee Training and Development (S8)		x	Textile Dust (E6)		x
Continuous Monitoring & Discipline Management (S11)		x	Fuel (E9)		x
OHSAS Training (S7)	x				
Ergonomics (S5)	x				
Social Responsibility (S10)	x				

In the next section, implications and discussions are described after Fuzzy DEMATEL method.

6 Implications and Discussion

By looking at the causal diagram for the social sustainability criteria, the company can find clues to make the right decisions and create new sustainability policies. According to the results, if the company aims to increase the sustainability performance at the warehouse level, the cause group should first examine and focus on those criteria.

Then, together with these studies, the effect group also will reach high performance levels. According to the examinations, S10, S5, S4, S12, and S7 have an influence on other criteria. These criteria affect the factors belonging to the effect group such as S8, S6, S11, S2, S1, S9, and S3. In general, the order of importance to be considered in the road map to be decided S9, S4, S2, S3, S12, S6, S1, S8, S11, S7, S5 and S10. The most important criterion for achieving the highest levels of performance in social sustainability is the social responsibility (S10) criterion. If a social sustainability is desired within the warehouse, social responsibility should be focused on and new ways should be followed.

Workers are sensitive to the local community's cultural, social and economic needs and social responsibility acts can make them feel more like a family with company. Especially corporate culture defines the effects between criteria. In general, the warehouse employs people living near the district where it is located, and especially the blue collars are affected by this situation. Because of that social responsibility (S10) can increase the employee satisfaction (S9) and workforce participation (S6) also can create better workforce stability and turnover (S1).

Since ergonomics are important for the safety, comfort and easy progress of the workers in the warehouse, the effect on other factors cannot be denied. It is particularly crucial for two criteria in effect group which are employee satisfaction (S9) and workforce stability and turnover (S1). Because if workers can move comfortably in the warehouse and do not have the risk of accidents, they will be satisfied with the work and their loyalty to the place they work will increase. But also, a layout created without considering the safety can cause major accidents inside the warehouse, so it can affect accidents and physical dangers (S12) which is in cause group.

Strong leadership (S4) has an impact on almost all of the criteria in effect group. Because the leaders are largely responsible for both employee's training and their motivation. At the same time, it affects workers' oversight, guidance and strong teamwork.

For the environmental sustainability criteria, according to the results of the method, it is known how much the criterion affects the sustainability. Particularly in terms of environmental sustainability, the relationship between the criteria and each other is more important. An improvement work done in any one affects others easily. Factors affecting the others, cause group, are E1, E5, E4 and E2. These criteria have important effects on other criteria. From the results, we can understand how important disposal (E1) is for sustainability. With the disposal of solid waste only, the environmental benefit is too great to be underestimated. Factors influenced by the cause group were seen as E1, E2, E9, E8 and E7. In order for the effect group to have a sustainable appearance, the cause group needs to be developed first. According to that, importance order is shown like E2, E4, E5, E7, E3, E1, E8, E6 and E9.

Disposal (E1), is located at the top of the cause group which means that it has the highest impact on other criteria according to Fuzzy DEMATEL implementation. To give an example to this, a decrease in its wastes and the disposal methods that they prefer, effects the criteria such as Reuse (E1), Recycling (E4). For instance, an increase in the disposal means that disposal has a significant impact on recycling likewise reuse. Solid Wastes (E5) has the second highest impact on other criteria. Solid wastes are very important for warehouse and waste treatment process is critical. It effects the other criteria and it has a close relationship with them. For instance, waste generated by the company effects criteria such as, Material (E7), Recovery (E3) and Reuse (E2). To specify one of these criteria, reusing items rather than throw away can have an impact on building an eco-friendly waste management. Due to this, effect of solid wastes on reuse can easily be seen. Recycling (E4) is the third highest in cause group. Its effects criteria such as Disposal (E1), Reuse (E2), Recovery (E3), Material (E7). As recycling can be counted as a disposal method, it can affect Reuse (E2) and Recovery (E3) as well. Recycling also effects material usage.

Material (E7), is located at the bottom of effect group which means that it is affected from other criteria the most. To illustrate its relationship between criteria, some examples could be given. For instance, Material (E7) is affected by Reuse (E2), Recycling (E4) and Solid wastes (E5). Such activities have a close relationship with Material (E7). Using renewable and reusable materials can have a significant impact to warehouse sustainability. Fuel (E9), is the second most effected criterion from other criteria. Fuel consumption is directly affected from almost every operation in the warehouse. Due to this, other impact of criteria on Fuel (E9) is significantly high. Electricity (E8), is the last criterion of the top three most effected criteria from the effect group. Amount of electricity consumed can change under the influence of many criteria.

Finally, some other recommendations can be given based on these results to improve social and environmental sustainability in warehouses. To start with lean principles can be applied to create minimal and useful layout in every sense and to reduce wastes. Also, workforce can be planned effectively, and inventory levels can be reduced at the same time, main processes (i.e. picking, sorting, repairing and packaging) can be simplified and standardized and, 5S can be used for process improvements. Moreover, circular economy can be adopted for to keep waste levels at a minimum and create sustainability. For warehouses, circularity brings the advantages of, reduced environmental wastes, misuse of labour force and ergonomics error, create efficient resource utilization, increased life cycle and economic benefits can be gained by reused, recycling, and remanufacturing activities. Furthermore, warehouses inventory management is critical issue. Systematic inventory management practices can be suggested to bring down the all-out stock out, backorder and salvage costs and don't consolidate sustainability criteria in the basic decision-making structure, various warehouse inventory systems, wireless barcode scanners and features such as RFID provides convenience in every way. This software is not limited to inventory management. Activities such as routing, packing, picking and sorting are more easily and technologically solved. They save time as well as saving labour and money. Also Enterprise Resource Planning (ERP) can be used for inventory management into a publicly accessible database. Thus, employees from different units can access information. The system can increase the functionality of picking, packing, receiving and shipping operations of the warehouse. Finally, reusable transportation packaging can be suggested for design of a reverse logistics system for textile warehouses. A closed-loop deposit-based system can be implemented by using collapsible plastic containers that can be rented as secondary packaging material.

7 Conclusion

Starting from the first phase of this study the core aim is to build a sustainable warehouse performance measurement system. In order to achieve this objective, several meetings are conducted with company experts and a literature review is made under related titles. Due to the field of study is a textile warehouse, after interviews and meetings performed with the company, main needs are decided as environmental and social sustainability. Literature review is specifically focused on environmental and social sustainability criteria. After criteria are examined and approved by company experts, to provide a road map, and to determine causal relationships and importance order between each criterion, Fuzzy DEMATEL method is implemented. Because of the difference between environmental and social sustainability criteria, Fuzzy DEMATEL implication is conducted separately under the light of the meetings performed with different experts from different departments. According to results that obtained from Fuzzy DEMATEL implication, methods that will provide sustainability are presented to company as recommendations. Within the implication study, performance metrics are determined for each criterion.

For future studies, as it is known in this study, only two of the three basic principles of sustainability were focused. In the following studies, it will be a great improvement to ensure the integrity of the subject by examining the economical sustainability within the warehouse. Also, case studies can be continued in order to work together with a few textile stores and give real-life examples. Working together with different warehouses is also a great benefit for multiplying criteria. For now, metrics are only for the company to be studied can be reproduced by examining various surveys and different warehouses.

References

1. Baker, P., Halim, Z.: An exploration of warehouse automation implementations: cost, service and flexibility issues. Supply Chain. Manag. Int. J. **12**(2), 129–138 (2007)
2. Frazelle, E.: Supply Chain Strategy: The Logistics of Supply Chain Management. McGrraw Hill, New York (2002)
3. Kazancoglu, Y., Ozturkoglu, Y.: Integrated framework of disassembly line balancing with Green and business objectives using a mixed MCDM. J. Clean. Prod. **191**, 179–191 (2018)
4. Hutchins, M.J., Sutherland, J.W.: An exploration of measures of social sustainability and their application to supply chain decisions. J. Clean. Prod. **16**(15), 1688–1698 (2008)
5. Seuring, S., Müller, M.: From a literature review to a conceptual framework for sustainable supply chain management. J. Clean. Prod. **16**(15), 1699–1710 (2008)
6. Dowton, S.B.: Imperatives in medical education and training in response to demands for a sustainable workforce. Med. J. Aust. **183**(11–12), 595–598 (2005)
7. Székely, F., Knirsch, M.: Responsible leadership and corporate social responsibility: metrics for sustainable performance. Eur. Manag. J. **23**(6), 628–647 (2005)
8. Smith, P.A., Sharicz, C.: The shift needed for sustainability. Learn. Organ. **18**(1), 73–86 (2011)
9. Baumgartner, R.J.: Organizational culture and leadership: preconditions for the development of a sustainable corporation. Sustain. Dev. **17**(2), 102–113 (2009)

10. Martínez-Jurado, P.J., Moyano-Fuentes, J.: Lean management, supply chain management and sustainability: a literature review. J. Clean. Prod. **85**, 134–150 (2014)
11. Muduli, K., Govindan, K., Barve, A., Kannan, D., Geng, Y.: Role of behavioural factors in green supply chain management implementation in Indian mining industries. Resour. Conserv. Recycl. **76**, 50–60 (2013)
12. Lee, S.M., Tae Kim, S., Choi, D.: Green supply chain management and organizational performance. Ind. Manag. Data Syst. **112**(8), 1148–1180 (2012)
13. King, A.M., Burgess, S.C.: The development of a remanufacturing platform design: a strategic response to the directive on waste electrical and electronic equipment. Proc. Inst. Mech. Eng. Part B J. Eng. Manuf. **219**(8), 623–631 (2005)
14. Gehin, A., Zwolinski, P., Brissaud, D.: A tool to implement sustainable end-of-life strategies in the product development phase. J. Clean. Prod. **16**(5), 566–576 (2008)
15. Rahman, S., Subramanian, N.: Factors for implementing end-of-life computer recycling operations in reverse supply chains. Int. J. Prod. Econ. **140**(1), 239–248 (2012)
16. Merli, R., Preziosi, M., Acampora, A.: How do scholars approach the circular economy? A systematic literature reviews. J. Clean. Prod. **178**, 703–722 (2018)
17. Jayaraman, V.: Production planning for closed-loop supply chains with product recovery and reuse: an analytical approach. Int. J. Prod. Econ. **44**(5), 981–998 (2006)
18. Suzuki, T., Ohashi, T., Hiroshige, Y., Ochiai, Y., Kitamura, T., Tachi, T., Hayashi, M.: U.S. Patent No. 6,226,617. Washington, DC: U.S. Patent and Trademark Office (2001)
19. Hetherington, K.: Accountability and disposal: visual impairment and the museum. Museum Soc. **1**(2), 104–115 (2003)
20. Williams, P.T.: Waste Treatment and Disposal. Wiley, Chichester (2005)
21. Ijomah, W.L.: A tool to improve training and operational effectiveness in remanufacturing. Int. J. Comput. Integr. Manuf. **21**(6), 676–701 (2008)
22. Shah, R.M., Hashim, R.: "Scientists Paradise": environmental sustainability and policy governance. Procedia Soc. Behav. Sci. **35**, 384–388 (2012)
23. Kemmer, S., Koskela, L.: Developing a lean model for production management of refurbishment projects. In: Proceedings of the 20th Annual Conference of the International Group for Lean Construction, San Diego, CA, USA, July 2012
24. Teixeira, J., Antunes, A.P., de Sousa, J.P.: Recyclable waste collection planning—a case study. Eur. J. Oper. Res. **158**(3), 543–554 (2004)
25. Dıez, C., Martınez, O., Calvo, L.F., Cara, J., Morán, A.: Pyrolysis of tyres. Influence of the final temperature of the process on emissions and the calorific value of the products recovered. Waste Manag. **24**(5), 463–469 (2004)
26. Kayhanian, M., Tchobanoglous, G.: Innovative two-stage process for the recovery of energy and compost from the organic fraction of municipal solid waste (MSW). Water Sci. Technol. **27**(2), 133–143 (1993)
27. Mrayyan, B., Hamdi, M.R.: Management approaches to integrated solid waste in industrialized zones in Jordan: a case of Zarqa City. Waste Manag. **26**(2), 195–205 (2006)
28. Willis, J.W., Cadou, C., Mitchell, M., Karagozian, A.R., Smith, O.I.: Destruction of liquid and gaseous waste surrogates in an acoustically excited dump combustor. Combustion and flame **99**(2), 280–287 (1994)
29. Kentish, S.E., Stevens, G.W.: Innovations in separations technology for the recycling and re-use of liquid waste streams. Chem. Eng. J. **84**(2), 149–159 (2001)
30. Shonnard, D.R., Hiew, D.S.: Comparative environmental assessments of VOC recovery and recycle design alternatives for a gaseous waste stream. Environ. Sci. Technol. **34**(24), 5222–5228 (2000)
31. LaGrega, M.D., Buckingham, P.L., Evans, J.C.: Hazardous waste. In: Hazardous Waste Management, pp. 1–39 (2001)

32. Saadah, Y.S.: Decreasing the CO2 emissions and the embodied energy in the UAE through the selection of sustainable building materials in the construction phase. Doctoral dissertation, The British University in Dubai (BUiD) (2008)
33. Menzies, G.F., Turan, S., Banfill, P.F.: Life-cycle assessment and embodied energy: a review. Proc. Inst. Civ. Eng. Constr. Mater. **160**(4), 135–143 (2007)
34. Rai, D., Sodagar, B., Fieldson, R., Hu, X.: Assessment of CO2 emissions reduction in a distribution warehouse. Energy **36**(4), 2271–2277 (2011)
35. Aronsson, H., Huge Brodin, M.: The environmental impact of changing logistics structures. Int. J. Logist. Manag. **17**(3), 394–415 (2006)
36. Rüdiger, D., Schön, A., Dobers, K.: Managing greenhouse gas emissions from warehousing and transshipment with environmental performance indicators. Transp. Res. Procedia **14**, 886–895 (2016)
37. Pan, J., Sun, Y., Li, W., Knight, J., Manthiram, A.: A green lead hydrometallurgical process based on a hydrogen-lead oxide fuel cell. Nat. Commun. **4**, 2178 (2013)
38. Tettelin, H., Masignani, V., Cieslewicz, M.J., Eisen, J.A., Peterson, S., Wessels, M.R., Paulsen, I.T., Nelson, K.E., Margarit, I., Read, T.D., Madoff, L.C.: Complete genome sequence and comparative genomic analysis of an emerging human pathogen, serotype V Streptococcus agalactiae. Proc. Natl. Acad. Sci. **99**(19), 12391–12396 (2002)
39. Piecyk, M., Browne, M., Whiteing, A., McKinnon, A. (eds.): Green Logistics: Improving the Environmental Sustainability of Logistics. Kogan Page Publishers, London (2015)
40. Stansfeld, S.A., Matheson, M.P.: Noise pollution: non-auditory effects on health. Br. Med. Bull. **68**(1), 243–257 (2003)
41. Pathak, V., Tripathi, B.D., Mishra, V.K.: Dynamics of traffic noise in a tropical city Varanasi and its abatement through vegetation. Environ. Monit. Assess. **146**(1–3), 67–75 (2008)
42. Francis, C.D., Kleist, N.J., Ortega, C.P., Cruz, A.: Noise pollution alters ecological services: enhanced pollination and disrupted seed dispersal. Proc. R. Soc. B Biol. Sci. **279**(1739), 2727–2735 (2012)
43. Marchant, H.K., Calosi, P., Spicer, J.I.: Short-term exposure to hypercapnia does not compromise feeding, acid–base balance or respiration of Patella vulgata but surprisingly is accompanied by radula damage. J. Mar. Biol. Assoc. U. K. **90**(7), 1379–1384 (2010)
44. Vincent, R., Bjarnason, S.G., Adamson, I.Y., Hedgecock, C., Kumarathasan, P., Guénette, J., Potvin, M., Goegan, P., Bouthillier, L.: Acute pulmonary toxicity of urban particulate matter and ozone. Am. J. Pathol. **151**(6), 1563 (1997)
45. Li, X., Wang, S., Duan, L., Hao, J., Li, C., Chen, Y., Yang, L.: Particulate and trace gas emissions from open burning of wheat straw and corn stover in China. Environ. Sci. Technol. **41**(17), 6052–6058 (2007)
46. Garbie, I.: Sustainability in Manufacturing Enterprises: Concepts, Analyses and Assessments for Industry 4.0. Springer, Cham (2016)
47. Alkaya, E., Demirer, G.N.: Sustainable textile production: a case study from a woven fabric manufacturing mill in Turkey. J. Clean. Prod. **65**, 595–603 (2014)
48. Dornfeld, D.A.: Moving towards green and sustainable manufacturing. Int. J. Precis. Eng. Manuf Green Technol. **1**(1), 63–66 (2014)
49. Veleva, V., Hart, M., Greiner, T., Crumbley, C.: Indicators of sustainable production. J. Clean. Prod. **9**(5), 447–452 (2001)
50. McIntyre, K., Smith, H., Henham, A., Pretlove, J.: Environmental performance indicators for integrated supply chains: the case of Xerox Ltd. Supply Chain Manag. Int. J. **3**(3), 149–156 (1998)
51. Dornfeld, D.A. (ed.): Green Manufacturing: Fundamentals and Applications. Springer, New York (2012)

52. Garetti, M., Taisch, M.: Sustainable manufacturing: trends and research challenges. Prod. Plan. Control. **23**(2–3), 83–104 (2012)
53. Sarhan, A.A., Sayuti, M., Hamdi, M.: Reduction of power and lubricant oil consumption in milling process using a new SiO 2 nanolubrication system. Int. J. Adv. Manuf. Technol. **63** (5–8), 505–512 (2012)
54. Guang Shi, V., Lenny Koh, S.C., Baldwin, J., Cucchiella, F.: Natural resource based green supply chain management. Supply Chain. Manag. Int. J. **17**(1), 54–67 (2012)
55. Fujii, H., Managi, S., Kaneko, S.: Decomposition analysis of air pollution abatement in China: empirical study for ten industrial sectors from 1998 to 2009. J. Clean. Prod. **59**, 22–31 (2013)
56. Gold, S., Seuring, S.: Supply chain and logistics issues of bio-energy production. J. Clean. Prod. **19**(1), 32–42 (2011)
57. Tan, H.X., Yeo, Z., Ng, R., Tjandra, T.B., Song, B.: A sustainability indicator framework for Singapore small and medium-sized manufacturing enterprises. Procedia CIRP **29**, 132–137 (2015)
58. Worrell, E., Price, L., Martin, N., Hendriks, C., Meida, L.O.: Carbon dioxide emissions from the global cement industry. Annu. Rev. Energy Environ. **26**(1), 303–329 (2001)
59. Salic, A., Zelic, B.: Introduction to environmental engineering. Phys. Sci. Rev. **3**, 10 (2018)
60. Aghazadeh, S.M.: Managing workforce diversity as an essential resource for improving organizational performance. Int. J. Prod. Perform. Manag. **53**(6), 521–531 (2004)
61. Kossek, E.E., Lobel, S.A., Brown, J.: Human resource strategies to manage workforce diversity. In: Handbook of Workplace Diversity, pp. 53–74 (2006)
62. DiTomaso, N., Post, C., Parks-Yancy, R.: Workforce diversity and inequality: Power, status, and numbers. Annu. Rev. Sociol. **33**, 473–501 (2007)
63. Bowen, W.G., Finegan, T.A.: The economics of Labor Force Participation, vol. 2054. Princeton University Press, Princeton (2015)
64. Carter, C.R.: Purchasing and social responsibility: a replication and extension. J. Supply Chain. Manag. **40**(3), 4–16 (2004)
65. Jones, P., Comfort, D., Hillier, D.: Corporate social responsibility and the UK construction industry. J. Corp. Real Estate **8**(3), 134–150 (2006)
66. O'Brien, D.: Integrating corporate social responsibility with competitive strategy. The Center for Corporate Citizenship at Boston College, pp. 3–23 (2001)
67. Labuschagne, C., Brent, A.C., Van Erck, R.P.: Assessing the sustainability performances of industries. J. Clean. Prod. **13**(4), 373–385 (2005)
68. Sarkis, J., Helms, M.M., Hervani, A.A.: Reverse logistics and social sustainability. Corp. Soc. Responsib. Environ. Manag. **17**(6), 337–354 (2010)
69. Min, H.: Examining sources of warehouse employee turnover. Int. J. Phys. Distrib. Logist. Manag. **37**(5), 375–388 (2007)
70. Sewell, G.: The discipline of teams: the control of team-based industrial work through electronic and peer surveillance. Adm. Sci. Q. **43**(2), 397–428 (1998)
71. Zald, M.N.: Spinning disciplines: critical management studies in the context of the transformation of management education. Organization **9**(3), 365–385 (2002)
72. Schäffer, U., Binder, C.: "Controlling" as an academic discipline: the development of management accounting and management control research in German-speaking countries between 1970 and 2003. Account. Hist. **13**(1), 33–74 (2008)
73. Spellman, F.R., Bieber, R.M.: Physical hazard control: preventing injuries in the workplace. Government Institutes (2011)
74. Berset, M., Semmer, N.K., Elfering, A., Amstad, F.T., Jacobshagen, N.: Work characteristics as predictors of physiological recovery on weekends. Scand. J. Work. Environ. Health **35**(3), 188–192 (2009)

75. Kossek, E., Lautsch, B.: CEO of Me: Creating a Life That Works in the Flexible Job Age. Pearson, Upper Saddle River (2007)
76. Turkle, S.: Alone Together: Why We Expect More from Technology and Less from Each Other. Basic Books, New York (2011)
77. Arscott, L.: Invited discussion-sustainable development in the oil and gas industry. J. Energy Resour. Technol. **126**, 1–5 (2004)
78. Mitropoulos, P., Abdelhamid, T.S., Howell, G.A.: Systems model of construction accident causation. J. Constr. Eng. Manag. **131**(7), 816–825 (2005)
79. Antonacopoulou, E.P.: Employee development through self-development in three retail banks. J. Pers. Rev. **29**(4), 491–508 (2000)
80. Sarkis, J.: The adoption of environmental and risk management practices: relationships to environmental performance. Ann. Oper. Res. **145**(1), 367–381 (2006)
81. Searcy, C., McCartney, D., Karapetrovic, S.: Sustainable development indicators for the transmission system of an electric utility. Corp. Soc. Responsib. Environ. Manag. **14**(3), 135–151 (2007)
82. Champathes, M.R.: Coaching for performance improvement: The coach model. Dev. Learn. Organ. **20**(2), 17–18 (2006)
83. Lee, C.H., Bruvold, N.T.: Creating value for employees: investment in employee development. Int. J. Hum. Resour. Manag. **14**(6), 981–1000 (2003)
84. Geurts, S.A.E., Beckers, D.G.J., Taris, T.W., Kompier, M.A.J., Smulders, P.G.W.: Worktime demands and work-family interference: does worktime control buffer the adverse effects of high demands? J. Bus. Ethics **84**, 229–241 (2009)
85. Grebner, S., Elfering, A., Semmer, N.K.: The success resource model of job stress. In: Perrewe, P.L., Ganster, D.C. (eds.) New Developments in Theoretical and Conceptual Approaches to Job Stress, vol. 8, pp. 61–108 (2010)
86. Newman, A., Thanacoody, R., Hui, W.: The impact of employee perceptions of training on organizational commitment and turnover intentions: a study of multinationals in the Chinese service sector. Int. J. Hum. Resour. Manag. **22**(8), 1765–1787 (2011)
87. Kossek, E., Baltes, B., Matthews, R.: How work-family research can finally have an impact on organizations. Ind. Organ. Psychol. Perspect. Sci. Pract. **4**, 352–369 (2011)
88. Kossek, E., Kaillaith, T., Kaillaith, P.: Achieving employee wellbeing in a changing work environment: an expert commentary on current scholarship. Int. J. Manpow. **33**(7), 738–753 (2012)
89. McMullan, A.: Supply chain management practices in Asia Pacific today. Int. J. Phys. Distrib. Logist. Manag. **26**(10), 79–95 (1996)
90. Genaidy, A.M., Sequeira, R., Rinder, M.M., A-Rehim, A.D.: Determinants of business sustainability: an ergonomics perspective. Ergonomics **52**(3), 273–301 (2009)
91. Ryan, B., Wilson, J.R.: Ergonomics in the development and implementation of organisational strategy for sustainability. Ergonomics **56**(3), 541–555 (2013)
92. Harik, R., El Hachem, W., Medini, K., Bernard, A.: Towards a holistic sustainability index for measuring sustainability of manufacturing companies. Int. J. Prod. Res. **53**(13), 4117–4139 (2015)
93. Cherrafi, A., Elfezazi, S., Govindan, K., Garza-Reyes, J.A., Benhida, K., Mokhlis, A.: A framework for the integration of Green and Lean Six Sigma for superior sustainability performance. Int. J. Prod. Res. **55**(15), 4481–4515 (2017)
94. Si, S.L., You, X.Y., Liu, H.C., Zhang, P.: DEMATEL technique: a systematic review of the state-of-the-art literature on methodologies and applications. Math. Probl. Eng. 1–33 (2018)

The Impacts of Foldable Containers, Street-Turn and Depot-Direct Strategies on Empty Container Repositioning Cost

Mehmet Serdar Erdoğan and Özgür Kabadurmuş(✉)

International Logistics Management, Yasar University, Izmir, Turkey
{mehmet.erdogan, ozgur.kabadurmus}@yasar.edu.tr

Abstract. In global logistics, the trade imbalances between the exporting and importing countries (and regions) cause the empty container repositioning problem. While export dominant countries face the difficulty to find empty containers to be loaded and shipped, import dominant countries have a surplus of empty containers and try to get rid of them because they are stacked at storage areas of the seaports. Transporting empty containers to the export dominant countries is as costly as transporting the loaded ones because they keep the same storage space and require the same time and handling operations. In this study, we formulated a novel Mixed-Integer Linear Programming (MILP) model considering a multi-period and multi-region shipping network to minimize the total cost for empty container repositioning operations. We investigated the impacts of foldable containers, street-turn and depot-direct strategies on the container repositioning cost. To test our proposed model, a hypothetical case study has been developed. The total costs of different strategies (i.e., street-turn, depot-direct, foldable containers) and their combinations are compared and the results are discussed.

Keywords: Empty container repositioning · Street-turn · Depot-direct · Foldable containers · Container purchasing

1 Introduction

The trade imbalances between countries resulted in empty container slack or surplus for companies. While the export dominant countries face difficulties to find empty containers to be loaded and shipped, the import dominant countries have a surplus of empty containers and try to get rid of them because they are stacked at storage areas of the seaports. To transport the containers to the export dominant countries, most of the containers go empty. Kuzmicz and Pesch [1] stated that transporting them empty does not has any added value, and more importantly, it is costly because empty containers keep the same storage space and require same time and expenditure with the loaded ones. On the other hand, holding them in the depot has also additional cost and the accumulation of containers may lead to storage problems.

Jahn and Schlingmeier [2] stated four causes of for empty container problem that are trade imbalance, seasonal demand, container type imbalance, and a large number of container owners. Epstein et al. [3] also indicated that inaccuracy in demand forecasting,

N. M. Durakbasa and M. G. Gençyılmaz (Eds.): ISPR 2019, LNME, pp. 811–825, 2020.
https://doi.org/10.1007/978-3-030-31343-2_67

lack of space in ships due to the higher priority of loaded containers, the difficulty of tracking containers and coordinating offices are some major problems in empty container balancing. According to Kuzmicz and Pesch [1], after a containerized cargo reached to an importer, they are emptied and moved with four different patterns. These four moving patterns are transportation to ports for global repositioning, transportation back to the owner, transportation to the depot, and transportation to a shipper located in a close region.

Transporting empty containers directly from consignees to local shipper is known as street-turn strategy. It has various advantages such as decreasing number of truck trips, cost, emission, and traffic congestion at the marine terminals. However, street-turn strategy is difficult to employ due to the mismatch of time, location, and container type. The same type of empty containers that a shipper requested must be available at the consignees at the right time. Apart from that, moving empty containers to off-dock depots and transporting them to the shippers when requested is the empty container repositioning strategy known as depot-direct. It creates a supply point of empty containers, reduces the congestion in the marine terminals and adds buffer capacity to marine terminals. Furthermore, foldable containers offer a solution to the empty container repositioning problem. They reduce the storage space up to 75%. Five folded containers can take the same space with a standard container. Therefore, it provides a significant storage advantage compared to the standard containers. However, they are vulnerable to damage and their production and maintenance costs are considerably high. Thus, their usage is not widespread.

In this study, we developed a mathematical model for an integrated shipping network to investigate the impact of street-turn, depot-direct, using foldable containers, and container purchasing strategies on empty container repositioning costs. The proposed model includes multiple ports, consignees, shippers, and periods along with standard and foldable containers. Purchasing, repositioning, holding, folding and unfolding costs are considered in the model. The model aims to fulfill the requests of all shippers in all period with minimum cost. The proposed model is formulated as a Mixed-Integer Linear Programming (MILP) model in IBM OPL CPLEX. Hypothetical data sets are generated to investigate how the street-turn, depot-direct, foldable containers, and container purchasing strategies affect the costs of repositioning, purchasing, holding and folding/unfolding.

This paper is organized as follows: Sect. 2 reviews the relevant literature. Section 3 presents the proposed model. Section 4 presents the results of the study. Lastly, Sect. 5 summarizes the concluding remarks and future work.

2 Literature Review

Empty container repositioning problem and strategies to handle the problem have been widely studied in the literature. Some of the studies have formulated different shipping networks and developed various solution approaches to make effective shipping decisions. Shintani et al. [4] designed a shipping network considering both loaded and empty containers. The objective of their problem is to maximize the company's profit related to revenue and cost of empty container positioning. Their model decides cruising speed, the number of ships to be used, the route, and the number of empty and loaded containers to be shipped. A Genetic Algorithm (GA) was applied to solve the model. Song and Dong [5] presented a cargo routing and empty container repositioning problem for a shipping network with multiple service routes and vessels. The objective of their model is to minimize the total cost of planning horizon that consists of the cost of container lifting, demand backordering, cost of waiting of the loaded containers at transshipment ports, inventory cost of empty containers, and transportation cost of empty containers. Monemi et al. [6] developed an integrated model for the shipping network design, fleet deployment, and empty container repositioning problems. Their model decides the ship types and quantities, routing of the loaded containers and empty container repositioning to maximize the profit that is determined according to revenue generated from demands, and cost of operations. Their problem was formulated as a MILP and a Branch, Cut and Benders algorithm was developed to solve it. In addition, Meng and Wang et al. [7] presented a liner shipping network design problem by considering hub-and-spoke, multi-port-calling operations, container handling times and empty container repositioning. The problem was formulated as a MILP and solved by using CPLEX.

Various studies investigated the impact of different strategies of empty container repositioning. Container substitution is a strategy that may also facilitate the solution of empty container repositioning problem. It is the practice of fulfilling the request of one type of containers with another type. Chang et al. [8] applied the container substitution strategy by developing a multi-commodity empty container substitution problem. They have developed a heuristic method to solve the problem and found that container substitution may reduce the cost of empty container interchange. Container sharing is another method for managing empty container repositioning problem. It refers to horizontal cooperation between container owner companies [9]. A company having a surplus of empty containers can share it with another company that lacks it. Zheng et al. [10] analyzed the empty container repositioning problem by considering the container sharing between companies. The results of their study showed that container sharing is a beneficial practice for companies and the perceived value of the containers at different ports is an important factor affecting the container sharing decisions. The model of their study was solved with a two-stage optimization technique. Furthermore, Zheng et al. [11] proposed a two-stage optimization method to decide perceived container leasing prices. Foldable container and container substitution strategies were considered in the study. Their study revealed that the cost related to a ship can be reduced by using foldable containers while container related costs can be reduced by using container substitution strategy. Moreover, Moon et al. [12] conducted a study to compare foldable containers with standard containers by formulating a mathematical model and a heuristic algorithm. Their study showed that foldable containers can be a good option to use in certain conditions, however, they cannot be widely used due to their high manufacturing and maintenance cost and their vulnerability to damage. Widespread usage of foldable containers depends on the reduction in their production cost in the future. Jula et al. [13] studied empty container traffic the Los Angeles and Long Beach. They presented an optimization method to minimize the cost of truck trips. They compared two different strategies for empty container repositioning that are street-turn and depot-direct. They considered multiple ports, however, they did not consider the movements between ports as we considered in our model. Their study found that depot-direct strategy is better if the time is an important factor for the company. If the transportation cost is the critical factor, street-turn strategy is better.

Apart from that, Choong et al. [14] formulated a mathematical model for empty container positioning problem and investigated the impact of length of the planning horizon on the empty container traffic. Their study concluded that a longer planning horizon may give better empty container distribution plans. Di Francesco et al. [15] emphasized the importance of uncertainty in the empty container repositioning problems. Their study assumed that the usage of historical data is useless and some parameters, such as the number of empty containers entering the network, and demand of empty containers, are unknown.

To the best of our knowledge, none of the studies have considered street-turn, depot-direct, and using foldable containers strategies simultaneously. In our model, we have incorporated these three strategies.

3 Problem Definition

In the proposed model, multiple ports, shippers, consignees and periods are considered. In addition, foldable and standard containers are distinguished in our model. Different numbers of consignees and shippers are located within the region of each port. Movements of empty containers from consignees to ports and shippers and from ports to shippers and other ports are allowed. Each consignee has a number available empty containers and each shipper has a number of empty container request in each period. The requests of the shippers can be fulfilled from both consignees and the depot located in the port of the same region. If the current supply of the consignee and the depot cannot satisfy the demand, the remaining containers must be purchased. Depots are assumed to be located at the port, and therefore, ports and depots are considered as the same place. Depots have a certain level of inventory in each period which is determined according to incoming and outgoing empty container flow. Note that each period is considered as a week in our model. Figure 1 illustrates the designed network for two ports, four consignees and four shippers.

The proposed model aims to satisfy empty container demand of all shippers for every period with minimum cost. The mathematical model is formulated as follows:

Sets:

C	Set of Consignees
CS	Set of Consignees and their associated Shippers with index of $(c, s) : c \in C, s \in S$
CP	Set of Consignees and their associated Ports with index of $(c, i) : c \in C, i \in P$
P	Set of Ports
S	Set of Shippers
SP	Set of Shippers and their associated Ports with index of $(s, i) : s \in S, i \in P$
N	Set of all Nodes: $C \cup P \cup S$
T	Set of Time Periods

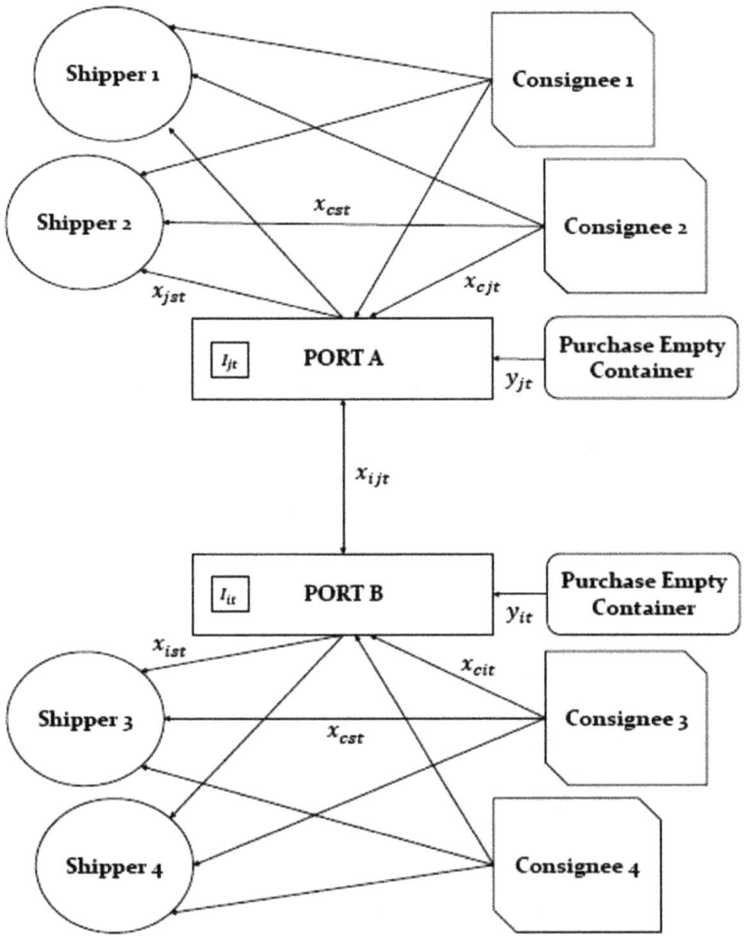

Fig. 1. An illustration of the designed network

Parameters:

A_{ct} Number of available standard empty containers in consignee $c \in C$ in period $t \in T$

A^F_{ct} Number of available foldable empty containers in consignee $c \in C$ in period $t \in T$

R_{st} Number of requested empty containers by shipper $s \in S$ in period $t \in T$

H_i Unit holding (storage) cost of a standard container at port $i \in P$

H^F_i Unit holding (storage) cost of a foldable container at port $i \in P$

P_i Unit purchasing price of a standard container at port $i \in P$

P^F_i Unit purchasing price of a foldable container at port $i \in P$

C_{ij} Unit repositioning cost of a standard container from node $i \in N$ to node $j \in N$

C^F_{ij} Unit repositioning cost of a foldable container from node $i \in N$ to port $j \in N$

L_i^F Unit folding cost of a foldable container at port $i \in P$
L_i^U Unit unfolding cost of a foldable container at port $i \in P$
v_{ij} Transportation time from port $i \in P$ to port $j \in P$
K_{ijt} Capacity that is available for repositioning empty containers from port $i \in P$ to port $j \in P$ in period $t \in T$
M Number of folded containers that can be packed in the size of a standard container

Decision Variables:
I_{it} Inventory level of standard containers at port $i \in P$ in period $t \in T$
I_{it}^F Inventory level of foldable containers at port $i \in P$ in period $t \in T$
x_{ijt} Number of standard containers to be transported from node $i \in N$ to node $j \in N$ in period $t \in T$
x_{ijt}^F Number of foldable containers to be transported from node $i \in N$ to node $j \in N$ in period $t \in T$
y_{it} Number of standard containers to be purchased at port $i \in P$ in period $t \in T$
y_{it}^F Number of foldable containers to be purchased at port $i \in P$ in period $t \in T$

Objective Function:

$$Min \sum_{i \in N} \sum_{i \in N: i \neq j} \sum_{t \in T} C_{ij} x_{ijt} + C_{ij}^F x_{ijt}^F + \sum_{i \in P} \sum_{t \in T} H_i I_{it} + H_i^F I_{it}^F + \sum_{i \in P} \sum_{t \in T} P_i y_{it} + P_i^F y_{it}^F + \sum_{i \in P} \sum_{j \in P} \sum_{t \in T} L_i^F x_{ijt}^F + L_i^U x_{ji(t-v_{ji})}^F \tag{1}$$

Subject to;

$$I_{it} = I_{i(t-1)} + y_{it} + \sum_{j \in P: i \neq j, t > v_{ji}} x_{ji(t-v_{ji})} + \sum_{(c,i) \in CP} x_{cit} - \sum_{j \in P: i \neq j, t \leq T - v_{ji}} x_{ijt} - \sum_{(s,i) \in SP} x_{ist} \quad \forall i \in P, t \in T \tag{2}$$

$$I_{it}^F = I_{i(t-1)}^F + y_{it}^F + \sum_{j \in P: i \neq j, t > v_{ji}} x_{ji(t-v_{ji})}^F + \sum_{(c,i) \in CP} x_{cit}^F - \sum_{j \in P: i \neq j, t \leq T - v_{ji}} x_{ijt}^F - \sum_{(s,i) \in SP} x_{ist}^F \quad \forall i \in P, t \in T \tag{3}$$

$$\sum_{(s,i) \in SP} (x_{ist} + x_{ist}^F) + \sum_{(c,s) \in CS} (x_{cst} + x_{cst}^F) = R_{st} \quad \forall s \in S, t \in T \tag{4}$$

$$\sum_{(c,i) \in CP} x_{cit} + \sum_{(c,s) \in CS} x_{cst} = A_{ct} \quad \forall c \in C, t \in T \tag{5}$$

$$\sum_{(c,i) \in CP} x_{cit}^F + \sum_{(c,s) \in CS} x_{cst}^F = A_{ct}^F \quad \forall c \in C, t \in T \tag{6}$$

$$x_{ijt} + \frac{x_{ijt}^F}{N} \leq K_{ijt} \quad \forall i \in P, j \in P, t \in T \tag{7}$$

$$I_{it}, I_{it}^F, x_{ijt}, x_{ijt}^F, y_{it} \text{ and } y_{it}^F \in \{0, 1, 2, \ldots\} \qquad \forall i \in N, j \in N, t \in T \qquad (8)$$

Objective function (1) minimizes the total cost which includes the costs of repositioning empty containers, the holding, the purchasing, and the folding/unfolding. Constraint (2) balances the inventory of standard containers while Constraint (3) balances the inventory of foldable containers. The number of empty containers going into the inventory includes the number of empty containers from the last period, the number of empty containers that are supplied from consignees in the current period, the number of containers that are repositioned from other ports in previous periods and scheduled to arrive in the current period, and the number of containers that are purchased in the current period. The number of empty containers going out of the inventory consists of the number of empty containers that are transported to other ports and the shippers. Constraint (4) ensures that requests of all shippers are fulfilled. Constraints (5) and (6) ensure that all available empty containers of consignees are used. Constraint (7) prevents the number of repositioned containers between ports to exceed shipping capacity. Constraint (8) is the boundary constraint of decision variables.

4 Computational Results

We formulated our model as Mixed-Integer Linear Programming (MILP) in IBM OPL CPLEX Optimization Studio. The model has been executed in a PC with Intel® Core™ i7-4510U CPU @2.00 GHz 2.60 GHz 8 GB RAM.

To test our proposed model, we have generated four different test scenarios in different sizes. Table 1 summarizes the four scenarios that we have generated. The data were generated hypothetically but in a realistic way. Purchasing cost of the standard containers at each port (P_i) is generated randomly between 4,000 and 5,000 while the purchasing cost of the foldable containers at each port (P_i^F) is generated randomly between 13,000 and 15,000. Unit holding cost of standard containers at each depot (H_i) is generated randomly between 20 and 30 while unit holding cost of foldable containers at each depot (H_i^F) is generated randomly between 4 and 7. The number of available standard containers at consignees (A_{ct}) is generated randomly between 0 and 180 while the number of available foldable containers at consignees (A_{ct}^F) is generated randomly between 0 and 40. The number of requested empty containers by shippers (R_{st}) is generated randomly between 40 and 120. Unit repositioning cost of standard containers between ports (C_{ij}) is generated randomly between 40 and 500, while unit repositioning cost of foldable containers between ports (C_{ij}^F) is generated randomly between 10 and 100. The available capacity of repositioning empty containers (K_{ijt}) is generated randomly between 200 and 500. Both unit folding and unfolding cost of foldable containers (L_i^F and L_i^U) are generated randomly between 80 and 140. Transportation time between ports (v_{ij}) is generated randomly between 0 and 2 periods. Lastly, the number of folded containers that can be packed in the size of a standard container (M) is choosen as 5.

Table 1. Summary of the scenarios

Scenario	|P|	|S|	|C|	|T|
1	6	12	12	6
2	6	12	12	12
3	6	24	24	6
3	6	24	24	12

To test the impact of each strategy, we applied a sensitivity analysis by combining different strategies. We created six different cases and applied all of these cases to each scenario. We have executed the model for 24 different cases in total (4 scenarios * 6 cases). Table 2 summarizes the cases that we considered.

Table 2. Summary of the cases

Case	Street-turn	Depot-direct	Foldable containers
A	✓	✓	✓
B	–	✓	✓
C	✓	–	✓
D	✓	✓	–
E	✓	–	–
F	–	✓	–

In case A, we considered all of the strategies while in cases B, C, and D, we did not consider street-turn, depot-direct, and foldable container strategies, respectively. In case E, we only considered street-turn strategy while in case F, we only considered depot-direct strategy.

Table 3 shows the results for all scenarios and cases. According to Table 3, foldable containers have not been purchased in any case. Since the purchasing cost of foldable containers is very high, leasing them rather than purchasing is a better decision. Furthermore, when depot-direct strategy is not allowed, the usage of foldable containers increases, while the usage of standard containers decreases. Since the holding cost of a foldable container is lower than a standard container, the model keeps more foldable containers at the depot. However, when there is no depot, using more foldable containers to fulfill the demand becomes a better choice. When street-turn strategy is not allowed, the costs related to standard containers increase dramatically. In that case, the model makes purchasing decision of standard containers to fulfill the demand since standard containers are cheaper.

Table 3. The results of all scenarios and cases

Scenario-case	Transportation cost of standard containers	Transportation cost of foldable containers	Holding cost of standard containers	Holding cost of foldable containers	Purchasing cost of standard containers	Purchasing cost of foldable containers	Folding/unfolding cost	Total cost
1-A	398,765	25,105	5,300	1,165	2,198,000	0	40,800	2,669,135
1-B	644,240	0	0	0	25,379,000	0	0	26,023,240
1-C	390,350	28,080	0	0	2,198,000	0	49,400	2,665,830
1-D	485,450	0	1,100	0	5,088,000	0	0	5,574,550
1-E	482,400	0	0	0	5,188,000	0	0	5,670,400
1-F	644,240	0	0	0	25,379,000	0	0	26,023,240
2-A	812,410	68,045	6,600	2,965	2,300,000	0	125,800	3,315,820
2-B	1,418,095	0	0	0	49,637,000	0	0	51,055,095
2-C	803,675	73,535	0	0	2,300,000	0	143,150	3,320,360
2-D	1,035,195	0	1,700	0	8,576,000	0	0	9,612,895
2-E	1,040,145	0	0	0	8,576,000	0	0	9,616,145
2-F	1,418,095	0	0	0	49,637,000	0	0	51,055,095
3-A	709,735	57,255	4,800	1,390	1,925,000	0	106,500	2,804,680
3-B	1,239,705	0	1,100	0	48,511,500	0	0	49,752,305
3-C	704,560	57,500	0	0	2,165,000	0	107,450	3,034,510
3-D	901,300	0	2,000	0	7,632,000	0	0	8,535,300
3-E	896,530	0	0	0	7,832,000	0	0	8,728,530
3-F	1,239,705	0	1,100	0	48,511,500	0	0	49,752,305
4-A	1,466,635	143,610	21,000	6,780	1,654,000	0	272,650	3,564,675
4-B	2,821,250	0	12,650	0	98,216,000	0	0	101,049,900
4-C	1,518,205	146,250	0	0	1,654,000	0	281,250	3,599,705
4-D	1,917,985	0	9,200	0	14,566,000	0	0	16,493,185
4-E	1,944,135	0	0	0	14,566,000	0	0	16,510,135
4-F	2,821,250	0	12,650	0	98,216,000	0	0	101,049,900

Figure 2 through Fig. 6 show the percent difference of all scenarios and cases in terms of transportation, holding, purchasing, folding/unfolding, and total costs compared to the base case of each scenario (i.e., Case A). Note that, the Case A is chosen as the base case since it allows all strategies.

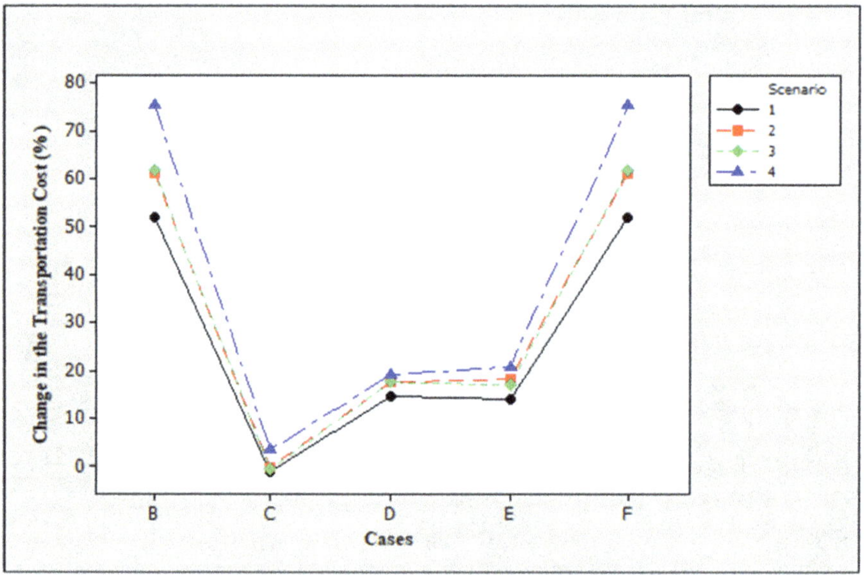

Fig. 2. The percent change in transportation cost of all scenarios and cases compared to their base case (Case A)

According to Fig. 2, transportation cost increases significantly when street-turn strategy is not allowed (Cases B and F) since containers have to be transported from the other ports. In addition, foldable containers have an important impact on transportation cost reduction since they keep less space and transporting them is cheaper. Depot-direct strategy does not have a significant impact on transportation cost.

According to Fig. 3, holding cost decreases when street-turn or foldable container strategies are not allowed. Since holding cost of standard containers is expensive, transporting them in the current period is a better decision than storing them in the depot. Therefore, a smaller level of total inventory is kept when foldable container strategy is not allowed (Cases D, E, and, F) and holding cost is decreased. When the street-turn strategy is not allowed (Cases B and F), the only way to get foldable containers is to purchase them. Since purchasing them is very costly, standard containers are used and they are not kept in the depot and the holding cost is decreased. Since there is no depot when depot-direct strategy was not allowed (Cases C and E), there is no holding cost.

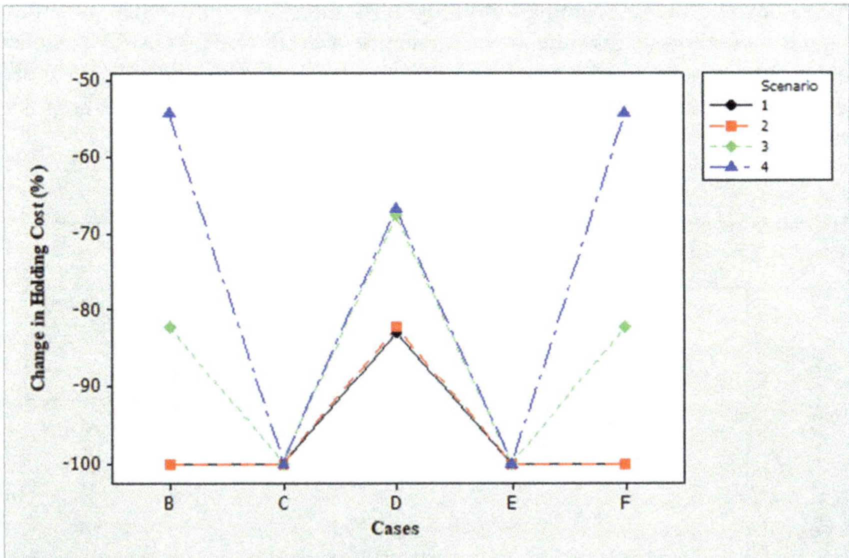

Fig. 3. The percent change in holding cost of all scenarios and cases compared to their base case (Case A)

According to Fig. 4, purchasing cost increases 10 to 60 times when the street-turn strategy is not allowed (Cases B and F) since only way obtain containers is to purchase them when there is no consignee. The absence of foldable containers also increases purchasing cost significantly (Cases D, E, and F). This is because the supply of foldable containers in the containers is not available. Depot-direct strategy has no impact on purchasing cost.

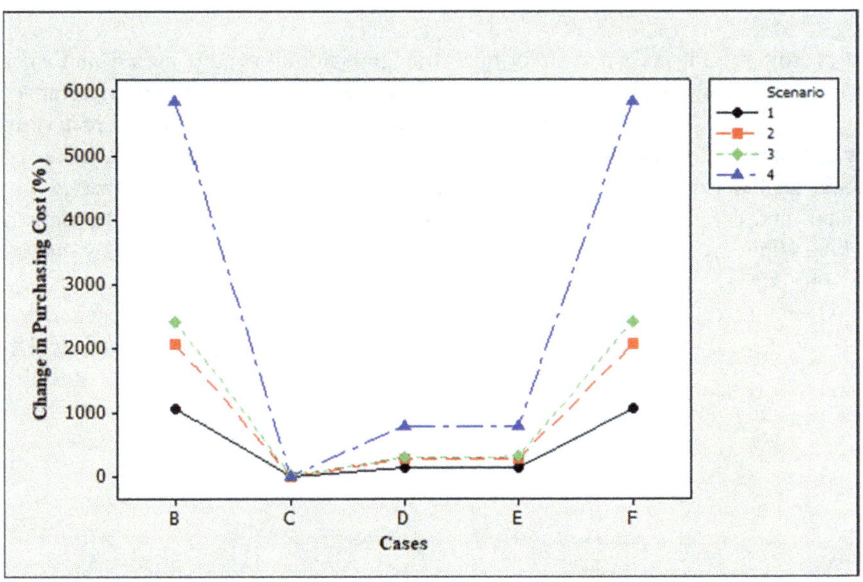

Fig. 4. The percent change in purchasing cost of all scenarios and cases compared to their base case (Case A)

According to Fig. 5, folding/unfolding cost increases only when depot-direct strategy is not allowed and the other strategies are allowed (Case C) since more foldable containers are used in transportation between ports rather than being stored in the depot. There is no folding/unfolding cost in the other cases since no foldable container is used.

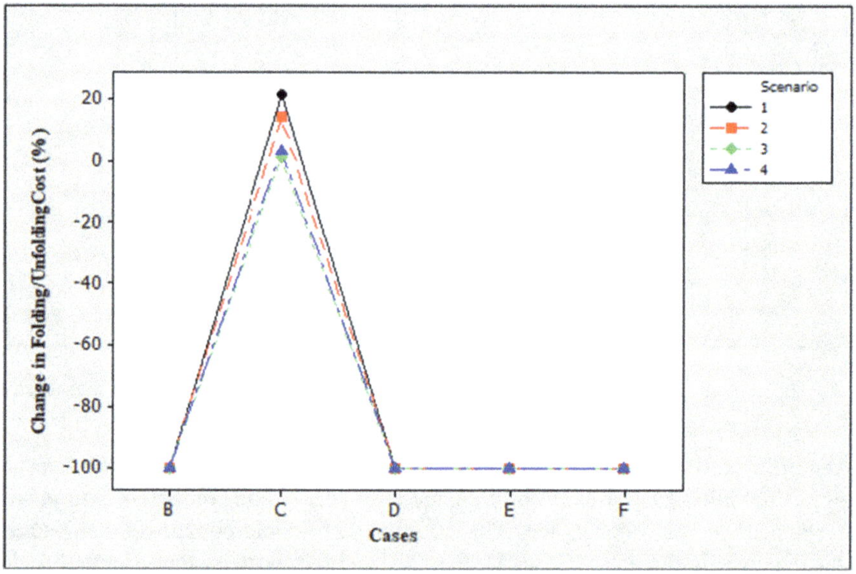

Fig. 5. The percent change in folding/unfolding cost of all scenarios and cases compared to their base case (Case A)

According to Fig. 6, when street-turn strategy is not allowed (Cases B and F), total cost increases dramatically. Not allowing foldable containers also has a high impact on the cost increase, however, not as much as street-turn strategy. Depot-direct strategy does not have a significant impact on cost. However, as we mentioned, it has other benefits such as decreasing congestion at terminals, adding buffer capacity to the terminal, and creating a supply point and by doing so responding the demand more quickly. While the least costly decision is to use all strategies simultaneously, not using street-turn strategy is the most costly one.

As for the solution times, the presented model was solved for all scenarios and cases very fast. Each run took less than one second to solve optimally in CPLEX.

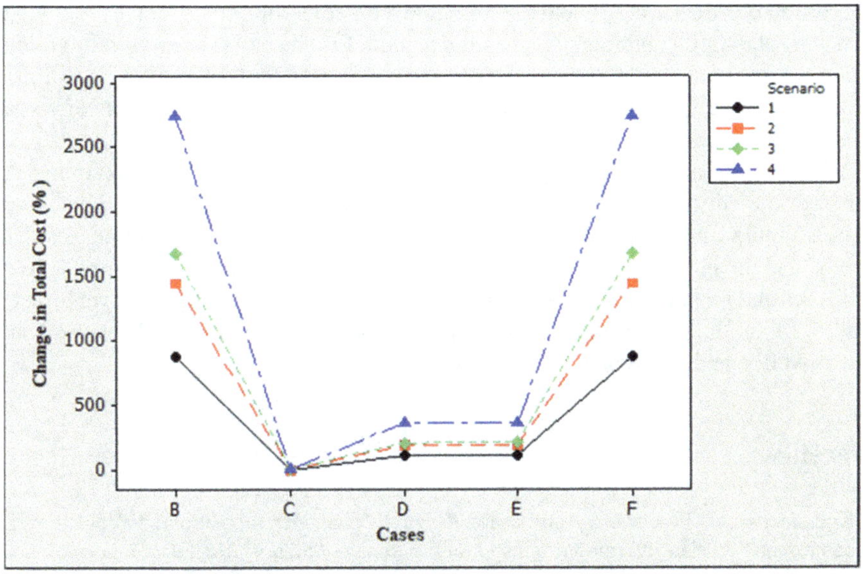

Fig. 6. The percent change in total cost of all scenarios and cases compared to their base case (Case A)

5 Conclusion and Further Research

Imbalance in the world trade makes repositioning of empty containers difficult. While an accumulation of empty containers occurs in import dominant regions, a lack of empty containers occurs in export dominant regions. Also, transporting containers empty is as costly as transporting them loaded since both of them keep the same storage space. On the other hand, storing containers at marine terminals has additional costs. To address empty container repositioning problem, various strategies exist such as street-turn (leasing containers from local consignees), depot-direct (transporting containers to off-dock depots), and using foldable containers (which reduces the storage space by %75).

In this study, we formulated a novel Mixed Integer Linear Programming model of a shipping network with multiple ports, shippers, consignees, and periods to compare the impacts of street-turn, depot-direct, and using foldable containers on empty container repositioning cost. We have generated hypothetical but realistic data sets to test the model and created four different scenarios in different sizes. We tested these four scenarios in six different cases that are allowing or not allowing some of the strategies.

Results of the study showed that street-turn and using foldable containers are very effective to reduce total cost while depot-direct strategy does not affect total cost significantly. Depot-direct strategy has other positive effects such as creating a supply point, decreasing the congestion in terminals, and adding buffer capacity to terminal. In comparison with the foldable container usage, the impact of street-turn on total cost is much higher. In addition, leasing foldable containers rather than purchasing them is a more profitable decision. If leasing foldable container option is not available then purchasing standard containers is a better decision. Furthermore, keeping more foldable containers in the depot and using relatively more standard containers to fulfill the demand is also a good strategy. This means that holding cost reduction of foldable containers is higher than transportation cost reduction.

As a limitation of this study, only hypothetical case studies are examined. For future work, the model proposed in this study can be tested in real applications. Further research should consider other strategies, such as container substitution and container sharing, and investigate the impacts of them on cost. Dynamic inventory levels of consignees and shippers can also be considered. In addition, shortage of inventories can be allowed in the model and a penalty cost can be included in the model since unsatisfied demand of shippers is common.

References

1. Kuzmicz, K.A., Pesch, E.: Approaches to empty container repositioning problems in the context of Eurasian intermodal transportation. Omega **85**, 194–213 (2018)
2. Jahn, C., Schlingmeier, J.: Cooperation in empty container logistics. Innov. Methods Logist. Supply Chain. Manag. **499**, 499–513 (2014)
3. Epstein, R., Neely, A., Weintraub, A., Valenzuela, F., Hurtado, S., Gonzalez, G., Beiza, A., Naveas, M., Infante, F., Alarcon, F., Angulo, G., Berner, C., Catalan, J., Gonzalez, C., Yung, D.: A strategic empty container logistics optimization in a major shipping company. Interfaces **42**(1), 5–16 (2012)
4. Shintani, K., Imai, A., Nishimura, E., Papadimitriou, S.: The container shipping network design problem with empty container repositioning. Transp. Res. Part E Logist. Transp. Rev. **43**(1), 39–59 (2007)
5. Song, D.P., Dong, J.X.: Cargo routing and empty container repositioning in multiple shipping service routes. Transp. Res. Part B Methodol. **46**(10), 1556–1575 (2012)
6. Monemi, R.N., Gelareh, S.: Network design, fleet deployment and empty repositioning in liner shipping. Transp. Res. Part E Logist. Transp. Rev. **108**, 60–79 (2017)
7. Meng, Q., Wang, X.: Intermodal hub-and-spoke network design: Incorporating multiple stakeholders and multi-type containers. Transp. Res. Part B Methodol. **45**(4), 724–742 (2011)
8. Chang, H., Jula, H., Chassiakos, A., Ioannou, P.: A heuristic solution for the empty container substitution problem. Transp. Res. Part E Logist. Transp. Rev. **44**(2), 203–216 (2008)
9. Furtado, P., Frayret, J.M.: Proposal sustainability assessment of resource sharing in intermodal freight transport with agent-based simulation. IFAC-PapersOnLine **48**(3), 436–441 (2015)
10. Zheng, J., Sun, Z., Gao, Z.: Empty container exchange among liner carriers. Transp. Res. Part E Logist. Transp. Rev. **83**, 158–169 (2015)
11. Zheng, J., Sun, Z., Zhang, F.: Measuring the perceived container leasing prices in liner shipping network design with empty container repositioning. Transp. Res. Part E Logist. Transp. Rev. **94**, 123–140 (2016)
12. Moon, I.K., Do-Ngoc, A.D., Hur, Y.S.: Positioning empty containers among multiple ports with leasing and purchasing considerations. OR Spectr. **32**(3), 765–786 (2010)
13. Jula, H., Chassiakos, A., Ioannou, P.: Port dynamic empty container reuse. Transp. Res. Part E Logist. Transp. Rev. **42**(1), 43–60 (2006)

14. Choong, S.T., Cole, M.H., Kutanoglu, E.: Empty container management for intermodal transportation networks. Transp. Res. Part E Logist. Transp. Rev. **38**(6), 423–438 (2002)
15. Di Francesco, M., Crainic, T.G., Zuddas, P.: The effect of multi-scenario policies on empty container repositioning. Transp. Res. Part E Logist. Transp. Rev. **45**(5), 758–770 (2009)

The Modification and the Line Balancing of an Assembly Line for the Inclusion of a New Product Model

Dilruba Telli[1], Yalçın Berberoğlu[1], Gizem Karadeniz[1], Doğangün Yılmaz Uzun[1], Elif Baykal[1], Damla Yüksel[1](✉), and Yiğit Kazançoğlu[2]

[1] Department of Industrial Engineering, Yasar University, Bornova, Turkey
dilruba.telli@gmail.com, yalcin.berberoglu@yahoo.com, gizemkaradeniz96@gmail.com, dogangunuzun@gmail.com, elifbaykal.yu@outlook.com, damla.yuksel@yasar.edu.tr
[2] Department of International Logistics Management, Yasar University, İzmir, Turkey
yigit.kazancoglu@yasar.edu.tr

Abstract. This study aims to modify the 5th lean assembly line in order to include a new product model in a combi boiler producer company. After reviewing customers' needs, company decides to produce a new product. New product model will be produced in the one of the existing lines in the plant. Most suitable line determined according to other product models job sequences' similarity with new product models'. After the analysis of the production process, the changes and possible constraints for the 5th assembly line are determined. After that, solution procedures are developed, the required calculations are made and implemented. The solutions are developed by various techniques and approaches such as Bubble diagram, Value Stream Mapping, Lean Production, Line Balancing. Particularly, mathematical model, which is solved by IBM ILOG Optimization Studio programmed comprehensive analysis with a simulation model, which is conducted by Arena Simulation software, are presented in this study. The outcomes of the solutions are simulated, compared and automatized. At the end, three options are provided for the 5th lean line to include a new product type while continuing the production process with the current products.

Keywords: Line balancing · Process design · Lean manufacturing · Mathematical modelling · Simulation · Value stream mapping

1 Introduction and Problem Definition

This study is conducted in one of the leading worldwide combi boiler producer company's plant, in Manisa, Turkey. The Company R&D Department started a new project based on considering user feedbacks due to the fierce competition. After developing the research, the company R&D Department came up with a new product which is more energy efficient and cost effective.

N. M. Durakbasa and M. G. Gençyılmaz (Eds.): ISPR 2019, LNME, pp. 826–838, 2020.
https://doi.org/10.1007/978-3-030-31343-2_68

This study presents a solution methodology for inclusion of the aforementioned new product to an existing line within a worldwide combi boiler producer company. The study covers line balancing of the 5th lean line. Various solution techniques have been applied to solve the problem. Initially, an extensive literature review is performed to find similar problems and their solution approaches. Mathematical models are used in order to optimize the system. The factory wants to integrate a new product type without effecting the existing products in the current production line. Creating a mathematical model is a key motivation for developing the existing line for the new product model. In this study, it is aimed to integrate the new device in production without affecting the current production. With this study, it is necessary to determine the material addresses of the new future family, determine the material feeding type, determine the box size for the materials, determine the cycle time, decide the number of operators, determine the work steps, rearrange the lay-out of the 5th lean line, and design the necessary fixtures and tools.

Combi devices are produced from various families in these lines. In the 5th line, 2 main family devices are produced. These are Model A and Model B devices. These devices are condensing combi devices. The cycle times of the Model A and Model B devices are 105 and 93 s, respectively. Material addresses, material box types and material feeding patterns of existing devices are evident. Likewise, the number of test stations and operators for these devices is also evident. Thanks to all these specific steps, work instructions are prepared and produced for the operators. Material addresses for new future devices, material box sizes, material feeding patterns, necessary fixture and equipment, work steps, number of operators, cycle time, lay-out work has not been done. As a result of all these studies at the end of this study, the 5th line product family can start production and reach the target with quality products without any loss in the targeted cycle time.

2 Literature Review

In a multi-model assembly line, it is aimed to minimize the cycle times and maximize the line efficiency while keeping the station and the number of employees constant for inclusion of a new product to the existing line. The related literature was covered accordingly. In order to minimize the number of employees; the collective priority scheme, methods of corrected task processing times, methods of inverse priority diagram; ant colony optimization and intuitive search; integer programming was used, respectively (Fokkert and Kok 1997; McMullen and Tarasewich 2007; Widyadana 2009). Ant Colony optimization has been used to minimize the number of stations for the given cycle time (Yagmahan 2011). Multi-integer programming, sequential weighting and computational algorithms to reduce cycle time; restriction method; genetic algorithm; The intuitive methods of the commodity were used, respectively (Karabatı and Sayın 2003; Yoosefelahi et al. 2012; Ramezanian and Ezztpanah 2014; Manavizadeh et al. 2015). In order to reduce time losses; sequential weight, largest candidate rule, trial-error method, Killbridge and Westers rule and many integer programming were used respectively (Zelter et al. 2013; Hwag and Katayama 2007). In order to increase the efficiency of the line, priority based genetic algorithm, improvement algorithm with genetic algorithm, roulette wheel and row weight were used (Hwang and Katayama 2007). Integer programming was used to minimize the number of stations (Akpınar and Baykasoglu 2014). Meta-heuristics and genetic algorithm are used to distribute the desired tasks equally to the stations by reducing the number of operators and stations (Raj et al. 2016). Genetic algorithm is used to balance the asynchronous assembly line with variable running times and to show the effectiveness of U-shaped lines against parallel lines with parallel workstations and buffers (Tiacci 2017). Meta-heuristics are used to minimize the total time required to produce a number of different product groups (Christensen et al. 2017). For a given cycle time, cyclic crossover method and genetic algorithm are used to group tasks at the minimum number of workstations, to maximize the assembly line and maximize efficiency (Sivasankaran and Shahabudeen 2017). All the above mentioned articles are summarized in Table 1 and the notations used are explained in Table 2.

Table 1. Literature table for line balancing

Author(s)/Date	Problem	Method/approach	Performance measure(s)													
			B	NS	NL	NT	NW	AT	P	C	IT	ST	TT	LE	LL	D
Fokkert and Kok (1997)	MiALB & MuALB	CPD, ATPTM & RPDM				x		x		x						
Karabatı and Sayın (2003)	MiALB	MIP, RPW & CA				x			x				x			
McMullen and Tarasewich (2007)	MoALB	ACO, SH		x			x			x						
Hwang and Katayama (2007)	U-shaped ALs	PGA, ASGA, RW & RPW		x						x			x			
Boysen et al. (2008)	AL	TMM		x						x				x		
GedeAgusWidyadana (2009)	U-type MoALB	MIP				x	x		x	x			x			
Fattahi et al. (2010)	MuMALB	MIP & ACO		x					x	x						
Yagmahan (2011)	B/S MiMALs	ACO				x			x	x						x
Yoosefelahi et al. (2012)	MIP	CM		x		x				x			x			
Al-zubaidy and Alrazaq (2013)	MuALB	RPW, LCR, TEM & KWR		x					x	x	x			x		
Zelter et al. (2013)	MiALB	MIP				x			x	x						
Akpinar and Baykasoglu (2014)	MiALB	MIP		x						x						
Ramezanian and Ezzatpanah (2014)	MiALB	GA		x					x	x			x	x		
Manavizadeh et al. (2015)	U-shaped MiALB	MH & EM		x					x	x			x			
Raj et al. (2016)	ALB	MH & GA							x						x	
Tiacci (2017)	U-shaped MiALB	GA		x				x		x						
Christensen et al. (2017)	MuALB	MH	x		x							x				
Sivasankaran and Shahabudeen (2017)	SMALB & MuALB	GA & CCM							x	x			x			

Table 2. Notations used in literature table

Notation	Refers to	Notation	Refers to
AT	Average time	ASGA	Amelioration Structure with a Genetic Algorithm
B	Batch size	ATPTM	The adjusted task processing times methods
C	Cycle time	CA	Computational Algorithms
D	Demand	CCM	Cyclic Crossover Method
IT	Idle time	CM	Constraint Method
LE	Line efficiency	CPD	Combined Precedence Diagram
LL	Line length	EA	Evolutionary Algorithm
NL	Number of lines	EM	Exact method
NS	Number of stations	GA	Genetic algorithm
NT	Number of tasks	KWR	Kilbridge and Westers Rule
NW	Number of workers	LCR	Largest Candidate Rule
P	Precedence relationship	MDGA	Multi decision genetic algorithm
ST	Setup time	MH	Meta Heuristic
TT	Task times	MIP	Multi Integer Programming
AL	Assembly Line	NV	Novel Heuristic
ALB	Assembly Line Balancing	PGA	Priority-based Genetic Algorithm
MiALB	Mixed-model assembly line balancing	RPDM	The reversed precedence diagram method
MuALB	Mixed-model assembly line balancing	RPW	Ranked Positional Weight
MuMALB	Multi-manned assembly line balancing	RW	Roulette Wheel
MoALB	Multi-objective assembly line balancing	SH	Search Heuristics
SMALB	Single line assembly line balancing	TEM	Trial and Error Method
ACO	Ant Colony Optimization	TMM	Time Measurement Methods

3 Line Balancing

Assembly lines are desired to be designed as lean production lines in the company. The loss of time and labor in the production environment should be eliminated or minimized. Assignment of tasks to stations needs to be optimized based on the desired performance criterion by considering the priority relationships between tasks are called assembly line balancing. These performance measures are usually the smallest number of stations or cycle time (Özcan and Peker 2007).

The most important performance measures in assembly line balancing is to minimize the number of workstations or cycle times with the same processing time on the line. The main work in line balancing is to distribute the total workload on the assembly line evenly to the stations in order to reduce the time reductions in the stations. For this purpose, the work elements are assigned to the workstations according to their priority relations with each other and their idle times. This problem is called the assembly line balancing problem in the literature (Kılınçcı 2004).

3.1 Calculation of Cycle Time and Number of Workstations

The cycle time and number of stations according to cycle time in the current 5th lean line is calculated as shown:

- One shift is 8 h.
- There is a half-hour meal break, two 10-min tea breaks, one 5-min sports time and 8-min TPM (Total Productive Maintenance) times.
- In this case, the production time can be calculated as follows; $8 * 60 - (30 + 2 * 10 + 5 + 8) = 417$ min.
- The forecasted demand values from February'19 to December'19 are already exists. Therefore, we have the whole data for 11 months. To calculate the cycle time with the forecasted demand values, the working time to be performed during the 11 months must be calculated previously.
- The working time for 11 months calculated as $417 * 60 * 22 * 11 * 3 =$ 18,164,520 s.
- Monthly-based forecasts from the customer are collected and the 11-month's total demand forecast is calculated.
- The factory allows 25% tolerance to the forecasted demand values to remain on the safe zone and to meet the customer's demands in unexpected conditions.
- In this case, the total forecasted demand values are multiplied by 1.25 to obtain the required inputs for the maximum cycle time calculation.
- Total forecasted demand = 63,084 units
- Total forecasted demand with 25% tolerance = 63,084 * 1.25 = 78,855 units
- Maximum cycle time = Total working time/Total forecast with 25% tolerance = 18,164,520/78,855
- Maximum cycle time = 230.35 s

The company proceed its production according to OEE (Overall Equipment Efficiency) calculations. Overall Equipment Effectiveness (OEE) is a measure of manufacturing operations performance and productivity, expressed in a percentage. OEE considers all losses, resulting in a measure of truly productive manufacturing time. It is calculated as:

- OEE = Availability × Performance × Quality
- Based on the calculations of the company, OEE is considered as 80%.
- Therefore, the real maximum cycle time is = 230.35 * 0.80 = 184.28 s

MTM is a time determination method. The method used to perform the process determines the time. MTM analysis provides the data in TMU (Time Measurement Unit). To convert the TMU unit to minute, it is multiplied by 0.0006. It is calculated based on cycle time calculation in seconds. Therefore, the result is multiplied by 60 and the second conversion is obtained. This conversion is need to perform the further analysis. In addition, the factory has a specified fatigue tolerance for the operators.

The tolerance value is 0.0421. Therefore, by multiplying the data that is provided in seconds by 1.0421, the time that the operator must perform for the process it obtained.

The minimum number of stations required for production is found by the ratio of the time of all works from MTM analysis to the time of indivisible work from MTM analysis. This ratio is rounded up to find the maximum number of stations.

The total time of all the works is 3,063.74 s. The time of indivisible works which are bottleneck from test stations is 136 s.

- Therefore, the minimum cycle time that is 3,063.74 s is divided by 136 s. Then, 22.52748335 is obtained. However, the number of stations must be integer thus 23 is obtained by rounding it up.
- At the end, the maximum of 23 stations to produce the production is obtained.

4 Mathematical Model

Two mathematical models are developed: minimization of number of stations and minimization of cycle time. First all the sets and indices, parameters and decision variables are given below. Then, both mathematical models are represented.

Sets and Indices

i operations $i \in I$
j stations $j \in J$

Parameters

I number of operations,
J number of stations,
N last assigned station,
T_i time of operation i (task time),
c cycle time,
P_i cluster of operations which should be done before start of operation *i*.
S_i cluster of operations which should not start before end of operation *i*.
LJ_i earliest station to assign operation *i*,
UJ_i latest station to assign operation *i*,
pr_i precedence operation *i* of operation *i*

Decision Variables

$x_{i,j}$ 1 if operation *i* assigned to station *j*, 0 otherwise

Model 1. Mathematical Model for Minimization of Number of Stations

$$\text{Min} \quad N; \tag{1}$$

Subject to;

$$\sum_{i=1}^{I} T_i * x_{i,j} \leq c \qquad i = 1, \ldots, I \tag{2}$$

$$\sum_{j=1}^{J} x_{i,j} \leq 1 \qquad j = 1, \ldots, J \tag{3}$$

$$\sum_{j=1}^{J} j * x_{i,j} \leq N \qquad i = 1, \ldots, I \tag{4}$$

$$\sum_{j=1}^{J} j * (x_{i,j} - x_{pr_i,j}) \leq J \qquad i = 1, \ldots, I \tag{5}$$

$$x_{i,j} = 0, \qquad \text{if } j < LJ_i \text{ and } j > UJ_i \tag{6}$$

$$LJ_i = \left\lceil \frac{T_i + \sum_{k \in P_i} T_k}{c} \right\rceil \tag{7}$$

$$UJ_i = J + 1 - \left\lceil \frac{T_i + \sum_{k \in S_i} T_k}{c} \right\rceil \tag{8}$$

$$x_{i,j} = 0 \text{ or } 1 \tag{9}$$

$$\text{if } j < LJ_i \text{ or } j > UJ_i \text{ then } \sum_{i=1}^{I} \sum_{j=1}^{J} x_{ij} = 0 \tag{10}$$

The objective (1) is to minimize the total number of stations. Constraint (2) provides the total time of all operations assigned to a station cannot exceed the cycle time. Constraint (3) provides that each operation should assign to exactly one station. Constraint (4) assures that the last assigned station number cannot exceed last station number. Constraint (5) prevents assigning operation i before its precedence. Constraint (6) provides operation i can assign at least LJ_i numbered station and at most UJ_i numbered station. (7) and (8) indicates calculations for precedence relationship. Constraint (7) displays the earliest station to assign for job i and constraint (8) presents the latest station to assign job i. P_i at constraint (7) stands for cluster of operations which should be done before start of operation i. In other words, operations that comes after operation i according to precedence relationship. S_i at constraint (8) stands for cluster of operations which should not start before end of operation i. It refers as the operation that comes before operation i according to precedence relationship. Constraint (10) refers to, j must be smaller than or equal to UJ_i and bigger than or equal to LJ_i, if not it is equal to 0.

By using the optimal number of stations obtained from model (i), the cycle time will be minimized in the second model as follows:

Model 2. Mathematical Model for Minimization of Cycle Time

$$\text{Min} \quad c; \tag{1}$$

Subject to;

$$\sum_{i=1}^{I} T_i * x_{i,j} \leq c \qquad i = 1, \ldots, I \tag{2}$$

$$\sum_{j=1}^{J} x_{i,j} \leq 1 \qquad j = 1, \ldots, J \tag{3}$$

$$\sum_{j=1}^{J} j * x_{i,j} \leq J \qquad i = 1, \ldots, I \tag{4}$$

$$\sum_{j=1}^{J} j * (x_{i,j} - x_{pr_i,j}) \leq J \qquad i = 1, \ldots, I \tag{5}$$

$$x_{i,j} = 0, \qquad \text{if } j < LJ_i \text{ and } j > UJ_i \tag{6}$$

$$LJ_i = \left\lceil \frac{T_i + \sum_{k \in P_i} T_k}{c} \right\rceil \tag{7}$$

$$UJ_i = J + 1 - \left\lceil \frac{T_i + \sum_{k \in S_i} T_k}{c} \right\rceil \tag{8}$$

$$x_{i,j} = 0 \text{ or } 1 \tag{9}$$

$$\text{if } j < LJ_i \text{ or } j > UJ_i \text{ then } \sum\nolimits_{i=1}^{I} \sum\nolimits_{j=1}^{J} x_{ij} = 0 \tag{10}$$

Our objective (1) is to minimize the cycle time. Constraint (2) provides that the total time of all operations assigned to a station cannot exceed the cycle time. Constraint (3) ensures that each operation must assign to exactly one station. Constraint (4) provides the last assigned station number cannot exceed the last station number. Constraint (5) prevents assigning operation *i* before its precedence. Constraint (6) provides operation i can assign to at least LJ_i numbered station and at most UJ_i numbered station. (7) and (8) indicates the calculations for precedence relationship. Constraint (7) displays the earliest station to assign for job *i* and constraint (8) shows the latest station to assign job *i*. P_i at constraint (7) stands for cluster of operations which should be done before start of operation i. In other words, operations that comes after operation i according to precedence relationship. S_i at constraint (8) stands for cluster of operations which should not start before end of operation i. It refers as the operation that comes before operation i according to precedence relationship. Constraint (10) refers to, j must be smaller than or equal to UJ_i and bigger than or equal to LJ_i, if not it is equal to 0.

5 Computational Results of the Mathematical Models

Both mathematical models have been run for each three-product model in IBM ILOG Optimization Studio version 12.8.0.0. Initially the mathematical model 1 is runned for all product type and then the station numbers are obtained for Model A, Model B and Model Cas 17, 15 and 19 respectively. According to the solution of the second model, the cycle times for Model A, Model B and Model Care obtained as 95.21, 91.74 and 175.49, respectively. (Used computer is MacBook Pro 13, Intel® Core™ i5-5257U @ 2.70 GHz, 8 GB DDR4.)

After obtaining the results of the Model 1 and Model 2 for all product model, station numbers are analysed to compare solutions of the different scenarios. Since the station number for Model B is 15, and the station number in the current system within the factory is 21, the further analysis is made from 15 to 21 station numbers. Therefore, the cycle times are calculated for each model, assuming station number varies from 15 to 21 and the results are presented in Table 3.

Table 3. Changeover times for different station numbers

	Number of stations						
	15	16	17	18	19	20	21
Changeover Time	113.64	104.17	102.08	90.27	83.76	72.88	77.18
Cycle time of Model A	116.52	111.03	103.57	100.96	95.21	89.53	85.00
Cycle time of Model B	115.60	109.22	103.39	98.54	91.74	89.00	84.20
Cycle time of New Product	229.24	213.39	205.47	188.81	175.50	161.88	161.38

As can be seen in Table 3, cycle time and the changeover time are calculated for each product model. Changeover time for each station number is calculated by subtracting the minimum cycle time from the maximum cycle time for each station. At the end, the minimum changeover time is found as 72.88 s sat the point where station number is equal to 20. According to the results provided in Table 3, three options are suggested for the company. If the company wants to decrease the station number, the station number can be chosen as 19. If the company wants to decrease the changeover time, the station number can be chosen as 20. However, for this situation, if they don't prefer to chance the station number, the station number can be chosen as 21 because of the very small changeover time difference between 20 and 21 station numbers.

The company where the study is conducted had clearly defined the upper and lower limits for the station number as from 19 stations to 21 stations. Therefore, to restrict the number of station and to analyze the system utilization, three options are used: 19 stations, 20 stations and 21 stations. Thus, in order to compare the average utilizations, a simulation model is created in Arena Simulation Software. For each station number, each model's average utilization is obtained for Model A, Model B and Model C, respectively. (In total, 9 simulation models are run.) After running the simulation model for all product models under different station number scenarios, the averages of the utilization are also obtained these are reported in Graph 1. According to Graph 1, the best result is given as the station number is 20.

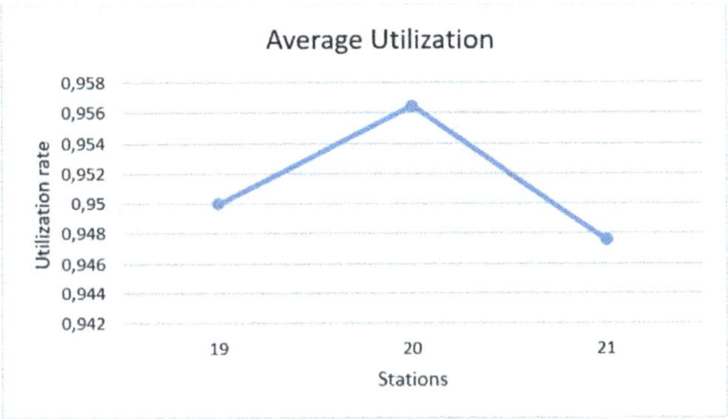

Graph 1. Average utilization for the line

6 Conclusion

As a conclusion, several methodologies are used for modification and line balancing of a new product. First, job sequence is determined from production of sample model as shown in Fig. 1. Operation times were found according to job sequence. Monthly demands are collected from customers for 11 months. Also 25% of total demand is added to forecasting value to meet customer demands in case of any changes. Total production time is also calculated. After these calculations, the company determines cycle time by using OEE value. According to OEE value, cycle time is calculated. Then, mathematical models are developed. By using IBM ILOG Optimization Studio, optimal station number is evaluated. According to number found, cycle time is updated. Optimal solutions within the framework of the mathematical models for three product models are determined. After these solutions, a simulation model is developed for the 5th lean line. The simulation is conducted based on different number of stations and cycle times, and then changeover time was calculated as an output. The simulation demonstrated utilization of each station.

Three options are presented as a result of this study. First option is (i) to minimize the station number, second one is (ii) to minimize the changeover time and third one is to (iii) keep station number constant. However these options are conflicting with each other. It means that, (i) to minimize the station number, 19 stations can be applied but this leads to decrease the fixed costs such as labor. On the other hand, the changeover time will be higher than the other options. (ii) To minimize the changeover time, 20 stations can be applied but this leads to decrease the utilization rate of workshop area usage in the 5th lean line since there are 21 stations on hand in the current system.

(iii) To keep the station number constant, 21 stations can still be processed but it is obvious that the utilization rate is the lowest one. To sum up, these options have benefits and disadvantages in different ways and can be prioritized based on managerial decisions.

Acknowledgment. This study is supported by TUBITAK (The Scientific and Technological Research Council of Turkey) in the program of "2209-B Undergraduate Research Projects for Industrial Applications Fellowship Program". We would like to thank our colleagues Cemal Berke Ersoy and Caner Samancı for their help and contribution to the study.

Appendix

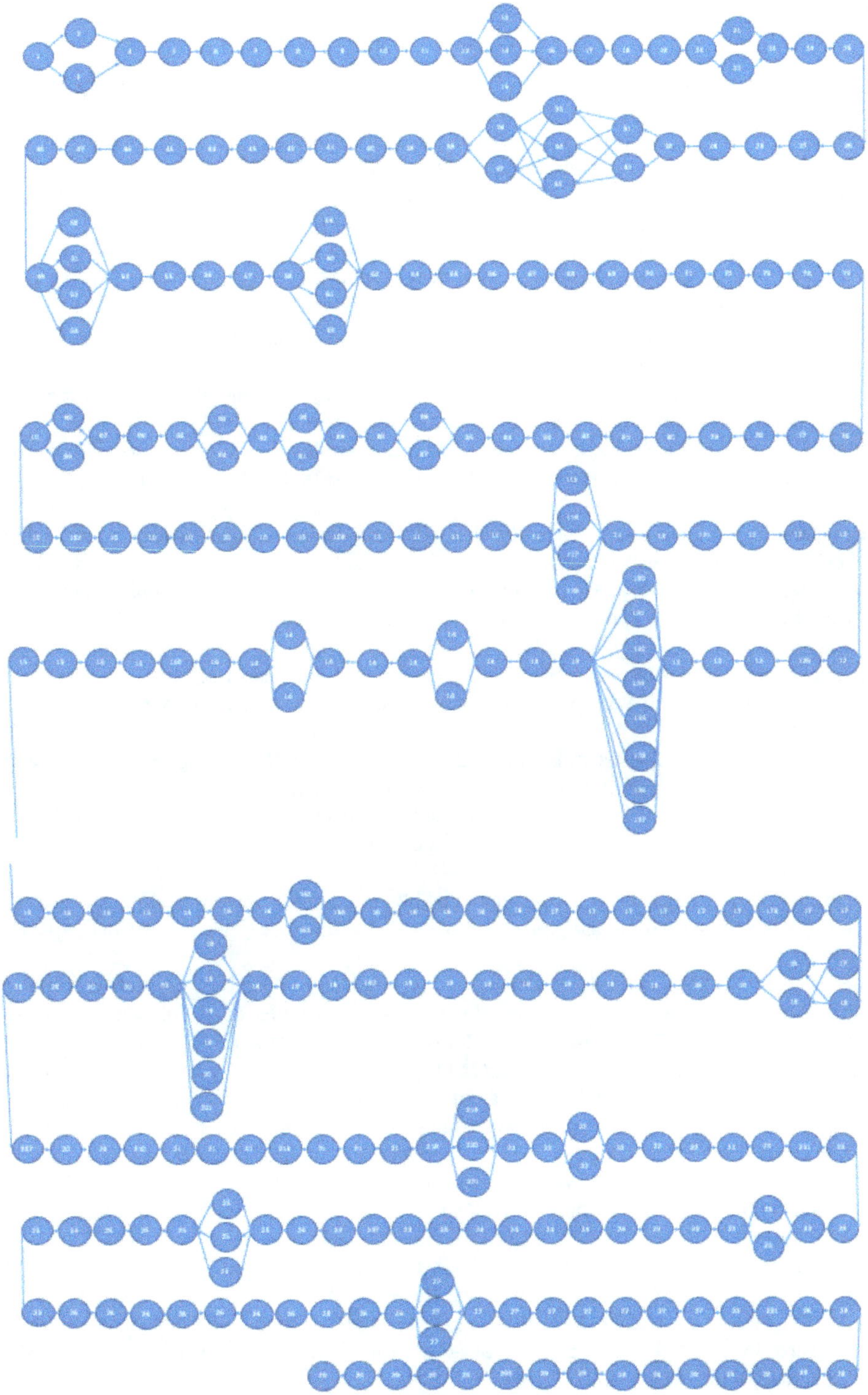

Fig. 1. Precedence relationship

References

van Zante-de Fokkert, J.I., de Kok, T.G.: The mixed and multi model line balancing problem: a comparison. Eur. J. Oper. Res. **100**(3), 399–412 (1997)

Karabatı, S., Sayın, S.: Assembly line balancing in a mixed-model sequencing environment with synchronous transfers. KocUniversity, Istanbul (2003)

McMullen, P.R., Tarasewich P.: Multi-objective assembly line balancing via a modified ant colony optimization technique **44**(1), 27-42 (2007). https://doi.org/10.1080/00207540500216748

Hwang, R., Katayama, H.: A multi-decision genetic approach for workload balancing of mixed-model U-shaped assembly line systems **47**(14), 3797–3822 (2007). https://doi.org/10.1080/00207540701851772

Boysen, N., Fliedner, M., Scholl, A.: Assembly line balancing: which model to use when? Universitat Hamburg, Germany (2007)

Widyadana, G.A.: Multi Objective Model for Balancing U-Type Assembly Line with Permanent and Temporary Workers. Petra Christian University, Indonesia (2009)

Fattahi, P., Roshani, A., Roshani, A.: A mathematical model and ant colony algorithm for multi-manned assembly line balancing problem **53**, 363–378 (2010). https://doi.org/10.1007/s00170-010-2832-y

Yagmahan, B.: Mixed-model assembly line balancing using a multi-objective ant colony optimization approach. Uludag University, Bursa (2011)

Yoosefelahi, A., Aminnayeri, M., Mosadegh, H., Ardakani, H.D.: Type II robotic assembly line balancing problem: an evolution strategies algorithm for a multi-objective model. Amirkabir University of Technology, Iran (2012)

Al-zubaidy, S.S., AbdAlrazaq, F.F.: Multi-model Production and Assembly Line Balancing (Caravans Production Workshop). Master's thesis. Journals of University of Babylon, Iraq (2013)

Zeltzer, L., Limère, V., Aghezzaf, E.-H., VanLandeghem, H.: Balancing mixed-model assembly lines in real world complex workstations. Ghent University, Belgium (2013)

Akpinar, Ş., Baykasoglu, A.: Modeling and solving mixed-model assembly line balancing problem with setups. Part I: a mixed integer linear programming model. Dokuz Eylul University, Izmir (2014)

Ramezanian, R., Ezzatpanah, A.: Modeling and solving multi-objective mixed-model assembly line balancing and worker assignment problem. K.N. Toosi University of Technology, Iran (2014)

Manavizadeh, N., Rabbani, M., Radmehr, F.: A new multi-objective approach in order to balancing and sequencing U-shaped mixed model assembly line problem: a proposed heuristic algorithm **79**, 415–425 (2015). https://doi.org/10.1007/s00170-015-6841-8

Raj, A.V., Mathew, J., Jose, P. and Sivan, G.: Optimization of Cycle Time in an Assembly Line Balancing Problem. St. Joseph's College of Engineering and Technology, Palai, India (2016)

Tiacci, L.: Mixed-model U-shaped assembly lines: balancing and comparing with straight lines with buffers and parallel workstations. Università degli Studi di Perugia, Italy (2017)

Christensen, M.K., Janardhanan, M.N., Nielsen, P.: Heuristics for solving a multi-model robotic assembly line balancing problem **5**(1), 410–424 (2017). https://doi.org/10.1080/21693277.2017.1403977

Sivasankaran, P., Shahabudeen, P.: Comparison of Single Model and Multi-Model Assembly Line Balancing Solutions. Anna University, India (2017)

Özcan, U., Peker, A.: Karışık Modelli U-Tipi Montaj Hatlarında Hat Dengeleme ve Model Sıralama Problemleri İçin Yeni Bir Sezgisel Yaklaşım. Gazi Üniversitesi Mühendislik-Mimarlık Fakültesi Dergisi **22**(2) (2007)

Kılınçcı, Ö.: A petri net approach for solving simple assembly line balancing problem.Dokuz Eylül Üniversitesi Mühendislik Fakültesi Fen ve Mühendislik Dergisi **6**(2), 1–15 (2004)

The Uniform Parallel Machine Scheduling Problem: A Case Study

Ege Duran, Gizem Görgülü, Ayben Pınar Kuruç, İpek Gülhan, Murat Can Doğruyol, Hande Öztop(✉), and Adalet Öner

Department of Industrial Engineering, Yaşar University, Izmir, Turkey
egeedurann@gmail.com, gizemgorgulu95@gmail.com, aybenkuruc@gmail.com, ipekglhn68@gmail.com, candogruyol@gmail.com, {hande.oztop, adalet.oner}@yasar.edu.tr

Abstract. In this study, the uniform parallel machine scheduling problem with non-common due dates and sequence-dependent setup times is addressed for a real-life problem in the dye house of a hood manufacturer company. The aim of this study is to create an efficient scheduling tool for the company, which minimizes lateness (earliness and tardiness) in the system and reduces the buffer stock caused by the lateness. A mathematical model is developed for the problem and optimal results are obtained for the small-sized instances. As the studied problem is NP-hard, three heuristic algorithms are also proposed to solve larger instances. The performance of the proposed algorithms is evaluated with a detailed computational experiment. Furthermore, a user-friendly decision support system (DSS) is developed using Excel VBA interface and proposed solution approaches are embedded in the DSS. The developed DSS enables users to make an efficient scheduling in very short computational time and provides the results with detailed schedule reports and Gantt charts. As this problem can be faced in various industrial areas, the proposed solution approaches can also be applied to different sectors and factories.

Keywords: Parallel machine scheduling · Lateness · Uniform machines · Sequence-dependent setup times · Heuristic algorithm · Decision support system

1 Introduction

This study considers a parallel machine scheduling problem with uniform machines, sequence-dependent setup times and non-common due dates (UPMSP-SDST-DD). The objective is to minimize total lateness in the system, which is the sum of the absolute differences between job completion times and their corresponding due dates. We consider a real-life UPMSP-SDST-DD, which is experienced in the dye house of a hood manufacturer company in Izmir, Turkey.

In the hood manufacturer company, production workflow of a hood follows as; mechanical section, shaping and assembling the raw material, dye house section, painting of the product, assembly section, and lastly, finalization of the product and

N. M. Durakbasa and M. G. Gençyılmaz (Eds.): ISPR 2019, LNME, pp. 839–850, 2020.
https://doi.org/10.1007/978-3-030-31343-2_69

packaging. Note that, there are many types of hoods with different features and production flows in the company. In general, initially, metal sheets are brought to the factory and enter the mechanical section to get their shapes. Shaping process is done with presses and laser machines according to hood type. Afterwards, shaped sheets are assembled and grouped by spot welding to form the products as semi-finished hoods. After this process, paintable semi-finished hoods are transferred to the dye house and painted, while non-paintable ones are directly transferred to the assembly line. In the assembly line, attachments such as engine groups, lamps, control groups and filters are assembled to the hoods. Finally, finished hoods are packaged and transferred to the warehouse.

In this study, we focus on the dye-house section of the hood manufacturing process. In the current production system, the scheduling of the jobs in the dye house is manually performed by the operators based on their experience without following a standard procedure. Therefore, high lateness and high buffer stocks are observed in the current system. In the dye house section, there are two dye houses. Initially, semi-finished products are hung on the hangers. Then, hangers move to the paint booth with the help of the conveyor. Painting robots spray products with powder paint in the paint booth. Then, powder applied parts enter the oven to be dried. Finally, after the painting process, the products are sent to the assembly line.

In the dye house section, there is a parallel machine scheduling problem, where orders indicate jobs and dye houses indicate machines. As the two dye houses have different conveyor speeds, there are uniform parallel machines. Furthermore, the system has sequence-dependent setup times due to the color changes, and starting time of the jobs in the assembly section determine the due dates of the jobs in dye house section. Note that, the due dates of the jobs are non-common. Consequently, the studied problem is classified as UPMSP-SDST-DD, where the aim is to develop a systematic and efficient scheduling tool to minimize lateness (earliness/tardiness) of the jobs and ensure that the products are ready at the designated time in the assembly section with minimum buffer stock. Then, the output of this study is a user-friendly decision support system which allows the users to line up the painted hoods for assembly section with minimum lateness and buffer stock.

The rest of the paper is organized as follows. A literature review on parallel machine scheduling is presented in Sect. 2. A mixed integer linear programming model (MILP) is presented in Sect. 3, as well as the formal problem definition. Heuristic algorithms are presented in Sect. 4. Section 5 presents the computational results. A user-friendly decision support system is also explained in Sect. 5. Finally, conclusions and future work suggestions are provided in Sect. 6.

2 Literature Review

There exists many studies that consider scheduling problems with earliness and tardiness. However, the literature on lateness criterion is rather limited when the job scheduling problem with multiple machines is considered. Most studies focus on sequence independent setup times, where they are assumed to be included in processing times. However, in our study we consider the uniform parallel machine scheduling problem in the dye house with sequence-dependent setup times and non-common due dates.

In the literature, two research papers were found to be much related to the studied problem. In the first study, the authors proposed a mathematical model for the job scheduling problem with non-identical parallel machines and sequence-dependent setup times to minimize the sum of weighted lateness [1]. In the second study, the authors proposed a mathematical model and a simulated annealing algorithm for the identical parallel machine scheduling problem with earliness-tardiness penalties and sequence-dependent setup times to minimize the total lateness of jobs [2]. In this study, we extend the mathematical model of [2] by introducing uniform machines to the problem.

Note that, the studied UPMSP-SDST-DD has already been proven as NP-hard, since the single machine version of the problem is known to be NP-hard [3]. Therefore, heuristic algorithms are generally addressed to solve large instances. Efficient constructive heuristic algorithms were presented to solve parallel machine scheduling problems in [4–6]. GTW (Genetic Time Warping) algorithm is also used in several studies [2, 7] to minimize lateness, which was originally developed by Garey et al. [3]. In this paper, we present three heuristic algorithms to solve UPMSP-SDST-DD inspiring from these heuristic algorithms [2, 3]. Furthermore, we develop a user friendly decision support system which is integrated with the proposed heuristic algorithms.

3 Problem Formulation

As mentioned before, there exists high lateness and high buffer stock in the system due to inefficient scheduling performed by the operators. Hence, the aim of this study is to schedule n jobs with sequence dependent setup times and non-common due dates on m uniform parallel machines, such a way that the total lateness in the system is minimized. It is assumed that all machines are eligible for all jobs and one job can be processed by only one machine at a time. Furthermore, there is no job preemption and all jobs are ready at the beginning. In this study, we extend the mathematical model of [2] by introducing uniform machines to the problem. The proposed MILP model for the UPMSP-SDST-DD is explained below:

Sets

N: set of jobs $\{1, 2, \ldots n\}$

M: set of machines $\{1, 2, \ldots m\}$

Parameters

d_i: due date of job i

p_{ik}: processing time of job i on machine k

s_{ij}: setup time when job j is processed immediately after job i

s_{0i}: initial setup time for job i

W: a large number

Decision Variables

c_i: completion time of job i

e_i: earliness of job i

t_i: tardiness of job i

x_{ijk}: $\begin{cases} 1 & \text{if job } j \text{ is performed immediately after job } i \text{ in the sequence on machine } k, \\ 0 & \text{otherwise.} \end{cases}$

x_{0ik}: $\begin{cases} 1 & \text{if job } i \text{ is the first job on machine } k, \\ 0 & \text{otherwise.} \end{cases}$

y_{ik}: $\begin{cases} 1 & \text{if job } i \text{ is assigned to machine } k, \\ 0 & \text{otherwise.} \end{cases}$

Mathematical Model

Minimize $\sum_{i=1}^{n} (e_i + t_i)$

Subject to

$$\sum_{i=1}^{n} x_{0ik} = 1 \qquad \forall k \in M \tag{1}$$

$$c_i + W(1 - x_{0ik}) \geq s_{0i} + p_{ik} \qquad \forall i \in N, \forall k \in M \tag{2}$$

$$\sum_{i=0:i \neq j}^{n} x_{ijk} = y_{jk} \qquad \forall j \in N, \forall k \in M \tag{3}$$

$$\sum_{j=1:i \neq j}^{n} x_{ijk} = y_{ik} \qquad \forall i \in N, \forall k \in M \tag{4}$$

$$c_j + W(1 - x_{ijk}) \geq c_i + s_{ij} + p_{jk} \qquad \forall i, j \in N : i \neq j, \quad \forall k \in M \tag{5}$$

$$c_i - d_i = t_i - e_i \qquad \forall i \in N \tag{6}$$

$$\sum_{k=1}^{m} y_{ik} = 1 \qquad \forall i \in N \tag{7}$$

$$c_i, t_i, e_i \geq 0 \qquad \forall i \in N \tag{8}$$

$$x_{ijk}, x_{0ik}, y_{ik} \in \{0, 1\} \qquad \forall i, j \in N : i \neq j, \quad \forall k \in M \tag{9}$$

Note that, a dummy job 0 is introduced for each machine, to handle the initial setup time for the real job assigned to the first position in the sequence on each machine. The objective function minimizes the total lateness (earliness/tardiness) of the jobs. The constraint set (1) guarantees that the dummy job 0 is placed at the beginning of the sequence before all real jobs on each machine. The constraint set (2) states that the finishing time of a first real job in the sequence on any machine must be greater than or equal to the sum of its processing time and initial setup time. The constraint set (3) says that if a real job is assigned to a machine, then it will be immediately preceded by one job. Similarly, the constraint set (4) ensures that if a real job is assigned to a machine then it must be succeeded by one job. The constraint sets (3) and (4) also verify that n real jobs are assigned to m machines. The constraint set (5) guarantees that the starting time of a real job j on a machine must be greater than or equal to the finishing time of the preceding job on that machine, regarding the sequence dependent setup time in between. The constraint set (6) states that the difference between completion time and the due date of a job is equal to the earliness or tardiness of the job. The constraint set (7) ensures that a real job should be assigned to exactly one machine. Finally, the constraint sets (8) and (9) define decision variables.

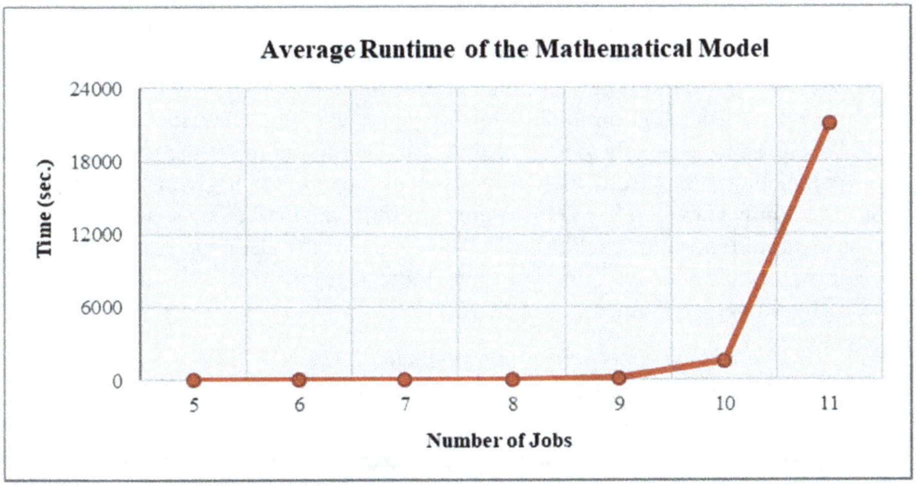

Fig. 1. Average runtime of the mathematical model

The proposed MILP model is solved for 35 instances with varying job sizes from 5 to 11, where there are 5 instances for each job size. Note that, there are two machines in each instance. The MILP formulation is coded on IBM ILOG CPLEX Optimization Studio and all instances are solved optimally with CPLEX 12.8. The average runtimes of the mathematical model for different job sizes are illustrated in Fig. 1. As shown in Fig. 1, the solution time of the mathematical model increases exponentially as the job size increases. This pattern is expected as the studied problem has already been proven as NP-hard [3]. Therefore, heuristic methods have been addressed in the following section.

4 Heuristic Algorithms

Three variants of heuristic algorithms are developed to solve the UPMSP-SDST-DD inspiring from the literature [2, 3, 7]. First heuristic algorithm is based on well-known "Earliest Due Date" (EDD) rule. Note that, EDD rule sort the jobs in increasing order of their due dates. The main steps of the Heuristic Algorithm 1 are explained in Fig. 2.

Heuristic Algorithm 1	
Step 1.	Sort the jobs according to EDD rule and form a job list.
Step 2.	Calculate the completion time of the first job in the list on each machine regarding the sequence-dependent setup times.
Step 3.	Calculate the lateness of the first job in the list on each machine using the completion time calculated at Step 2.
Step 4.	Schedule the first job in the list to the machine with the minimum lateness value, in case of an equality schedule the first job in the list to the machine with the minimum processing time.
Step 5.	Update list by removing the scheduled job from the list.
Step 6.	If the job list is not empty return to Step 2.

Fig. 2. Heuristic Algorithm 1

As second and third heuristic approaches, we employ the efficient heuristic algorithm proposed by [2]. Note that, the algorithm of Radhakrishnan and Ventura [2] employs GTW algorithm [3] in their procedure. In this study, the GTW algorithm is not included in Heuristic Algorithm 2, while it is included in Heuristic Algorithm 3. The GTW algorithm is an efficient algorithm that aims to minimize the total deviations of job completion times from their respective due dates, for a given job sequence. While scheduling a new job, the GTW algorithm shifts the previously scheduled jobs in some steps to minimize the total lateness. The details of the Heuristic Algorithm 2 and Heuristic Algorithm 3 are explained in Figs. 3 and 4, respectively. The notations used in these algorithms are defined as below [2]:

EC_{jk} = *earliest completion time of job j on machine k*
ES_k = *earliest available starting position on machine k*
$last_k$ = *current last job on machine k*

$s_{(last_k)j}$: *setup time when job j is processed immediately after job $last_k$ on machine k*

$$EC_{jk} = Max\{ES_k + s_{(last_k)j} + p_{jk}, d_j\}$$

Heuristic Algorithm 2	
Step 1	Let job list = $\{1,2,\ldots,n\}$.
Step 2	Calculate EC_{jk} for each job-machine pair and find the minimum job-machine pair ($Min\{EC_{jk}\}$)
Step 3	Schedule the selected job to the selected machine, in case of an equality schedule the job to the machine with the least number of scheduled jobs.
Step 4	Update the number of scheduled jobs and the job sequence on the selected machine.
Step 5	Update ES_k and $last_k$ for the selected machine.
Step 6	Remove the scheduled job from the job list and return to Step 2.

Fig. 3. Heuristic Algorithm 2

Heuristic Algorithm 3	
Step 1	Let job list = $\{1,2,\ldots,n\}$.
Step 2	Calculate EC_{jk} for each job-machine pair and find the minimum job-machine pair ($Min\{EC_{jk}\}$)
Step 3	Schedule the selected job to the selected machine, in case of an equality schedule the job to the machine with the least number of scheduled jobs.
Step 4	Update the number of scheduled jobs and the job sequence on the selected machine.
Step 5	Apply GTW algorithm to the job sequence of the selected machine.
Step 6	Update ES_k and $last_k$ for the selected machine.
Step 7	Remove the scheduled job from the job list and return to Step 2.

Fig. 4. Heuristic Algorithm 3

In order to improve solution quality, additional swap operations within and between machines are also applied to the results obtained from the aforementioned three heuristic algorithms. Note that, a swap move interchanges the positions of previously assigned two jobs within a machine or different machines.

5 Computational Results

The computational results of the proposed heuristic algorithms are presented in this section. In order to evaluate the performance of the proposed heuristic methods, 35 instances with varying job sizes from 5 to 11 are developed, where there are 5 instances for each job size. Note that, there are two machines in each instance. The proposed MILP model in Sect. 3 is coded on IBM ILOG CPLEX Optimization Studio and all instances are solved optimally with CPLEX 12.8 on a Core 2.40 GHz computer. All proposed heuristic algorithms are coded in Microsoft Visual Basic for Applications (VBA) Software and all instances are solved by three heuristic algorithms within few seconds on a Core 2.40 GHz computer. The relative percent deviations (RPD) from the optimal results are reported for each heuristic algorithm as below:

$$RPD = \frac{Heuristic\ Solution - Optimal\ Solution}{Optimal\ Solution} \times 100 \tag{10}$$

Table 1 reports the RPD values from the optimal results for the proposed heuristic algorithms. As shown in the Table 1, Heuristic Algorithm 1 finds the optimal result for only one instance over 35 instances, while Heuristic Algorithm 2 and 3 find the optimal results for the two and four instances respectively. The results show that Heuristic Algorithm 3, which includes GTW algorithm [3], has the minimum average RPD from the optimal results with 14.37%.

Table 1. Comparison of the heuristic algorithms without swap operations

RPD (%)					
Job size	Instance	Heuristic Algorithm 1	Heuristic Algorithm 2	Heuristic Algorithm 3	Best result
5	1	73.68	31.58	15.79	15.79
	2	21.95	34.15	7.32	7.32
	3	0.00	0.00	0.00	0.00
	4	2.70	8.11	3.60	2.70
	5	96.36	12.73	12.73	12.73
6	1	172.73	27.27	27.27	27.27
	2	3.45	3.45	0.00	0.00
	3	11.48	13.93	9.02	9.02
	4	8.57	5.71	5.71	5.71
	5	5.31	27.43	24.78	5.31

(*continued*)

Table 1. *(continued)*

RPD (%)					
Job size	Instance	Heuristic Algorithm 1	Heuristic Algorithm 2	Heuristic Algorithm 3	Best result
7	1	126.09	71.74	52.17	52.17
	2	77.94	53.68	19.85	19.85
	3	43.14	31.37	13.73	13.73
	4	10.90	2.84	0.00	0.00
	5	41.35	25.00	21.15	21.15
8	1	63.04	42.39	29.35	29.35
	2	67.52	16.24	0.85	0.85
	3	4.57	20.30	10.66	4.57
	4	191.89	0.00	54.05	0.00
	5	31.02	25.74	23.76	23.76
9	1	60.97	29.87	26.21	26.21
	2	32.35	30.72	10.46	10.46
	3	18.72	32.42	0.00	0.00
	4	82.21	54.33	6.73	6.73
	5	175.00	19.64	23.21	19.64
10	1	22.32	33.68	11.58	11.58
	2	46.43	43.25	7.54	7.54
	3	7.18	34.45	18.66	7.18
	4	20.98	38.84	20.54	20.54
	5	16.67	11.05	6.55	6.55
11	1	13.99	33.23	6.52	6.52
	2	28.57	18.62	3.83	3.83
	3	15.83	8.32	3.00	3.00
	4	16.63	24.89	2.41	2.41
	5	25.84	52.15	23.92	23.92
	Average	**46.78**	**26.26**	**14.37**	**11.64**

As mentioned in Sect. 4, to improve the results of the heuristic algorithms, additional swap operations are implemented within and between machines randomly. The RPD values of the three heuristics after the swap operations are shown in Table 2 for 35 instances. As shown in the Table 2, Heuristic Algorithm 1 finds the optimal results for only four instances over 35 instances, while Heuristic Algorithm 2 and 3 find the optimal results for the 10 and 12 instances respectively.

Table 2. Comparison of the heuristic algorithms after swap operations

RPD (%)					
Job size	Instance	Heuristic Algorithm 1	Heuristic Algorithm 2	Heuristic Algorithm 3	Best result
5	1	2.63	18.42	10.53	2.63
	2	0.00	0.00	0.00	0.00
	3	0.00	0.00	0.00	0.00
	4	0.00	2.70	0.00	0.00
	5	114.55	12.73	12.73	12.73
6	1	127.27	9.09	9.09	9.09
	2	3.45	3.45	0.00	0.00
	3	3.28	7.38	3.28	3.28
	4	0.95	0.00	0.00	0.00
	5	5.31	1.77	1.77	1.77
7	1	30.43	28.26	21.74	21.74
	2	77.94	8.82	7.35	7.35
	3	13.73	13.73	13.73	13.73
	4	10.90	0.00	0.00	0.00
	5	7.69	7.69	7.69	7.69
8	1	13.04	0.00	0.00	0.00
	2	0.85	0.00	0.00	0.00
	3	0.00	0.00	0.00	0.00
	4	51.35	0.00	51.35	0.00
	5	10.23	13.20	9.90	9.90
9	1	72.73	100.00	45.45	45.45
	2	3.27	3.92	3.27	3.27
	3	5.94	7.31	0.00	0.00
	4	6.73	8.65	6.73	6.73
	5	45.71	10.36	15.36	10.36
10	1	1.26	5.26	1.26	1.26
	2	11.11	0.40	4.76	0.40
	3	0.48	5.74	0.48	0.48
	4	1.26	5.26	1.26	1.26
	5	3.75	0.00	0.00	0.00
11	1	2.23	2.23	2.70	2.23
	2	6.38	6.38	3.83	3.83
	3	1.77	0.00	0.00	0.00
	4	1.72	1.61	2.18	1.61
	5	11.00	5.26	13.40	5.26
	Average	**18.54**	**8.27**	**7.14**	**4.92**

From the results after swap operations, it can be said that Heuristic Algorithm 3 still performs better than the other two heuristic algorithms with an average RPD value of 7.14%. Note that, swap operations have improved the performance of the Heuristic Algorithm 3 by 7.23% on average. Also, the average best result deviation decreased from 11.64% to 4.92%. Therefore, Heuristic Algorithm 3 which uses GTW algorithm is selected as a final solution approach with the implementation of swap local search operations. Consequently, Heuristic Algorithm 3 is integrated in the Decision Support System to provide an efficient scheduling tool.

The proposed Heuristic Algorithm 3 is coded using Excel VBA to create a user-friendly Decision Support System (DSS). The DSS takes the due dates, the processing times and the setup times of the jobs to be painted in the dye house as inputs from the user. The home page of the developed DSS is presented in Fig. 5. Then the DSS reports the schedules obtained by Heuristic Algorithm 3 with detailed lists and Gantt Charts. Report screens of the DSS are provided in Figs. 6 and 7. Note that, the necessary input data can be automatically extracted from the related files or manually edited to provide flexibility for the users. In addition, it is possible to edit, delete and add all the data entered by the user. Furthermore, reports can be transferred via e-mail to expedite and facilitate communication within the factory. It is also possible to print out the detailed reports from the system.

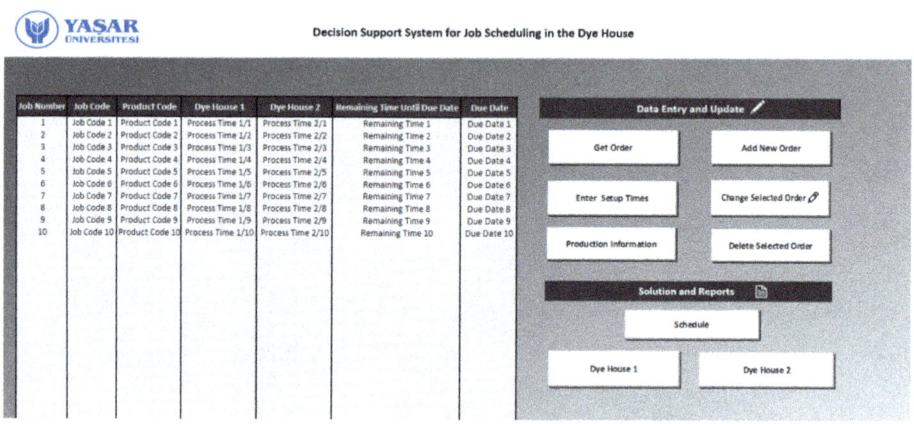

Fig. 5. DSS main screen

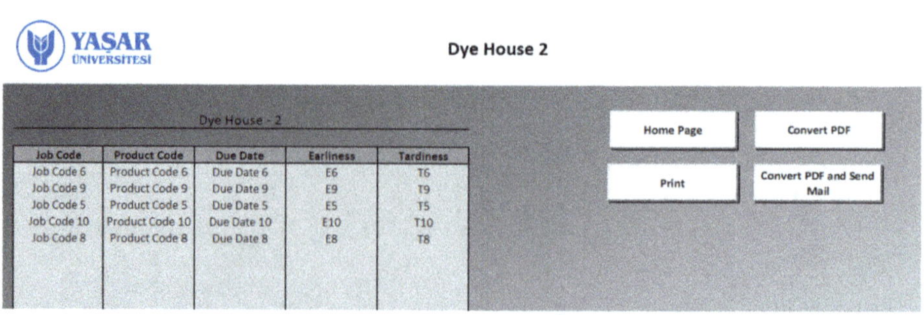

Fig. 6. DSS report (job list)

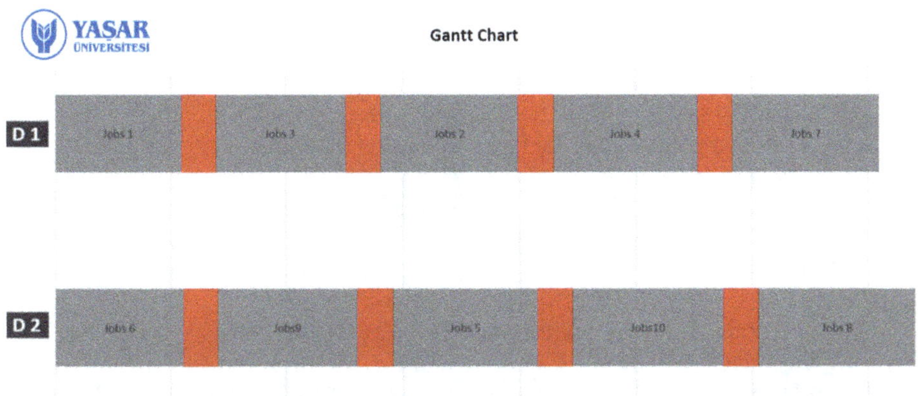

Fig. 7. DSS report (Gantt chart)

6 Conclusions

In this paper, a real-life parallel machine scheduling problem in the dye house of a hood manufacturer company is addressed. A mathematical model, three heuristic algorithms and a user friendly decision support system are developed for the studied real-life problem of the company. Initially, a general system analysis and a micro system analysis were conducted to have a better understanding of the factory's current system. The problem in the dye house is determined as a parallel machine scheduling problem with uniform machines, sequence-dependent setup times and non-common due dates (UPMSP-SDST-DD). After the detailed literature review, a mathematical model is developed for the problem. Toy examples with different job sizes from 5 to 11 were solved with IBM ILOG CPLEX to check the validity of the model formulation. Results have shown that the runtime of the mathematical model increases exponentially as the job size increases. Note that, the studied problem has already been proven as NP-hard [3]. Hence, three different heuristic algorithms are developed. In the proposed heuristic algorithms, additional swap operations are also implemented within/between machines to improve the solution quality. The performance of the proposed heuristic algorithms is evaluated using the optimal results obtained from mathematical model. The computational results show that the Heuristic Algorithm 3 which involves GTW [3] algorithm gives the best RPD values from the optimal solutions. Therefore, Heuristic Algorithm 3 with swap operations is chosen as the ultimate solution approach for the studied problem.

Consequently, the Heuristic Algorithm 3 is integrated into the Decision Support System (DSS) to provide a user-friendly and an efficient scheduling system to the company. The DSS takes the due dates, the processing times and the setup times of the jobs to be painted in the dye house as inputs and reports the schedules with detailed lists and Gantt Charts. Note that, the data can be automatically taken from the required files or manually edited by the user, to ensure the flexibility. In addition, it is possible to change/edit the data entered by the user and reports can be transferred via e-mail to facilitate the communication within the factory. Since the studied (UPMSP-SDST-DD) can be faced in various areas, the proposed solution approaches can also be applied to different sectors and companies.

Acknowledgment. This work cannot be completed without the assistance of Kaan Özsümbül and Batuhan Uğuz. We are thankful for their contribution. Furthermore, we are grateful to the company for their co-operation.

References

1. Heady, R.B., Zhu, Z.: Minimizing the sum of job earliness and tardiness in a multimachine system. Int. J. Prod. Res. **36**(6), 1619–1632 (1998)
2. Radhakrishnan, S., Ventura, J.A.: Simulated annealing for parallel machine scheduling with earliness-tardiness penalties and sequence-dependent set-up times. Int. J. Prod. Res. **38**(10), 2233–2252 (2000)
3. Garey, M.R., Tarjan, R.E., Wilfong, G.T.: One-processor scheduling with symmetric earliness and tardiness penalties. Math. Oper. Res. **13**(2), 330–348 (1988)
4. Kim, D.W., Kim, K.H., Jang, W., Chen, F.F.: Unrelated parallel machine scheduling with setup times using simulated annealing. Robot. Comput. Integr. Manuf. **18**(3–4), 223–231 (2002)
5. Lee, J.H., Yu, J.M., Lee, D.H.: A tabu search algorithm for unrelated parallel machine scheduling with sequence-and machine-dependent setups: minimizing total tardiness. Int. J. Adv. Manuf. Technol. **69**(9–12), 2081–2089 (2013)
6. Lee, Y.H., Pinedo, M.: Scheduling jobs on parallel machines with sequence-dependent setup times. Eur. J. Oper. Res. **100**(3), 464–474 (1997)
7. Hassin, R., Shani, M.: Machine scheduling with earliness, tardiness and non-execution penalties. Comput. Oper. Res. **32**(3), 683–705 (2005)

Author Index

N. M. Durakbasa and M. G. Gençyılmaz (Eds.): ISPR 2019, LNME, pp. 851–853, 2020.
https://doi.org/10.1007/978-3-030-31343-2